Real Analysis with Economic Applications

Real Analysis with Economic Applications

Efe A. Ok

PRINCETON UNIVERSITY PRESS | PRINCETON AND OXFORD

Published by Princeton University Press, 41 William Street,
Princeton, New Jersey 08540

In the United Kingdom: Princeton University Press, 3 Market Place,
Woodstock, Oxfordshire OX20 1SY

Library of Congress Cataloging-in-Publication Data

Ok, Efe A.
Real analysis with economic applications / Efe A. Ok.
p. cm.
Includes bibliographical references and index.
ISBN-13: 978-0-691-11768-3 (alk. paper)
ISBN-10: 0-691-11768-3 (alk. paper)
1. Economics, Mathematical. 2. Mathematical analysis. I. Title.

HB135.O45 2007
330.1′51—dc22 2006049378

British Library Cataloging-in-Publication Data is available

This book has been composed in Scala and Scala Sans

Printed on acid-free paper. ∞

pup.princeton.edu

Printed in the United States of America

10 9 8 7 6

Mathematics is very much like poetry. . . . What makes a good poem—a great poem—is that there is a large amount of thought expressed in very few words. In this sense formulas like

$$e^{\pi i} + 1 = 0 \quad \text{or} \quad \int_{-\infty}^{\infty} e^{-x^2} dx = \sqrt{\pi}$$

are poems.

—*Lipman Bers*

Contents

Preface

This is primarily a textbook on mathematical analysis for graduate students in economics. While there are a large number of excellent textbooks on this broad topic in the mathematics literature, most of these texts are overly advanced relative to the needs of the vast majority of economics students and concentrate on various topics that are not readily helpful for studying economic theory. Moreover, it seems that most economics students lack the time or courage to enroll in a math course at the graduate level. Sometimes this is not even for bad reasons, for only few math departments offer classes that are designed for the particular needs of economists. Unfortunately, more often than not, the consequent lack of mathematical background creates problems for the students at a later stage of their education, since an exceedingly large fraction of economic theory is impenetrable without some rigorous background in real analysis. The present text aims at providing a remedy for this inconvenient situation.

My treatment is rigorous yet selective. I prove a good number of results here, so the reader will have plenty of opportunity to sharpen his or her understanding of the "theorem-proof" duality and to work through a variety of "deep" theorems of mathematical analysis. However, I take many shortcuts. For instance, I avoid complex numbers at all cost, assume compactness of things when one could get away with separability, introduce topological and topological linear concepts only via metrics or norms, and so on. My objective is not to report even the main theorems in their most general form but rather to give a good idea to the student why these are true, or, even more important, why one should suspect that they must be true even before they are proved. But the shortcuts are not overly extensive in the sense that the main results covered here possess a good degree of applicability, especially for mainstream economics. Indeed, the purely mathematical development of the text is put to good use through several applications that provide concise introductions to a variety of topics from economic theory. Among these topics are individual decision theory, cooperative and noncooperative game theory, welfare economics, information theory, general equilibrium and finance, and intertemporal economics.

An obvious dimension that differentiates this text from various books on real analysis pertains to the choice of topics. I place much more emphasis on topics that are immediately relevant for economic theory and omit some standard themes of real analysis that are of secondary importance for economists. In particular, unlike most treatments of mathematical analysis found in the literature, I present quite a bit on order theory, convex analysis, optimization, linear and nonlinear correspondences, dynamic programming, and calculus of variations. Moreover, apart from direct applications to economic theory, the exposition includes quite a few fixed point theorems, along with a leisurely introduction to differential calculus in Banach spaces. (Indeed, the second half of the book can be thought of as providing a modest introduction to geometric (non)linear analysis.) However, because they play only a minor role in modern economic theory, I do not discuss topics such as Fourier analysis, Hilbert spaces, and spectral theory in this book.

While I assume here that the student is familiar with the notion of proof—this goal must be achieved during the first semester of a graduate economics program—I also spend quite a bit of time telling the reader why things are proved the way they are, especially in the earlier part of each chapter. At various points there are visible attempts to help the reader "see" a theorem (either by discussing informally the plan of attack or by providing a false-proof), in addition to confirming its validity by means of a formal proof. Moreover, whenever possible I have tried to avoid rabbit-out-of-the-hat proofs and rather give rigorous arguments that explain the situation that is being analyzed. Longer proofs are thus often accompanied by footnotes that describe the basic ideas in more heuristic terms, reminiscent of how one would "teach" the proof in the classroom.[1] This way the material is hopefully presented at a level that is readable for most second- or third-semester graduate students in economics and advanced undergraduates in mathematics while still preserving the aura of a serious analysis course. Having said this, however, I should note that the exposition gets less restrained toward the end of each chapter, and the analysis is presented without being overly pedantic. This goes especially for the starred sections, which cover more advanced material than the rest of the text.

[1] In keeping with this, I have written most of the footnotes in the first person singular pronoun, while using exclusively the first person plural pronoun in the body of the text.

The basic approach is, of course, primarily that of a textbook rather than a reference. But the reader will still find here the careful yet unproved statements of a good number of "difficult" theorems that fit well with the overall development; some examples include Blumberg's Theorem, noncontractibility of the sphere, Rademacher's Theorem on the differentiability of Lipschitz continuous functions, Motzkin's Theorem, and Reny's Theorem on the existence of the Nash equilibrium. At the very least, this should hint to the student what might be expected in a higher-level course. Furthermore, some of these results are widely used in economic theory, so it is desirable that the students begin at this stage developing a precursory understanding of them. To this end, I discuss some of these results at length, talk about their applications, and at times give proofs for special cases. It is worth noting that the general exposition relies on a select few of these results.

Last but not least, it is my sincere hope that the present treatment provides glimpses of the strength of abstract reasoning, whether it comes from applied mathematical analysis or from pure analysis. I have tried hard to strike a balance in this regard. Overall, I put far more emphasis on the applicability of the main theorems relative to their generalizations or strongest formulations, only rarely mention if something can be achieved without invoking the Axiom of Choice, and use the method of proof by contradiction more frequently than a "purist" might like to see. On the other hand, by means of various remarks, exercises, and the starred sections, I touch on a few topics that carry more of a pure mathematician's emphasis. (Some examples here include the characterization of metric spaces with the Banach fixed point property, the converse of Weierstrass' Theorem, various characterizations of infinite-dimensional normed linear spaces, and so on.) This reflects my full agreement with the following wise words of Tom Körner:

> A good mathematician can look at a problem in more than one way. In particular, a good mathematician will "think like a pure mathematician when doing pure mathematics and like an applied mathematician when doing applied mathematics." (Great mathematicians think like themselves when doing mathematics.)[2]

[2] Little is lost in translation if one adapts this quote for economists. You decide:

> A good economist can look at a problem in more than one way. In particular, a good economist will "think like a pure theorist when doing pure economic theory and like an applied theorist when doing applied theory." (Great economists think like themselves when doing economics.)

On the Structure of the Text

This book consists of four parts:

I. Set Theory (Chapters A and B)

II. Analysis on Metric Spaces (Chapters C, D, and E)

III. Analysis on Linear Spaces (Chapters F, G, and H)

IV. Analysis on Metric/Normed Linear Spaces (Chapters I, J, and K)

Part I provides an elementary yet fairly comprehensive overview of (intuitive) set theory. Covering the fundamental notions of sets, relations, functions, real sequences, basic calculus, and countability, this part is a prerequisite for the rest of the text. It also introduces the Axiom of Choice and some of its equivalent formulations, and sketches a brief introduction to order theory. Among the most notable theorems covered here are Tarski's Fixed Point Theorem and Szpilrajn's Extension Theorem.

Part II is (almost) a standard course on real analysis on metric spaces. It studies at length the topological properties of separability and compactness and the uniform property of completeness, along with the theory of continuous functions and correspondences, in the context of metric spaces. I also talk about the elements of fixed point theory (in Euclidean spaces) and introduce the theories of stationary dynamic programming and Nash equilibrium. Among the most notable theorems covered here are the Contraction Mapping Principle, the Stone-Weierstrass Theorem, the Tietze Extension Theorem, Berge's Maximum Theorem, the fixed point theorems of Brouwer and Kakutani, and Michael's Selection Theorem.

Part III begins with an extensive review of some linear algebraic concepts (such as linear spaces, bases and dimension, and linear operators), then proceeds to convex analysis. A purely linear algebraic treatment of both the analytic and geometric forms of the Hahn-Banach Theorem is given here, along with several economic applications that range from individual decision theory to financial economics. Among the most notable theorems covered are the Hahn-Banach Extension Theorem, the Krein-Rutman Theorem, and the Dieudonné Separation Theorem.

Part IV can be considered a primer on geometric linear and nonlinear analysis. Since I wish to avoid the consideration of general topology in this text, the entire discussion is couched within metric and/or normed linear spaces. The results on the extension of linear functionals and the separation by hyperplanes are sharpened in this context, an introduction to infinite-dimensional convex analysis is outlined, and the fixed point theory developed earlier within Euclidean spaces is carried into the realm of normed linear spaces. The final chapter considers differential calculus and optimization in Banach spaces and, by way of an application, provides an introductory but rigorous approach to calculus of variations. Among the most notable theorems covered here are the Separating Hyperplane Theorem, the Uniform Boundedness Principle, the Glicksberg-Fan Fixed Point Theorem, Schauder's Fixed Point Theorems, and the Krein-Milman Theorem.

On the Exercises

As in most mathematics textbooks, the exercises provided throughout the text are integral to the ongoing exposition and hence appear within the main body of various sections. Many of them appear after the introduction of a particularly important concept to make the reader better acquainted with that concept. Others are given after a major theorem in order to illustrate how to apply the associated result or the method of proof that yielded it.

Some of the exercises look like this:

EXERCISE 6

Such exercises are "must do" ones that will be used in the material that follows them. Other exercises look like

EXERCISE 6

Such exercises aim to complement the exposition at a basic level and provide practice ground for students to improve their understanding of the related topic. Some even suggest directions for further study.[3]

While most of the exercises in this book are quite doable (with a reasonable amount of suffering), some are challenging (these are starred), and

[3] While quite a few of these exercises are original, several of them come from the problem sets of my teachers, Tosun Terzioglu, Peter Lax, and Oscar Rothaus.

some are for the very best students (these are double-starred). Hints and partial solutions are provided for about one-third of the exercises at the end of the book.[4] All in all—and this will be abundantly clear early on—the guiding philosophy behind this text strongly subscribes to the view that there is only one way of learning mathematics, and that is learning by doing. In his preface, Chae (1995) uses the following beautiful Asian proverb to drive this point home:

> I hear, and I forget;
> I see, and I remember;
> I do, and I understand.

This recipe, I submit, should also be tried by those who wish to have some fun throughout the following 700-some pages.

On Measure Theory and Integration

This text omits the theory of measure and Lebesgue integration in its entirety. These topics are covered in a forthcoming companion volume, *Probability Theory with Economic Applications.*

On Alternative Uses of the Text

This book is intended to serve as a textbook for a number of different courses, and also for independent study.

- *A second graduate course on mathematics for economists.* Such a course would use Chapter A for review and cover the first section of Chapter B, along with pretty much all of Chapters C, D, and E. This should take about one-half to two-thirds of a semester, depending on how long one wishes to spend on the applications of dynamic programming and game theory. The remaining part of the semester may then be used to go deeper into a variety of fields, such as convex analysis (Chapters F, G, and H and parts of Chapters I and J), introductory linear analysis

[4] To the student: Please work on the exercises as hard as you can, before seeking out these hints. This is for your own good. Believe it or not, you'll thank me later.

(Chapters F through J), or introductory nonlinear analysis and fixed point theory (parts of the Chapters F, G, and I, and Chapters J and K). Alternatively, one may alter the focus and offer a little course in probability theory, whose coverage may now be accelerated. (For what it's worth, this is how I taught from the text at New York University for about six years, with some success.)

- *A first graduate course on mathematics for economists.* Unless the math preparation of the class is extraordinary, this text would not serve well as a primary textbook for this sort of course. However, it may be useful for complementary reading on a good number of topics that are traditionally covered in a first math-for-economists course, especially if the instructor wishes to touch on infinite-dimensional matters as well. (For examples, the earlier parts of Chapters C, D, and E complement the standard coverage of real analysis within $\mathbb{R}^n$, Chapter C spends quite a bit of time on the Contraction Mapping Theorem and its applications, Chapter E provides extensive coverage of matters related to correspondences, and Chapters F and G investigate linear spaces, operators, and basic convex analysis and include numerous separating and supporting hyperplane theorems of varying generality.)

- *An advanced undergraduate or graduate-level course on real analysis for mathematics students.* While my topic selection is dictated by the needs of modern economic theory, the present book is foremost a mathematics book. It is therefore duly suitable to be used for a course on mathematical analysis at the senior undergraduate or first-year graduate level. Especially if the instructor wishes to emphasize the fixed point theory and some economic applications (regarding, say, individual decision theory), it may well help organize a full-fledged math course.

- *Independent study.* One of the major objectives of this book is to provide the student with a glimpse of what lies behind the horizon of the standard mathematics that is covered in the first year of most graduate economics programs. Good portions of Chapters G through K, for instance, are relatively advanced and hence may be deemed unsuitable for the courses mentioned above. Yet I have tried to make these chapters accessible to the student who needs to learn the related material but finds it difficult to follow the standard texts on linear and nonlinear functional analysis. It may eventually be necessary to study matters from more advanced treatments, but the coverage in this book may perhaps ease the pain by building a bridge between advanced texts and a standard math-for-economists course.

On Related Textbooks

A few words on how this text fares in comparison with other related textbooks are in order. It will become abundantly clear early on that my treatment is a good deal more advanced than that in the excellent introductory book by Simon and Blume (1994) and the slightly more advanced text by de la Fuente (1999). Although the topics of Simon and Blume (1994) are prerequisites for the present course, de la Fuente (1999) dovetails with my treatment. On the other hand, my treatment for the most part is equally as advanced as the popular treatise by Stokey and Lucas (1989), which is sometimes taught as a second math course for economists. Most of what is assumed to be known in the latter reference is covered here. So, after finishing the present course, the student who wishes to take an introductory class on the theory of dynamic programming and discrete stochastic systems would be able to read this book at a considerably accelerated pace. Similarly, after the present course, advanced texts such as Mas-Colell (1989), Duffie (1996), and Becker and Boyd (1997) should be within reach.

Within the mathematics folklore, this book would be viewed as a continuation of a first mathematical analysis course, which is usually taught after or along with advanced calculus. In that sense, it is more advanced than the expositions of Rudin (1976), Ross (1980), and Körner (2004), and is roughly at the same level as Kolmogorov and Fomin (1970), Haaser and Sullivan (1991), and Carothers (2000). The widely popular Royden (1994) and Folland (1999) overlap in coverage quite a bit with this book as well, but those treatises are a bit more advanced. Finally, a related text that is exceedingly more advanced than this one is Aliprantis and Border (1999). That book covers an amazing plethora of topics from functional analysis and should serve as a useful advanced reference book for any student of economic theory.

Errors

Although I desperately tried to avoid them, a number of errors have surely managed to sneak past me. I can only hope they are not substantial. The errors that I have identified after publication of the book will be posted on my web page, http://homepages.nyu.edu/~eo1/books.html. Please do not hesitate to inform me by email of the ones you find—my email address is efe.ok@nyu.edu.

Acknowledgments

Many economists and mathematicians have contributed significantly to this book. My good students Sophie Bade, Boğaçhan Çelen, Juan Dubra, Andrei Gomberg, Yusufcan Masatlioglu, Francesc Ortega, Onur Özgür, Liz Potamites, Gil Riella, Maher Said, and Hilary Sarneski-Hayes carefully read substantial parts of the manuscript and identified several errors. All the figures in the text are drawn kindly, and with painstaking care, by Boğaçhan Çelen—I don't think I could have completed this book without his phenomenally generous help.

In addition, numerous comments and corrections I received from Jose Apesteguia, Jean-Pierre Benoît, Alberto Bisin, Kim Border, David Dillenberger, Victor Klee, Peter Lax, Claude Lemaréchal, Jing Li, Massimo Marinacci, Tapan Mitra, Louise Nirenberg, Debraj Ray, Ennio Stacchetti, and Srinivasa Varadhan shaped the structure of the text considerably. I had long discussions about the final product especially with Jean-Pierre Benoît and Ennio Stacchetti. I also have to note that my mathematical upbringing, and hence the making of this book, owe much to the many discussions I had with Tapan Mitra at Cornell by his famous little blackboard. Finally, I am grateful to Tim Sullivan and Jill Harris of Princeton University Press for expertly guiding me through the production process and to Marjorie Pannell for correcting my ubiquitous abuse of English grammar.

At the end of the day, however, my greatest debt is to my students, who have unduly suffered the preliminary stages of this text. I can only hope that I was able to teach them as much as they taught me.

Efe A. Ok
New York, 2006

Prerequisites

This text is intended primarily for an audience that has taken at least one mathematics-for-economists type of course at the graduate level or an advanced calculus course with proofs. Consequently, it is assumed that the reader is familiar with the basic methods of calculus, linear algebra, and nonlinear (static) optimization that would be covered in such a course. For completeness purposes, a relatively comprehensive review of the basic theory of real sequences, functions, and ordinary calculus is provided in Chapter A. In fact, many facts concerning real functions are re-proved later in the book in a more general context. Nevertheless, having a good understanding of real-to-real functions often helps in developing an intuition about things in more abstract settings. Finally, while most students come across metric spaces by the end of the first semester of their graduate education in economics, I do not assume any prior knowledge of this topic here.

To judge things for yourself, check if you have some feeling for the following facts:

- Every monotonic sequence of real numbers in a closed and bounded interval converges in that interval.
- Every concave function defined on an open interval is continuous and quasiconcave.
- Every differentiable function on $\mathbb{R}$ is continuous, but not conversely.
- Every continuous real function defined on a closed and bounded interval attains its maximum.
- A set of vectors that spans $\mathbb{R}^n$ has at least n vectors.
- A linear function defined on $\mathbb{R}^n$ is continuous.

- The (Riemann) integral of every continuous function defined on a closed and bounded interval equals a finite number.
- The Fundamental Theorem of Calculus
- The Mean Value Theorem

If you have certainly seen these results before, and if you can sketch a quick (informal) argument regarding the validity of about half of them, you are well prepared to read this book. (All of these results, or substantial generalizations of them, are proved in the text.)

The economic applications covered here are foundational for the large part, so they do not require any sophisticated economic training. However, you will probably appreciate the importance of these applications better if you have taken at least one graduate course in microeconomic theory.

Basic Conventions

- The frequently used phrase "if and only if" is often abbreviated in the text as "iff."

- Roughly speaking, I label a major result as a *theorem*, a result less significant than a theorem, but still of interest, as a *proposition*, a more or less immediate consequence of a theorem or proposition as a *corollary*, a result whose main utility derives from its aid in deriving a theorem or proposition as a *lemma*, and finally, certain auxiliary results as *claims*, *facts*, or *observations*.

- Throughout this text, n stands for an arbitrary positive integer. This symbol will correspond almost exclusively to the (algebraic) dimension of a Euclidean space, hence the notation $\mathbb{R}^n$. If $x \in \mathbb{R}^n$, then it is understood that x_i is the real number that corresponds to the ith coordinate of x, that is, $x = (x_1, \ldots, x_n)$.

- I use the notation $\subset$ in the strict sense. That is, implicit in the statement $A \subset B$ is that $A \neq B$. The "subsethood" relation in the weak sense is denoted by $\subseteq$.

- Throughout this text, the symbol $\square$ symbolizes the ending of a particular discussion, be it an example, an observation, or a remark. The symbol $\|$ ends a claim within a proof of a theorem, proposition, and so on, while $\blacksquare$ ends the proof itself.

- For any symbols $\clubsuit$ and $\heartsuit$, the expressions $\clubsuit := \heartsuit$ and $\heartsuit =: \clubsuit$ mean that $\clubsuit$ is defined by $\heartsuit$. (This is the so-called Pascal notation.)

- Although the chapters are labeled by Latin letters (A, B, etc.), the sections and subsections are all identified conventionally by positive integers. Consider the following sentence:

  By Proposition 4, the conclusion of Corollary B.1 would be valid here, so by using the observation noted in Example D.3.[2], we find that the solution to the problem mentioned at the end of Section J.7 exists.

Here Proposition 4 refers to the proposition numbered 4 in the chapter that this sentence is taken from. Corollary B.1 is the first corollary in Chapter B, Example D.3.[2] refers to part 2 of Example 3 in Chapter D, and Section J.7 is the seventh section of Chapter J. (*Theorem.* The chapter from which this sentence is taken cannot be any one of the Chapters B, D, and J.)

- The rest of the notation and conventions that I adopt throughout the text are explained in Chapter A.

Part I
SET THEORY

CHAPTER A

Preliminaries of Real Analysis

A principal objective of this largely rudimentary chapter is to introduce the basic set-theoretical nomenclature that we adopt throughout the text. We start with an intuitive discussion of the notion of set, and then introduce the basic operations on sets, Cartesian products, and binary relations. After a quick excursion to order theory (in which the only relatively advanced topic that we cover is the completion of a partial order), functions are introduced as special cases of binary relations and sequences as special cases of functions. Our coverage of abstract set theory concludes with a brief discussion of the Axiom of Choice and the proof of Szpilrajn's Theorem on the completion of a partial order.

We assume here that the reader is familiar with the elementary properties of the real numbers and thus provide only a heuristic discussion of the basic number systems. No construction for the integers is given, in particular. After a short elaboration on ordered fields and the Completeness Axiom, we note without proof that the rational numbers form an ordered field and the real numbers form a complete ordered field. The related discussion is intended to be read more quickly than anywhere else in the text.

We next turn to real sequences. These we discuss relatively thoroughly because of the important role they play in real analysis. In particular, even though our coverage will serve only as a review for most readers, we study here the monotonic sequences and subsequential limits with some care, and prove a few useful results, such as the Bolzano-Weierstrass Theorem and Dirichlet's Rearrangement Theorem. These results will be used freely in the remainder of the book.

The final section of the chapter is nothing more than a swift refresher on the analysis of real functions. First we recall some basic definitions, and then, very quickly, we go over the concepts of limits and continuity of real functions defined on the real line. We then review the elementary theory of differentiation for single-variable functions, mostly through exercises. The primer we present on Riemann integration is a bit more leisurely.

In particular, we give a complete proof of the Fundamental Theorem of Calculus, which is used in the remainder of the book freely. We invoke our calculus review also to outline a basic analysis of exponential and logarithmic real functions. These maps are used in many examples throughout the book. The chapter concludes with a brief discussion of the theory of concave functions on the real line.

1 Elements of Set Theory

1.1 Sets

Intuitively speaking, a "set" is a collection of objects.[1] The distinguishing feature of a set is that whereas it may contain numerous objects, it is nevertheless conceived as a single entity. In the words of Georg Cantor, the great founder of abstract set theory, "a set is a Many which allows itself to be thought of as a One." It is amazing how much follows from this simple idea.

The objects that a set S contains are called the "elements" (or "members") of S. Clearly, to know S, it is necessary and sufficient to know all elements of S. The principal concept of set theory, then, is the relation of "being an element/member of." The universally accepted symbol for this relation is $\in$; that is, $x \in S$ (or $S \ni x$) means that "x is an element of S" (also read "x is a member of S," or "x is contained in S," or "x belongs to S," or "x is in S," or "S includes x," etc.). We often write $x, y \in S$ to denote that both $x \in S$ and $y \in S$ hold. For any natural number m, a statement like $x_1, \ldots, x_m \in S$ (or equivalently, $x_i \in S$, $i = 1, \ldots, m$) is understood analogously. If $x \in S$ is a false statement, then we write $x \notin S$, and read "x is not an element of S."

If the sets A and B have exactly the same elements, that is, $x \in A$ iff $x \in B$, then we say that A and B are identical, and write $A = B$; otherwise we write $A \neq B$.[2] (So, for instance, $\{x, y\} = \{y, x\}$, $\{x, x\} = \{x\}$, and $\{\{x\}\} \neq \{x\}$.) If every member of A is also a member of B, then we say that A is a **subset** of B (also read "A is a set in B" or "A is contained in B") and write $A \subseteq B$ (or $B \supseteq A$). Clearly, $A = B$ holds iff both $A \subseteq B$ and $B \subseteq A$ hold. If $A \subseteq B$

[1] The notion of "object" is left undefined, that is, it can be given any meaning. All I demand of our "objects" is that they be logically *distinguishable*. That is, if x and y are two objects, $x = y$ and $x \neq y$ cannot hold simultaneously, and the statement "either $x = y$ or $x \neq y$" is a tautology.

[2] *Reminder.* iff = if and only if.

but $A \neq B$, then A is said to be a **proper subset** of B, and we denote this situation by writing $A \subset B$ (or $B \supset A$).

For any set S that contains finitely many elements (in which case we say S is **finite**), we denote by $|S|$ the total number of elements that S contains, and refer to this number as the **cardinality** of S. We say that S is a **singleton** if $|S| = 1$. If S contains infinitely many elements (in which case we say S is **infinite**), then we write $|S| = \infty$. Obviously, we have $|A| \leq |B|$ whenever $A \subseteq B$, and if $A \subset B$ and $|A| < \infty$, then $|A| < |B|$.

We sometimes specify a set by enumerating its elements. For instance, $\{x, y, z\}$ is the set that consists of the objects x, y, and z. The contents of the sets $\{x_1, \ldots, x_m\}$ and $\{x_1, x_2, \ldots\}$ are similarly described. For example, the set $\mathbb{N}$ of positive integers can be written as $\{1, 2, \ldots\}$. Alternatively, one may describe a set S as a collection of all objects x that satisfy a given property P. If $P(x)$ stands for the (logical) statement "x satisfies the property P," then we can write $S = \{x : P(x) \text{ is a true statement}\}$, or simply $S = \{x : P(x)\}$. If A is a set and B is the set that contains all elements x of A such that $P(x)$ is true, we write $B = \{x \in A : P(x)\}$. For instance, where $\mathbb{R}$ is the set of all real numbers, the collection of all real numbers greater than or equal to 3 can be written as $\{x \in \mathbb{R} : x \geq 3\}$.

The symbol $\emptyset$ denotes the **empty set**, that is, the set that contains no elements (i.e., $|\emptyset| = 0$). Formally speaking, we can define $\emptyset$ as the set $\{x : x \neq x\}$, for this description entails that $x \in \emptyset$ is a false statement for any object x. Consequently, we write

$$\emptyset := \{x : x \neq x\},$$

meaning that the symbol on the left-hand side is *defined* by that on the right-hand side.[3] Clearly, we have $\emptyset \subseteq S$ for any set S, which in particular implies that $\emptyset$ is unique. (Why?) If $S \neq \emptyset$, we say that S is **nonempty**. For instance, $\{\emptyset\}$ is a nonempty set. Indeed, $\{\emptyset\} \neq \emptyset$—the former, after all, is a set of sets that contains the empty set, while $\emptyset$ contains nothing. (An empty box is not the same thing as nothing!)

We define the class of all subsets of a given set S as

$$2^S := \{T : T \subseteq S\},$$

[3] Recall my notational convention: For any symbols ♣ and ♡, either one of the expressions ♣ := ♡ and ♡ =: ♣ means that ♣ is defined by ♡.

which is called the **power set** of S. (The choice of notation is motivated by the fact that the power set of a set that contains m elements has exactly 2^m elements.) For instance, $2^{\emptyset} = \{\emptyset\}$, $2^{2^{\emptyset}} = \{\emptyset, \{\emptyset\}\}$, and $2^{2^{2^{\emptyset}}} = \{\emptyset, \{\emptyset\}, \{\{\emptyset\}\}, \{\emptyset, \{\emptyset\}\}\}$, and so on.

NOTATION. Throughout this text, the class of all nonempty *finite* subsets of any given set S is denoted by $\mathcal{P}(S)$, that is,

$$\mathcal{P}(S) := \{T : T \subseteq S \text{ and } 0 < |T| < \infty\}.$$

Of course, if S is finite, then $\mathcal{P}(S) = 2^S \backslash \{\emptyset\}$.

Given any two sets A and B, by $A \cup B$ we mean the set $\{x : x \in A \text{ or } x \in B\}$, which is called the **union** of A and B. The **intersection** of A and B, denoted as $A \cap B$, is defined as the set $\{x : x \in A \text{ and } x \in B\}$. If $A \cap B = \emptyset$, we say that A and B are **disjoint**. Obviously, if $A \subseteq B$, then $A \cup B = B$ and $A \cap B = A$. In particular, $\emptyset \cup S = S$ and $\emptyset \cap S = \emptyset$ for any set S.

Taking unions and intersections are *commutative* operations in the sense that

$$A \cap B = B \cap A \quad \text{and} \quad A \cup B = B \cup A$$

for any sets A and B. They are also *associative*, that is,

$$A \cap (B \cap C) = (A \cap B) \cap C \quad \text{and} \quad A \cup (B \cup C) = (A \cup B) \cup C,$$

and *distributive*, that is,

$$A \cap (B \cup C) = (A \cap B) \cup (A \cap C) \quad \text{and} \quad A \cup (B \cap C) = (A \cup B) \cap (A \cup C),$$

for any sets A, B, and C.

EXERCISE 1 Prove the commutative, associative, and distributive laws of set theory stated above.

EXERCISE 2 Given any two sets A and B, by $A \backslash B$—the **difference** between A and B—we mean the set $\{x : x \in A \text{ and } x \notin B\}$.

(a) Show that $S \backslash \emptyset = S$, $S \backslash S = \emptyset$, and $\emptyset \backslash S = \emptyset$ for any set S.

(b) Show that $A \backslash B = B \backslash A$ iff $A = B$ for any sets A and B.

(c) (*De Morgan Laws*) Prove: For any sets A, B, and C,

$$A \backslash (B \cup C) = (A \backslash B) \cap (A \backslash C) \quad \text{and} \quad A \backslash (B \cap C) = (A \backslash B) \cup (A \backslash C).$$

Throughout this text we use the terms "class" or "family" only to refer to a nonempty collection of sets. So if $\mathcal{A}$ is a class, we understand that $\mathcal{A} \neq \emptyset$ and that any member $A \in \mathcal{A}$ is a set (which may itself be a collection of sets). The union of all members of this class, denoted as $\cup\mathcal{A}$, or $\cup\{A : A \in \mathcal{A}\}$, or $\cup_{A\in\mathcal{A}}A$, is defined as the set $\{x : x \in A$ for some $A \in \mathcal{A}\}$. Similarly, the intersection of all sets in $\mathcal{A}$, denoted as $\cap\mathcal{A}$, or $\cap\{A : A \in \mathcal{A}\}$, or $\cap_{A\in\mathcal{A}}A$, is defined as the set $\{x : x \in A$ for each $A \in \mathcal{A}\}$.

A common way of specifying a class $\mathcal{A}$ of sets is by designating a set I as a set of indices and by defining $\mathcal{A} := \{A_i : i \in I\}$. In this case, $\cup\mathcal{A}$ may be denoted as $\cup_{i\in I}A_i$. If $I = \{k, k+1, \ldots, K\}$ for some integers k and K with $k < K$, then we often write $\cup_{i=k}^{K}A_i$ (or $A_k \cup \cdots \cup A_K$) for $\cup_{i\in I}A_i$. Similarly, if $I = \{k, k+1, \ldots\}$ for some integer k, then we may write $\cup_{i=k}^{\infty}A_i$ (or $A_k \cup A_{k+1} \cup \cdots$) for $\cup_{i\in I}A_i$. Furthermore, for brevity, we frequently denote $\cup_{i=1}^{K}A_i$ as $\cup^K A_i$, and $\cup_{i=1}^{\infty}A_i$ as $\cup^\infty A_i$, throughout the text. Similar notational conventions apply to intersections of sets as well.

Warning. The symbols $\cup\emptyset$ and $\cap\emptyset$ are left *undefined* (in much the same way that the symbol $\frac{0}{0}$ is undefined in number theory).

Exercise 3 Let A be a set and $\mathcal{B}$ a class of sets. Prove that

$$A \cap \cup\mathcal{B} = \bigcup\{A \cap B : B \in \mathcal{B}\} \quad \text{and} \quad A \cup \cap\mathcal{B} = \bigcap\{A \cup B : B \in \mathcal{B}\},$$

while

$$A \backslash \cup\mathcal{B} = \bigcap\{A \backslash B : B \in \mathcal{B}\} \quad \text{and} \quad A \backslash \cap\mathcal{B} = \bigcup\{A \backslash B : B \in \mathcal{B}\}.$$

A word of caution may be in order before we proceed further. While duly intuitive, the "set theory" we have outlined so far provides us with no demarcation criterion for identifying what exactly constitutes a set. This may suggest that one is completely free in deeming any given collection of objects a set. But in fact, this would be a pretty bad idea that would entail serious foundational difficulties. The best known example of such difficulties was given by Bertrand Russell in 1902 when he asked if the set of all objects that are not members of *themselves* is a set: Is $S := \{x : x \notin x\}$ a set?[4] There is

[4] While a bit unorthodox, $x \in x$ may well be a statement that is true for some objects. For instance, the collection of all sets that I have mentioned in my life, say x, is a set that I have just mentioned, so $x \in x$. But the collection of all cheesecakes I have eaten in my life, say y, is not a cheesecake, so $y \notin y$.

nothing in our intuitive discussion above that forces us to conclude that S is not a set; it is a collection of objects (sets in this case) that is considered as a single entity. But we cannot accept S as a set, for if we do, we have to be able to answer the question, Is $S \in S$? If the answer is yes, then $S \in S$, but this implies $S \notin S$ by definition of S. If the answer is no, then $S \notin S$, but this implies $S \in S$ by definition of S. That is, we have a contradictory state of affairs no matter what! This is the so-called *Russell's paradox*, which started a severe foundational crisis for mathematics that eventually led to a complete axiomatization of set theory in the early twentieth century.[5]

Roughly speaking, this paradox would arise only if we allowed "unduly large" collections to be qualified as sets. In particular, it will not cause any harm for the mathematical analysis that will concern us here, precisely because in all of our discussions, we will *fix* a universal set of objects, say X, and consider sets like $\{x \in X : P(x)\}$, where $P(x)$ is an unambiguous logical statement in terms of x. (We will also have occasion to work with sets of such sets, and sets of sets of such sets, and so on.) Once such a domain X is fixed, Russell's paradox cannot arise. Why, you may ask, can't we have the same problem with the set $S := \{x \in X : x \notin x\}$? No, because now we can answer the question: Is $S \in S$? The answer is no! The statement $S \in S$ is false, simply because $S \notin X$. (For, if $S \in X$ was the case, then we would end up with the contradiction $S \in S$ iff $S \notin S$.)

So when the context is clear (that is, when a universe of objects is fixed), and when we define our sets as just explained, Russell's paradox will not be a threat against the resulting set theory. But can there be any other paradoxes? Well, there is really not an easy answer to this. To even discuss the matter unambiguously, we must leave our intuitive understanding of the notion of set and address the problem through a completely axiomatic approach (in which we would leave the expression "$x \in S$" undefined and give meaning to it *only* through axioms). This is, of course, not at all the place to do this. Moreover, the "intuitive" set theory that we covered here is more than enough for the mathematical analysis to come. We thus leave this topic by

[5] Russell's paradox is a classic example of the dangers of using *self-referential* statements carelessly. Another example of this form is the ancient *paradox of the liar*: "Everything I say is false." This statement can be declared neither true nor false! To get a sense of some other kinds of paradoxes and the way axiomatic set theory avoids them, you might want to read the popular account of Rucker (1995).

referring the reader who wishes to get a broader introduction to abstract set theory to Chapter 1 of Schechter (1997) or Marek and Mycielski (2001); both of these expositions provide nice introductory overviews of axiomatic set theory. If you want to dig deeper, then try the first three chapters of Enderton (1977).

1.2 Relations

An **ordered pair** is an ordered list (a, b) consisting of two objects a and b. This list is *ordered* in the sense that, as a defining feature of the notion of ordered pair, we assume the following: For any two ordered pairs (a, b) and (a', b'), we have $(a, b) = (a', b')$ iff $a = a'$ and $b = b'$.[6]

The **(Cartesian) product** of two nonempty sets A and B, denoted as $A \times B$, is defined as the set of all ordered pairs (a, b) where a comes from A and b comes from B. That is,

$$A \times B := \{(a, b) : a \in A \text{ and } b \in B\}.$$

As a notational convention, we often write A^2 for $A \times A$. It is easily seen that taking the Cartesian product of two sets is not a commutative operation. Indeed, for any two *distinct* objects a and b, we have $\{a\} \times \{b\} = \{(a, b)\} \neq \{(b, a)\} = \{b\} \times \{a\}$. Formally speaking, it is not associative either, for $(a, (b, c))$ is not the same thing as $((a, b), c)$. Yet there is a natural correspondence between the elements of $A \times (B \times C)$ and $(A \times B) \times C$, so one can really think of these two sets as the same, thereby rendering the status of the set $A \times B \times C$ unambiguous.[7] This prompts us to define an ***n*-vector**

[6] This defines the notion of ordered pair as a new "primitive" for our set theory, but in fact, this is not really necessary. One can define an ordered pair by using only the concept of "set" as $(a, b) := \{\{a\}, \{a, b\}\}$. With this definition, which is due to Kazimierz Kuratowski, one can *prove* that, for any two ordered pairs (a, b) and (a', b'), we have $(a, b) = (a', b')$ iff $a = a'$ and $b = b'$. The "if" part of the claim is trivial. To prove the "only if" part, observe that $(a, b) = (a', b')$ entails that either $\{a\} = \{a'\}$ or $\{a\} = \{a', b'\}$. But the latter equality may hold only if $a = a' = b'$, so we have $a = a'$ in all contingencies. Therefore, $(a, b) = (a', b')$ entails that either $\{a, b\} = \{a\}$ or $\{a, b\} = \{a, b'\}$. The latter case is possible only if $b = b'$, while the former possibility arises only if $a = b$. But if $a = b$, then we have $\{\{a\}\} = (a, b) = (a, b') = \{\{a\}, \{a, b'\}\}$, which holds only if $\{a\} = \{a, b'\}$, that is, $b = a = b'$.

Quiz. (Wiener) Show that we would also have $(a, b) = (a', b')$ iff $a = a'$ and $b = b'$, if we instead defined (a, b) as $\{\{\emptyset, \{a\}\}, \{\{b\}\}\}$.

[7] What is this "natural" correspondence?

(for any natural number n) as a list $(a_1, \ldots, a_n)$, with the understanding that $(a_1, \ldots, a_n) = (a'_1, \ldots, a'_n)$ iff $a_i = a'_i$ for each $i = 1, \ldots, n$. The **(Cartesian) product** of n sets $A_1, \ldots, A_n$, is then defined as

$$A_1 \times \ldots \times A_n := \{(a_1, \ldots, a_n) : a_i \in A_i,\ i = 1, \ldots, n\}.$$

We often write $\mathsf{X}^n A_i$ to denote $A_1 \times \cdots \times A_n$, and refer to $\mathsf{X}^n A_i$ as the ***n*-fold product** of $A_1, \ldots, A_n$. If $A_i = S$ for each i, we then write S^n for $A_1 \times \ldots \times A_n$, that is, $S^n := \mathsf{X}^n S$.

EXERCISE 4 For any sets A, B, and C, prove that

$$A \times (B \cap C) = (A \times B) \cap (A \times C) \quad \text{and} \quad A \times (B \cup C) = (A \times B) \cup (A \times C).$$

Let X and Y be two nonempty sets. A subset R of $X \times Y$ is called a *(binary)* **relation from X to Y**. If $X = Y$, that is, if R is a relation from X to X, we simply say that it is a **relation on X**. Put differently, R is a relation on X iff $R \subseteq X^2$. If $(x, y) \in R$, then we think of R as associating the object x with y, and if $\{(x, y), (y, x)\} \cap R = \emptyset$, we understand that there is no connection between x and y as envisaged by R. In concert with this interpretation, we adopt the convention of writing xRy instead of $(x, y) \in R$ throughout this text.

DEFINITION

A relation R on a nonempty set X is said to be **reflexive** if xRx for each $x \in X$, and **complete** if either xRy or yRx holds for each $x, y \in X$. It is said to be **symmetric** if, for any $x, y \in X$, xRy implies yRx, and **antisymmetric** if, for any $x, y \in X$, xRy and yRx imply $x = y$. Finally, we say that R is **transitive** if xRy and yRz imply xRz for any $x, y, z \in X$.

The interpretations of these properties are straightforward, so we do not elaborate on them here. But note: While every complete relation is reflexive, there are no other logical implications between these properties.

EXERCISE 5 Let X be a nonempty set, and R a relation on X. The **inverse** of R is defined as the relation $R^{-1} := \{(x, y) \in X^2 : yRx\}$.

(a) If R is symmetric, does R^{-1} have to be also symmetric? Antisymmetric? Transitive?
(b) Show that R is symmetric iff $R = R^{-1}$.
(c) If R_1 and R_2 are two relations on X, the **composition** of R_1 and R_2 is the relation $R_2 \circ R_1 := \{(x, y) \in X^2 : xR_1z \text{ and } zR_2y \text{ for some } z \in X\}$. Show that R is transitive iff $R \circ R \subseteq R$.

EXERCISE 6 A relation R on a nonempty set X is called **circular** if xRz and zRy imply yRx for every $x, y, z \in X$. Prove that R is reflexive and circular iff it is reflexive, symmetric, and transitive.

EXERCISE 7[H] Let R be a reflexive relation on a nonempty set X. The **asymmetric part** of R is defined as the relation P_R on X as xP_Ry iff xRy but not yRx. The relation $I_R := R \backslash P_R$ on X is then called the **symmetric part** of R.
(a) Show that I_R is reflexive and symmetric.
(b) Show that P_R is neither reflexive nor symmetric.
(c) Show that if R is transitive, so are P_R and I_R.

EXERCISE 8 Let R be a relation on a nonempty set X. Let $R_0 = R$, and for each positive integer m, define the relation R_m on X by xR_my iff there exist $z_1, \ldots, z_m \in X$ such that xRz_1, $z_1Rz_2, \ldots,$ and z_mRy. The relation $tr(R) := R_0 \cup R_1 \cup \cdots$ is called the **transitive closure** of R. Show that $tr(R)$ is transitive, and if R' is a transitive relation with $R \subseteq R'$, then $tr(R) \subseteq R'$.

1.3 Equivalence Relations

In mathematical analysis, one often needs to "identify" two distinct objects when they possess a particular property of interest. Naturally, such an identification scheme should satisfy certain consistency conditions. For instance, if x is identified with y, then y must be identified with x. Similarly, if x and y are deemed identical, and so are y and z, then x and z should be identified. Such considerations lead us to the notion of *equivalence relation.*

Definition

A relation $\sim$ on a nonempty set X is called an **equivalence relation** if it is reflexive, symmetric, and transitive. For any $x \in X$, the **equivalence class** of x relative to $\sim$ is defined as the set

$$[x]_\sim := \{y \in X : y \sim x\}.$$

The class of all equivalence classes relative to $\sim$, denoted as $X/_\sim$, is called the **quotient set** of X relative to $\sim$, that is,

$$X/_\sim := \{[x]_\sim : x \in X\}.$$

Let X denote the set of all people in the world. "Being a sibling of" is an equivalence relation on X (provided that we adopt the convention of saying that any person is a sibling of himself). The equivalence class of a person relative to this relation is the set of all of his or her siblings. On the other hand, you would probably agree that "being in love with" is not an equivalence relation on X. Here are some more examples (that fit better with the "serious" tone of this course).

Example 1

[1] For any nonempty set X, the **diagonal relation** $D_X := \{(x, x) : x \in X\}$ is the smallest equivalence relation that can be defined on X (in the sense that if R is any other equivalence relation on X, we have $D_X \subseteq R$). Clearly, $[x]_{D_X} = \{x\}$ for each $x \in X$.[8] At the other extreme is X^2 which is the largest equivalence relation that can be defined on X. We have $[x]_{X^2} = X$ for each $x \in X$.

[2] By Exercise 7, the symmetric part of any reflexive and transitive relation on a nonempty set is an equivalence relation.

[3] Let $X := \{(a, b) : a, b \in \{1, 2, \ldots\}\}$, and define the relation $\sim$ on X by $(a, b) \sim (c, d)$ iff $ad = bc$. It is readily verified that $\sim$ is an equivalence relation on X, and that $[(a, b)]_\sim = \left\{(c, d) \in X : \frac{c}{d} = \frac{a}{b}\right\}$ for each $(a, b) \in X$.

[4] Let $X := \{\ldots, -1, 0, 1, \ldots\}$, and define the relation $\sim$ on X by $x \sim y$ iff $\frac{1}{2}(x - y) \in X$. It is easily checked that $\sim$ is an equivalence relation

[8] I say an equally suiting name for D_X is the "equality relation." What do you think?

on X. Moreover, for any integer x, we have $x \sim y$ iff $y = x - 2m$ for some $m \in X$, and hence $[x]_\sim$ equals the set of all even integers if x is even, and that of all odd integers if x is odd. □

One typically uses an equivalence relation to simplify a situation in a way that all things that are indistinguishable from a particular perspective are put together in a set and treated as if they were a single entity. For instance, suppose that for some reason we are interested in the signs of people. Then, any two individuals who are of the same sign can be thought of as "identical," so instead of the set of all people in the world, we would rather work with the set of all Capricorns, all Virgos, and so on. But the set of all Capricorns is of course none other than the equivalence class of any given Capricorn person relative to the equivalence relation of "being of the same sign." So when someone says "a Capricorn is. . . ," then one is really referring to a whole class of people. The equivalence relation of "being of the same sign" divides the world into twelve equivalence classes, and we can then talk "as if" there were only twelve individuals in our context of reference.

To take another example, ask yourself how you would define the set of positive rational numbers, given the set of natural numbers $\mathbb{N} := \{1, 2, \ldots\}$ and the operation of "multiplication." Well, you may say, a positive rational number is the ratio of two natural numbers. But wait, what is a "ratio"? Let us be a bit more careful about this. A better way of looking at things is to say that a positive rational number is an ordered pair $(a, b) \in \mathbb{N}^2$, although in daily practice, we write $\frac{a}{b}$ instead of (a, b). Yet we don't want to say that each ordered pair in $\mathbb{N}^2$ is a distinct rational number. (We would like to think of $\frac{1}{2}$ and $\frac{2}{4}$ as the same number, for instance.) So we "identify" all those ordered pairs that we wish to associate with a single rational number by using the equivalence relation $\sim$ introduced in Example 1.[3], and then define a rational number simply as an equivalence class $[(a, b)]_\sim$. Of course, when we talk about rational numbers in daily practice, we simply talk of a fraction like $\frac{1}{2}$, not $[(1, 2)]_\sim$, even though, formally speaking, what we really mean is $[(1, 2)]_\sim$. The equality $\frac{1}{2} = \frac{2}{4}$ is obvious, precisely because the rational numbers are constructed as equivalence classes such that $(2, 4) \in [(1, 2)]_\sim$.

This discussion suggests that an equivalence relation can be used to decompose a grand set of interest into subsets such that the members of the same subset are thought of as "identical," while the members of distinct subsets are viewed as "distinct." Let us now formalize this intuition. By a **partition** of a nonempty set X, we mean a class of pairwise disjoint, nonempty subsets of X whose union is X. That is, $\mathcal{A}$ is a partition of X iff $\mathcal{A} \subseteq 2^X \backslash \{\emptyset\}$, $\cup\mathcal{A} = X$ and $A \cap B = \emptyset$ for every distinct A and B in $\mathcal{A}$. The next result says that the set of equivalence classes induced by any equivalence relation on a set is a partition of that set.

PROPOSITION 1

For any equivalence relation $\sim$ on a nonempty set X, the quotient set $X/_\sim$ is a partition of X.

PROOF

Take any nonempty set X and an equivalence relation $\sim$ on X. Since $\sim$ is reflexive, we have $x \in [x]_\sim$ for each $x \in X$. Thus any member of $X/_\sim$ is nonempty, and $\cup\{[x]_\sim : x \in X\} = X$. Now suppose that $[x]_\sim \cap [y]_\sim \neq \emptyset$ for some $x, y \in X$. We wish to show that $[x]_\sim = [y]_\sim$. Observe first that $[x]_\sim \cap [y]_\sim \neq \emptyset$ implies $x \sim y$. (Indeed, if $z \in [x]_\sim \cap [y]_\sim$, then $x \sim z$ and $z \sim y$ by symmetry of $\sim$, so we get $x \sim y$ by transitivity of $\sim$.) This implies that $[x]_\sim \subseteq [y]_\sim$, because if $w \in [x]_\sim$, then $w \sim x$ (by symmetry of $\sim$), and hence $w \sim y$ by transitivity of $\sim$. The converse containment is proved analogously. ■

The following exercise shows that the converse of Proposition 1 also holds. Thus the notions of equivalence relation and partition are really two different ways of looking at the same thing.

EXERCISE 9 Let $\mathcal{A}$ be a partition of a nonempty set X, and consider the relation $\sim$ on X defined by $x \sim y$ iff $\{x, y\} \subseteq A$ for some $A \in \mathcal{A}$. Prove that $\sim$ is an equivalence relation on X.

1.4 Order Relations

Transitivity property is the defining feature of any *order* relation. Such relations are given various names depending on the properties they possess in addition to transitivity.

Definition

A relation $\succsim$ on a nonempty set X is called a **preorder** on X if it is transitive and reflexive. It is said to be a **partial order** on X if it is an antisymmetric preorder on X. Finally, $\succsim$ is called a **linear order** on X if it is a partial order on X that is complete.

By a **preordered set** we mean a list $(X, \succsim)$ where X is a nonempty set and $\succsim$ is a preorder on X. If $\succsim$ is a partial order on X, then $(X, \succsim)$ is called a **poset** (short for *partially ordered set*), and if $\succsim$ is a linear order on X, then $(X, \succsim)$ is called either a **chain** or a **loset** (short for *linearly ordered set*).

It is convenient to talk as if a preordered set $(X, \succsim)$ were indeed a set when referring to properties that apply only to X. For instance, by a "finite preordered set," we understand a preordered set $(X, \succsim)$ with $|X| < \infty$. Or, when we say that Y is a subset of the preordered set $(X, \succsim)$, we mean simply that $Y \subseteq X$. A similar convention applies to posets and losets as well.

Notation. Let $(X, \succsim)$ be a preordered set. Unless otherwise is stated explicitly, we denote by $\succ$ the asymmetric part of $\succsim$ and by $\sim$ the symmetric part of $\succsim$ (Exercise 7).

The main distinction between a preorder and a partial order is that the former may have a large symmetric part, while the symmetric part of the latter must equal the diagonal relation. As we shall see, however, in most applications this distinction is immaterial.

Example 2

[1] For any nonempty set X, the diagonal relation $D_X := \{(x, x) : x \in X\}$ is a partial order on X. In fact, this relation is the only partial order on X that is also an equivalence relation. (Why?) The relation X^2, on the other hand, is a complete preorder, which is not antisymmetric unless X is a singleton.

[2] For any nonempty set X, the equality relation $=$ and the subsethood relation $\supseteq$ are partial orders on 2^X. The equality relation is not linear, and $\supseteq$ is not linear unless X is a singleton.

[3] $(\mathbb{R}^n, \geq)$ is a poset for any positive integer n, where $\geq$ is defined coordinatewise, that is, $(x_1, \ldots, x_n) \geq (y_1, \ldots, y_n)$ iff $x_i \geq y_i$ for each

$i = 1, \ldots, n$. When we talk of $\mathbb{R}^n$ without specifying explicitly an alternative order, we always have in mind this partial order (which is sometimes called the **natural** (or **canonical**) **order** of $\mathbb{R}^n$). Of course, $(\mathbb{R}, \geq)$ is a loset.

[4] Take any positive integer n and preordered sets $(X_i, \succsim_i)$, $i = 1, \ldots, n$. The **product** of the preordered sets $(X_i, \succsim_i)$, denoted as $\boxtimes^n (X_i, \succsim_i)$, is the preordered set $(X, \succsim)$ with $X := \mathsf{X}^n X_i$ and

$$(x_1, \ldots, x_n) \succsim (y_1, \ldots, y_n) \quad \text{iff} \quad x_i \succsim_i y_i \quad \text{for all } i = 1, \ldots, n.$$

In particular, $(\mathbb{R}^n, \geq) = \boxtimes^n (\mathbb{R}, \geq)$. □

Example 3

In individual choice theory, a **preference relation** $\succsim$ on a nonempty alternative set X is defined as a preorder on X. Here the reflexivity is a trivial condition to require, and transitivity is viewed as a fundamental rationality postulate. (We will talk more about this in Section B.4.) The **strict preference relation** $\succ$ is defined as the asymmetric part of $\succsim$ (Exercise 7). This relation is transitive but not reflexive. The **indifference relation** $\sim$ is then defined as the symmetric part of $\succsim$, and is easily checked to be an equivalence relation on X. For any $x \in X$, the equivalence class $[x]_\sim$ is called in this context the **indifference class** of x, and is simply a generalization of the familiar concept of "*the indifference curve that passes through* x." In particular, Proposition 1 says that no two distinct indifference sets can have a point in common. (This is the gist of the fact that "distinct indifference curves cannot cross!")

In social choice theory, one often works with multiple (complete) preference relations on a given alternative set X. For instance, suppose that there are n individuals in the population, and $\succsim_i$ stands for the preference relation of the ith individual. The **Pareto dominance** relation $\succsim$ on X is defined as $x \succsim y$ iff $x \succsim_i y$ for each $i = 1, \ldots, n$. This relation is a preorder on X in general, and a partial order on X if each $\succsim_i$ is antisymmetric. □

Let $(X, \succsim)$ be a preordered set. By an **extension** of $\succsim$ we understand a preorder $\trianglerighteq$ on X such that $\succsim \subseteq \trianglerighteq$ and $\succ \subseteq \vartriangleright$, where $\vartriangleright$ is the asymmetric part of $\trianglerighteq$. Intuitively speaking, an extension of a preorder is "more complete" than the original relation in the sense that it allows one to compare more elements, but it certainly agrees exactly with the original relation when the latter applies. If $\trianglerighteq$ is a partial order, then it is an extension of $\succsim$ iff $\succsim \subseteq \trianglerighteq$. (Why?)

A fundamental result of order theory says that every partial order can be extended to a linear order, that is, for every poset $(X, \succsim)$ there is a loset $(X, \trianglerighteq)$ with $\succsim \subseteq \trianglerighteq$. Although it is possible to prove this by mathematical induction when X is finite, the proof in the general case is built on a relatively advanced method that we will cover later in the course. Relegating its proof to Section 1.7, we only state here the result for future reference.[9]

Szpilrajn's Theorem

Every partial order on a nonempty set X can be extended to a linear order on X.

A natural question is whether the same result holds for preorders as well. The answer is yes, and the proof follows easily from Szpilrajn's Theorem by means of a standard method.

Corollary 1

Let $(X, \succsim)$ be a preordered set. There exists a complete preorder on X that extends $\succsim$.

Proof

Let $\sim$ denote the symmetric part of $\succsim$, which is an equivalence relation. Then $(X/_\sim, \succsim^*)$ is a poset where $\succsim^*$ is defined on $X/_\sim$ by

$$[x]_\sim \succsim^* [y]_\sim \quad \text{if and only if} \quad x \succsim y.$$

By Szpilrajn's Theorem, there exists a linear order $\trianglerighteq^*$ on $X/_\sim$ such that $\succsim^* \subseteq \trianglerighteq^*$. We define $\trianglerighteq$ on X by

$$x \trianglerighteq y \quad \text{if and only if} \quad [x]_\sim \trianglerighteq^* [y]_\sim.$$

It is easily checked that $\trianglerighteq$ is a complete preorder on X with $\succsim \subseteq \trianglerighteq$ and $\succ \subseteq \triangleright$, where $\succ$ and $\triangleright$ are the asymmetric parts of $\succsim$ and $\trianglerighteq$, respectively. ■

[9] For an extensive introduction to the theory of linear extensions of posets, see Bonnet and Pouzet (1982).

EXERCISE 10 Let $(X, \succsim)$ be a preordered set, and define $\mathcal{L}(\succsim)$ as the set of all complete preorders that extend $\succsim$. Prove that $\succsim = \cap\mathcal{L}(\succsim)$. (Where do you use Szpilrajn's Theorem in the argument?)

EXERCISE 11 Let $(X, \succsim)$ be a finite preordered set. Taking $\mathcal{L}(\succsim)$ as in the previous exercise, we define $dim(X, \succsim)$ as the smallest positive integer k such that $\succsim = R_1 \cap \cdots \cap R_k$ for some $R_i \in \mathcal{L}(\succsim)$, $i = 1, \ldots, k$.

(a) Show that $dim(X, \succsim) \leq \left|X^2\right|$.

(b) What is $dim(X, D_X)$? What is $dim(X, X^2)$?

(c) For any positive integer n, show that $dim(\boxtimes^n(X_i, \succsim_i)) = n$, where $(X_i, \succsim_i)$ is a loset with $|X_i| \geq 2$ for each $i = 1, \ldots, n$.

(d) Prove or disprove: $dim(2^X, \supseteq) = |X|$.

DEFINITION

Let $(X, \succsim)$ be a preordered set, and $\emptyset \neq Y \subseteq X$. An element x of Y is said to be $\succsim$**-maximal** in Y if there is no $y \in Y$ with $y \succ x$, and $\succsim$**-minimal** in Y if there is no $y \in Y$ with $x \succ y$. If $x \succsim y$ for all $y \in Y$, then x is called the $\succsim$**-maximum** of Y, and if $y \succsim x$ for all $y \in Y$, then x is called the $\succsim$**-minimum** of Y.

Obviously, for any preordered set $(X, \succsim)$, every $\succsim$-maximum of a nonempty subset of X is $\succsim$-maximal in that set. Also note that if $(X, \succsim)$ is a poset, then there can be at most one $\succsim$-maximum of any $Y \in 2^X \backslash \{\emptyset\}$.

EXAMPLE 4

[1] Let X be any nonempty set, and $\emptyset \neq Y \subseteq X$. Every element of Y is both D_X-maximal and D_X-minimal in Y. Unless it is a singleton, Y has neither a D_X-maximum nor a D_X-minimum element. On the other hand, every element of Y is both X^2-maximum and X^2-minimum of Y.

[2] Given any nonempty set X, consider the poset $(2^X, \supseteq)$, and take any nonempty $\mathcal{A} \subseteq 2^X$. The class $\mathcal{A}$ has a $\supseteq$-maximum iff $\cup\mathcal{A} \in \mathcal{A}$, and it has a $\supseteq$-minimum iff $\cap\mathcal{A} \in \mathcal{A}$. In particular, the $\supseteq$-maximum of 2^X is X and the $\supseteq$-minimum of 2^X is $\emptyset$.

[3] (*Choice Correspondences*) Given a preference relation $\succsim$ on an alternative set X (Example 3) and a nonempty subset S of X, we define the "set of choices from S" for an individual whose preference relation is $\succsim$

as the set of all $\succsim$-maximal elements in S. That is, denoting this set as $C_{\succsim}(S)$, we have

$$C_{\succsim}(S) := \{x \in S : y \succ x \text{ for no } y \in S\}.$$

Evidently, if S is a finite set, then $C_{\succsim}(S)$ is nonempty. (Proof?) Moreover, if S is finite and $\succsim$ is complete, then there exists at least one $\succsim$-maximum element in S. The finiteness requirement cannot be omitted in this statement, but as we shall see throughout this book, there are various ways in which it can be substantially weakened. □

EXERCISE 12

(a) Which subsets of the set of positive integers have a $\geq$-minimum? Which ones have a $\geq$-maximum?

(b) If a set in a poset $(X, \succsim)$ has a unique $\succsim$-maximal element, does that element have to be a $\succsim$-maximum of the set?

(c) Which subsets of a poset $(X, \succsim)$ possess an element that is *both* $\succsim$-maximum *and* $\succsim$-minimum?

(d) Give an example of an infinite set in $\mathbb{R}^2$ that contains a unique $\geq$-maximal element that is also the unique $\geq$-minimal element of the set.

EXERCISE 13[H] Let $\succsim$ be a complete relation on a nonempty set X, and S a nonempty finite subset of X. Define

$$c_{\succsim}(S) := \{x \in S : x \succsim y \text{ for all } y \in S\}.$$

(a) Show that $c_{\succsim}(S) \neq \emptyset$ if $\succsim$ is transitive.

(b) We say that $\succsim$ is **acyclic** if there does not exist a positive integer k such that $x_1, \ldots, x_k \in X$ and $x_1 \succ \cdots \succ x_k \succ x_1$. Show that every transitive relation is acyclic, but not conversely.

(c) Show that $c_{\succsim}(S) \neq \emptyset$ if $\succsim$ is acyclic.

(d) Show that if $c_{\succsim}(T) \neq \emptyset$ for every finite $T \in 2^X \backslash \{\emptyset\}$, then $\succsim$ must be acyclic.

EXERCISE 14[H] Let $(X, \succsim)$ be a poset, and take any $Y \in 2^X \backslash \{\emptyset\}$ that has a $\succsim$-maximal element, say x^*. Prove that $\succsim$ can be extended to a linear order $\trianglerighteq$ on X such that x^* is $\trianglerighteq$-maximal in Y.

EXERCISE 15 Let $(X, \succsim)$ be a poset. For any $Y \subseteq X$, an element x in X is said to be an **$\succsim$-upper bound** for Y if $x \succsim y$ for all $y \in Y$; a **$\succsim$-lower bound** for Y is defined similarly. The **$\succsim$-supremum** of Y, denoted $\sup_{\succsim} Y$, is defined as the $\succsim$-minimum of the set of all $\succsim$-upper bounds for Y, that is, $\sup_{\succsim} Y$ is an $\succsim$-upper bound for Y and has the property that $z \succsim \sup_{\succsim} Y$ for any $\succsim$-upper bound z for Y. The **$\succsim$-infimum** of Y, denoted as $\inf_{\succsim} Y$, is defined analogously.

(a) Prove that there can be only one $\succsim$-supremum and only one $\succsim$-infimum of any subset of X.

(b) Show that $x \succsim y$ iff $\sup_{\succsim}\{x, y\} = x$ and $\inf_{\succsim}\{x, y\} = y$, for any $x, y \in X$.

(c) Show that if $\sup_{\succsim} X \in X$ (that is, if $\sup_{\succsim} X$ exists), then $\inf_{\succsim} \emptyset = \sup_{\succsim} X$.

(d) If $\succsim$ is the diagonal relation on X, and x and y are any two distinct members of X, does $\sup_{\succsim}\{x, y\}$ exist?

(e) If $X := \{x, y, z, w\}$ and $\succsim := \{(z, x), (z, y), (w, x), (w, y)\}$, does $\sup_{\succsim}\{x, y\}$ exist?

EXERCISE 16[H] Let $(X, \succsim)$ be a poset. If $\sup_{\succsim}\{x, y\}$ and $\inf_{\succsim}\{x, y\}$ exist for all $x, y \in X$, then we say that $(X, \succsim)$ is a **lattice.** If $\sup_{\succsim} Y$ and $\inf_{\succsim} Y$ exist for all $Y \in 2^X$, then $(X, \succsim)$ is called a **complete lattice.**

(a) Show that every complete lattice has an upper and a lower bound.

(b) Show that if X is finite and $(X, \succsim)$ is a lattice, then $(X, \succsim)$ is a complete lattice.

(c) Give an example of a lattice which is not complete.

(d) Prove that $(2^X, \supseteq)$ is a complete lattice.

(e) Let $\mathcal{X}$ be a nonempty subset of 2^X such that $X \in \mathcal{X}$ and $\cap\mathcal{A} \in \mathcal{X}$ for any (nonempty) class $\mathcal{A} \subseteq \mathcal{X}$. Prove that $(\mathcal{X}, \supseteq)$ is a complete lattice.

1.5 Functions

Intuitively, we think of a *function* as a rule that transforms the objects in a given set to those of another. Although this is not a formal definition—what is a "rule"?—we may now use the notion of binary relation to formalize the idea. Let X and Y be any two nonempty sets. By a **function f that maps X into** Y, denoted as $f : X \to Y$, we mean a relation $f \in X \times Y$ such that

(i) for every $x \in X$, there exists a $y \in Y$ such that $x f y$;

(ii) for every $y, z \in Y$ with $x f y$ and $x f z$, we have $y = z$.

Here X is called the **domain** of f and Y the **codomain** of f. The **range** of f is, on the other hand, defined as

$$f(X) := \{y \in Y : xf\,y \quad \text{for some } x \in X\}.$$

The set of all functions that map X into Y is denoted by Y^X. For instance, $\{0,1\}^X$ is the set of all functions on X whose values are either 0 or 1, and $\mathbb{R}^{[0,1]}$ is the set of all real-valued functions on $[0,1]$. The notation $f \in Y^X$ will be used interchangeably with the expression $f : X \to Y$ throughout this course. Similarly, the term **map** is used interchangeably with the term "function."

Although our definition of a function may look a bit strange at first, it is hardly anything other than a set-theoretic formulation of the concept we use in daily discourse. After all, we want a *function* f that maps X into Y to assign each member of X to a member of Y, right? Our definition says simply that one can think of f as a set of ordered pairs, so "$(x,y) \in f$" means "x is mapped to y by f." Put differently, all that f "does" is completely identified by the set $\{(x,f(x)) \in X \times Y : x \in X\}$, which is what f "is." The familiar notation $f(x) = y$ (which we shall also adopt in the rest of the exposition) is then nothing but an alternative way of expressing $xf\,y$. When $f(x) = y$, we refer to y as the **image** (or **value**) **of** x **under** f. Condition (i) says that every element in the domain X of f has an image under f in the codomain Y. In turn, condition (ii) states that no element in the domain of f can have more than one image under f.

Some authors adhere to the intuitive definition of a function as a "rule" that transforms one set into another and refer to the set of all ordered pairs $(x,f(x))$ as the **graph** of the function. Denoting this set by $Gr(f)$, then, we can write

$$Gr(f) := \{(x,f(x)) \in X \times Y : x \in X\}.$$

According to the formal definition of a function, f and $Gr(f)$ are the same thing. So long as we keep this connection in mind, there is no danger in thinking of a function as a "rule" in the intuitive way. In particular, we say that two functions f and g are equal if they have the same graph, or equivalently, if they have the same domain and codomain, and $f(x) = g(x)$ for all $x \in X$. In this case, we simply write $f = g$.

If its range equals its codomain, that is, if $f(X) = Y$, then one says that f maps X **onto** Y, and refers to it as a **surjection** (or as a **surjective** function/map). If f maps distinct points in its domain to distinct points in its codomain, that is, if $x \neq y$ implies $f(x) \neq f(y)$ for all $x, y \in X$, then we say that f is an **injection** (or a **one-to-one** or **injective** function/map). Finally, if f is both injective and surjective, then it is called a **bijection** (or a **bijective** function/map). For instance, if $X := \{1, \ldots, 10\}$, then $f := \{(1,2), (2,3), \ldots, (10,1)\}$ is a bijection in X^X, while $g \in X^X$, defined as $g(x) := 3$ for all $x \in X$, is neither an injection nor a surjection. When considered as a map in $(\{0\} \cup X)^X$, f is an injection but not a surjection.

WARNING. Every injective function can be viewed as a bijection, provided that one views the codomain of the function as its range. Indeed, if $f : X \to Y$ is an injection, then the map $f : X \to Z$ is a bijection, where $Z := f(X)$. This is usually expressed as saying that $f : X \to f(X)$ is a bijection.

Before we consider some examples, let us note that a common way of defining a particular function in a given context is to describe the domain and codomain of that function and the image of a generic point in the domain. So one would say something like, "let $f : X \to Y$ be defined by $f(x) := \ldots$" or "consider the function $f \in Y^X$ defined by $f(x) := \ldots$." For example, by the function $f : \mathbb{R} \to \mathbb{R}_+$ defined by $f(t) := t^2$, we mean the surjection that transforms every real number t to the nonnegative real number t^2. Since the domain of the function is understood from the expression $f : X \to Y$ (or $f \in Y^X$), it is redundant to add the phrase "for all $x \in X$" after the expression "$f(x) := \ldots$," although sometimes we may do so for clarity. Alternatively, when the codomain of the function is clear, a phrase like "the map $x \mapsto f(x)$ on X" is commonly used. For instance, one may refer to the quadratic function mentioned above unambiguously as "the map $t \mapsto t^2$ on $\mathbb{R}$."

EXAMPLE 5

In the following examples, X and Y stand for arbitrary nonempty sets.

[1] A **constant function** is one that assigns the same value to every element of its domain, that is, $f \in Y^X$ is constant iff there exists a $y \in Y$ such that $f(x) = y$ for all $x \in X$. (Formally speaking, this constant function is the set $X \times \{y\}$.) Obviously, $f(X) = \{y\}$ in this case, so a constant function is not surjective unless its codomain is a singleton, and it is not injective unless its domain is a singleton.

[2] A function whose domain and codomain are identical, that is, a function in X^X, is called a **self-map** on X. An important example of a self-map is the **identity function** on X. This function is denoted as id_X, and it is defined as $\mathrm{id}_X(x) := x$ for all $x \in X$. Obviously, id_X is a bijection, and formally speaking, it is none other than the diagonal relation D_X.

[3] Let $S \subseteq X$. The function that maps X into $\{0, 1\}$ such that every member of S is assigned to 1 and all the other elements of X are assigned to zero is called the **indicator function of S in X**. This function is denoted as 1_S (assuming that the domain X is understood from the context). By definition, we have

$$1_S(x) := \begin{cases} 1, & \text{if } x \in S \\ 0, & \text{if } x \in X \backslash S \end{cases}.$$

You can check that, for every $A, B \subseteq X$, we have $1_{A \cup B} + 1_{A \cap B} = 1_A + 1_B$ and $1_{A \cap B} = 1_A 1_B$. □

The following examples point to some commonly used methods of obtaining new functions from a given set of functions.

Example 6

In the following examples, X, Y, Z, and W stand for arbitrary nonempty sets.

[1] Let $Z \subseteq X \subseteq W$, and $f \in Y^X$. By the **restriction** of f to Z, denoted as $f|_Z$, we mean the function $f|_Z \in Y^Z$ defined by $f|_Z(z) := f(z)$. By an **extension** of f to W, on the other hand, we mean a function $f^* \in Y^W$ with $f^*|_X = f$, that is, $f^*(x) = f(x)$ for all $x \in X$. If f is injective, so must $f|_Z$, but surjectivity of f does not entail that of $f|_Z$. Of course, if f is not injective, $f|_Z$ may still turn out to be injective (e.g., $t \mapsto t^2$ is not injective on $\mathbb{R}$, but it is so on $\mathbb{R}_+$).

[2] Sometimes it is possible to extend a given function by *combining* it with another function. For instance, we can combine any $f \in Y^X$ and $g \in W^Z$ to obtain the function $h : X \cup Z \to Y \cup W$ defined by

$$h(t) := \begin{cases} f(t), & \text{if } t \in X \\ g(t), & \text{if } t \in Z \end{cases},$$

provided that $X \cap Z = \emptyset$, or $X \cap Z \neq \emptyset$ and $f|_{X \cap Z} = g|_{X \cap Z}$. Note that this method of combining functions does *not* work if $f(t) \neq g(t)$ for some $t \in X \cap Z$. For, in that case h would not be well-defined as a function. (What would be the image of t under h?)

[3] A function $f \in X^{X \times Y}$ defined by $f(x, y) := x$ is called the **projection from** $X \times Y$ **onto** X.[10] (The projection from $X \times Y$ onto Y is similarly defined.) Obviously, $f(X \times Y) = X$, that is, f is necessarily surjective. It is not injective unless Y is a singleton.

[4] Given functions $f : X \to Z$ and $g : Z \to Y$, we define the **composition** of f and g as the function $g \circ f : X \to Y$ by $g \circ f\,(x) := g(f(x))$. (For easier reading, we often write $(g \circ f)(x)$ instead of $g \circ f\,(x)$.) This definition accords with the way we defined the composition of two relations (Exercise 5). Indeed, we have $(g \circ f)(x) = \{(x, y) : x f z \text{ and } z g y \text{ for some } z \in Z\}$.

Obviously, $\mathrm{id}_Z \circ f = f = f \circ \mathrm{id}_X$. Even when $X = Y = Z$, the operation of taking compositions is not commutative. For instance, if the self-maps f and g on $\mathbb{R}$ are defined by $f(t) := 2$ and $g(t) := t^2$, respectively, then $(g \circ f)(t) = 4$ and $(f \circ g)(t) = 2$ for any real number t. The composition operation is, however, associative. That is, $h \circ (g \circ f) = (h \circ g) \circ f$ for all $f \in Y^X$, $g \in Z^Y$ and $h \in W^Z$. □

EXERCISE 17 Let $\sim$ be an equivalence relation on a nonempty set X. Show that the map $x \mapsto [x]_\sim$ on X (called the **quotient map**) is a surjection on X which is injective iff $\sim = D_X$.

EXERCISE 18[H] (*A Factorization Theorem*) Let X and Y be two nonempty sets. Prove: For any function $f : X \to Y$, there exists a nonempty set Z, a surjection $g : X \to Z$, and an injection $h : Z \to Y$ such that $f = h \circ g$.

EXERCISE 19 Let X, Y, and Z be nonempty sets, and consider any $f, g \in Y^X$ and $u, v \in Z^Y$. Prove:

(a) If f is surjective and $u \circ f = v \circ f$, then $u = v$.
(b) If u is injective and $u \circ f = u \circ g$, then $f = g$.
(c) If f and u are injective (respectively, surjective), then so is $u \circ f$.

[10] Strictly speaking, I should write $f((x, y))$ instead of $f(x, y)$, but that's just splitting hairs.

EXERCISE 20[H] Show that there is no surjection of the form $f : X \to 2^X$ for any nonempty set X.

For any given nonempty sets X and Y, the (*direct*) **image** of a set $A \subseteq X$ under $f \in Y^X$, denoted $f(A)$, is defined as the collection of all elements y in Y with $y = f(x)$ for some $x \in A$. That is,

$$f(A) := \{f(x) : x \in A\}.$$

The range of f is thus the image of its entire domain: $f(X) = \{f(x) : x \in X\}$. (*Note.* If $f(A) = B$, then one says that "f maps A onto B.")

The **inverse image** of a set B in Y under f, denoted as $f^{-1}(B)$, is defined as the set of all x in X whose images under f belong to B, that is,

$$f^{-1}(B) := \{x \in X : f(x) \in B\}.$$

By convention, we write $f^{-1}(y)$ for $f^{-1}(\{y\})$, that is,

$$f^{-1}(y) := \{x \in X : f(x) = y\} \quad \text{for any } y \in Y.$$

Obviously, $f^{-1}(y)$ is a singleton for each $y \in Y$ iff f is an injection. For instance, if f stands for the map $t \mapsto t^2$ on $\mathbb{R}$, then $f^{-1}(1) = \{-1, 1\}$, whereas $f|_{\mathbb{R}_+}^{-1}(1) = \{1\}$.

The issue of whether or not one can express the image (or the inverse image) of a union/intersection of a collection of sets as the union/intersection of the images (inverse images) of each set in the collection arises quite often in mathematical analysis. The following exercise summarizes the situation in this regard.

EXERCISE 21 Let X and Y be nonempty sets and $f \in Y^X$. Prove that, for any (nonempty) classes $\mathcal{A} \subseteq 2^X$ and $\mathcal{B} \subseteq 2^Y$, we have

$$f(\cup\mathcal{A}) = \bigcup\{f(A) : A \in \mathcal{A}\} \quad \text{and} \quad f(\cap\mathcal{A}) \subseteq \bigcap\{f(A) : A \in \mathcal{A}\},$$

whereas

$$f^{-1}(\cup\mathcal{B}) = \bigcup\{f^{-1}(B) : B \in \mathcal{B}\} \quad \text{and} \quad f^{-1}(\cap\mathcal{B}) = \bigcap\{f^{-1}(B) : B \in \mathcal{B}\}.$$

A general rule that surfaces from this exercise is that inverse images are quite well-behaved with respect to the operations of taking unions and intersections, while the same cannot be said for direct images in the case of taking intersections. Indeed, for any $f \in Y^X$, we have $f(A \cap B) \supseteq f(A) \cap f(B)$ for all $A, B \subseteq X$ if, *and only if*, f is injective.[11] The "if" part of this assertion is trivial. The "only if" part follows from the observation that, if the claim was not true, then, for any distinct $x, y \in X$ with $f(x) = f(y)$, we would find $\emptyset = f(\emptyset) = f(\{x\} \cap \{y\}) = f(\{x\}) \cap f(\{y\}) = \{f(x)\}$, which is absurd.

Finally, we turn to the problem of *inverting* a function. For any function $f \in Y^X$, let us define the set

$$f^{-1} := \{(y, x) \in Y \times X : x f y\}$$

which is none other than the inverse of f viewed as a relation (Exercise 5). This relation simply *reverses* the map f in the sense that if x is mapped to y by f, then f^{-1} maps y back to x. Now, f^{-1} may or may not be a function. If it is, we say that f is **invertible** and f^{-1} is the **inverse** of f. For instance, $f : \mathbb{R} \to \mathbb{R}_+$ defined by $f(t) := t^2$ is not invertible (since $(1, 1) \in f^{-1}$ *and* $(1, -1) \in f^{-1}$, that is, 1 does not have a unique image under f^{-1}), whereas $f|_{\mathbb{R}_+}$ is invertible and $f|_{\mathbb{R}_+}^{-1}(t) = \sqrt{t}$ for all $t \in \mathbb{R}$.

The following result gives a simple characterization of invertible functions.

PROPOSITION 2
Let X and Y be two nonempty sets. A function $f \in Y^X$ is invertible if, and only if, it is a bijection.

EXERCISE 22 Prove Proposition 2.

By using the composition operation defined in Example 6.[4], we can give another useful characterization of invertible functions.

PROPOSITION 3
Let X and Y be two nonempty sets. A function $f \in Y^X$ is invertible if, and only if, there exists a function $g \in X^Y$ such that $g \circ f = \mathrm{id}_X$ and $f \circ g = \mathrm{id}_Y$.

[11] Of course, this does not mean that $f(A \cap B) = f(A) \cap f(B)$ can never hold for a function that is not one-to-one. It only means that, for any such function f, we can always find nonempty sets A and B in the domain of f such that $f(A \cap B) \supseteq f(A) \cap f(B)$ is false.

Proof

The "only if" part is readily obtained upon choosing $g := f^{-1}$. To prove the "if" part, suppose there exists a $g \in X^Y$ with $g \circ f = \mathrm{id}_X$ and $f \circ g = \mathrm{id}_Y$, and note that, by Proposition 2, it is enough to show that f is a bijection. To verify the injectivity of f, pick any $x, y \in X$ with $f(x) = f(y)$, and observe that

$$x = \mathrm{id}_X(x) = (g \circ f)(x) = g(f(x)) = g(f(y)) = (g \circ f)(y) = \mathrm{id}_X(y) = y.$$

To see the surjectivity of f, take any $y \in Y$ and define $x := g(y)$. Then we have

$$f(x) = f(g(y)) = (f \circ g)(y) = \mathrm{id}_Y(y) = y,$$

which proves $Y \subseteq f(X)$. Since the converse containment is trivial, we are done. ∎

1.6 Sequences, Vectors, and Matrices

By a *sequence* in a given nonempty set X, we intuitively mean an ordered array of the form $(x_1, x_2, \ldots)$ where each term x_i of the sequence is a member of X. (Throughout this text we denote such a sequence by (x_m), but note that some books prefer instead the notation $(x_m)_{m=1}^{\infty}$.) As in the case of ordered pairs, one could introduce the notion of a sequence as a new object to our set theory, but again there is really no need to do so. Intuitively, we understand from the notation $(x_1, x_2, \ldots)$ that the ith term in the array is x_i. But then we can think of this array as a function that maps the set $\mathbb{N}$ of positive integers into X in the sense that it tells us that "the ith term in the array is x_i" by mapping i to x_i. With this definition, our intuitive understanding of the ordered array $(x_1, x_2, \ldots)$ is formally captured by the function $\{(i, x_i) : i = 1, 2, \ldots\} = f$. Thus, we define a **sequence** in a nonempty set X as any function $f : \mathbb{N} \to X$, and *represent* this function as $(x_1, x_2, \ldots)$ where $x_i := f(i)$ for each $i \in \mathbb{N}$. Consequently, the set of all sequences in X is equal to $X^{\mathbb{N}}$. As is common, however, we denote this set as X^{∞} throughout the text.

By a *subsequence* of a sequence $(x_m) \in X^{\infty}$, we mean a sequence that is made up of the terms of (x_m) that appear in the subsequence in the same order they appear in (x_m). That is, a subsequence of (x_m) is of the form $(x_{m_1}, x_{m_2}, \ldots)$, where (m_k) is a sequence in $\mathbb{N}$ such that $m_1 < m_2 < \cdots$. (We denote this subsequence as (x_{m_k}).) Once again, we use the notion of function to formalize this definition. Strictly speaking, a **subsequence** of a sequence $f \in X^{\mathbb{N}}$ is a function of the form $f \circ \sigma$, where $\sigma : \mathbb{N} \to \mathbb{N}$ is strictly increasing (that is, $\sigma(k) < \sigma(l)$ for any $k, l \in \mathbb{N}$ with $k < l$). We *represent* this function as the array $(x_{m_1}, x_{m_2}, \ldots)$ with the understanding that $m_k = \sigma(k)$ and $x_{m_k} = f(m_k)$ for each $k = 1, 2, \ldots$. For instance, $(x_{m_k}) := (1, \frac{1}{3}, \frac{1}{5}, \ldots)$ is a subsequence of $(x_m) := (\frac{1}{m}) \in \mathbb{R}^\infty$. Here (x_m) is a representation for the function $f \in \mathbb{R}^{\mathbb{N}}$, which is defined by $f(i) := \frac{1}{i}$, and (x_{m_k}) is a representation of the map $f \circ \sigma$, where $\sigma(k) := 2k - 1$ for each $k \in \mathbb{N}$.

By a *double sequence* in X, we mean an infinite matrix each term of which is a member of X. Formally, a **double sequence** is a function $f \in X^{\mathbb{N}\times\mathbb{N}}$. As in the case of sequences, we *represent* this function as (x_{kl}), with the understanding that $x_{kl} := f(k, l)$. The set of all double sequences in X equals $X^{\mathbb{N}\times\mathbb{N}}$, but it is customary to denote this set as $X^{\infty\times\infty}$. We note that one can always view (in more than one way) a double sequence in X as a sequence of sequences in X, that is, as a sequence in X^∞. For instance, we can think of (x_{kl}) as $((x_{1l}), (x_{2l}), \ldots)$ or as $((x_{k1}), (x_{k2}), \ldots)$.

The basic idea of viewing a string of objects as a particular function also applies to *finite* strings, of course. For instance, how about $X^{\{1,\ldots,n\}}$, where X is a nonempty set and n is some positive integer? The preceding discussion shows that this function space is none other than the set $\{(x_1, \ldots, x_n) : x_i \in X, i = 1, \ldots, n\}$. Thus we may define an ***n*-vector** in X as a function $f : \{1, \ldots, n\} \to X$, and *represent* this function as $(x_1, \ldots, x_n)$ where $x_i := f(i)$ for each $i = 1, \ldots, n$. (Check that $(x_1, \ldots, x_n) = (x_1', \ldots, x_n')$ iff $x_i = x_i'$ for each $i = 1, \ldots, n$, so everything is in concert with the way we defined n-vectors in Section 1.2.) The ***n*-fold product** of X is then defined as $X^{\{1,\ldots,n\}}$, but is denoted as X^n. (So $\mathbb{R}^n = \mathbb{R}^{\{1,\ldots,n\}}$. This makes sense, no?) The main lesson is that everything that is said about arbitrary functions also applies to sequences and vectors.

Finally, for any positive integers m and n, by an $m \times n$ **matrix** (read "m by n matrix") in a nonempty set X, we mean a function $f : \{1, \ldots, m\} \times \{1, \ldots, n\} \to X$. We *represent* this function as $[a_{ij}]_{m\times n}$, with the understanding that $a_{ij} := f(i, j)$ for each $i = 1, \ldots, m$ and $j = 1, \ldots, n$. (As you know, one often views a matrix like $[a_{ij}]_{m\times n}$ as a rectangular array with m rows and n columns in which a_{ij} appears in the ith row and jth column.)

The set of all $m \times n$ matrices in X is $X^{\{1,\ldots,m\}\times\{1,\ldots,n\}}$, but it is much better to denote this set as $X^{m\times n}$. Needless to say, both $X^{1\times n}$ and $X^{n\times 1}$ can be identified with X^n. (Wait, what does this mean?)

1.7* A Glimpse of Advanced Set Theory: The Axiom of Choice

We now turn to a problem that we have so far conveniently avoided: How do we define the Cartesian product of *infinitely* many nonempty sets? Intuitively speaking, the Cartesian product of all members of a class $\mathcal{A}$ of sets is the set of all collections each of which contains one and only one element of each member of $\mathcal{A}$. That is, a member of this product is really a function on $\mathcal{A}$ that selects a single element from each set in $\mathcal{A}$. The question is simple to state: Does there exist such a function?

If $|\mathcal{A}| < \infty$, then the answer would obviously be yes, because we can *construct* such a function by choosing an element from each set in $\mathcal{A}$ one by one. But when $\mathcal{A}$ contains infinitely many sets, then this method does not readily work, so we need to *prove* that such a function exists.

To get a sense of this, suppose $\mathcal{A} := \{A_1, A_2, \ldots\}$, where $\emptyset \neq A_i \subseteq \mathbb{N}$ for each $i = 1, 2, \ldots$. Then we're okay. We can define $f : \mathcal{A} \to \cup\mathcal{A}$ by $f(A) :=$ the smallest element of A – this well-defines f as a map that selects one element from each member of $\mathcal{A}$ *simultaneously*. Or, if each A_i is a bounded interval in $\mathbb{R}$, then again we're fine. This time we can define f, say, as follows: $f(A) :=$ the midpoint of A. But what if all we knew was that each A_i consists of real numbers? Or worse, what if we were not told anything about the contents of $\mathcal{A}$? You see, in general, we can't write down a formula, or an algorithm, the application of which yields such a function. Then how do you know that such a thing exists in the first place?[12]

[12] But, how about the following algorithm? Start with A_1, and pick any a_1 in A_1. Now move to A_2 and pick any $a_2 \in A_2$. Continue this way, and define $g : \mathcal{A} \to \cup\mathcal{A}$ by $g(A_i) = a_i$, $i = 1, 2, \ldots$. Aren't we done? No, we are not! The function at hand is not well-defined—its definition does not tell me *exactly* which member of A_{27} is assigned to $g(A_{27})$—this is very much unlike how I defined f above in the case where each A_i was contained in $\mathbb{N}$ (or was a bounded interval).

Perhaps you are still not quite comfortable about this. You might think that f is well-defined, it's just that it is defined *recursively*. Let me try to illustrate the problem by means of a concrete example. Take any infinite set S, and ask yourself if you can define an injection f from $\mathbb{N}$ into S. Sure, you might say, "recursion" is again the name of the game. Let $f(1)$ be any member a_1 of S. Then let $f(2)$ be any member of $S\backslash\{a_1\}$, $f(3)$ any member $S\backslash\{a_1, a_2\}$, and so on. Since $S\backslash T \neq \emptyset$ for any finite $T \subset S$, this well-defines f, recursively, as an injection from $\mathbb{N}$ into S. Wrong! If this were the case, on the basis of the knowledge of $f(1), \ldots, f(26)$, I would know the value of f at 27. The "definition" of f doesn't do that—it just points to some *arbitrary* member of A_{27}—so it is not a proper definition at all.

(*Note.* As "obvious" as it might seem, the proposition "for any infinite set S, there is an injection in $S^{\mathbb{N}}$," cannot be proved within the standard realm of set theory.)

In fact, it turns out that the problem of "finding an $f : \mathcal{A} \to \cup\mathcal{A}$ for *any* given class $\mathcal{A}$ of sets" cannot be settled in one way or another by means of the standard axioms of set theory.[13] The status of our question is thus a bit odd, it is *undecidable.*

To make things a bit more precise, let us state formally the property that we are after.

The Axiom of Choice. *For any (nonempty) class $\mathcal{A}$ of sets, there exists a function $f : \mathcal{A} \to \cup\mathcal{A}$ such that $f(A) \in A$ for each $A \in \mathcal{A}$.*

One can reword this in a few other ways.

EXERCISE 23 Prove that the Axiom of Choice is equivalent to the following statements.

(i) For any nonempty set S, there exists a function $f : 2^S\backslash\{\emptyset\} \to S$ such that $f(A) \in A$ for each $\emptyset \neq A \subseteq S$.

(ii) (*Zermelo's Postulate*) If $\mathcal{A}$ is a (nonempty) class of sets such that $A \cap B = \emptyset$ for each distinct $A, B \in \mathcal{A}$, then there exists a set S such that $|S \cap A| = 1$ for every $A \in \mathcal{A}$.

(iii) For any nonempty sets X and Y, and any relation R from X into Y, there is a function $f : Z \to Y$ with $\emptyset \neq Z \subseteq X$ and $f \subseteq R$. (That is: *Every relation contains a function.*)

The first thing to note about the Axiom of Choice is that it cannot be disproved by using the standard axioms of set theory. Provided that these axioms are consistent (that is, no contradiction may be logically deduced from them), adjoining the Axiom of Choice to these axioms yields again a consistent set of axioms. This raises the possibility that perhaps the Axiom of Choice can be deduced as a "theorem" from the standard axioms. The second thing to know about the Axiom of Choice is that this is false, that is, the Axiom of Choice is not provable from the standard axioms of set theory.[14]

[13] For brevity, I am again being imprecise about this *standard* set of axioms (called the *Zermelo-Fraenkel-Skolem* axioms). For the present discussion, nothing will be lost if you just think of these as the formal properties needed to "construct" the set theory we outlined intuitively earlier. It is fair to say that these axioms have an unproblematic standing in mathematics.

[14] These results are of extreme importance for the foundations of the entire field of mathematics. The first one was proved by Kurt Gödel in 1939 and the second one by Paul Cohen in 1963.

We are then at a crossroads. We must either reject the validity of the Axiom of Choice and confine ourselves to the conclusions that can be reached only on the basis of the standard axioms of set theory, or alternatively, adjoin the Axiom of Choice to the standard axioms to obtain a richer set theory that is able to yield certain results that could not have been proved within the confines of the standard axioms. Most analysts follow the second route. However, it is fair to say that the status of the Axiom of Choice is in general viewed as less appealing than the standard axioms, so one often makes it explicit if this axiom is a prerequisite for a particular theorem to be proved. Given our applied interests, we will be more relaxed about this matter and mention the (implicit) use of the Axiom of Choice in our arguments only rarely.

As an immediate application of the Axiom of Choice, we now define the **Cartesian product** of an arbitrary (nonempty) class $\mathcal{A}$ of sets as the set of all $f : \mathcal{A} \to \cup\mathcal{A}$ with $f(A) \in A$ for each $A \in \mathcal{A}$. We denote this set by $\mathsf{X}\mathcal{A}$, and note that $\mathsf{X}\mathcal{A} \neq \emptyset$ because of the Axiom of Choice. If $\mathcal{A} = \{A_i : i \in I\}$, where I is an index set, then we write $\mathsf{X}_{i\in I} A_i$ for $\mathsf{X}\mathcal{A}$. Clearly, $\mathsf{X}_{i\in I} A_i$ is the set of all maps $f : I \to \cup\{A_i : i \in I\}$ with $f(i) \in A_i$ for each $i \in I$. It is easily checked that this definition is consistent with the definition of the Cartesian product of finitely many sets given earlier.

There are a few equivalent versions of the Axiom of Choice that are often more convenient to use in applications than the original statement of the axiom. To state the most widely used version, let us first agree on some terminology. For any poset $(X, \succsim)$, by a "poset in $(X, \succsim)$," we mean a poset like $(Y, \succsim \cap Y^2)$ with $Y \subseteq X$, but we denote this poset more succinctly as $(Y, \succsim)$. By an *upper bound* for such a poset, we mean an element x of X with $x \succsim y$ for all $y \in Y$ (Exercise 15).

> ZORN'S LEMMA
> *If every loset in a given poset has an upper bound, then that poset must have a maximal element.*

Although this is a less intuitive statement than the Axiom of Choice (no?), it can in fact be shown to be equivalent to the Axiom of Choice.[15] (That is, we can deduce Zorn's Lemma from the standard axioms *and* the Axiom of Choice, and we can prove the Axiom of Choice by using the standard axioms

[15] For a proof, see Enderton (1977, pp. 151–153) or Kelley (1955, pp. 32–35).

and Zorn's Lemma.) Since we take the Axiom of Choice as "true" in this text, therefore, we must also accept the validity of Zorn's Lemma.

We conclude this discussion by means of two quick applications that illustrate how Zorn's Lemma is used in practice. We will see some other applications in later chapters.

Let us first prove the following fact:

> The Hausdorff Maximal Principle
> *There exists a $\supseteq$-maximal loset in every poset.*

Proof

Let $(X, \succsim)$ be a poset, and

$$\mathcal{L}(X, \succsim) := \{Z \subseteq X : (Z, \succsim) \text{ is a loset}\}.$$

(Observe that $\mathcal{L}(X, \succsim) \neq \emptyset$ by reflexivity of $\succsim$.) We wish to show that there is a $\supseteq$-maximal element of $\mathcal{L}(X, \succsim)$. This will follow from Zorn's Lemma, if we can show that every loset in the poset $(\mathcal{L}(X, \succsim), \supseteq)$ has an upper bound, that is, for any $\mathcal{A} \subseteq \mathcal{L}(X, \succsim)$ such that $(\mathcal{A}, \supseteq)$ is a loset, there is a member of $\mathcal{L}(X, \succsim)$ that contains $\mathcal{A}$. To establish that this is indeed the case, take any such $\mathcal{A}$, and let $Y := \cup\mathcal{A}$. Then $\succsim$ is a complete relation on Y, because, since $\supseteq$ linearly orders $\mathcal{A}$, for any $x, y \in Y$ we must have $x, y \in A$ for some $A \in \mathcal{A}$ (why?), and hence, given that $(A, \succsim)$ is a loset, we have either $x \succsim y$ or $y \succsim x$. Therefore, $(Y, \succsim)$ is a loset, that is, $Y \in \mathcal{L}(X, \succsim)$. But it is obvious that $Y \supseteq A$ for any $A \in \mathcal{A}$. ■

In fact, the Hausdorff Maximal Principle is equivalent to the Axiom of Choice.

Exercise 24 Prove Zorn's Lemma assuming the validity of the Hausdorff Maximal Principle.

As another application of Zorn's Lemma, we prove Szpilrajn's Theorem.[16] Our proof uses the Hausdorff Maximal Principle, but you now know that this is equivalent to invoking Zorn's Lemma or the Axiom of Choice.

[16] In case you are wondering, Szpilrajn's Theorem is *not* equivalent to the Axiom of Choice.

Proof of Szpilrajn's Theorem

Let $\succsim$ be a partial order on a nonempty set X. Let $\mathcal{T}_X$ be the set of all partial orders on X that extend $\succsim$. Clearly, $(\mathcal{T}_X, \supseteq)$ is a poset, so by the Hausdorff Maximal Principle, it has a maximal loset, say, $(\mathcal{A}, \supseteq)$. Define $\succsim^* := \cup\mathcal{A}$. Since $(\mathcal{A}, \supseteq)$ is a loset, $\succsim^*$ is a partial order on X that extends $\succsim$. (Why?) $\succsim^*$ is in fact complete. To see this, suppose we can find some $x, y \in X$ with neither $x \succsim^* y$ nor $y \succsim^* x$. Then the transitive closure of $\succsim^* \cup \{(x, y)\}$ is a member of $\mathcal{T}_X$ that contains $\succsim^*$ as a proper subset (Exercise 8). (Why exactly?) This contradicts the fact that $(\mathcal{A}, \supseteq)$ is a maximal loset within $(\mathcal{T}_X, \supseteq)$. (Why?) Thus $\succsim^*$ is a linear order, and we are done. ■

2 Real Numbers

This course assumes that the reader has a basic understanding of the real numbers, so our discussion here will be brief and duly heuristic. In particular, we will not even attempt to give a *construction* of the set $\mathbb{R}$ of real numbers. Instead we will mention some axioms that $\mathbb{R}$ satisfies, and focus on certain properties that $\mathbb{R}$ possesses. Some books on real analysis give a fuller view of the construction of $\mathbb{R}$, some talk about it even less than we do. If you are really curious about this, it's best if you consult a book that specializes in this sort of a thing. (Try, for instance, Chapters 4 and 5 of Enderton (1977).)

2.1 Ordered Fields

In this subsection we talk briefly about a few topics in abstract algebra that will facilitate our discussion of real numbers.

Definition

Let X be any nonempty set. We refer to a function of the form $\bullet : X \times X \to X$ as a **binary operation** on X, and write $x \bullet y$ instead of $\bullet(x, y)$ for any $x, y \in X$.

For instance, the usual addition and multiplication operations $+$ and $\cdot$ are binary operations on the set $\mathbb{N}$ of natural numbers. The subtraction operation is, on the other hand, not a binary operation on $\mathbb{N}$ (e.g., $1+(-2) \notin \mathbb{N}$), but it is a binary operation on the set of all integers.

Definition

Let X be any nonempty set, let $+$ and $\cdot$ be two binary operations on X, and let us agree to write xy for $x \cdot y$ for simplicity. The list $(X, +, \cdot)$ is called a **field** if the following properties are satisfied:

(i) (*Commutativity*) $x + y = y + x$ and $xy = yx$ for all $x, y \in X$;

(ii) (*Associativity*) $(x + y) + z = x + (y + z)$ and $(xy)z = x(yz)$ for all $x, y, z \in X$;[17]

(iii) (*Distributivity*) $x(y + z) = xy + xz$ for all $x, y, z \in X$;

(iv) (*Existence of Identity Elements*) There exist elements 0 and 1 in X such that $0 + x = x = x + 0$ and $1x = x = x1$ for all $x \in X$;

(v) (*Existence of Inverse Elements*) For each $x \in X$ there exists an element $-x$ in X (the *additive inverse* of x) such that $x + -x = 0 = -x + x$, and for each $x \in X\backslash\{0\}$ there exists an element x^{-1} in X (the *multiplicative inverse* of x) such that $xx^{-1} = 1 = x^{-1}x$.

A field $(X, +, \cdot)$ is an algebraic structure that envisions two binary operations, $+$ and $\cdot$, on the set X in a way that makes a satisfactory *arithmetic* possible. In particular, given the $+$ and $\cdot$ operations, we can define the two other (inverse) operations $-$ and $/$ by $x - y := x + -y$ and $x/y := xy^{-1}$, the latter provided that $y \neq 0$. (Strictly speaking, the division operation $/$ is not a binary operation; for instance, $1/0$ is not defined in X.)

Pretty much the entire arithmetic that we are familiar with in the context of $\mathbb{R}$ can be performed within an arbitrary field. To illustrate this, let us establish a few arithmetic laws that you may recall from high school algebra. In particular, let us show that

$$x+y = x+z \quad \text{iff} \quad y = z, \qquad -(-x) = x \quad \text{and} \quad -(x+y) = -x+-y \quad (1)$$

in any field $(X, +, \cdot)$. The first claim is a *cancellation law*, which is readily proved by observing that, for any $w \in X$, we have $w = 0+w = (-x+x)+w = -x + (x + w)$. Thus, $x + y = x + z$ implies $y = -x + (x + y) = z$, and we're done. As an immediate corollary of this cancellation law, we find that

[17] Throughout this exposition, (w) is the same thing as w, for any $w \in X$. For instance, $(x + y)$ corresponds to $x + y$, and $(-x)$ corresponds to $-x$. The brackets are used at times only for clarity.

the additive inverse of each element in X is unique. (The same holds for the multiplicative inverses as well. *Quiz.* Prove!) On the other hand, the second claim in (1) is true because

$$x = x+0 = x+(-x+-(-x)) = (x+-x)+-(-x) = 0+-(-x) = -(-x).$$

Finally, given that the additive inverse of $x+y$ is unique, the last claim in (1) follows from the following argument:

$$\begin{aligned}(x+y)+(-x+-y) &= (x+y)+(-y+-x)\\ &= x+(y+(-y+-x))\\ &= x+((y+-y)+-x)\\ &= x+(0+-x)\\ &= x+-x\\ &= 0.\end{aligned}$$

(*Quiz.* Prove that $-1x = -x$ in any field. *Hint.* There is something to be proved here!)

EXERCISE 25 (*Rules of Exponentiation*) Let $(X,+,\cdot)$ be a field. For any $x \in X$, we define $x^0 := 1$, and for any positive integer k, we let $x^k := x^{k-1}x$ and $x^{-k} := (x^k)^{-1}$. For any integers i and j, prove that $x^i x^j = x^{i+j}$ and $(x^i)^j = x^{ij}$ for any $x \in X$, and $x^i/x^j = x^{i-j}$ and $(y/x)^i = y^i/x^i$ for any $x \in X\backslash\{0\}$.

Although a field provides a rich environment for doing arithmetic, it lacks structure for *ordering* things. We introduce such a structure next.

DEFINITION

The list $(X,+,\cdot,\geq)$ is called an **ordered field** if $(X,+,\cdot)$ is a field, and if $\geq$ is a partial order on X that is compatible with the operations $+$ and $\cdot$ in the sense that $x \geq y$ implies $x+z \geq y+z$ for any $x,y,z \in X$, and $xz \geq yz$ for any $x,y,z \in X$ with $z \geq 0$. We note that the expressions $x \geq y$ and $y \leq x$ are identical. The same goes also for the expressions $x > y$ and

$y < x$.[18] We also adopt the following notation:

$$X_+ := \{x \in X : x \geq 0\} \quad \text{and} \quad X_{++} := \{x \in X : x > 0\},$$

and

$$X_- := \{x \in X : x \leq 0\} \quad \text{and} \quad X_{--} := \{x \in X : x < 0\}.$$

An ordered field is a rich algebraic system within which many algebraic properties of real numbers can be established. This is of course not the place to get into a thorough algebraic analysis, but we should consider at least one example to give you an idea about how this can be done.

EXAMPLE 7

(*The Triangle Inequality*) Let $(X, +, \cdot, \geq)$ be an ordered field. The function $|\cdot| : X \to X$ defined by

$$|x| := \begin{cases} x, & \text{if } x \geq 0 \\ -x, & \text{if } x < 0 \end{cases}$$

is called the **absolute value** function.[19] The following is called the **triangle inequality**:

$$|x + y| \leq |x| + |y| \qquad \text{for all } x, y \in X.$$

You have surely seen this inequality in the case of real numbers. The point is that it is valid within *any* ordered field, so the only properties responsible for it are the ordered field axioms.

We divide the argument into five easy steps. All x and y that appear in these steps are arbitrary elements of X.

(a) $|x| \geq x$. *Proof.* If $x \geq 0$, then $|x| = x$ by definition. If $0 > x$, on the other hand, we have

$$|x| = -x = 0 + -x \geq x + -x = 0 \geq x.$$

(b) $x \geq 0$ implies $-x \leq 0$, and $x \leq 0$ implies $-x \geq 0$. *Proof.* If $x \geq 0$, then

$$0 = x + -x \geq 0 + -x = -x.$$

[18] Naturally, $x > y$ means that x and y are distinct members of X with $x \geq y$. That is, $>$ is the asymmetric part of $\geq$.

[19] We owe the notation $|x|$ to Karl Weierstrass. Before Weierstrass's famous 1858 lectures, there was apparently no unity on denoting the absolute value function. For instance, Bernhard Bolzano would write $\pm x$!

The second claim is proved analogously.

(c) $x \geq -|x|$. *Proof.* If $x \geq 0$, then $x \geq 0 \geq -x = -|x|$ where the second inequality follows from (b). If $0 > x$, then $-|x| = -(-x) = x$ by (1).

(d) $x \geq y$ implies $-y \geq -x$. *Proof.* Exercise.

(e) $|x+y| \leq |x| + |y|$. *Proof.* Applying (a) twice,

$$|x| + |y| \geq x + |y| = |y| + x \geq y + x = x + y.$$

Similarly, by using (c) twice,

$$x + y \geq -|x| + -|y| = (|x| + |y|)$$

where we used the third claim in (1) to get the final equality. By (d), therefore, $|x| + |y| \geq -(x+y)$, and we are done. □

EXERCISE 26 Let $(X, +, \cdot, \geq)$ be an ordered field. Prove:

$$|xy| = |x|\,|y| \quad \text{and} \quad |x-y| \geq \big||x| - |y|\big| \quad \text{for all } x, y \in X.$$

2.2 Natural Numbers, Integers, and Rationals

As you already know, we denote the set of all *natural numbers* by $\mathbb{N}$, that is, $\mathbb{N} := \{1, 2, 3, \ldots\}$. Among the properties that this system satisfies, a particularly interesting one that we wish to mention is the following:

THE PRINCIPLE OF MATHEMATICAL INDUCTION
If S is a subset of $\mathbb{N}$ such that $1 \in S$, and $i + 1 \in S$ whenever $i \in S$, then $S = \mathbb{N}$.

This property is actually one of the main *axioms* that are commonly used to construct the natural numbers.[20] It is frequently employed when giving

[20] Roughly speaking, the standard construction goes as follows. One postulates that $\mathbb{N}$ is a set with a linear order, called the *successor relation*, which specifies an immediate successor for each member of $\mathbb{N}$. If $i \in \mathbb{N}$, then the immediate successor of i is denoted as $i + 1$. Then, $\mathbb{N}$ is the set that is characterized by the Principle of Mathematical Induction and the following three axioms: (i) there is an element 1 in $\mathbb{N}$ that is not a successor of any other element in $\mathbb{N}$; (ii) if $i \in \mathbb{N}$, then $i + 1 \in \mathbb{N}$; and (iii) if i and j have the same successor, then $i = j$. Along with the Principle of Mathematical Induction, these properties are known as the *Peano axioms* (in honor of Giuseppe Peano (1858–1932), who first formulated these postulates and laid out an axiomatic foundation for the integers). The binary operations $+$ and $\cdot$ are defined via the successor relation, and behave "well" because of these axioms.

a recursive definition (as in Exercise 25), or when proving infinitely many propositions by recursion. Suppose $P_1, P_2, \ldots$ are logical statements. If we can prove that P_1 is true, and then show that the validity of P_{i+1} would in fact follow from the validity of P_i (i being arbitrarily fixed in $\mathbb{N}$), then we may invoke the Principle of Mathematical Induction to conclude that each proposition in the string $P_1, P_2, \ldots$ is true. For instance, suppose we wish to prove that

$$1 + \frac{1}{2} + \frac{1}{4} + \cdots + \frac{1}{2^i} = 2 - \frac{1}{2^i} \quad \text{for each } i \in \mathbb{N}. \tag{2}$$

Then we first check if the claim holds for $i = 1$. Since $1 + \frac{1}{2} = 2 - \frac{1}{2}$, this is indeed the case. On the other hand, if we assume that the claim is true for an arbitrarily fixed $i \in \mathbb{N}$ (the *induction hypothesis*), then we see that the claim is true for $i + 1$, because

$$\begin{aligned} 1 + \frac{1}{2} + \frac{1}{4} + \cdots + \frac{1}{2^{i+1}} &= \left(1 + \frac{1}{2} + \frac{1}{4} + \cdots + \frac{1}{2^i}\right) + \frac{1}{2^{i+1}} \\ &= 2 - \frac{1}{2^i} + \frac{1}{2^{i+1}} \quad \text{(by the induction hypothesis)} \\ &= 2 - \frac{1}{2^{i+1}}. \end{aligned}$$

Thus, by the Principle of Mathematical Induction, we conclude that (2) holds. We shall use this principle numerous times throughout the text. Here is another example.

Exercise 27 Let $(X, +, \cdot, \geq)$ be an ordered field. Use the Principle of Mathematical Induction to prove the following generalization of the triangle inequality: For any $m \in \mathbb{N}$,

$$|x_1 + \cdots + x_m| \leq |x_1| + \cdots + |x_m| \qquad \text{for all } x_1, \ldots, x_m \in X.$$

Adjoining to $\mathbb{N}$ an element to serve as the additive identity, namely the *zero*, we obtain the set of all *nonnegative integers*, which is denoted as $\mathbb{Z}_+$. In turn, adjoining to $\mathbb{Z}_+$ the set $\{-1, -2, \ldots\}$ of all *negative integers* (whose construction would mimic that of $\mathbb{N}$), we obtain the set $\mathbb{Z}$ of all *integers*. In the process, the binary operations $+$ and $\cdot$ are suitably extended from $\mathbb{N}$ to $\mathbb{Z}$ so that they become binary operations on $\mathbb{Z}$ that satisfy all of the field axioms except the existence of multiplicative inverse elements.

Unfortunately, the nonexistence of multiplicative inverses is a serious problem. For instance, while an equation like $2x = 1$ makes sense in $\mathbb{Z}$, it cannot possibly be solved in $\mathbb{Z}$. To be able to solve such linear equations, we need to extend $\mathbb{Z}$ to a *field*. Doing this (in the minimal way) leads us to the set $\mathbb{Q}$ of all *rational numbers*, which can be thought of as the collection of all fractions $\frac{m}{n}$ with $m, n \in \mathbb{Z}$ and $n \neq 0$. The operations $+$ and $\cdot$ are extended to $\mathbb{Q}$ in the natural way (so that, for instance, the additive and multiplicative inverses of $\frac{m}{n}$ are $-\frac{m}{n}$ and $\frac{n}{m}$, respectively, provided that $m, n \neq 0$). Moreover, the standard order $\geq$ on $\mathbb{Z}$ (which is deduced from the *successor relation* that leads to the construction of $\mathbb{N}$) is also extended to $\mathbb{Q}$ in the straightforward manner.[21] The resulting algebraic system, which we denote simply as $\mathbb{Q}$ instead of the fastidious $(\mathbb{Q}, +, \cdot, \geq)$, is significantly richer than $\mathbb{Z}$. In particular, the following is true.

PROPOSITION 4

$\mathbb{Q}$ is an ordered field.

Since we did not give a formal construction of $\mathbb{Q}$, we cannot prove this fact here.[22] But it is certainly good to know that all algebraic properties of an ordered field are possessed by $\mathbb{Q}$. For instance, thanks to Proposition 4, Example 7, and Exercise 25, the triangle inequality and the standard rules of exponentiation are valid in $\mathbb{Q}$.

2.3 Real Numbers

Although it is far superior to that of $\mathbb{Z}$, the structure of $\mathbb{Q}$ is nevertheless not strong enough to deal with many worldly matters. For instance, if we take a square with sides having length one, and attempt to compute the length

[21] [*Only for the formalists*] These definitions are meaningful only insofar as one knows the operation of "division" (and we don't, since the binary operation / is not defined on $\mathbb{Z}$). As noted in Section 1.3, the proper approach is to define $\mathbb{Q}$ as the set of equivalence classes $[(m, n)]_\sim$ where the equivalence relation $\sim$ is defined on $\mathbb{Z} \times (\mathbb{Z}\backslash\{0\})$ by $(m, n) \sim (k, l)$ iff $ml = nk$. The addition and multiplication operations on $\mathbb{Q}$ are then defined as $[(m, n)]_\sim + [(k, l)]_\sim = [(ml + nk, nl)]_\sim$ and $[(m, n)]_\sim[(k, l)]_\sim = [(mk, nl)]_\sim$. Finally, the linear order $\geq$ on $\mathbb{Q}$ is defined via the ordering of integers as follows: $[(m, n)]_\sim \geq [(k, l)]_\sim$ iff $ml \geq nk$.

[22] If you followed the previous footnote, you should be able to supply a proof, *assuming* the usual properties of $\mathbb{Z}$.

r of its diagonal, we would be in trouble if we were to use only the rational numbers. After all, we know from planar geometry (from the Pythagorean Theorem, to be exact) that r must satisfy the equation $r^2 = 2$. The trouble is that no rational number is equal to the task. Suppose that $r^2 = 2$ holds for some $r \in \mathbb{Q}$. We may then write $r = \frac{m}{n}$ for some integers $m, n \in \mathbb{Z}$ with $n \neq 0$. Moreover, we can assume that m and n do not have a common factor. (Right?) Then $m^2 = 2n^2$, from which we conclude that m^2 is an even integer. But this is possible only if m is an even integer itself. (Why?) Hence we may write $m = 2k$ for some $k \in \mathbb{Z}$. Then we have $2n^2 = m^2 = 4k^2$ so that $n^2 = 2k^2$, that is, n^2 is an even integer. But then n is even, which means 2 is a common factor of both m and n, a contradiction.

This observation is easily generalized:

EXERCISE 28 Prove: If a is a positive integer such that $a \neq b^2$ for any $b \in \mathbb{Z}$, then there is no rational number r such that $r^2 = a$.[23]

Here is another way of looking at the problem above. There are certainly two rational numbers p and q such that $p^2 > 2 > q^2$, but now we know that there is no $r \in \mathbb{Q}$ with $r^2 = 2$. It is as if there were a "hole" in the set of rational numbers. Intuitively speaking, then, we wish to *complete* $\mathbb{Q}$ by filling up its holes with "new" numbers. And, lo and behold, doing this leads us to the set $\mathbb{R}$ of real numbers. (*Note.* Any member of the set $\mathbb{R}\backslash\mathbb{Q}$ is said to be an **irrational number**.)

This is not the place to get into the formal details of how such a completion would be carried out, so we will leave things at this fairy tale level. However, we remark that, during this completion, the operations of addition and multiplication are extended to $\mathbb{R}$ in such a way as to make it a field. Similarly, the order $\geq$ is extended from $\mathbb{Q}$ to $\mathbb{R}$ nicely, so a great many algebraic properties of $\mathbb{Q}$ are inherited by $\mathbb{R}$.

[23] This fact provides us with lots of real numbers that are not rational, e.g., $\sqrt{2}, \sqrt{3}, \sqrt{5}, \sqrt{6}, \ldots$, etc. There are many other *irrational* numbers. (Indeed, there is a sense in which there are more of such numbers than of rational numbers.) However, it is often difficult to prove the irrationality of a number. For instance, while the problem of incommensurability of the circumference and the diameter of a circle was studied since the time of Aristotle, it was not until 1766 that a complete proof of the irrationality of π was given. Fortunately, elementary proofs of the fact that $\pi \notin \mathbb{Q}$ are since then formulated. If you are curious about this issue, you might want to take a look at Chapter 6 of Aigner and Ziegler (1999), where a brief and self-contained treatment of several such results (e.g., $\pi^2 \notin \mathbb{Q}$ and $e \notin \mathbb{Q}$) is given.

> **Proposition 5**
> *$\mathbb{R}$ is an ordered field.*

Notation. Given Propositions 4 and 5, it is natural to adopt the notations $\mathbb{Q}_+, \mathbb{Q}_{++}, \mathbb{Q}_-$, and $\mathbb{Q}_{--}$ to denote, respectively, the nonnegative, positive, nonpositive, and negative subsets of $\mathbb{Q}$, and similarly for $\mathbb{R}_+, \mathbb{R}_{++}, \mathbb{R}_-$, and $\mathbb{R}_{--}$.

There are, of course, many properties that $\mathbb{R}$ satisfies but $\mathbb{Q}$ does not. To make this point clearly, let us restate the order-theoretic properties given in Exercise 15 for the special case of $\mathbb{R}$. A set $S \subseteq \mathbb{R}$ is said to be **bounded from above** if it has an $\geq$-upper bound, that is, if there is a real number a such that $a \geq s$ for all $s \in S$. In what follows, we shall refer to an $\geq$-upper bound (or the $\geq$-maximum, etc.) of a set in $\mathbb{R}$ simply as an upper bound (or the maximum, etc.) of that set. Moreover, we will denote the $\geq$-supremum of a set $S \subseteq \mathbb{R}$ by $\sup S$. That is, $s^* = \sup S$ iff s^* is an upper bound of S, and $a \geq s^*$ holds for all upper bounds a of S. (The number $\sup S$ is often called the **least upper bound** of S.) The lower bounds of S and $\inf S$ are defined dually. (The number $\inf S$ is called the **greatest lower bound** of S.)

The main difference between $\mathbb{Q}$ and $\mathbb{R}$ is captured by the following property:

> **The Completeness Axiom**
> *Every nonempty subset S of $\mathbb{R}$ that is bounded from above has a supremum in $\mathbb{R}$. That is, if $\emptyset \neq S \subseteq \mathbb{R}$ is bounded from above, then there exists a real number s^* such that $s^* = \sup S$.*

It is indeed this property that distinguishes $\mathbb{R}$ from $\mathbb{Q}$. For instance, $S := \{q \in \mathbb{Q} : q^2 < 2\}$ is obviously a set in $\mathbb{Q}$ that is bounded from above. Yet $\sup S$ does *not* exist in $\mathbb{Q}$, as we will prove shortly. But $\sup S$ exists in $\mathbb{R}$ by the Completeness Axiom (or, as is usually said, by the *completeness of the reals*), and of course, $\sup S = \sqrt{2}$. (This is not entirely trivial; we will prove it shortly.) In an intuitive sense, therefore, $\mathbb{R}$ is obtained from $\mathbb{Q}$ by filling the "holes" in $\mathbb{Q}$ to obtain an ordered field that satisfies the Completeness Axiom. We thus say that $\mathbb{R}$ is a **complete ordered field**.[24]

In the rest of this section, we explore some important consequences of the completeness of the reals. Let us first warm up with an elementary exercise that tells us why we did not need to assume anything about the *greatest lower bound* of a set when stating the Completeness Axiom.

EXERCISE 29[H] Prove: If $\emptyset \neq S \subseteq \mathbb{R}$ and there exists an $a \in \mathbb{R}$ with $a \leq s$ for all $s \in S$, then $\inf S \in \mathbb{R}$.

Here is a result that shows how powerful the Completeness Axiom really is.

PROPOSITION 6

(a) *(The Archimedean Property) For any* $(a, b) \in \mathbb{R}_{++} \times \mathbb{R}$, *there exists an* $m \in \mathbb{N}$ *such that* $b < ma$.

(b) *For any* $a, b \in \mathbb{R}$ *such that* $a < b$, *there exists a* $q \in \mathbb{Q}$ *such that* $a < q < b$.[25]

PROOF

(a) This is an immediate consequence of the completeness of $\mathbb{R}$. Indeed, if the claim was not true, then there would exist a real number $a > 0$ such that $\{ma : m \in \mathbb{N}\}$ is bounded from above. But then $s = \sup\{ma : m \in \mathbb{N}\}$ would be a real number, and hence $a > 0$ would imply that $s - a$ is not an upper bound of $\{ma : m \in \mathbb{N}\}$, that is, there exists an $m^* \in \mathbb{N}$ such that $s < (m^* + 1)a$, which is not possible in view of the choice of s.

24 Actually, one can say a bit more in this junction. $\mathbb{R}$ is not only "a" complete ordered field, it is in fact "the" complete ordered field. To say this properly, let us agree to call an ordered field $(X, \oplus, \odot, \succcurlyeq)$ *complete* if $\sup_{\succcurlyeq} S \in X$ for any $S \in 2^X \backslash \{\emptyset\}$ that has an $\succcurlyeq$-upper bound in X. It turns out that any such ordered field is equivalent to $\mathbb{R}$ up to relabeling. That is, for any complete ordered field $(X, \oplus, \odot, \succcurlyeq)$, there exists a bijection $f : X \to \mathbb{R}$ such that $f(x \oplus y) = f(x) + f(y)$, $f(x \odot y) = f(x)f(y)$, and $x \succcurlyeq y$ iff $f(x) \geq f(y)$. (This is *the Isomorphism Theorem*. McShane and Botts (1959) prove this as Theorem 6.1 (of Chapter 1) in their classic treatment of real analysis (reprinted by Dover in 2005).)

25 We thus say that the rationals are *order-dense* in the reals.

(b) Take any $a, b \in \mathbb{R}$ with $b - a > 0$. By the Archimedean Property, there exists an $m \in \mathbb{N}$ such that $m(b-a) > 1$, that is, $mb > ma + 1$. Define $n := \min\{k \in \mathbb{Z} : k > ma\}$.[26] Then $ma < n \leq 1 + ma < mb$ (why?), so letting $q := \frac{n}{m}$ completes the proof. ∎

EXERCISE 30[H] Show that, for any $a, b \in \mathbb{R}$ with $a < b$, there exists a $c \in \mathbb{R}\backslash\mathbb{Q}$ such that $a < c < b$.

We will make use of Proposition 6.(b) (and hence the Archimedean Property, and hence the Completeness Axiom) on many occasions. Here is a quick illustration. Let $S := \{q \in \mathbb{Q} : q < 1\}$. What is $\sup S$? The natural guess is, of course, that it is 1. Let us prove this formally. First of all, note that S is bounded from above (by 1, in particular), so by the Completeness Axiom, we know that $\sup S$ is a real number. Thus, if $1 \neq \sup S$, then by definition of $\sup S$, we must have $1 > \sup S$. But then by Proposition 6.(b), there exists a $q \in \mathbb{Q}$ such that $1 > q > \sup S$. Yet the latter inequality is impossible, since $q \in S$ and $\sup S$ is an upper bound of S. Hence, $1 = \sup S$.

One can similarly compute the sup and inf of other sets, although the calculations are bound to be a bit tedious at this primitive stage of the development. For instance, let us show that

$$\sup\{q \in \mathbb{Q} : q^2 < 2\} = \sqrt{2}.$$

That is, where $S := \{q \in \mathbb{Q} : q^2 < 2\}$, we wish to show that $\sup S$ is a real number the square of which equals 2. Notice first that S is a nonempty set that is bounded from above, so the Completeness Axiom ensures that $s := \sup S$ is real number. Suppose we have $s^2 > 2$. Then $s^2 - 2 > 0$, so by the Archimedean Property there exists an $m \in \mathbb{N}$ such that $m(s^2 - 2) > 2s$. Then

$$\left(s - \frac{1}{m}\right)^2 = s^2 - \frac{2s}{m} + \frac{1}{m^2} > s^2 - (s^2 - 2) = 2,$$

which means that $\left(s - \frac{1}{m}\right)^2 > q^2$ for all $q \in S$. But then $s - \frac{1}{m}$ is an upper bound for S, contradicting that s is the *smallest* upper bound for S. It follows that we have $s^2 \leq 2$. Good, let us now look at what happens if we have $s^2 < 2$.

[26] By the Archimedean Property, there must exist a $k \in \mathbb{N}$ such that $k > ma$, so n is well-defined.

In that case we use again the Archimedean Property to find an $m \in \mathbb{N}$ such that $m(2-s^2) > 4s$ and $m > \frac{1}{2s}$. Then

$$\left(s+\frac{1}{m}\right)^2 = s^2 + \frac{2s}{m} + \frac{1}{m^2} < s^2 + \frac{2s}{m} + \frac{2s}{m} < s^2 + (2-s^2) = 2.$$

But, by Proposition 6.(b), there exists a $q \in \mathbb{Q}$ with $s < q < s + \frac{1}{m}$. It follows that $s < q \in S$, which is impossible, since s is an upper bound for S. Conclusion: $s^2 = 2$. Put differently, the equation $x^2 = 2$ has a solution in $\mathbb{R}$, thanks to the Completeness Axiom, while it does not have a solution in $\mathbb{Q}$.

EXERCISE 31 Let S be a nonempty subset of $\mathbb{R}$ that is bounded from above. Show that $s^* = \sup S$ iff both of the following two conditions hold:

(i) $s^* \geq s$ for all $s \in S$;

(ii) for any $\varepsilon > 0$, there exists an $s \in S$ such that $s > s^* - \varepsilon$.

EXERCISE 32 Let A and B be two nonempty subsets of $\mathbb{R}$ that are bounded from above. Show that $A \subseteq B$ implies $\sup A \leq \sup B$, and that

$$\sup\{a+b : (a,b) \in A \times B\} = \sup A + \sup B.$$

Moreover, if $c \geq a$ for all $a \in A$, then $c \geq \sup A$.

EXERCISE 33 Let $S \subseteq \mathbb{R}$ be a nonempty set that is bounded from below. Prove that $\inf S = -\sup(-S)$, where $-S := \{-s \in \mathbb{R} : s \in S\}$. Use this result to state and prove the versions of the results reported in Exercises 31 and 32 for nonempty subsets of $\mathbb{R}$ that are bounded from below.

2.4 Intervals and $\overline{\mathbb{R}}$

For any real numbers a and b with $a < b$, the open interval (a,b) is defined as $(a,b) := \{t \in \mathbb{R} : a < t < b\}$, and the semiopen intervals $(a,b]$ and $[a,b)$ are defined as $(a,b] := (a,b) \cup \{b\}$ and $[a,b) := \{a\} \cup (a,b)$, respectively.[27] Finally, the closed interval $[a,b]$ is defined as $[a,b] := \{t \in \mathbb{R} : a \leq t \leq b\}$. Any one of these intervals is said to be **bounded** and of **length** $b-a$. Any

[27] The French tradition is to denote these sets as $]a,b[$, $]a,b]$ and $[a,b[$, respectively. While this convention has the advantage of avoiding use of the same notation for ordered pairs and open intervals, it is not commonly adopted in the literature.

one of them is called **nondegenerate** if $b - a > 0$. In this book, when we write (a, b) or $(a, b]$ or $[a, b)$, we always mean that these intervals are nondegenerate. (We allow for $a = b$ when we write $[a, b]$, however.) We also adopt the following standard notation for **unbounded intervals**: $(a, \infty) := \{t \in \mathbb{R} : t > a\}$ and $[a, \infty) := \{a\} \cup (a, \infty)$. The unbounded intervals $(-\infty, b)$ and $(-\infty, b]$ are defined similarly. By an **open interval**, we mean an interval of the form (a, b), (a, ∞), $(-\infty, b)$, or $\mathbb{R}$; the **closed intervals** are defined similarly.

We have $\sup(-\infty, b) = \sup(a, b) = \sup(a, b] = b$ and $\inf(a, \infty) = \inf(a, b) = \inf[a, \infty) = a$. The Completeness Axiom says that every nonempty subset S of $\mathbb{R}$ that fits in an interval of finite length has both an inf and a sup. Conversely, if S does not fit in any interval of the form $(-\infty, b)$, then $\sup S$ does not exist (i.e., $\sup S \notin \mathbb{R}$). We sometimes indicate that this is the case by writing $\sup S = \infty$, but this is only a notational convention since ∞ is not a real number. (The statement $\inf S = -\infty$ is interpreted similarly.)

It will be convenient on occasion to work with a trivial extension of $\mathbb{R}$ that is obtained by adjoining to $\mathbb{R}$ the symbols $-\infty$ and ∞. The resulting set is called the set of **extended real numbers** and is denoted by $\overline{\mathbb{R}}$. By definition, $\overline{\mathbb{R}} := \mathbb{R} \cup \{-\infty, \infty\}$. We *extend* the linear order $\geq$ of $\mathbb{R}$ to $\overline{\mathbb{R}}$ by letting

$$\infty > -\infty \quad \text{and} \quad \infty > t > -\infty \quad \text{for all } t \in \mathbb{R}, \tag{3}$$

and hence view $\overline{\mathbb{R}}$ itself as a loset. Interestingly, $\overline{\mathbb{R}}$ satisfies the Completeness Axiom. In fact, a major advantage of $\overline{\mathbb{R}}$ is that *every set S in $\overline{\mathbb{R}}$ has a $\geq$-infimum and a $\geq$-supremum.* (Just as in $\mathbb{R}$, we denote these extended real numbers as $\inf S$ and $\sup S$, respectively.) For, if $S \subseteq \overline{\mathbb{R}}$ and $\sup S \notin \mathbb{R}$, then (3) implies that $\sup S = \infty$, and similarly for $\inf S$.[28] In this sense, the supremum (infimum) of a set is quite a different notion than the maximum (minimum) of a set. Recall that, for any set S in $\mathbb{R}$, the **maximum** of S, denoted as $\max S$, is defined to be the number $s^* \in S$, with $s^* \geq s$ for all $s \in S$. (The **minimum** of S, denoted as $\min S$, is defined dually.) Clearly, $\sup(0, 1) = 1$ but $\max(0, 1)$ does not exist. Of course, if S is finite, then both $\max S$ and $\min S$ exist. In general, we have $\sup S = \max S$ and $\inf S = \min S$, provided that $\max S$ and $\min S$ exist.

[28] Even $\sup \emptyset$ is well-defined in $\overline{\mathbb{R}}$. *Quiz.* $\sup \emptyset =$? (*Hint.* $\inf \emptyset > \sup \emptyset$!)

The interval notation introduced above extends readily to $\overline{\mathbb{R}}$. For instance, for any extended real $a > -\infty$, the semiopen interval $[-\infty, a)$ stands for the set $\{t \in \overline{\mathbb{R}} : -\infty \leq t < a\}$. Other types of intervals in $\overline{\mathbb{R}}$ are defined similarly. Clearly, $\min[-\infty, a) = \inf[-\infty, a) = -\infty$ and $\max[-\infty, a) = \sup[-\infty, a) = a$.

Finally, we extend the standard operations of addition and multiplication to $\overline{\mathbb{R}}$ by means of the following definitions: For any $t \in \mathbb{R}$,

$$t + \infty := \infty + t := \infty, \quad t + -\infty := -\infty + t := -\infty, \quad \infty + \infty := \infty,$$

$$-\infty + -\infty := -\infty, \quad t.\infty := \infty.t := \begin{cases} \infty, & \text{if } 0 < t \leq \infty \\ -\infty, & \text{if } -\infty \leq t < 0 \end{cases},$$

and

$$t(-\infty) := (-\infty)t := \begin{cases} -\infty, & \text{if } 0 < t \leq \infty \\ \infty, & \text{if } -\infty \leq t < 0 \end{cases}.$$

WARNING. The expressions $\infty + (-\infty)$, $-\infty + \infty$, $\infty \cdot 0$, and $0 \cdot \infty$ are left undefined, so $\overline{\mathbb{R}}$ cannot be considered a field.

EXERCISE 34 Letting $|t| := t$ for all $t \in [0, \infty]$, and $|t| := -t$ for all $t \in [-\infty, 0)$, show that $|a + b| \leq |a| + |b|$ for all $a, b \in \overline{\mathbb{R}}$ with $a + b \in \overline{\mathbb{R}}$. Also show that $|ab| = |a|\,|b|$ for all $a, b \in \overline{\mathbb{R}} \setminus \{0\}$.

3 Real Sequences

3.1 Convergent Sequences

By a **real sequence**, we mean a sequence in $\mathbb{R}$. The set of all real sequences is thus $\mathbb{R}^{\mathbb{N}}$, but recall that we denote this set instead by $\mathbb{R}^\infty$. We think of a sequence $(x_m) \in \mathbb{R}^\infty$ as *convergent* if there is a real number x such that the later terms of the sequence get arbitrarily close to x. Put precisely, (x_m) is said to **converge** to x if, for each $\varepsilon > 0$, there exists a real number M (that may depend on ε) such that $|x_m - x| < \varepsilon$ for all $m \in \mathbb{N}$ with $m \geq M$.[29]

[29] By the Archimedean Property, we can always choose M to be a natural number, and write "for all $m = M, M + 1, \ldots$" instead of "for all $m \in \mathbb{N}$ with $m \geq M$" in this definition. Since the fact that each m must be a natural number is clear from the context, one often writes simply "for all $m \geq M$" instead of either of these expressions (whether or not $M \in \mathbb{N}$).

In this case, we say that (x_m) is **convergent**, and x is the **limit** of (x_m). We describe this situation by writing $\lim_{m\to\infty} x_m = x$, or $\lim x_m = x$, or simply, $x_m \to x$ (as $m \to \infty$). In words, $x_m \to x$ means that, no matter how small $\varepsilon > 0$ is, all but finitely many terms of the sequence (x_m) are contained in the open interval $(x - \varepsilon, x + \varepsilon)$.[30]

A sequence that does not converge to a real number is called **divergent**. If, for every real number y, there exists an $M \in \mathbb{R}$ with $x_m \geq y$ for each $m \geq M$, then we say that (x_m) **diverges** (or **converges**) **to** ∞, or that "the limit of (x_m) is ∞," and write either $x_m \to \infty$ or $\lim x_m = \infty$. We say that (x_m) **diverges** (or **converges**) **to** $-\infty$, or that "the limit of (x_m) is $-\infty$," and write $x_m \to -\infty$ or $\lim x_m = -\infty$, if $-x_m \to \infty$. (See Figure 1.)

The idea is that the *tail* of a convergent real sequence approximates the limit of the sequence to any desired degree of accuracy. Some initial (finitely many) terms of the sequence may be quite apart from its limit point, but *eventually* all terms of the sequence accumulate around this limit. For instance, the real sequence $(\frac{1}{m})$ and $(y_m) := (1, 2, \ldots, 100, 1, \frac{1}{2}, \frac{1}{3}, \ldots)$ have the same long-run behavior—they both converge to 0—even though their first few terms are quite different from each other. The initial terms of the sequence have no say on the behavior of the tail of the sequence.

To see this more clearly, let us show formally that $\frac{1}{m} \to 0$. To this end, pick an arbitrary $\varepsilon > 0$, and ask if there is an $M \in \mathbb{R}$ large enough to guarantee that $\left|\frac{1}{m} - 0\right| = \left|\frac{1}{m}\right| < \varepsilon$ for all $m \geq M$. In this simple example, the choice is clear. By choosing M to be a number strictly greater than $\frac{1}{\varepsilon}$, we get the desired inequality straightaway. The point is that we can prove that $y_m \to 0$ analogously, except that we need to choose our threshold M larger in this case, meaning that we need to wait a bit longer (in fact, for 100 more "periods") for the terms of (y_m) to enter and never leave the interval $(0, \varepsilon)$.

For another example, note that $((-1)^m)$ and (m) are divergent real sequences. While there is no real number a such that all but finitely many terms of $((-1)^m)$ belong to $(a - \frac{1}{2}, a + \frac{1}{2})$, we have $\lim m = \infty$ by the

[30] While the "idea" of convergence of a sequence was around for some time, we owe this precise formulation to Augustin-Louis Cauchy (1789–1857). It would not be an exaggeration to say that Cauchy is responsible for the emergence of what is called real analysis today. (The same goes for complex analysis too, as a matter of fact.) Just to give you an idea, let me note that it was Cauchy who proved the Fundamental Theorem of Calculus (in 1822) as we know it today (although for uniformly continuous functions). Cauchy published 789 mathematical articles in his lifetime.

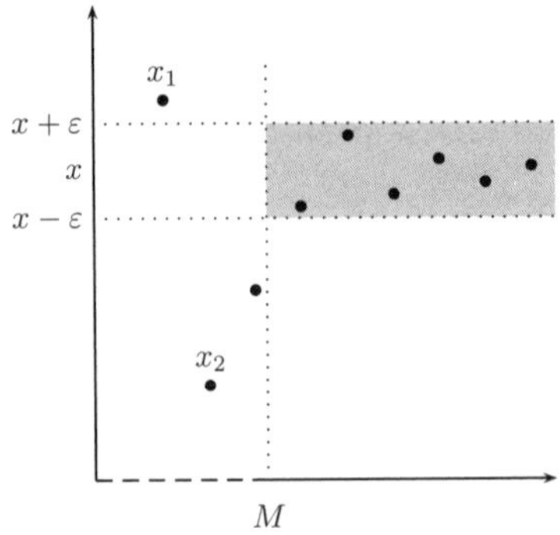

(x_m) converges to x.

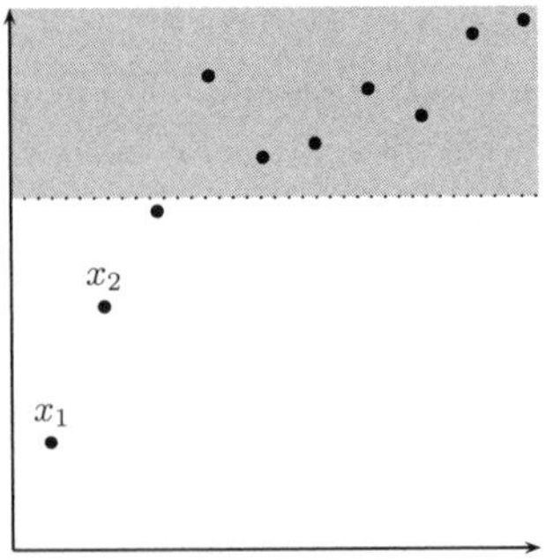

(x_m) diverges to ∞.

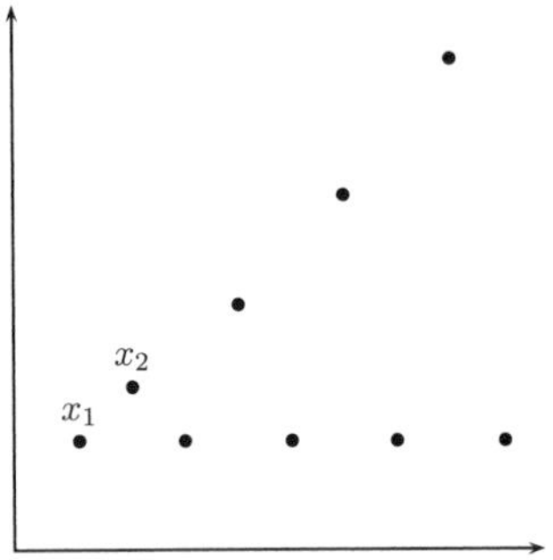

(x_m) is divergent but
it does not diverge to ∞.

FIGURE 1

Archimedean Property. Also note that $\lim a^m = 0$ for any real number a with $|a| < 1$.[31] The following example is also very useful.

> LEMMA 1
> *For any real number $x \in \mathbb{R}$, there exists a sequence (q_m) of rational numbers and (p_m) of irrational numbers such that $q_m \to x$ and $p_m \to x$.*

[31] *Quiz.* Prove this! *Hint.* Use the Principle of Mathematical Induction to obtain first the *Bernoulli Inequality*: $(1+t)^m \geq 1 + mt$ for any $(t, m) \in \mathbb{R} \times \mathbb{N}$. This inequality will make the proof very easy.

PROOF

Take any $x \in \mathbb{R}$, and use Proposition 6.(b) to choose a $q_m \in (x, x + \frac{1}{m})$ for each $m \in \mathbb{N}$. For any $\varepsilon > 0$, by choosing any real number $M > \frac{1}{\varepsilon}$, we find that $|q_m - x| < \frac{1}{m} < \varepsilon$ for all $m \geq M$. Recalling Exercise 30, the second assertion is proved analogously. ■

A real sequence cannot have more than one limit. For, if (x_m) is a convergent real sequence such that $x_m \to x$ and $x_m \to y$ with $x \neq y$, then by choosing $\varepsilon := \frac{1}{2}|x - y|$, we can find an $M > 0$ large enough to guarantee that $|x_m - x| < \frac{\varepsilon}{2}$ and $|x_m - y| < \frac{\varepsilon}{2}$ for all $m \geq M$. Thanks to the triangle inequality, this yields the following contradiction:

$$|x - y| \leq |x - x_m| + |x_m - y| < \frac{\varepsilon}{2} + \frac{\varepsilon}{2} = \varepsilon = \frac{1}{2}|x - y|.$$

Here is another simple illustration of how one works with convergent real sequences in practice. Suppose we are given a sequence $(x_m) \in \mathbb{R}^\infty$ with $x_m \to x \in \mathbb{R}$. We wish to show that if b is real number with $x_m \leq b$ for all m, then we have $x \leq b$. The idea is that if $x > b$ were the case, then, since the terms of (x_m) get eventually very close to x, we would have $x_m > b$ for m large enough. To say this formally, let $\varepsilon := x - b > 0$, and note that there exists an $M \in \mathbb{R}$ such that $|x_m - x| < \varepsilon$ for all $m \geq M$, so $x_M > x - \varepsilon = b$, which contradicts our main hypothesis. Amending this argument only slightly, we can state a more general fact: *For any* $-\infty \leq a < b \leq \infty$, *and convergent* $(x_m) \in [a, b]^\infty$, *we have* $\lim x_m \in [a, b]$.[32]

The following exercises may help you recall some other common tricks that come up when playing with convergent sequences.

EXERCISE 35 Let (x_m) and (y_m) be two real sequences such that $x_m \to x$ and $y_m \to y$ for some real numbers x and y. Prove:

(a) $|x_m| \to |x|$;

(b) $x_m + y_m \to x + y$;

(c) $x_m y_m \to xy$;

(d) $\frac{1}{x_m} \to \frac{1}{x}$, provided that $x, x_m \neq 0$ for each m.

[32] *Reminder.* For any nonempty subset S of $\mathbb{R}$, "$(x_m) \in S^\infty$" means that (x_m) is a real sequence such that $x_m \in S$ for each m. (Recall Section 1.6.)

EXERCISE 36[H] Let (x_m), (y_m) and (z_m) be real sequences such that $x_m \leq y_m \leq z_m$ for each m. Show that if $\lim x_m = \lim z_m = a$, then $y_m \to a$.

3.2 Monotonic Sequences

We say that a real sequence (x_m) is **bounded from above** if $\{x_1, x_2, \ldots\}$ is bounded from above, that is, if there exists a real number K with $x_m \leq K$ for all $m = 1, 2, \ldots$. By the Completeness Axiom, this is equivalent to saying that

$$\sup\{x_m : m \in \mathbb{N}\} < \infty.$$

Dually, (x_m) is said to be **bounded from below** if $\{x_1, x_2, \ldots\}$ is bounded from below, that is, if $\inf\{x_m : m \in \mathbb{N}\} > -\infty$. Finally, (x_m) is called **bounded** if it is bounded from both above and below, that is,

$$\sup\{|x_m| : m \in \mathbb{N}\} < \infty.$$

Boundedness is a property all convergent real sequences share. For, if all but finitely (say, M) many terms of a sequence are at most some $\varepsilon > 0$ away from a fixed number x, then this sequence is bounded either by $|x| + \varepsilon$ or by the largest of the first M terms (in absolute value). This is almost a proof, but let us write things out precisely anyway.

PROPOSITION 7
Every convergent real sequence is bounded.

PROOF
Take any $(x_m) \in \mathbb{R}^\infty$ with $x_m \to x$ for some real number x. Then there must exist a natural number M such that $|x_m - x| < 1$, and hence $|x_m| < |x| + 1$, for all $m \geq M$. But then $|x_m| \leq \max\{|x| + 1, |x_1|, \ldots, |x_M|\}$ for all $m \in \mathbb{N}$. ■

The converse of Proposition 7 does not hold, of course. (Think of the sequence $((-1)^m)$, for instance.) However, there is one very important class of bounded sequences that always converge.

Definition

A real sequence (x_m) is said to be **increasing** if $x_m \leq x_{m+1}$ for each $m \in \mathbb{N}$, and **strictly increasing** if $x_m < x_{m+1}$ for each $m \in \mathbb{N}$. It is said to be **(strictly) decreasing** if $(-x_m)$ is (strictly) increasing. Finally, a real sequence which is either increasing or decreasing is referred to as a **monotonic** sequence.[33] If (x_m) is increasing and converges to $x \in \overline{\mathbb{R}}$, then we write $x_m \nearrow x$, and if it is decreasing and converges to $x \in \overline{\mathbb{R}}$, we write $x_m \searrow x$.

The following fact attests to the importance of monotonic sequences. We owe it to the Completeness Axiom.

Proposition 8

Every increasing (decreasing) real sequence that is bounded from above (below) converges.

Proof

Let $(x_m) \in \mathbb{R}^\infty$ be an increasing sequence which is bounded from above, and let $S := \{x_1, x_2, \ldots\}$. By the Completeness Axiom, $x := \sup S \in \mathbb{R}$. We claim that $x_m \nearrow x$. To show this, pick an arbitrary $\varepsilon > 0$. Since x is the least upper bound of S, $x - \varepsilon$ cannot be an upper bound of S, so $x_M > x - \varepsilon$ for some $M \in \mathbb{N}$. Since (x_m) is increasing, we must then have $x \geq x_m \geq x_M > x - \varepsilon$, so $|x_m - x| < \varepsilon$, for all $m \geq M$. The proof of the second claim is analogous. ■

Proposition 8 is an extremely useful observation. For one thing, monotonic sequences are not terribly hard to come by. In fact, within every real sequence there is one!

Proposition 9

Every real sequence has a monotonic subsequence.

[33] That is, an increasing (decreasing) real sequence is an increasing (decreasing) real function on $\mathbb{N}$. Never forget that a real sequence is just a special kind of a real function.

PROOF
(Thurston) Take any $(x_m) \in \mathbb{R}^\infty$ and define $S_m := \{x_m, x_{m+1}, \ldots\}$ for each $m \in \mathbb{N}$. If there is no maximum element in S_1, then it is easy to see that (x_m) has a monotonic subsequence. (Let $x_{m_1} := x_1$, let x_{m_2} be the first term in the sequence $(x_2, x_3, \ldots)$ greater than x_1, let x_{m_3} be the first term in the sequence $(x_{m_2+1}, x_{m_2+2}, \ldots)$ greater than x_{m_2}, and so on.) By the same logic, if for any $m \in \mathbb{N}$ there is no maximum element in S_m, then we are done. Assume, then, max S_m exists for each $m \in \mathbb{N}$. Now define the subsequence (x_{m_k}) recursively as follows:

$$x_{m_1} := \max S_1, \qquad x_{m_2} := \max S_{m_1+1}, \qquad x_{m_3} := \max S_{m_2+1}, \qquad \ldots.$$

Clearly, (x_{m_k}) is decreasing. ■

Putting the last two observations together, we get the following famous result as an immediate corollary.

THE BOLZANO-WEIERSTRASS THEOREM.[34]
Every bounded real sequence has a convergent subsequence.

EXERCISE 37 Show that every unbounded real sequence has a subsequence that diverges to either ∞ or $-\infty$.

EXERCISE 38[H] Let S be a nonempty bounded subset of $\mathbb{R}$. Show that there is an increasing sequence $(x_m) \in S^\infty$ such that $x_m \nearrow \sup S$, and a decreasing sequence $(y_m) \in S^\infty$ such that $y_m \searrow \inf S$.

EXERCISE 39 For any real number x and $(x_m) \in \mathbb{R}^\infty$, show that $x_m \to x$ iff every subsequence of (x_m)has itself a subsequence that converges to x.

[34] Bernhard Bolzano (1781–1848) was one of the early founders of real analysis. Much of his work was found too unorthodox by his contemporaries and so was ignored. The depth of his discoveries was understood only after his death, after a good number of them were rediscovered and brought to light by Karl Weierstrass (1815–1897). The Bolzano-Weierstrass Theorem is perhaps best viewed as an outcome of an intertemporal (in fact, intergenerational) collaboration.

Exercise 40[H] (*The Cauchy Criterion*) We say that an $(x_m) \in \mathbb{R}^\infty$ is a **real Cauchy sequence** if, for any $\varepsilon > 0$, there exists an $M \in \mathbb{R}$ such that $|x_k - x_l| < \varepsilon$ for all $k, l \geq M$.

(a) Show that every real Cauchy sequence is bounded.

(b) Show that every real Cauchy sequence converges.

A double real sequence $(x_{kl}) \in \mathbb{R}^{\infty \times \infty}$ is said to **converge** to $x \in \mathbb{R}$, denoted as $x_{kl} \to x$, if, for each $\varepsilon > 0$, there exists a real number M (that may depend on ε) such that $|x_{kl} - x| < \varepsilon$ for all $k, l \geq M$. The following exercise tells us when one can conclude that (x_{kl}) converges by looking at the behavior of (x_{kl}) first as $k \to \infty$ and then as $l \to \infty$ (or vice versa).

Exercise 41[H] (*The Moore-Osgood Theorem*) Take any $(x_{kl}) \in \mathbb{R}^{\infty \times \infty}$ such that there exist $(y_k) \in \mathbb{R}^\infty$ and $(z_l) \in \mathbb{R}^\infty$ such that

(i) for any $\varepsilon > 0$, there exists an $L \in \mathbb{N}$ such that $\left|x_{kl} - y_k\right| < \varepsilon$ for all $k \geq 1$ and $l \geq L$; and

(ii) for any $\varepsilon > 0$ and $l \in \mathbb{N}$, there exists a $K_l \in \mathbb{N}$ such that $|x_{kl} - z_l| < \varepsilon$ for all $k \geq K_l$.

(a) Prove that there exists an $x \in \mathbb{R}$ such that $x_{kl} \to x$ and

$$\lim_{k\to\infty} \lim_{l\to\infty} x_{kl} = x = \lim_{l\to\infty} \lim_{k\to\infty} x_{kl}. \tag{4}$$

(b) Check if (4) holds for the double sequence $\left(\frac{kl}{k^2+l^2}\right)$. What goes wrong?

3.3 Subsequential Limits

Any subsequence of a convergent real sequence converges to the limit of the mother sequence. (Why?) What is more, even if the mother sequence is divergent, it may still possess a convergent subsequence (as in the Bolzano-Weierstrass Theorem). This suggests that we can get at least some information about the long-run behavior of a sequence by studying those points to which at least one subsequence of the sequence converges. Given any $(x_m) \in \mathbb{R}^\infty$, we say that $x \in \overline{\mathbb{R}}$ is a **subsequential limit** of (x_m) if there exists a subsequence (x_{m_k}) with $x_{m_k} \to x$ (as $k \to \infty$). For instance, -1 and 1 are the only subsequential limits of $((-1)^m)$, and -1, 1 and ∞ are the only subsequential limits of the sequence (x_m) where $x_m = -1$ for each odd m not divisible by 3, $x_m = 1$ for each even m, and $x_m = m$ for each odd m divisible by 3.

If x is a subsequential limit of (x_m), then we understand that (x_m) visits the interval $(x - \varepsilon, x + \varepsilon)$ *infinitely often*, no matter how small $\varepsilon > 0$ is. It is in this sense that subsequential limits give us asymptotic information about the long-run behavior of a real sequence. Of particular interest in this regard are the largest and smallest subsequential limits of a real sequence. These are called the **limit superior** (abbreviated as lim sup) and **limit inferior** (abbreviated as lim inf) of a real sequence.

Definition

For any $x \in \mathbb{R}$ and $(x_m) \in \mathbb{R}^\infty$, we write $\limsup x_m = x$ if

(i) for any $\varepsilon > 0$, there exists an $M > 0$ such that $x_m < x + \varepsilon$ for all $m \geq M$,

(ii) for any $\varepsilon > 0$ and $m \in \mathbb{N}$, there exists an integer $k > m$ such that $x_k > x - \varepsilon$.

We write $\limsup x_m = \infty$ if ∞ is a subsequential limit of (x_m); and $\limsup x_m = -\infty$ if $x_m \to -\infty$. The expression $\liminf x_m$ is defined dually (or by letting $\liminf x_m := -\limsup(-x_m)$).

If $\limsup x_m = x \in \mathbb{R}$, we understand that all but finitely many terms of the sequence are smaller than $x + \varepsilon$, no matter how small $\varepsilon > 0$ is. (Such a sequence is thus bounded from above, but it need not be bounded from below.) If $x = \lim x_m$ was the case, we could say in addition to this that all but finitely many terms of (x_m) are also larger than $x - \varepsilon$, no matter how small $\varepsilon > 0$ is. When $x = \limsup x_m$, however, all we can say in this regard is that infinitely many terms of (x_m) are larger than $x - \varepsilon$, no matter how small $\varepsilon > 0$ is. That is, if $x = \limsup x_m$, then the terms of the sequence (x_m) need not accumulate around x; it is just that all but finitely many of them are in $(-\infty, x + \varepsilon)$, and infinitely many of them are in $(x - \varepsilon, x + \varepsilon)$, no matter how small $\varepsilon > 0$ is. (See Figure 2.) The expression $\liminf x_m = x$ is similarly interpreted. For instance, $\lim(-1)^m$ does not exist, but $\limsup(-1)^m = 1$ and $\liminf(-1)^m = -1$.

It is easy to see that any real sequence (x_m) has a monotonic subsequence (x_{m_k}) such that $x_{m_k} \to \limsup x_m$. (For, $\limsup x_m$ is a subsequential limit of (x_m) (why?), so the claim obtains upon applying Proposition 9 to a

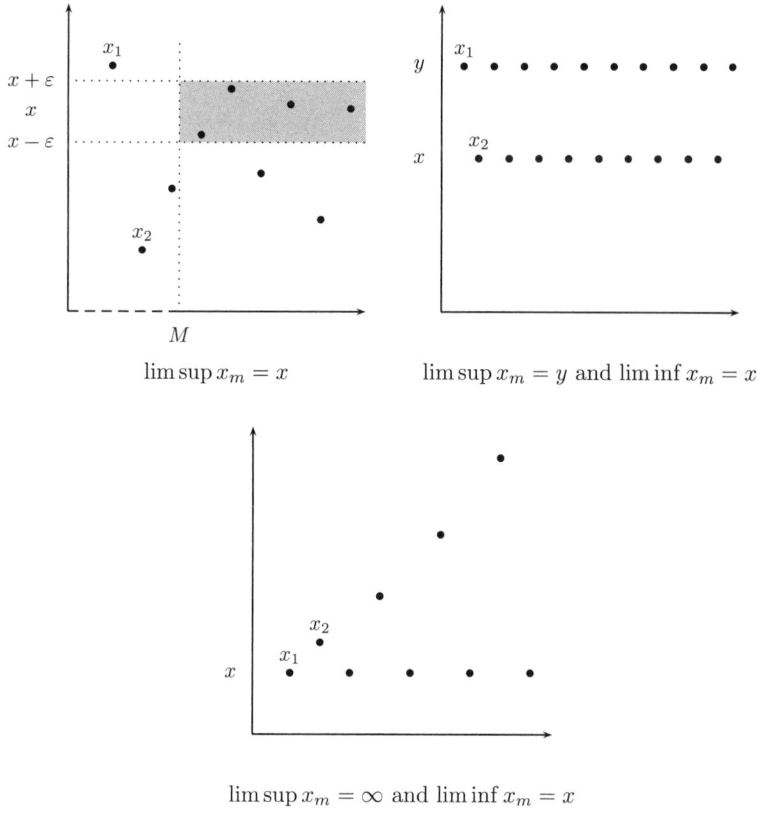
x_1
$x + \varepsilon$
x
$x - \varepsilon$
x_2
M
$\limsup x_m = x$
y
x_1
x
x_2
$\limsup x_m = y$ and $\liminf x_m = x$
x_2
x_1
x
$\limsup x_m = \infty$ and $\liminf x_m = x$

Figure 2

subsequence of (x_m) that converges to $\limsup x_m$.) Of course, the analogous claim is true for $\liminf x_m$ as well. It also follows readily from the definitions that, for any $(x_m) \in \mathbb{R}^\infty$,

$$\liminf x_m \leq \limsup x_m,$$

and

$$(x_m) \text{ is convergent} \quad \text{iff} \quad \liminf x_m = \limsup x_m.$$

(Right?) Thus, to prove that a real sequence (x_m) converges, it is enough to show that $\liminf x_m \geq \limsup x_m$, which is sometimes easier than adopting the direct approach. The following exercises outline some other facts concerning the lim sup and lim inf of real sequences. If you're not already familiar with these concepts, it is advisable that you work through these before proceeding further.

EXERCISE 42 Let (x_m) be a real sequence and $x \in \overline{\mathbb{R}}$. Show that the following statements are equivalent:

(i) $\limsup x_m = x$.

(ii) x is the largest subsequential limit of (x_m).

(iii) $x = \inf\{\sup\{x_m, x_{m+1}, \ldots\} : m = 1, 2, \ldots\}$.

State and prove the analogous result for the lim inf of (x_m).

A Corollary of Exercise 42. The lim sup and lim inf of any real sequence exist in $\overline{\mathbb{R}}$.

EXERCISE 43[H] Prove: For any bounded real sequences (x_m) and (y_m), we have

$$\begin{aligned}\liminf x_m + \liminf y_m &\leq \liminf(x_m + y_m)\\ &\leq \limsup(x_m + y_m)\\ &\leq \limsup x_m + \limsup y_m.\end{aligned}$$

Also, give an example for which *all* of these inequalities hold strictly.

EXERCISE 44 Prove: For any $x \geq 0$ and $(x_m), (y_m) \in \mathbb{R}^\infty$ with $x_m \to x$, we have $\limsup x_m y_m = x \limsup y_m$.

3.4 Infinite Series

Let (x_m) be a real sequence. We define

$$\sum_{i=1}^{m} x_i := x_1 + \cdots + x_m \quad \text{and} \quad \sum_{i=k}^{m} x_i := \sum_{i=1}^{m-k+1} x_{i+k-1}$$

for any $m \in \mathbb{N}$ and $k \in \{1, \ldots, m\}$.[35] For simplicity, however, we often write $\sum^m x_i$ for $\sum_{i=1}^m x_i$ within the text. For any nonempty finite subset S of $\mathbb{N}$, we write $\sum_{i \in S} x_i$ to denote the sum of all terms of (x_m) the indices of which belong to S.[36]

[35] There is no ambiguity in the definition of $\sum_{i=1}^m x_i$, precisely because the addition operation on $\mathbb{R}$ is associative.

[36] Formally speaking, $\sum_{i \in S} x_i := \sum_{i=1}^{|S|} x_{\sigma(i)}$, where σ is *any* bijection from $\{1, \ldots, |S|\}$ onto S.

Convention. For any $(x_m) \in \mathbb{R}^\infty$, we let $\sum_{i\in\emptyset} x_i := 0$ in this text. This is nothing more than a notational convention.

By an **infinite series**, we mean a real sequence of the form $(\sum^m x_i)$ for some $(x_m) \in \mathbb{R}^\infty$. When the limit of this sequence exists in $\overline{\mathbb{R}}$, we denote it as $\sum_{i=1}^\infty x_i$, but, again, we write $\sum^\infty x_i$ for $\sum_{i=1}^\infty x_i$ within the text. That is,

$$\sum_{i=1}^{\infty} x_i = \lim_{m\to\infty} \sum_{i=1}^{m} x_i,$$

provided that $(\sum^m x_i)$ converges in $\overline{\mathbb{R}}$. Similarly,

$$\sum_{i=k}^{\infty} x_i = \sum_{i=1}^{\infty} x_{i+k-1}, \quad k = 1, 2, \ldots,$$

provided that the right-hand side is well-defined. We say that an infinite series $(\sum^m x_i)$ is **convergent** if it has a finite limit (i.e., $\sum^\infty x_i \in \mathbb{R}$). In this case, with a standard abuse of terminology, we say that "the series $\sum^\infty x_i$ is convergent." If $(\sum^m x_i)$ diverges to ∞ or $-\infty$, that is, $\sum^\infty x_i \in \{-\infty, \infty\}$, then we say that the series is **divergent**. With the same abuse of terminology, we say then that "the series $\sum^\infty x_i$ is divergent."

Warning. In the present terminology, $\sum^\infty x_i$ may not be well-defined. For instance, the infinite series $(\sum^m (-1)^i)$ does not have a limit, so the notation $\sum^\infty (-1)^i$ is meaningless. Before dealing with an object like $\sum^\infty x_i$ in practice, you should first make sure that it is well-defined.

It is useful to note that the convergence of $\sum^\infty x_i$ implies $\lim x_m = 0$, but not conversely. For,

$$\lim_{m\to\infty} x_m = \lim_{m\to\infty} \left(\sum_{i=1}^{m+1} x_i - \sum_{i=1}^{m} x_i \right) = \lim_{m\to\infty} \sum_{i=1}^{m+1} x_i - \lim_{m\to\infty} \sum_{i=1}^{m} x_i = 0$$

where, given that $(\sum^m x_i)$ is convergent, the second equality follows from Exercise 35.(b). (What about the third equality?) The series $\sum^\infty \frac{1}{i}$, on the other hand, diverges to ∞, so the converse of this observation does not hold in general.[37]

Here are a few examples that come up frequently in applications.

[37] Consider the sequence $(y_m) := \left(\frac{1}{2}, \frac{1}{4}, \frac{1}{4}, \frac{1}{8}, \frac{1}{8}, \frac{1}{8}, \frac{1}{8}, \frac{1}{16}, \frac{1}{16}, \ldots\right)$, and check that $\sum^\infty \frac{1}{i} \geq \sum^\infty y_i = \infty$.

EXAMPLE 8

[1] $\sum^{\infty} \frac{1}{i^{\alpha}}$ *converges iff* $\alpha > 1$. This is easily proved by calculus: For any $m \in \mathbb{N}$ and $\alpha > 1$,

$$\sum_{i=1}^{m} \frac{1}{i^{\alpha}} \leq 1 + \int_{1}^{m} \frac{1}{t^{\alpha}} dt = 1 + \frac{1}{\alpha-1}\left(1 - \frac{1}{m^{\alpha-1}}\right) < 1 + \frac{1}{\alpha-1} = \frac{\alpha}{\alpha-1}.$$

(Draw a picture to see why the first inequality is true.) Thus, $\sum^{\infty} \frac{1}{i^{\alpha}} \leq \frac{\alpha}{\alpha-1}$ whenever $\alpha > 1$. Conversely, $\sum^{\infty} \frac{1}{i^{\alpha}} \geq \sum^{\infty} \frac{1}{i} = \infty$ for any $\alpha \leq 1$.

[2] $\sum^{\infty} \frac{1}{2^i} = 1$. For, by using (2), we have $\lim \sum^{m} \frac{1}{2^i} = \lim \left(1 - \frac{1}{2^i}\right) = 1$. The next example generalizes this useful observation.

[3] $\sum^{\infty} \delta^i = \frac{\delta}{1-\delta}$ for any $-1 < \delta < 1$. To prove this, observe that

$$(1 + \delta + \cdots + \delta^m)(1 - \delta) = 1 - \delta^{m+1}, \qquad m = 1, 2, \ldots$$

so that, for any $\delta \neq 1$, we have

$$\sum_{i=1}^{\infty} \delta^i = \lim_{m\to\infty} \sum_{i=1}^{m} \delta^i = \lim_{m\to\infty} \frac{1-\delta^{m+1}}{1-\delta} - 1 = \frac{\delta - \lim \delta^{m+1}}{1-\delta}.$$

But when $|\delta| < 1$, we have $\lim \delta^{m+1} = 0$ (as you were asked to prove about ten pages ago), and hence the claim. □

EXERCISE 45[H] For any infinite series $(\sum^m x_i)$, prove:
(a) If $\sum^{\infty} x_i$ converges, then $\lim_{k\to\infty} \sum_{i=k}^{\infty} x_i = 0$;
(b) If $\sum^{\infty} x_i$ converges, $\left|\sum_{i=k}^{\infty} x_i\right| \leq \sum_{i=k}^{\infty} |x_i|$ for any $k \in \mathbb{N}$.

EXERCISE 46 Prove: If $(x_m) \in \mathbb{R}^{\infty}$ is a decreasing sequence such that $\sum^{\infty} x_i$ converges, then $mx_m \to 0$.

EXERCISE 47 Let $0! := 1$, and define $m! := ((m-1)!)\, m$ for any $m \in \mathbb{N}$. Prove that $\lim \left(1 + \frac{1}{m}\right)^m = 1 + \sum^{\infty} \frac{1}{i!}$. (*Note.* The common value of these expressions equals the real number $e = 2.71\ldots$. Can you show that e is irrational, by the way?)

*EXERCISE 48[H] Let (x_m) be a real sequence, and $s_j := \sum^j x_i, j = 1, 2, \ldots$.
(a) Give an example to show that $(\frac{1}{m} \sum^m s_i)$ may converge even if $\sum^{\infty} x_i$ is not convergent.
(b) Show that if $\sum^{\infty} x_i$ is convergent, then $\lim \frac{1}{m} \sum^m s_i = \sum^{\infty} x_i$.

*EXERCISE 49 Prove *Tannery's Theorem*: Take any $(x_{kl}) \in \mathbb{R}^{\infty\times\infty}$ such that $\sum_{j=1}^{\infty} x_{kj}$ converges for each k and $(x_{1l}, x_{2l}, \ldots)$ converges for each l. If there exists a real sequence $(K_1, K_2, \ldots)$ such that $|x_{kl}| \leq K_l$ for each l, and $\sum^{\infty} K_i$ converges, then

$$\lim_{k\to\infty} \sum_{j=1}^{\infty} x_{kj} = \sum_{j=1}^{\infty} \lim_{k\to\infty} x_{kj}.$$

3.5 Rearrangement of Infinite Series

An issue that arises frequently in practice concerns the *rearrangement* of an infinite series. The question is if, and when, one can sum the terms of a given real sequence in different orders and obtain the same number in result (as it would be the case for any n-vector). Let's consider an example that points to the fact that the problem is not trivial. Fix any $\alpha \geq 1$. It is easily checked that $\sum^{\infty} \frac{1}{2i-1} = \infty$, so there must exist a smallest natural number $m_1 \geq 2$ with

$$\sum_{i=1}^{m_1} \frac{1}{2i-1} > \alpha.$$

Due to the choice of m_1, we have

$$\sum_{i=1}^{m_1} \frac{1}{2i-1} - \frac{1}{3} \leq \sum_{i=1}^{m_1} \frac{1}{2i-1} - \frac{1}{2m_1-1} \leq \sum_{i=1}^{m_1-1} \frac{1}{2i-1} \leq \alpha.$$

(Why?) Now let m_2 be the smallest number in $\{m_1+1, m_1+2, \ldots\}$ such that

$$\sum_{i=1}^{m_1} \frac{1}{2i-1} - \frac{1}{3} + \sum_{i=m_1+1}^{m_2} \frac{1}{2i-1} > \alpha$$

which implies

$$\sum_{i=1}^{m_1} \frac{1}{2i-1} - \frac{1}{3} + \sum_{i=m_1+1}^{m_2} \frac{1}{2i-1} - \frac{1}{9} \leq \alpha.$$

Continuing this way inductively, we obtain the sequence

$$(x_m) := \left(1, \tfrac{1}{3}, \ldots, \tfrac{1}{2m_1-1}, -\tfrac{1}{2}, \tfrac{1}{2m_1+1}, \ldots, \tfrac{1}{2m_2-1}, -\tfrac{1}{4}, \ldots\right).$$

The upshot is that we have $\sum^{\infty} x_m = \alpha$. (Check this!) Yet the sequence (x_m) is none other than the rearrangement of the sequence $\left(\frac{(-1)^{m+1}}{m}\right)$, so the series $\sum^\infty x_m$ is equal to the series $\sum^\infty \frac{(-1)^{i+1}}{i}$, except that it is summed in a different order. If such a rearrangement does not affect the value of the series, then we must conclude that $\sum^\infty \frac{(-1)^{i+1}}{i} = \alpha$. But this is absurd, for $\alpha \geq 1$ is completely arbitrary here; for instance, our conclusion yields $1 = \sum^\infty \frac{(-1)^{i+1}}{i} = 2$.[38]

This example tells us that one has to be careful in rearranging a given infinite series. Fortunately, there would be no problem in this regard if all terms of the series were nonnegative (or all were nonpositive). This simple fact is established next.

PROPOSITION 10

For any given $(x_m) \in \mathbb{R}_+^\infty$ *and bijection* $\sigma : \mathbb{N} \to \mathbb{N}$, *we have* $\sum^\infty x_i = \sum^\infty x_{\sigma(i)}$.

PROOF

Since σ is bijective, for any given $m \in \mathbb{N}$ there exist integers K_m and L_m such that $K_m \geq L_m \geq m$ and $\{1, \ldots, m\} \subseteq \{\sigma(1), \ldots, \sigma(L_m)\} \subseteq \{1, \ldots, K_m\}$. So, by nonnegativity, $\sum^m x_i \leq \sum^{L_m} x_{\sigma(i)} \leq \sum^{K_m} x_i$. Letting $m \to \infty$ yields the claim. ■

The following result gives another condition that is sufficient for any rearrangement of an infinite series to converge to the same limit as the original series. This result is often invoked when Proposition 10 does not apply because the series at hand may have terms that alternate in sign.

DIRICHLET'S REARRANGEMENT THEOREM

For any given $(x_m) \in \mathbb{R}^\infty$ *and any bijection* $\sigma : \mathbb{N} \to \mathbb{N}$, *we have* $\sum^\infty x_i = \sum^\infty x_{\sigma(i)}$, *provided that* $\sum^\infty |x_i|$ *converges.*

[38] This is not an idle example. According to a theorem of Bernhard Riemann that was published (posthumously) in 1867, for any convergent infinite series $\sum^\infty x_i$ such that $\sum^\infty |x_i| = \infty$ (such a series is called *conditionally convergent*), and any $\alpha \in \overline{\mathbb{R}}$, there exists a bijection $\sigma : \mathbb{N} \to \mathbb{N}$ such that $\sum^\infty x_{\sigma(i)} = \alpha$. (The proof is analogous to the one I gave above to show that the series $\sum^\infty \frac{(-1)^{i+1}}{i}$ can be rearranged to converge to *any* number.)

Bernhard Riemann (1826–1865) is a towering figure in mathematics. Argued by some to be the best mathematician who ever lived, in his short lifetime he revolutionized numerous subjects, ranging from complex and real analysis to geometry and mathematical physics. There are many books about the life and genius of this great man; I would recommend Laugwitz (1999) for an engaging account.

Proof

Note first that both $\sum^\infty x_i$ and $\sum^\infty x_{\sigma(i)}$ converge. (For instance, $\left(\sum^m x_{\sigma(i)}\right)$ is convergent, because $\sum^m x_{\sigma(i)} \leq \sum^m \left|x_{\sigma(i)}\right| \leq \sum^\infty |x_i|$ for any $m \in \mathbb{N}$.) Define $s_m := \sum^m x_i$ and $t_m := \sum^m x_{\sigma(i)}$ for each m, and let $s := \sum^\infty x_i$ and $t := \sum^\infty x_{\sigma(i)}$. We wish to show that $s = t$. For any $\varepsilon > 0$, we can clearly find an $M \in \mathbb{N}$ such that $\sum_{i=M}^\infty |x_i| < \frac{\varepsilon}{3}$ and $\sum_{i=M}^\infty \left|x_{\sigma(i)}\right| < \frac{\varepsilon}{3}$ (Exercise 45). Now choose $K \in \mathbb{N}$ large enough to guarantee that $\{1, \ldots, M\} \subseteq \{\sigma(1), \ldots, \sigma(K)\}$. Then, for any positive integer $k > K$, we have $\sigma(k) > M$, so letting $S_k := \{i \in \{1, \ldots, k\} : \sigma(i) > M\}$, we have

$$|t_k - s_M| = \left|x_{\sigma(1)} + \cdots + x_{\sigma(k)} - x_1 - \cdots - x_M\right|$$
$$\leq \sum_{i \in S_k} \left|x_{\sigma(i)}\right| \leq \sum_{i=k+1}^{\infty} |x_i| < \tfrac{\varepsilon}{3}.$$

(Recall Exercise 27.) But then, for any $k > K$,

$$|t - s| \leq |t - t_k| + |t_k - s_M| + |s_M - s| < \tfrac{\varepsilon}{3} + \tfrac{\varepsilon}{3} + \tfrac{\varepsilon}{3} = \varepsilon.$$

Since $\varepsilon > 0$ is arbitrary here, this proves that $s = t$. ∎

3.6 Infinite Products

Let (x_m) be a real sequence. We define

$$\prod_{i=1}^{m} x_i := x_1 \cdots x_m \qquad \text{for any } m = 1, 2, \ldots,$$

but write $\prod^m x_i$ for $\prod_{i=1}^m x_i$ within the text. By an **infinite product**, we mean a real sequence of the form $(\prod^m x_i)$ for some $(x_m) \in \mathbb{R}^\infty$. When the limit of this sequence exists in $\overline{\mathbb{R}}$, we denote it by $\prod_{i=1}^\infty x_i$. (But again, we often write $\prod^\infty x_i$ for $\prod_{i=1}^\infty x_i$ to simplify the notation.) That is,

$$\prod_{i=1}^{\infty} x_i := \lim_{m\to\infty} \prod_{i=1}^{m} x_i,$$

provided that $(\prod^m x_i)$ converges in $\overline{\mathbb{R}}$. We say that $(\prod^m x_i)$ (or, abusing terminology, $\prod^\infty x_i$) is **convergent** if $\lim \prod^m x_i \in \mathbb{R}$. If $(\prod^m x_i)$ diverges to ∞ or $-\infty$, that is, $\prod^\infty x_i \in \{-\infty, \infty\}$, then we say that the infinite product (or, $\prod^\infty x_i$) is **divergent**.

EXERCISE 50 For any $(x_m) \in \mathbb{R}^\infty$, prove the following statements.

(a) If there is an $0 < \varepsilon < 1$ such that $0 \leq |x_m| < 1 - \varepsilon$ for all but finitely many m, then $\prod^\infty x_i = 0$. (Can we take $\varepsilon = 0$ in this claim?)

(b) If $\prod^\infty x_i$ converges to a positive number, then $x_m \to 1$.

(c) If $x_m \geq 0$ for each m, then $\prod^\infty (1 + x_i)$ converges iff $\sum^\infty x_i$ converges.

4 Real Functions

This section is a refresher on the theory of real functions on $\mathbb{R}$. Because you are familiar with the elements of this theory, we go as fast as possible. Most of the proofs are either left as exercises or given only in brief sketches.

4.1 Basic Definitions

By a **real function** (or a **real-valued function**) on a nonempty set T, we mean an element of $\mathbb{R}^T$. If $f \in \mathbb{R}^T$ equals the real number a everywhere, that is, if $f(t) = a$ for all $t \in T$, then we write $f = a$. If $f \neq a$, it follows that $f(t) \neq a$ for *some* $t \in T$. Similarly, if $f, g \in \mathbb{R}^T$ are such that $f(t) \geq g(t)$ for all $t \in T$, we write $f \geq g$. If $f \geq g$ but not $g \geq f$, we then write $f > g$. If, on the other hand, $f(t) > g(t)$ for all $t \in T$, then we write $f \gg g$. The expressions $f \leq g$, $f < g$ and $f \ll g$ are understood similarly. Note that $\geq$ is a partial order on $\mathbb{R}^T$ which is linear iff $|T| = 1$.

We define the addition and multiplication of real functions by using the binary operations $+$ and $\cdot$ *pointwise*. That is, for any $f, g \in \mathbb{R}^T$, we define $f + g$ and $fg \in \mathbb{R}^T$ as the real functions on T with

$$(f + g)(t) := f(t) + g(t) \quad \text{and} \quad (fg)(t) := f(t)g(t) \quad \text{for all } t \in T.$$

Similarly, for any $a \in \mathbb{R}$, the map $af \in \mathbb{R}^T$ is defined by $(af)(t) := af(t)$. The subtraction operation is then defined on $\mathbb{R}^T$ in the straightforward way: $f - g := f + (-1)g$ for each $f, g \in \mathbb{R}^T$. Provided that $g(t) \neq 0$ for all $t \in T$, we also define $\frac{f}{g} \in \mathbb{R}^T$ by $\left(\frac{f}{g}\right)(t) := \frac{f(t)}{g(t)}$.

REMARK 1. Let $n \in \mathbb{N}$. By setting $T := \{1, \ldots, n\}$, we see that the definitions above also tell us how we order, add and multiply vectors in $\mathbb{R}^n$. In particular,

$\geq$ is none other than the natural order of $\mathbb{R}^n$. Moreover, for any $\lambda \in \mathbb{R}$ and real n-vectors $x := (x_1, \ldots, x_n)$ and $y := (y_1, \ldots, y_n)$, we have

$$x + y = (x_1 + y_1, \ldots, x_n + y_n) \quad \text{and} \quad \lambda x = (\lambda x_1, \ldots, \lambda x_n).$$

Of course, these are the natural addition and scalar multiplication operations in $\mathbb{R}^n$; when we talk about $\mathbb{R}^n$ we always have these operation in mind. In particular, these operations are used to define the **line segment between** x **and** y algebraically as $\{\lambda x + (1 - \lambda)y : 0 \leq \lambda \leq 1\}$.

Similar remarks apply to real matrices and sequences. Indeed, given any positive integers m and n, by setting $T := \{1, \ldots, m\} \times \{1, \ldots, n\}$, we obtain the definitions for ordering, summing and multiplying by a real number the members of $\mathbb{R}^{m\times n}$. Similarly, by setting $T := \mathbb{N}$, we find out about the situation for $\mathbb{R}^\infty$. For instance, for any real number λ and any matrices $[a_{ij}]_{m\times n}$ and $[b_{ij}]_{m\times n}$, we have

$$[a_{ij}]_{m\times n} + [b_{ij}]_{m\times n} = [a_{ij} + b_{ij}]_{m\times n} \quad \text{and} \quad \lambda[a_{ij}]_{m\times n} = [\lambda a_{ij}]_{m\times n}.$$

Similarly, for any $\lambda \in \mathbb{R}$ and any $(x_m), (y_m) \in \mathbb{R}^\infty$, we have $(x_m) + (y_m) = (x_m + y_m)$ and $\lambda(x_m) = (\lambda x_m)$, while $(x_m) \geq (0, 0, \ldots)$ means that $x_m \geq 0$ for each m. □

When $|T| \geq 2$, $(\mathbb{R}^T, +, \cdot)$ is not a field, because not every map in $\mathbb{R}^T$ has a multiplicative inverse. (What is the inverse of the map that equals 0 at a given point and 1 elsewhere, for instance?) Nevertheless, $\mathbb{R}^T$ has a pretty rich algebraic structure. In particular, it is a partially ordered linear space (see Chapter F).

When the domain of a real function is a poset, we can talk about how this map affects the ordering of things in its domain. Of particular interest in this regard is the concept of a *monotonic* function defined on a subset of $\mathbb{R}^n$, $n \in \mathbb{N}$. (Of course, we think of $\mathbb{R}^n$ as a poset with respect to its natural order (Example 2.[3]).) For any $\emptyset \neq T \subseteq \mathbb{R}^n$, the map $f \in \mathbb{R}^T$ is said to be **increasing** if, for any $x, y \in T$, $x \geq y$ implies $f(x) \geq f(y)$, and **strictly increasing** if, for any $x, y \in T$, $x > y$ implies $f(x) > f(y)$. (An obvious example of an increasing real function that is not strictly increasing is a constant function on $\mathbb{R}$.) We say that $f \in \mathbb{R}^T$ is **decreasing** or **strictly decreasing** if $-f$ is increasing or strictly increasing, respectively. By a **monotonic** function in $\mathbb{R}^T$, we understand a map in $\mathbb{R}^T$ that is either increasing or decreasing.

We say that $f \in \mathbb{R}^T$ is **bounded** if there is an $M \in \mathbb{R}$ such that $|f(t)| \leq M$ for all $t \in T$. Note that, given any $-\infty < a \leq b < \infty$, every monotonic function in $\mathbb{R}^{[a,b]}$ is bounded. Indeed, for any such function, we have $|f(t)| \leq \max\{|f(a)|, |f(b)|\}$ for all $a \leq t \leq b$. It is also easily seen that a strictly increasing function f in $\mathbb{R}^{[a,b]}$ is injective. Thus $f : [a,b] \to f([a,b])$ is then a bijection, and hence it is invertible (Proposition 2). Moreover, the inverse of f is itself a strictly increasing function on $f([a,b])$. Similar observations hold for strictly decreasing functions.

4.2 Limits, Continuity, and Differentiation

Let T be a nonempty subset of $\mathbb{R}$, and $f \in \mathbb{R}^T$. If x is an extended real number that is the limit of at least one decreasing sequence in $T\backslash\{x\}$, then we say that $y \in \overline{\mathbb{R}}$ is the **right-limit** of f at x, and write $f(x+) = y$, provided that $f(x_m) \to y$ for every sequence (x_m) in $T\backslash\{x\}$ with $x_m \searrow x$. (Notice that f does not have to be defined at x.) The **left-limit** of f at x, denoted as $f(x-)$, is defined analogously. Finally, if x is an extended real number that is the limit of at least one sequence in $T\backslash\{x\}$, we say that y is the **limit** of f at x, and write

$$\lim_{t \to x} f(t) = y,$$

provided that $f(x_m) \to y$ for every sequence (x_m) in $T\backslash\{x\}$ with $x_m \to x$.[39] Equivalently, for any such x, we have $\lim_{t\to x} f(t) = y$ iff, for each $\varepsilon > 0$, we can find a $\delta > 0$ (which may depend on x and ε) such that $|y - f(t)| < \varepsilon$ for all $t \in T\backslash\{x\}$ with $|x - t| < \delta$. (Why?) In particular, when T is an open interval and $x \in T$, we have $\lim_{t\to x} f(t) = y$ iff $f(x+) = y = f(x-)$.

Let $x \in T$. If there is no sequence (x_m) in $T\backslash\{x\}$ with $x_m \to x$ (so x is an *isolated* point of T), or if there is such a sequence and $\lim_{t\to x} f(t) = f(x)$, we say that f is **continuous at** x. Intuitively, this means that f maps the points nearby x to points that are close to $f(x)$. For any nonempty subset S of T, if f is continuous at each $x \in S$, then it is said to be **continuous on** S. If $S = T$ here, then we simply say that f is **continuous**. The set of all continuous functions on T is denoted by $\mathbf{C}(T)$. (But if $T := [a,b]$ for

[39] WARNING. The limit of a function may fail to exist at every point on its domain. (Check the limits of the indicator function of $\mathbb{Q}$ in $\mathbb{R}$, for instance.)

some $a, b \in \mathbb{R}$ with $a \leq b$, then we write $\mathbf{C}[a, b]$ instead of $\mathbf{C}([a, b])$. It is obvious that if $f \in \mathbb{R}^T$ is continuous, then so is $f|_S$ for any $S \in 2^T \backslash \{\emptyset\}$. Put differently, continuity of a real function implies its continuity on any nonempty subset of its domain.

A function $f \in \mathbb{R}^T$ is said to be **uniformly continuous on** $S \subseteq T$ if for each $\varepsilon > 0$, there exists a $\delta > 0$ such that $|f(x) - f(y)| < \varepsilon$ for *all* $x, y \in S$ with $|x - y| < \delta$. If $S = T$ here, then we say that f is **uniformly continuous**. While continuity is a "local" phenomenon, uniform continuity is a "global" property that says that whenever *any* two points in the domain of the function are close to each other, so should the values of the function at these points.

It is obvious that if $f \in \mathbb{R}^T$ is uniformly continuous, then it is continuous. (Yes?) The converse is easily seen to be false. For instance, $f : (0, 1) \to \mathbb{R}$ defined by $f(t) := \frac{1}{t}$ is continuous, but not uniformly continuous. There is, however, one important case in which uniform continuity and continuity coincide.

Proposition 11

(Heine) *Let T be any subset of $\mathbb{R}$ that contains the closed interval $[a, b]$, and take any $f \in \mathbb{R}^T$. Then f is continuous on $[a, b]$ if, and only if, it is uniformly continuous on $[a, b]$.*

Proof

To derive a contradiction, assume that f is continuous on $[a, b]$, but not uniformly so. Then there exists an $\varepsilon > 0$ such that we can find two sequences (x_m) and (y_m) in $[a, b]$ with

$$|x_m - y_m| < \frac{1}{m} \quad \text{and} \quad |f(x_m) - f(y_m)| \geq \varepsilon, \qquad m = 1, 2, \ldots \tag{5}$$

(Why?) By the Bolzano-Weierstrass Theorem, there exists a convergent subsequence (x_{m_k}) of (x_m). Let $x := \lim x_{m_k}$, and note that $a \leq x \leq b$, so f is continuous at x. Then, since the first part of (5) guarantees that $\lim y_{m_k} = x$, we have $\lim f(x_{m_k}) = f(x) = \lim f(y_{m_k})$, which, of course, entails $|f(x_M) - f(y_M)| < \varepsilon$ for some $M \in \mathbb{N}$ large enough, contradicting the second part of (5). ∎

Back to our review. Let x be a real number that is the limit of at least one sequence in $T\backslash\{x\}$. If the limits of $f, g \in \mathbb{R}^T$ at x exist and are finite, then we have

$$\lim_{t\to x}\big(f(t)+g(t)\big) = \lim_{t\to x} f(t) + \lim_{t\to x} g(t) \qquad \text{and}$$

$$\lim_{t\to x} f(t)g(t) = \lim_{t\to x} f(t) \lim_{t\to x} g(t). \tag{6}$$

(If $\lim_{t\to x} f(t) = \infty$, then these formulas remain valid provided that $\lim_{t\to x} g(t) \neq -\infty$ and $\lim_{t\to x} g(t) \neq 0$, respectively.) Moreover, we have

$$\lim_{t\to x} \frac{f(t)}{g(t)} = \frac{\lim_{t\to x} f(t)}{\lim_{t\to x} g(t)},$$

provided that $g(t) \neq 0$ for all $t \in T$ and $\lim_{t\to x} g(t) \neq 0$. (*Proof.* These assertions follow readily from those of Exercise 35.) It follows that $\mathbf{C}(T)$ is closed under addition and multiplication. More generally, if $f, g \in \mathbb{R}^T$ are continuous at $x \in T$, then $f + g$ and fg are continuous at x. (Of course, provided that it is well-defined on T, the same also goes for $\frac{f}{g}$.)

In passing, we note that, for any $m \in \mathbb{Z}_+$, a **polynomial of degree** m on T is a real function $f \in \mathbb{R}^T$ with

$$f(t) = a_0 + a_1 t + a_2 t^2 + \cdots + a_m t^m \qquad \text{for all } t \in T$$

for some $a_0, \ldots, a_m \in \mathbb{R}$ such that $a_m \neq 0$ if $m > 0$. The set of all polynomials (of any degree) on T is denoted as $\mathbf{P}(T)$, but again, if T is an interval of the form $[a, b]$, we write $\mathbf{P}[a, b]$ instead of $\mathbf{P}([a, b])$.

Clearly, $\mathbf{P}(T)$ is closed under addition and multiplication. Moreover, since any constant function on T, along with id_T, is continuous, and $\mathbf{C}(T)$ is closed under addition and multiplication, a straightforward application of the Principle of Mathematical Induction shows that $\mathbf{P}(T) \subseteq \mathbf{C}(T)$.

The following exercises aim to substantiate this brief review. We take up the theory of continuous functions in a much more general setting in Chapter D, where, you will be happy to know, the exposition will proceed under the speed limit.

EXERCISE 51 Let S and T be two nonempty subsets of $\mathbb{R}$, and take any $(f, g) \in \mathbb{R}^T \times \mathbb{R}^S$ with $f(T) \subseteq S$. Show that if f is continuous at $x \in T$ and g is continuous at $f(x)$, then $g \circ f$ is continuous at x.

Exercise 52 For any given $-\infty < a < b < \infty$, let $f \in \mathbb{R}^{(a,b)}$ be a continuous bijection. Show that f^{-1} is a continuous bijection defined on $f((a,b))$.

Exercise 53[H] (*"Baby" Weierstrass' Theorem*) Show that for any $a, b \in \mathbb{R}$ with $a \leq b$, and any $f \in \mathbf{C}[a,b]$, there exist $x, y \in [a,b]$ such that $f(x) \geq f(t) \geq f(y)$ for all $t \in [a,b]$.

Exercise 54[H] (*"Baby" Intermediate Value Theorem*) Let I be any interval and $a, b \in I$. Prove: If $f \in \mathbf{C}[a,b]$ and $f(a) < f(b)$, then $(f(a), f(b)) \subseteq f((a,b))$.

Let I be a nondegenerate interval, and take any $f \in \mathbb{R}^I$. For any given $x \in I$, we define the **difference-quotient map** $Q_{f,x} : I \backslash \{x\} \to \mathbb{R}$ by

$$Q_{f,x}(t) := \frac{f(t) - f(x)}{t - x}.$$

If the right-limit of this map at x exists as a real number, that is, $Q_{f,x}(x+) \in \mathbb{R}$, then f is said to be **right-differentiable at** x. In this case, the number $Q_{f,x}(x+)$ is called the **right-derivative** of f at x, and is denoted by $f'_+(x)$. Similarly, if $Q_{f,x}(x-) \in \mathbb{R}$, then f is said to be **left-differentiable at** x, and the **left-derivative** of f at x, denoted by $f'_-(x)$, is defined as the number $Q_{f,x}(x-)$. If x is the left end point of I and $f'_+(x)$ exists, or if x is the right end point of I and $f'_-(x)$ exists, or if x is not an end point of I and f is both right- and left-differentiable at x with $f'_+(x) = f'_-(x)$, then we say that f is **differentiable at** x. In the first case $f'_+(x)$, in the second case $f'_-(x)$, and in the third case the common value of $f'_+(x)$ and $f'_-(x)$ is denoted as either $f'(x)$ or $\frac{d}{dt}f(x)$. As you know, when it exists, the number $f'(x)$ is called the **derivative of** f **at** x. It is readily checked that f is differentiable at x iff

$$\lim_{t \to x} \frac{f(t) - f(x)}{t - x} \in \mathbb{R},$$

in which case $f'(x)$ equals precisely to this number. If J is an interval contained in I, and f is differentiable at each $x \in J$, then we say that f is **differentiable on** J. If $J = I$ here, then we simply say that f is **differentiable**. In this case the **derivative** of f is defined as the function $f' : I \to \mathbb{R}$ that maps each $x \in I$ to the derivative of f at x. (If f' is differentiable, then f is said to be **twice differentiable**, and the **second derivative** of f is defined as the function $f'' : I \to \mathbb{R}$ that maps each $x \in I$ to the derivative of f' at x.)

Similarly, if f is right-differentiable at each $x \in I$, then it is said to be **right-differentiable**, and in this case we define the **right-derivative of** f as a real function on I that maps every $x \in I$ to $f'_+(x)$. Naturally, this function is denoted as f'_+. Left-differentiability of f and the function f'_- are analogously defined.

The following exercises recall a few basic facts about the differentiation of real functions on the real line.

EXERCISE 55 Let I be an open interval and take any $f \in \mathbb{R}^I$.

(a) Show that if $f \in \mathbb{R}^I$ is differentiable then it is continuous.

(b) Show that if $f, g \in \mathbb{R}^I$ are differentiable and $\alpha \in \mathbb{R}$, then $\alpha f + g$ and fg are differentiable.

(c) Show that every $f \in \mathbf{P}(I)$ is differentiable.

(d) (*The Chain Rule*) Let $f \in \mathbb{R}^I$ be differentiable and $f(I)$ an open interval. Show that if $g \in \mathbb{R}^{f(I)}$ is differentiable, then so is $g \circ f$ and $(g \circ f)' = (g' \circ f)f'$.

For any $-\infty < a < b < \infty$ and $f \in \mathbf{C}[a, b]$, the definition above maintains that the derivatives of f at a and at b are $f'_+(a)$ and $f'_-(b)$, respectively. Thus, f being differentiable means that $f|_{[a,b)}$ is right-differentiable, $f|_{(a,b]}$ is left-differentiable, and $f'_+(x) = f'_-(x)$ for each $a < x < b$. If $f' \in \mathbf{C}[a, b]$, then we say that f is **continuously differentiable**—the class of all such real functions is denoted by $\mathbf{C}^1[a, b]$. If, further, $f' \in \mathbf{C}^1[a, b]$, then we say that f is **twice continuously differentiable**, and denote the class of all such maps by $\mathbf{C}^2[a, b]$. We define the classes $\mathbf{C}^3[a, b]$, $\mathbf{C}^4[a, b]$, etc., inductively. In turn, for any positive integer k, we let $\mathbf{C}^k[a, \infty)$ stand for the class of all $f \in \mathbf{C}[a, \infty)$ such that $f|_{[a,b]} \in \mathbf{C}^k[a, b]$ for every $b > a$. (The classes $\mathbf{C}^k(-\infty, b]$ and $\mathbf{C}^k(\mathbb{R})$ are defined analogously.)

Let f be differentiable on the bounded open interval (a, b). If f assumes its maximum at some $x \in (a, b)$, that is, $f(x) \geq f(t)$ for all $a < t < b$, then a fairly obvious argument shows that the derivative of f must vanish at x, that is, $f'(x) = 0$. (*Proof.* If $f'(x) > 0$ (or < 0), then we could find a small enough $\varepsilon > 0$ (< 0, respectively) such that $x + \varepsilon \in (a, b)$ and $f(x+\varepsilon) > f(x)$, contradicting that f assumes its maximum at x.) Of course, the same would be true if f assumed instead its minimum at x. (*Proof.* Just apply the previous observation to $-f$.) Combining these observations with the "baby" Weierstrass Theorem of Exercise 53 yields the following simple but very useful result.

Rolle's Theorem
Let $-\infty < a < b < \infty$ and $f \in \mathbf{C}[a, b]$. If f is differentiable on (a, b) and $f(a) = f(b)$, then $f'(c) = 0$ for some $c \in (a, b)$.

Proof
Since f is continuous, the "baby" Weierstrass Theorem (Exercise 53) implies that there exist $a \leq x, y \leq b$ such that $f(y) \leq f(t) \leq f(x)$ for all $a \leq t \leq b$. Now assume that f is differentiable on (a, b), and $f(a) = f(b)$. If $\{x, y\} \subseteq \{a, b\}$, then f must be a constant function, and hence $f'(t) = 0$ for all $a \leq t \leq b$. If this is not the case, then either $x \in (a, b)$ or $y \in (a, b)$. In the former case we have $f'(x) = 0$ (because f assumes its maximum at x), and in the latter case $f'(y) = 0$. ∎

There are many important consequences of this result. The following exercise recounts some of them.

Exercise 56[H] Let $-\infty < a < b < \infty$, and take any $f \in \mathbf{C}[a, b]$ that is differentiable on (a, b).

(a) Prove the *Mean Value Theorem*: There exists a $c \in (a, b)$ such that $f(b) - f(a) = f'(c)(b - a)$.

(b) Show that if $f' = 0$, then f is a constant function.

(c) Show that if $f' \geq 0$, then f is increasing, and if $f' > 0$, then it is strictly increasing.

We shall revisit the theory of differentiation in Chapter K in a much broader context and use it to give a potent introduction to optimization theory.

4.3 Riemann Integration

Throughout this section we work mostly with two arbitrarily fixed real numbers a and b, with $a \leq b$. For any $m \in \mathbb{N}$, we denote by $[a_0, \ldots, a_m]$ the set

$$\{[a_0, a_1], [a_1, a_2], \ldots, [a_{m-1}, a_m]\} \quad \text{where} \quad a = a_0 < \cdots < a_m = b,$$

provided that $a < b$. In this case, we refer to $[a_0, \ldots, a_m]$ as a **dissection** of $[a, b]$, and we denote the class of all dissections of $[a, b]$ by $\mathcal{D}[a, b]$. By convention, we let $\mathcal{D}[a, b] := \{\{a\}\}$ when $a = b$.

For any $\mathbf{a} := [a_0, \ldots, a_m]$ and $\mathbf{b} := [b_0, \ldots, b_k]$ in $\mathcal{D}[a, b]$, we write $\mathbf{a} \Cup \mathbf{b}$ for the dissection $[c_0, \ldots, c_l] \in \mathcal{D}[a, b]$ where $\{c_0, \ldots, c_l\} = \{a_0, \ldots, a_m\} \cup \{b_0, \ldots, b_k\}$. Moreover, we say that $\mathbf{b}$ is **finer than a** if $\{a_0, \ldots, a_m\} \subseteq \{b_0, \ldots, b_k\}$. Evidently, $\mathbf{a} \Cup \mathbf{b} = \mathbf{b}$ iff $\mathbf{b}$ is finer than $\mathbf{a}$.

Now let $f \in \mathbb{R}^{[a,b]}$ be any bounded function. For any $\mathbf{a} := [a_0, \ldots, a_m] \in \mathcal{D}[a, b]$, we define

$$K_{f,\mathbf{a}}(i) := \sup\{f(t) : a_{i-1} \leq t \leq a_i\} \quad \text{and}$$
$$k_{f,\mathbf{a}}(i) := \inf\{f(t) : a_{i-1} \leq t \leq a_i\}$$

for each $i = 1, \ldots, m$. (Thanks to the Completeness Axiom, everything is well-defined here.) By the **a-upper Riemann sum** of f, we mean the number

$$R_{\mathbf{a}}(f) := \sum_{i=1}^{m} K_{f,\mathbf{a}}(i) \left(a_i - a_{i-1}\right),$$

and by the **a-lower Riemann sum** of f, we mean

$$r_{\mathbf{a}}(f) := \sum_{i=1}^{m} k_{f,\mathbf{a}}(i) \left(a_i - a_{i-1}\right).$$

Clearly, $R_{\mathbf{a}}(f)$ decreases, and $r_{\mathbf{a}}(f)$ increases, as $\mathbf{a}$ becomes finer, while we always have $R_{\mathbf{a}}(f) \geq r_{\mathbf{a}}(f)$. Moreover—and this is important—

$$R(f) := \inf\{R_{\mathbf{a}}(f) : \mathbf{a} \in \mathcal{D}[a, b]\} \geq \sup\{r_{\mathbf{a}}(f) : \mathbf{a} \in \mathcal{D}[a, b]\} =: r(f).$$

($R(f)$ and $r(f)$ are called the **upper** and **lower Riemann integrals** of f, respectively.) This is not entirely obvious. Make sure you prove it before proceeding any farther.[40]

[40] *Hint.* Otherwise we would have $R_{\mathbf{a}}(f) < r_{\mathbf{b}}(f)$ for some $\mathbf{a}, \mathbf{b} \in \mathcal{D}[a, b]$. Compare $R_{\mathbf{a} \Cup \mathbf{b}}(f)$ and $r_{\mathbf{a} \Cup \mathbf{b}}(f)$.

Definition

Let $f \in \mathbb{R}^{[a,b]}$ be a bounded function. If $R(f) = r(f)$, then f is said to be **Riemann integrable**, and the number

$$\int_a^b f(t)dt := R(f)$$

is called the **Riemann integral** of f.[41] In this case, we also define

$$\int_b^a f(t)dt := -R(f).$$

Finally, if $g \in \mathbb{R}^{[a,\infty)}$ is a bounded function, then we define the **improper Riemann integral** of g as

$$\int_a^\infty g(t)dt := \lim_{b\to\infty} R(g|_{[a,b]}),$$

provided that $g|_{[a,b]}$ is Riemann integrable for each $b > a$, and the limit on the right-hand side exists (in $\overline{\mathbb{R}}$). (For any bounded $g \in \mathbb{R}^{(-\infty,a]}$, the improper Riemann integral $\int_{-\infty}^a g(t)dt$ is analogously defined.)

As you surely recall, the geometric motivation behind this formulation relates to the calculation of the area under the graph of f on the interval $[a,b]$. (When $f \geq 0$, the intuition becomes clearer.) Informally put, we approximate the area that we wish to compute from above (by an upper Riemann sum) and from below (by a lower one), and by choosing finer and finer dissections, we check if these two approximations converge to the same real number. If they do, then we call the common limit the Riemann integral of f. If they don't, then $R(f) > r(f)$, and we say that f is *not* Riemann integrable.

Almost immediate from the definitions is the following simple but very useful result.

Proposition 12

If $f \in \mathbb{R}^{[a,b]}$ is bounded and Riemann integrable, then

$$\left|\int_a^b f(t)dt\right| \leq (b-a)\sup\{|f(t)| : a \leq t \leq b\}.$$

[41] Of course, t acts as a "dummy variable" here—the expressions $\int_a^b f(t)dt$, $\int_a^b f(x)dx$ and $\int_a^b f(\omega)d\omega$ all denote the same number. (For this reason, some authors prefer to write $\int_a^b f$ for the Riemann integral of f.)

EXERCISE 57[H] Let $\alpha \in \mathbb{R}$ and let $f, g \in \mathbb{R}^{[a,b]}$ be bounded functions. Show that if f and g are Riemann integrable, then so is $\alpha f + g$, and we have

$$\int_a^b (\alpha f + g)(t)dt = \alpha \int_a^b f(t)dt + \int_a^b g(t)dt.$$

EXERCISE 58 Take any $c \in [a, b]$ and let $f \in \mathbb{R}^{[a,b]}$ be a bounded function. Show that if f is Riemann integrable, then so is $f|_{[a,c]}$ and $f|_{[c,b]}$, and we have

$$\int_a^b f(t)dt = \int_a^c f(t)dt + \int_c^b f(t)dt.$$

(Here $\int_a^c f(t)dt$ stands for $\int_a^c f|_{[a,c]}(t)dt$, and similarly for $\int_c^b f(t)dt$.)

EXERCISE 59 Prove Proposition 12.

EXERCISE 60 Let $f \in \mathbb{R}^{[a,b]}$ be a bounded function, and define $f^+, f^- \in \mathbb{R}^{[a,b]}$ by

$$f^+(t) := \max\{f(t), 0\} \qquad \text{and} \qquad f^-(t) := \max\{-f(t), 0\}.$$

(a) Verify that $f = f^+ - f^-$ and $|f| = f^+ + f^-$.
(b) Verify that $R_{\mathbf{a}}(f) - r_{\mathbf{a}}(f) \geq R_{\mathbf{a}}(f^+) - r_{\mathbf{a}}(f^+) \geq 0$ for any $\mathbf{a} \in \mathcal{D}[a, b]$, and state and prove a similar result for f^-.
(c) Show that if f is Riemann integrable, then so are f^+ and f^-.
(d) Show that if f is Riemann integrable, then so is $|f|$, and

$$\left| \int_a^b f(t)dt \right| \leq \int_a^b |f(t)|\, dt.$$

(Here, as usual, we write $|f(t)|$ for $|f|(t)$.)

An important issue in the theory of integration concerns the identification of Riemann integrable functions. Fortunately, we don't have to spend much time on this matter. The main integrability result that we need in the sequel is quite elementary.

PROPOSITION 13
Any $f \in \mathbf{C}[a, b]$ is Riemann integrable.

PROOF
We assume $a < b$, for otherwise the claim is obvious. Take any $f \in \mathbf{C}[a, b]$, and fix an arbitrary $\varepsilon > 0$. By Proposition 11, f is uniformly continuous on

$[a,b]$. Thus, there exists a $\delta > 0$ such that $|f(t) - f(t')| < \frac{\varepsilon}{b-a}$ for *all* t, t' in $[a,b]$ with $|t - t'| < \delta$. Then, for any dissection $\mathbf{a} := [a_0, \ldots, a_m]$ of $[a,b]$ with $|a_i - a_{i-1}| < \delta$ for each $i = 1, \ldots, m$, we have

$$R_{\mathbf{a}}(f) - r_{\mathbf{a}}(f) = \sum_{i=1}^{m} (K_{f,\mathbf{a}}(i) - k_{f,\mathbf{a}}(i))(a_i - a_{i-1}) < \tfrac{\varepsilon}{b-a} \sum_{i=1}^{m} (a_i - a_{i-1}) = \varepsilon.$$

Since $R_{\mathbf{a}}(f) \geq R(f) \geq r(f) \geq r_{\mathbf{a}}(f)$, it follows that $|R(f) - r(f)| < \varepsilon$. Since $\varepsilon > 0$ is arbitrary here, we are done. ■

EXERCISE 61[H] If $f \in \mathbf{C}[a,b]$ and $f \geq 0$, then $\int_a^b f(t)dt = 0$ implies $f = 0$.

EXERCISE 62 Let f be a bounded real map on $[a,b]$ which is continuous at all but finitely many points of $[a,b]$. Prove that f is Riemann integrable.

We conclude with a (slightly simplified) statement of the Fundamental Theorem of Calculus, which you should carry with yourself at all times. As you might recall, this result makes precise in what way one can think of the "differentiation" and "integration" as inverse operations. Its importance cannot be overemphasized.

THE FUNDAMENTAL THEOREM OF CALCULUS

For any $f \in \mathbf{C}[a,b]$ and $F \in \mathbb{R}^{[a,b]}$, we have

$$F(x) = F(a) + \int_a^x f(t)dt \quad \text{for all } a \leq x \leq b, \tag{7}$$

if, and only if, $F \in \mathbf{C}^1[a,b]$ and $F' = f$.

PROOF

Take any $f \in \mathbf{C}[a,b]$ and $F \in \mathbb{R}^{[a,b]}$ such that (7) holds. Consider any $a \leq x < b$, and let ε be a fixed but arbitrary positive number. Since f is continuous at x, there exists a $\delta > 0$ such that $|f(t) - f(x)| < \varepsilon$ for any $a < t < b$ with $|t - x| < \delta$. Thus, for any $x < y < b$ with $y - x < \delta$, we have

$$\left| \frac{F(y) - F(x)}{y - x} - f(x) \right| \leq \frac{1}{y - x} \int_x^y |f(t) - f(x)| \, dt \leq \varepsilon$$

by Exercise 58 and Proposition 12. It follows that $F|_{[a,b)}$ is right-differentiable and $F'_+(x) = f(x)$ for each $a \leq x < b$. Moreover, an analogous argument would show that $F|_{(a,b]}$ is left-differentiable and $F'_-(x) = f(x)$ for each $a < x \leq b$. Conclusion: F is differentiable and $F' = f$.

Conversely, take any $f \in \mathbf{C}[a,b]$ and $F \in \mathbf{C}^1[a,b]$ such that $F' = f$. We wish to show that (7) holds. Fix any $a \leq x \leq b$, and let $\varepsilon > 0$. It is easy to see that, since f is Riemann integrable (Proposition 13), there exists a dissection $\mathbf{a} := [a_0, \ldots, a_m]$ in $\mathcal{D}[a,x]$ such that $R_\mathbf{a}(f) - r_\mathbf{a}(f) < \varepsilon$. (Yes?) By the Mean Value Theorem (Exercise 56), for each $i = 1, \ldots, m$ there exists an $x_i \in (a_{i-1}, a_i)$ with $F(a_i) - F(a_{i-1}) = f(x_i)(a_i - a_{i-1})$. It follows that

$$F(x) - F(a) = \sum_{i=1}^{m}(F(a_i) - F(a_{i-1})) = \sum_{i=1}^{m} f(x_i)(a_i - a_{i-1}),$$

and hence $R_\mathbf{a}(f) \geq F(x) - F(a) \geq r_\mathbf{a}(f)$. Since $R_\mathbf{a}(f) - r_\mathbf{a}(f) < \varepsilon$ and $R_\mathbf{a}(f) \geq \int_a^x f(t)dt \geq r_\mathbf{a}(f)$, therefore,

$$\left| \int_a^x f(t)dt - (F(x) - F(a)) \right| < \varepsilon.$$

Since $\varepsilon > 0$ is arbitrary here, the theorem is proved. ■

REMARK 2. In the statement of the Fundamental Theorem of Calculus, we may replace (7) by

$$F(x) = F(b) - \int_x^b f(t)dt \qquad \text{for all } a \leq x \leq b.$$

The proof goes through (almost) verbatim. □

EXERCISE 63 (*Integration by Parts Formula*) Prove: If $f, g \in \mathbf{C}^1[a,b]$, then

$$\int_a^b f(t)g'(t)dt = f(b)g(b) - f(a)g(a) - \int_a^b f'(t)g(t)dt.$$

4.4 Exponential, Logarithmic, and Trigonometric Functions

Other than the polynomials, we use only four types of special real functions in this book: the exponential, the logarithmic, and the two most basic trigonometric functions. The rigorous development of these functions from scratch is a tedious task that we do not wish to get into here. Instead, by using integral calculus, we introduce these functions here at a far quicker pace.

Let us begin with the **logarithmic function**: We define the map $x \mapsto \ln x$ on $\mathbb{R}_{++}$ by

$$\ln x := \int_1^x \tfrac{1}{t}dt.$$

This map is easily checked to be strictly increasing and continuous, and of course, $\ln 1 = 0$. (Verify!) By the Fundamental Theorem of Calculus (and Remark 2), the logarithmic function is differentiable, and we have $\frac{d}{dx}\ln x = \frac{1}{x}$ for any $x > 0$. Two other important properties of this function are:

$$\ln xy = \ln x + \ln y \quad \text{and} \quad \ln \tfrac{x}{y} = \ln x - \ln y \tag{8}$$

for any $x, y > 0$. To prove the first assertion, fix any $y > 0$, and define $f : \mathbb{R}_{++} \to \mathbb{R}$ by $f(x) := \ln xy - \ln x - \ln y$. Observe that f is differentiable, and $f'(x) = 0$ for all $x > 0$ by the Chain Rule. Since $f(1) = 0$, it follows that $f(x) = 0$ for all $x > 0$. (Verify this by using Exercise 56.) To prove the second claim in (8), on the other hand, set $x = \frac{1}{y}$ in the first equation of (8) to find $\ln \frac{1}{y} = -\ln y$ for any $y > 0$. Using this fact and the first equation of (8) again, we obtain $\ln \frac{x}{y} = \ln x + \ln \frac{1}{y} = \ln x - \ln y$ for any $x, y > 0$.

Finally, we note that

$$\lim_{x\to 0} \ln x = -\infty \quad \text{and} \quad \lim_{x\to\infty} \ln x = \infty. \tag{9}$$

(See Figure 3.) Let us prove the second assertion here, the proof of the first claim being analogous. Take any $(x_m) \in \mathbb{R}^\infty_{++}$ with $x_m \to \infty$. Clearly, there exists a strictly increasing sequence (m_k) of natural numbers such that $x_{m_k} \geq 2^k$ for all $k = 1, 2, \ldots$. Since $x \mapsto \ln x$ is increasing, we thus have $\ln x_{m_k} \geq \ln 2^k = k \ln 2$. (The final equality follows from the Principle of Mathematical Induction and the first equation in (8).) Since $\ln 2 > 0$, it is obvious that $k \ln 2 \to \infty$ as $k \to \infty$. It follows that the strictly increasing sequence $(\ln x_m)$ has a subsequence that diverges to ∞, which is possible only if $\ln x_m \to \infty$. (Why?)

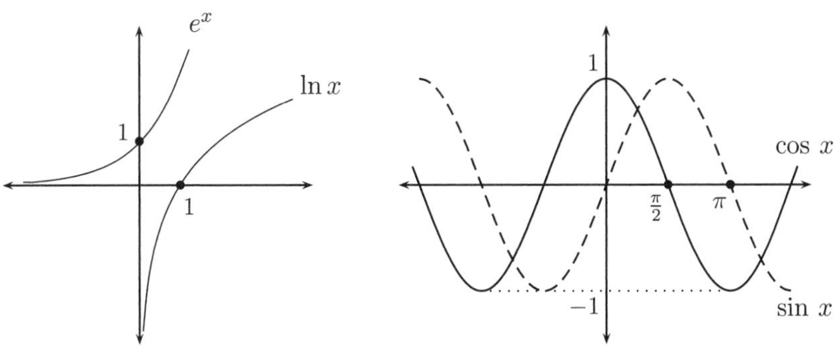

FIGURE 3

Since the logarithmic function is continuous, (9) and an appeal to the "baby" Intermediate Value Theorem (Exercise 54) entail that the range of this map is the entire $\mathbb{R}$. Since it is also strictly increasing, the logarithmic function is invertible, with its inverse function being a strictly increasing map from $\mathbb{R}$ onto $\mathbb{R}_{++}$. The latter map, denoted as $x \mapsto e^x$, is called the **exponential function**. (See Figure 3.) By definition, we have

$$\ln e^x = x \quad \text{and} \quad e^{\ln x} = x \quad \text{for all } x > 0.$$

Of course, the real number e^1 is denoted as e.[42]

The following property of the exponential function is basic:

$$e^{x+y} = e^x e^y \quad \text{for all } x, y > 0. \tag{10}$$

Indeed, by (8),

$$\ln e^x e^y = \ln e^x + \ln e^y = x + y = \ln e^{x+y}$$

for all $x, y > 0$, so, since the logarithmic function is injective, we get (10).[43]

Finally, let us show that the exponential function is differentiable, and compute its derivative. Since the derivative of the logarithmic function at 1 equals 1, we have $\lim_{y\to 1} \frac{\ln y}{y-1} = 1$, which implies that $\lim_{y\to 1} \frac{y-1}{\ln y} = 1$. (Why?) Then, since $e^{x_m} \to 1$ for any real sequence (x_m) with $x_m \to 0$, we have $\frac{e^{x_m}-1}{x_m} \to 1$. (Why?) Since (x_m) is arbitrary here, we thus have $\lim_{\varepsilon\to 0} \frac{e^\varepsilon - 1}{\varepsilon} = 1$. It follows that $x \mapsto e^x$ is differentiable at 0, and its derivative equals 1 there. Therefore, by (10),

$$\frac{d}{dx} e^x = \lim_{\varepsilon\to 0} \frac{e^{x+\varepsilon} - e^x}{\varepsilon} = e^x \lim_{\varepsilon\to 0} \frac{e^\varepsilon - 1}{\varepsilon} = e^x, \qquad -\infty < x < \infty.$$

We conclude that the exponential map is differentiable, and the derivative of this function is equal to the exponential function itself.

Among the trigonometric functions, we only need to introduce the *sine* and the *cosine* functions, and we will do this again by using integral calculus. Let us define first the real number π by the equation

$$\pi := 2 \int_0^1 \frac{1}{\sqrt{1-t^2}} dt,$$

[42] Therefore, e is the (unique) real number with the property $\int_1^e \frac{1}{t} dt = 1$, but of course, there are various other ways of defining the number e (Exercise 47).

[43] By the way, do you think there is another increasing map f on $\mathbb{R}$ with $f(1) = e$ and $f(x+y) = f(x)f(y)$ for any $x, y \in \mathbb{R}$? (This question will be answered in Chapter D.)

that is, we define π as the area of the circle with radius 1. Now define the function $f \in \mathbb{R}^{(-1,1)}$ by

$$f(x) := \int_0^x \frac{1}{\sqrt{1-t^2}}dt \qquad \text{for any } x \geq 0$$

and

$$f(x) := -\int_x^0 \frac{1}{\sqrt{1-t^2}}dt \qquad \text{for any } x < 0.$$

This map is a bijection from $(-1, 1)$ onto $(-\frac{\pi}{2}, \frac{\pi}{2})$. (Why?) We define the map $x \mapsto \sin x$ on $(-\frac{\pi}{2}, \frac{\pi}{2})$ as the inverse of f, and then extend it to $[-\frac{\pi}{2}, \frac{\pi}{2}]$ by setting $\sin \frac{-\pi}{2} := -1$ and $\sin \frac{\pi}{2} = 1$. (How does this definition relate to the geometry behind the sine function?) Finally, the **sine function** is defined on the entire $\mathbb{R}$ by requiring the following periodicity: $\sin(x + \pi) = -\sin x$ for all $x \in \mathbb{R}$. It is easy to see that this function is an *odd* function, that is, $\sin(-x) = -\sin x$ for any $x \in \mathbb{R}$ (Figure 3).

Now define the map $x \mapsto \cos x$ on $[-\frac{\pi}{2}, \frac{\pi}{2}]$ by $\cos x := \sqrt{1 - (\sin x)^2}$, and then extend it to $\mathbb{R}$ by requiring the same periodicity with the sine function: $\cos(x + \pi) = -\cos x$ for any $x \in \mathbb{R}$. The resulting map is called the **cosine function.** This is an *even* function, that is, $\cos(-x) = \cos x$ for any $x \in \mathbb{R}$, and we have $\cos 0 = 1$ and $\cos \frac{\pi}{2} = 0 = \cos \frac{-\pi}{2}$ (Figure 3).

Exercise 64 Show that the sine and cosine functions are differentiable, and $\frac{d}{dx} \sin x = \cos x$ and $\frac{d}{dx} \cos x = -\sin x$ for all $x \in \mathbb{R}$.

Exercise 65 Prove: $\lim_{x\to 0} \frac{\sin x}{x} = 1$.

4.5 Concave and Convex Functions

Let $n \in \mathbb{N}$, and recall that a subset T of $\mathbb{R}^n$ is said to be **convex** if the line segment connecting any two elements of T lies entirely within T, that is, $\lambda x + (1 - \lambda)y \in T$ for all $x, y \in T$ and $0 \leq \lambda \leq 1$. Given any such nonempty set T, a function $\varphi \in \mathbb{R}^T$ is called **concave** if

$$\varphi(\lambda x+(1-\lambda)y) \geq \lambda\varphi(x)+(1-\lambda)\varphi(y) \qquad \text{for any } x, y \in T \text{ and } 0 \leq \lambda \leq 1,$$

and **strictly concave** if this inequality holds strictly for any distinct $x, y \in T$ and $0 < \lambda < 1$. The definitions of convex and strictly convex functions are obtained by reversing these inequalities. Equivalently, φ is called **(strictly) convex** if $-\varphi$ is (strictly) concave. (This observation allows us to convert any property that a concave function may possess into a property for convex functions in a straightforward manner.) Finally, φ is said to be **affine** if it is both concave and convex.

If φ and ψ are concave functions in $\mathbb{R}^T$, and $\alpha \geq 0$, then $\alpha\varphi + \psi$ is a concave function in $\mathbb{R}^T$. Similarly, if S is an interval with $\varphi(T) \subseteq S$, and $\varphi \in \mathbb{R}^T$ and $\psi \in \mathbb{R}^S$ are concave, then so is $\psi \circ \varphi$. The following exercises provide two further examples of functional operations that preserve the concavity of real functions.

EXERCISE 66 For any given $n \in \mathbb{N}$, let T be a nonempty convex subset of $\mathbb{R}^n$ and $\mathcal{F}$ a (nonempty) class of concave functions in $\mathbb{R}^T$. Show that if $\inf\{\varphi(x) : \varphi \in \mathcal{F}\} > -\infty$ for all $x \in T$, then the map $x \mapsto \inf\{\varphi(x) : \varphi \in \mathcal{F}\}$ is a concave function in $\mathbb{R}^T$.

EXERCISE 67 For any given $n \in \mathbb{N}$, let T be a nonempty convex subset of $\mathbb{R}^n$ and (φ_m) a sequence of concave functions in $\mathbb{R}^T$. Show that if $\lim \varphi_m(x) \in \mathbb{R}$ for each $x \in T$, then the map $x \mapsto \lim \varphi_m(x)$ is a concave function in $\mathbb{R}^T$.

We now specialize to concave functions defined on an open interval $I \subseteq \mathbb{R}$. The first thing to note about such a function is that it is continuous. In fact, we can prove a stronger result with the aid of the following useful observation about the boundedness of concave functions defined on a bounded interval.

LEMMA 2
For any given $-\infty < a \leq b < \infty$, if $f \in \mathbb{R}^{[a,b]}$ is concave (or convex), then

$$\inf f([a,b]) > -\infty \quad \textit{and} \quad \sup f([a,b]) < \infty.$$

PROOF
Let f be a concave real map on $[a,b]$. Obviously, for any $a \leq t \leq b$, we have $t = \lambda_t a + (1-\lambda_t)b$ for some $0 \leq \lambda_t \leq 1$, whereas $f(\lambda_t a + (1-\lambda_t)b) \geq \min\{f(a), f(b)\}$ by concavity. It follows that $\inf f([a,b]) > -\infty$.

The proof of the second claim is trickier. Let us denote the midpoint $\frac{a+b}{2}$ of the interval $[a,b]$ by M, and fix an arbitrary $a \leq t \leq b$. Note that there is a real number c_t such that $|c_t| \leq \frac{b-a}{2}$ and $t = M + c_t$. (Simply define c_t by the latter equation.) Then, $M - c_t$ belongs to $[a, b]$, so, by concavity,

$$\begin{aligned} f(M) &= f\left(\tfrac{1}{2}(M + c_t) + \tfrac{1}{2}(M - c_t)\right) \\ &\geq \tfrac{1}{2}f(M + c_t) + \tfrac{1}{2}f(M - c_t) \\ &= \tfrac{1}{2}f(t) + \tfrac{1}{2}f(M - c_t), \end{aligned}$$

so $f(t) \leq 2f(M) - \inf f([a, b]) < \infty$. Since t was chosen arbitrarily in $[a, b]$, this proves that $\sup f([a, b]) < \infty$. ■

Here is the main conclusion we wish to derive from this observation.

Proposition 14

Let I be an open interval and $f \in \mathbb{R}^I$. If f is concave (or convex), then for every $a, b \in \mathbb{R}$ with $a \leq b$ and $[a, b] \subset I$, there exists a $K > 0$ such that

$$|f(x) - f(y)| \leq K|x - y| \qquad \textit{for all } a \leq x, y \leq b.$$

Proof

Since I is open, there exists an $\varepsilon > 0$ such that $[a-\varepsilon, b+\varepsilon] \subseteq I$. Let $a' := a-\varepsilon$ and $b' := b + \varepsilon$. Assume that f is concave, and let $\alpha := \inf f([a', b'])$ and $\beta := \sup f([a', b'])$. By Lemma 2, α and β are real numbers. Moreover, if $\alpha = \beta$, then f is constant, so all becomes trivial. We thus assume that $\beta > \alpha$.

For any distinct $x, y \in [a, b]$, let

$$z := y + \varepsilon\frac{y - x}{|y - x|} \qquad \text{and} \qquad \lambda := \frac{|y - x|}{\varepsilon + |y - x|}.$$

Then $a' \leq z \leq b'$ and $y = \lambda z + (1 - \lambda)x$—we defined z the way we did in order to satisfy these two properties. Hence, by concavity of f, $f(y) \geq \lambda(f(z) - f(x)) + f(x)$, that is,

$$\begin{aligned} f(x) - f(y) &\leq \lambda(f(x) - f(z)) \\ &\leq \lambda(\beta - \alpha) \\ &= \frac{\beta - \alpha}{\varepsilon + |y - x|}|y - x| \\ &< \frac{\beta - \alpha}{\varepsilon}|y - x|. \end{aligned}$$

Interchanging the roles of x and y in this argument and letting $K := \frac{\beta-\alpha}{\varepsilon}$ complete the proof. ■

COROLLARY 2
Let I be an open interval and $f \in \mathbb{R}^I$. If f is concave (or convex), then it is continuous.

EXERCISE 68[H] Show that every concave function on an open interval is both right-differentiable and left-differentiable. (Of course, such a map need not be differentiable.)

In passing, we recall that, provided that f is a differentiable real map on an open interval I, then it is (strictly) concave iff f' is (strictly) decreasing. (Thus $x \mapsto \ln x$ is a concave map on $\mathbb{R}_{++}$, and $x \mapsto e^x$ is a convex map on $\mathbb{R}$.) Provided that f is twice differentiable, it is concave iff $f'' \leq 0$, while $f'' < 0$ implies the strict concavity of f. (The converse of the latter statement is false; for instance, the derivative of the strictly concave function $x \mapsto x^2$ vanishes at 0.) These are elementary properties, and they can easily be proved by using the Mean Value Theorem (Exercise 56). We will not, however, lose more time on this matter here.

This is all we need in terms of concave and convex real functions on the real line. In later chapters we will revisit the notion of concavity in much broader contexts. For now, we conclude by noting that a great reference that specializes on the theory of concave and convex functions is Roberts and Varberg (1973). That book certainly deserves a nice spot on the bookshelves of any economic theorist.

4.6 Quasiconcave and Quasiconvex Functions

With T being a nonempty convex subset of $\mathbb{R}^n$, $n \in \mathbb{N}$, we say that a function $\varphi \in \mathbb{R}^T$ is **quasiconcave** if

$$\varphi(\lambda x + (1-\lambda)y) \geq \min\{\varphi(x), \varphi(y)\} \qquad \text{for any } x, y \in T \text{ and } 0 \leq \lambda \leq 1,$$

and **strictly quasiconcave** if this inequality holds strictly for any distinct $x, y \in T$ and $0 < \lambda < 1$. (φ is called **(strictly) quasiconvex** if $-\varphi$ is (strictly) quasiconcave.) It is easy to show that φ is quasiconcave iff $\varphi^{-1}([a, \infty))$ is a convex set for any $a \in \mathbb{R}$. It is also plain that every concave function in $\mathbb{R}^T$ is quasiconcave, but not conversely.[44]

Quasiconcavity plays an important role in optimization theory, and it is often invoked to establish the uniqueness of a solution for a maximization problem. Indeed, if $f \in \mathbb{R}^T$ is strictly quasiconcave, and there exists an $x \in T$ with $f(x) = \max f(T)$, then x must be the only element of T with this property. For, if $x \neq y = \max f(T)$, then $f(x) = f(y)$, so $f(\frac{x}{2} + \frac{y}{2}) > f(x) = \max f(T)$ by strict quasiconcavity. Since $x, y \in T$ and T is convex, this is impossible.

EXERCISE 69 Give an example of two quasiconcave functions on the real line the sum of which is not quasiconcave.

EXERCISE 70 Let I be an interval, $f \in \mathbb{R}^I$, and let $g \in \mathbb{R}^{f(I)}$ be a strictly increasing function. Show that if f is quasiconcave, then so is $g \circ f$. Would $g \circ f$ be necessarily concave if f was concave?

[44] If $\emptyset \neq T \subseteq \mathbb{R}$, then every monotonic function in $\mathbb{R}^T$ is quasiconcave, but of course, not every monotonic function in $\mathbb{R}^T$ is concave.

Chapter B

Countability

This chapter is about Cantor's countability theory, which is a standard prerequisite for elementary real analysis. Our treatment is incomplete in that we cover only those results that are immediately relevant for the present course. In particular, we have little to say here about cardinality theory and ordinal numbers.[1] We shall, however, cover two relatively advanced topics here, namely, the theory of order isomorphisms and the Schröder-Bernstein Theorem on the "equivalence" of infinite sets. If you are familiar with countable and uncountable sets, you might want to skip Section 1 and jump directly to the discussion of these two topics. The former one is put to good use in the last section of the chapter, which provides an introduction to ordinal utility theory, a topic we shall revisit a few more times later. The Schröder-Bernstein Theorem is, in turn, proved via Tarski's Fixed Point Theorem, the first of the many fixed point theorems that are discussed in this book. This theorem should certainly be included in the toolkit of an economic theorist, for it has recently found a number of important applications in game theory.

1 Countable and Uncountable Sets

In this section we revisit set theory and begin to sketch a systematic method of thinking about the "size" of any given set. The issue is not problematic in the case of finite sets, for we can simply count the number of members of a given finite set and use this number as a measure of its size. Thus, quite simply, one finite set is "more crowded" than another if it contains more elements than the other. But how can one extend this method to the case of infinite sets? Or, how can we decide whether or not a given infinite set

[1] For a rigorous introduction to these topics, and to set theory in general, you should consult outlets such as Halmos (1960), Kaplansky (1977), Enderton (1977), or Devlin (1993). My favorite by far is Enderton (1977).

is "more crowded" than another such set? Clearly, things get trickier with infinite sets. One may even think at first that any two infinite sets are equally crowded, or even that the question is meaningless. There is, however, an intuitive way to approach the problem of ranking the sizes of infinite sets, and it is this approach that the current section is about. As a first pass, we will talk about the sense in which an infinite set can be thought of as "small." We will then apply our discussion to compare the sizes of the sets of integers, rational numbers, and real numbers. Various other applications will be given as we proceed.[2]

Let us begin by defining the two fundamental concepts that will play a major role in what follows.

> DEFINITION
> A set X is called **countably infinite** if there exists a bijection f that maps X onto the set $\mathbb{N}$ of natural numbers. X is called **countable** if it is either finite or countably infinite. X is called **uncountable** if it is not countable.

So, quite intuitively, we can "count" the members of a countable set just as we could count the natural numbers. An infinite set like $X = \{x_1, x_2, \ldots\}$ is thus countable, for we can put the members of X into one-to-one correspondence with the natural numbers:

$$x_1 \longleftrightarrow 1 \qquad x_2 \longleftrightarrow 2 \qquad \cdots \qquad x_m \longleftrightarrow m \qquad \cdots$$

Conversely, if X is countably infinite, then it can be enumerated as $X = \{x_1, x_2, \ldots\}$. To see this formally, let f be a bijection from X onto $\mathbb{N}$. Then f is invertible (Proposition A.2), and the inverse function f^{-1} is a bijection

[2] We owe our ability to compare the sizes of infinite sets to the great German mathematician Georg Cantor (1845–1818). Although the notion of infinity had been debated in philosophy for over two thousand years, it was Cantor who provided a precise manner in which infinity can be understood, studied, and even manipulated. It is not an exaggeration to say that Cantor's ideas were decades ahead of his contemporaries'. His strength did not stem from his capability to do hard proofs but from his ability to think "outside the box." Most mathematicians of his cohort, including some eminent figures (such as Klein, Kronecker, and Poincaré), found Cantor's theory of infinite sets nonsensical, but this theory was later found to provide a sound foundation for much of mathematics at large. After Cantor, writes James in Remarkable Mathematicians, "mathematics was never to be the same again" (2002, p. 214). There are not a lot of people in history about whom one can say something like this.

There are many references that detail Cantor's life and work. My favorite is Dauben (1980), from which one learns some mathematics as well.

from $\mathbb{N}$ onto X. But this means that we must have $X = \{f^{-1}(1), f^{-1}(2), \ldots\}$. Thus, if we let $x_i = f^{-1}(i)$, $i = 1, 2, \ldots$, we may write $X = \{x_1, x_2, \ldots\}$—nice and easy.

REMARK 1. Let X be a countable subset of $\mathbb{R}$. Can we meaningfully talk about the sum of the elements of X? Well, it depends! If X is finite, there is of course no problem; the sum of the elements of X, denoted $\sum_{x \in X} x$, is well-defined.[3] When X is countably infinite, we need to be a bit more careful. The natural inclination is to take an enumeration $\{x_1, x_2, \ldots\}$ of X, and define $\sum_{x \in X} x$ as $\sum^{\infty} x_i$, provided that $\sum^{\infty} x_i \in \overline{\mathbb{R}}$. Unfortunately, this need not well-define $\sum_{x \in X} x$—what if we used a different enumeration of X? You see, the issue is none other than the rearrangement of infinite series that we discussed in Section A.3.5. The right definition is: $\sum_{x \in X} x := \sum^{\infty} x_i$, where $\{x_1, x_2, \ldots\}$ is any enumeration of X, *provided that* $\sum^{\infty} x_i \in \overline{\mathbb{R}}$ *and* $\sum^{\infty} x_i$ *is invariant under rearrangements.* (If either of the latter two conditions fails, we say that $\sum_{x \in X} x$ is *undefined*.) In particular, thanks to Proposition A.10, $\sum_{x \in X} x$ is well-defined when $X \subseteq \mathbb{R}_+$. □

Let us look at a few examples of countable sets. $\mathbb{N}$ is countable, as it is evident from the definition of countability. It is also easy to see that any subset of $\mathbb{N}$ is countable. For instance, the set of all even natural numbers is countably infinite, because we can put this set into one-to-one correspondence with $\mathbb{N}$:

$$2 \longleftrightarrow 1 \quad 4 \longleftrightarrow 2 \quad \cdots \quad 2m \longleftrightarrow m \quad \cdots$$

Similarly, the set of all prime numbers is countable.[4] So, in a sense, there are as many even (or prime) numbers as there are natural numbers.[5] Since

[3] Recall that we handle the exceptional case $X = \emptyset$ by convention: $\sum_{x \in \emptyset} x = 0$.

[4] This set is, in fact, countably infinite, for according to a celebrated theorem of Euclid, there are infinitely many prime numbers. Euclid's proof of this theorem is almost as famous as the theorem itself: Suppose there are finitely many primes, say, $x_1, \ldots, x_m$, and show that $1 + \prod^m x_i$ is prime. (One needs to use in the argument the fact that every integer $k > 1$ is either prime or a product of primes, but this can easily be proved by using the Principle of Mathematical Induction.)

[5] This observation is popularized by means of the following anectode. One night, countably infinitely many passengers came to *Hilbert's Hotel*, each looking for a room. Now, Hilbert's Hotel did contain countably infinitely many rooms, but that night all of the rooms were occupied. This was no problem for Hilbert. He asked everybody staying in the hotel to come down to the lobby, and reallocated each of them using only the even-numbered rooms. This way all of the newcomers could be accommodated in the odd-numbered rooms.

the set of all positive even integers is a proper subset of $\mathbb{N}$, this may appear counterintuitive at first. Nonetheless, this sort of a thing is really in the nature of the notion of "infinity," and it simply tells us that one has to develop a new kind of "intuition" to deal with infinite sets. The following result is a step in this direction.

Proposition 1
Every subset of a countable set is countable.

Proof
Let X be a countable set, and take any subset S of X. If S is finite, there is nothing to prove, so say $|S| = \infty$. Then X must be countably infinite, and thus we can enumerate it as $X = \{x_1, x_2, \ldots\}$. Now define the self-map f on $\mathbb{N}$ inductively as follows:

$$\begin{aligned} f(1) &= \min\{i \in \mathbb{N} : x_i \in S\} \\ f(2) &= \min\{i \in \mathbb{N} : i > f(1) \text{ and } x_i \in S\} \\ f(3) &= \min\{i \in \mathbb{N} : i > f(2) \text{ and } x_i \in S\} \\ & \quad \ldots . \end{aligned}$$

(So $x_{f(1)}$ is the first term of the sequence $(x_1, x_2, \ldots)$ that belongs to S, $x_{f(2)}$ is the second, and so on.) Now consider $g \in S^{\mathbb{N}}$ defined by $g(i) := x_{f(i)}$. Since g is a bijection (isn't it?), we are done. ∎

Exercise 1 Show that if B is a countable set, and if there exists an injection from a set A into B, then A must be countable.

Exercise 2^{H} Show that every infinite set has a countably infinite subset.

Proposition 1 shows how one can obtain new countable sets by "shrinking" a countable set. It is also possible to "expand" a countable set in order to obtain another countable set. For instance, $\mathbb{Z}_+$ is countable, for $f : i \mapsto i+1$ defines a bijection from $\mathbb{N} \cup \{0\}$ onto $\mathbb{N}$.[6] Similarly, the set $\mathbb{Z}$ of all integers is

(Here Hilbert refers to David Hilbert, one of the most prominent figures in the history of mathematics. We will meet him later in the course.)

[6] What would Hilbert do if a new customer came to his hotel when all rooms were occupied?

countable. Indeed, the map $f : \mathbb{N} \to \mathbb{Z}\backslash\{0\}$, defined by $f(i) := \frac{i+1}{2}$ if i is odd and by $f(i) := -\frac{i}{2}$ if i is even, is a bijection. More generally, we have the following useful result.

PROPOSITION 2
A countable union of countable sets is countable.

PROOF
Let X_i be a countable set for each $i \in \mathbb{N}$. A moment's reflection will convince you that we would be done if we could show that $X := X_1 \cup X_2 \cup \cdots$ is a countable set. (But make sure you are really convinced.) It is without loss of generality to assume that $X_i \cap X_j = \emptyset$ for every distinct i and j. (Right?) Since each X_i is countable, we may enumerate it as $X_i = \{x_1^i, x_2^i, \ldots\}$ (but note that X_i may be finite). Now define the mapping $f : X \to \mathbb{N}^2$ by $f(x_k^i) := (i, k)$. (Is f well-defined? Why?) Clearly, f is injective, and hence, by Exercise 1, we may conclude that X is countable, provided that $\mathbb{N}^2$ is countable. The proof is thus complete in view of the next exercise.[7] ■

EXERCISE 3 Counting the members of $A \times B$ as indicated in Figure 1, show that if A and B are countable sets, then so is $A \times B$.

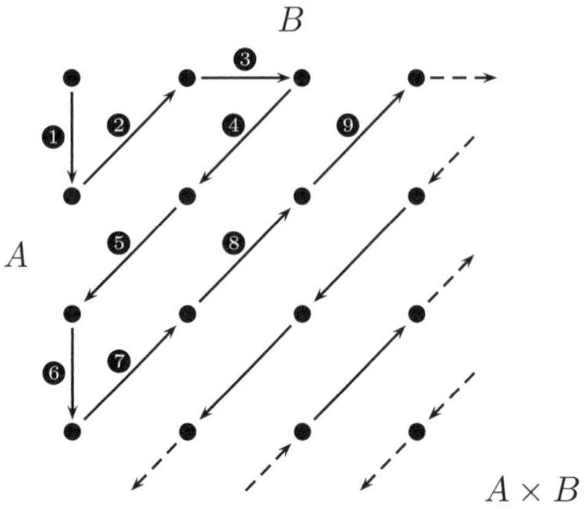

FIGURE 1

[7] Hidden in the proof is the Axiom of Choice. *Quiz.* Where is it? (*Hint.* This is subtle. Let $\mathcal{A}_i$ be the set of all enumerations of X_i (that is, the set of all bijections from $\mathbb{N}$ (or a finite set) onto X_i), $i = 1, 2, \ldots$. True, I know that each $\mathcal{A}_i$ is nonempty, because each X_i is countable. But in the proof I work with an element of $X^{\infty}\mathcal{A}_i$, don't I?)

Exercise 4 (*Another Proof for Proposition* 2) Using the notation of the proof of Proposition 2, define $g \in \mathbb{N}^X$ by $g(x_i^k) := 2^i 3^k$. Show that g is injective, and invoke Exercise 1 to conclude that X is countable.

An important implication of these results concerns the countability of the set $\mathbb{Q}$ of all rational numbers. Indeed, we have $\mathbb{Q} = \cup\{X_n : n \in \mathbb{N}\}$ where $X_n = \{\frac{m}{n} : m \in \mathbb{Z}\}$ for all $n \in \mathbb{N}$. But, obviously, there is an injection from each X_n into $\mathbb{Z}$, so, since $\mathbb{Z}$ is countable, so is each X_n (Exercise 1). Therefore, by Proposition 2, we find that $\mathbb{Q}$ must be countably infinite. (By Proposition 1, $\mathbb{Q} \cap I$ is also countably infinite for any interval I in $\mathbb{R}$.)

Corollary 1
(Cantor) $\mathbb{Q}$ is countable.

So, there are as many rational numbers as there are natural numbers! This is truly a "deep" observation. It tells us that $\mathbb{Q}$ *can be enumerated* as $\{q_1, q_2, \ldots\}$, while we do not at all know how to "construct" such an enumeration. It suffices to note that Cantor himself was fascinated by this theorem, about which he is often quoted as saying, "I see it, but I don't believe it."[8]

The countability of $\mathbb{Q}$ has profound implications. For instance, it implies that any (nondegenerate) interval partition of $\mathbb{R}$ must be countable, an observation that is certainly worth keeping in mind. This is because in any given nondegenerate interval, there exists at least one rational number (in fact, there are infinitely many of them). Thus, if there existed uncountably many nonoverlapping intervals in $\mathbb{R}$, we could deduce that $\mathbb{Q}$ contains an uncountable subset, which contradicts Proposition 1. In fact, we can say something a little stronger in this regard.

Proposition 3
Let $\mathcal{I}$ be a set of nondegenerate intervals in $\mathbb{R}$ such that $|I \cap J| \leq 1$ for any $I, J \in \mathcal{I}$. Then $\mathcal{I}$ is countable.

[8] To be more precise, however, I should note that Cantor made this remark about the possibility of constructing a bijection between a given line and a plane. While the spirit of this is similar to the argument that yields the countability of $\mathbb{Q}$, it is considerably deeper. See Dauben (1980) for a detailed historical account of the matter.

Proof

Given the nondegeneracy hypothesis, for each $I \in \mathcal{I}$, there exist real numbers $a_I < b_I$ with $(a_I, b_I) \subseteq I$, so by Proposition A.6, we can find a $q_I \in \mathbb{Q}$ with $q_I \in I$. Define $f : \mathcal{I} \to \mathbb{Q}$ by $f(I) := q_I$. Since any two members of $\mathcal{I}$ overlap at most at one point, f is injective. (Yes?) Thus the claim follows from the countability of $\mathbb{Q}$ and Exercise 1. ■

So much for countable sets. What of uncountable ones? As you might guess, there are plenty of them. Foremost, our beloved $\mathbb{R}$ is uncountable. Indeed, one cannot enumerate $\mathbb{R}$ as $\{x_1, x_2, \ldots\}$, or what is the same thing, one cannot exhaust all real numbers by counting them as one counts the natural numbers. Put differently, $\mathbb{R}$ is "more crowded" than any countable set.

Proposition 4

(Cantor) $\mathbb{R}$ *is uncountable.*

There are various ways of proving this result. The proof we present here has the advantage of making transparent the reliance of the result on the Completeness Axiom. It is based on the following useful fact.

Cantor's Nested Interval Lemma

Let $I_m := [a_m, b_m]$ be a closed interval for each $m \in \mathbb{N}$. If $I_1 \supseteq I_2 \supseteq \cdots$, then $\cap^\infty I_i \neq \emptyset$. If in addition, $b_m - a_m \to 0$, then $\cap^\infty I_i$ is a singleton.

Proof

$I_1 \supseteq I_2 \supseteq \cdots$ implies that (a_m) is a bounded and increasing sequence while (b_m) is a bounded and decreasing sequence. By Proposition A.8, both of these sequences converge (and this we owe to the Completeness Axiom), so $\lim a_m = a$ and $\lim b_m = b$ for some real numbers a and b. It is easy to see that $a_m \leq a \leq b \leq b_m$ for all m, and that $[a, b] = \cap^\infty I_i$. (Check!) The second claim follows from the fact that $b - a = \lim b_m - \lim a_m = \lim(b_m - a_m)$. ■

We can now prove the uncountability of the reals by means of a method that we will later use on a few other occasions as well. (*Note.* This method is sometimes referred to as "butterfly hunting." You'll see why in a second.)

PROOF OF PROPOSITION 4

To derive a contradiction, suppose that $[0,1] = \{a_1, a_2, \ldots\} =: I_0$. We wish to find a real number (a "butterfly") that is not equal to a_i for any $i \in \mathbb{N}$.

Divide I_0 into any three closed nondegenerate intervals (say, $[0, 1/3]$, $[1/3, 2/3]$ and $[2/3, 1]$). Clearly, a_1 does not belong to at least one of these three intervals, call any one such interval I_1. (There has to be a "butterfly" in I_1.) Now play the same game with I_1, that is, divide I_1 into three closed nondegenerate intervals, and observe that a_2 does not belong to at least one of these subintervals, say I_2. (There has to be a "butterfly" in I_2.) Continuing this way, we obtain a nested sequence (I_m) of closed intervals, so Cantor's Nested Interval Lemma yields $\cap^\infty I_i \neq \emptyset$. (Aha! We caught our "butterfly" in $\cap^\infty I_i$.) But, by construction, $a_i \notin I_i$ for each i, so we must have $\cap^\infty I_i = \emptyset$, a contradiction. ■

While the theory of countability is full of surprising results, it is extremely useful and its basics are not very difficult to master. However, to develop a good intuition about the theory, one needs to play around with a good number of countable and uncountable sets. The following exercises provide an opportunity to do precisely this.

EXERCISE 5^{H}

(a) Show that $\mathbb{N}^m$ is countable for any $m \in \mathbb{N}$.

(b) Prove or disprove: $\mathbb{N}^\infty$ is countable.

EXERCISE 6 Let A and B be two sets such that A is countable, B is uncountable, and $A \subseteq B$. Can $B \backslash A$ be countable? Is the set of all irrational numbers countable?

EXERCISE 7^{H} Let A and B be two sets such that A is countable and B is uncountable. Show that $A \cup B$ is uncountable. In fact, show that there is a bijection from B onto $A \cup B$.

EXERCISE 8^{H} For any self-map f on $\mathbb{R}$, we say that $x \in \mathbb{R}$ is a **point of discontinuity** of f if f is not continuous at x. Show that if f is monotonic, then it can have at most countably many points of discontinuity. Give an example of a real function on $\mathbb{R}$ that has uncountably many points of discontinuity.

*Exercise 9[H] Let I be an open interval and $f \in \mathbb{R}^I$. Let $\mathbf{D}_f$ stand for the set of all $x \in I$ such that f is differentiable at x. Prove that if f is concave, then $I \backslash \mathbf{D}_f$ is countable.

Exercise 10[H] For any sets A and B, we write $A \succ_{\text{card}} B$ if there exists an injection from B into A but there does not exist an injection from A into B.

(a) Show that $A \succ_{\text{card}} B$ need not hold even if B is a proper subset of A.
(b) Show that $\mathbb{R} \succ_{\text{card}} \mathbb{Q}$.
(c) Show that $2^A \succ_{\text{card}} A$ for any nonempty set A.
(d) Show that $2^{\mathbb{N}}$ is uncountable.
(e) (*Cantor's Paradox*) Let $X = \{x : x \text{ is a set}\}$. Use part (c) and the fact that $2^X \subseteq X$ to establish that X cannot be considered as a "set."

2 Losets and $\mathbb{Q}$

The set of rational numbers is not only countable, it is also ordered in a natural way (Proposition A.4). These two properties of $\mathbb{Q}$ combine nicely in various applications of real analysis. It may thus be a good idea to see what kinds of sets we can in general not only *count* like $\mathbb{Q}$ but also *order* like $\mathbb{Q}$. In fact, we will see in Section 4 that this issue is closely related to a fundamental problem in decision theory. In this section, therefore, we will make its statement precise, and then outline Cantor's solution for it.

Let us first see in what way we can relate a given linearly ordered countable set to the set of rational numbers.

Proposition 5

(Cantor) *Let X be a countable set and $\succsim$ a linear order on X. There exists a function $f : X \to \mathbb{Q}$ such that*

$$f(x) \geq f(y) \quad \textit{if and only if} \quad x \succsim y$$

for any $x, y \in X$.

Proof

The claim is trivial when X is finite, so we assume that X is countably infinite. Owing to their countability, we may enumerate X and $\mathbb{Q}$ as

$$X = \{x_1, x_2, \ldots\} \quad \text{and} \quad \mathbb{Q} = \{q_1, q_2, \ldots\}.$$

We construct the function $f \in \mathbb{Q}^X$ as follows. First let $f(x_1) := q_1$. If $x_1 \succsim x_2$ ($x_2 \succsim x_1$), then set $f(x_2)$ as the first element (with respect to the subscripts) in $\{q_2, \ldots\}$ such that $q_1 \geq f(x_2)$ ($f(x_2) \geq q_1$, respectively). Proceeding inductively, for any $m = 2,3,\ldots$, we set $f(x_m)$ as the first element of $\{q_1, \ldots\} \backslash \{f(x_1), \ldots, f(x_{m-1})\}$, which has the same order relation (with respect to $\geq$) to $f(x_1), \ldots, f(x_{m-1})$ as x_m has to $x_1, \ldots, x_{m-1}$ (with respect to $\succsim$). (Why is f well-defined?) It follows readily from this construction that, for any $x, y \in X$, we have $f(x) \geq f(y)$ iff $x \succsim y$. ■

Is it then true that the order-theoretic properties of any countable loset are identical to those of the rational numbers? No! While, according to Proposition 5, we can *embed* the loset $(\mathbb{N}, \geq)$ in $(\mathbb{Q}, \geq)$ in an order-preserving manner, it is not true that $(\mathbb{N}, \geq)$ and $(\mathbb{Q}, \geq)$ are identical with respect to all order-theoretic properties. For instance, although $\mathbb{N}$ has a $\geq$-minimum, $\mathbb{Q}$ does not have a $\geq$-minimum. The problem is that we cannot *embed* $(\mathbb{Q}, \geq)$ back in $(\mathbb{N}, \geq)$. Formally speaking, these two losets are not *order-isomorphic.*

DEFINITION

Let $(X, \succsim_X)$ and $(Y, \succsim_Y)$ be two posets. A map $f \in Y^X$ is said to be **order-preserving** (or **isotonic**), provided that

$$f(x) \succsim_Y f(y) \qquad \text{if and only if} \qquad x \succsim_X y$$

for any $x, y \in X$. If $f \in Y^X$ is an order-preserving injection, then it is called an **order-embedding** from X into Y. If such an f exists, then we say that $(X, \succsim_X)$ can be **order-embedded in** $(Y, \succsim_Y)$. Finally, if f is an order-preserving bijection, then it is called an **order-isomorphism.** If such an f exists, $(X, \succsim_X)$ and $(Y, \succsim_Y)$ are said to be **order-isomorphic.**

If two posets are order-isomorphic, they are indistinguishable from each other insofar as their order-theoretic properties are concerned; one can simply be thought of as the relabeling of the other in this regard. In concert with this view, "being order-isomorphic" acts as an equivalence relation on any given class of posets. (Proof?) In this sense, the order structures of any two finite losets of the same size are identical, but those of two infinite losets may be different from each other. Consider, for instance, $(\mathbb{Z}_+, \geq)$ and $(\mathbb{Z}_-, \geq)$; the $\geq$-minimum of the former is the $\geq$-maximum of the latter. Similarly, the posets $(\{1,3,7,\ldots\}, \geq)$ and $(\{2,4,5,\ldots\}, \geq)$ are order-theoretically identical, but $(\mathbb{N}, \geq)$ and $(\mathbb{Q}, \geq)$ are not. In fact, Proposition 5 says that every countable loset can be order-embedded in $(\mathbb{Q}, \geq)$ but of course, not every loset is order-isomorphic to $(\mathbb{Q}, \geq)$. Here are some more examples.

EXERCISE 11 Let $A := \left\{\frac{1}{2}, \frac{1}{3}, \frac{1}{4}, \ldots\right\}$ and $B := \left\{\frac{3}{2}, \frac{5}{3}, \frac{7}{4}, \ldots\right\}$. Prove:
(a) $(A, \geq)$ and $(\mathbb{N}, \geq)$ are not order-isomorphic, but $(A, \geq)$ and $(A \cup \{1\}, \geq)$ are.
(b) $(A, \geq)$ and $(A \cup B, \geq)$ are not order-isomorphic.

EXERCISE 12 Let $(X, \succsim_X)$ and $(Y, \succsim_Y)$ be two order-isomorphic posets. Show that X has a $\succsim_X$-maximum iff Y has a $\succsim_Y$-maximum. (The same also applies to the minimum elements, of course.)

EXERCISE 13[H] Prove that any poset $(S, \succsim)$ is order-isomorphic to $(X, \supseteq)$ for some class X of sets.

It is now time to identify those countable losets whose order structures are indistinguishable from that of $\mathbb{Q}$. The following property turns out to be crucial for this purpose.

DEFINITION
Let $(X, \succsim)$ be a preordered set and $S \subseteq X$. If, for any $x, y \in X$ such that $x \succ y$, there exists an element s of S such that $x \succ s \succ y$, we say that S is **$\succsim$-dense** (or, **order-dense**) **in** X. If $S = X$ here, we simply say that X is **$\succsim$-dense** (or, **order-dense**).

PROPOSITION 6
(Cantor) *Let $(X, \succsim_X)$ and $(Y, \succsim_Y)$ be two countable losets with neither maximum nor minimum elements (with respect to their respective linear orders). If both $(X, \succsim_X)$ and $(Y, \succsim_Y)$ are order-dense, then they are order-isomorphic.*

EXERCISE 14[H] Prove Proposition 6.

A special case of Proposition 6 solves the motivating problem of this section by characterizing the losets that are order-isomorphic to $\mathbb{Q}$. This finding is important enough to deserve separate mention, so we state it as a corollary below. In Section 4, we will see how important this observation is for utility theory.

Corollary 2
Any countable and order-dense loset with neither maximum nor minimum elements is order-isomorphic to $\mathbb{Q}$ *(and, therefore, to* $\mathbb{Q} \cap (0,1)$*).*

3 Some More Advanced Set Theory

3.1 The Cardinality Ordering

We now turn to the problem of comparing the "size" of two infinite sets. We have already had a head start on this front in Section 1. That section taught us that $\mathbb{N}$ and $\mathbb{Q}$ are "equally crowded" (Corollary 1), whereas $\mathbb{R}$ is "more crowded" than $\mathbb{N}$ (Proposition 4). This is because we can put all members of $\mathbb{N}$ into a one-to-one correspondence with all members of $\mathbb{Q}$, whereas mapping each natural number to a real number can never exhaust the entire $\mathbb{R}$. If we wish to generalize this reasoning to compare the "size" of any two sets, then we arrive at the following notion.

Definition
Let A and B be any two sets. We say that A is **cardinally larger than** B, denoted $A \succsim_{\text{card}} B$, if there exists an injection f from B into A. If, on the other hand, we can find a bijection from A onto B, then we say that A and B are **cardinally equivalent**, and denote this by $A \sim_{\text{card}} B$.

So, a set is countably infinite iff it is cardinally equivalent to $\mathbb{N}$. Moreover, we learned in Section 1 that $\mathbb{N} \sim_{\text{card}} \mathbb{N}^2 \sim_{\text{card}} \mathbb{Q}$, while $[0,1] \succsim_{\text{card}} \mathbb{N}$ but not $\mathbb{N} \succsim_{\text{card}} [0,1]$. Similarly, $2^S \succsim_{\text{card}} S$, but not conversely, for any nonempty set S (Exercise 10). For another example, we note that $2^{\mathbb{N}} \sim_{\text{card}} \{0,1\}^\infty$, that is, the class of all subsets of the natural numbers is cardinally equivalent to the set of all 0–1 sequences. Indeed, the map $f : 2^{\mathbb{N}} \to \{0,1\}^\infty$ defined by $f(S) := (x_m^S)$, where $x_m^S = 1$ if $m \in S$ and $x_m^S = 0$ otherwise, is a bijection. Here are some other examples.

EXERCISE 15[H] Prove the following facts:

(a) $\mathbb{R} \sim_{\text{card}} \mathbb{R}\backslash\mathbb{N}$.

(b) $[0,1] \sim_{\text{card}} 2^{\mathbb{N}}$.

(c) $\mathbb{R} \sim_{\text{card}} \mathbb{R}^2$.

(d) $\mathbb{R} \sim_{\text{card}} \mathbb{R}^{\infty}$.

EXERCISE 16[H] Prove: A set S is infinite iff there is a set $T \subset S$ with $T \sim_{\text{card}} S$.

It is readily verified that $A \succcurlyeq_{\text{card}} B$ and $B \succcurlyeq_{\text{card}} C$ imply $A \succcurlyeq_{\text{card}} C$ for any sets A, B, and C, that is, $\succcurlyeq_{\text{card}}$ is a preorder on any given class of sets. But there is still a potential problem with taking the relation $\succcurlyeq_{\text{card}}$ as a basis for comparing the "sizes" of two sets. This is because at the moment, we do not know what to make of the case $A \succcurlyeq_{\text{card}} B$ and $B \succcurlyeq_{\text{card}} A$. Intuitively, we would like to say in this case that A and B are equally crowded, or formally, we would like to declare that A is cardinally equivalent to B. But, given our definitions, this is not at all an obvious conclusion. What is more, if it didn't hold, we would be in serious trouble. Assume for a moment that we can find a set S such that $S \succcurlyeq_{\text{card}} \mathbb{N}$ and $\mathbb{N} \succcurlyeq_{\text{card}} S$, but not $S \sim_{\text{card}} \mathbb{N}$. According to our interpretation, we would like to say in this case that the "size" of S is neither larger nor smaller than $\mathbb{N}$, but $S \succcurlyeq_{\text{card}} \mathbb{N}$ and not $S \sim_{\text{card}} \mathbb{N}$ entail that S is an uncountable set! Obviously, this would be a problematic situation that would forbid thinking of uncountable sets as being much more "crowded" than countable ones. But of course, such a problem never arises.

> THE SCHRÖDER-BERNSTEIN THEOREM[9]
> *For any two sets A and B such that $A \succcurlyeq_{\text{card}} B$ and $B \succcurlyeq_{\text{card}} A$, we have $A \sim_{\text{card}} B$.*

In the rest of this section we provide a proof of this important theorem. Our proof is based on the following result, which is of interest in and of

[9] This result is sometimes referred to as Bernstein's Theorem, or the Cantor-Bernstein Theorem. This is because Cantor conjectured the result publicly in 1897 but was unable to prove it. Cantor's conjecture was proved that year by then 19-year-old Felix Bernstein. This proof was never published; it was popularized instead by an 1898 book of Émile Borel. In the same year, Friedrich Schröder published an independent proof of the fact (but his argument contained a relatively minor error, which he corrected in 1911).

itself. It is the first of many fixed point theorems that you will encounter in this text.

> **Tarski's Fixed Point Theorem**
> *Let $(X, \succsim)$ be a poset such that*
>
> (i) *if $S \in 2^X \backslash \{\emptyset\}$ has an $\succsim$-upper bound in X, then $\sup_{\succsim}(S) \in X$;*
>
> (ii) *X has a $\succsim$-maximum and a $\succsim$-minimum.*
>
> *If f is a self-map on X such that $x \succsim y$ implies $f(x) \succsim f(y)$ for any $x, y \in X$, then it has a fixed point, that is, $f(x) = x$ for some $x \in X$.*[10]

Proof

Let f be a self-map on X such that $f(x) \succsim f(y)$ whenever $x \succsim y$. Define $S := \{z \in X : f(z) \succsim z\}$, and observe that S is nonempty and bounded from above by (ii). (Why nonempty?) By (i), on the other hand, $x := \sup_{\succsim} S$ exists. Observe that, for any $y \in S$, we have $f(y) \succsim y$ and $x \succsim y$. The latter expression implies that $f(x) \succsim f(y)$, so combining this with the former, we find $f(x) \succsim y$ for any $y \in S$. Thus $f(x)$ is an $\succsim$-upper bound for S in X, which implies $f(x) \succsim x$. In turn, this yields $f(f(x)) \succsim f(x)$, which means that $f(x) \in S$. But then, since $x = \sup_{\succsim} S$, we have $x \succsim f(x)$. Since $\succsim$ is antisymmetric, we must then have $f(x) = x$. ■

This theorem is surprisingly general. It guarantees the existence of a fixed point by using the order structure at hand and nothing more. For instance, it implies, as an immediate corollary, the following beautiful result (which is, of course, not valid for decreasing functions; see Figure 2).

> **Corollary 3**
> *Every increasing self-map on $[0,1]$ has a fixed point.*

Or how about the following? *If $n \in \mathbb{N}$, $\varphi_i : [0,1]^n \to [0,1]$ is an increasing function, $i = 1, \ldots, n$, and Φ is a self-map on $[0,1]^n$ defined by*

[10] A special case of this result is the famous *Knaster-Tarski Fixed Point Theorem*: *Every order-preserving self-map on a complete lattice has a fixed point.* (See Exercise A.16 to recall the definition of a complete lattice.)

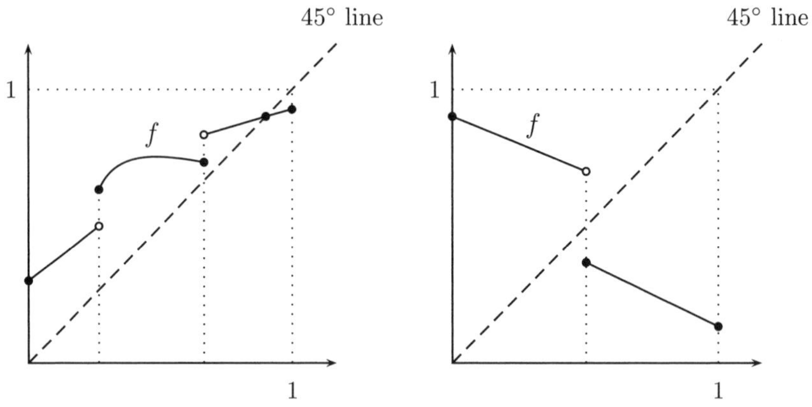

FIGURE 2

$\Phi(t) := (\varphi_1(t), \ldots, \varphi_n(t))$, *then* Φ *has a fixed point.* Noticing the similarity of these results with those obtained from the Brouwer Fixed Point Theorem (which you have probably seen somewhere before; if not, don't worry: we'll talk about it quite a bit in Section D.8) should give you an idea about how useful Tarski's Fixed Point Theorem is. Indeed, when one lacks the continuity of the involved individuals' objective functions but has rather the monotonicity of them, this theorem does the work of the Brouwer Fixed Point Theorem to ensure the existence of equilibrium in a variety of economic models.[11]

EXERCISE 17 Let $(X, \succsim)$ and f satisfy the hypotheses of Tarski's Fixed Point Theorem. Let $Fix(f)$ stand for the set of all fixed points of f on X. Prove or disprove: $(Fix(f), \succsim)$ is a poset with a $\succsim$-maximum and $\succsim$-minimum element.

Let us now turn back to our primary goal of proving the Schröder-Bernstein Theorem. It turns out that a clever application of Tarski's Fixed Point Theorem delivers this result on a silver platter.[12] We will first derive the following intermediate step, which makes the logic of the proof transparent.

[11] If you don't know what I am talking about here, that's just fine. The importance of fixed point theory will become crystal clear as you learn more economics. This text will help too; almost every chapter henceforth contains a bit of fixed point theory and its applications.

[12] I learned this method of proof from Gleason (1991), which is also used by Carothers (2000). There are many alternative proofs. See, for instance, Halmos (1960), Cox (1968), or Kaplansky (1977).

Banach's Decomposition Theorem

Let X and Y be any two nonempty sets, and take any $f \in Y^X$ and $g \in X^Y$. Then X can be written as the union of two disjoint sets X_1 and X_2 and Y can be written as the union of two disjoint sets Y_1 and Y_2 such that $f(X_1) = Y_1$ and $g(Y_2) = X_2$.

Proof

Clearly, $(2^X, \supseteq)$ is a poset that has a $\supseteq$-maximum and a $\supseteq$-minimum. Moreover, for any (nonempty) class $\mathcal{A} \in 2^X$, we have $\sup_{\supseteq} \mathcal{A} = \cup\mathcal{A}$. Now define the self-map Φ on 2^X by

$$\Phi(S) := X \backslash g(Y \backslash f(S)),$$

and observe that $B \supseteq A$ implies $\Phi(B) \supseteq \Phi(A)$ for any $A, B \subseteq X$. We may thus apply Tarski's Fixed Point Theorem to conclude that there exists a set $S \subseteq X$ such that $\Phi(S) = S$. Then $g(Y \backslash f(S)) = X \backslash S$, and hence defining $X_1 := S, X_2 := X \backslash S, Y_1 := f(S)$, and $Y_2 := Y \backslash f(S)$ completes the proof.[13] ■

Now the stage is set for settling the score, for the Schröder-Bernstein Theorem is an almost immediate corollary of Banach's Decomposition Theorem.

Proof of the Schröder-Bernstein Theorem

Let A and B be two nonempty sets, and take any two injections $f : A \to B$ and $g : B \to A$. By Banach's Decomposition Theorem, there exist a doubleton partition $\{A_1, A_2\}$ of A and a doubleton partition $\{B_1, B_2\}$ of B such that $F := f|_{A_1}$ is a bijection from A_1 onto B_1 and $G := g|_{B_2}$ is a bijection from B_2 onto A_2. But then $h : A \to B$, defined by

$$h(x) := \begin{cases} F(x), & \text{if } x \in A_1 \\ G^{-1}(x), & \text{if } x \in A_2 \end{cases}$$

is a bijection, and we are done. ■

[13] Whoa! Talk about a rabbit-out-of-the-hat proof! What is going on here? Let me explain. Suppose the assertion is true, so there exists such a decomposition of X and Y. Then X is such a set that when you subtract the image $f(X_1)$ from Y, you end up with a set the image of which under g is $X \backslash X_1$, that is, $g(Y \backslash f(X_1)) = X \backslash X_1$. But the converse is also true, that is, if we can get our hands on an $X_1 \subseteq X$ with $g(Y \backslash f(X_1)) = X \backslash X_1$, then we are done. That is, the assertion is true iff we can find a (possibly empty) subset S of X with $g(Y \backslash f(S)) = X \backslash S$, that is, a fixed point of Φ! And how do you find such a fixed point? Well, with this little structure at hand, Alfred Tarski should be the first person we should ask help from.

This proof is a beautiful illustration of the power of fixed point theorems. Like most proofs that are based on a fixed point argument, however, it is not constructive. That is, it does not tell us how to map A onto B in a one-to-one manner, it just says that such a thing is possible to do. If this worries you at all, then you will be pleased to hear that there are other ways of proving the Schröder-Bernstein Theorem, and some of these actually "construct" a tangible bijection between the involved sets. Two such proofs are reported in Halmos (1960) and Kaplansky (1977).

3.2* The Well-Ordering Principle

We now turn to the concept of well-ordering. Since this notion will play a limited role in this book, however, our exposition will be brief.

DEFINITION
Let $(X, \succsim)$ be a poset. The relation $\succsim$ is said to be a **well-ordering** relation if every nonempty subset of X has a $\succsim$-minimum. In this case, X is said to be **well-ordered** by $\succsim$, and $(X, \succsim)$ is called a **well-ordered set**, or shortly a **woset**.

Well-ordered sets are in general very useful because they possess an inductive structure. In particular, for any such set we have the following result, which is much in the same spirit as the Principle of Mathematical Induction.

EXERCISE 18 (*The Principle of Transfinite Induction*) Let $(X, \succsim)$ be a woset. Prove: If S is a nonempty subset of X such that, for any $x \in X$, $\{y \in X : x \succ y\} \subseteq S$ implies $x \in S$, then $S = X$.

It is obvious that the set of all natural numbers is well-ordered by the usual order $\geq$. On the other hand, $(2^{\{1,2\}}, \supseteq)$ is not a woset since the set $\{\{1\}, \{2\}\}$ does not have a $\supseteq$-minimum. Similarly, $([0, 1], \geq)$ is not a woset. However, a fundamental theorem (or axiom, if you will) of set theory says that, on any given set, we can always define a partial order that well-orders that set. So, for instance, we can well-order $2^{\{1,2\}}$, say, by using the partial order $\succsim$ on $2^{\{1,2\}}$ that satisfies $\{1, 2\} \succ \{2\} \succ \{1\} \succ \emptyset$. (Notice that $\{\{1\}, \{2\}\}$ does have a $\succsim$-minimum, which is $\{1\}$.) To give another example, consider the set $\{1, 2\} \times \mathbb{N}$. We can well-order this set by using the partial order $\succsim$ on $\{1, 2\} \times \mathbb{N}$ with

$$\cdots \succ (2, 2) \succ (2, 1) \succ \cdots \succ (1, 2) \succ (1, 1).$$

Much less trivially, thanks to the Axiom of Choice, it is possible to make $[0, 1]$ a woset as well. (Think about it! How can this be?) In fact, we have the following sensational result.

> **The Well-Ordering Principle**
> (Zermelo) *Every set can be well-ordered.*

This is one of the most surprising consequences of the Axiom of Choice.[14] Our interest in the Well-Ordering Principle here stems from the fact that this principle allows us to answer the following question: Can we always rank the cardinalities of any two sets? The answer is affirmative, and, when one is equipped with the Well-Ordering Principle, not difficult to prove.

Exercise 19 Show that, for any two sets A and B, we have either $A \succcurlyeq_{\text{card}} B$ or $B \succcurlyeq_{\text{card}} A$. Conclude that $\succcurlyeq_{\text{card}}$ acts as a complete preorder on any class of sets.

Exercise 20 Prove that the diagonal relation D_X on any nonempty set X (Example A.1.[1]) can be written as the intersection of two linear orders on X.

4 Application: Ordinal Utility Theory

In this section we outline the basic framework of ordinal utility theory, and show that the notion of countability plays an important role in this theory. We shall revisit this topic in Chapters C and D and present a number of classic results that lie at the foundation of economic theory.

[14] The Well-Ordering Principle is in fact equivalent to the Axiom of Choice (given the standard axioms of set theory), even though the former seems impossible and the latter self-evident.

4.1 Preference Relations

Throughout this section the set X is viewed as a nonempty set of outcomes (or prizes, or alternatives). As noted in Example A.3, by a **preference relation** $\succsim$ on X, we mean a preorder on X. As you already know well, the preference relation of an individual contains all the information that concerns her tastes about the outcomes in X. If $x \succsim y$ holds, we understand that the individual with preference relation $\succsim$ views the alternative x to be at least as good as the alternative y. Induced from $\succsim$ are the **strict preference relation** $\succ$ on X, and the **indifference relation** $\sim$ on X, defined by $x \succ y$ iff $x \succsim y$ but not $y \succsim x$, and $x \sim y$ iff $x \succsim y$ and $y \succsim x$, respectively. Both of these relations are readily checked to be transitive (Exercise A.7). Furthermore, it is easy to verify that if either $x \succ y$ and $y \succsim z$, or $x \succsim y$ and $y \succ z$, then $x \succ z$.

While reflexivity is trivial, the transitivity requirement for a preference relation is an immediate reflection of the hypothesis of rationality that is almost always imposed on economic agents. A good illustration of this is through what is called the "money pump" argument. Suppose that the *strict* preference relation of a certain individual on X is given by a nontransitive binary relation R. Then we may find a triplet $(x, y, z) \in X^3$ such that x is strictly better than y and y is weakly better than z, while the agent either cannot compare x and z, or finds z at least as good as x. Now suppose that the individual owns x to begin with. Then it is quite conceivable that she may exchange x with z, barring any status quo bias. But she likes y better than z, so it shouldn't be too difficult to convince her to trade z with y. What is more, she would certainly be willing to pay at least a small amount of money to exchange y with x. So, after all this trade, the agent ends up where she started, but she is now a bit poorer. In fact, repeating this argument, we may extract arbitrarily large amounts of money from her—something a "rational" individual should not allow. Thus, so the argument concludes, there is good reason to view transitivity (of at least that of one's strict preference relation) as a basic tenet of rationality. We subscribe to this position in what follows.[15]

[15] This said, however, be warned that transitivity of $\sim$ is a somewhat problematic postulate. The problem is that an individual may be indifferent between two alternatives just because she fails to perceive a difference between them. The now classic argument goes as follows: "I'm indifferent between a cup of coffee with no sugar and the same cup of coffee with one grain of sugar added. I'm also indifferent between the latter and the same cup of coffee with

Another property commonly imposed on a preference relation is that of *completeness* (i.e., the requirement that either $x \succsim y$ or $y \succsim x$ holds for every $x, y \in X$). This property is, however, less appealing than transitivity on intuitive grounds in that it appears much less linked to the notion of rationality. Why can't a rational individual be indecisive on occasion? Can't a rational person ever say "I don't know"? (You might want to read Aumann (1962) and Mandler (2005) for illuminating discussions about the conceptual distinction between the notion of rationality and the Completeness Axiom, and Eliaz and Ok (2006) to see how one may behaviorally distinguish between indifference and indecisiveness in general.) At the very least, there are instances in which a decision maker is actually composed of several agents, each with a possibly distinct objective function. For instance, in coalitional bargaining games, it may be quite natural to specify the preferences of each coalition by means of a vector of utility functions (one for each member of the coalition), and this requires one to view the preference relation of each coalition as an incomplete preference relation. The same reasoning applies to social choice problems: the most commonly used social welfare ordering in economics, the Pareto dominance (Example A.3), is an incomplete preorder (unless the society consists of a single individual). At any rate, at least some of the main results of individual choice theory can be obtained without imposing the completeness property. Thus, in this book, we choose not to impose this requirement on a preference relation at the onset. The only price to pay for this is the need to qualify a preference relation with the adjective "complete" in the statements of some of the subsequent results.

Before proceeding further, let us define the following bit of terminology, which will be used repeatedly in what follows.

---

another grain of sugar added; I simply cannot tell the difference. And this, ad infinitum. But then transitivity would require that I be indifferent between a cup of coffee with no sugar and the same cup with an enormous amount of sugar added in, and that I am not." Moreover, there are other appealing reasons for studying nontransitive preferences, especially in sequential (intertemporal) choice contexts (cf. Fishburn (1991) and Ok and Masatlioglu (2005)). Nevertheless, it is almost universal practice to work with transitive preferences, and this is not the place to depart from this tradition. In what follows, therefore, I will work only with such preferences, and will, furthermore, ignore the issues related to the description of choice problems (the so-called *framing effects*) and the computational limitations of decision makers (the *bounded rationality* theory).

Definition
Let $\succsim$ be a preference relation on X. For any $x \in X$, the **weak** and **strict upper $\succsim$-contour sets** of x are the sets defined as

$$U_{\succsim}(x) := \{y \in X : y \succsim x\} \quad \text{and} \quad U_{\succ}(x) := \{y \in X : y \succ x\},$$

respectively. The **weak** and **strict lower $\succsim$-contour sets** of x are defined analogously:

$$L_{\succsim}(x) := \{y \in X : x \succsim y\} \quad \text{and} \quad L_{\succ}(x) := \{y \in X : x \succ y\}.$$

4.2 Utility Representation of Complete Preference Relations

While the preference relation of an agent contains all the information that concerns her tastes, this is really not the most convenient way of summarizing this information. Maximizing a binary relation (while a well-defined matter) is a much less friendly exercise than maximizing a function. Thus, it would be quite useful if we knew how and when one could find a real function that attaches to an alternative x a (strictly) higher value than an alternative y iff x is ranked (strictly) above y by a given preference relation. As you will surely recall, such a function is called the *utility function* of the individual who possesses this preference relation. A fundamental question in the theory of individual choice is therefore the following: What sort of preference relations can be described by means of a utility function?

We begin by formally defining what it means to "describe a preference relation by means of a utility function."

Definition
Let X be a nonempty set, and $\succsim$ a preference relation on X. For any $\emptyset \neq S \subseteq X$, we say that $u \in \mathbb{R}^S$ **represents $\succsim$ on** S, if u is an order-preserving function, that is, if

$$x \succsim y \quad \text{if and only if} \quad u(x) \geq u(y)$$

for any $x, y \in S$. If u represents $\succsim$ on X, we simply say that u **represents** $\succsim$. If such a function exists, then $\succsim$ is said to be **representable**, and u is called a **utility function** for $\succsim$.

Thus, if u represents $\succsim$, and $u(x) > u(y)$, we understand that x is strictly preferred to y by an agent with the preference relation $\succsim$. (Notice that if u represents $\succsim$, then $\succsim$ is complete, and $u(x) > u(y)$ iff $x \succ y$, and $u(x) = u(y)$ iff $x \sim y$, for any $x, y \in X$.) It is commonplace to say in this case that the agent derives more utility from obtaining alternative x than y; hence the term *utility function*. However, one should be careful in adopting this interpretation, for a utility function that represents a preference relation $\succsim$ is *not* unique. Therefore, a utility function cannot be thought of as measuring the "util" content of an alternative. It is rather *ordinal* in the sense that if u represents $\succsim$, then so does $f \circ u$ for any strictly increasing self-map f on $\mathbb{R}$. More formally, we say that *an ordinal utility function is unique up to strictly increasing transformations.*

Proposition 7
Let X be a nonempty set, and let $u \in \mathbb{R}^X$ represent the preference relation $\succsim$ on X. Then, $v \in \mathbb{R}^X$ represents $\succsim$ if, and only if, there exists a strictly increasing function $f \in \mathbb{R}^{u(X)}$ such that $v = f \circ u$.

Exercise 21 Prove Proposition 7.

Exercise 22[H] Let $\succsim$ be a complete preference relation on a nonempty set X, and let $\emptyset \neq B \subseteq A \subseteq X$. If $u \in [0,1]^A$ represents $\succsim$ on A and $v \in [0,1]^B$ represents $\succsim$ on B, then there exists an extension of v that represents $\succsim$ on A. True or false?

We now proceed to the analysis of preference relations that actually admit a representation by a utility function. It is instructive to begin with the trivial case in which X is a finite set. A moment's reflection will be enough to convince yourself that any complete preference relation $\succsim$ on X is representable in this case. Indeed, if $|X| < \infty$, then all we have to do is to find the set of least preferred elements of X (which exist by finiteness), say S, and assign the utility value 1 to any member of S. Next we choose the least preferred elements in $X \backslash S$ and assign the utility value 2 to any such element. Continuing this way, we eventually exhaust X (since X is finite) and hence obtain a representation of $\succsim$ as we sought. (Hidden in the argument is the Principle of Mathematical Induction, right?)

In fact, Proposition 5 brings us exceedingly close to concluding that any complete preference relation $\succsim$ on X is representable whenever X is a countable set. The only reason why we cannot conclude this immediately is because this proposition is proved for linear orders, while a preference relation need not be antisymmetric (that is, its indifference classes need not be singletons). However, this is not a serious difficulty; all we have to do is to make sure we assign the same utility value to all members that belong to the same indifference class. (We used the same trick when proving Corollary A.1, remember?)

PROPOSITION 8
Let X be a nonempty countable set, and $\succsim$ a complete preference relation on X. Then $\succsim$ is representable.

PROOF
Recall that $\sim$ is an equivalence relation, so the quotient set $X/_\sim$ is well-defined. Define the linear order $\succsim^*$ on $X/_\sim$ by $[x]_\sim \succsim^* [y]_\sim$ iff $x \succsim y$. (Why is $\succsim^*$ well-defined?) By Proposition 5, there exists a function $f : X/_\sim \to \mathbb{Q}$ that represents $\succsim^*$. But then $u \in \mathbb{R}^X$, defined by $u(x) := f([x]_\sim)$, represents $\succsim$. ■

This proposition also paves way toward the following interesting result, which applies to a rich class of preference relations.

PROPOSITION 9
(Birkhoff) *Let X be a nonempty set, and $\succsim$ a complete preference relation on X. If X contains a countable $\succsim$-dense subset, then $\succsim$ can be represented by a utility function $u \in [0,1]^X$.*

PROOF
If $\succ = \emptyset$, then it is enough to take u as any constant function, so we may assume $\succ \neq \emptyset$ to concentrate on the nontrivial case. Assume that there is a countable $\succsim$-dense set Y in X. By Proposition 8, there exists a $w \in \mathbb{R}^Y$ such that $w(x) \geq w(y)$ iff $x \succsim y$ for any $x, y \in Y$. Clearly, the function $v \in [0,1]^Y$, defined by $v(x) := \frac{1}{2}\left(\frac{w(x)}{1-|w(x)|} + 1\right)$, also represents $\succsim$ on Y. (Why?)

Now take any $x \in X$ with $L_\succ(x) \neq \emptyset$, and define

$$\alpha_x := \sup\{v(t) : t \in L_\succ(x) \cap Y\}.$$

By $\succsim$-denseness of Y and boundedness of v, α_x is well-defined for any $x \in X$. (Why?) Define next the function $u \in [0,1]^X$ by

$$u(x) := \begin{cases} 1, & \text{if } U_{\succ}(x) = \emptyset \\ 0, & \text{if } L_{\succ}(x) = \emptyset \\ \alpha_x, & \text{otherwise} \end{cases}.$$

$\succ \neq \emptyset$ implies that u is well-defined. (Why?) The rest of the proof is to check that u actually represents $\succsim$ on X.[16] We leave verifying this as an exercise, but just to get you going, let's show that $x \succ y$ implies $u(x) > u(y)$ for any $x, y \in X$ with $L_{\succ}(y) \neq \emptyset$. Since Y is $\succsim$-dense in X, there must exist $z_1, z_2 \in Y$ such that $x \succ z_1 \succ z_2 \succ y$. Since $z_1 \in L_{\succ}(x) \cap Y$ and v represents $\succsim$ on Y, we have $\alpha_x \geq v(z_1) > v(z_2)$. On the other hand, since $v(z_2) > v(t)$ for all $t \in L_{\succ}(y) \cap Y$ (why?), we also have $v(z_2) \geq \alpha_y$. Combining these observations yields $u(x) > u(y)$. ■

EXERCISE 23 Complete the proof of Proposition 9.

EXERCISE 24 (*Another Proof for Proposition* 9) Assume the hypotheses of Proposition 9, and let Y be a countable $\succsim$-dense subset of X. Enumerate Y as $\{y_1, y_2, \ldots\}$, and define $\mathcal{L}(x) := \{m \in \mathbb{N} : x \succ y_m\}$. Finally, define $u \in [0,1]^X$ by $u(x) := 0$ if $\mathcal{L}(x) = \emptyset$, and $u(x) := \sum_{i \in \mathcal{L}(x)} \frac{1}{2^i}$ otherwise. Show that u represents $\succsim$.

The next exercise provides a generalization of Proposition 9 by offering an alternative denseness condition that is actually *necessary* and sufficient for the representability of a linear order.

*EXERCISE 25^{H} Let X be a nonempty set and $\succsim$ a linear order on X. Then, $\succsim$ is representable iff X contains a countable set Y such that, for each $x, y \in X \backslash Y$ with $x \succ y$, there exists a $z \in Y$ such that $x \succ z \succ y$.

The characterization result given in Exercise 25 can be used to identify certain preference relations that are not representable by a utility function. Here is a standard example of such a relation.

[16] Alternative representations may be obtained by replacing the role of α_x in this proof with $\lambda\alpha_x + (1-\lambda)\inf\{v(y) : y \in U_{\succ}(x) \cap Y\}$ for any $\lambda \in [0,1]$.

EXAMPLE 1

(*The Lexicographic Preference Relation*) Consider the linear order $\succsim_{\text{lex}}$ on $\mathbb{R}^2$ defined as $x \succsim_{\text{lex}} y$ iff either $x_1 > y_1$ or $x_1 = y_1$ and $x_2 \geq y_2$. (This relation is called the **lexicographic order**.) We write $x \succ_{\text{lex}} y$ whenever $x \succsim_{\text{lex}} y$ and $x \neq y$. For instance, $(1, 3) \succ_{\text{lex}} (0, 4)$ and $(1, 3) \succ_{\text{lex}} (1, 2)$. As you probably recall from a microeconomics class you've taken before, $\succsim_{\text{lex}}$ is not representable by a utility function. We now provide two proofs of this fact.

FIRST PROOF

Suppose that there exists a set $Y \subset \mathbb{R}^2$ such that, for all $x, y \in \mathbb{R}^2 \backslash Y$ with $x \succ_{\text{lex}} y$, we have $x \succ_{\text{lex}} z \succ_{\text{lex}} y$ for some $z \in Y$. We shall show that Y must then be uncountable, which is enough to conclude that $\succsim_{\text{lex}}$ is not representable in view of Exercise 25. Take any real number a, and pick any $z^a \in Y$ such that $(a, 1) \succ_{\text{lex}} z^a$ and $z^a \succ_{\text{lex}} (a, 0)$. Clearly, we must have $z_1^a = a$ and $z_2^a \in (0, 1)$. But then $z^a \neq z^b$ for any $a \neq b$, while $\{z^a : a \in \mathbb{R}\} \subseteq Y$. It follows that Y is not countable (Proposition 1).

SECOND PROOF

Let $u : \mathbb{R}^2 \to \mathbb{R}$ represent $\succsim_{\text{lex}}$. Then, for any $a \in \mathbb{R}$, we have $u(a, a + 1) > u(a, a)$ so that $I(a) := (u(a, a), u(a, a + 1))$ is a nondegenerate interval in $\mathbb{R}$. Moreover, $I(a) \cap I(b) = \emptyset$ for any $a \neq b$, for we have

$$u(b, b) > u(a, a + 1) \quad \text{whenever} \quad b > a,$$

and

$$u(b, b + 1) < u(a, a) \quad \text{whenever} \quad b < a.$$

Therefore, the map $a \mapsto I(a)$ is an injection from $\mathbb{R}$ into $\{I(a) : a \in \mathbb{R}\}$. But since $\{I(a) : a \in \mathbb{R}\}$ is countable (Proposition 3), this entails that $\mathbb{R}$ is countable (Proposition 1), a contradiction. □

The class of all preference relations that do not possess a utility representation is recently characterized by Beardon et al. (2002). As the next example illustrates, this class includes nonlexicographic preferences as well.[17]

[17] For another example in an interesting economic context, see Basu and Mitra (2003).

EXERCISE 26 (Dubra-Echenique) Let X be a nonempty set, and let $\mathcal{P}(X)$ denote the class of all partitions of X. Assume that $\succsim$ is a complete preference relation on $\mathcal{P}(X)$ such that $\mathcal{B} \succ \mathcal{A}$ holds if for every $B \in \mathcal{B}$ there is an $A \in \mathcal{A}$ with $B \subseteq A$.[18] Prove:

(a) If $X = [0, 1]$, then $\succsim$ is not representable by a utility function.

(b) If X is uncountable, then $\succsim$ is not representable by a utility function.

4.3* Utility Representation of Incomplete Preference Relations

As noted earlier, there is a conceptual advantage in taking a (possibly incomplete) preorder as the primitive of analysis in the theory of rational choice. Yet an incomplete preorder cannot be represented by a utility function—if it did, it would not be incomplete. Thus it seems that the analytical scope of adopting such a point of view is perforce rather limited. Fortunately, however, it is still possible to provide a utility representation for an incomplete preference relation, provided that we suitably generalize the notion of a "utility function."

To begin with, let us note that while it is obviously not possible to find a $u \in \mathbb{R}^X$ for an incomplete preorder $\succsim$ such that $x \succsim y$ iff $u(x) \geq u(y)$ for any $x, y \in X$, there may nevertheless exist a real function u on X such that

$$x \succ y \text{ implies } u(x) > u(y) \quad \text{and} \quad x \sim y \text{ implies } u(x) = u(y)$$

for any $x, y \in X$. Among others, this approach was explored first by Richter (1966) and Peleg (1970). We shall thus refer to such a real function u as a **Richter-Peleg utility function** for $\succsim$. Recall that Szpilrajn's Theorem guarantees that $\succsim$ can be extended to a complete preference relation (Corollary A.1). If $\succsim$ has a Richter-Peleg utility function u, then, and only then, $\succsim$ can be extended to a complete preference relation that is represented by u in the ordinary sense.

The following result shows that Proposition 9 can be extended to the case of incomplete preorders if one is willing to accept this particular notion of utility representation. (You should contrast the associated proofs.)

[18] Szpilrajn's Theorem assures us that there exists such a preference relation. (Why?)

LEMMA 1
(Richter) *Let X be a nonempty set, and $\succsim$ a preference relation on X. If X contains a countable $\succsim$-dense subset, then there exists a Richter-Peleg utility function for $\succsim$.*

PROOF
(Peleg) We will prove the claim assuming that $\succsim$ is a partial order; the extension to the case of preorders is carried out as in the proof of Proposition 8. Obviously, if $\succ = \emptyset$, then there is nothing to prove. So let $\succ \neq \emptyset$, and assume that X contains a countable $\succsim$-dense set—let's call this set Y. Clearly, there must then exist $a, b \in Y$ such that $b \succ a$. Thus $\{(a,b)_{\succsim} : a, b \in Y \text{ and } b \succ a\}$ is a countably infinite set, where $(a,b)_{\succsim} := \{x \in X : b \succ x \succ a\}$. (Why?) We enumerate this set as $\{(a_1,b_1)_{\succsim}, (a_2,b_2)_{\succsim}, \ldots\}$. Each $(a_i,b_i)_{\succsim} \cap Y$ is partially ordered by $\succsim$ so that, by the Hausdorff Maximal Principle, it contains a $\supseteq$-maximal loset, say $(Z_i, \succsim)$. By $\succsim$-denseness of Y, Z_i has neither a $\succsim$-maximum nor a $\succsim$-minimum. Moreover, by its maximality, it is $\succsim$-dense in itself. By Corollary 2, therefore, there exists a bijection $f_i : Z_i \to (0,1) \cap \mathbb{Q}$ such that $x \succsim y$ iff $f_i(x) \geq f_i(y)$ for any $x, y \in Z_i$. Now define the map $\varphi_i \in [0,1]^X$ by

$$\varphi_i(x) := \begin{cases} \sup\{f_i(t) : x \succ t \in Z_i\}, & \text{if } L_{\succ}(x) \cap Z_i \neq \emptyset \\ 0, & \text{otherwise.} \end{cases}$$

Clearly, we have $\varphi_i(x) = 0$ for all $x \in L_{\succsim}(a_i)$, and $\varphi_i(x) = 1$ for all $x \in U_{\succsim}(b_i)$. Using this observation and the definition of f_i, one can show that for any $x, y \in X$ with $x \succ y$, we have $\varphi_i(x) \geq \varphi_i(y)$. (Verify!) To complete the proof, then, define

$$u(x) := \sum_{i=1}^{\infty} \frac{\varphi_i(x)}{2^i} \qquad \text{for all } x \in X.$$

(Since the range of each φ_i is contained in $[0,1]$ and $\sum^{\infty} \frac{1}{2^i} = 1$ (Example 8. [2]), u is well-defined.) Notice that, for any $x, y \in X$ with $x \succ y$, there exists a $j \in \mathbb{N}$ with $x \succ b_j \succ a_j \succ y$ so that $\varphi_j(x) = 1 > 0 = \varphi_j(y)$. Since $x \succ y$ implies $\varphi_i(x) \geq \varphi_i(y)$ for every i, therefore, we find that $x \succ y$ implies $u(x) > u(y)$, and the proof is complete. ∎

Unfortunately, the Richter-Peleg formulation of utility representation has a serious shortcoming in that it may result in a substantial information loss.

Indeed, one cannot recover the original preference relation $\succsim$ from a Richter-Peleg utility function u; the information contained in u is strictly less than that contained in $\succsim$. All we can deduce from the statement $u(x) > u(y)$ is that it is not the case that y is strictly better than x for the subject individual. We cannot tell if this agent actually likes x better than y, or that she is unable to rank x and y. (That is, we are unable to capture the "indecisiveness" of a decision maker by using a Richter-Peleg utility function.) The problem is, of course, due to the fact that the range of a real function is completely ordered, while its domain is not. One way of overcoming this problem is by using a poset-valued utility function, or better, by using a *set of* real-valued utility functions in the following way.

> Definition
>
> Let X be a nonempty set, and $\succsim$ a preference relation on X. We say that the set $\mathcal{U} \subseteq \mathbb{R}^X$ **represents** $\succsim$, if
>
> $$x \succsim y \quad \text{if and only if} \quad u(x) \geq u(y) \quad \text{for all } u \in \mathcal{U},$$
>
> for any $x, y \in X$.

Here are some immediate examples.

Example 2

[1] Let $n \in \{2, 3, \ldots\}$. Obviously, we cannot represent the partial order $\geq$ on $\mathbb{R}^n$ by a single utility function, but we can represent it by a (finite) set of utility functions. Indeed, defining $u_i(x) := x_i$ for each $x \in \mathbb{R}^n$, $i = 1, \ldots, n$, we find

$$x \geq y \quad \text{iff} \quad u_i(x) \geq u_i(y) \quad i = 1, \ldots, n.$$

for any $x, y \in \mathbb{R}^n$.

[2] $Z := \{1, \ldots, m\}$, $m \geq 2$ and let $\mathcal{L}_Z$ stand for the set of all probability distributions (lotteries) on Z, that is, $\mathcal{L}_Z := \{(p_1, \ldots, p_m) \in \mathbb{R}^m_+ : \sum^m p_i = 1\}$. Then the **first-order stochastic dominance ordering** on $\mathcal{L}_Z$, denoted by $\succcurlyeq_{\text{FSD}}$, is defined as follows:

$$p \succcurlyeq_{\text{FSD}} q \quad \text{iff} \quad \sum_{i=1}^{k} p_i \leq \sum_{i=1}^{k} q_i \quad k = 1, \ldots, m-1.$$

(Here we interpret m as the best prize, $m-1$ as the second best prize, and so on.) The partial order $\succcurlyeq_{\text{FSD}}$ on $\mathcal{L}_Z$ is represented by the set $\{u_1, \ldots, u_{m-1}\}$ of real functions on $\mathcal{L}_Z$, where $u_k(p) := -\sum^k p_i$, $k = 1, \ldots, m-1$.

[3] For any nonempty set X, the diagonal relation D_X can be represented by a set of two utility functions. This follows from Exercise 20 and Proposition 8. (*Note.* If X is countable, we do not have to use the Axiom of Choice to prove this fact.) □

EXERCISE 27 Let $X := \{x \in \mathbb{R}^2 : x_1^2 + x_2^2 = 1 \text{ and } x_1 \neq 0\}$ and define the partial order $\succsim$ on X as

$$x \succsim y \quad \text{iff} \quad x_1 y_1 > 0 \quad \text{and} \quad x_2 \geq y_2.$$

Show that $\succsim$ can be represented by a set of two continuous real functions on X.

EXERCISE 28 Prove: If there exists a countable set of bounded utility functions that represent a preference relation, then there is a Richter-Peleg utility function for that preference relation.

EXERCISE 29 Prove: If there exists a countable set of utility functions that represents a complete preference relation, then this relation is representable in the ordinary sense.

EXERCISE 30^{H} Let $\succsim$ be a reflexive partial order on $\mathbb{R}$ such that $x \succ y$ iff $x > y + 1$ for any $x, y \in \mathbb{R}$. Is there a $\mathcal{U} \subseteq \mathbb{R}^{\mathbb{R}}$ that represents $\succsim$?

EXERCISE 31 Define the partial order $\succsim$ on $\mathbb{N} \times \mathbb{R}$ by $(m, x) \succsim (n, y)$ iff $m = n$ and $x \geq y$. Is there a $\mathcal{U} \subseteq \mathbb{R}^{\mathbb{N} \times \mathbb{R}}$ that represents $\succsim$?

Suppose $\succsim$ is represented by $\{u, v\} \subseteq \mathbb{R}^X$. One interpretation we can give to this situation is that the individual with the preference relation $\succsim$ is a person who deems two dimensions relevant for comparing the alternatives in X. (Think of a potential graduate student who compares the graduate schools that she is admitted in according to the amount of financial aid they provide and the reputation of their programs.) Her preferences over the first dimension are represented by the utility function u, and the second by v. She then judges the value of an alternative x on the basis of its performance on both dimensions, that is, by the 2-vector $(u(x), v(x))$, and prefers this alternative to $y \in X$ iff x performs better than y on both of the dimensions, that is, $(u(x), v(x)) \geq (u(y), v(y))$. The utility representation notion we advance above is thus closely related to decision making with multiple objectives.[19]

At any rate, this formulation generalizes the usual utility representation notion we studied above. Moreover, it does not cause any information loss; the potential incompleteness of $\succsim$ is fully reflected in the set $\mathcal{U}$ that represents $\succsim$. Finally, it makes working with incomplete preference relations analytically less difficult, for it is often easier to manipulate vector-valued functions than preorders.

So, what sort of preference relations can be represented by means of a *set of* utility functions? It turns out that the answer is not very difficult, especially if one is prepared to adopt our usual order-denseness requirement. In what follows, given any preference relation $\succsim$ on a nonempty set X, we write $x \bowtie y$ when x and y are $\succsim$*-incomparable*, that is, when neither $x \succsim y$ nor $y \succsim x$ holds. Observe that this defines $\bowtie$ as an irreflexive binary relation on X.

PROPOSITION 10[20]
Let X be a nonempty set, and $\succsim$ a preference relation on X. If X contains a countable $\succsim$-dense subset, then there exists a nonempty set $\mathcal{U} \subseteq \mathbb{R}^X$ that represents $\succsim$.

[19] Of course, this is an "as if" interpretation. The primitive of the model is $\succsim$, so when $\mathcal{U}$ represents $\succsim$, one may only think "as if" each member of $\mathcal{U}$ measures (completely) how the agent feels about a particular dimension of the alternatives.

Let me elaborate on this a bit. Suppose the agent indeed attributes two dimensions to the alternatives and ranks the first one with respect to u and the second with respect to v, but she can compare some of the alternatives even in the absence of dominance in both alternatives. More concretely, suppose $\succsim$ is given as: $x \succsim y$ iff $U_\alpha(x) \geq U_\alpha(y)$ for all $\alpha \in [\frac{1}{3}, \frac{2}{3}]$, where $U_\alpha := \alpha u + (1-\alpha)v$ for all real α. Then, $\succsim$ is represented by $\mathcal{U} := \{U_\alpha : \frac{1}{3} \leq \alpha \leq \frac{2}{3}\}$, and while we may interpret "as if" the agent views every member of $\mathcal{U}$ measuring the value of a dimension of an alternative (so it is "as if" there are uncountably many dimensions for her), we in fact know here that each member of $\mathcal{U}$ corresponds instead to a potential aggregation of the values of the actual dimensions relevant to the problem.

[20] A special case of this result was obtained in Ok (2002b), where you can also find some results that guarantee the *finiteness* of the representing set of utility functions.

PROOF

Once again we will prove the claim assuming that $\succsim$ is a partial order; the extension to the case of preorders is straightforward. Assume that X contains a countable $\succsim$-dense subset, and let $\mathcal{U}$ be the collection of all $u \in \mathbb{R}^X$ such that $x \succ y$ implies $u(x) > u(y)$ for any $x, y \in X$. By Lemma 1, $\mathcal{U}$ is nonempty. We wish to show that $\mathcal{U}$ actually represents $\succsim$. Evidently, this means that, for any $x, y \in X$ with $x \bowtie y$, there exist at least two functions u and v in $\mathcal{U}$ such that $u(x) > u(y)$ and $v(y) > v(x)$. (Why?)

Fix any $x^*, y^* \in X$ with $x^* \bowtie y^*$, and pick an arbitrary $w_o \in \mathcal{U}$. Define $w \in [0,1]^X$ by $w(z) := \frac{1}{2}\left(\frac{w_o(z)}{1-|w_o(z)|} + 1\right)$, and note that w is also a Richter-Peleg utility function for $\succsim$. (We have also used the same trick when proving Proposition 9, remember?) Now let

$$Y := \{z \in X : z \succ x^* \quad \text{or} \quad z \succ y^*\},$$

and define $u, v \in \mathbb{R}^X$ as

$$u(z) := \begin{cases} w(z) + 4, & \text{if } z \in Y \\ 3, & \text{if } z = x^* \\ 2, & \text{if } z = y^* \\ w(z), & \text{otherwise} \end{cases} \qquad \text{and}$$

$$v(z) := \begin{cases} w(z) + 4, & \text{if } z \in Y \\ 2, & \text{if } z = x^* \\ 3, & \text{if } z = y^* \\ w(z), & \text{otherwise} \end{cases}.$$

We leave it for you to verify that both u and v are Richter-Peleg utility functions for $\succsim$. Thus $u, v \in \mathcal{U}$ and we have $u(x^*) > u(y^*)$, while $v(y^*) > v(x^*)$. ■

EXERCISE 32 Complete the proof of Proposition 10.

EXERCISE 33[H] Define the partial order $\succsim$ on $\mathbb{R}_+$ as: $x \succsim y$ iff $x \in \mathbb{Q}_+$ and $y \in \mathbb{R}_+ \backslash \mathbb{Q}$. Show that there is no countable $\succsim$-dense subset of $\mathbb{R}_+$, but there exists a $\mathcal{U} \subseteq \mathbb{R}^{\mathbb{R}_+}$ with $|\mathcal{U}| = 2$ that represents $\succsim$.

All this is nice, but it is clear that we need stronger utility representation results for applications. For instance, at present we have no way of even speaking about representing a preference relation by a *continuous* utility function (or a set of such functions). In fact, a lot can be said about this issue, but this requires us first to go through a number of topics in real analysis. We will come back to this problem in Chapters C and D when we are better prepared to tackle it.

PART II

ANALYSIS ON METRIC SPACES

Chapter C

Metric Spaces

This chapter provides a self-contained review of the basic theory of metric spaces. Chances are good that you are familiar with the rudiments of this theory, so our exposition starts a bit faster than usual. We slow down when we get to the "real stuff"—the analysis of the properties of connectedness, separability, compactness, and completeness for metric spaces.

Connectedness is a geometric property that will be of limited use in this course. Consequently, its discussion here is quite brief. All we do is identify the connected subsets of $\mathbb{R}$ and prepare for the Intermediate Value Theorem, which will be given in the next chapter. Our treatment of separability is also relatively short, even though this concept will be important for us later on. Because separability usually makes an appearance only in relatively advanced contexts, we will study this property in greater detail later.

Utility theory, which we sketched out in Section B.4, can be taken to the next level with the help of an elementary investigation of connected and separable metric spaces. As a brief application, therefore, we formulate here the "metric" versions of some of the utility representation theorems that were proved in that section. The story is brought to its conclusion in Chapter D.

The bulk of this chapter is devoted to the analysis of metric spaces that are either compact or complete. A good understanding of these two properties is essential for real analysis and optimization theory, so we spend quite a bit of time studying them. In particular, we consider several examples, give two proofs of the Heine-Borel Theorem for good measure, and discuss why closed and bounded spaces need not be compact in general. Totally bounded sets, the sequential characterization of compactness, and the relationship between compactness and completeness are also studied with care.

Most of the results established in this chapter are relatively preliminary observations whose main purpose is to create good grounds to derive a number of deeper facts in later chapters. But there is one major exception—the

Banach Fixed Point Theorem. Although this theorem is elementary and has an amazingly simple proof, the result is of substantial interest and has numerous applications. We thus explore it here at length. In particular, we consider some of the variants of this celebrated theorem and show how it can be used to prove the "existence" of a solution to certain types of functional equations. As a major application, we prove here both the local and global versions of the fundamental existence theorem of Emile Picard for differential equations. Two major generalizations of the Banach Fixed Point Theorem, along with further applications, will be considered in subsequent chapters.[1]

1 Basic Notions

Recall that we think of a real function f on $\mathbb{R}$ as continuous at a given point $x \in \mathbb{R}$ iff the image of a point (under f) that is close to x is itself close to $f(x)$. So, for instance, the indicator function $\mathbf{1}_{\{\frac{1}{2}\}}$ on $\mathbb{R}$ is not continuous at $\frac{1}{2}$, because points that are arbitrarily close to $\frac{1}{2}$ are not mapped by this function to points that are arbitrarily close to its value at $\frac{1}{2}$. On the other hand, this function is continuous at every other point in its domain.

It is crucial to understand at the outset that this "geometric" way of thinking about continuity depends intrinsically on the notion of distance between two points on the real line. Although there is an obvious measure of distance in $\mathbb{R}$, this observation is important precisely because it paves the way toward thinking about the continuity of functions defined on more complicated sets on which the meaning of the term "close" is not transparent. As a prerequisite for a suitably general analysis of continuous functions, therefore, we need to elaborate on the notion of distance between two elements of an arbitrary set. This is precisely what we intend to do in this section.

[1] Among the excellent introductory references for the analysis of metric spaces are Sutherland (1975), Rudin (1976), Kaplansky (1977), and Haaser and Sullivan (1991). Of the more recent expositions, my personal favorite is Carothers (2000). The first part of that beautifully written book not only provides a much broader perspective of metric spaces than I am able to do here, it also covers additional topics (such as compactification and completion of metric spaces, and category-type theorems), and sheds light on the historical development of the material. For a more advanced but still very readable account, I refer you to Royden (1994), which is a classic text on real analysis.

1.1 Metric Spaces: Definition and Examples

We begin with the formal definition of a metric space.

DEFINITION
Let X be a nonempty set. A function $d : X \times X \to \mathbb{R}_+$ that satisfies the following properties is called a **distance function** (or a **metric**) on X: For any $x, y, z \in X$,

(i) $d(x, y) = 0$ if and only if $x = y$;

(ii) (*Symmetry*) $d(x, y) = d(y, x)$;

(iii) (*Triangle Inequality*) $d(x, y) \leq d(x, z) + d(z, y)$.

If d is a distance function on X, we say that (X, d) is a **metric space**, and refer to the elements of X as **points** in (X, d). If d satisfies (ii) and (iii), and $d(x, x) = 0$ for any $x \in X$, then we say that d is a **semimetric** on X, and (X, d) is a **semimetric space**.

Recall that we think of the distance between two points x and y on the real line as $|x - y|$. Thus the map $(x, y) \mapsto |x - y|$ serves as a function that tells us how much apart any two elements of $\mathbb{R}$ are from each other. Among others, this function satisfies properties (i)–(iii) of the definition above (Example A.7). By way of abstraction, the notion of distance function is built *only* on these three properties. It is remarkable that these properties are strong enough to introduce to an arbitrary nonempty set a geometry rich enough to build a satisfactory theory of continuous functions.[2]

NOTATION. When the (semi)metric under consideration is apparent from the context, it is customary to dispense with the notation (X, d) and refer to X as a metric space. We also adhere to this convention here (and spare the notation d for a generic metric on X). But when we feel that there is a danger of confusion, or we endow X with a particular metric d, then we shall revert back to the more descriptive notation (X, d).

[2] The concept of metric space was first introduced in the 1906 dissertation of Maurice Fréchet (1878–1973). (We owe the term "metric space" to Felix Hausdorff, however.) Considered as one of the major founders of modern real (and functional) analysis, Fréchet is also the mathematician who first introduced the abstract formulation of compactness and completeness properties (see Dieudonné (1981) and Taylor (1982)).

Convention. We often talk as if a metric space (X, d) were indeed a set when referring to properties that apply only to X. For instance, when we say that Y is a subset of the metric space (X, d), we mean simply that $Y \subseteq X$.

Let us look at some standard examples of metric spaces.

Example 1

[1] Let X be any nonempty set. A trivial way of making X a metric space is to use the metric $d : X \times X \to \mathbb{R}_+$, which is defined by

$$d(x, y) := \begin{cases} 1, & x \neq y \\ 0, & x = y \end{cases}.$$

It is easy to check that (X, d) is indeed a metric space. Here d is called the **discrete metric** on X, and (X, d) is called a **discrete space**.

[2] Let $X := \{x \in \mathbb{R}^2 : x_1^2 + x_2^2 = 1\}$, and define $d \in \mathbb{R}^{X \times X}$ by letting $d(x, y)$ be the length of the shorter arc in X that join x and y. It is easy to see that this defines d as a metric on X, and thus (X, d) is a metric space.

[3] Given any $n \in \mathbb{N}$, there are various ways of metrizing $\mathbb{R}^n$. Indeed, $(\mathbb{R}^n, d_p)$ is a metric space for each $1 \leq p \leq \infty$, where $d_p : \mathbb{R}^n \times \mathbb{R}^n \to \mathbb{R}_+$ is defined by

$$d_p(x, y) := \left(\sum_{i=1}^{n} |x_i - y_i|^p \right)^{\frac{1}{p}} \quad \text{for } 1 \leq p < \infty,$$

and

$$d_p(x, y) := \max\{|x_i - y_i| : i = 1, \ldots, n\} \quad \text{for } p = \infty.$$

It is easy to see that each d_p satisfies the first two axioms of being a distance function. The verification of the triangle inequality in the case of $p \in [1, \infty)$ is, on the other hand, not a trivial matter. Rather, it follows from the following celebrated result of Hermann Minkowski:

Minkowski's Inequality 1

For any $n \in \mathbb{N}$, $a_i, b_i \in \mathbb{R}$, $i = 1, \ldots, n$, and any $1 \leq p < \infty$,

$$\left(\sum_{i=1}^{n} |a_i + b_i|^p \right)^{\frac{1}{p}} \leq \left(\sum_{i=1}^{n} |a_i|^p \right)^{\frac{1}{p}} + \left(\sum_{i=1}^{n} |b_i|^p \right)^{\frac{1}{p}}.$$

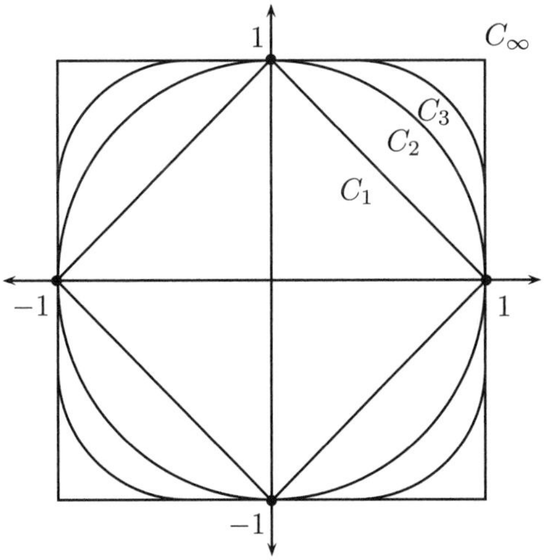

Figure 1

To be able to move faster, we postpone the proof of this important inequality to the end of this subsection. You are invited at this point, however, to show that $(\mathbb{R}^n, d_p)$ is *not* a metric space for $p < 1$.

It may be instructive to examine the geometry of the unit "circle"

$$C_p := \{x \in \mathbb{R}^2 : d_p(\mathbf{0}, x) = 1\}$$

for various choices of p. (Here $\mathbf{0}$ stands for the 2-vector $(0, 0)$.) This is done in Figure 1, which suggests that the sets C_p in some sense "converges" to the set C_∞. Indeed, for every $x, y \in \mathbb{R}^2$, we have $d_m(x, y) \to d_\infty(x, y)$. (Proof?)

The space $(\mathbb{R}^n, d_2)$ is called the ***n*-dimensional Euclidean space** in analysis. When we refer to $\mathbb{R}^n$ in the sequel without specifying a particular metric, you should understand that we view this set as metrized by the metric d_2. That is to say, the notation $\mathbb{R}^n$ is spared for the n-dimensional Euclidean space in what follows. If we wish to endow $\mathbb{R}^n$ with a metric different than d_2, we will be explicit about it.

Notation. Throughout this text, we denote the metric space $(\mathbb{R}^n, d_p)$ as $\mathbb{R}^{n,p}$ for any $1 \le p \le \infty$. However, we almost always use the notation $\mathbb{R}^n$ instead of $\mathbb{R}^{n,2}$.

Before leaving this example, let's see how we can metrize an extended Euclidean space, say $\overline{\mathbb{R}}$. For this purpose, we define the function $f : \overline{\mathbb{R}} \to [-1,1]$ by $f(-\infty) := -1$, $f(\infty) := 1$, and $f(t) := \frac{t}{1+|t|}$ for all $t \in \mathbb{R}$. The standard metric d^* on $\overline{\mathbb{R}}$ is then defined by

$$d^*(x,y) := \left|f(x) - f(y)\right|.$$

The important thing to observe here is that this makes $\overline{\mathbb{R}}$ a metric space that is essentially identical to $[-1,1]$. This is because f is a bijection from $\overline{\mathbb{R}}$ onto $[-1,1]$ that leaves the distance between any two points intact: $d^*(x,y) = d_1(f(x),f(y))$ for any $x,y \in \overline{\mathbb{R}}$. So, we should expect that any metric property that is true in $[-1,1]$ is also true in $\overline{\mathbb{R}}$.[3]

[4] For any $1 \le p < \infty$, we define

$$\ell^p := \left\{ (x_m) \in \mathbb{R}^\infty : \sum_{i=1}^{\infty} |x_i|^p < \infty \right\}.$$

This set is metrized by means of the metric $d_p : \ell^p \times \ell^p \to \mathbb{R}_+$ with

$$d_p\left((x_m), (y_m)\right) := \left(\sum_{i=1}^{\infty} |x_i - y_i|^p \right)^{\frac{1}{p}}.$$

(When we speak of ℓ^p as a metric space, we always have this metric in mind!) Of course, we have to check that d_p is well-defined as a real-valued function, and that it satisfies the triangle inequality. These facts follow readily from the following generalization of Minkowski's Inequality 1:

> MINKOWSKI'S INEQUALITY 2
> *For any* $(x_m), (y_m) \in \mathbb{R}^\infty$ *and* $1 \le p < \infty$,
>
> $$\left(\sum_{i=1}^{\infty} |x_i + y_i|^p \right)^{\frac{1}{p}} \le \left(\sum_{i=1}^{\infty} |x_i|^p \right)^{\frac{1}{p}} + \left(\sum_{i=1}^{\infty} |y_i|^p \right)^{\frac{1}{p}}. \qquad (1)$$

We will prove this inequality at the end of this subsection. You should assume its validity for now, and verify that d_p is a metric on ℓ^p for any $1 \le p < \infty$.

[3] This point may be somewhat vague right now. It will become clearer bit by bit as we move on.

By ℓ^∞, we mean the set of all bounded real sequences, that is,

$$\ell^\infty := \left\{(x_m) \in \mathbb{R}^\infty : \sup\{|x_m| : m \in \mathbb{N}\} < \infty\right\}.$$

It is implicitly understood that this set is endowed with the metric $d_\infty : \ell^\infty \times \ell^\infty \to \mathbb{R}_+$ with

$$d_\infty\left((x_m), (y_m)\right) := \sup\{\left|x_m - y_m\right| : m \in \mathbb{N}\}.$$

That d_∞ is indeed a metric will be verified below. This metric is called the **sup-metric** on the set of all bounded real sequences.
Before we leave this example, let us stress that any ℓ^p space is smaller than the set of all real sequences $\mathbb{R}^\infty$, since the members of such a space are real sequences that are either bounded or that satisfy some form of a summability condition (that ensures that d_p is real-valued). Indeed, no d_p defines a distance function on the entire $\mathbb{R}^\infty$. (Why?) But this does not mean that we cannot metrize the set of all real sequences in a useful way. We can, and we will, later in this chapter.

[5] Let T be any nonempty set. By $\mathbf{B}(T)$ we mean the set of all bounded real functions defined on T, that is,

$$\mathbf{B}(T) := \left\{f \in \mathbb{R}^T : \sup\{\left|f(x)\right| : x \in T\} < \infty\right\}.$$

We will always think of this space as metrized by the **sup-metric** $d_\infty : \mathbf{B}(T) \times \mathbf{B}(T) \to \mathbb{R}_+$, which is defined by

$$d_\infty(f, g) := \sup\{\left|f(x) - g(x)\right| : x \in T\}.$$

It is easy to see that d_∞ is real-valued. Indeed, for any $f, g \in \mathbf{B}(T)$,

$$d_\infty(f, g) \le \sup\{\left|f(x)\right| : x \in T\} + \sup\{\left|g(x)\right| : x \in T\} < \infty.$$

It is also readily checked that d_∞ satisfies the first two requirements of being a distance function. As for the triangle inequality, all we need is to invoke the corresponding property of the absolute value function (Example A.7). After all, if $f, g, h \in \mathbf{B}(T)$, then

$$\begin{aligned}\left|f(x) - g(x)\right| &\le \left|f(x) - h(x)\right| + \left|h(x) - g(x)\right| \\ &\le \sup\{\left|f(y) - h(y)\right| : y \in T\} \\ &\quad + \sup\{\left|h(y) - g(y)\right| : y \in T\} \\ &= d_\infty(f, h) + d_\infty(h, g)\end{aligned}$$

for *any* $x \in T$, so

$$d_\infty(f,g) = \sup\{|f(x) - g(x)| : x \in T\} \leq d_\infty(f,h) + d_\infty(h,g).$$

Given that a sequence and/or an n-vector can always be thought of as special functions (Section A.1.6), it is plain that $\mathbf{B}(\{1,\ldots,n\})$ coincides with $\mathbb{R}^{n,\infty}$ (for any $n \in \mathbb{N}$) while $\mathbf{B}(\mathbb{N})$ coincides with ℓ^∞. (Right?) Therefore, the inequality we just established proves in one stroke that both $\mathbb{R}^{n,\infty}$ and ℓ^∞ are metric spaces. □

REMARK 1. Distance functions need not bounded. However, given any metric space (X,d), we can always find a bounded metric d' on X that orders the distances between points in the space *ordinally* the same way as the original metric. (That is, such a metric d' satisfies: $d(x,y) \geq d(z,w)$ iff $d'(x,y) \geq d'(z,w)$ for all $x,y,z,w \in X$.) Indeed, $d' := \frac{d}{1+d}$ is such a distance function. (*Note.* We have $0 \leq d' \leq 1$.) As we proceed further, it will become clear that there is a good sense in which (X,d) and $(X, \frac{d}{1+d})$ can be thought of as "equivalent" in terms of certain characteristics and not so in terms of others.[4] □

If X is a metric space (with metric d) and $\emptyset \neq Y \subset X$, we can view Y as a metric space in its own right by using the distance function induced by d on Y. More precisely, we make Y a metric space by means of the distance function $d|_{Y\times Y}$. We then say that $(Y, d|_{Y\times Y})$, or simply Y, is a **metric subspace** of X. For instance, we think of any interval, say $[0,1]$, as a metric subspace of $\mathbb{R}$; this means simply that the distance between any two elements x and y of $[0,1]$ is calculated by viewing x and y as points in $\mathbb{R}$: $d_1(x,y) = |x-y|$. Of course, we can also think of $[0,1]$ as a metric subspace of $\mathbb{R}^2$. Formally, we would do this by "identifying" $[0,1]$ with the set $[0,1] \times \{0\}$ (or with $\{0\} \times [0,1]$, or with $[0,1] \times \{47\}$, etc.) and considering $[0,1] \times \{0\}$ as a metric subspace of $\mathbb{R}^2$. This would render the distance between x and y equal to, again, $|x-y|$ (for, $d_2((x,0),(y,0)) = |x-y|$).[5]

[4] This is definitely a good point to keep in mind. When there is a "natural" unbounded metric on the space that you are working with, but for some reason you need a bounded metric, you can always modify the original metric to get a bounded and ordinally equivalent metric on your space. (More on this in Section 1.5 below.) Sometimes—and we will encounter such an instance later—this little trick does wonders.

[5] *Quiz.* Is the metric space given in Example 1. [2] a metric subspace of $\mathbb{R}^2$?

Convention. Throughout this book, when we consider a nonempty subset S of a Euclidean space $\mathbb{R}^n$ as a metric space without explicitly mentioning a particular metric, you should understand that we view S as a metric subspace of $\mathbb{R}^n$.

Example 2

[1] For any positive integer n, we may think of $\mathbb{R}^n$ as a metric subspace of $\mathbb{R}^{n+1}$ by identifying it with the subset $\mathbb{R}^n \times \{0\}$ of $\mathbb{R}^{n+1}$. By induction, therefore, $\mathbb{R}^n$ can be thought of as a metric subspace of $\mathbb{R}^m$ for any $m, n \in \mathbb{N}$ with $m > n$.

[2] Let $-\infty < a < b < \infty$, and consider the metric space $\mathbf{B}[a,b]$ as it is introduced in Example 1.[5]. Recall that every continuous function on $[a,b]$ is bounded (Exercise A.53), and hence $\mathbf{C}[a,b] \subseteq \mathbf{B}[a,b]$. Consequently, we can consider $\mathbf{C}[a,b]$ as a metric subspace of $\mathbf{B}[a,b]$. Indeed, throughout this text, whenever we talk about $\mathbf{C}[a,b]$ as a metric space, we think of the distance between any f and g in $\mathbf{C}[a,b]$ as $d_\infty(f,g)$, unless otherwise is explicitly mentioned.

[3] Let $-\infty < a < b < \infty$, and recall that we denote the set of all continuously differentiable functions on $[a,b]$ by $\mathbf{C}^1[a,b]$ (Section A.4.2). The metric that is used for this space is usually not the sup-metric. That is, we do not define $\mathbf{C}^1[a,b]$ as a metric subspace of $\mathbf{B}[a,b]$. (There are a good reasons for this, but we'll get to them later.) Instead, $\mathbf{C}^1[a,b]$ is commonly metrized by means of the distance function $d_{\infty,\infty} : \mathbf{C}^1[a,b] \times \mathbf{C}^1[a,b] \to \mathbb{R}_+$ defined by

$$d_{\infty,\infty}(f,g) := d_\infty(f,g) + d_\infty(f',g').$$

It is this metric that we have in mind when talking about $\mathbf{C}^1[a,b]$ as a metric space. □

Exercise 1[H] If (X,d) and (X,ρ) are metric spaces, is $(X, \max\{d,\rho\})$ necessarily a metric space? How about $(X, \min\{d,\rho\})$?

Exercise 2 For any metric space X, show that $|d(x,y) - d(y,z)| \leq d(x,z)$ for all $x, y, z \in X$.

Exercise 3 For any semimetric space X, define the binary relation $\approx$ on X by $x \approx y$ iff $d(x,y) = 0$. Now define $[x] := \{y \in X : x \approx y\}$ for all

$x \in X$, and let $\mathcal{X} := \{[x] : x \in X\}$. Finally, define $D : \mathcal{X}^2 \to \mathbb{R}_+$ by $D([x],[y]) = d(x,y)$.

(a) Show that $\approx$ is an equivalence relation on X.

(b) Prove that $(\mathcal{X}, D)$ is a metric space.

EXERCISE 4 Show that $(\mathbf{C}^1[0,1], d_{\infty,\infty})$ is a metric space.

EXERCISE 5 Let (X,d) be a metric space and $f : \mathbb{R}_+ \to \mathbb{R}$ be a concave and strictly increasing function with $f(0) = 0$. Show that $(X, f \circ d)$ is a metric space.

The final order of business in this subsection is to prove Minkowski's Inequalities, which we invoked above to verify that $\mathbb{R}^{n,p}$ and ℓ^p are metric spaces for any $1 \leq p < \infty$. Since the first one is a special case of the second (yes?), all we need is to establish Minkowski's Inequality 2.

PROOF OF MINKOWSKI'S INEQUALITY 2

Take any $(x_m), (y_m) \in \mathbb{R}^\infty$ and fix any $1 \leq p < \infty$. If either $\sum^\infty |x_i|^p = \infty$ or $\sum^\infty |y_i|^p = \infty$, then (1) becomes trivial, so we assume that $\sum^\infty |x_i|^p < \infty$ and $\sum^\infty |y_i|^p < \infty$. (1) is also trivially true if either (x_m) or (y_m) equals $(0,0,\ldots)$, so we focus on the case where both $\alpha := \left(\sum^\infty |x_i|^p\right)^{\frac{1}{p}}$ and $\beta := \left(\sum^\infty |y_i|^p\right)^{\frac{1}{p}}$ are positive real numbers.

Define the real sequences $(\hat{x}_m)$ or $(\hat{y}_m)$ by $\hat{x}_m := \frac{1}{\alpha}|x_m|$ and $\hat{y}_m := \frac{1}{\beta}|y_m|$. (Notice that $\sum^\infty |\hat{x}_i|^p = 1 = \sum^\infty |\hat{y}_i|^p$.) Using the triangle inequality for the absolute value function (Example A.7) and the fact that $t \mapsto t^p$ is an increasing map on $\mathbb{R}_+$, we find

$$\begin{aligned}|x_i + y_i|^p \leq (|x_i| + |y_i|)^p &= (\alpha|\hat{x}_i| + \beta|\hat{y}_i|)^p \\ &= (\alpha+\beta)^p\left(\tfrac{\alpha}{\alpha+\beta}|\hat{x}_i| + \tfrac{\beta}{\alpha+\beta}|\hat{y}_i|\right)^p\end{aligned}$$

for each $i = 1, 2, \ldots$. But since $t \mapsto t^p$ is a convex map on $\mathbb{R}_+$, we have

$$\left(\tfrac{\alpha}{\alpha+\beta}|\hat{x}_i| + \tfrac{\beta}{\alpha+\beta}|\hat{y}_i|\right)^p \leq \tfrac{\alpha}{\alpha+\beta}|\hat{x}_i|^p + \tfrac{\beta}{\alpha+\beta}|\hat{y}_i|^p, \qquad i = 1, 2, \ldots,$$

and hence

$$|x_i + y_i|^p \leq (\alpha+\beta)^p\left(\tfrac{\alpha}{\alpha+\beta}|\hat{x}_i|^p + \tfrac{\beta}{\alpha+\beta}|\hat{y}_i|^p\right), \qquad i = 1, 2, \ldots.$$

Summing over i, then,

$$\sum_{i=1}^{\infty} |x_i + y_i|^p \leq (\alpha+\beta)^p \left(\tfrac{\alpha}{\alpha+\beta} \sum_{i=1}^{\infty} |\hat{x}_i|^p + \tfrac{\beta}{\alpha+\beta} \sum_{i=1}^{\infty} |\hat{y}_i|^p \right)$$

$$= (\alpha+\beta)^p \left(\tfrac{\alpha}{\alpha+\beta} + \tfrac{\beta}{\alpha+\beta} \right)^{\frac{1}{p}}.$$

Thus $\sum^{\infty} |x_i + y_i|^p \leq (\alpha+\beta)^p$, which is equivalent to (1). ■

We conclude by noting that (1) holds as an equality (for any given $1 < p < \infty$) iff either $x = (0, 0, \ldots)$ or $y = \lambda x$ for some $\lambda \geq 0$. The proof is left as an exercise.

1.2 Open and Closed Sets

We now review a number of fundamental concepts regarding metric spaces.

Definition

Let X be a metric (or a semimetric) space. For any $x \in X$ and $\varepsilon > 0$, we define the **ε-neighborhood of x in X** as the set

$$N_{\varepsilon,X}(x) := \{y \in X : d(x,y) < \varepsilon\}.$$

In turn, a **neighborhood of x in X** is any subset of X that contains at least one ε-neighborhood of x in X.

The first thing that you should note about the ε-neighborhood of a point x in a (semi)metric space is that such a set is never empty, for it contains x. Second, make sure you understand that this notion is based on *four* primitives. Obviously, the ε-neighborhood of x in a metric space X depends on ε and x. But it also depends on the set X and the distance function d used to metrize this set. For instance, the 1-neighborhood of $\mathbf{0} := (0,0)$ in $\mathbb{R}^2$ is $\{(x_1, x_2) \in \mathbb{R}^2 : x_1^2 + x_2^2 < 1\}$, whereas the 1-neighborhood of $\mathbf{0}$ in $\mathbb{R} \times \{0\}$ (viewed as a metric subspace of $\mathbb{R}^2$) is $\{(x_1, 0) \in \mathbb{R}^2 : -1 < x_1 < 1\}$. Similarly, the 1-neighborhood of $\mathbf{0}$ in $\mathbb{R}^2$ is distinct from that in $\mathbb{R}^{2,p}$ for $p \neq 2$.

The notion of ε-neighborhoods plays a major role in real analysis mainly through the following definition.

Definition

A subset S of X is said to be **open in** X (or an **open subset of** X) if, for each $x \in S$, there exists an $\varepsilon > 0$ such that $N_{\varepsilon,X}(x) \subseteq S$. A subset S of X is said to be **closed in** X (or a **closed subset of** X) if $X \backslash S$ is open in X.

Because an ε-neighborhood of a point is inherently connected to the underlying metric space, which sets are open and which sets are closed depends on the metric under consideration. Please keep in mind that changing the metric on a given set, or concentrating on a metric subspace of the original metric space, would in general yield different classes of open (and hence closed) sets.

Definition

Let X be a metric space and $S \subseteq X$. The largest open set in X that is contained in S (that is, the $\supseteq$-maximum of the class of all open subsets of X contained in S) is called the **interior** of S (*relative to* X) and is denoted by $int_X(S)$. On the other hand, the **closure** of S (*relative to* X), denoted by $cl_X(S)$, is defined as the smallest closed set in X that contains S (that is, the $\supseteq$-minimum of the class of all closed subsets of X that contain S). The **boundary** of S (*relative to* X), denoted by $bd_X(S)$, is defined as

$$bd_X(S) := cl_X(S) \backslash int_X(S).$$

Let X be a metric space and Y a metric subspace of X. For any subset S of Y, we may think of the interior of S as lying in X or in Y. (And yes, these may well be quite different!) It is for this reason that we use the notation $int_X(S)$, instead of $int(S)$, to mean the interior of S relative to the metric space X. However, if there is only one metric space under consideration, or the context leaves no room for confusion, we may, and will, simply write $int(S)$ to denote the interior of S relative to the appropriate space. (The same comments apply to the closure and boundary operators as well.)

Example 3

[1] In any metric space X, the sets X and $\varnothing$ are both open and closed. (Sets that are both open and closed are sometimes called **clopen** in

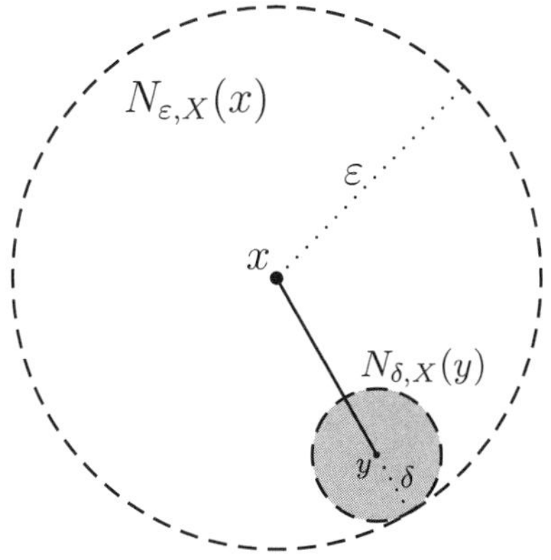

Figure 2

analysis.) On the other hand, for any $x \in X$ and $\varepsilon > 0$, the set $N_{\varepsilon,X}(x)$ is open and the set $\{x\}$ is closed.

To prove that $N_{\varepsilon,X}(x)$ is open, take any $y \in N_{\varepsilon,X}(x)$ and define $\delta := \varepsilon - d(x,y) > 0$. We have $N_{\delta,X}(y) \subseteq N_{\varepsilon,X}(x)$ because, by the triangle inequality,

$$d(x,z) \leq d(x,y) + d(y,z) < d(x,y) + \varepsilon - d(x,y) = \varepsilon$$

for any $z \in N_{\delta,X}(y)$. (See Figure 2 to see the intuition of the argument.)

To prove that $\{x\}$ is closed, we need to show that $X\backslash\{x\}$ is open. If $X = \{x\}$, there is nothing to prove. (Yes?) On the other hand, if there exists a $y \in X\backslash\{x\}$, we then have $N_{\varepsilon,X}(y) \subseteq X\backslash\{x\}$, where $\varepsilon := d(x,y)$. It follows that $X\backslash\{x\}$ is open and $\{x\}$ is closed.

[2] Any subset S of a nonempty set X is open with respect to the discrete metric. For, if $x \in S \subseteq X$, then we have $N_{\frac{1}{2},X}(x) = \{x\} \subseteq S$, where the discrete metric is used in computing $N_{\frac{1}{2},X}(x)$. Thus: *Any subset of a discrete space is clopen.*

[3] It is possible for a set in a metric space to be neither open nor closed. In $\mathbb{R}$, for instance, $(0,1)$ is open, $[0,1]$ is closed, and $[0,1)$ is neither open nor closed. But observe that the structure of the mother metric space is crucial for the validity of these statements. For instance, the set $[0,1)$ is open when considered as a set in the metric space $\mathbb{R}_+$. (Indeed, *relative to this metric subspace of* $\mathbb{R}$, 0 belongs to the interior of

$[0, 1)$, and the boundary of $[0, 1)$ equals $\{1\}$.) More generally, the following fact is true:

> EXERCISE 6 Given any metric space X, let Y be a metric subspace of X, and take any $S \subseteq Y$. Show that S is open in Y iff $S = O \cap Y$ for some open subset O of X, and it is closed in Y iff $S = C \cap Y$ for some closed subset C of X.

WARNING. Given any metric space X, let Y be a metric subspace of X, and take any $U \subseteq Y$. An immediate application of Exercise 6 shows that

$$U \text{ is open in } X \qquad \text{only if} \qquad U \text{ is open in } Y.$$

(Yes?) But the converse is false! For instance, $(0, 1) \times \{0\}$ is an open subset of $[0, 1] \times \{0\}$, but it is not an open subset of $[0, 1]^2$. Yet, if the metric subspace under consideration is an open subset of the mother metric space, all goes well. Put precisely, *provided that Y is open in X*,

$$U \text{ is open in } X \qquad \text{iff} \qquad U \text{ is open in } Y.$$

(*Proof.* If U is open in Y, then, by Exercise 6, there is an open subset O of X such that $U = O \cap Y$, so if Y is open in X as well, U must be open in X (because the intersection of two open sets is open).)

Similar remarks apply to closed sets as well, of course. If $C \subseteq Y$, then

$$C \text{ is closed in } X \qquad \text{only if} \qquad C \text{ is closed in } Y,$$

and, *provided that Y is closed in X*,

$$C \text{ is closed in } X \qquad \text{iff} \qquad C \text{ is closed in } Y.$$

(Proofs?)

[4] How do we know that $int_X(S)$ is well-defined for any subset S of a metric space X? (Perhaps the class $\{O \in 2^X : O \text{ is open in } X \text{ and } O \subseteq S\}$ does not have a $\supseteq$-maximum, that is, there is no *largest* open subset of X that is contained in S!) The reason is that the union of any collection of open sets is open in a metric space. (Yes?) Thus $int_X(S)$ is well-defined, since, thanks to this property, we have

$$int_X(S) = \bigcup\{O \in \mathcal{O}_X : O \subseteq S\},$$

where $\mathcal{O}_X$ is the class of all open subsets of X.[6,7] By contrast, the intersection of any collection of closed subsets of X is closed (why?), and hence $cl_X(S)$ is well-defined for any $S \in 2^X$:

$$cl_X(S) = \bigcap\{C \in \mathcal{C}_X : S \subseteq C\},$$

where $\mathcal{C}_X$ is the class of all closed subsets of X.

WARNING. While the intersection of a *finite* collection of open sets is open (why?), an arbitrary intersection of open sets need not be open in general. For instance, $(-1, 1) \cap \left(-\frac{1}{2}, \frac{1}{2}\right) \cap \cdots = \{0\}$ is not an open subset of $\mathbb{R}$. Similarly, the union of a *finite* collection of closed sets is closed, but an arbitrary union of closed sets need not be closed.

[5] It is obvious that a set S in a metric space X is closed iff $cl_X(S) = S$. (Is it?) Similarly, S is open iff $int_X(S) = S$. Also observe that, for any subset S of X, we have $x \in bd_X(S)$ iff $S \cap N_{\varepsilon,X}(x)$ and $(X\backslash S) \cap N_{\varepsilon,X}(x)$ are nonempty for any $\varepsilon > 0$. (Proofs?)[8]

[6] A set is open in a given metric space (X, d) iff it is open in $\left(X, \frac{d}{1+d}\right)$. (Recall Remark 1.) So, in terms of their open set structures, these two metric spaces are identical (even though the "distance" between any two points in X would be assessed differently by d and $\frac{d}{1+d}$).

[7] In contrast to metric spaces, semimetric spaces may have very few open (thus closed) sets. For instance, (X, d_o) is a semimetric space if $d_o(x, y) := 0$ for all $x, y \in X$. This space is not a metric space unless $|X| = 1$, and the only open and/or closed sets in it are $\emptyset$ and X. One may thus view such a space as a polar opposite of a discrete space—it is called an **indiscrete space**. □

EXERCISE 7[H] Can you find a metric on $\mathbb{N}$ such that $\emptyset \neq S \subseteq \mathbb{N}$ is open iff $\mathbb{N}\backslash S$ is finite?

EXERCISE 8 Prove all the assertions made in Examples 3.[4] and 3.[5].

EXERCISE 9[H] Show that, for any subset S of a metric space X, the interior of $bd_X(S)$ equals S iff $S = \emptyset$. Find an example of a metric space X that contains a nonempty set S with $int_X(bd_X(S)) \supseteq S$.

6 This presumes that $\{O \in \mathcal{O}_X : O \subseteq S\} \neq \emptyset$; how do I know that this is the case?
7 *Corollary.* $x \in int_X(S)$ iff there exists an $\varepsilon > 0$ such that $N_{\varepsilon,X}(x) \subseteq S$.
8 *Quiz.* What is the boundary of the unit "circle" C_∞ defined in Example 1.[3]?

Exercise 10[H] Given a metric space X, let Y be a metric subspace of X, and $S \subseteq X$. Show that

$$int_X(S) \cap Y \subseteq int_Y(S \cap Y) \quad \text{and} \quad cl_X(S) \cap Y \supseteq cl_Y(S \cap Y),$$

and give examples to show that the converse containments do not hold in general. Also prove that

$$int_X(S) \cap Y = int_Y(S \cap Y),$$

provided that Y is open in X. Similarly, $cl_X(S) \cap Y = cl_Y(S \cap Y)$ holds if $Y$is closed in X.

Exercise 11 Let S be a closed subset of a metric space X, and $x \in X \backslash S$. Show that there exists an open subset O of X such that $S \subseteq O$ and $x \in X \backslash O$.

1.3 Convergent Sequences

The notion of closedness (and hence openness) of a set in a metric space can be characterized by means of the sequences that live in that space. Since this characterization often simplifies things considerably, it will be a good idea to provide it here before proceeding further into the theory of metric spaces. Let us first recall what it means for a sequence to converge in a metric space.

Definition
Let X be a metric (or a semimetric) space, $x \in X$, and $(x^m) \in X^\infty$.[9] We say that (x^m) **converges to** x if, for each $\varepsilon > 0$, there exists a real number M (that may depend on ε) such that $d(x^m, x) < \varepsilon$ for all $m \geq M$. (*Note.* This is the same thing as saying $d(x^m, x) \to 0$.) In this case, we say that (x^m) **converges in** X, or that it is **convergent in** X, we refer to x as the **limit of** (x^m), and write either $x^m \to x$ or $\lim x^m = x$.

[9] In this book, as a notational convention, I denote a generic sequence in a given (abstract) metric space X mostly by (x^m), (y^m), etc. (This convention becomes particularly useful when, for instance, the terms of (x^m) are themselves sequences.) The generic real (or extended real) sequences are denoted as (x_m), (y_m), etc., and generic sequences of real functions are denoted as (f_m), (g_m), etc.

A sequence (x^m) in a metric space X can converge to at most one limit. Indeed, if both $x^m \to x$ and $x^m \to y$ held true, then, by symmetry and the triangle inequality, we would have

$$d(x,y) \leq d(x,x^m) + d(x^m,y) \qquad \text{for all } m = 1,2,\ldots$$

while

$$d(x,x^m) + d(x^m,y) \to 0,$$

which implies $x = y$. (Why?)[10] The situation is different in the case of semimetric spaces.

Exercise 12^{H} Show that a convergent sequence may have more than one limit in a semimetric space.

To recap, a sequence (x^m) in a metric space X converges to a point x in this space, if, for any $\varepsilon > 0$, all but finitely many terms of the sequence (x^m) belong to $N_{\varepsilon,X}(x)$. One way of thinking about this intuitively is viewing the sequence (x^m) as "staying in $N_{\varepsilon,X}(x)$ eventually," no matter how small ε is. Equivalently, we have $x^m \to x$ iff, for every open neighborhood O of x in X, there exists a real number M such that $x^m \in O$ for all $m \geq M$. So, for instance, the sequence $(1, \frac{1}{2}, \frac{1}{3}, \ldots) \in \mathbb{R}_+^\infty$ converges to zero, because, for any open neighborhood O of 0 in $\mathbb{R}_+$, there exists an $M \in \mathbb{N}$ such that $\frac{1}{m} \in O$ for all $m \geq M$. (Yes?)

Example 4

[1] A sequence (x^m) is convergent in a discrete space iff it is eventually constant (that is, there exists an $M \in \mathbb{N}$ such that $x^M = x^{M+1} = \cdots$.)

[2] A constant sequence in any metric space is convergent.

[3] Take any $n \in \mathbb{N}$, and let $(x^m) = ((x_1^m, \ldots, x_n^m))$ be a sequence in $\mathbb{R}^n$. It is easy to show that $x^m \to (x_1, \ldots, x_n)$ iff (x_i^m) converges to x_i for each $i = 1, \ldots, n$. (Prove!)

[10] I could write the argument more compactly as $d(x,y) \leq d(x,x^m) + d(x^m,y) \to 0$. I will use this sort of a shorthand expression quite frequently in what follows.

[4] Consider the following real sequences: For each $m \in \mathbb{N}$,

$$x^m := \left(0,\ldots,0,\tfrac{1}{m},0,\ldots\right), \qquad y^m := (0,\ldots,0,1,0,\ldots), \quad \text{and}$$
$$z^m := \left(\tfrac{1}{m},\ldots,\tfrac{1}{m},0,\ldots\right),$$

where the only nonzero term of the sequences x^m and y^m is the mth one, and all but the first m terms of z^m are zero. Since $d_p((x^m),(0,0,\ldots)) = \frac{1}{m}$ for each m, we have $(x^m) \to (0,0,\ldots)$ in ℓ^p for any $1 \le p \le \infty$. (Don't forget that here (x^m) is a sequence of real sequences.) In contrast, it is easily checked that the sequence $(y^1, y^2, \ldots)$ is not convergent in any ℓ^p space. On the other hand, we have $d_\infty(z^m,(0,0,\ldots)) = \frac{1}{m} \to 0$, so (z^m) converges to $(0,0,\ldots)$ in ℓ^∞. Yet $d_1(z^m,(0,0,\ldots)) = 1$ for each m, so (z^m) does not converge to $(0,0,\ldots)$ in ℓ^1. Is (z^m) convergent in any ℓ^p, $1 < p < \infty$?

[5] Let $f_m \in \mathbf{B}[0,1]$ be defined by $f_m(t) := t^m$, $m = 1,2,\ldots$. Since $f_m(t) \to 0$ for all $t \in [0,1)$ and $f_m(1) = 1$ for all m, it may at first seem plausible that $f_m \to f$ where $f(t) = 0$ for all $0 \le t < 1$ and $f(1) = 1$. But this is false, because $d_\infty(f_m, f) = 1$ for all $m \in \mathbb{N}$. This example shows again how detrimental the choice of metric may be in studying the convergence of sequences. (We will come back to this example in due course.) □

EXERCISE 13 For any given metric space (X,d), show that, for any $(x^m) \in X^\infty$ and $x \in X$, we have $x^m \to x$ in (X,d) iff $x^m \to x$ in $\left(X, \frac{d}{1+d}\right)$.

1.4 Sequential Characterization of Closed Sets

Here is the sequential characterization of closed sets we promised above.

PROPOSITION 1
A set S in a metric space X is closed if, and only if, every sequence in S that converges in X converges to a point in S.

PROOF
Let S be a closed subset of X, and take any $(x^m) \in S^\infty$ with $x^m \to x$ for some $x \in X$. If $x \in X \backslash S$, then we can find an $\varepsilon > 0$ with $N_{\varepsilon,X}(x) \subseteq X \backslash S$, because $X \backslash S$ is open in X. But since $d(x^m, x) \to 0$, there must exist a large enough

$M \in \mathbb{N}$ such that $x^M \in N_{\varepsilon,X}(x)$, contradicting that all terms of the sequence (x^m) lies in S. Conversely, suppose that S is not closed in X. Then $X\backslash S$ is not open, so we can find an $x \in X\backslash S$ such that every ε-neighborhood around x intersects S. Thus, for any $m = 1, 2, \ldots$, there is an $x^m \in N_{\frac{1}{m},X}(x) \cap S$. (Formally, we invoke the Axiom of Choice here.) But then $(x^m) \in S^\infty$ and $\lim x^m = x$, and yet $x \notin S$. Thus if S was not closed, there would exist at least one sequence in S that converges to a point outside of S. ■

To understand what this result says (or better, what it does not say), consider the metric space $(0, 1)$ and ask yourself if $(0, 1)$ is a closed subset of this space. A common mistake is to answer this question in the negative, and use Proposition 1 to suggest a proof. The fallacious argument goes as follows: "By Proposition 1, the interval $(0, 1)$ cannot be closed, because the sequence $\left(1, \frac{1}{2}, \frac{1}{3}, \ldots\right)$ in $(0, 1)$ converges to 0, a point that is outside of $(0, 1)$." The problem with this argument is that it works with a *non*-convergent sequence in $(0, 1)$. Indeed, the sequence $\left(1, \frac{1}{2}, \frac{1}{3}, \ldots\right)$ does not converge anywhere in the space $(0, 1)$. After all, the only possible limit for this sequence is 0, but 0 does not live in the mother space. (*Note.* $\left(1, \frac{1}{2}, \frac{1}{3}, \ldots\right)$ would be convergent, for instance, if our mother space was $[0, 1)$.) In fact, any convergent sequence in $(0, 1)$ must converge in $(0, 1)$ (because of the funny structure of this metric space), and therefore we must conclude that $(0, 1)$ is closed as a subset of itself, which is, of course, a triviality (Example 3.[1]). This observation points once again to the fact that the metric properties of sets (such as the convergence of sequences) depend crucially on the structure of the mother metric space under consideration.

Exercise 14[H] Prove that, for any subset S of a metric space X, the following statements are equivalent:

(i) $x \in cl_X(S)$.
(ii) Every open neighborhood of x in X intersects S.
(iii) There exists a sequence in S that converges to x.

Exercise 15 Let (x^m) be a sequence in a metric space X. We say that $x \in X$ is a **cluster point** of (x^m) if, for each $\varepsilon > 0$, $N_{\varepsilon,X}(x)$ contains infinitely many terms of (x^m).

(a) Show that any convergent sequence has exactly one cluster point. (*Note.* The converse is not true; consider, for instance, $(1, 0, 2, 0, 3, 0, \ldots) \in \mathbb{R}^\infty$.)

(b) For any $k \in \mathbb{Z}_+$, give an example of a sequence (in some metric space) with exactly k many cluster points.
(c) Show that x is a cluster point of (x^m) iff there is a subsequence of (x^m) that converges to x.

1.5 Equivalence of Metrics

When would endowing a given nonempty set X with two different metrics d and D yield metric spaces that could reasonably be considered as "equivalent"? While this query is rather vague at present, we can still come up with a few benchmark responses to it. For instance, it makes perfect sense to view two metric spaces of the form (X, d) and $(X, 2d)$ as "identical." The second space simply equals a version of the former space in which the distances are measured in a different scale. Fine, how about (X, d) and $(X, \sqrt{d})$?[11] This comparison seems a bit more subtle. While d and $\sqrt{d}$ are *ordinally* identical, one is not a simple rescaling of the other, so the metric spaces they induce may in principle look rather different from certain points of view. At the very least, it seems that the connection between (X, d) and $(X, 2d)$ is tighter than that between (X, d) and $(X, \sqrt{d})$, even though we would not expect the properties of the latter two spaces to be vastly different from each other.

Let us be more precise now.

> DEFINITION
> Let d and D be two metrics on a nonempty set X, and denote the classes of all open subsets of X with respect to d and D as $\mathcal{O}(d)$ and $\mathcal{O}(D)$, respectively. We say that d and D (and/or (X, d) and (X, D)) are **equivalent** if $\mathcal{O}(d) = \mathcal{O}(D)$, and that they are **strongly equivalent** if
>
> $$\alpha d \leq D \leq \beta d$$
>
> for some real numbers $\alpha, \beta \geq 0$.

As we proceed further in the course, it will become clear that the class of all open subsets of a given metric space determines a good deal of the properties of this space (at least with respect to the basic questions concerning

[11] If d is a metric on X, then so is $\sqrt{d}$. Why?

real analysis). Consequently, if two metrics on a given nonempty set generate precisely the same class of open sets, then the resulting metric spaces are bound to look "identical" from a variety of viewpoints. For instance, the classes of all closed subsets of two equivalent metric spaces are the same. Moreover, if a sequence in one metric space converges, then it also does so in any equivalent metric space. (Why?)

The reasons for considering two strongly equivalent spaces as "identical" are even more compelling. Notice first that any two strongly equivalent metrics are equivalent. (Why?) To show that the converse is false, we use the following concept.

> Definition
>
> A subset S of a metric space X is called **bounded** (*in* X) if there exists an $\varepsilon > 0$ such that $S \subseteq N_{\varepsilon,X}(x)$ for some $x \in S$. If S is not bounded, then it is said to be **unbounded**.

It is clear that if (X, d) and (X, D) are strongly equivalent metric spaces, then a subset S of X is bounded in (X, d) iff it is bounded in (X, D). (Yes?) By contrast, boundedness is not a property that is invariant under equivalence of metrics. Indeed, if (X, d) is an *un*bounded metric space, then $(X, \frac{d}{1+d})$ is a bounded metric space, whereas d and $\frac{d}{1+d}$ are equivalent. (Recall Example 3.[6]). Thus, d and $\frac{d}{1+d}$ are equivalent metrics on X that are not strongly equivalent.

In fact, strong equivalence of metrics is substantially more demanding than their equivalence. We used the boundedness property here only for illustrative purposes. Two equivalent metric spaces (X, d) and (X, D) need not be strongly equivalent, even if X is rendered bounded by both of these metrics. For instance, $([0, 1], d_1)$ and $([0, 1], \sqrt{d_1})$ are equivalent (bounded) metric spaces.[12] Yet d_1 and $\sqrt{d_1}$ are not strongly equivalent metrics on $[0, 1]$. (*Proof.* There is no $\alpha > 0$ such that $\alpha\sqrt{d_1(x, y)} \leq d_1(x, y)$ for all $0 \leq x, y \leq 1$.)

[12] More generally, the following is true for any metric space (X, d): *If* $f : \mathbb{R}_+ \to \mathbb{R}$ *is strictly increasing, continuous, and subadditive, then* (X, d) *and* $(X, f \circ d)$ *are equivalent metric spaces.* Proof? (*Note.* Here subadditivity of f means that $f(a + b) \leq f(a) + f(b)$ for all $a, b \geq 0$.)

To give another example, let $n \in \mathbb{N}$, and take any $1 \leq p \leq \infty$. Clearly, for any metric d_p on $\mathbb{R}^n$ we have

$$d_\infty(x, y) \leq d_p(x, y) \leq n d_\infty(x, y) \qquad \text{for all } x, y \in \mathbb{R}^n.$$

Thus, $\mathbb{R}^{n,p}$ is strongly equivalent to $\mathbb{R}^{n,\infty}$ for any $1 \leq p < \infty$. Since "being strongly equivalent to" is an equivalence relation on the class of all metrics on $\mathbb{R}^n$, we conclude: *d_p and d_q are strongly equivalent metrics on $\mathbb{R}^n$ for any $1 \leq p, q \leq \infty$.*[13]

2 Connectedness and Separability

The notion of metric space alone is too general to be useful in applications. Indeed, some metric spaces can be quite ill-behaved (e.g., discrete spaces), so we need to "find" those spaces that possess certain regularity properties. We consider two such properties in this section. The first of these, connectedness, gives one a glimpse of how one would study the geometry of an arbitrary metric space. The second one, separability, identifies those metric spaces that have relatively "few" open sets. While they are important in other contexts, these properties play a limited role in this book. We thus proceed at somewhat of a quick pace here.

2.1 Connected Metric Spaces

Intuitively speaking, a connected subset of a metric space is one that cannot be partitioned into two (or more) separate pieces; rather, it is in one whole piece. In $\mathbb{R}$, for instance, we like to think of $(0, 1)$ as connected and $[0, 1] \cup [2, 3)$ as disconnected. The definition below formalizes this simple geometric intuition.

[13] Due to this fact, it simply does not matter which d_p metric is used to metrize $\mathbb{R}^n$, *for the purposes of this text.* I do not, of course, claim that all properties of interest are shared by $\mathbb{R}^{n,p}$ and $\mathbb{R}^{n,q}$ for any $p, q \geq 1$. If we were interested in the shape of the unit circles, for example, then there is no way we would view $\mathbb{R}^{n,1}$ and $\mathbb{R}^n$ as identical (see Figure 1).

Definition

We say that a metric space X is **connected** if there do not exist two nonempty and disjoint open subsets O and U of X such that $O \cup U = X$. In turn, a subset S of X is said to be **connected in** X (or a **connected subset of** X) if S is a connected metric subspace of X. (So, $S \subseteq X$ is connected in X iff it cannot be written as a disjoint union of two nonempty sets that are open *in* S.)

The following simple result provides an interesting characterization of connected metric spaces.

Proposition 2

Let X be a metric space. Then, X is connected if, and only if, the only clopen subsets of X are $\emptyset$ and X.

Proof

If $S \notin \{\emptyset, X\}$ is a clopen subset of X, then X cannot be connected since $X = S \cup (X \backslash S)$. Conversely, assume that X is not connected. In this case we can find nonempty and disjoint open subsets O and U of X such that $O \cup U = X$. But then $U = X \backslash O$ so that O must be both open and closed. Since $O \notin \{\emptyset, X\}$, this proves the claim. ∎

So, for instance, a discrete space is not connected unless it contains only one element, because any subset of the discrete space is clopen. Similarly, $\mathbb{Q}$ is not connected in $\mathbb{R}$, for $\mathbb{Q} = (-\infty, \sqrt{2}) \cup (\sqrt{2}, \infty)$. In fact, the only connected subsets of $\mathbb{R}$ are the intervals, as we show next.

Example 5

We claim that any interval I is connected in $\mathbb{R}$. To derive a contradiction, suppose we could write $I = O \cup U$ for nonempty and disjoint open subsets O and U of I. Pick any $a \in O$ and $b \in U$, and let $a < b$ without loss of generality. Define $c := \sup\{t \in O : t < b\}$, and note that $a \leq c \leq b$ and hence $c \in I$ (because I is an interval). If $c \in O$, then $c \neq b$ (since $O \cap U = \emptyset$), so $c < b$. But, since O is open, there exists an $\varepsilon > 0$ such that $b > c + \varepsilon \in O$, which contradicts the choice of c. (Why?) If, on the other hand, $c \notin O$, then $a < c \in U$ (since $I = O \cup U$). Then, given that U is open, $(c - \varepsilon, c) \subseteq U$, and hence $(c - \varepsilon, c) \cap O = \emptyset$, which means that $\sup\{t \in O : t < b\} \leq c - \varepsilon$, contradiction.

Conversely, let I be a nonempty connected set in $\mathbb{R}$, but assume that I is not an interval. The latter hypothesis implies that there exist two points a and b in I such that $(a, b)\backslash I \neq \emptyset$. (Why?) Pick any $c \in (a, b)\backslash I$, and define $O := I \cap (-\infty, c)$ and $U := I \cap (c, \infty)$. Observe that O and U are nonempty and disjoint open subsets of I (Exercise 6) while $O \cup U = I$, contradicting the connectedness of I. Conclusion: *A nonempty subset of* $\mathbb{R}$ *is connected iff it is an interval.* □

We won't elaborate on the importance of the notion of connectedness just yet. This is best seen when one considers the properties of continuous functions defined on connected metric spaces, and thus we relegate further discussion of connectedness to the next chapter (Section D.2).

EXERCISE 16 Show that the closure of any connected subset of a metric space is connected.

EXERCISE 17 Show that if $\mathcal{S}$ is a finite (nonempty) class of connected subsets of a metric space such that $\cap\mathcal{S} \neq \emptyset$, then $\cup\mathcal{S}$ must be connected.

EXERCISE 18 For any given $n \in \{2, 3, \ldots\}$, prove that a convex subset of $\mathbb{R}^n$ is necessarily connected, but not conversely.

*EXERCISE 19[H] Show that a metric space that has countably many points is connected iff it contains only one point.

2.2 Separable Metric Spaces

DEFINITION

Let X be a metric space and $Y \subseteq X$. If $cl_X(Y) = X$, then Y is said to be **dense in** X (or a **dense subset of** X). In turn, X is said to be **separable** if it contains a countable dense set.

Intuitively speaking, one may think of a separable metric space as a space that is "not very large." After all, in such a space, there is a countable set that is "almost" equal to the entire space.

Thanks to Exercise 14, it is readily observed that a set $Y \subseteq X$ is dense in a metric space X iff any point in the grand space X can be approached by means of a sequence that is contained entirely in Y. So, X is a separable metric space if, and only if, it contains a countable set Y such that $x \in X$ iff there exists a $(y_m) \in Y^\infty$ with $y_m \to x$. This characterization of separability often proves useful in proofs. For instance, it allows us to use Lemma A.1 to conclude that $\mathbb{Q}$ is dense in $\mathbb{R}$. (By the way, $bd_{\mathbb{R}}(\mathbb{Q}) = \mathbb{R}$, right?) Thus, $\mathbb{R}$ is a separable metric space. This fact allows us to find other examples of separable spaces.

Example 6

[1] *$\mathbb{R}^n$ is separable, $n = 1, 2, \ldots$.* This is because any $x \in \mathbb{R}^n$ can be approached by a sequence in $\mathbb{R}^n$ all components of which are rational; that is, $\mathbb{Q}^n$ is dense in $\mathbb{R}^n$. ($\mathbb{Q}^n$ is countable, right?) In fact, $\mathbb{R}^{n,p}$ is separable for any $1 \leq p \leq \infty$. (Why?)

[2] A discrete space is separable iff it is countable.

*[3] *Any metric subspace of a separable metric space is separable.*[14] To prove this, let X be any metric space and Y a countable dense subset of X. Take any metric subspace Z of X that we wish to prove to be separable. Define

$$Y_m := \{y \in Y : N_{\frac{1}{m},X}(y) \cap Z \neq \emptyset\}.$$

(*Note.* Each Y_m is nonempty, thanks to the denseness of Y.) Now pick an arbitrary $z^m(y) \in N_{\frac{1}{m},X}(y) \cap Z$ for each $m \in \mathbb{N}$ and $y \in Y_m$, and define

$$W := \{z^m(y) : y \in Y_m \text{ and } m \in \mathbb{N}\}.$$

Clearly, W is a countable subset of Z. (Yes?) Now take any $z \in Z$. By denseness of Y, for each $m \in \mathbb{N}$ we can find a $y^m \in Y$ with $d(z, y^m) < \frac{1}{m}$. So $z \in N_{\frac{1}{m},X}(y^m) \cap Z$, and hence $y^m \in Y_m$. Therefore,

$$d(z, z^m(y^m)) \leq d(z, y^m) + d(y^m, z^m(y^m)) < \tfrac{1}{m} + \tfrac{1}{m} = \tfrac{2}{m}.$$

It follows that any element of Z is in fact a limit of some sequence in W, that is, W is dense in Z.

[14] The proof is a bit subtle; it may be a good idea to skip it in the first reading.

[4] $\mathbf{C}[a,b]$ *is separable for any* $-\infty < a \leq b < \infty$. (Recall that $\mathbf{C}[a,b]$ is endowed with the sup-metric (Example 2.[2]).) An elementary proof of this claim could be given by using the piecewise linear functions with kinks occurring at rational points with rational values. Since the construction is a bit tedious, we omit the details here.

The separability of $\mathbf{C}[a,b]$ can also be seen as a corollary of the following famous result of Karl Weierstrass (which he proved in 1885).

THE WEIERSTRASS APPROXIMATION THEOREM

The set of all polynomials defined on $[a,b]$ *is dense in* $\mathbf{C}[a,b]$. *That is,*

$$cl_{\mathbf{C}[a,b]}(\mathbf{P}[a,b]) = \mathbf{C}[a,b].$$

We shall later obtain this theorem as a special case of a substantially more general result (in Section D.6.4). What is important for us at present is the observation that the separability of $\mathbf{C}[a,b]$ follows readily from this theorem and the fact that the set of all polynomials on $[a,b]$ with rational coefficients is dense in $\mathbf{P}[a,b]$.[15] You should not have much difficulty in supplying the details of the argument.

[5] ℓ^p *is separable for any* $1 \leq p < \infty$. It may be tempting to suggest $\ell^p \cap \mathbb{Q}^\infty$ as a candidate for a countable dense subset of ℓ^p. But this wouldn't work, because, while this set is indeed dense in ℓ^p, it is not countable (since $\mathbb{Q}^\infty$ is uncountable). Instead we consider the set of those sequences in $\ell^p \cap \mathbb{Q}^\infty$ all but finitely many components of which are zero.[16] It is easy to check that this set, call it Y, is countable. (Proof?) To verify that Y is dense, take any sequence $(x_m) \in \ell^p$ and fix any $\varepsilon > 0$. Since $\sum^\infty |x_i|^p < \infty$, there must exist an $M \in \mathbb{N}$ such that $\sum_{i=M+1}^\infty |x_i|^p < \frac{\varepsilon^p}{2}$.[17] Moreover,

[15] Observe that the cardinality of this set is equal to that of $\mathbb{Q} \cup \mathbb{Q}^2 \cup \cdots$ (and not of $\mathbb{Q}^\infty$). Since a countable union of countable sets is countable (Proposition B.2), the set of all polynomials on $[a,b]$ with rational coefficients is thus countable.

[16] *Idea of proof.* Since (the pth power of) a sequence in ℓ^p is summable, the tail of any such sequence must be "very close" to the zero sequence. (*Warning.* This would not be the case in ℓ^∞.) But since $\mathbb{Q}$ is dense in $\mathbb{R}$, we can approximate the earlier (finitely many) terms of such a sequence by using the rational numbers. Thus, it seems, all we need are those sequences of rational numbers whose all but finitely many terms are zero.

[17] Why do we have $\lim_{M\to\infty} \sum_{i=M+1}^\infty |x_i|^p = 0$? If you solved Exercise A.45, you know the answer. Otherwise, note that since $(\sum^M |x_i|^p)$ converges to some number as $M \to \infty$, say a, we can find, for any $\delta > 0$, a large enough $M \in \mathbb{N}$ such that $\sum^M |x_i|^p \geq a - \delta$, that is, $\sum_{i=M+1}^\infty |x_i|^p \leq \delta$.

since $\mathbb{Q}$ is dense in $\mathbb{R}$, we can find a rational r_i such that $|r_i - x_i|^p < \frac{\varepsilon^p}{2M}$ for each $i = 1, \ldots, M$. But then

$$d_p\left((x_m), (r_1, \ldots, r_M, 0, 0, \ldots)\right) = \left(\sum_{i=1}^{M} |r_i - x_i|^p + \sum_{i=M+1}^{\infty} |x_i|^p\right)^{\frac{1}{p}}$$

$$< \left(\frac{\varepsilon^p}{2} + \frac{\varepsilon^p}{2}\right)^{\frac{1}{p}}$$

$$= \varepsilon.$$

Thus any element of ℓ^p can be approached by a sequence (of sequences) in Y. □

One major reason for why separable metric spaces are useful is that in such spaces all open sets can be described in terms of a countable set of open sets. For instance, all open subsets of $\mathbb{R}^n$ can be expressed by using the members of the countable collection $\{N_{\varepsilon,\mathbb{R}^n}(x) : x \in \mathbb{Q}^n \text{ and } \varepsilon \in \mathbb{Q}_{++}\}$. Indeed, by using the denseness of $\mathbb{Q}$ in $\mathbb{R}$, it can easily be verified that

$$U = \bigcup \left\{N_{\varepsilon,\mathbb{R}^n}(x) \in 2^U : x \in \mathbb{Q}^n \text{ and } \varepsilon \in \mathbb{Q}_{++}\right\}$$

for any open subset U of $\mathbb{R}^n$. Thus, there is sense in which all open subsets of $\mathbb{R}^n$ are *generated* by a given countable collection of open sets in $\mathbb{R}^n$. The following result generalizes this observation to the case of an arbitrary separable metric space.

PROPOSITION 3

Let X be metric space. If X is separable, then there exists a countable class $\mathcal{O}$ of open subsets of X such that

$$U = \bigcup \{O \in \mathcal{O} : O \subseteq U\}$$

for any open subset U of X.

PROOF

Assume that X is separable. Pick any countable dense subset Y of X, and define

$$\mathcal{O} := \{N_{\varepsilon,X}(y) : y \in Y \text{ and } \varepsilon \in \mathbb{Q}_{++}\}.$$

$\mathcal{O}$ is countable since it has the same cardinality with $Y \times \mathbb{Q}_{++}$. Now pick any open subset U of X, and take any $x \in U$. We claim that $x \in O$ for some $O \in \mathcal{O}$ with $O \subseteq U$. Since U is open in X, there exists an $\varepsilon \in \mathbb{Q}_{++}$ such that $N_{\varepsilon,X}(x) \subseteq U$. But since $cl_X(Y) = X$, there must exist a $y \in Y$ with $d(x,y) < \frac{\varepsilon}{2}$. But then $x \in N_{\frac{\varepsilon}{2},X}(y) \subseteq N_{\varepsilon,X}(x) \subseteq U$. (Yes?) This proves that $U \subseteq \bigcup\{O \in \mathcal{O} : O \subseteq U\}$. The converse containment is obvious. ■

Keep in mind that many interesting properties of a metric space are defined through the notion of an open set. For instance, if we know all the open subsets of a metric space, then we know all the closed and/or connected sets in this space. (Why?) Similarly, if we know all the open subsets of a metric space, then we know which sequences in this space are convergent and which ones are not. (Why?) Consequently, a lot can be learned about the general structure of a metric space by studying the class of all open subsets of this space. But the significance of this observation would be limited, if, in a sense, we had "too many" open sets lying around. In the case of a separable metric space this is not the case, for such a space has *only* a countable number of open sets "that matter" in the sense that all other open subsets of this space can be described using only these (countably many) open sets. This is the gist of Proposition 3, the importance of which will become clearer when you see it in action in the following subsection.[18]

Now you may ask, given this motivation, why don't we forget about separability, and deal instead with spaces that satisfy the conclusion of Proposition 3 directly? Good question! The answer is given in the next exercise.

EXERCISE 20 Prove the converse of Proposition 3.

EXERCISE 21

(a) Let X be a metric space such that there exists an $\varepsilon > 0$ and an uncountable set $S \subseteq X$ such that $d(x,y) > \varepsilon$ for all distinct $x, y \in S$. Show that X cannot be separable.

(b) Show that ℓ^∞ is not a separable metric space.

[18] By the way, this proposition also shows that if d and D are equivalent metrics on a nonempty set X, then (X,d) is a separable iff (X,D) is separable. (The same also goes for the connectedness property, of course.)

Exercise 22[H] Show that $\mathbb{R}$ is cardinally larger than any separable metric space. Also show that $\mathbf{C}[0,1]$ is cardinally equivalent to $\mathbb{R}$.[19]

2.3 Applications to Utility Theory

To give a quick application, we now go back to the decision theoretic setting described in Section B.4, and see how one may be able to improve the utility representation results we have obtained there in the case where the alternative space X has a metric structure. We need the following terminology which builds on the definition introduced at the end of Section B.4.1.

Definition

Let X be a metric space and $\succsim$ a preference relation on X. We say that $\succsim$ is **upper semicontinuous** if $L_{\succ}(x)$ is an open subset of X for each $x \in X$, and that $\succsim$ is **lower semicontinuous** if $U_{\succ}(x)$ is an open subset of X for each x. In turn, $\succsim$ is called **continuous** if it is both upper and lower semicontinuous.

Intuitively speaking, if $\succsim$ is an upper semicontinuous preference relation on a metric space X, and if $x \succ y$, then an alternative z that is "very close" to y should also be deemed strictly worse than x. Put differently, if the sequence $(y^m) \in X^\infty$ converges to y, then there exists an $M \in \mathbb{R}$ such that $x \succ y^m$ for each $m \geq M$. (Notice how the "metric" in question, which is a purely mathematical term, and the preference relation, which is a psychological concept, are linked tightly by the notion of semicontinuity.) Lower semicontinuity is interpreted similarly.

If $\succsim$ is a *complete* preference relation, then $\succsim$ is upper semicontinuous iff $U_{\succsim}(x)$ is closed for all $x \in X$, and it is lower semicontinuous iff $L_{\succsim}(x)$ is closed for all $x \in X$. (Why?) So, given any $x, y \in X$, if (y^m) is a sequence in X with $y^m \to y$ and $y^m \succsim x$ for each $m = 1, 2, \ldots$, then we have $y \succsim x$, provided that $\succsim$ is a complete and upper semicontinuous preference relation on X.[20]

[19] This exercise presumes familiarity with Section B.3.1.

[20] But if $\succsim$ is not complete, then this conclusion need not be true; upper semicontinuity does not, in general, imply that $U_{\succsim}(x)$ is a closed subset of X for each $x \in X$. Besides, as you are asked to demonstrate in Exercise 25, it is often simply impossible to demand the closedness of all weak upper and weak lower $\succsim$-contour sets from an incomplete and continuous preference relation $\succsim$.

As we shall elaborate further in the next chapter, finding a utility representation for a continuous and complete preference relation is a relatively easy matter, provided that the metric space X (of alternatives) under consideration is well-behaved. To illustrate this point, let's consider the case where X is both connected and separable. Let Y be a countable dense subset of X, and pick any $x, y \in X$ with $x \succ y$. (If there is no such x and y in X, then any constant function would represent $\succsim$.) Since X is connected, and $L_\succ(x)$ and $U_\succ(y)$ are open in X, we must have $L_\succ(x) \cap U_\succ(y) \neq \emptyset$. (Why?) But then $L_\succ(x) \cap U_\succ(y)$ is a nonempty open subset of X, and hence, since a dense set intersects every nonempty open set (why?), we must have $Y \cap L_\succ(x) \cap U_\succ(y) \neq \emptyset$. This proves that Y is $\succsim$-dense in X (Section B.2). Applying Proposition B.9, therefore, we find: *Every continuous complete preference relation on a connected and separable metric space can be represented by a utility function.*

This is very nice already, but we can do much better. By means of a different argument, we can show that we don't in fact need the connectedness hypothesis in this statement, and continuity can be relaxed to upper (or lower) semicontinuity in it. This argument, due to Trout Rader, illustrates a powerful technique that is frequently used in utility representation exercises. (We have in fact already used this technique when proving Lemma B.1.)

RADER'S UTILITY REPRESENTATION THEOREM 1

Let X be a separable metric space, and $\succsim$ a complete preference relation on X. If $\succsim$ is upper semicontinuous, then it can be represented by a utility function $u \in [0,1]^X$.

PROOF

Since X is a separable metric space, there must exist a countable collection $\mathcal{O}$ of open subsets of X such that $U = \bigcup\{O \in \mathcal{O} : O \subseteq U\}$ for any open set U in X (Proposition 3). Since $\mathcal{O}$ is countable, we may enumerate it as $\mathcal{O} = \{O_1, O_2, \ldots\}$. Now let

$$M(x) := \{i \in \mathbb{N} : O_i \subseteq L_\succ(x)\}$$

and define $u \in [0,1]^X$ by

$$u(x) := \sum_{i \in M(x)} \frac{1}{2^i}.$$

Notice that

$$x \succsim y \text{ iff } L_{\succ}(x) \supseteq L_{\succ}(y) \text{ only if } M(x) \supseteq M(y) \text{ only if } u(x) \geq u(y)$$

for any $x, y \in X$.[21] Therefore, we will be done if we can show that $M(x) \supseteq M(y)$ implies $L_{\succ}(x) \supseteq L_{\succ}(y)$ for any $x, y \in X$. (Why?) To this end, take any x and y in X such that $L_{\succ}(x) \supseteq L_{\succ}(y)$ is false, that is, there is a $z \in L_{\succ}(y)\backslash L_{\succ}(x)$. Then, since $L_{\succ}(y)$ is an open subset of X by hypothesis, there is an $O \in \mathcal{O}$ that contains z and that is contained in $L_{\succ}(y)$. (Why?) Since $z \notin L_{\succ}(x)$, O is not contained in $L_{\succ}(x)$, which proves that $M(x) \supseteq M(y)$ is false. Therefore $M(x) \supseteq M(y)$ implies $L_{\succ}(x) \supseteq L_{\succ}(y)$, and we are done.[22] ■

EXERCISE 23 Is $\geq$ an upper semicontinuous preference relation on $\mathbb{R}^2_+$? How about $\succsim_{\text{lex}}$ of Example B.1?

EXERCISE 24 Let $X := \mathbb{R}$, and show that the utility function constructed in the proof of Rader's Utility Representation Theorem 1 need not be continuous.

*EXERCISE 25[H] (Schmeidler) Let X be a connected metric space and $\succsim$ a preference relation on X with $\succ \neq \emptyset$. Prove that if $\succsim$ is continuous, and $U_{\succsim}(x)$ and $L_{\succsim}(x)$ are closed in X for each $x \in X$, then $\succsim$ must be complete.

We will turn to issues related to the problem of representing a complete preference relation by means of a *continuous* utility function in Section D.5.2.

3 Compactness

We now come to one of the most fundamental concepts of real analysis, and one that plays an important role in optimization theory: compactness. Our immediate task is to outline a basic analysis of those metric spaces that possess this property. Plenty of applications will be given later.

[21] Recall my convention about summing over the empty set: $\sum_{i\in\emptyset}(\text{whatever}) = 0$. So, if $M(x) = \emptyset$, we have $u(x) = 0$ here.

[22] Note the slick use of separability and upper semicontinuity together in the argument. A good idea to better understand what is going on here is to check how the argument would fail in the case of the lexicographic order (on $\mathbb{R}^2$).

3.1 Basic Definitions and the Heine-Borel Theorem

Definition

Let X be a metric space and $S \subseteq X$. A class $\mathcal{O}$ of subsets of X is said to **cover** S if $S \subseteq \cup\mathcal{O}$. If all members of such a class $\mathcal{O}$ are open in X, then we say that $\mathcal{O}$ is an **open cover** of S.

Here comes the definition of the compactness property. If it seems a bit unnatural to you, that's okay. Part of our task later on will be to explain why in fact this is such a fundamental property.

Definition

A metric space X is said to be **compact** if every open cover of X has a finite subset that also covers X. A subset S of X is said to be **compact in** X (or a **compact subset of** X) if every open cover of S has a finite subset that also covers S.

Warning. Compactness of a subset S of X means, by definition, the following: If $\mathcal{O}$ is a class of open sets *in* X such that $S \subseteq \cup\mathcal{O}$, then there is a finite $\mathcal{U} \subseteq \mathcal{O}$ with $S \subseteq \cup\mathcal{U}$. But what if we regard S as a metric subspace of X? In that case S is a metric space in its own right, and hence its compactness means the following: If $\mathcal{O}$ is a class of open sets *in* S such that $S \subseteq \cup\mathcal{O}$, then there is a finite $\mathcal{U} \subseteq \mathcal{O}$ with $S \subseteq \cup\mathcal{U}$. Thus, "$S$ is a compact subset of X" and "S is a compact metric subspace of X" are distinct statements. Fortunately, this is only academic, for these two statements are in fact equivalent. That is, S is compact iff every open cover of S (with sets open *in* S) has a finite subset that covers S.[23] Thus, for any subset S of X, the phrase "S is compact" is unambiguous.

(An immediate implication: Compactness is a property that is invariant under equivalence of metrics.)

(Another immediate implication: If S is a compact subset of a metric space Y and Y is a metric subspace of X, then S is compact in X.)

As a first pass, let us examine a space that is not compact, namely, the open interval $(0, 1)$. Consider the collection $\mathcal{O} := \{(\frac{1}{i}, 1) : i = 1, 2, \ldots\}$

[23] There is something to be proved here. Please recall Exercise 6, and supply a proof.

and observe that $(0,1) = (\frac{1}{2},1) \cup (\frac{1}{3},1) \cup \cdots$, that is, $\mathcal{O}$ is an open cover of $(0,1)$. Does $\mathcal{O}$ have a finite subset that covers $(0,1)$? No, because the greatest lower bound of any finite subset of $\mathcal{O}$ is bounded away from 0, so no such subset can possibly cover $(0,1)$ entirely. Therefore, we conclude that $(0,1)$ is not a compact subset of $\mathbb{R}$ (or that $(0,1)$ is not a compact metric space).[24]

For a positive example, note that a *finite* subset of any metric space is necessarily compact. Much less trivially, $[0,1]$ is a compact subset of $\mathbb{R}$. Let us prove this claim by means of a bisection argument that parallels the one we gave in Section B.1 to establish the uncountability of $\mathbb{R}$. (This is the method of "butterfly hunting," remember?) Suppose there exists an open cover $\mathcal{O}$ of $[0,1]$ no finite subset of which covers $[0,1]$.[25] Then, either $[0,\frac{1}{2}]$ or $[\frac{1}{2},1]$ is not covered by any finite subset of $\mathcal{O}$. (Why?) Pick any one of these intervals with this property, call it $[a_1,b_1]$. (The "butterfly" must be in $[a_1,b_1]$.) Then, either $[a_1,\frac{1}{2}(b_1+a_1)]$ or $[\frac{1}{2}(b_1+a_1),b_1]$ is not covered by any finite subset of $\mathcal{O}$. Pick any one of these intervals with this property, call it $[a_2,b_2]$. (The "butterfly" must be in $[a_2,b_2]$.) Continuing this way inductively, we obtain two sequences (a_m) and (b_m) in $[0,1]$ such that

(i) $a_m \leq a_{m+1} < b_{m+1} \leq b_m$,

(ii) $b_m - a_m = \frac{1}{2^m}$,

(iii) $[a_m,b_m]$ is not covered by any finite subset of $\mathcal{O}$,

for each $m = 1,2,\ldots$. Clearly, (i) and (ii) allow us to invoke Cantor's Nested Interval Lemma to find a real number c with $\{c\} = \cap^\infty[a_i,b_i]$. Now take any $O \in \mathcal{O}$ which contains c. Since O is open and $\lim a_m = \lim b_m = c$, we must have $[a_m,b_m] \subset O$ for m large enough. (Right?) But this contradicts condition (iii). Conclusion: $[0,1]$ is a compact subset of $\mathbb{R}$. (Unambiguously, we can also say that "$[0,1]$ is compact," where it is understood that $[0,1]$ is metrized in the usual way.)

This observation kindly generalizes to the case of any Euclidean space.

[24] *Quiz.* How about $\mathbb{R}$?

[25] Again, it doesn't matter whether the elements of $\mathcal{O}$ are open in the mother metric space $\mathbb{R}$ or in $[0,1]$.

The Heine-Borel Theorem[26]

For any $-\infty < a < b < \infty$, the n-dimensional cube $[a, b]^n$ is compact.

The following exercise sketches a proof for this result using the multi-dimensional analogue of the bisection argument given above. In Section 4 we will give an alternative proof.

Exercise 26 Let $n \in \mathbb{N}$, and take any real numbers a and b with $b > a$. Assume that $\mathcal{O}$ is an open cover of $[a, b]^n$ no finite subset of which covers $[a, b]^n$.

(a) Bisect $[a, b]^n$ into 2^n equal cubes by planes parallel to its faces. At least one of these cubes is not covered by any finite subset of $\mathcal{O}$, call it C_1. Proceed inductively to obtain a sequence (C_m) of cubes in $\mathbb{R}^n$ such that, for each $m \in \mathbb{N}$, we have (i) $C_{m+1} \subset C_m$, (ii) the length of an edge of C_m is $\frac{1}{2^m}(b - a)$, and (iii) C_m is not covered by any finite subset of $\mathcal{O}$.

(b) Use part (a) to prove the Heine-Borel Theorem.

The following simple fact helps us find other examples of compact sets.

Proposition 4

Any closed subset of a compact metric space X is compact.

Proof

Let S be a closed subset of X. If $\mathcal{O}$ is an open cover of S (with sets open *in* X), then $\mathcal{O} \cup \{X \backslash S\}$ is an open cover of X. Since X is compact, there exists a finite subset of $\mathcal{O} \cup \{X \backslash S\}$, say $\mathcal{O}'$, that covers X. Then $\mathcal{O}' \backslash \{X \backslash S\}$ is a finite subset of $\mathcal{O}$ that covers S. ■

By the Heine-Borel Theorem and Proposition 4, we may conclude that any n-dimensional prism $[a_1, b_1] \times \cdots \times [a_n, b_n]$ is a compact subset of $\mathbb{R}^n$. (Why?) More can, and will, be said, of course.

[26] Eduard Heine used the basic idea behind this result in 1872 (when proving Proposition A.11), but the exact formulation (with a slightly weaker definition of compactness) was given by Émile Borel in 1895. (We owe the modern formulation of the result to Henri Lebesgue.)

Exercise 27 Let $\mathcal{A}$ be a finite (nonempty) class of compact subsets of a metric space X. Is $\cup\mathcal{A}$ necessarily compact? What if $\mathcal{A}$ were not finite?

Exercise 28 [H]

(a) Show that a separable metric space need not be compact, but a compact metric space X is separable.

(b) Show that a connected metric space need not be compact, and that a compact metric space need not be connected.

3.2 Compactness as a Finite Structure

What is compactness good for? Well, the basic idea is that compactness is some sort of a generalization of the notion of finiteness. To clarify what we mean by this, let us ask the following question: What sort of sets are bounded? (Please go back and check the definition of boundedness given in Section 1.5.) The most obvious example (after the empty set) would be a nonempty finite set. Indeed, for any $k \in \mathbb{N}$ and points $x^1, \ldots, x^k$ in a given metric space X, the set $\{x^1, \ldots, x^k\}$ is bounded in X. Indeed, for ε large enough – any $\varepsilon > \max\{d(x^1, x^i) : i = 1, \ldots, k\}$ would do—we have $\{x^1, \ldots, x^k\} \subseteq N_{\varepsilon,X}(x^1)$.

Now take an arbitrary nonempty compact subset S of a metric space. Is this set bounded? An immediate temptation is to apply the previous argument by fixing an arbitrary $x \in S$ and checking if $S \subseteq N_{\varepsilon,X}(x)$ holds for large ε. But how large should ε be? We can't simply choose $\varepsilon > \sup\{d(x,y) : y \in S\}$ anymore, since we do not know if $\sup\{d(x,y) : y \in S\} < \infty$ at the outset. If S was finite, we would be okay, but all we have right now is its compactness. But in fact this is all we need! The key observation is that $\{N_{m,X}(x) : m = 1, 2, \ldots\}$ is an open cover of S (for any fixed $x \in S$). So, by compactness of S, there must exist finitely many $N_{m_1,X}(x), \ldots, N_{m_l,X}(x)$ such that $S \subseteq \cup^l N_{m_i,X}(x)$. Thus $S \subseteq N_{\varepsilon,X}(x)$ for $\varepsilon := \max\{m_1, \ldots, m_l\}$, which means that S is a bounded subset of X. (Another way of saying this is: *Every compact metric space is bounded.*)

This is, then, the power of compactness: *providing a finite structure for infinite sets.* In many problems where finiteness makes life easier (such as in optimization problems), compactness does the same thing.

The following examples, the first one of which may perhaps be a bit closer to home, illustrate this point further.

EXAMPLE 7

(*Choice Correspondences*) Let X be a metric space—interpret it as the set of *all* choice alternatives. For any nonempty subset S of X—interpreted as the set of all *feasible* alternatives—we defined

$$C_{\succsim}(S) := \{x \in S : y \succ x \text{ for no } y \in S\}$$

in Example A.4.[3] as the "set of choices from S" of an individual with the preference relation $\succsim$ on X. This set is viewed as the collection of all feasible alternatives that the individual deems as "optimal" according to her tastes.

We ask: Is there an optimal choice available to the subject individual? That is, do we have $C_{\succsim}(S) \neq \emptyset$? If S is finite, this is no pickle, for then the transitivity of $\succsim$ ensures that $C_{\succsim}(S) \neq \emptyset$. (Why?) Unfortunately, finiteness is a very demanding requirement that is not met in most applications. So, we need to go beyond finiteness here, and we may do so by recalling that "compactness" is finiteness in disguise. This leads us to focus on *compact* feasible sets instead of the finite ones. Consequently, we take the set of all **choice problems** in this abstract setting as the class of all nonempty compact subsets of X.[27]

The question now becomes: Is $C_{\succsim}(S) \neq \emptyset$ for all choice problems? The answer is still no, not necessarily. (Why?) But we are now close to identifying an interesting class of preference relations that induce nonempty choice sets. Recall that we say that $\succsim$ is upper semicontinuous if $L_{\succ}(x)$ is an open subset of X for each $x \in X$ (Section 2.3). Our claim is: *If $\succsim$ is upper semicontinuous, then $C_{\succsim}(S)$ is a nonempty compact set for any choice problem S in X.*[28]

To prove this, pick any nonempty compact set S in X, and suppose that $C_{\succsim}(S) = \emptyset$. By upper semicontinuity, this implies that $\{L_{\succ}(x) : x \in S\}$ is an open cover of S. (Why?) Now use the compactness of S to find a finite subset T of S such that $\{L_{\succ}(x) : x \in T\}$ also covers S. (Note how compactness is doing the dirty work for us here.) By transitivity of $\succsim$, there must exist a $\succsim$-maximal element of T, say x^*. But since $x^* \in S \backslash L_{\succ}(x^*)$, we must have $x^* \in L_{\succ}(x)$ for some $x \in T \backslash \{x^*\}$, which contradicts the $\succsim$-maximality of x^* in T.

27 This abstraction should not bother you. For instance, the standard consumer problem is a special case of the abstract model we consider here. As we shall see, many dynamic economic problems too are "choice problems" in the sense just described.

28 Notice that we do not assume here that $\succsim$ is a complete preference relation. Under this assumption, we could actually guarantee the existence of a $\succsim$-maximum element in any nonempty compact subset S of X.

To prove the compactness of $C_{\succsim}(S)$, observe that $S \backslash L_{\succ}(x)$ is a closed subset of S for each $x \in S$. Thus $\cap\{S \backslash L_{\succ}(x) : x \in S\}$ is compact, because it is a closed subset of the compact set S (Proposition 4). But $C_{\succsim}(S) = \cap\{S \backslash L_{\succ}(x) : x \in S\}$, no? □

EXAMPLE 8

A (nonempty) class $\mathcal{A}$ of sets is said to have the **finite intersection property** if $\cap\mathcal{B} \neq \emptyset$ for any finite (nonempty) subclass $\mathcal{B}$ of $\mathcal{A}$. Suppose that X is a metric space and $\mathcal{A}$ is a class of closed subsets of X. Question: If $\mathcal{A}$ has the finite intersection property, does it follow that $\cap\mathcal{A} \neq \emptyset$?[29]

Well, the answer depends on the structure of X. If X is finite, the answer is obviously yes, for then $\mathcal{A}$ is itself a finite class. Perhaps, then, the answer would also be affirmative when X is compact. (After all, the present author keeps saying for some reason that compactness provides a finite structure for infinite sets.) And yes, this is exactly what happens. Suppose X is compact but $\cap\mathcal{A} = \emptyset$. Then $X = X \backslash (\cap\mathcal{A}) = \cup\{X \backslash A : A \in \mathcal{A}\}$, so $\{X \backslash A : A \in \mathcal{A}\}$ is an open cover of X. Since X is compact, then, there exists a finite subset $\mathcal{B}$ of $\mathcal{A}$ such that $\cup\{X \backslash A : A \in \mathcal{B}\} = X$. But this implies $\cap\mathcal{B} = \emptyset$, which contradicts $\mathcal{A}$ having the finite intersection property.

We proved: *A class of closed subsets of a compact metric space that has the finite intersection property has a nonempty intersection.* Not impressed? Well then, please combine this fact with the Heine-Borel Theorem, write down what you get, and go back and compare it with Cantor's Nested Interval Lemma. □

EXERCISE 29[H]

(a) Prove or disprove: If (S_m) is a sequence of nonempty compact subsets of a metric space such that $S_1 \supseteq S_2 \supseteq \cdots$, then $\cap^{\infty} S_i \neq \emptyset$.

(b) Prove or disprove: If (S_m) is a sequence of nonempty closed and bounded subsets of a metric space such that $S_1 \supseteq S_2 \supseteq \cdots$, then $\cap^{\infty} S_i \neq \emptyset$.

[29] This is not a silly abstract question. We have already encountered two instances in which our ability to answer this sort of a question proved useful: recall how we proved the uncountability of $\mathbb{R}$ (Section B.1) and the Heine-Borel Theorem.

EXERCISE 30^{H} Let T be any metric space, and define

$$X := \{f \in \mathbb{R}^T : \{x \in T : |f(x)| \geq \varepsilon\} \text{ is compact for any } \varepsilon > 0\}.$$

Show that $X \subseteq \mathbf{B}(T)$.

EXERCISE 31 Let X be a metric space such that $\bigcap \mathcal{A} \neq \emptyset$ for any (nonempty) class $\mathcal{A}$ of closed subsets of X that has the finite intersection property. Prove that X is compact.

3.3 Closed and Bounded Sets

Here is one further insight into the nature of compact sets.

PROPOSITION 5
Any compact subset of a metric space X is closed and bounded.

PROOF
Let S be a compact subset of X. We have already seen that every compact metric space is bounded, so here we only need to prove that S is closed in X. If $S = X$ there is nothing to prove, so assume otherwise, and pick any $x \in X \backslash S$. Clearly, for any $y \in S$, we can find an $\varepsilon_y > 0$ such that $N_{\varepsilon_y,X}(x) \cap N_{\varepsilon_y,X}(y) = \emptyset$. (Choose, for instance, $\varepsilon_y := \frac{1}{2}d(x,y)$.) Since the collection $\{N_{\varepsilon_y,X}(y) : y \in S\}$ is an open cover of S, there must exist a finite $T \subseteq S$ such that $\{N_{\varepsilon_y,X}(y) : y \in T\}$ also covers S. Define $\varepsilon := \min\{\varepsilon_y : y \in T\}$ and observe that $N_{\varepsilon,X}(x) \subseteq X \backslash S$.[30] Thus $X \backslash S$ must be open. ■

EXERCISE 32 Show that a subset of a metric space X is closed iff its intersection with every compact subset of X is closed.

The following important result shows that the converse of Proposition 5 also holds in any Euclidean space, and hence provides an interesting characterization of compactness for such spaces. This result is also called the Heine-Borel Theorem in the real analysis folklore.

[30] Note again how this proof utilizes the finite structure provided by compactness.

> **Theorem 1**
> *Given any $n \in \mathbb{N}$, a subset of $\mathbb{R}^n$ is compact if, and only if, it is closed and bounded.*[31]

Proof
Thanks to Proposition 5, all we need to show is that a closed and bounded subset S of $\mathbb{R}^n$ is compact. By boundedness, we can find an $\varepsilon > 0$ such that $S \subseteq N_{\varepsilon,\mathbb{R}^n}(x)$ for some $x \in S$. Therefore, S must be a closed subset of a cube $[a, b]^n$.[32] But $[a, b]^n$ is compact by the Heine-Borel Theorem, and hence S must be compact by Proposition 4. ■

A common mistake is to "think of" any closed and bounded set in a metric space as compact. This is mostly due to the fact that some textbooks that focus exclusively on Euclidean spaces define compactness as the combination of the properties of closedness and boundedness. It is important to note that while, by Theorem 1, this is justified when the metric space under consideration is $\mathbb{R}^n$, a closed and bounded set need not be compact in an arbitrary metric space. That is, compactness is, in general, a stronger property than closedness and boundedness put together; it often introduces significantly more structure to the analysis. Since this is an important point that is often missed, we illustrate it here by several examples.

Example 9

[1] If X is any infinite set and d is the discrete metric, then (X, d) cannot be a compact metric space. For instance, $\{N_{1,X}(x) : x \in X\}$ is an open cover of X that does not have a finite subset that covers X. However, X is obviously closed and bounded (since $X = N_{2,X}(x)$ for any $x \in X$).

[2] $(0, 1)$ is a closed and bounded metric space that is not compact. (Here we view $(0, 1)$ as a metric subspace of $\mathbb{R}$, of course.)

[3] Let $\mathbf{e}^1 := (1, 0, 0, \ldots)$, $\mathbf{e}^2 := (0, 1, 0, \ldots)$, etc. We claim that $S := \{(\mathbf{e}^m) : m \in \mathbb{N}\}$ is a closed and bounded subset of ℓ^2 that is not compact. Indeed, any convergent sequence in S must be eventually constant, and

31 *Quiz.* This theorem is valid in any $\mathbb{R}^{n,p}$, $1 \leq p \leq \infty$. Why?

32 This is an innocent shortcut. You may pick $a = \min\{x_i - \varepsilon : i = 1, \ldots, n\}$ and $b = \max\{x_i + \varepsilon : i = 1, \ldots, n\}$ for concreteness.

hence by Proposition 1, S is closed in ℓ^2. Furthermore, it is easily checked that $S \subset N_{\varepsilon,\ell^2}(\mathbf{e}^1)$ for any $\varepsilon > \sqrt{2}$, and hence S is also bounded. But no finite subset of the open cover $\{N_{\sqrt{2},\ell^2}(\mathbf{e}^m) : m \in \mathbb{N}\}$ of S can possibly cover S, so S is not compact.

[4] The major part of this example is contained in the next exercise, which you should solve after reading what follows.

EXERCISE 33 Find an example of a closed and bounded subset of $\mathbf{C}[0,1]$ that is not compact.

It is often quite hard to prove the compactness of a set in a given function space. In the case of $\mathbf{C}[0,1]$, however, there is a very nice characterization of compact sets. A subset $\mathcal{F}$ of $\mathbf{C}[0,1]$ is said to be **equicontinuous at** $x \in [0,1]$ if, for any given $\varepsilon > 0$, there exists a $\delta > 0$ such that $|f(x) - f(y)| < \varepsilon$ for all $y \in [0,1]$ with $|x - y| < \delta$ *and all* $f \in \mathcal{F}$. The collection $\mathcal{F}$ is called **equicontinuous** if it is equicontinuous at every $x \in [0,1]$. (Some say that if $\mathcal{F} \subseteq \mathbf{C}[0,1]$ is equicontinuous, then all members of $\mathcal{F}$ are "equally continuous.")

To give an example, define $f_\alpha \in \mathbf{C}[0,1]$ by $f_\alpha(t) = \alpha t$, and consider the set $\mathcal{F}_K := \{f_\alpha : \alpha \in [0,K)\}$ where K is a strictly positive extended real number. It is easy to verify that $\mathcal{F}_K$ is equicontinuous for any given $K \in \mathbb{R}_{++}$. To see this, pick any $x \in [0,1]$ and $\varepsilon > 0$. Define next $\delta := \frac{\varepsilon}{K}$, and observe that, for any $\alpha \in [0,K)$, we have $|f(x) - f(y)| = \alpha |x - y| < K |x - y| < \varepsilon$ for any $y \in [0,1]$ with $|x - y| < \delta$. Since x is arbitrary here, we may conclude that $\mathcal{F}_K$ is equicontinuous.[33] On the other hand, $\mathcal{F}_\infty$ is *not* equicontinuous. (Prove this!)

The notion of equicontinuity is noteworthy because of the following result.

THE "BABY" ARZELÀ-ASCOLI THEOREM
The closure of a set $\mathcal{F}$ in $\mathbf{C}[0,1]$ is compact if, and only if, $\mathcal{F}$ is bounded and equicontinuous. In particular, a subset of $\mathbf{C}[0,1]$ is compact if, and only if, it is closed, bounded and equicontinuous.

[33] More generally, let $\mathcal{F}$ be any set of differentiable functions on $[0,1]$ such that there exists a real number $K > 0$ with $|f'(t)| \leq K$ for all $t \in [0,1]$ and all $f \in \mathcal{F}$. An easy application of the Mean Value Theorem shows that $\mathcal{F}$ is equicontinuous. But is $\mathcal{F} := \{f_m : m \in \mathbb{N}\}$, where $f_m \in \mathbf{C}[0,1]$ is defined by $f_m(t) := t^m$, an equicontinuous family?

We omit the proof for the time being; a substantially more general version of this theorem will be proved in Chapter D.

Now go back to Exercise 33 and solve it. You now know that you should look for a subset of $\mathbf{C}[0,1]$ that is not equicontinuous. □

4 Sequential Compactness

Recall that we can characterize the closedness property in terms of convergent sequences (Proposition 1). Given that every compact space is closed, it makes sense to ask if it is possible to characterize compactness in the same terms as well. The answer turns out to be affirmative, although the associated characterization is more subtle than that of closed sets. Nevertheless, it is very useful in applications, so it is worth spending some time and energy to understand it well.

Our aim in this section is to prove the following important theorem.

THEOREM 2
A subset S of a metric space X is compact if, and only if, every sequence in S has a subsequence that converges to a point in S.

A set that satisfies the convergence property mentioned in this theorem is said to be **sequentially compact**. So, Theorem 2 tells us that the properties of compactness and sequential compactness are equivalent for any given metric space. In particular, when working within a compact metric space, even though you may not be able to prove the convergence of a particular sequence that you are interested in, you can always pick a convergent subsequence of this sequence. You might be surprised how often this solves the problem.

EXAMPLE 10

[1] Here is an alternative proof of the fact that compactness implies closedness. Let S be a compact subset of a metric space X, and let $(x^m) \in S^\infty$ converge somewhere in X. By Theorem 2, (x^m) must have a subsequence that converges to a point x in S. But any subsequence of a convergent sequence must converge to the limit of the entire sequence, and hence $x^m \to x$. By Proposition 1, then, S must be closed in X.

[2] Here is an alternative proof of Proposition 4. Let S be a closed subset of a compact metric space X, and $(x^m) \in S^\infty$. Since X is sequentially compact by Theorem 2, (x^m) must have a subsequence that converges somewhere in X. But S is closed in X, so by Proposition 1, the limit of this subsequence must belong to S. Thus S is sequentially compact, and hence compact.

[3] We have earlier established the compactness of $[0, 1]$ by means of a nested interval argument. This argument extends to the multidimensional case as well (Exercise 26), but becomes a bit harder to follow. By contrast, we can give a very quick proof by using Theorem 2. First of all, notice that the compactness of $[0, 1]$ is an immediate consequence of the Bolzano-Weierstrass Theorem and Theorem 2. To prove the same for $[0, 1]^n$ for any $n \in \mathbb{N}$, take any sequence $(x^m) = ((x_1^m, \ldots, x_n^m))$ in $[0, 1]^n$. Then (x_1^m) is a real sequence in $[0, 1]$. So, by Theorem 2, (x_1^m) has a convergent subsequence in $[0, 1]$, call it $(x_1^{m_k})$. Now observe that $(x_2^{m_k})$ must have a convergent subsequence in $[0, 1]$. Continuing this way, we can obtain a subsequence of (x^m) that converges in $[0, 1]^n$. This proves that $[0, 1]^n$ is sequentially compact. Thus, by Theorem 2, $[0, 1]^n$ is compact.

[4] $\mathbb{R}$ is obviously not compact, but how about $\overline{\mathbb{R}}$? (Recall the last paragraph of Example 1. [3].) $\overline{\mathbb{R}}$ is trivially closed, and it is also bounded (because we are using the bounded metric d^* on it). While, of course, this is no guarantee for the compactness of $\overline{\mathbb{R}}$, d^* actually does make $\overline{\mathbb{R}}$ a compact metric space. Sequential compactness provides an easy way of seeing this. If (x_m) is a sequence in $\overline{\mathbb{R}}$, then it has a monotonic subsequence (x_{m_k}) by Proposition A.9, and since $-\infty \leq x_{m_k} \leq \infty$ for each k, this subsequence must converge in $\overline{\mathbb{R}}$. (Why?) Thus $\overline{\mathbb{R}}$ is sequentially compact, and hence, compact. □

EXERCISE 34 For any given $n \in \mathbb{N}$, let (X_i, d_i) be a metric space, $i = 1, \ldots, n$. Let $X := \mathsf{X}^n X_i$, and define the map $d : X \times X \to \mathbb{R}_+$ by

$$d(x, y) := \sum_{i=1}^{n} d_i(x_i, y_i).$$

Prove that (X, d) is a metric space, and that it is compact if each (X_i, d_i) is compact.

Exercise 35[H] Recall Example 2.[3], and show that any compact subset of $\mathbf{C}^1[0,1]$ (relative to the $d_{\infty,\infty}$ metric) is compact in $\mathbf{C}[0,1]$. More generally, if S is a bounded subset of $\mathbf{C}^1[0,1]$, then it is *relatively compact* in $\mathbf{C}[0,1]$, that is, $cl_{\mathbf{C}[0,1]}(S)$ is a compact subset of $\mathbf{C}[0,1]$.

We hope that these examples have convinced you of the usefulness of sequential compactness. If not, don't worry; there will be numerous other occasions to use Theorem 2 in the next chapter. We now turn to the proof of Theorem 2. Since the "only if" part of the theorem is relatively easy, we deal with it first.

Let S be a compact subset of X, and $(x^m) \in S^\infty$. Suppose that (x^m) does not have a subsequence that converges in S. In that case, $T := \{x^1, x^2, \ldots\}$ must be a closed subset of S.[34] Since S is compact, then, T is a compact subset of S (Proposition 5). Since (x^m) lacks a convergent subsequence, for any $m \in \mathbb{N}$ there exists an $\varepsilon_m > 0$ such that $N_{\varepsilon_m,X}(x^m) \cap \{x^1, x^2, \ldots\} = \{x^m\}$. (Why?) But $\{N_{\varepsilon_m,X}(x^m) : m \in \mathbb{N}\}$ is an open cover of T, and by compactness of T, it has a finite subset that also covers T. It follows that T is a finite set, which means that at least one term of (x^m) must be repeated infinitely often in the sequence, that is, there is a constant subsequence of (x^m) (which is trivially convergent in S), a contradiction.

Let us now move on to the "if" part of Theorem 2. We need some preparation for this. First of all, let us agree to call a set S in a metric space X **totally bounded** (or **precompact**) if, for any $\varepsilon > 0$, there exists a finite subset T of S such that $S \subseteq \cup\{N_{\varepsilon,X}(x) : x \in T\}$. The following exercise shows that this property is strictly more demanding than boundedness.

Exercise 36

(a) Show that every totally bounded subset of a metric space is bounded.

(b) Show that an infinite discrete space is bounded but not totally bounded.

(c) Give an example of a bounded set in ℓ^∞ that is not totally bounded.

(d) Prove that a subset of $\mathbb{R}^n$ is bounded iff it is totally bounded.

It turns out that there is a fundamental link between compactness and total boundedness. Foremost we have the following useful observation.

34 *Quiz.* Why?

LEMMA 1
Every sequentially compact subset of a metric space X is totally bounded.

PROOF
Suppose the claim is not true, that is, there is a sequentially compact subset S of X with the following property: there exists an $\varepsilon > 0$ such that $\{N_{\varepsilon,X}(x) : x \in T\}$ does not cover S for *any* finite $T \subseteq S$. To derive a contradiction, we wish to construct a sequence in S with no convergent subsequence. Begin by picking an $x^1 \in S$ arbitrarily. By hypothesis, we cannot have $S \subseteq N_{\varepsilon,X}(x^1)$, so there must exist an $x^2 \in S$ such that $d(x^1, x^2) \geq \varepsilon$. Again, $S \subseteq N_{\varepsilon,X}(x^1) \cup N_{\varepsilon,X}(x^2)$ cannot hold, so we can find an $x^3 \in S$ such that $d(x^i, x^3) \geq \varepsilon$, $i = 1, 2$. Proceeding inductively, we obtain a sequence $(x^m) \in S^\infty$ such that $d(x^i, x^j) \geq \varepsilon$ for any distinct $i, j \in \mathbb{N}$.[35] Since S is sequentially compact, there must exist a convergent subsequence, say (x^{m_k}), of (x^m). But this is impossible, since $\lim x^{m_k} = x$ implies that

$$d(x^{m_k}, x^{m_l}) \leq d(x^{m_k}, x) + d(x, x^{m_l}) < \varepsilon$$

for large enough (distinct) k and l. ∎

The following is the second step of our proof of Theorem 2.

LEMMA 2
Let S be a sequentially compact subset of a metric space X and $\mathcal{O}$ an open cover of S. Then there exists an $\varepsilon > 0$ such that, for any $x \in S$, we can find an $O_x \in \mathcal{O}$ with $N_{\varepsilon,X}(x) \subseteq O_x$.

PROOF
Assume that we cannot find such an $\varepsilon > 0$. Then, for any $m \in \mathbb{N}$, there exists an $x^m \in S$ such that $N_{\frac{1}{m},X}(x^m)$ is not contained in any member of $\mathcal{O}$. (Otherwise we would choose $\varepsilon = \frac{1}{m}$.) By sequential compactness of S, we can find a subsequence of (x^m) that converges to an $x \in S$. Relabeling if necessary, let us denote this subsequence again by (x^m). Since $\mathcal{O}$ is an open cover of S, we have $x \in O$ for some $O \in \mathcal{O}$. Given that O is open, we have $N_{\varepsilon,X}(x) \subseteq O$ for $\varepsilon > 0$ small enough. Since $x^m \to x$, there exists an $M \in \mathbb{R}$

[35] "Proceeding inductively" is a bit of wishful phrase here. We owe the present fact to the good graces of the Axiom of Choice, to be sure.

such that $x^m \in N_{\frac{\varepsilon}{2},X}(x)$ for all $m \geq M$.[36] Now pick any $m \geq \max\{M, \frac{2}{\varepsilon}\}$, and observe that we have $N_{\frac{1}{m},X}(x^m) \subseteq N_{\varepsilon,X}(x)$, since, for any $y \in N_{\frac{1}{m},X}(x^m)$,

$$d(x, y) \leq d(x, x^m) + d(x^m, y) < \tfrac{\varepsilon}{2} + \tfrac{1}{m} < \varepsilon.$$

But then $N_{\frac{1}{m},X}(x^m) \subseteq O \in \mathcal{O}$, which contradicts the construction of (x^m). ∎

All we need do now is to put these two facts together.

PROOF OF THEOREM 2
Since the "only if" part was proved above, we only need to show that sequential compactness implies compactness. To this end, let S be a sequentially compact subset of X and $\mathcal{O}$ an open cover of S. By using first Lemma 2 and then Lemma 1, we can find an $\varepsilon > 0$ and a finite $T \subseteq S$ with the following properties:

(i) $S \subseteq \cup\{N_{\varepsilon,X}(x) : x \in T\}$; and

(ii) for each $x \in T$, there exists an $O_x \in \mathcal{O}$ with $N_{\varepsilon,X}(x) \subseteq O_x$.

But then $\{O_x : x \in T\}$ is a finite subset of $\mathcal{O}$ that covers S, and we are done. ∎

5 Completeness

We now come to the final major property that we will consider in this chapter for a metric space, the property of *completeness*. A quick review of *Cauchy sequences* is a prerequisite for this, so we start with that.

5.1 Cauchy Sequences

Intuitively speaking, by a Cauchy sequence we mean a sequence the terms of which eventually get arbitrarily close to one another. This idea is formalized as follows.

[36] Why $\frac{\varepsilon}{2}$? Because I wish to get $N_{\frac{1}{m},X}(x^m) \subseteq N_{\varepsilon,X}(x)$ for large m. If I chose M such that $x^m \in N_{\varepsilon,X}(x)$ for each $m \geq M$, then I would not be able to guarantee fitting $N_{\frac{1}{m},X}(x^m)$ in $N_{\varepsilon,X}(x)$ even for large m. With this $\frac{\varepsilon}{2}$ trick, things go orderly.

> **DEFINITION**
> A sequence (x^m) in a metric space X is called a **Cauchy sequence** if for any $\varepsilon > 0$, there exists an $M \in \mathbb{R}$ (which may depend on ε) such that $d(x^k, x^l) < \varepsilon$ for all $k, l \geq M$.

For instance, $(1, \frac{1}{2}, \frac{1}{3}, \ldots)$ is a Cauchy sequence in $\mathbb{R}$, for the terms of this sequence get closer and closer toward its tail. More formally, $(1, \frac{1}{2}, \frac{1}{3}, \ldots)$ is Cauchy because

$$\left|\frac{1}{k} - \frac{1}{l}\right| \leq \left|\frac{1}{k}\right| + \left|\frac{1}{l}\right| \to 0 \qquad (\text{as } k, l \to \infty).$$

Still more formally—although the previous argument was fine, really—this sequence is Cauchy because, for any $\varepsilon > 0$, there exist M_1 and M_2 in $\mathbb{R}$ such that $\left|\frac{1}{k}\right| < \frac{\varepsilon}{2}$ and $\left|\frac{1}{l}\right| < \frac{\varepsilon}{2}$ whenever $k \geq M_1$ and $l \geq M_2$. Thus, by the triangle inequality, $\left|\frac{1}{k} - \frac{1}{l}\right| < \varepsilon$ for all $k, l \geq \max\{M_1, M_2\}$.

As another example, note that $(-1, 1, -1, 1, \ldots)$ is not a Cauchy sequence in $\mathbb{R}$, for $\left|(-1)^m - (-1)^{m+1}\right| = 2$ for all $m \in \mathbb{N}$.

WARNING. For any sequence (x^m) in a metric space X, the condition that *consecutive* terms of the sequence get closer and closer, that is, $d(x^m, x^{m+1}) \to 0$, is a necessary but not sufficient condition for (x^m) to be Cauchy. The proof of the first claim is easy. To verify the second claim, consider the real sequence (x_m), where $x_1 = 1$ and $x_m = 1 + \cdots + \frac{1}{m-1}$ for $m = 2, 3, \ldots$. Similarly, $(\ln 1, \ln 2, \ldots)$ is a divergent real sequence, but $\ln(m+1) - \ln m = \ln(1 + \frac{1}{m}) \to 0$.

EXERCISE 37 Prove: If (x^m) is a sequence in a metric space X such that $\sum^\infty d(x^i, x^{i+1}) < \infty$, then it is Cauchy.

The first thing to note about Cauchy sequences is that they are *bounded*. That is, the set of all terms of a Cauchy sequence in a metric space is a bounded subset of that space. Indeed, if (x^m) is a Cauchy sequence in a metric space X, then, by choosing an integer $M \geq 2$ such that $d(x^k, x^l) < 1$ for all $k, l \geq M$, we obtain $\{x^m : m \in \mathbb{N}\} \subseteq N_{\delta,X}(x^M)$, where $\delta := \max\{1, d(x^1, x^M), \ldots, d(x^{M-1}, x^M)\}$. Moreover, any convergent sequence (x^m) in X is Cauchy, because we have

$$d(x^k, x^l) \leq d(x^k, \lim x^m) + d(\lim x^m, x^l) \to 0 \qquad (\text{as } k, l \to \infty).$$

On the other hand, a Cauchy sequence need not converge. For example, consider $(1, \frac{1}{2}, \frac{1}{3}, \ldots)$ as a sequence in the metric space $(0, 1]$ (not in $\mathbb{R}$). As we have shown above, this sequence is Cauchy in $(0, 1]$, but $(\frac{1}{m})$ does not converge *in this space* (while it does in $[0, 1]$ or $\mathbb{R}$). Thus, we conclude, *a Cauchy sequence need not be convergent.* Yet if we know that (x^m) is Cauchy and that it has a convergent subsequence, say (x^{m_k}), then we can conclude that (x^m) converges. For, $x^{m_k} \to x$ implies

$$d(x^m, x) \leq d(x^m, x^{m_k}) + d(x^{m_k}, x) \to 0 \quad \text{(as } m, k \to \infty).$$

We summarize these properties of Cauchy sequences for future reference.

PROPOSITION 6

Let (x^m) be a sequence in a metric space X.

(a) *If (x^m) is convergent, then it is Cauchy.*

(b) *If (x^m) is Cauchy, then $\{x^1, x^2, \ldots\}$ is bounded, but (x^m) need not converge in X.*

(c) *If (x^m) is Cauchy and has a subsequence that converges in X, then it converges in X as well.*

5.2 Complete Metric Spaces: Definition and Examples

Suppose that we are given a sequence (x^m) in some metric space, and we need to check if this sequence converges. Doing this directly requires us to "guess" a candidate limit x for the sequence, and then to show that we actually have $x^m \to x$ (or that this never holds for any choice of x). But guessing is not always the most efficient way, or even a feasible way, of dealing with this problem. An alternative and sometimes unambiguously superior method is to check whether or not the sequence at hand is Cauchy. If it is not Cauchy, then it cannot be convergent by Proposition 6, and we are done. What if (x^m) turned out to be Cauchy, however? Proposition 6 would not settle the score in this case, but, depending on the structure of the mother space we are working with, we may still have learned something. Indeed, if we knew somehow that in our metric space all Cauchy sequences converge, then we would be done again. In such a space a sequence is convergent iff it is Cauchy, and hence convergence can always be tested by using the "Cauchyness" condition. As we develop the theory further, it will become apparent that you should really think of such metric spaces as rather friendly environments. At any rate, most metric spaces that one deals with in practice do possess this property; it is about time that we give them a name.

> DEFINITION
> A metric space X is said to be **complete** if every Cauchy sequence in X converges to a point in X.

We have just seen that $(0, 1]$ is not complete. But note well: this conclusion depends on the metric we use. For instance, if we chose not to view $(0, 1]$ as a metric subspace of $\mathbb{R}$, and endowed it instead with, say, the discrete metric, then the resulting space would be complete. Indeed, if (x_m) is a Cauchy sequence in a discrete space X, then there must exist an $M \in \mathbb{R}$ such that $x_m = x_{m+1} = \cdots$ for all $m \geq M$ (why?), which implies that (x_m) converges in X. Another example of an incomplete metric space is $\mathbb{Q}$ (viewed as a metric subspace of $\mathbb{R}$). Indeed, since $\mathbb{Q}$ is dense in $\mathbb{R}$, for any $x \in \mathbb{R} \backslash \mathbb{Q}$ we can find an $(x_m) \in \mathbb{Q}^\infty$ with $\lim x_m = x$. Then, (x_m) is Cauchy, but it does not converge in $\mathbb{Q}$.

Here are some less trivial examples.

EXAMPLE 11

[1] $\mathbb{R}$ *is complete*. This statement is essentially equivalent to the Completeness Axiom.[37] Since we take the Completeness Axiom as a primitive in this book, however, all we need to show here is that this axiom implies that *all real Cauchy sequences converge* (Exercise A.40). And this is true, for if (x_m) is a real Cauchy sequence, then it is bounded (Proposition 6), so by the Bolzano-Weierstrass Theorem (which is based on the Completeness Axiom), (x_m) must have a subsequence that converges in $\mathbb{R}$, and hence it is itself convergent (Proposition 6). Thus, $\mathbb{R}$ *is complete*.

[2] For any $n \in \mathbb{N}$, let (X_i, d_i) be a complete metric space, $i = 1, \ldots, n$, and let (X, d) be the (product) metric space defined in Exercise 34. We

[37] More precisely, when extending $\mathbb{Q}$ to $\mathbb{R}$, we can use the Archimedean Property (Proposition A.6) and the property that "all Cauchy sequences converge" instead of the Completeness Axiom. It can be shown that the resulting ordered field would satisfy the Completeness Axiom, and would thus coincide with the real number system.

claim that (X, d) is complete. Let $(x^m) = ((x_1^m, \ldots, x_n^m))$ be a Cauchy sequence in X. Then, for each $\varepsilon > 0$, there exists an $M \in \mathbb{R}$ such that $d_i(x_i^k, x_i^l) \leq d(x^k, x^l) < \varepsilon$ for all $k, l \geq M$, which means that $(x_i^1, x_i^2, \ldots)$ is a Cauchy sequence in (X_i, d_i), $i = 1, \ldots, n$. Since (X_i, d_i) is complete, $x_i^m \to x_i$ for some $x_i \in X_i$, $i = 1, \ldots, n$. But then,

$$d(x^m, (x_1, \ldots, x_n)) = \sum_{i=1}^{n} d_i(x_i^m, x_i) \to 0,$$

so we may conclude that (x^m) converges in (X, d). Thus: *A finite product of complete metric spaces is complete.*

[3] $\mathbb{R}^{n,1}$ *is complete.* This is an immediate consequence of the previous two observations. In fact, the metric d_1 does not play any special role here. The argument given in [2] can easily be used to establish that any $\mathbb{R}^{n,p}$ is complete. (Do you see this?)

[4] ℓ^p *is complete,* $1 \leq p < \infty$.[38] Now for this we will have to work harder. Fix any $1 \leq p < \infty$, and let $(x^m) = ((x_1^m, x_2^m, \ldots))$ be a Cauchy sequence (of sequences) in ℓ^p. As in [2] above, we can show easily that (x_i^m) must be a real Cauchy sequence (for each i). (Verify!) So, by completeness of $\mathbb{R}$, we have $x_i^m \to x_i$ for some $x_i \in \mathbb{R}$, $i = 1, 2, \ldots$. Then, the natural claim is that $(x^m) \to (x_1, x_2, \ldots) \in \ell^p$. (Notice that there are *two* things to prove here.)

Since (x^m) is Cauchy, given any $\varepsilon > 0$, we can find a positive integer M such that, for any $r \in \mathbb{N}$,

$$\sum_{i=1}^{r} \left| x_i^k - x_i^l \right|^p \leq \left(d_p((x^k), (x^l)) \right)^p < \varepsilon^p \qquad \text{for all } k, l \geq M.$$

Now keep r and $l \geq M$ fixed, and let $k \to \infty$ to get

$$\sum_{i=1}^{r} \left| x_i - x_i^l \right|^p < \varepsilon^p \qquad \text{for all } r \in \mathbb{N} \text{ and } l \geq M.^{39}$$

[38] This is a special case of the famous *Riesz-Fischer Theorem.*

[39] I'm cheating a bit here. Indeed, I'm invoking the continuity of the map $(t_1, \ldots, t_r) \mapsto \sum^r |t_i - \alpha_i|^p$ on $\mathbb{R}^r$ for any given real numbers $\alpha_1, \ldots, \alpha_r$—this, shamelessly, even though I have not yet talked about the continuity of maps defined on an arbitrary Euclidean space. But I trust you know that all is kosher. If you have doubts, take a quick look at Example D.2.[4] and Proposition D.1.

But this inequality implies

$$\sum_{i=1}^{\infty} \left|x_i - x_i^l\right|^p \leq \varepsilon^p \qquad \text{for all } l \geq M.$$

So, since $\varepsilon > 0$ is arbitrary here, we will be done as soon as we show that $(x_1, x_2, \ldots) \in \ell^p$. (Yes?) But, since $x^M \in \ell^p$, we have $\sum_{i=1}^{\infty} |x_i^M|^p < \infty$, so

$$\begin{aligned}\sum_{i=1}^{\infty} |x_i|^p &= \sum_{i=1}^{\infty} |(x_i - x_i^M) + x_i^M|^p \\ &\leq \sum_{i=1}^{\infty} |x_i - x_i^M|^p + \sum_{i=1}^{\infty} |x_i^M|^p \\ &\leq \varepsilon^p + \sum_{i=1}^{\infty} |x_i^M|^p,\end{aligned}$$

and it follows that $(x_1, x_2, \ldots) \in \ell^p$. We are done: *$\ell^p$ is complete.*

[5] $\mathbf{B}(T)$ *is complete for any nonempty set* T. Let (f_m) be a Cauchy sequence in $\mathbf{B}(T)$. What is a plausible candidate for the limit of (f_m) (with respect to d_∞)? It is plain that, for each $x \in T$, $(f_m(x))$ is a Cauchy sequence in $\mathbb{R}$, so we have $f_m(x) \to f(x)$ for some $f(x) \in \mathbb{R}$. Thus the pointwise limit of (f_m), call it $f \in \mathbb{R}^T$, exists. Naturally, this map is a good candidate to be the limit of (f_m). (Be careful here. We have $f_m(x) \to f(x)$ for each x, but this does *not* imply in general that $d_\infty(f_m, f) \to 0$ (Example 4.[5]). But, in this particular case, where we have the additional information that (f_m) is Cauchy, we can prove that $d_\infty(f_m, f) \to 0$.) We first need to show that $f \in \mathbf{B}(T)$. To this end, fix any $\varepsilon > 0$, and use the "Cauchyness" of (f_m) to find an $M \in \mathbb{R}$ (which may of course depend on ε) such that $d_\infty(f_k, f_l) < \varepsilon$ for all $k, l \geq M$. Then, for any $l \geq M$,

$$\left|f(x) - f_l(x)\right| = \lim_{k\to\infty} \left|f_k(x) - f_l(x)\right| \leq \lim_{k\to\infty} d_\infty(f_k, f_l) \leq \varepsilon \qquad \text{for all } x \in T \tag{2}$$

(where we used the continuity of the absolute value function). So we have $\left|f(x)\right| \leq \left|f_l(x)\right| + \varepsilon$ for all $x \in T$, and hence, since $f_l \in \mathbf{B}(T)$, we have $f \in \mathbf{B}(T)$. Moreover, (2) gives us also that $d_\infty(f, f_l) \leq \varepsilon$ for all $l \geq M$.

Since ε is arbitrary here, we may thus conclude that $d_\infty(f, f_m) \to 0$, as we sought.

[6] The completeness property is invariant under the strong equivalence of metrics. That is, if (X, d) and (X, D) are strongly equivalent, then (X, d) is complete iff (X, D) is also complete. (Why?) It is important to note that the equivalence of metrics is not sufficient for this result. That is, it is possible that two metrics spaces are equivalent, and yet one is complete and the other is not. (In this sense, completeness is a different kind of property than connectedness, separability, or compactness.) For instance, $(\mathbb{N}, d_1)$ and $(\mathbb{N}, D)$, where $D : \mathbb{N}^2 \to \mathbb{R}_+$ is defined by $D(i,j) := \left|\frac{1}{i} - \frac{1}{j}\right|$, are equivalent metric spaces, yet $(\mathbb{N}, d_1)$ is complete, while $(\mathbb{N}, D)$ is not. □

EXERCISE 38 Let $\mathcal{A}$ be a finite (nonempty) class of complete metric subspaces of a given metric space X. Is $\cup\mathcal{A}$ complete metric subspace of X? What if $\mathcal{A}$ was not finite? How about $\cap\mathcal{A}$?

EXERCISE 39 Prove that $\mathbb{R}^{n,p}$ is complete for any $(n, p) \in \mathbb{N} \times [1, \infty]$.

EXERCISE 40[H] Prove that ℓ^∞ and $\mathbf{C}[0, 1]$ are complete metric spaces.

5.3 Completeness versus Closedness

Recall that $(0, 1]$ is not a complete metric space, while $[0, 1]$ is a complete metric subspace of $\mathbb{R}$. This suggests a tight connection between the closedness of a set and its completeness as a metric subspace. Indeed, a complete subspace of a metric space is closed. We even have a partial converse of this fact.

PROPOSITION 7
Let X be a metric space, and Y a metric subspace of X. If Y is complete, then it is closed in X. Conversely, if Y is closed in X and X is complete, then Y is complete.

PROOF
Let Y be complete, and take any $(x^m) \in Y^\infty$ that converges in X. Then (x^m) is Cauchy (Proposition 6), and thus $\lim x^m \in Y$. It follows from Proposition 1 that Y is closed.

To prove the second claim, assume that X is complete and Y is closed in X. If (x^m) is a Cauchy sequence in Y, then by completeness of X, it must converge in X. But since Y is closed, $\lim x^m$ must belong to Y (Proposition 1). It follows that Y is complete. ∎

COROLLARY 1
A metric subspace of a complete metric space X is complete if, and only if, it is closed in X.

This is a useful observation that lets us obtain other complete metric spaces from the ones we already know to be complete. A classic example is the space $\mathbf{C}[0,1]$, which is a metric subspace of $\mathbf{B}[0,1]$ (Example 2.[2]). While one can verify the completeness of this space directly (as you were asked to do in Exercise 40), an easier way of doing this is to verify that $\mathbf{C}[0,1]$ is a closed subset of $\mathbf{B}[0,1]$, thanks to Proposition 7 and Example 11.[5]. In fact, as we shall see later, this method yields a good deal more.

EXERCISE 41 [H] Show that every compact metric space is complete.

EXERCISE 42 [H] Let X denote the set of all continuous real functions on $[0,1]$ and define $d_1 \in \mathbb{R}^{X\times X}$ by

$$d_1(f,g) := \int_0^1 |f(t) - g(t)|\, dt.$$

(a) Prove that (X, d_1) is a metric space.
(b) (Sutherland) Consider the sequence (f_m) of continuous functions on $[0,1]$ with

$$f_m(t) := \begin{cases} 0, & \text{if } 0 \leq t \leq \frac{1}{2} - \frac{1}{m} \\ m(t + \frac{1}{m} - \frac{1}{2}), & \text{if } \frac{1}{2} - \frac{1}{m} < t \leq \frac{1}{2} \\ 1, & \text{if } \frac{1}{2} < t \leq 1 \end{cases}$$

(Figure 3). Verify that (f_m) is a Cauchy sequence in (X, d_1).
(i) Find what is wrong with the following argument: "$f_m(t) \to f(t)$ for each $t \in [0,1]$, where f equals 0 on $[0, \frac{1}{2})$ and 1 on $[\frac{1}{2}, 1]$. Thus the only possible limit of (f_m), which is f, is not continuous on $[0,1]$. So (X, d_1) is not complete."

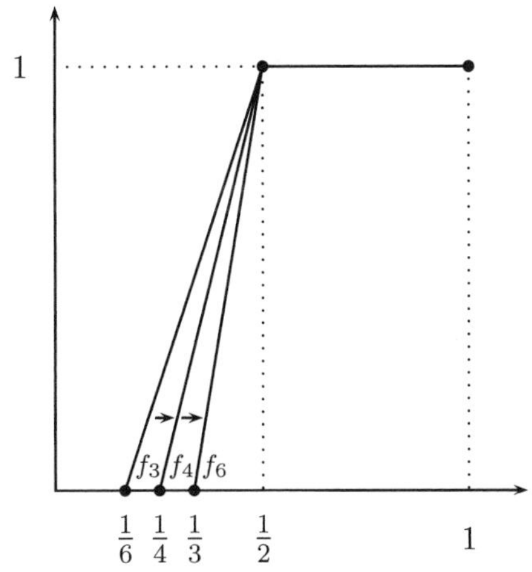

FIGURE 3

(ii) Find what is wrong with the following alternative argument: "Define f as in part (i). We have $d_1(f_m, f) = \frac{1}{2m} \to 0$. Thus $f_m \to f$ in the d_1 metric. But f is not continuous on $[0, 1]$, so (X, d_1) is not complete."

(iii) Assume that (f_m) converges in (X, d_1), and derive a contradiction. Conclude that (X, d_1) is not complete.

Note. The previous exercise points to one of the main reasons why the sup-metric is more useful than d_1 in metrizing the set of all continuous real functions on $[0, 1]$. The reason why $d_{\infty,\infty}$ is more appropriate than d_∞ for metrizing $\mathbf{C}^1[0, 1]$ (Example 2.[3]) is similar.

EXERCISE 43 H

(a) Prove: For any sequence (h_m) in $\mathbf{C}[0, 1]$, and $h \in \mathbf{C}[0, 1]$, if $d_\infty(h_m, h) \to 0$, then $\int_0^1 h_m(t)dt \to \int_0^1 h(t)dt$.

(b) Let (f_m) be a sequence in $\mathbf{C}^1[0, 1]$, and $f, g \in \mathbf{C}[0, 1]$. Use part (a) and the Fundamental Theorem of Calculus to show that if $d_\infty(f_m, f) \to 0$ and $d_\infty(f'_m, g) \to 0$, then $f \in \mathbf{C}^1[0, 1]$ and $f' = g$.

(c) Show that $\mathbf{C}^1[0, 1]$ is a complete metric space, but $(\mathbf{C}^1[0, 1], d_\infty)$ is not complete.

We define the **diameter** of any bounded subset S of a metric space X as

$$diam(S) := \sup\{d(x, y) : x, y \in S\}.$$

This concept allows us to state the following important generalization of Cantor's Nested Interval Lemma.

THE CANTOR-FRÉCHET INTERSECTION THEOREM[40]
A metric space X is complete if, and only if, for any sequence (S_m) of nonempty closed subsets of X with

$$S_1 \supseteq S_2 \supseteq \cdots \quad \text{and} \quad diam(S_m) \to 0, \tag{3}$$

we have $\cap^\infty S_i \neq \emptyset$ (so that $|\cap^\infty S_i| = 1$).

EXERCISE 44[H] Prove the Cantor-Fréchet Intersection Theorem.

EXERCISE 45 Give an example of
(a) a sequence (S_m) of nonempty closed subsets of $(0, 1)$ such that (3) holds and $\cap^\infty S_i = \emptyset$;
(b) a sequence (S_m) of nonempty subsets of $\mathbb{R}$ such that (3) holds and $\cap^\infty S_i = \emptyset$;
(c) a sequence (S_m) of nonempty closed subsets of $\mathbb{R}$ such that $S_1 \supseteq S_2 \supseteq \cdots$ and $\cap^\infty S_i = \emptyset$.

EXERCISE 46[H] Let X be a complete metric space, and denote by $\mathcal{B}_X$ the class of all nonempty bounded subsets of X. The **Kuratowski measure of noncompactness** on X is the function $\kappa : \mathcal{B}_X \to \mathbb{R}_+$ defined by $\kappa(S) := \inf\{\varepsilon > 0 : \text{there is a finite } \mathcal{A} \subseteq \mathcal{B}_X \text{ such that } S \subseteq \cup\mathcal{A} \text{ and } diam(A) \leq \varepsilon$ for each $A \in \mathcal{A}\}$. Prove:
(a) $\kappa(S) = 0$ iff S is totally bounded.
(b) $\kappa(S) = \kappa(cl_X(S))$ for all $S \in \mathcal{B}_X$.
(c) (*The Generalized Cantor-Fréchet Intersection Theorem*) If (S_m) is a sequence of nonempty closed and bounded subsets of X with $S_1 \supseteq S_2 \supseteq \cdots$ and $\lim \kappa(S_m) = 0$, then $\cap^\infty S_i$ is a nonempty compact set.

[40] Most textbooks refer to this result as the *Cantor Intersection Theorem*, presumably because the crux of it is already contained in Cantor's Nested Interval Lemma. However, the generalization noted here is quite substantial, and was first proved by Maurice Fréchet in his famous 1906 dissertation.

5.4 Completeness versus Compactness

While every compact metric space is complete (Exercise 41), it is obvious that completeness does not imply compactness. $\mathbb{R}$, for instance, is complete but not compact. In fact, even a complete *and bounded* metric space need not be compact. For example, a discrete space that contains infinitely many points is bounded and complete, but it is not compact. Similarly, metrizing $\mathbb{R}$ by means of the metric $(x, y) \mapsto \frac{|x-y|}{1+|x-y|}$ yields a bounded and complete metric space that is not compact. However, the following important result shows that the gap between completeness and compactness disappears if we strengthen the boundedness hypothesis here to *total boundedness.* Put differently, the converse of Lemma 1 holds in complete metric spaces.

THEOREM 3
A metric space is compact if, and only if, it is complete and totally bounded.

PROOF
The "only if" part follows readily from Theorem 2, Lemma 1, and Proposition 6. (How?) To prove the "if" part, take any complete and totally bounded metric space X, and let $\mathcal{O}$ be an arbitrary open cover of X. The plan is to proceed as in the proof of the Heine-Borel Theorem. (You remember the method of "butterfly hunting," don't you?) To derive a contradiction, assume that no finite subset of $\mathcal{O}$ covers X. Since X is totally bounded, it can be covered by finitely many nonempty closed sets of diameter at most 1. Therefore, at least one of these closed sets, say S_1, cannot be covered by finitely many elements of $\mathcal{O}$. Obviously, S_1 itself is totally bounded, being a subset of a totally bounded set. Then it can be covered by finitely many nonempty closed sets of diameter at most $\frac{1}{2}$, and at least one of these sets, say S_2, cannot be covered by finitely many elements of $\mathcal{O}$. Continuing inductively, we obtain a sequence (S_m) of nonempty closed subsets of X such that $S_1 \supseteq S_2 \supseteq \cdots$ and $diam(S_m) \leq \frac{1}{m}$ for each m, whereas no S_m can be covered by finitely many sets in $\mathcal{O}$. By the Cantor-Fréchet Intersection Theorem, then, there exists an x in $\cap^\infty S_i$.[41] Then $x \in O$ for some $O \in \mathcal{O}$ (because $\mathcal{O}$ covers X). Since O is open, there exists an $\varepsilon > 0$ such that $N_{\varepsilon,X}(x) \subseteq O$. But for any

[41] Or, for a direct proof, pick any $x^m \in S_m$ for each m, and note that (x^m) is a Cauchy sequence. So, since X is complete, (x^m) converges to some $x \in X$. Since each S_m is closed and $x^k \in S_m$ for all $k \geq m$, we have $x \in \cap^\infty S_i$.

$m > \frac{1}{\varepsilon}$, we have $diam(S_m) < \varepsilon$ and $x \in S_m$, so $S_m \subseteq N_{\varepsilon,X}(x)$. Thus S_m is covered by an element of $\mathcal{O}$, contradiction. ■

Combining this result with Corollary 1 leads us to the following wisdom:

A closed subset of a complete metric space is compact iff it is totally bounded.

More often than not, it is in this form that Theorem 3 is used in practice.

Exercise 47[H] Show that a subset S of a metric space X is totally bounded iff every sequence in S has a Cauchy subsequence.

6 Fixed Point Theory I

Certain types of self-maps defined on a complete metric space possess a very desirable fixed point property.[42] This section is devoted to the analysis of such maps and this important property.

6.1 Contractions

Definition
Let X be any metric space. A self-map Φ on X is said to be a **contraction** (or a **contractive self-map**) if there exists a real number $0 < K < 1$ such that

$$d(\Phi(x), \Phi(y)) \leq Kd(x,y) \qquad \text{for all } x, y \in X.$$

In this case, the infimum of the set of all such K is called the **contraction coefficient** of Φ.

For instance, $f_\alpha \in \mathbb{R}^{\mathbb{R}}$, defined by $f_\alpha(t) := \alpha t$, is a contraction (with the contraction coefficient $|\alpha|$) iff $|\alpha| < 1$. More generally, a differentiable real function on a nonempty open subset O of $\mathbb{R}$ is a contraction, provided that the absolute value of its derivative at any point in O is bounded by a fixed

42 *Reminder.* A *self-map* is a function whose domain and codomain are identical.

number strictly smaller than 1. We will make repeated use of this simple fact in what follows.[43]

Don't be fooled by the simplicity of these observations. In general, it may be quite difficult to check whether or not a given self-map is contractive. In most applications, however, one works with self-maps that are defined on relatively well-behaved metric spaces. For instance, self-maps defined on certain types of metric subspaces of the space of bounded functions are usually easy to manipulate. In particular, we have the following simple criterion for checking the contraction property for such maps. In the literature on dynamic programming, this result is sometimes referred to as *Blackwell's Contraction Lemma.*

LEMMA 3

(Blackwell) *Let T be a nonempty set, and X a nonempty subset of $\mathbf{B}(T)$ that is closed under addition by positive constant functions.*[44] *Assume that Φ is an increasing self-map on X.*[45] *If there exists a $0 < \delta < 1$ such that*

$$\Phi(f+\alpha) \leq \Phi(f) + \delta\alpha \qquad \textit{for all } (f,\alpha) \in X \times \mathbb{R}_+,$$

then Φ is a contraction.

PROOF

For any $f, g \in X$ we have $f(x) - g(x) \leq |f(x) - g(x)| \leq d_\infty(f,g)$ for all $x \in T$. Thus $f \leq g + d_\infty(f,g)$, and hence, by our hypotheses,

$$\Phi(f) \leq \Phi(g + d_\infty(f,g)) \leq \Phi(g) + \delta d_\infty(f,g)$$

for some $\delta \in (0,1)$. Interchanging the roles of f and g, we find $|\Phi(f) - \Phi(g)| \leq \delta d_\infty(f,g)$, which implies that $d_\infty(\Phi(f), \Phi(g)) \leq \delta d_\infty(f,g)$. Since f and g are arbitrary in X, this proves that Φ is a contraction. ■

[43] Formally, the claim is that, for any nonempty open subset O of $\mathbb{R}$, a differentiable function $f : O \to O$ with $\sup\{|f'(t)| : t \in O\} < 1$ is a contraction. (*Quiz.* Prove this by using the Mean Value Theorem (Exercise A.56).)

WARNING. Notice that the requirement on the derivative of f is stronger than the statement $-1 < f' < 1$. The latter statement is in fact not strong enough to guarantee that f is a contraction. (Right?)

[44] That is, $f \in X$ implies $f + \alpha \in X$ for any $\alpha > 0$.

[45] That is, $\Phi(X) \subseteq X$, and we have $\Phi(f) \geq \Phi(g)$ for any $f, g \in X$ with $f \geq g$.

Let us now look at some examples.

Example 12

[1] Let φ be any continuous function on $[0,1]^2$. Take any $0 < \delta < 1$ and define $\Phi : \mathbf{C}[0,1] \to \mathbb{R}^{[0,1]}$ as:

$$\Phi(f)(x) := \max\{\varphi(x,y) + \delta f(y) : 0 \leq y \leq 1\} \qquad \text{for all } 0 \leq x \leq 1.$$

This is actually a very important function, the variations of which play a crucial role in dynamic programming. (We will talk more about it later in Example 13.[1] and in Section E.4.) By a result we will prove in Section E.3 (*the Maximum Theorem*), Φ maps a continuous function to a continuous function, that is, it is a self-map on $\mathbf{C}[0,1]$. Moreover, a straightforward application of Lemma 3 shows that Φ is a contraction. (Verify!)

[2] Fix any $h \in \mathbf{C}[0,1]$, and take a continuous function of the form $\varphi : [0,1]^2 \to \mathbb{R}_+$. To focus on the nontrivial case, we assume that

$$M_\varphi := \max\{\varphi(x,y) : 0 \leq x, y \leq 1\} > 0.$$

Now let $\lambda \in \mathbb{R}$ and define $\Phi : \mathbf{C}[0,1] \to \mathbb{R}^{[0,1]}$ as:

$$\Phi(f)(x) := h(x) + \lambda \int_0^1 \varphi(x,t) f(t)dt \qquad \text{for all } 0 \leq x \leq 1.$$

It follows from the Riemann integration theory that $\Phi(f)$ is a continuous function on $[0,1]$ for any $f \in \mathbf{C}[0,1]$, that is, Φ is a self-map on $\mathbf{C}[0,1]$.[46] Furthermore, *provided that* $\lambda \geq 0$, we have $\Phi(f) \geq \Phi(g)$ for any $f, g \in \mathbf{C}[0,1]$ with $f \geq g$, that is, Φ is increasing. Moreover, for any $x \in [0,1]$ and $f \in \mathbf{C}[0,1]$, Proposition A.12 implies

$$\Phi(f+\alpha)(x) = \Phi(f)(x) + \left(\lambda \int_0^1 \varphi(x,t)dt\right)\alpha \leq \Phi(f)(x) + \lambda M_\varphi \alpha.$$

So, it follows from Lemma 3 that Φ is a contraction, *provided that* $0 \leq \lambda < \frac{1}{M_\varphi}$.

[46] I am sure you can easily prove this. There is a slight mischief here in that I did not yet talk about the continuity of a real function defined on $[0,1]^2$. But of course, you are familiar with this notion, and thus need not wait until the next chapter to supply a proof. (All you need from the Riemann integration theory is contained in Section A.4.3.)

In this case we can actually do a bit better if we try to determine conditions on λ that would make Φ a contraction directly. Indeed, for any $f, g \in \mathbf{C}[0,1]$, Proposition A.12 implies

$$d_\infty(\Phi(f), \Phi(g)) = \sup_{t\in[0,1]} \left| \lambda \int_0^1 \varphi(x,t)(f(t) - g(t))dt \right| \leq |\lambda| \, M_\varphi d_\infty(f,g).$$

Thus Φ is a contraction whenever $-\frac{1}{M_\varphi} < \lambda < \frac{1}{M_\varphi}$. □

Please note that the notion of contraction is intimately linked with the distance function underlying the metric space. It is possible that a self-map Φ on (X, d) is not a contraction, while it might be a contraction if we instead metrized X with some other metric (which may even be strongly equivalent to d). Thus, in applications, when you need a certain self-map to be a contraction, but know that this is not the case under the "natural" metric of your space, all hope is not lost. By a suitable remetrization of your space, you can "make" your self-map a contraction. In Section 7.2, we shall encounter a major instance in which this trick does wonders.

EXERCISE 48 Let $n \in \mathbb{N}$. For any $n \times n$ matrix $\mathbf{A} := [a_{ij}]_{n\times n}$ (Section A.1.6) and any $x \in \mathbb{R}^n$, we write $\mathbf{A}x$ for the n-vector $(\sum^n a_{1j}x_j, \ldots, \sum^n a_{nj}x_j)$. Give an example of an $n \times n$ matrix $\mathbf{A}$ such that the map $x \mapsto \mathbf{A}x$ is a contraction when $\mathbb{R}^n$ is metrized by d_1, but not so when it is metrized by d_∞.

6.2 The Banach Fixed Point Theorem

So, why are contractions of interest? Because, when it is defined on a complete metric space, a contraction must map a point to itself, that is, it must have a fixed point. In fact, more is true: it must have a *unique* fixed point. This is almost too good to be true. (You may not see any reason to cheer yet, but read on.) It is actually rather easy to give an intuition of this fact in the case of real functions—a geometric description of things is provided in Figure 4. The idea is to pick *any* point x in the domain of a given contraction $f \in \mathbb{R}^\mathbb{R}$, and observe the behavior of the sequence $(x, f(x), f(f(x)), \ldots)$. The contraction property tells us that this is a Cauchy sequence, and since $\mathbb{R}$ is complete, it must converge. But can it converge anywhere but *the* fixed point of f?

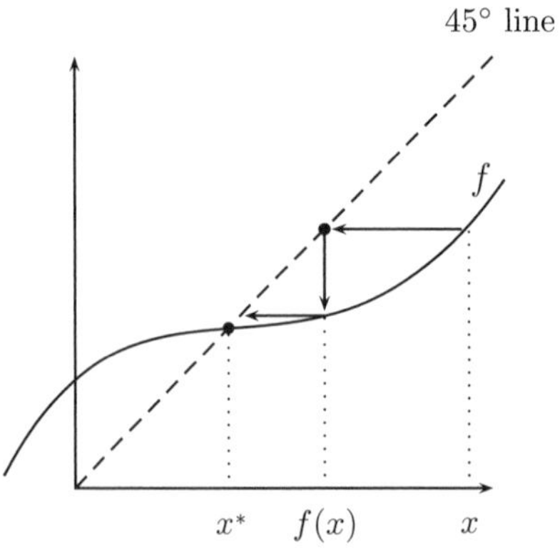

f is a contraction with the unique fixed point x^*.

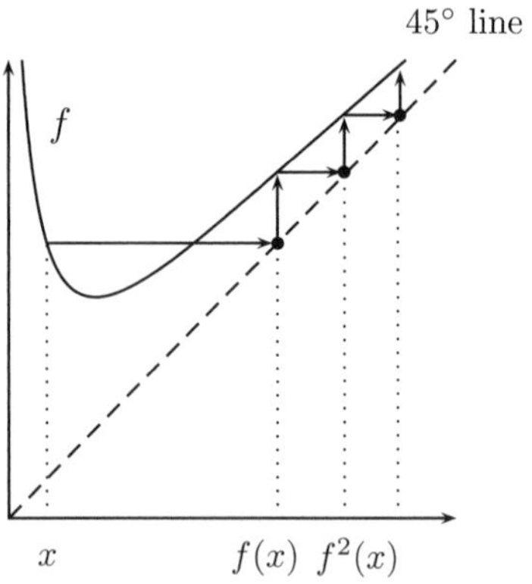

f is not a contraction, and it has no fixed points.

FIGURE 4

We now generalize this heuristic argument to the case of contractions defined on an arbitrary complete metric space. It should be noted that the method of proof used here, called the *method of successive approximations*, is almost as important as the main result itself.[47] At the very least, it gives a good idea about why the assertion is true.

> THE BANACH FIXED POINT THEOREM[48]
> *Let X be a complete metric space. If $\Phi \in X^X$ is a contraction, then there exists a unique $x^* \in X$ such that $\Phi(x^*) = x^*$.*

PROOF

Assume that $\Phi \in X^X$ is a contraction. Let us first prove the existence of a fixed point for Φ. Pick any $x^0 \in X$ and define $(x^m) \in X^\infty$ recursively as $x^{m+1} := \Phi(x^m)$, $m = 0, 1, \ldots$. We claim that this sequence is Cauchy. To see this, let K be the contraction coefficient of Φ, and notice that, by the

[47] This method dates back apparently even to the days of Cauchy.

[48] This theorem was proved by Stefan Banach, the great founder of the Polish school of functional analysis, in his 1920 dissertation. You will also come across several other results that also bear the name of this extraordinary man who is said to have never been able to bring himself to write down his ideas in the form of a paper (this includes his dissertation), but instead dictated them to his assistants and later edited the outcome. For this reason it is feared that a great many discoveries of Banach may have been lost for lack of proper recording.

Principle of Mathematical Induction, we have $d(x^{m+1}, x^m) \leq K^m d(x^1, x^0)$ for all $m = 1, 2, \ldots$, where d is the metric of X. Thus, for any $k > l+1$,

$$\begin{aligned} d(x^k, x^l) &\leq d(x^k, x^{k-1}) + \cdots + d(x^{l+1}, x^l) \\ &\leq (K^{k-1} + \cdots + K^l) d(x^1, x^0) \\ &= \frac{K^l(1-K^{k-l})}{1-K} d(x^1, x^0) \end{aligned}$$

so that $d(x^k, x^l) < \frac{K^l}{1-K} d(x^1, x^0)$. That (x^m) is Cauchy follows readily from this inequality. (Right?)

Since X is complete, (x^m) must then converge to a point in X, say x^*. (This is our fixed point.) Then, for any $\varepsilon > 0$, there must exist an $M \in \mathbb{N}$ such that $d(x^*, x^m) < \frac{\varepsilon}{2}$ for all $m = M, M+1, \ldots$, and hence

$$\begin{aligned} d(\Phi(x^*), x^*) &\leq d(\Phi(x^*), x^{m+1}) + d(x^{m+1}, x^*) \\ &= d(\Phi(x^*), \Phi(x^m)) + d(x^{m+1}, x^*) \\ &\leq K d(x^*, x^m) + d(x^{m+1}, x^*) \\ &< \tfrac{\varepsilon}{2} + \tfrac{\varepsilon}{2}. \end{aligned}$$

Since $\varepsilon > 0$ is arbitrary here, we must then have $d(\Phi(x^*), x^*) = 0$, which is possible only if $\Phi(x^*) = x^*$.

To prove the uniqueness assertion, observe that if $x \in X$ was another fixed point of Φ, we would then have $d(x, x^*) = d(\Phi(x), \Phi(x^*)) \leq K d(x, x^*)$, which is possible only if $x = x^*$ (since $K < 1$). ■

EXERCISE 49 Let X be a complete metric space, and Φ a surjective self-map on X such that there exists a real number $\alpha > 1$ with

$$d(\Phi(x), \Phi(y)) \geq \alpha d(x, y) \qquad \text{for all } x, y \in X.$$

Prove that Φ has a unique fixed point.

EXERCISE 50 [H] Let X be a metric space. A self-map Φ on X is said to be a **pseudocontraction** if $d(\Phi(x), \Phi(y)) < d(x, y)$ holds for all distinct $x, y \in X$. The present exercise compares this property with that of being a contraction.

(a) Show that the word "contraction" cannot be replaced with "pseudocontraction" in the statement of the Banach Fixed Point Theorem.

(b) If $\Phi \in X^X$ is a pseudocontraction, then $(d(\Phi^{m+1}(x), \Phi^m(x)))$ is a decreasing sequence, and hence converges. (Here $\Phi^1 := \Phi$ and $\Phi^{m+1} := \Phi \circ \Phi^m$, $m = 1, 2, \ldots$.) Use this to show that if $(\Phi^m(x))$ has a convergent subsequence, then $d(x^*, \Phi(x^*)) = d(\Phi(x^*), \Phi^2(x^*))$ for some $x^* \in X$.

(c) Prove *Edelstein's Fixed Point Theorem*: If X is a compact metric space and $\Phi \in X^X$ a pseudocontraction, then there exists a unique $x^* \in X$ such that $\Phi(x^*) = x^*$. Moreover, we have $\lim \Phi^m(x) = x^*$ for any $x \in X$.

EXERCISE 51[H] Let X be a complete metric space, and (Φ_m) a sequence of contractive self-maps on X such that $\sup\{K_m : m \in \mathbb{N}\} < 1$, where K_m is the contraction coefficient of Φ_m, $m = 1, 2, \ldots$. By the Banach Fixed Point Theorem, Φ_m has a unique fixed point, say x_m. Show that if

$$\sup\{d(\Phi_m(x), \Phi(x)) : x \in X\} \to 0$$

for some $\Phi \in X^X$, then Φ is a contraction with the unique fixed point $\lim x_m$. (Would this be true if all we knew was that $d(\Phi_m(x), \Phi(x)) \to 0$ for every $x \in X$?)

The Banach Fixed Point Theorem, also called **the Contraction Mapping Theorem**, is often used to establish that a unique solution to a given equation exists. To illustrate this, suppose that we are interested in solving an equation like

$$x = g(x) + a,$$

where $(g, a) \in \mathbb{R}^{\mathbb{R}} \times \mathbb{R}$. If g is nonlinear, it may not be a trivial matter to determine if a solution to this equation exists. But if we know that g is a contraction, then we are assured that there exists a unique solution to our equation. Indeed, defining the self-map f on $\mathbb{R}$ by $f(t) := g(t) + a$, it is readily observed that f would in this case be a contraction as well. Moreover, x solves our equation iff $x = f(x)$, that is, x is a fixed point of f. But, by the Banach Fixed Point Theorem, f has a unique fixed point. Therefore, our equation must have a unique solution. For example, there is a unique $x \in \mathbb{R}$ such that $x = \frac{1}{2}\cos x + 1$, because the map $t \mapsto \frac{1}{2}\cos t$ is a contraction on $\mathbb{R}$. (Right?)

Alternatively, we may be interested in solving an equation like

$$g(x) = \mathbf{0},$$

where g is a self-map on a complete metric subspace X of a Euclidean space $\mathbb{R}^n$ such that $\mathbf{0} \in X$. (Here $\mathbf{0}$ is the n-vector $(0, \ldots, 0)$, of course.) An indirect way of checking if there is a unique solution to this equation is to check if $\mathrm{id}_X - g$ is a contraction. In fact, there is quite a bit of leeway in how we may choose to reduce the problem of solving this equation to a fixed point problem. For instance, if h is an injective self-map on X with $h(\mathbf{0}) = \mathbf{0}$, then $x \in X$ is a fixed point of $\mathrm{id}_X - h \circ g$ iff $g(x) = \mathbf{0}$.

To illustrate, take any $0 < a < b$, and suppose we wish to find out if there is a nonnegative real number x such that $g(x) = 0$, where g is a differentiable self-map on $\mathbb{R}_+$ with $a \leq g' \leq b$. Clearly, $\mathrm{id}_{\mathbb{R}_+} - g$ need not be a contraction. But $\mathrm{id}_{\mathbb{R}_+} - \frac{1}{a+b} g$ is a contraction, so by the Banach Fixed Point Theorem, we may conclude that there exists a unique $x \geq 0$ with $g(x) = 0$. (We will encounter more interesting examples of this sort in Section 7.)

6.3* Generalizations of the Banach Fixed Point Theorem

The literature provides numerous generalizations of the Banach Fixed Point Theorem.[49] Some of these establish the claim in a more general context than complete metric spaces, while others apply to some noncontractive maps as well. In this section we note a few generalizations of the latter type. Two further generalizations will be considered in Chapters D and E.

Let us first recall the notion of *iteration* of a self-map.

> DEFINITION
> Let X be a nonempty set and Φ a self-map on X. We let $\Phi^1 := \Phi$, and define $\Phi^{m+1} := \Phi \circ \Phi^m$ for any $m = 1, 2, \ldots$. The self-map Φ^m is called the m**th iteration** of Φ.

[49] To get an overall view of the related field (which is called *metric fixed point theory*), I should mention that Smart (1974) and Dugundji and Granas (1982) are classic references, while Goebel and Kirk (1990) and Agarwal, Meehan, and O'Regan (2001) provide more recent accounts. (I would start with Smart (1974), if I were you.)

Here is an immediate reason why iterations are of great interest in the present context.

COROLLARY 2
Let X be a complete metric space. For any self-map Φ on X such that Φ^m is a contraction for some $m \in \mathbb{N}$, there exists a unique $x^* \in X$ such that $\Phi(x^*) = x^*$.

PROOF
By the Banach Fixed Point Theorem, Φ^m has a unique fixed point in X, say x^*. Then $\Phi^m(x^*) = x^*$, so

$$\Phi(x^*) = \Phi(\Phi^m(x^*)) = \Phi^{m+1}(x^*) = \Phi^m(\Phi(x^*)),$$

that is, $\Phi(x^*)$ is a fixed point of Φ^m. Since Φ^m has a unique fixed point, however, we must have $\Phi(x^*) = x^*$. The uniqueness part is an easy exercise. ∎

This observation sometimes does the job of the Banach Fixed Point Theorem when the latter is not applicable. For instance, suppose we wish to find out if the equation $x = \cos x$ has a unique real solution. Since $t \mapsto \cos t$ is not a contraction (why not?), we cannot use the Banach Fixed Point Theorem to answer this question. But it is relatively easy to verify that $t \mapsto \cos(\cos(t))$ is actually a contraction. (Differentiate and see!) Thus, by Corollary 2, the said equation has a unique real solution.

As another example, consider the self-map Φ on $\mathbf{C}[0, 1]$ defined by

$$\Phi(f)(x) := \int_0^x f(t)dt \qquad \text{for all } 0 \leq x \leq 1.$$

It is easily checked that Φ is not a contraction. Yet Φ^2 is a contraction, because, for any $0 \leq y \leq 1$, we have

$$\begin{aligned}\left|\Phi^2(f)(y) - \Phi^2(g)(y)\right| &\leq \int_0^y \int_0^x \left|f(t) - g(t)\right| dtdx \\ &\leq d_\infty(f, g) \int_0^y xdx \\ &\leq \tfrac{1}{2} d_\infty(f, g).\end{aligned}$$

Therefore, by Corollary 2, Φ has a unique fixed point. (What is this fixed point?)

A significant generalization of Corollary 2 is recently obtained by Merryfield and Stein, Jr. (2002). Although the (combinatorial) techniques needed to prove this generalization are beyond the scope of the present course, the statement of this powerful result is easy enough.

The Generalized Banach Fixed Point Theorem 1 (Merryfield-Stein Jr.) *Let X be a complete metric space and Φ a self-map on X. If there exists a $(K, M) \in (0, 1) \times \mathbb{N}$ such that*

$$\min\{d(\Phi^m(x), \Phi^m(y)) : m = 1, \ldots, M\} \leq Kd(x, y) \qquad \textit{for all } x, y \in X,$$

then there exists an $x^* \in X$ such that $\Phi(x^*) = x^*$.[50]

Another important generalization of the Banach Fixed Point Theorem is the following.

The Generalized Banach Fixed Point Theorem 2 (Matkowski) *Let X be a complete metric space, and $f : \mathbb{R}_+ \to \mathbb{R}_+$ an increasing function with $\lim f^m(t) = 0$ for all $t \geq 0$. If $\Phi \in X^X$ satisfies*

$$d(\Phi(x), \Phi(y)) \leq f(d(x, y)) \qquad \textit{for all } x, y \in X,$$

then there exists a unique $x^* \in X$ such that $\Phi(x^*) = x^*$.[51]

Defining $f : \mathbb{R}_+ \to \mathbb{R}_+$ by $f(t) := Kt$, where K is an arbitrary, but fixed, number in $(0, 1)$, and checking the statements of the associated results, it becomes clear that the above result is a genuine generalization of the Banach Fixed Point Theorem. Its proof is relegated to the next chapter. (See Example D.3.[5].)

[50] If you know what it means for a function defined on a metric space to be continuous, then you can readily verify that any self-map that satisfies the requirement of Corollary 2 must be continuous. It is thus remarkable that the Generalized Banach Fixed Point Theorem 1 applies to self-maps that are not necessarily continuous. Indeed, Jachymski, Schroder, and Stein Jr. (1999) provide an example that shows that a self-map that satisfies the hypothesis of the theorem (with $M = 3$) need not be continuous.

[51] For a closely related result in which the monotonicity of f is replaced by a continuity property, see Boyd and Wong (1969).

*REMARK 2 (*A Characterization of Completeness*) Does the Banach Fixed Point Theorem hold in a metric space that is not complete? Interestingly, the answer to this question is no, that is, this theorem is essentially a characterization of the completeness of a metric space. To make things precise, let us define the following property for a metric space X:

THE BANACH FIXED POINT PROPERTY. Every contraction $\Phi \in S^S$, where S is a nonempty closed subset of X, has a fixed point.

From Corollary 1 and the Banach Fixed Point Theorem, we know that every complete metric space possesses the Banach Fixed Point Property. In fact, completeness is not only sufficient but also necessary for a space to have this property.[52] This is because the terms of a nonconvergent Cauchy sequence must constitute a closed set, and we can always define a contraction on this set without a fixed point. This idea is formalized below.

CLAIM. A metric space is complete iff it has the Banach Fixed Point Property.

PROOF OF CLAIM
Suppose X is a metric space in which there exists a Cauchy sequence (x^m) that does not converge. We may assume that all terms of (x^m) are distinct. (Otherwise, we would work with a subsequence that has this property). Define $\theta \in \mathbb{R}_+^X$ by

$$\theta(x) := \inf\{d(x, x^m) : x^m \neq x \text{ for each } m \in \mathbb{N}\}.$$

Clearly, if $\theta(x) = 0$, then there exists a subsequence of $(d(x, x^m))$ that converges to 0, and hence, there exists a subsequence of (x^m) that converges to x, contradicting that (x^m) is a Cauchy sequence that does not converge in X (Proposition 6). Thus we have $\theta(x) > 0$ for all $x \in X$. Now pick any $0 < K < 1$, and construct the subsequence (x^{m_r}) recursively as follows: $x^{m_1} = x^1$ and $x^{m_{r+1}}$ is the first (or any) term after x^{m_r} such that $d(x^k, x^l) \leq K\theta(x^{m_r})$ for all $k, l \geq m_{r+1}$. (We can find such a subsequence because (x^m) is Cauchy and $\theta > 0$.) Now define $S := \{x^{m_r} : r \in \mathbb{N}\}$, and note that S is a closed subset of X (since (x^m) does not have any convergent subsequence). Define next the self-map Φ on S by $\Phi(x^{m_r}) := x^{m_{r+1}}$. This

[52] To my knowledge, Hu (1967) was the first to make note of this fact.

function is easily checked to be a contraction without a fixed point. (Please check this—the logic of the argument will then get quite clear.) Conclusion: X does not have the Banach Fixed Point Property. || □

7 Applications to Functional Equations

7.1 Solutions of Functional Equations

This section provides several examples that illustrate how the Banach Fixed Point Theorem can be used to prove the existence and uniqueness of a solution to a functional equation. Given our work in the previous section, the first two examples require almost no effort.

EXAMPLE 13

[1] Adopting the definitions and notation of Example 12.[1], consider the following equation:

$$f(x) = \max\{\varphi(x, y) + \delta f(y) : 0 \leq y \leq 1\} \quad \text{for all } 0 \leq x \leq 1. \tag{4}$$

This equation is called **Bellman's functional equation** (and is a special case of a more general form that we shall study in Section E.4). By a "solution" to this equation, we mean a function $f \in \mathbf{C}[0,1]$ that satisfies (4). Put differently, f solves this equation iff $f = \Phi(f)$, where the self-map Φ is defined on $\mathbf{C}[0,1]$ as in Example 12.[1]. But we already know that Φ is a contraction and $\mathbf{C}[0,1]$ is a complete metric space (Exercise 40). Thus, it follows from the Banach Fixed Point Theorem that our functional equation has a unique solution.

[2] Adopting the definitions and notation of Example 12.[2], consider the following equation:

$$f(x) = h(x) + \lambda \int_0^1 \varphi(x, t) f(t) dt \quad \text{for all } 0 \leq x \leq 1. \tag{5}$$

This equation is called **Fredholm's second (linear) integral equation.**

Question: Under what conditions on h, φ and λ does Fredholm's second integral equation admit a solution (that is, there exists an $f \in \mathbf{C}[0,1]$ that satisfies (5))?

Now if $\varphi = 0$, then all is trivial; in that case, setting $f := h$ yields the unique solution to (5). The situation is much more interesting when $\varphi > 0$. After all, in that case too there is a unique solution to Fredholm's second integral equation, at least when λ *is small enough*. Indeed, a solution to (5) exists iff the self-map Φ, defined on $\mathbf{C}[0,1]$ as in Example 12.[2], has a fixed point. But we have seen in that example that Φ is a contraction whenever $|\lambda| < \frac{1}{M_\varphi}$, where $M_\varphi := \max \varphi([0,1]^2)$. Thus, the Banach Fixed Point Theorem tells us that, for any $-\frac{1}{M_\varphi} < \lambda < \frac{1}{M_\varphi}$, there is a unique $f \in \mathbf{C}[0,1]$ that satisfies (5).[53] □

EXERCISE 52^H For any $n \in \mathbb{N}$, let f_i be a contraction on $\mathbb{R}$, $i = 1, \ldots, n$. Show that there exists a unique $(x_1, \ldots, x_n) \in \mathbb{R}^n$ such that $x_1 = f_1(x_n)$ and $x_i = f_i(x_{i-1})$ for all $i = 2, \ldots, n$.

EXERCISE 53^H Let $n \in \mathbb{N}$, and take an $n \times n$ matrix $\mathbf{A} := [a_{ij}]_{n \times n}$. Recall that, for any $x \in \mathbb{R}^n$, we write $\mathbf{A}x$ for the n-vector $(\sum^n a_{1j}x_j, \ldots, \sum^n a_{nj}x_j)$. Let f be a differentiable self-map on $\mathbb{R}$ such that $s := \sup\{|f'(t)| : t \in \mathbb{R}\} < \infty$, and define the self-map F on $\mathbb{R}^n$ by $F(x) := (f(x_1), \ldots, f(x_n))$. Finally, fix an n-vector w. We are interested in solving the following system of nonlinear equations:

$$z = \mathbf{A}F(z) + w. \tag{6}$$

(a) Show that there is a unique $z \in \mathbb{R}^n$ that satisfies (6) if $\max\{\sum_{j=1}^n |a_{ij}| : i = 1, \ldots, n\} < \frac{1}{s}$.
(b) Show that there is a unique $z \in \mathbb{R}^n$ that satisfies (6) if $\sum_{i=1}^n \sum_{j=1}^n |a_{ij}| < \frac{1}{s^2}$.

EXERCISE 54 Let $(a_{kl}) \in \mathbb{R}^{\infty \times \infty}$ be a double real sequence such that $\sup\{\sum_{j=1}^\infty |a_{ij}| : i \in \mathbb{N}\} < 1$. Show that a bounded real sequence (x_m) satisfies

$$x_i = \sum_{j=1}^\infty a_{ij}x_j, \qquad i = 1, 2, \ldots,$$

iff $(x_m) = (0, 0, \ldots)$.

[53] Wow! How would you try to solve this problem if you didn't know the Banach Fixed Point Theorem?

EXERCISE 55 Let h be a self-map on $\mathbb{R}_+$, and let $H : \mathbb{R}_+ \times \mathbb{R} \to \mathbb{R}$ be a bounded function such that there exists a $K \in (0, 1)$ with

$$\left|H(x, y) - H(x, z)\right| < K \left|y - z\right| \quad \text{for all } x \geq 0 \text{ and } y, z \in \mathbb{R}.$$

Prove that there exists a unique bounded function $f : \mathbb{R}_+ \to \mathbb{R}$ such that

$$f(x) = H(x, f(h(x))) \quad \text{for all } x \geq 0.$$

EXERCISE 56[H] Prove that there is a unique $f \in \mathbf{C}[0, 1]$ such that $f \geq 0$ and

$$f(x) = 1 + \tfrac{3}{4} \ln \left(1 + \int_0^x f(t)dt\right) \qquad \text{for all } 0 \leq x \leq 1.$$

EXERCISE 57 Prove that there is a unique continuous and bounded real function f on $\mathbb{R}_+$ such that

$$f(x) = 1 + \int_0^x e^{-t^2} f(xt)dt \qquad \text{for all } x \geq 0.$$

EXERCISE 58 (*The Linear Volterra Integral Equation*) Let λ be an arbitrary real number, and consider the equation

$$f(x) = h(x) + \lambda \int_0^x \varphi(x, t) f(t)dt \qquad \text{for all } 0 \leq x \leq 1,$$

where $h \in \mathbf{C}[0, 1]$ and $\varphi \in \mathbf{C}([0, 1]^2)$.

(a) Using the Banach Fixed Point Theorem, show that there exists a unique $f \in \mathbf{C}[0, 1]$ that satisfies this equation, provided that λ is small.

(b) Using Corollary 2, show that, for any $\lambda \in \mathbb{R}$, there exists a unique $f_\lambda \in \mathbf{C}[0, 1]$ that satisfies this equation.

*EXERCISE 59 (Nirenberg) Consider the integro-differential equation

$$4f'(x) + \sin f(x) + \int_{x/2}^x (1 + f(t)^2) \sin t dt = 0 \qquad \text{for all } 0 \leq x \leq 1.$$

Prove that there exists a unique $f \in \mathbf{C}^1[0, 1]$ that satisfies this equation, $f(0) = 0$, and $\left|f\right| \leq 1$.

The Banach Fixed Point Theorem plays an important role also in the theory of differential equations.

EXAMPLE 14

Let $H : \mathbb{R} \to \mathbb{R}$ be a continuous function such that, for some $\alpha > 0$,

$$\left|H(y) - H(z)\right| \leq \alpha \left|y - z\right| \tag{7}$$

for any real numbers y and z. (You might recall that such a function is called *Lipschitz continuous*.) Fix any x_0, y_0 in $\mathbb{R}$. We wish to prove the following.

CLAIM. The boundary condition $y_0 = f(x_0)$ and the differential equation $f'(x) = H(f(x))$ are simultaneously satisfied on some interval $I_\delta := [x_0 - \delta, x_0 + \delta]$ (with $\delta > 0$) by a unique differentiable $f : I_\delta \to \mathbb{R}$.

In this sense, we say that the said differential equation (with the boundary condition) has a unique *local solution*.[54]

How would one prove such a thing? The main idea behind the proof comes from two relatively elementary observations. First, we note that a differentiable $f : I_\delta \to \mathbb{R}$ satisfies $y_0 = f(x_0)$ and $f'(x) = H(f(x))$ for all $x \in I_\delta$ iff $f \in \mathbf{C}^1(I_\delta)$ and

$$f(x) = y_0 + \int_{x_0}^{x} H(f(t))dt, \qquad x_0 - \delta \leq x \leq x_0 + \delta, \tag{8}$$

and this by the Fundamental Theorem of Calculus. (Yes?) Second, we observe that we may be able to solve this integral equation by using the same method we used in Example 13.[2], that is, by using the Banach Fixed Point Theorem. It turns out that this strategy works out beautifully here.

Define the self-map Φ on $\mathbf{C}(I_\delta)$ by

$$\Phi(f)(x) := y_0 + \int_{x_0}^{x} H(f(t))dt, \qquad x_0 - \delta \leq x \leq x_0 + \delta,$$

[54] As we shall see in Chapter J, (7) can actually be relaxed to ordinary continuity if one is interested only in the *existence* of a local solution. For *uniqueness*, however, this requirement is crucial. Indeed, both of the functions $f_1(x) := 0$ and $f_2(x) := x^2$ (defined on $\mathbb{R}_+$) satisfy the differential equation $f'(x) = 2\sqrt{f(x)}$ for all $x \geq 0$ along with the boundary equation $f(0) = 0$. (This doesn't violate the claim I just stated, because the map $t \mapsto 2\sqrt{t}$ is not Lipschitz continuous on any interval of the form $[0, a]$.)

and observe that solving (8) is tantamount to finding a fixed point of Φ. (*Note.* The fact that $\Phi(\mathbf{C}(I_\delta)) \subseteq \mathbf{C}(I_\delta)$ follows, again, from the Fundamental Theorem of Calculus.) To this end, we wish to show that δ can be chosen small enough to ensure that Φ is a contraction. (This is why the solution we will obtain will only be a *local* one.[55]) In fact, finding a condition on δ that would make Φ a contraction is easy. For any $f, g \in \mathbf{C}(I_\delta)$, we have

$$\left|\Phi(f)(x) - \Phi(g)(x)\right| = \left|\int_{x_0}^{x} (H(f(t)) - H(g(t)))dt\right| \leq \int_{x_0}^{x} \left|H(f(t)) - H(g(t))\right| dt$$

for each $x \in I_\delta$ (Exercise A.60). Therefore, by (7), for any $x \in I_\delta$,

$$\left|\Phi(f)(x) - \Phi(g)(x)\right| \leq \alpha \int_{x_0}^{x} \left|f(t) - g(t)\right| dt \leq \alpha d_\infty(f, g)\,|x - x_0| \leq \alpha\delta d_\infty(f, g),$$

which yields $d_\infty(\Phi(f), \Phi(g)) \leq \alpha\delta d_\infty(f, g)$. Thus, Φ is a contraction whenever $0 < \delta < \frac{1}{\alpha}$. Conclusion: There is a unique function $f \in \mathbf{C}(I_\delta)$ that satisfies (8) and the boundary condition $y_0 = f(x_0)$, provided that $0 < \delta < \frac{1}{\alpha}$. As noted earlier, this is all we needed to prove our main claim. □

7.2 Picard's Existence Theorems

The observation noted in Example 14 generalizes to a major theorem on the existence and uniqueness of a solution of a general ordinary differential equation. This theorem is of foundational importance for the theory of differential equations, and nicely illustrates the power of real analysis as we have covered it so far. It was proved by Emile Picard in 1890.

[55] What of a *global* solution to our differential equation (with the boundary condition)? It exists and it is unique, believe it or not. Read on!

Picard's Existence Theorem

(The Local Version) *Let* $-\infty < a < b < \infty$, *and* $(x_0, y_0) \in (a, b)^2$. *If* $H : [a, b]^2 \to \mathbb{R}$ *is continuous, and*

$$\left|H(x, y) - H(x, z)\right| \leq \alpha \left|y - z\right|, \qquad a \leq x, y, z \leq b,$$

for some $\alpha > 0$, *then there is a* $\delta > 0$ *such that there exists a unique differentiable real function* f *on* $[x_0 - \delta, x_0 + \delta]$ *with* $y_0 = f(x_0)$ *and*

$$f'(x) = H(x, f(x)), \qquad x_0 - \delta \leq x \leq x_0 + \delta.$$

Since the proof of this theorem parallels the argument given in Example 14 closely, we leave it here as an exercise.

Exercise 60 Prove Picard's Existence Theorem. Here is some help.

(a) Let $I_\delta := [x_0 - \delta, x_0 + \delta]$ for any $\delta > 0$, and reduce the problem to finding a $\delta > 0$ and an $f \in \mathbf{C}(I_\delta)$ such that

$$f(x) = y_0 + \int_{x_0}^{x} H(t, f(t))dt$$

for all $x \in I_\delta$.

(b) Let $\beta := \max\{\left|H(x, y)\right| : a \leq x, y \leq b\}$. (You may assume that β is well-defined.) Choose $\delta > 0$ such that

$$\delta < \tfrac{1}{\alpha}, \qquad I_\delta \subseteq [a, b] \qquad \text{and} \qquad [y_0 - \beta\delta, y_0 + \beta\delta] \subseteq [a, b].$$

Let C_δ denote the set of all continuous real maps f defined on I_δ that satisfy $\left|f(x) - y_0\right| \leq \beta\delta$. Prove that (C_δ, d_∞) is a complete metric space.

(c) Define the operator $\Phi : C_\delta \to \mathbb{R}^{I_\delta}$ by

$$\Phi(f)(x) := y_0 + \int_{x_0}^{x} H(t, f(t))dt.$$

First show that Φ is a self-map, and then verify that it is a contraction.

(d) Use step (a) and the Banach Fixed Point Theorem to complete the proof.

As impressive as it is, there is a major shortcoming to Picard's Existence Theorem. This result assures the existence and uniqueness of a solution only *locally*. Since in applications one often needs to solve a differential equation globally, this theorem provides us with only partial comfort. Yet the following result shows there is little reason to worry.

Picard's Existence Theorem

(The Global Version) *Let* $-\infty < a < b < \infty$, *and* $(x_0, y_0) \in [a, b] \times \mathbb{R}$. *If* $H : [a, b] \times \mathbb{R} \to \mathbb{R}$ *is continuous, and*

$$\left|H(x, y) - H(x, z)\right| \leq \alpha \left|y - z\right|, \qquad a \leq x \leq b, \ -\infty < y, z < \infty,$$

for some $\alpha > 0$, *then there exists a unique differentiable real function* f *on* $[a, b]$ *with* $y_0 = f(x_0)$ *and*

$$f'(x) = H(x, f(x)), \qquad a \leq x \leq b.$$

Once again, the problem here is to prove that there is a unique $f \in \mathbf{C}[a, b]$ such that

$$f(x) = y_0 + \int_{x_0}^{x} H(t, f(t))dt$$

for all $a \leq x \leq b$. Unfortunately, however, our usual contraction mapping argument is not readily applicable, because we have no assurance here that the self-map Φ on $\mathbf{C}[a, b]$, defined by

$$\Phi(f)(x) := y_0 + \int_{x_0}^{x} H(t, f(t))dt, \tag{9}$$

is a contraction, that is, we don't know if it satisfies

$$d_\infty(\Phi(f), \Phi(g)) \leq K d_\infty(f, g)$$

for some $0 < K < 1$. (The "idea" of Exercise 60 was to argue that Φ is a contraction on $\mathbf{C}[x_0 - \delta, x_0 + \delta]$ for some "small" $\delta > 0$.) But do we have to metrize the set of all continuous real maps on $[a, b]$ by the sup-metric? Can't we find a metric on $\mathbf{C}[a, b]$ that would make it a complete metric space and Φ a contraction? After all, the Banach Fixed Point Theorem doesn't care which metric one uses on the subject space so long as we have completeness and the contraction property. Aha!

It turns out that our problem is easily solved if we metrized the set of all continuous real maps on $[a, b]$ by a weighted version of the sup-metric. In particular, we may use the map $D : \mathbf{C}[a, b] \times \mathbf{C}[a, b] \to \mathbb{R}_+$ defined by

$$D(f, g) := \sup \left\{ e^{-\alpha(t-x_0)} \left| f(t) - g(t) \right| : a \le t \le b \right\}.$$

OBSERVATION 1. D is a metric on $\mathbf{C}[a, b]$ such that

$$e^{-\alpha(b-x_0)} d_\infty(f, g) \le D(f, g) \le e^{-\alpha(a-x_0)} d_\infty(f, g) \quad \text{for all } f, g \in \mathbf{C}[a, b]. \tag{10}$$

That D is indeed a distance function is proved by imitating the argument we gave in Example 1.[5]. The proof of (10) is also very easy—just recall that $t \mapsto e^{-\alpha(t-x_0)}$ is a strictly decreasing map on $\mathbb{R}$. Thus D is a metric on $\mathbf{C}[a, b]$ that is strongly equivalent to the sup-metric. It follows that if we metrize $\mathbf{C}[a, b]$ by D, we obtain a complete metric space (Example 11.[6].) Let us denote this metric space by X.

OBSERVATION 2. The map $\Phi : X \to \mathbb{R}^X$ defined by (9) is a contractive self-map.

PROOF
That Φ is a self-map on X follows from the Fundamental Theorem of Calculus. To see the contraction property, notice that, for any $a \le x \le b$, we have

$$\begin{aligned} & e^{-\alpha(x-x_0)} \left| \Phi(f)(x) - \Phi(g)(x) \right| \\ & \quad \le e^{-\alpha(x-x_0)} \int_{x_0}^{x} \alpha \left| f(t) - g(t) \right| e^{\alpha(t-x_0)} e^{-\alpha(t-x_0)} dt \\ & \quad \le e^{-\alpha(x-x_0)} D(f, g) \int_{x_0}^{x} \alpha e^{\alpha(t-x_0)} dt. \end{aligned}$$

But a little calculus shows that $\int_{x_0}^{x} \alpha e^{\alpha(t-x_0)} dt = e^{\alpha(x-x_0)} - 1$, so

$$\begin{aligned} e^{-\alpha(x-x_0)} \left| \Phi(f)(x) - \Phi(g)(x) \right| & \le e^{-\alpha(x-x_0)} D(f, g) \left(e^{\alpha(x-x_0)} - 1 \right) \\ & = (1 - e^{-\alpha(x-x_0)}) D(f, g) \\ & \le (1 - e^{-\alpha(b-x_0)}) D(f, g), \end{aligned}$$

for any $x \in [a, b]$. Therefore, we conclude that

$$D(\Phi(f), \Phi(g)) \leq (1 - e^{-\alpha(b-x_0)})D(f,g).$$

Since $0 \leq 1 - e^{-\alpha(b-x_0)} < 1$, we are done. □

This is it! We now know that the map Φ defined by (9) is a contractive self-map on X, whereas X is a complete metric space. By the Banach Fixed Point Theorem Φ has a unique fixed point, and the global version of Picard's Existence Theorem is proved.[56]

Exercise 61 Let $H : \mathbb{R} \to \mathbb{R}$ be a continuous function such that

$$\left|H(y) - H(z)\right| \leq \alpha \left|y - z\right|, \qquad -\infty < y, z < \infty,$$

for some $\alpha > 0$. Prove that, given any real numbers x_0 and y_0, there exists a unique differentiable $f : \mathbb{R} \to \mathbb{R}$ such that $y_0 = f(x_0)$ and $f'(x) = H(f(x))$ for all $x \in \mathbb{R}$. (Compare with Example 14.)

Exercise 62 (Körner) Let $H : \mathbb{R}^2 \to \mathbb{R}$ be a continuous function such that, for any $c > 0$, there exists an $\alpha_c \geq 0$ with

$$\left|H(x, y) - H(x, z)\right| \leq \alpha_c \left|y - z\right|, \qquad -c \leq x \leq c, \ -\infty < y, z < \infty.$$

Prove that, given any real numbers x_0 and y_0, there exists a unique differentiable $f : \mathbb{R} \to \mathbb{R}$ such that $y_0 = f(x_0)$ and $f'(x) = H(x, f(x))$ for all $x \in \mathbb{R}$.

[56] Hopefully, this discussion shows you how powerful real analysis can be at times. In case you are wondering how on earth one could come up with a metric like D, let me tell you. Note first that, for any continuous and strictly decreasing $\vartheta : [a, b] \to \mathbb{R}_{++}$, the real map D_ϑ defined on $\mathbf{C}[a, b] \times \mathbf{C}[a, b]$ by

$$D_\vartheta(f, g) := \sup\left\{\vartheta(t)\left|f(t) - g(t)\right| : a \leq t \leq b\right\},$$

is a metric that is strongly equivalent to d_∞. (We have $\vartheta(b)d_\infty \leq D_\vartheta \leq \vartheta(a)d_\infty$.) Thus, for any such ϑ, metrizing $\mathbf{C}[a, b]$ with D_ϑ yields a complete metric space. It remains to choose ϑ in order to ensure that Φ is a contraction. Go through the argument given in the proof of Observation 2 as long as you can with an arbitrary ϑ. You will then realize that the ϑ that you want to pick should satisfy $\int_{x_0}^{x} \frac{\vartheta(x)}{\vartheta(t)} dt < \frac{1}{\alpha}$ for every $a \leq x \leq b$. The integral on the left-hand side tells me that if ϑ is of the exponential form, then I may be able to evaluate it. So why not try the map $t \mapsto e^{pt+q}$ for appropriately chosen real numbers p and q? The rest is easy.

8 Products of Metric Spaces

In this section we consider the issue of metrizing the Cartesian product of finitely, or countably infinitely, many nonempty sets, when each of these sets is a metric space in its own right. This issue will arise in a variety of contexts later on, so it makes sense to settle it here once and for all.

8.1 Finite Products

Take any $n \in \mathbb{N}$, and let (X_i, d_i) be a metric space, $i = 1, \ldots, n$. We wish to make the n-fold product $X := \mathsf{X}^n X_i$ a metric space in a way that is *consistent* with the metrics d_i in the sense that the metric imposed on X agrees with each d_i on any set of the form $\{(x^1, \ldots, x^n) \in X : x^i \in S_i\}$ where $\emptyset \neq S_i \subseteq X_i$ and x^j is an arbitrary point in X_j, $j \neq i$. To be more precise, suppose we decide to endow X with a metric ρ. For any given $x \in X$, this metric induces a metric $D_1^{x,\rho}$ on X_1 as follows:

$$D_1^{x,\rho}(u, v) := \rho((u, x^2, \ldots, x^n), (v, x^2, \ldots, x^n)).$$

If we were to view ρ consistent with d_1, we would certainly wish to have the property that $D_1^{x,\rho} = d_1$ for any $x \in X$. (Yes?) And, of course, the analogous consistency condition should hold for the rest of the d_is as well.

There are many metrics that satisfy this consistency condition. For instance, the following would do nicely:

$$\rho((x^1, \ldots, x^n), (y^1, \ldots, y^n)) := \sum_{i=1}^{n} d_i(x^i, y^i). \tag{11}$$

Or we can let

$$\rho((x^1, \ldots, x^n), (y^1, \ldots, y^n)) := \left(\sum_{i=1}^{n} \left(d_i(x^i, y^i)\right)^p\right)^{\frac{1}{p}} \tag{12}$$

for an arbitrarily fixed $p \geq 1$, or choose

$$\rho((x^1, \ldots, x^n), (y^1, \ldots, y^n)) := \max\{d_i(x^i, y^i) : i = 1, \ldots, n\}. \tag{13}$$

(Any one of these define ρ as a metric on X. Why?) Notice that in the case of any of these definitions (11), (12), and (13), we have $D_1^{x,\rho} = d_1$ for any $x \in X$. Moreover, for most practical purposes (and certainly all purposes of this text), it is simply a matter of convenience to choose among the above alternatives, because any two of these definitions yield strongly equivalent metrics on X. (Why?) Due to its simplicity, we designate (11) as our default choice.

DEFINITION
Take any $n \in \mathbb{N}$ and metric spaces (X_i, d_i), $i = 1, \ldots, n$. Let $X := \mathsf{X}^n X_i$. We call the map $\rho : X \times X \to \mathbb{R}_+$ defined by (11) the **product metric** on X, and refer to (X, ρ) as the **product** of the metric spaces (X_i, d_i). We denote this space as $\mathsf{X}^n(X_i, d_i)$, and refer to (X_i, d_i) as the ith **coordinate space** of $\mathsf{X}^n(X_i, d_i)$.

So, according to this definition, $\mathsf{X}^n(\mathbb{R}, d_1) = \mathbb{R}^{n,1}$. But, as noted above, this is nothing more than a convention for the purposes of this text, for, as you know, $\mathbb{R}^{n,1}$ is strongly equivalent to $\mathbb{R}^{n,p}$ for any $1 \leq p \leq \infty$.

What sort of properties would a product metric space inherit from its coordinate spaces? A timely question, to be sure. We can in fact answer it pretty easily too, because it is evident that a sequence $(x_1^m, \ldots, x_n^m)$ in the product of n metric spaces converges to the point $(x_1, \ldots, x_n)$ iff $x_i^m \to x_i$ for each $i = 1, \ldots, n$. (Yes?) Still, let us postpone this matter for the moment, because we are about to tackle it in a more general setup.

8.2 Countably Infinite Products

Life gets slightly more complicated when one is interested in the product of countably infinitely many metric spaces (X_i, d_i), $i = 1, 2, \ldots$, for none of the alternatives (11), (12), or (13) we considered above have straightforward extensions to this case, owing to obvious summability problems. (For instance, $((x_m), (y_m)) \to \sum^\infty |x_i - y_i|$ does not work as a metric on $\mathbb{R}^\infty$.) Indeed, metrizing $\mathsf{X}^\infty X_i$ in a way that satisfies the consistency property proposed in the previous subsection is not possible in general. But we can still satisfy a slightly weaker consistency condition by using the following definition.

DEFINITION

Let (X_i, d_i) be a metric space, $i = 1, 2, \ldots$, and let $X := \mathsf{X}^\infty X_i$. We call the map $\rho : X \times X \to \mathbb{R}_+$ defined by

$$\rho(x, y) := \sum_{i=1}^{\infty} \tfrac{1}{2^i} \min\{1, d_i(x^i, y^i)\}$$

the **product metric** on X, and refer to (X, ρ) as the **product** of the metric spaces (X_i, d_i).[57] We denote this space as $\mathsf{X}^\infty(X_i, d_i)$, and refer to (X_i, d_i) as the ***i*th coordinate space** of $\mathsf{X}^\infty(X_i, d_i)$.

Fix an arbitrary sequence $((X_m, d_m))$ of metric spaces. Two important observations about the product of these spaces are in order.

OBSERVATION 3. $\mathsf{X}^\infty(X_i, d_i)$ is a bounded metric space.

PROOF

Let $X := \mathsf{X}^\infty X_i$, and observe that the product metric ρ on X is well-defined, because for any $x, y \in X$, the sequence $(\sum^m \frac{1}{2^i} \min\{1, d_i(x^i, y^i)\})$, being an increasing real sequence that is bounded above (by 1), converges. Moreover, ρ is a bounded function: $0 \leq \rho \leq \sum^\infty \frac{1}{2^i} = 1$ (Example A.8.[2]). The fact that ρ is a distance function is straightforward. □

OBSERVATION 4. For any $n \in \mathbb{N}$, the space $\mathsf{X}^\infty(X_i, d_i)$ is consistent with $\mathsf{X}^n(X_i, d_i)$ in the following sense: If ρ_n stands for the map on $(\mathsf{X}^n X_i) \times (\mathsf{X}^n X_i)$ defined by

$$\rho_n((x^1, \ldots, x^n), (y^1, \ldots, y^n))$$
$$:= \rho\left((x^1, \ldots, x^n, z^{n+1}, z^{n+2}, \ldots), (y^1, \ldots, y^n, z^{n+1}, z^{n+2}, \ldots)\right),$$

where $z^i \in X_i$ is arbitrarily fixed, $i = n+1, n+2, \ldots$, then ρ_n is a metric equivalent to the product metric on $\mathsf{X}^n X_i$.

PROOF

Exercise. □

[57] The metric ρ is sometimes called the *Fréchet metric* in the current literature. (*Note.* Actually, Fréchet was only interested in metrizing $\mathbb{R}^\infty$, and to this end he introduced the map $((x_m), (y_m)) \mapsto \sum^\infty \frac{1}{i!}\left(\frac{|x_i - y_i|}{1+|x_i - y_i|}\right)$ on $\mathbb{R}^\infty \times \mathbb{R}^\infty$. Of course, it is hardly a leap to apply the idea behind this metric to metrize an arbitrary countably infinite product of metric spaces.)

As we shall see, this fact allows us to deduce many results about the products of finitely many metric spaces from those established for the case of countably infinite products. (See Exercise 63.)

It is useful to see under which circumstances a sequence in $\mathsf{X}^\infty(X_i, d_i)$ converges. First, note that if the sequence (x^m) in $\mathsf{X}^\infty(X_i, d_i)$ converges to some x in $\mathsf{X}^\infty X_i$, then we must have $d_i(x_i^m, x_i) \to 0$ for each i. (Yes?) That is, a sequence in a product metric space converges only if each coordinate of the sequence converges in the space that it lies. We next show that the converse is also true.[58]

> PROPOSITION 8
>
> *Let (X_i, d_i) be a metric space, $i = 1, 2, \ldots$, $x \in \mathsf{X}^\infty X_i$, and (x^m) a sequence in $\mathsf{X}^\infty(X_i, d_i)$. We have:*
>
> $$x^m \to x \quad \textit{if and only if} \quad x_i^m \to x_i \textit{ for each } i = 1, 2, \ldots.$$

PROOF

We only need to prove the "if" part of the proposition. So assume that $x_i^m \to x_i$ for each i, and fix any $\varepsilon > 0$. We can obviously choose a positive integer $k \in \mathbb{N}$ large enough to guarantee that $\sum_{i=k+1}^\infty \frac{1}{2^i} < \frac{\varepsilon}{2}$ (Exercise A.45). Since $d_i(x_i^m, x_i) \to 0$ for each i, we can also choose an $M > 0$ large enough so that

$$d_i(x_i^m, x_i) < \frac{\varepsilon}{2\left(\frac{1}{2} + \cdots + \frac{1}{2^k}\right)} \qquad \text{for all } m \geq M \text{ and } i = 1, \ldots, k.$$

But then, for each $m \geq M$,

$$\sum_{i=1}^k \tfrac{1}{2^i} \min\{1, d_i(x_i^m, x_i)\} < \tfrac{\varepsilon}{2} \qquad \text{and} \qquad \sum_{i=k+1}^\infty \tfrac{1}{2^i} \min\{1, d_i(x_i^m, x_i)\} < \tfrac{\varepsilon}{2},$$

and hence $\rho((x^m), x) < \varepsilon$. Since $\varepsilon > 0$ is arbitrary here, we may conclude that $x^m \to x$, as we sought. ■

EXAMPLE 15

(*The Hilbert Cube*) For any $a, b \in \mathbb{R}$ with $a < b$, consider the set $[a, b]^\infty$ of all real sequences all terms of which lie between a and b. When $[a, b]^\infty$

[58] For this reason, some authors refer to the product metric as the metric of *coordinatewise convergence*.

is metrized by the product metric, the resulting metric space is called the **Hilbert cube**. In this book we reserve the notation $[a,b]^\infty$ for the Hilbert cube; that is, by default, we view $[a,b]^\infty$ as $\mathsf{X}^\infty([a,b],d_1)$.

Please note that the Hilbert cube is distinct from $([a,b]^\infty, d_\infty)$. For any sequence (x^m) in $[a,b]^\infty$, $d_\infty(x^m,x) \to 0$ implies $x_i^m \to x_i$, and hence, by Proposition 8, we have $\lim \rho(x^m,x) = 0$, where ρ is the product metric on $[a,b]^\infty$. The converse implication, however, does not hold. To see this, consider the sequence (x^m) of sequences where

$$x^m = \left(a + \frac{b-a}{m}, a + \frac{b-a}{m-1}, \ldots, a + \frac{b-a}{2}, b, 0, 0, \ldots\right), \qquad m = 2, 3, \ldots,$$

and check that while $x_i^m \to a$ for each i, we have $d_\infty(x^m, (a, a, \ldots)) = b - a$ for all m. Thus $\rho(x^m, (a, a, \ldots)) \to 0$ but (x^m) does not converge in $([a,b]^\infty, d_\infty)$. We conclude: *A convergent sequence in $([a,b]^\infty, d_\infty)$ is also convergent in $[a,b]^\infty$, but not conversely.* This implies that any set which is closed in $[a,b]^\infty$ is also closed in $([a,b]^\infty, d_\infty)$, but not conversely. (Why?) That is, there are fewer closed (and hence open) sets in the Hilbert cube $[a,b]^\infty$ than in $([a,b]^\infty, d_\infty)$. □

CONVENTION. In this book, whenever we speak of the set $\mathbb{R}^\infty$ of all real sequences as a metric space without mentioning a particular distance function, it is implicitly understood that $\mathbb{R}^\infty$ is metrized by the product metric. Put differently, unless it is explicitly stated otherwise, we identify $\mathbb{R}^\infty$ with $\mathsf{X}^\infty(\mathbb{R}, d_1)$.

The following is the main result of this section. In a nutshell, it says simply that a product space inherits the major metric space properties we have considered in this chapter from its constituent coordinate spaces.

THEOREM 4
Let (X_i, d_i) be a metric space, $i = 1, 2, \ldots$. Then:

(a) *If each (X_i, d_i) is separable, so is $\mathsf{X}^\infty(X_i, d_i)$.*

(b) *If each (X_i, d_i) is connected, so is $\mathsf{X}^\infty(X_i, d_i)$.*

(c) *If each (X_i, d_i) is compact, so is $\mathsf{X}^\infty(X_i, d_i)$.*

(d) *If each (X_i, d_i) is complete, so is $\mathsf{X}^\infty(X_i, d_i)$.*

Proof

(a) Let Y_i be a countable dense set in X_i, $i = 1, 2, \ldots$. Fix any $x_*^i \in X_i$ for each i, and define

$$Y := \bigcup_{k=1}^{\infty} \left\{ (x^1, \ldots, x^k, x_*^{k+1}, x_*^{k+2}, \ldots) : x_i \in Y_i, \;\; i = 1, \ldots, k \right\},$$

which is a countable set, being a countable union of countable sets.[59] We wish to show that Y is dense in X. To see this, take any $z \in X$ and fix any $\varepsilon > 0$. Since $\sum^{\infty} \frac{1}{2^i}$ converges, there is a $k \in \mathbb{N}$ large enough so that $\sum_{i=k+1}^{\infty} \frac{1}{2^i} \leq \frac{\varepsilon}{2}$. Moreover, by denseness of Y_i in X_i, there must exist an $x^i \in Y_i$ such that $d_i(x^i, z^i) \leq \frac{\varepsilon}{2k}$, $i = 1, \ldots, k$. Then $y := (x^1, \ldots, x^k, x_*^{k+1}, x_*^{k+2}, \ldots) \in Y$, and we have

$$\rho(y, z) \leq \sum_{i=1}^{k} d_i(x^i, z^i) + \sum_{i=k+1}^{\infty} \tfrac{1}{2^i} \min\{1, d_i(x_*^i, z^i)\} \leq \tfrac{\varepsilon}{2} + \tfrac{\varepsilon}{2} = \varepsilon.$$

Since $\varepsilon > 0$ is arbitrary here, we are done.

(b) This is left as an exercise.

(c) In view of Theorem 2, it is enough to prove that $\mathsf{X}^{\infty}(X_i, d_i)$ is sequentially compact whenever each X_i is sequentially compact. To this end, pick any sequence $(x^m) = ((x_1^1, x_2^1, \ldots), (x_1^2, x_2^2, \ldots), \ldots)$ in $\mathsf{X}^{\infty} X_i$. By sequential compactness of X_1, the sequence (x_1^m) must have a subsequence that converges in X_1. Denote this subsequence by $(x_1^{m_k^1})$ and write

$$x_1^{m_k^1} \to x_1 \quad \text{as } (k \to \infty)$$

where $x_1 \in X_1$. Now consider the sequence $(x_2^{m_k^1})$, which must have a convergent subsequence in X_2. Denote this subsequence by $(x_2^{m_k^2})$ and write

$$x_2^{m_k^2} \to x_2 \quad \text{as } (k \to \infty)$$

[59] Observe that defining $Y := \mathsf{X}^{\infty} Y_i$ would not work here because the infinite product of countable sets need not be countable. (For instance, $\mathbb{N}^{\infty}$ is an uncountable set—there is a bijection between this set and $\mathbb{R}$. Think of the decimal (or binary) expansion of real numbers!) This is why we define Y here via a truncation trick; we have earlier used a similar argument in Example 6.[5].

where $x_2 \in X_2$. Proceeding this way, and letting $(m_k^0) := (1, 2, \ldots)$, we find, for each $t \in \mathbb{N}$, a strictly increasing sequence (m_k^t) in $\mathbb{N}$ and an $(x_1, x_2, \ldots) \in X_1 \times X_2 \times \cdots$ such that

(i) (m_k^t) is a subsequence of (m_k^{t-1}),

(ii) $x_t^{m_k^t} \to x_t \quad \text{as } (k \to \infty)$.

Now define the sequence (x^{m_k}) in X by

$$x^{m_k} = \left(x_1^{m_k^k}, x_2^{m_k^k}, \ldots \right).$$

(Formally we invoke the Axiom of Choice here.) Clearly, (x^{m_k}) is a subsequence of (x^m). Moreover, by (i) and (ii) we have $x_i^{m_k^k} \to x_i$ (as $k \to \infty$) for each i. Thus, by Proposition 8, $x^{m_k} \to (x_1, x_2, \ldots)$, and we conclude that $\mathsf{X}^\infty(X_i, d_i)$ is sequentially compact.

(d) If (x^m) is Cauchy in $\mathsf{X}^\infty(X_i, d_i)$, then (x_i^m) must be Cauchy in (X_i, d_i) for each i. (Why?) But then using the completeness of each (X_i, d_i), along with Proposition 8, yields the claim. ■

The most penetrating statement in Theorem 4 is certainly the one that pertains to the compactness of a product space. You should note that the method of proof used here extends the simple one that we used in Example 10.[3] to the countably infinite case. It is called the Cantor's **diagonal method**, and is used frequently in mathematical analysis to extract a convergent subsequence from a double sequence (with the help of the Axiom of Choice). We will have occasion to use this technique again in the sequel.

Let us now conclude with a few exercises.

EXERCISE 63^{H} The product of finitely many compact metric spaces is compact. Obtain this result as an immediate consequence of Observation 4 and Theorem 4. Prove the same claim with respect to the completeness property as well. (Theorem 4 is not of immediate help for this latter case. Why?)

EXERCISE 64 Let (X_i, d_i) be a discrete space, $i = 1, 2, \ldots$.

(a) Show that $\mathsf{X}^n(X_i, d_i)$ is a discrete space for any $n \in \mathbb{N}$.

(b) Is $\mathsf{X}^\infty(X_i, d_i)$ a discrete space?

(c) Give an example of a convergent sequence in $\mathsf{X}^\infty(X_i, d_i)$ that does not converge relative to the discrete metric.

EXERCISE 65[H] Let (X_i, d_i) be a metric space, and O_i an open subset of X_i, $i = 1, 2, \ldots$. Is $\mathsf{X}^\infty O_i$ necessarily open in $\mathsf{X}^\infty(X_i, d_i)$?

EXERCISE 66 Let (X_i, d_i) be a metric space, $i = 1, 2, \ldots$, and O an open subset of $\mathsf{X}^\infty(X_i, d_i)$. Prove that $x \in O$ iff there exist an $m \in \mathbb{N}$ and open subsets O_i of X_i, $i = 1, \ldots, m$, such that $(x^1, \ldots, x^m) \in \mathsf{X}^m O_i$ and $\mathsf{X}^m O_i \times X_{m+1} \times X_{m+1} \times \cdots \subseteq O$.

EXERCISE 67 Compute the interior of the Hilbert cube $[0, 1]^\infty$.

EXERCISE 68[H] Let $\mathbf{c}_0$ denote the set of all real sequences only finitely many terms of which are nonzero. What is the closure of $\mathbf{c}_0$ in $\mathbb{R}^\infty$? What is the closure of $\mathbf{c}_0$ in ℓ^∞?

EXERCISE 69[H] Prove Theorem 4.(b).

EXERCISE 70 Prove or disprove: The product of countably infinitely many totally bounded metric spaces is totally bounded.

Chapter D

Continuity I

This chapter provides a basic introduction to the theory of functions in general, and that of continuous maps between two metric spaces in particular. Many of the results that you have seen in your earlier studies in terms of real functions on $\mathbb{R}$ are derived here in the context of metric spaces. Examples include the Intermediate Value Theorem, Weierstrass' Theorem, and the basic results on uniform convergence (such as those about the interchangeability of limits and Dini's Theorem). We also introduce and lay out a basic analysis of a few concepts that may be new to you, such as stronger notions of continuity (e.g., uniform, Lipschitz and Hölder continuity), weaker notions of continuity (e.g., upper and lower semicontinuity), homeomorphisms, and isometries.

This chapter addresses at least four topics that are often not covered in standard courses on real analysis but that nevertheless see good playing time in various branches of economic theory. In particular, and as applications of the main body of the chapter, we study Caristi's famous generalization of the Banach Fixed Point Theorem, the characterization of additive continuous maps on Euclidean spaces, and de Finetti's theorem on the representation of additive preorders. We also revisit the problem of representing a preference relation by a utility function and discuss two of the best-known results of utility theory, namely, the Debreu and Rader Utility Representation Theorems. This completes our coverage of ordinal utility theory; we will be able to take up issues related to cardinal utility only in the second part of the book.

The pace of the chapter is leisurely for the most part, and our treatment is fairly elementary. Toward the end, however, we study two topics that may be considered relatively advanced. (These may be omitted in the first reading.) First, we discuss Marshall Stone's important generalization of the Weierstrass Approximation Theorem. We prove this landmark result and consider a few of its applications, such as the proof of the separability of the set of all continuous real functions defined on a compact metric space. Second, we explore the problem of extending a given continuous function defined on a subset of a metric space to the entire space. The fundamental result in this regard is the Tietze Extension Theorem. We prove this theorem and then supplement it with further extension results, this time for functions that are either uniformly or Lipschitz continuous.[1] The final section of the chapter contains our next major trip into fixed point theory. It provides a preliminary discussion of the fixed point property and retracts, and then goes on to discuss the Brouwer Fixed Point Theorem and some of its applications.

1 Continuity of Functions

1.1 Definitions and Examples

Recall that a function $f \in \mathbb{R}^{[0,1]}$ is continuous if, for any $x \in [0,1]$, the images of points nearby x under f are close to $f(x)$. In conceptual terms, where we think of f as transforming inputs into outputs, this property can be thought of as ensuring that a small perturbation in the input entails only a small perturbation in the output.

It is easy to generalize the definition of continuity so that it applies to functions defined on arbitrary metric spaces. If (X, d_X) and (Y, d_Y) are two metric spaces and $f \in Y^X$ is any function, then, for any $x \in X$, the statement "the images of points nearby x under f are close to $f(x)$" can be formalized as follows: However small an $\varepsilon > 0$ one picks, if y is a point in X that is sufficiently close to x (closer than some $\delta > 0$), then the distance between $f(x)$ and $f(y)$ is bound to be smaller than ε. Here goes the formal definition.

[1] This concludes our introduction to the classical theory of functions. We barely touch on approximation theory here, and omit matters related to differentiation altogether, other than one isolated instance. For those who wish to get a more complete introduction to the basic theory of real functions, a standard recommendation would be Rudin (1976) or Marsden and Hoffman (1993) at the entry level, and Apostol (1974) or Royden (1986) or Haaser and Sullivan (1991) at the intermediate level. Again, my personal favorites are Körner (2003) at the introductory (but pleasantly challenging) level and Carothers (2000) at the intermediate level.

DEFINITION

Let (X, d_X) and (Y, d_Y) be two metric spaces. We say that the map $f : X \to Y$ is **continuous at** $x \in X$ if, for any $\varepsilon > 0$, there exists a $\delta > 0$ (which may depend on both ε and x) such that

$$d_X(x, y) < \delta \quad \text{implies} \quad d_Y(f(x), f(y)) < \varepsilon$$

for each $y \in X$, that is,

$$d_Y(f(x), f(y)) < \varepsilon \quad \text{for all } y \in N_{\delta,X}(x).$$

Put differently, f is continuous at x if, for any $\varepsilon > 0$, there exists a $\delta > 0$ such that

$$f(N_{\delta,X}(x)) \subseteq N_{\varepsilon,Y}(f(x)).$$

If f is not continuous at x, then it is said to be **discontinuous at** x. For any nonempty $S \subseteq X$, we say that f is **continuous on** S, if it is continuous at every $x \in S$. In turn, f is said to be **continuous**, if it is continuous on X.

Let's look at some simple examples. Consider the self-map f on $\mathbb{R}_{++}$ defined by $f(t) := \frac{1}{t}$. Of course you already know that f is continuous, but if only for practice, let us give a rigorous proof anyway. Fix any $x > 0$ and notice that, for any $y > 0$, we have

$$\left| f(x) - f(y) \right| = \left| \tfrac{1}{x} - \tfrac{1}{y} \right| = \tfrac{|x-y|}{xy}.$$

Now take any $\varepsilon > 0$. We wish to find a $\delta > 0$ such that $\frac{|x-y|}{xy} < \varepsilon$ for any $y \in \mathbb{R}_{++}$ with $\left| x - y \right| < \delta$. Since δ is allowed to depend both on x and ε, this is not difficult. Indeed, we have $\frac{|x-y|}{xy} < \frac{\delta}{x(x-\delta)}$ for any $\delta \in (0, x)$ and $y > 0$ with $\left| x - y \right| < \delta$. But $\frac{\delta}{x(x-\delta)} < \varepsilon$ if $\delta < \frac{\varepsilon x^2}{1+\varepsilon x}$, so by choosing any such $\delta > 0$ (which is necessarily smaller than x), we find $f(N_{\delta,\mathbb{R}_{++}}(x)) \subseteq N_{\varepsilon,\mathbb{R}_{++}}(f(x))$. (Notice that δ depends on both ε and x.) Since $x > 0$ is arbitrary in this observation, we may conclude that f is continuous.

Consider next the function $\varphi : \ell^\infty \to \mathbb{R}_+$ defined by

$$\varphi((x_m)) := \sup\{|x_m| : m \in \mathbb{N}\}.$$

For any bounded real sequences (x_m) and (y_m), we have

$$\begin{aligned}\left|\varphi((x_m)) - \varphi((y_m))\right| &= \left|\sup_{m\in\mathbb{N}} |x_m| - \sup_{m\in\mathbb{N}} |y_m|\right| \\ &\leq \sup_{m\in\mathbb{N}} \left|x_m - y_m\right| \\ &= d_\infty((x_m),(y_m)).\end{aligned}$$

(Why the inequality?) Therefore, for any $\varepsilon > 0$ and any $(x_m) \in \ell^\infty$, we have $\varphi(N_{\varepsilon,\ell^\infty}((x_m))) \subseteq N_{\varepsilon,\mathbb{R}_+}(\varphi((x_m)))$. Conclusion: φ is continuous.

A similar argument would show that the function $L \in \mathbb{R}^{\mathbf{C}[0,1]}$ defined by

$$L(f) := \int_0^1 f(t)dt$$

is also continuous. Indeed, an immediate application of Proposition A.12 yields

$$\left|L(f) - L(g)\right| \leq d_\infty(f,h) \qquad \text{for all } f,g \in \mathbf{C}[0,1].$$

This is more than enough to conclude that L is continuous.[2]

To give a simple example of a discontinuous function, consider $f := \mathbf{1}_{\mathbb{R}_{++}}$, the indicator function of $\mathbb{R}_{++}$ in $\mathbb{R}$ (Example A.5.[3]). This function is discontinuous at 0, because, for any $\delta > 0$, we have $f(N_{\delta,\mathbb{R}}(0)) = \{0,1\}$, while $N_{\frac{1}{2},\mathbb{R}}(f(0)) = (-\frac{1}{2}, \frac{1}{2})$. Thus there is no $\delta > 0$ for which $f(N_{\delta,\mathbb{R}}(0)) \subseteq N_{\frac{1}{2},\mathbb{R}}(f(0))$, whence f is not continuous at 0. In words, the image of a point arbitrarily close to 0 under f is not necessarily arbitrarily close to $f(0)$; this is the source of discontinuity of f at 0. As less trivial examples of discontinuous functions on $\mathbb{R}$, consider the maps $\mathbf{1}_{\mathbb{Q}}$ and $g(t) := \begin{cases} t, & \text{if } t \in \mathbb{Q} \\ -t, & \text{if } t \in \mathbb{R}\backslash\mathbb{Q} \end{cases}$. You can check that $\mathbf{1}_{\mathbb{Q}}$ is discontinuous at every point in $\mathbb{R}$ while g is discontinuous at every point in $\mathbb{R}$ but at 0.

It is crucial to understand that the continuity of a function that maps a metric space to another depends intrinsically on the involved metrics. Suppose we are given two metric spaces, (X, d_X) and (Y, d_Y), and $f \in Y^X$

[2] *Quiz.* Define the self-map L on $\mathbf{C}[0,1]$ by $L(f)(x) := \int_0^x f(t)dt$, and show that L is continuous.

is continuous. Now suppose we were to endow X and/or Y with distance functions other than d_X and/or d_Y. Would f still be continuous in this new setting? The answer is no, not necessarily! For instance, consider $f := \mathbf{1}_{\mathbb{R}_{++}}$ which we have just seen to be discontinuous at 0. This conclusion is valid conditional on the fact that we use (implicitly) the standard metric d_1 on the domain of f. Suppose that we instead use the discrete metric on $\mathbb{R}$ (Example C.1.[1]). In this case, denoting the resulting metric space by X, we would have $f(N_{\frac{1}{2},X}(0)) = \{f(0)\} = \{0\} \subseteq N_{\varepsilon,\mathbb{R}}(f(0))$ for any $\varepsilon > 0$, and hence we would conclude that f is continuous at 0. The moral of the story is that continuity of a function is conditional on the distance functions used to metrize the domain and codomain of the function.

NOTATION. Some authors prefer to write $f : (X, d_X) \to (Y, d_Y)$ to make it clear that the continuity properties of f depend on both d_X and d_Y. Since it leads to somewhat cumbersome notation, we shall mostly refrain from doing this, but it is advisable that you view the notation $f : X \to Y$ (or $f \in Y^X$) as $f : (X, d_X) \to (Y, d_Y)$ throughout this chapter. After a while, this will become automatic anyway.

NOTATION. The symbols X, Y, and Z are used in this chapter only to denote arbitrary metric spaces. Generically speaking, we denote the metric on X simply by d, whereas the metrics on Y and Z are denoted more explicitly as d_Y and d_Z.

EXAMPLE 1

[1] The identity function id_X on a metric space X is continuous, for we have $\mathrm{id}_X(N_{\varepsilon,X}(x)) = N_{\varepsilon,X}(x) = N_{\varepsilon,X}(\mathrm{id}_X(x))$ for all $x \in X$ and $\varepsilon > 0$. Similarly, it is easily checked that a constant function on any metric space is continuous.

[2] For any given metric space Y, if X is a discrete space, then any $f \in Y^X$ must be continuous, because in this case we have $f(N_{\frac{1}{2},X}(x)) = \{f(x)\} \subseteq N_{\varepsilon,Y}(f(x))$ for any $\varepsilon > 0$. Thus: *Any function defined on a discrete space is continuous.*

[3] Let S be any nonempty subset of a metric space X. The distance between S and a point $x \in X$ is defined as

$$d(x, S) := \inf\{d(x, z) : z \in S\}.$$

Thus the function $f \in \mathbb{R}_+^X$ defined by $f(x) := d(x, S)$ measures the distance of any given point in X from the set S in terms of the metric d. For self-consistency, it is desirable that this function be continuous. This is indeed the case, because, for each $x, y \in X$, the triangle inequality yields

$$f(x) = d(x, S) \leq \inf\{d(x, y) + d(y, z) : z \in S\} = d(x, y) + f(y),$$

and similarly, $f(y) \leq d(y, x) + f(x)$. Thus, $|f(x) - f(y)| \leq d(x, y)$ for all $x, y \in X$, and it follows that f is continuous.

[4] Given two metric spaces X and Y, if $f \in Y^X$ is continuous and S is a metric subspace of X, then $f|_S$ is a continuous function. The converse is, of course, false. For instance, while the indicator function of $\mathbb{R}_{++}$ in $\mathbb{R}$ is discontinuous at 0, the restriction of this function to $\mathbb{R}_{++}$ is trivially continuous *on this metric subspace.*[3]

[5] (*On the Continuity of Concave Functions*) In Section A.4.5 we showed that any concave real function defined on an open interval must be continuous. This fact generalizes to the case of real functions defined on a Euclidean space: *For every $n \in \mathbb{N}$, any concave (or convex) function defined on an open subset of $\mathbb{R}^n$ is continuous.*[4] □

EXERCISE 1 Let X be any metric space, and $\varphi \in \mathbb{R}^X$.

(a) Show that if φ is continuous, then the sets $\{x : \varphi(x) \geq \alpha\}$ and $\{x : \varphi(x) \leq \alpha\}$ are closed in X for *any* real number α. Also show that the continuity of φ is necessary for this conclusion to hold.

(b) Prove that if φ is continuous and $\varphi(x) > 0$ for some $x \in X$, then there exists an open subset O of X such that $\varphi(y) > 0$ for all $y \in O$.

EXERCISE 2[H] Let A and B be two nonempty closed subsets of a metric space X with $A \cap B = \emptyset$. Define $\varphi \in \mathbb{R}_+^X$ and $\psi \in \mathbb{R}^X$ by $\varphi(x) := d(x, A)$

[3] It may be worthwhile to pursue this matter a little further. While we have seen earlier that a real function on $\mathbb{R}$ can well be discontinuous everywhere, it turns out that any such function is continuous on some dense subspace S of $\mathbb{R}$. That is—and this is the famous *Blumberg's Theorem*—for any $f \in \mathbb{R}^{\mathbb{R}}$ there exists a dense subset S of $\mathbb{R}$ such that $f|_S$ is a continuous member of $\mathbb{R}^S$.

[4] The proof of this claim is a bit harder than that of Proposition A.14, so I don't want to get into it here. A substantially more general result will be proved later (in Section I.2.4).

and $\psi(x) := d(x, A) - d(x, B)$, respectively. Prove:

(a) $A = \{x \in X : \varphi(x) = 0\}$, so we have $d(x, A) > 0$ for all $x \in X \backslash A$.

(b) ψ is continuous, so $\{x \in X : \psi(x) < 0\}$ and $\{x \in X : \psi(x) > 0\}$ are open.

(c) There exist disjoint open sets O and U in X such that $A \subseteq O$ and $B \subseteq U$. (Compare with Exercise C.11.)

EXERCISE 3 Let X be a metric space, and for any $n \in \mathbb{N}$, define the map $\rho : X^n \times X^n \to \mathbb{R}_+$ by

$$\rho((x_1, \ldots, x_n), (y_1, \ldots, y_n)) := \left(\sum_{i=1}^{n} d(x_i, y_i)^2 \right)^{\frac{1}{2}}.$$

Now metrize $X^n \times X^n$ by the product metric, and show that ρ is a continuous function on the resulting metric space.

EXERCISE 4^{H} Let $((X_m, d_m))$ be a sequence of metric spaces, and let X stand for the product of all (X_i, d_i) s. Is the function $f : X \to X_i$ defined by $f(x_1, x_2, \ldots) := x_i$ continuous?

EXAMPLE 2

[1] (*Composition of Continuous Functions*) For any metric spaces X, Y, and Z, let $f : X \to Y$ and $g : f(X) \to Z$ be continuous functions. Then, we claim, $h := g \circ f$ is a continuous function on X. (Here we obviously consider $f(X)$ as a subspace of Y. So, a special case of this claim is the case in which g is continuous on the entire Y.) To see this, take any $x \in X$ and $\varepsilon > 0$. Since g is continuous at $f(x)$, we can find a $\delta' > 0$ such that

$$g(N_{\delta', f(X)}(f(x))) \subseteq N_{\varepsilon, Z}(g(f(x))) = N_{\varepsilon, Z}(h(x)).$$

But since f is continuous at x, there exists a $\delta > 0$ with $f(N_{\delta, X}(x)) \subseteq N_{\delta', f(X)}(f(x))$ so that

$$h(N_{\delta, X}(x)) = g(f(N_{\delta, X}(x))) \subseteq g(N_{\delta', f(X)}(f(x))).$$

Combining these two observations, we find $h(N_{\delta, X}(x)) \subseteq N_{\varepsilon, Z}(h(x))$, as we sought. In words: *The composition of two continuous functions is continuous.*

[2] For any given $n \in \mathbb{N}$, take any metric spaces (X_i, d_i), $i = 1, \ldots, n$, and let (X, ρ) be the product of these spaces. (As usual, we abbreviate a

point like $(x^1, \ldots, x^n)$ in X by x.) We define the ***i*th projection map** $\pi_i : X \to X_i$ by $\pi_i(x^1, \ldots, x^n) := x^i$. It is easily seen that π_i is a continuous function. Indeed, we have

$$d_i(\pi_i(x), \pi_i(y)) = d_i(x^i, y^i) \leq \sum_{j=1}^{n} d_j(x^j, y^j) = \rho(x, y),$$

for any $x, y \in X$ and $i = 1, \ldots, n$.

[3] For any given $m \in \mathbb{N}$, take any metric spaces Y_i, $i = 1, \ldots, m$, and let Y be the product of these spaces. Now take any other metric space X, and consider any maps $f_i : X \to Y_i$, $i = 1, \ldots, m$. Let us define the function $f : X \to Y$ by $f(x) := (f_1(x), \ldots, f_m(x))$. (Here each f_i is referred to as a **component map** of f.) Now if f is continuous, then, by observations [1] and [2] above, $f_i = \pi_i \circ f$ is continuous. Conversely, if each f_i is continuous, then f must be continuous as well. (Proof?)

A special case of this observation is the following fact, which you must have seen before in some calculus course: For any $m, n \in \mathbb{N}$, if $\Phi : \mathbb{R}^n \to \mathbb{R}^m$ is defined as $\Phi(x) := (\varphi_1(x), \ldots, \varphi_m(x))$ for some real maps φ_i on $\mathbb{R}^n$, $i = 1, \ldots, m$, then Φ is continuous iff each φ_i is continuous.

[4] Fix any $n \in \mathbb{N}$ and any metric space X. Let $\varphi_i \in \mathbb{R}^X$ be a continuous map, $i = 1, \ldots, n$, and pick any continuous $F : \mathbb{R}^n \to \mathbb{R}$. We wish to show that the map $\psi \in \mathbb{R}^X$ defined by $\psi(x) := F(\varphi_1(x), \ldots, \varphi_n(x))$ is continuous. To this end, define $\varphi : X \to \mathbb{R}^n$ by $\varphi(x) := (\varphi_1(x), \ldots, \varphi_n(x))$, and observe that $\psi = F \circ \varphi$. Applying the observations [1] and [3] above, therefore, we find that ψ is continuous.[5]

The following situation obtains as a special case of this observation. If X is the product of the metric spaces $(X_1, d_1), \ldots, (X_n, d_n)$, and $F : \mathbb{R}^n \to \mathbb{R}$ and $\phi_i : X_i \to \mathbb{R}$ are continuous, $i = 1, \ldots, n$, then $\psi \in \mathbb{R}^X$ defined by $\psi(x^1, \ldots, x^n) := F(\phi_1(x^1), \ldots, \phi_n(x^n))$ is a continuous function. (*Proof.* By the findings of [1] and [2], the map $\varphi_i := \phi_i \circ \pi_i$ is continuous (for each i). Now apply what we have found in the previous paragraph.) □

[5] It does not matter which of the metrics d_p is used here to metrize $\mathbb{R}^n$. Why? (*Hint.* Think about the strong equivalence of metrics.)

EXERCISE 5 Given any $n \in \mathbb{N}$, let X be a metric space, and $\varphi_i \in \mathbb{R}^X$ a continuous map, $i = 1, \ldots, n$. Show that $|\varphi_1|$, $\sum^n \varphi_i$, $\prod^n \varphi_i$, $\max\{\varphi_1, \ldots, \varphi_n\}$ and $\min\{\varphi_1, \ldots, \varphi_n\}$ are continuous real functions on X.

EXERCISE 6 For any $n \in \mathbb{N}$, a function $\varphi : \mathbb{R}^n \to \mathbb{R}$ is called a (*multivariate*) **polynomial** if there exist real numbers $\alpha_{i_1,\ldots,i_n}$ such that

$$\varphi(t_1, \ldots, t_n) = \sum \alpha_{i_1,\ldots,i_n} \prod_{j=1}^{n} t_j^{i_j} \qquad \text{for all } (t_1, \ldots, t_n) \in \mathbb{R}^n,$$

where the sum runs through a finite set of n-tuples of indices $(i_1, \ldots, i_n) \in \mathbb{N}^n$. Prove that any polynomial is continuous.

1.2 Uniform Continuity

The notion of continuity is an inherently local one. If $f \in Y^X$ is continuous, we know that, for any $x \in X$, "the images of points nearby x under f are close to $f(x)$," but we do not know if the word "nearby" in this statement depends on x or not. A global property would allow us to say something like this: "Give me any $\varepsilon > 0$, and I can give you a $\delta > 0$ such that, *for any point* $x \in X$, the images of points at most δ-away from x under f are at most ε-away from $f(x)$." This property says something about the behavior of f on its entire domain, not only in certain neighborhoods of the points in its domain. It is called *uniform continuity.*

DEFINITION
Let X and Y be two metric spaces. We say that a function $f \in Y^X$ is **uniformly continuous** if, for all $\varepsilon > 0$, there exists a $\delta > 0$ (which may depend on ε) such that $f(N_{\delta,X}(x)) \subseteq N_{\varepsilon,Y}(f(x))$ for all $x \in X$.

Obviously, a uniformly continuous function is continuous. On the other hand, a continuous function need not be uniformly continuous. For instance, consider the continuous function $f : \mathbb{R}_{++} \to \mathbb{R}_{++}$ defined by $f(t) := \frac{1}{t}$. Intuitively, you might sense that this function is not uniformly continuous. It has a relatively peculiar behavior near 0; it is continuous, but the nature of its continuity at 1 and at 0.0001 seems quite different. In a manner of speaking, the closer we are to 0, the harder it gets to verify that

f is continuous (in the sense that, in our ε-δ definition, for a given $\varepsilon > 0$, we need to choose smaller $\delta > 0$). If f were uniformly continuous, this would not be the case.

To demonstrate that $t \mapsto \frac{1}{t}$ is not uniformly continuous on $\mathbb{R}_{++}$ formally, choose $\varepsilon = 1$, and ask yourself if we can find a $\delta > 0$ such that $f(N_{\delta,\mathbb{R}_{++}}(x)) \subseteq N_{1,\mathbb{R}}(f(x))$ for *all* $x > 0$. The question is, does there exist a $\delta > 0$ such that, for any $x > 0$, we have $\left|\frac{1}{x} - \frac{1}{y}\right| < 1$ (i.e., $|x - y| < xy$) whenever $y > 0$ satisfies $|x - y| < \delta$? It is plain that the answer is no. For instance, if we choose $y = x + \frac{\delta}{2}$, then we need to have $\frac{\delta}{2} < x(x + \frac{\delta}{2})$ for *all* $x > 0$. Obviously, no $\delta > 0$ is equal to this task, no matter how small. (If δ were allowed to depend on x, there would be no problem, of course. After all, $t \mapsto \frac{1}{t}$ is a continuous map on $\mathbb{R}_{++}$.)

WARNING. A continuous map from a metric space X into another metric space Y remains continuous if we remetrize X by a metric equivalent to d_X, and similarly for Y. This is *not* true for uniform continuity. Remetrizing the domain of a uniformly continuous map f with an equivalent metric may render f not uniformly continuous. (Can you give an example to illustrate this?) Remetrization with *strongly* equivalent metrics, however, leaves uniformly continuous maps uniformly continuous. (Why?)

EXERCISE 7[H]

(a) Show that the real map $t \mapsto \frac{1}{t}$ is uniformly continuous on $[a, \infty)$ for any $a > 0$.

(b) Show that the real map $t \mapsto t^2$ is not uniformly continuous on $\mathbb{R}$.

(c) Is the map $f : \ell^2 \to \ell^1$ defined by $f((x_m)) := (\frac{1}{m}x_m)$ uniformly continuous?

EXERCISE 8[H] Let φ and ψ be uniformly continuous bounded real functions on a metric space X. Show that $\varphi\psi$ is uniformly continuous. What if the boundedness condition did not hold?

Why should you care about uniform continuity? There are plenty of reasons for this, and we shall encounter many of them later. In the meantime, chew on the following.

EXERCISE 9 Let X and Y be metric spaces and $f \in Y^X$. Show that if $(x_m) \in X^\infty$ is Cauchy and f is uniformly continuous, then $(f(x_m)) \in Y^\infty$ is Cauchy. Would this be true if f was only known to be continuous?

1.3 Other Continuity Concepts

Ordinary continuity and uniform continuity are the most commonly used continuity properties in practice. However, sometimes one needs to work with other kinds of continuity conditions that demand more regularity from a function. For any $\alpha > 0$, a function $f \in Y^X$ is said to be **α-Hölder continuous** if there exists a $K > 0$ such that

$$d_Y(f(x), f(y)) \leq Kd(x,y)^\alpha \qquad \text{for all } x, y \in X.$$

(Recall that we denote the metric of X by d.) It is called **Hölder continuous** if it is α-Hölder continuous for some $\alpha > 0$, and **Lipschitz continuous** if it is 1-Hölder continuous, that is, if there exists a $K > 0$ such that

$$d_Y(f(x), f(y)) \leq Kd(x,y) \qquad \text{for all } x, y \in X.$$

(The smallest such K is called the **Lipschitz constant of** f.) On the other hand, as you'd surely guess, it is called a **contraction** (or a **contractive map**) if there exists a $0 < K < 1$ such that

$$d_Y(f(x), f(y)) \leq Kd(x,y) \qquad \text{for all } x, y \in X,$$

and **nonexpansive** if

$$d_Y(f(x), f(y)) \leq d(x,y) \qquad \text{for all } x, y \in X.$$

(The latter two definitions generalize the corresponding ones given in Section C.6.1, which applied only to self-maps.)

We have already seen some examples of nonexpansive and Lipschitz continuous functions. For instance, we have shown in Section 1.1 that the functions $\varphi \in \mathbb{R}_+^{\ell^\infty}$ and $L \in \mathbb{R}_+^{\mathbf{C}[0,1]}$ defined by $\varphi((x_m)) := \sup\{|x_m| : m \in \mathbb{N}\}$ and $L(f) := \int_0^1 f(t)dt$ are nonexpansive. Similarly, we have seen that the map $x \mapsto d(x, S)$ on any metric space X (with S being a nonempty set in X) is nonexpansive. (*Quiz.* Is any of these maps a contraction?) Finally, we also know that the restriction of any concave function on $\mathbb{R}$ to a compact interval is Lipschitz continuous (Proposition A.14), but it does not have to be nonexpansive.

It is often easy to check whether a differentiable self-map f on $\mathbb{R}$ is Lipschitz continuous or not. Indeed, any such f is Lipschitz continuous if its derivative is bounded, it is nonexpansive if $\sup\{|f'(t)| : t \in \mathbb{R}\} \leq 1$, and it is a contraction if $\sup\{|f'(t)| : t \in \mathbb{R}\} \leq K < 1$ for some real number K.

These observations are straightforward consequences of the Mean Value Theorem (Exercise A.56).

For future reference, let us explicitly state the logical connections between all of the continuity properties we introduced so far.

$$\boxed{\text{contraction property}} \Longrightarrow \boxed{\text{nonexpansiveness}} \Longrightarrow \boxed{\text{Lipschitz continuity}} \Longrightarrow \boxed{\text{Hölder continuity}} \Longrightarrow \boxed{\text{uniform continuity}} \Longrightarrow \boxed{\text{continuity}} \tag{1}$$

The converse of any one of these implications is false. As an example, let us show that Hölder continuity does not imply Lipschitz continuity. Consider the function $f \in \mathbb{R}_+^{[0,1]}$ defined by $f(t) := \sqrt{t}$. This function is $\frac{1}{2}$-Hölder continuous, because

$$\left|f(x) - f(y)\right| = \left|\sqrt{x} - \sqrt{y}\right| \leq \sqrt{|x-y|} \quad \text{for all } 0 \leq x, y \leq 1.$$

(The proof of the claimed inequality is elementary.) On the other hand, f is not Lipschitz continuous because, for any $K > 0$, we have $|f(x) - f(0)| = \sqrt{x} > Kx$ for any $0 < x < \frac{1}{K^2}$.

EXERCISE 10 Prove (1) and provide examples to show that the converse of any of the implications in (1) is false in general.

EXERCISE 11 Let X be a metric space, $\varphi \in \mathbb{R}^X$, $\alpha > 0$, and $\lambda \in \mathbb{R}$.
(a) Show that if φ and ψ are α-Hölder continuous, then so is $\lambda\varphi + \psi$.
(b) Prove or disprove: If φ and ψ are nonexpansive, then so is $\lambda\varphi + \psi$.
(c) Prove or disprove: If φ and ψ are Hölder continuous, then so is $\lambda\varphi + \psi$.

EXERCISE 12 For any $0 < \alpha < \beta \leq 1$, show that if $f \in \mathbb{R}^{[0,1]}$ is β-Hölder continuous, then it is also α -Hölder continuous.

EXERCISE 13 Let Y be a metric space and $\alpha > 1$. Show that $F \in Y^{\mathbb{R}}$ is α-Hölder continuous iff it is a constant function.

1.4* Remarks on the Differentiability of Real Functions

We noted earlier that a monotonic function on $\mathbb{R}$ can have at most countably many discontinuity points (Exercise B.8). In fact, one can also say quite a bit about the differentiability of such a function. Let us agree to say that a set S in $\mathbb{R}$ is **null** if, for all $\varepsilon > 0$, there exist countably many intervals such that (i) S is contained in the union of these intervals, and (ii) the sum of the lengths of these intervals is at most ε. For instance, $\mathbb{Q}$ (or any countable set) is null.[6] Clearly, one should intuitively think of null sets as being very "small" (although—and this is important—such sets need not be countable). We therefore say that a property holds **almost everywhere** if it holds on $\mathbb{R}\backslash S$ for some null subset S of $\mathbb{R}$. For instance, we can say that a monotonic function on $\mathbb{R}$ is continuous almost everywhere (but again, Exercise B.8 says something stronger than this).[7]

One of the main results of the theory of real functions concerns the differentiability of monotonic functions; it establishes that any such real function on $\mathbb{R}$ is differentiable almost everywhere. This is:

> Lebesgue's Theorem
> *Any monotonic $f : \mathbb{R} \to \mathbb{R}$ is differentiable almost everywhere.*

Put differently, the set of points on which a monotonic real function on $\mathbb{R}$ fails to be differentiable is null. Since we will not need this result in the sequel, its lengthy proof is omitted here.[8]

Lebesgue's Theorem shows that the differentiability properties of continuous functions are in general quite distinct from those of monotonic functions. Indeed, a continuous function need not possess derivatives *anywhere.*[9]

[6] *Quiz.* Show that the union of countably many null sets is null.

[7] There is a lot of stuff here that I don't want to get into right now. All I expect you to do is to get an intuitive feeling for the idea that if something is true *almost everywhere*, then it is true everywhere but on a negligibly small set.

[8] The modern proof of this result is based on a result called the Rising Sun Lemma (due to Frigyes Riesz). An easily accessible account is given in Riesz and Nagy (1990, pp. 5–9), but this result can be found in essentially any graduate textbook on real analysis.

[9] As explained by Boyer and Merzbach (1989, p. 577), this surprising fact was first established by Bernhard Bolzano in 1834. It became commonly known, however, only after an example to this effect was produced by Karl Weierstrass. Since then many other

However, the situation is quite different for Lipschitz continuous functions. Indeed, if $f \in \mathbb{R}^{\mathbb{R}}$ is Lipschitz continuous with Lipschitz constant $K > 0$, then the function $g \in \mathbb{R}^{\mathbb{R}}$ defined by $g(x) := f(t) + Kt$ is increasing. Thus g, and hence f, are differentiable almost everywhere by Lebesgue's Theorem.

> **Rademacher's Theorem**
> *Any Lipschitz continuous function $f\colon \mathbb{R} \to \mathbb{R}$ is differentiable almost everywhere.*[10]

1.5 A Fundamental Characterization of Continuity

Let us now turn back to the investigation of functions that are continuous in the ordinary sense. First, a characterization theorem.

> **Proposition 1**
> *For any metric spaces X and Y, and $f \in Y^X$, the following statements are equivalent:*
>
> (a) *f is continuous.*
>
> (b) *For every open subset O of Y, the set $f^{-1}(O)$ is open in X.*
>
> (c) *For every closed subset S of Y, the set $f^{-1}(S)$ is closed in X.*
>
> (d) *For any $x \in X$ and $(x^m) \in X^\infty$, $x^m \to x$ implies $f(x^m) \to f(x)$.*[11]

---

examples have been devised. For instance, consider the real map f defined on $\mathbb{R}$ by $f(t) := \sum^\infty \left|10^i t - [10^i t]\right| / 10^i$, where $[10^i t]$ stands for an integer closest to $10^i t$. (This example is due to Bartel van der Waerden.)

Quiz. Prove that f is continuous, but $f'(t)$ does not exist for any real number t.

[10] This result remains true for functions that map $\mathbb{R}^n$ to $\mathbb{R}^m$ as well, provided that we suitably extend the notion of "almost everywhere" to $\mathbb{R}^n$. For an elementary proof of this in the case $n = m$, consult Zajicek (1992). The general case is quite complicated and is treated in, say, Federer (1996, Section 3.1).

[11] If $x^m \to x$ implies $f(x^m) \to f(x)$ for any sequence $(x^m) \in X^\infty$ then we are assured that f is continuous at x. Thus the sequential characterization of continuity applies *locally* as well. We can also formulate the "local" version of (b) as: "The inverse image of any open neighborhood of $f(x)$ (under f) is open in X." It is easy to check that this statement holds iff f is continuous at x.

Proof

(a) ⇒ *(b)*. Take any open $O \subseteq Y$ and any x in $f^{-1}(O)$. Then $f(x) \in O$, so, since O is open, there exists an $\varepsilon > 0$ such that $N_{\varepsilon,Y}(f(x)) \subseteq O$. But, by continuity of f at x, we can find a $\delta > 0$ such that $f(N_{\delta,X}(x)) \subseteq N_{\varepsilon,Y}(f(x))$ so that $N_{\delta,X}(x) \subseteq f^{-1}(O)$. Since x is arbitrary in $f^{-1}(O)$, this means that $f^{-1}(O)$ is open.

(b) ⇔ *(c)*. If S is a closed subset of Y, then $Y \backslash S$ is open in Y, so that (b) implies that $f^{-1}(Y \backslash S)$ is open in X. Since $X \backslash f^{-1}(S) = f^{-1}(Y \backslash S)$, this means that $f^{-1}(S)$ is closed. That (c) implies (b) is shown analogously.

(b) ⇒ *(d)*. Take any $x \in X$ and any sequence $(x^m) \in X^\infty$ with $x^m \to x$. Fix an arbitrary $\varepsilon > 0$. We wish to show that the terms of the sequence $(f(x^m))$ belong to $N_{\varepsilon,Y}(f(x))$ eventually. But $x \in f^{-1}(N_{\varepsilon,Y}(f(x)))$, and since $N_{\varepsilon,Y}(f(x))$ is open (Example C.3.[1]), (b) implies that $f^{-1}(N_{\varepsilon,Y}(f(x)))$ is also open. So, $\lim x^m = x$ implies that there exists an $M > 0$ such that $x^m \in f^{-1}(N_{\varepsilon,Y}(f(x)))$ for all $m \geq M$. Thus $f(x^m) \in N_{\varepsilon,Y}(f(x))$ for all $m \geq M$, and we are done.

(d) ⇒ *(a)*. Take any $x \in X$ and $\varepsilon > 0$. We wish to find a $\delta > 0$ such that $f(N_{\delta,X}(x)) \subseteq N_{\varepsilon,Y}(f(x))$. To derive a contradiction, suppose that such a δ does not exist. Then we can find a sequence (y^m) in Y such that $y^m \in f(N_{\frac{1}{m},X}(x)) \backslash N_{\varepsilon,Y}(f(x))$ for each $m \geq 1$. (Formally, we invoke the Axiom of Choice here.) Clearly, $y^m = f(x^m)$ for some $x^m \in N_{\frac{1}{m},X}(x)$ for each $m = 1, 2, \ldots$. But it is obvious that $x^m \to x$, so, by (d), $y^m \to f(x)$. This implies that there exists an $M > 0$ such that $y^m \in N_{\varepsilon,X}(f(x))$ for all $m \geq M$, contradicting the choice of y^m. ■

Proposition 1 provides four different viewpoints of continuity. Depending on the nature of the problem at hand, any one of these viewpoints may prove more useful than the others.

Example 3

In the following examples, X and Y stand for arbitrary metric spaces. (X is complete in the last example, however.)

[1] If $f \in Y^X$ is an open injection, that is, if f is an injection that maps every open subset of X onto an open subset of Y, then f^{-1} is a continuous function on $f(X)$. This fact follows immediately from the open set characterization of continuity.

[2] As you were asked to prove in Exercise 1, the set $\{x \in X : \varphi(x) \geq \alpha\}$ is closed for any continuous $\varphi \in \mathbb{R}^X$ and any $\alpha \in \mathbb{R}$. Since $\{x : \varphi(x) \geq \alpha\} = \varphi^{-1}([\alpha, \infty))$, this is proved in one line by using the closed set characterization of continuity.

[3] Suppose that you were asked to prove the following observation: *A continuous function is determined by its values on a dense set.* That is, if f and g are continuous functions in Y^X, and $f|_S = g|_S$ with S being a dense set in X, then we must have $f = g$. With the sequential characterization of continuity, this problem is easily solved: Since S is dense in X, for any $x \in X$ there exists a sequence $(x^m) \in S^\infty$ such that $x^m \to x$ so that $f(x) = \lim f(x^m) = \lim g(x^m) = g(x)$.

[4] Let φ and ψ be real functions on X. Then, $\varphi + \psi$, $|\varphi|$ and $\varphi\psi$ are continuous, while $\frac{\varphi}{\psi}$ is continuous provided that it is well-defined. These claims are straightforward consequences of the sequential characterization of continuity and Exercise A.35.

*[5] (*Proof of the Generalized Banach Fixed Point Theorem* 2) We adopt the notation used in the statement of this theorem in Section C.6.3. The first step of the proof is to observe that Φ must be continuous. Indeed, since $f^m(t) \to 0$ for all $t \geq 0$, we have either $f(t) = 0$ for some $t > 0$ or $\{f(t) : t > 0\}$ contains arbitrarily small numbers. (Why?) In either case, using monotonicity of f and the inequality $d(\Phi(x), \Phi(y)) \leq f(d(x, y))$, which is valid for all $x, y \in X$, we find that Φ must be continuous. The second (and main) step of the proof is to show that $(\Phi^m(x))$ is a Cauchy sequence for any $x \in X$. (Recall the method of successive approximations!) We leave this step as a (somewhat challenging) exercise to the reader. In the final step of the proof, we use the completeness of X to ensure that $(\Phi^m(x))$ converges to an element of X, say y. But then, by continuity of Φ and Proposition 1, we have $\Phi^{m+1}(x) = \Phi(\Phi^m(x)) \to \Phi(y)$. But of course, $\Phi^{m+1}(x) \to y$ by definition of y, and hence we conclude that $y = \Phi(y)$. This establishes that Φ has a fixed point. The uniqueness of this point follows from the observation that, for any two fixed points y and z, we have $d(y, z) = d(\Phi^m(y), \Phi^m(z)) \leq f^m(d(x, z)) \to 0$. ∎

Since the sequential characterization of continuity is frequently used in applications, we illustrate its use with two further examples.

Exercise 14 Prove: For any two metric spaces X and Y, if X is separable and there exists a continuous surjection $f \in Y^X$, then Y is also separable.

Exercise 15^H For any two metric spaces X and Y, show that $f \in Y^X$ is continuous iff it is continuous on every compact subset of X.

In the rest of this chapter, we use Proposition 1 mostly without giving explicit reference to it. You should thus begin regarding the properties stated in this result as alternative "definitions" of continuity.

1.6 Homeomorphisms

If $f : X \to Y$ is a bijection from a metric space X onto another metric space Y, and both f and f^{-1} are continuous, then it is called a **homeomorphism** between X and Y. If there exists such a bijection, then we say that X and Y are **homeomorphic** (or that "Y is homeomorphic to X"). If f is not necessarily surjective, but $f : X \to f(X)$ is a homeomorphism, then f is called an **embedding** (from X into Y). If there exists such an injection, we say that "X can be embedded in Y."

Two homeomorphic spaces are indistinguishable from each other insofar as their neighborhood structures are concerned. If X and Y are homeomorphic, then corresponding to each open set O in X there is an open set $f(O)$ in Y, and conversely, corresponding to each open set U in Y there exists an open set $f^{-1}(U)$ in X. (*Proof.* Apply Proposition 1.) Therefore, loosely speaking, Y possesses any property that X possesses so long as this property is defined in terms of open sets. (Such a property is called a *topological property.*[12])

For instance, if X is a connected metric space and Y is homeomorphic to X, then Y must also be connected. (You may choose to prove this now—it is easy—or wait for Proposition 2.) Put differently, connectedness is a topological property. The same goes for separability and compactness as well.

[12] Formally, a property for metric spaces is referred to as a **topological property** if it is invariant under any homeomorphism, that is, whenever this property is true for X, it must also be true for any other metric space that is homeomorphic to X.

(See Exercise 14 and Proposition 3). So, for example, neither $(0,1)\backslash\{\frac{1}{2}\}$ nor $[0,1]$ can be homeomorphic to $\mathbb{R}_+$. On the other hand, $[0,1)$ is homeomorphic to $\mathbb{R}_+$. (Indeed, $t \mapsto \frac{t}{1-t}$ is a homeomorphism between $[0,1)$ and $\mathbb{R}_+$.)

As another example, note that if d and D are equivalent metrics on a nonempty set X, then (X,d) and (X,D) are necessarily homeomorphic. In fact, the identity map id_X constitutes a homeomorphism between these two spaces. It follows that $\mathbb{R}$ and $(\mathbb{R},d')$ are homeomorphic, where $d'(x,y) := \frac{|x-y|}{1+|x-y|}$. Similarly, $\mathbb{R}^{n,p}$ and $\mathbb{R}^{n,q}$ are homeomorphic for any $n \in \mathbb{N}$ and $1 \leq p,q \leq \infty$.

The fact that $\mathbb{R}_+$ and $[0,1)$ are homeomorphic shows that neither completeness nor boundedness is preserved by a homeomorphism; these are *not* topological properties.[13] Thus, there are important senses in which two homeomorphic metric spaces may be of different character. If, however, $f \in Y^X$ is a homeomorphism that preserves the distance between any two points, that is, if

$$d_Y(f(x), f(y)) = d(x,y) \qquad \text{for all } x,y \in X,$$

then we may conclude that the spaces (X,d) and (Y,d_Y) are indistinguishable as metric spaces—one is merely a relabeling of the other. In this case, X and Y are said to be **isometric**, and we say that f is an **isometry** between them. For instance, $\overline{\mathbb{R}}$ and $[-1,1]$ are isometric (Example C.1.[3]).

EXERCISE 16[H]

(a) Let $X := (-\infty,0) \cup [1,\infty)$ and define $f \in \mathbb{R}^X$ by $f(t) := t$ if $t < 0$, and $f(t) := t-1$ if $t \geq 1$. Show that f is a continuous bijection that is not a homeomorphism.

(b) Show that $(0,1)$ and $\mathbb{R}$ are homeomorphic but not isometric.

(c) Any isometry is uniformly continuous. Why?

(d) Take any $1 \leq p \leq \infty$, and define the **right-shift** and **left-shift operators** on ℓ^p as the self-maps R and L with $R(x_1,x_2,\ldots) := (0,x_1,x_2,\ldots)$ and $L(x_1,x_2,\ldots) := (x_2,x_3,\ldots)$, respectively. Show that R is an isometry. How about L?

[13] If only to drive this point home, note that *every* metric space is homeomorphic to a bounded metric space. (There is nothing puzzling about this. Go back and read Remark C.1 again.)

EXERCISE 17 For any real numbers a and b with $a < b$, show that $\mathbf{C}[0,1]$ and $\mathbf{C}[a,b]$ are isometric.

EXERCISE 18 Let d and D be two metrics on a nonempty set X.

(a) Show that d and D are equivalent iff id_X is an homeomorphism.

(b) (Carothers) (X,d) and (X,D) may be homeomorphic, even if d and D are not equivalent. For instance, let $X := \{0, 1, \frac{1}{2}, \frac{1}{3}, \ldots\}$, $d := d_1$, and define D on X^2 as follows: $D(x,1) := x$ and $D(x,0) := 1 - x$ if $x \in X\backslash\{0,1\}$, while $D(x,y) := |x - y|$ otherwise. Use part (a) to show d and D are not equivalent, but (X,d) and (X,D) are homeomorphic.

EXERCISE 19^{H} Two metric spaces X and Y are said to be **uniformly homeomorphic** if there exists a bijection $f \in Y^X$ such that both f and f^{-1} are uniformly continuous. Show that if X and Y are uniformly homeomorphic and X is complete, then Y must be complete as well.

EXERCISE 20^{H} Let X and Y be two metric spaces, and $f \in Y^X$. Prove:

(a) f is continuous iff $f(cl_X(S)) \subseteq cl_Y(f(S))$ for any $S \in 2^X$.

(b) If f is injective, then f is a homeomorphism iff $f(cl_X(S)) = cl_Y(f(S))$ for any $S \in 2^X$.

EXERCISE 21 By Theorem C.4, the Hilbert cube $[0,1]^\infty$ is separable. In fact, there is a sense in which $[0,1]^\infty$ "includes" all separable metric spaces: *Every separable metric space X can be embedded in the Hilbert cube $[0,1]^\infty$*. This can be proved as follows: Let $Y = \{y^1, y^2, \ldots\}$ be a countable dense set in X, define d' as in Remark C.1, and define $f : X \to [0,1]^\infty$ by $f(x) := (d'(y^1,x), d'(y^2,x), \ldots)$. Now show that f is a homeomorphism from X onto $f(X)$.

2 Continuity and Connectedness

Our aim in this section is to investigate the properties of continuous functions defined on an arbitrary connected metric space. Let us begin by making note of the following simple characterization of the connectedness property in terms of continuous functions.

EXERCISE 22 Prove that a metric space X is connected iff there does not exist a continuous surjection $f \in \{0, 1\}^X$.[14]

The following simple but very useful result assures us that a continuous image of a connected set is connected.

PROPOSITION 2
Let X and Y be two metric spaces, and $f \in Y^X$ a continuous function. If X is connected, then $f(X)$ is a connected subset of Y.

PROOF
If $f(X)$ is not connected in Y, then we can find two nonempty and disjoint open subsets O and U of $f(X)$ such that $O \cup U = f(X)$. But then, by Proposition 1, $f^{-1}(O)$ and $f^{-1}(U)$ are nonempty open subsets of X. Moreover, these sets are disjoint, and we have $f^{-1}(O) \cup f^{-1}(U) = f^{-1}(O \cup U) = X$. (Why?) Conclusion: X is not connected. ∎

In the previous chapter we were unable to provide many examples of connected sets. We characterized the class of all connected sets in $\mathbb{R}$ (i.e., the class of all intervals), but that was about it. But now we can use Proposition 2, along with the fact that an interval is connected, to find many other connected sets. For instance, we can now show that any semicircle in $\mathbb{R}^2$ is connected. Consider the following two semicircles:

$$A := \{(x_1, x_2) : x_1^2 + x_2^2 = 1, x_2 \geq 0\} \quad \text{and}$$
$$B := \{(x_1, x_2) : x_1^2 + x_2^2 = 1, x_2 \leq 0\}.$$

Let us define the map $F : [-1, 1] \to \mathbb{R}^2$ by $F(t) := (t, \sqrt{1 - t^2})$. This map is continuous on $[-1, 1]$ (Example 2.[3]) so that, by Proposition 2, $F([-1, 1]) = A$ is connected in $\mathbb{R}^2$. Similarly, B is connected, and using these two observations together, one sees that the unit circle is connected in $\mathbb{R}^2$ (Exercise C.17). More generally, a **path** in $\mathbb{R}^n$, $n \in \mathbb{N}$, is defined as any set of the form $\{(f_1(t), \ldots, f_n(t)) : t \in I\}$, where I is an interval and f_i is a continuous real function on I, $i = 1, \ldots, n$. By Example 2.[3] and Proposition 2, we may now conclude: *Every path in $\mathbb{R}^n$ is connected.* (Compare with Example C.5.)

[14] There is no ambiguity in this problem since there is in effect only one way of metrizing the set $\{0, 1\}$.

EXERCISE 23 A metric space X is called **path-connected** if, for any $x, y \in X$, there exists a continuous function $F \in X^{[0,1]}$ with $F(0) = x$ and $F(1) = y$. Show that every path-connected space is connected.[15]

It is now time to see what connectedness can do for you. Recall that every continuous real function on $\mathbb{R}$ has the intermediate value property (Exercise A.54), a very useful fact that would be covered in any introductory calculus course. As an immediate corollary of Proposition 2, we now obtain a substantial generalization of this result—apparently, any connected metric space would do as well as $\mathbb{R}$ on this score.

THE INTERMEDIATE VALUE THEOREM

Let X be a connected metric space and $\varphi \in \mathbb{R}^X$ a continuous function. If $\varphi(x) \leq \alpha \leq \varphi(y)$ for some $x, y \in X$, then there exists a $z \in X$ such that $\varphi(z) = \alpha$.

PROOF

By Proposition 2, $\varphi(X)$ is connected in $\mathbb{R}$. Hence $\varphi(X)$ must be an interval (Example C.5). ■

COROLLARY 1

Given any $-\infty < a \leq b < \infty$, any continuous self-map f on $[a, b]$ has a fixed point.

PROOF

Take any continuous self-map f on $[a, b]$, and define $g \in \mathbb{R}^{[a,b]}$ by $g(t) := f(t) - t$ (Figure 1). Obviously, g is continuous. Moreover, $g(t) > 0$ cannot be true for all $a \leq t \leq b$. (Why?) Similarly, we cannot have $g < 0$. Therefore, either $g(x) = 0$ for some $a \leq x \leq b$, or we have $g(x) < 0 < g(y)$ for some $a \leq x, y \leq b$. In the former case x is a fixed point of f, while in the latter case we can apply the Intermediate Value Theorem to complete the proof. ■

[15] A connected space need not be path-connected. (An example of such a space is the following metric subspace of $\mathbb{R}^2$: $(\{0\} \times [-1, 1]) \cup \{(x, \sin(\frac{1}{x})) : 0 < x \leq 1]\}$. This space is sometimes referred to as the topologist's sine curve; plot it for yourself.) However, every connected *open* set in $\mathbb{R}^n$ is path-connected. (Talk is cheap. Proofs?)

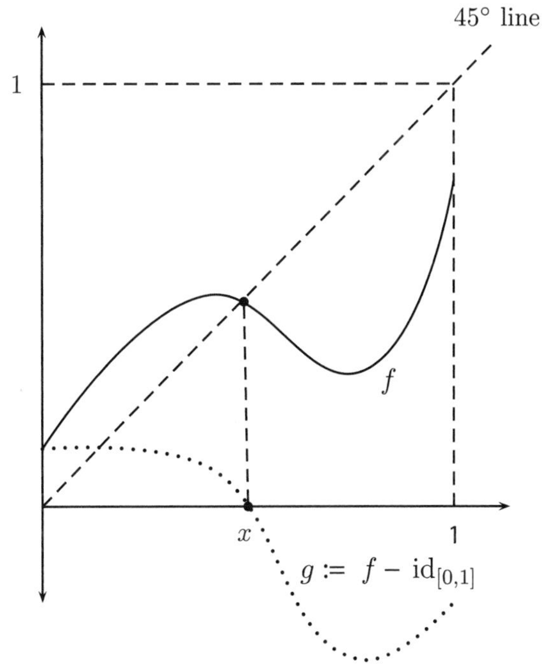

Figure 1

Proposition 2 and/or the Intermediate Value Theorem can sometimes be used to prove that a function is *not* continuous. For instance, Proposition 2 yields readily that there does not exist a continuous function that maps $[0, 1]$ onto $[0, 1]\backslash\{\frac{1}{2}\}$.

To give a less trivial example, let us show that *no injection of the form* $\varphi : \mathbb{R}^2 \to \mathbb{R}$ *can be continuous.* Indeed, if a real map φ on $\mathbb{R}^2$ is injective, then its values at the points $(-1, 0)$ and $(1, 0)$ must be distinct, say $\varphi(-1, 0) < \varphi(1, 0)$. Then, since the semicircles $A := \{(x_1, x_2) : x_1^2 + x_2^2 = 1, x_2 \geq 0\}$ and $B := \{(x_1, x_2) : x_1^2 + x_2^2 = 1, x_2 \leq 0\}$ are both connected in $\mathbb{R}^2$, if φ were continuous, the Intermediate Value Theorem would yield $[\varphi(-1, 0), \varphi(1, 0)] \subseteq \varphi(A)$ and $[\varphi(-1, 0), \varphi(1, 0)] \subseteq \varphi(B)$. This would, of course, contradict the injectivity of φ.

We conclude with two more examples of this nature.

Exercise 24[H] Show that there is no continuous self-map f on $\mathbb{R}$ with $f(\mathbb{Q}) \subseteq \mathbb{R}\backslash\mathbb{Q}$ and $f(\mathbb{R}\backslash\mathbb{Q}) \subseteq \mathbb{Q}$.

Exercise 25[H] Show that $\mathbb{R}$ and $\mathbb{R}^n$ are homeomorphic iff $n = 1$.

3 Continuity and Compactness

3.1 Continuous Image of a Compact Set

We now turn to the investigation of the properties of continuous functions defined on compact metric spaces. A fundamental observation in this regard is the following: Continuous image of a compact set is compact.

> **PROPOSITION 3**
> *Let X and Y be two metric spaces, and $f \in Y^X$ a continuous function. If S is a compact subset of X, then $f(S)$ is a compact subset of Y.*

PROOF
Take any compact subset S of X, and let $\mathcal{O}$ be an open cover of $f(S)$. Then $f^{-1}(O)$ is open for each $O \in \mathcal{O}$ (Proposition 1), and we have

$$\bigcup\{f^{-1}(O) : O \in \mathcal{O}\} = f^{-1}(\cup\mathcal{O}) \supseteq f^{-1}(f(S)) \supseteq S.$$

That is, $\{f^{-1}(O) : O \in \mathcal{O}\}$ is an open cover of S. If S is compact, then there exists a finite subset of $\mathcal{U}$ of $\mathcal{O}$ such that $\{f^{-1}(U) : U \in \mathcal{U}\}$ covers S. But then

$$S \subseteq \bigcup\{f^{-1}(U) : U \in \mathcal{U}\} = f^{-1}(\cup\mathcal{U})$$

so that $f(S) \subseteq \cup\mathcal{U}$. That is, $\mathcal{U}$ is a finite subset of $\mathcal{O}$ that covers $f(S)$. ■

EXERCISE 26 Give an alternative proof of Proposition 3 by using Theorem C.2.

Proposition 3 is a very important observation that has many implications. First of all, it gives us a useful sufficient condition for the inverse of an invertible continuous function to be continuous.

> **THE HOMEOMORPHISM THEOREM**
> *If X is a compact metric space and $f \in Y^X$ is a continuous bijection, then f is a homeomorphism.*

PROOF
Let X be a compact metric space and $f \in Y^X$ a continuous bijection. We wish to show that for any nonempty closed subset S of $X, f(S)$ is closed in Y.

(By Propositions A.2 and 1, this means that f^{-1} is a continuous function. Right?)

Indeed, the closedness of any nonempty $S \subseteq X$ implies that S is a compact metric subspace of X (Proposition C.4). Thus, in this case, Proposition 3 ensures that $f(S)$ is a compact subset of Y, and it follows that $f(S)$ is closed in Y (Proposition C.5). ■

WARNING. The compactness hypothesis is essential for the validity of the Homeomorphism Theorem. Please go back and reexamine Exercise 16.(a) if you have any doubts about this.

A second corollary of Proposition 3 concerns the real functions.

COROLLARY 2
Any continuous real function φ defined on a compact metric space X is bounded.

3.2 The Local-to-Global Method

Since any compact set is bounded (Proposition C.5), Corollary 2 is really an immediate consequence of Proposition 3. But to highlight the crucial role that compactness plays here, we would like to produce a direct proof that does not rely on Proposition 3. Observe first that continuity readily delivers us the boundedness of φ in a neighborhood of each point in X. Indeed, by continuity, there exists a $\delta(x) > 0$ such that $\left|\varphi(x) - \varphi(y)\right| < 1$ for all $y \in N_{\delta(x),X}(x)$, which means that

$$\sup\{|\varphi(y)| : y \in N_{\delta(x),X}(x)\} \leq 1 + |\varphi(x)|.$$

But this is only a "local" observation, it does not provide a uniform bound for φ. In fact, continuity alone is not strong enough to give us a uniform bound for φ. (What is $\sup\left\{\frac{1}{t} : 0 < t < 1\right\}$?) It is precisely to find such a uniform bound that we need compactness. After all, as argued before, compactness provides a finite structure for infinite sets, thereby allowing us to extend a "local" observation to a "global" one. So cover X with $\{N_{\delta(x),X}(x) : x \in X\}$ and use its compactness to find a finite subset S of X such that $\{N_{\delta(x),X}(x) : x \in S\}$ covers X. This finite cover delivers us the uniform bound that we were looking for:

$$\sup\{|\varphi(x)| : x \in X\} < 1 + \max\{|\varphi(x)| : x \in S\}.$$

And there follows Corollary 2, just as another illustration of the mighty powers of compactness.

There is actually a "method" here—let us agree to call it the **local-to-global method**—that will be useful to us in some other, less trivial instances. For any given metric space X, suppose Λ is a property (that may or may not be satisfied by the open subsets of X) such that

($\blacktriangle$) Λ is satisfied by an open neighborhood of every point in X; and

($\triangledown$) if Λ is satisfied by the open subsets O and U of X, then it is also satisfied by $O \cup U$.

Then, if X is compact, it must possess the property Λ. (You see, by compactness, we can deduce that Λ is satisfied *globally* from the knowledge that it is satisfied locally.) Why is this? Well, by ($\blacktriangle$), for every $x \in X$ there is an $\varepsilon(x) > 0$ such that $N_{\varepsilon(x),X}(x)$ satisfies property Λ. But $\{N_{\varepsilon(x),X}(x) : x \in X\}$ is an open cover of X, so compactness of X entails that there exists a finite subset S of X such that $\{N_{\varepsilon(x),X}(x) : x \in S\}$ also covers X. But, by ($\triangledown$), property Λ is satisfied by $\cup\{N_{\varepsilon(x),X}(x) : x \in S\} = X$, and we are done. (Lesson: Compactness is your friend!)

The proof we sketched for Corollary 2 above used the local-to-global method in disguise. The property of interest there was the *boundedness* of $\varphi \in \mathbf{C}(X)$. Say that an open set O of X satisfies the property Λ if $\varphi|_O$ is bounded. The local-to-global method says that, given that X is compact, all we need is to show that this property is such that ($\blacktriangle$) and ($\triangledown$) are true. In this case ($\triangledown$) is trivial, so the only thing we need to do is to verify ($\blacktriangle$), but this is easily seen to hold true, thanks to the continuity of φ. It is as simple as this.[16]

Sometimes one needs to adopt some sort of a variation of the local-to-global method to extend a local fact to a global one by means of compactness. A case in point is provided by the fact that continuity becomes equivalent to uniform continuity on a compact metric space. This is a substantial generalization of Proposition A.11, and it is a result that you should always keep in mind. We leave its proof as an exercise here. You can either adapt the proof we gave for Proposition A.11, or apply a modification of the

[16] *Quiz.* Prove: If X is a compact metric space and $\mathcal{O}$ is a cover of X such that every point of X has a neighborhood that intersects only *finitely* many members of $\mathcal{O}$, then $\mathcal{O}$ is a finite set. (*Hint.* This is a showcase for the local-to-global method.)

local-to-global method. (Please give a clean proof here—this is a must-do exercise!)

EXERCISE 27 [H] For any two metric spaces X and Y, prove that if X is compact and $f : X \to Y$ is continuous, then f is uniformly continuous.

3.3 Weierstrass' Theorem

It is easy to improve on Corollary 2. Indeed, Proposition 3 tells us that $\varphi(X)$ is a closed and bounded subset of $\mathbb{R}$ whenever $\varphi \in \mathbb{R}^X$ is continuous and X is compact. (Why?) But any such set contains its sup and inf. (Yes?) Therefore, we have the following result, which is foundational for optimization theory.

WEIERSTRASS' THEOREM
If X is a compact metric space and $\varphi \in \mathbb{R}^X$ is a continuous function, then there exist $x, y \in X$ with $\varphi(x) = \sup \varphi(X)$ and $\varphi(y) = \inf \varphi(X)$.

Here are some applications.

EXAMPLE 4

[1] Let $c > 0$ and let $f : \mathbb{R}_+ \to \mathbb{R}$ be any increasing and concave function such that $f(0) = 0$. Let us think of f as modeling the production technology of a given firm that produces a single good, that is, $f(t)$ is interpreted as the amount of outcome produced by a firm upon employing some t level of inputs. Suppose the firm operates under a constant marginal cost $c > 0$, and the market price of its product is \$1. Then, the problem of the firm would simply be to maximize the map $t \mapsto f(t) - ct$ on $\mathbb{R}_+$. This problem may or may not have a solution in general, but if $f(x_0) < cx_0$ for some x_0 (which is a very weak condition that is almost always satisfied in economic models), there exists a solution. For, under these assumptions, one can show that $\sup\{f(t) - ct : t \geq 0\} = \max\{f(t) - ct : 0 \leq t \leq x_0\}$ by Weierstrass' Theorem. (Why, exactly?)

[2] The canonical (static) individual choice problem in economics is of the following form:

$$\text{Maximize } u(x) \quad \text{such that} \quad x \in \mathbb{R}^n_+ \text{ and } px \leq \iota, \tag{2}$$

where $n \in \mathbb{N}$, and px stands for the *inner product* of the n-vectors p and x, that is, $px := \sum^n p_i x_i$. Here $\mathbb{R}^n_+$ plays the role of the consumption space—there are n goods that the consumer can consume—while $u : \mathbb{R}^n_+ \to \mathbb{R}$ stands for the utility function of the subject individual. Finally, $p_i \geq 0$ stands for the price of good i, and $\iota \geq 0$ is the monetary income of the agent.

Is there a solution to (2)? Yes, if u is continuous and $p_i > 0$ for each i. Indeed, the question is none other than if $u(x) = \sup u(X)$ for some $x \in X$, where $X := \{x \in \mathbb{R}^n_+ : px \leq \iota\}$ is the budget set of the individual. If $p_i > 0$ for each i, X is obviously bounded. Moreover, by using Example 2.[4] and Proposition 1, it is easily seen to be closed. (Verify!) By Theorem C.1, then, X is compact, and thus by an appeal to Weierstrass' Theorem, we may conclude that there is a solution to (2), provided that u is continuous and $p_i > 0$, $i = 1, \ldots, n$. □

EXAMPLE 5

(*The Projection Operator*) Fix any $n \in \mathbb{N}$. Let S be a nonempty closed subset of $\mathbb{R}^n$, and let d stand for any one of the metrics d_p defined in Example C.1.[3]. We claim that

$$d(x, S) = \min\{d(x, y) : y \in S\} < \infty \qquad \text{for all } x \in \mathbb{R}^n.$$

To see this, observe that the claim is trivial for $x \in S$, so take any $x \notin S$. Now fix any $y \in S$, and define $\varepsilon := d(x, y)$. (Since we do not know if S is compact, we can't immediately apply Weierstrass' Theorem. We shall thus first use a little trick; see Figure 2.) Define $T := \{w \in \mathbb{R}^n : d(x, w) \leq \varepsilon\}$, which is easily checked to be closed and bounded, and hence compact (Theorem C.1). Since $y \in T$, we have $S \cap T \neq \emptyset$. Moreover, $d(x, y) \leq d(x, z)$ for any $z \in S \backslash T$. But since T is compact and S is closed, $S \cap T$ is compact (Proposition C.4), so by Example 1.[3] (which ensures that $z \mapsto d(x, z)$ is a continuous map on $\mathbb{R}^n$) and by Weierstrass' Theorem, there must exist a $y^* \in S \cap T$ such that $d(x, y^*) \leq d(x, z)$ for all $z \in S \cap T$. Conclusion: $d(x, S) = d(x, y^*)$.

While we are on this topic, let us note that we can do much better in the case of Euclidean spaces (where the underlying metric is d_2): *Given any point x and a nonempty closed and convex set S in a Euclidean space, there is a unique point in S that is closest to x.* We have just established the existence assertion. The uniqueness claim, on the other hand, follows from

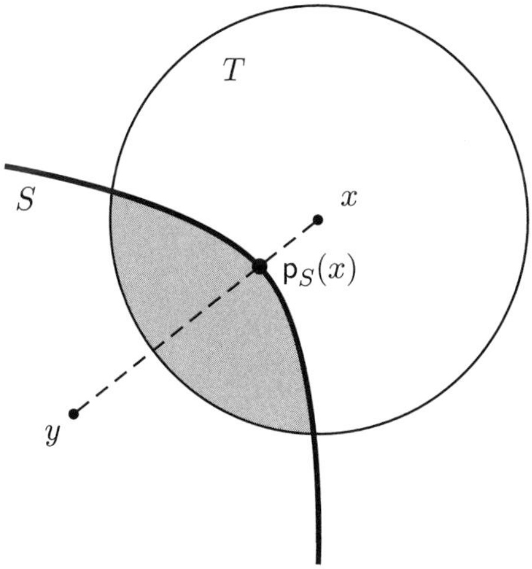

Figure 2

the simple observation that $d_2(x,\cdot)^2$ is a strictly convex function on $\mathbb{R}^n$ so that $d_2(x,y^*) = d_2(x,z)$ and $y^* \neq z$ imply

$$d_2\left(x, \tfrac{1}{2}(y^* + z)\right)^2 < d_2(x,y^*)^2 = d_2(x,S)^2,$$

which is impossible since $\frac{1}{2}(y+z) \in S$ by convexity of S.[17]

Let's dig a little deeper. Define the function $\mathsf{p}_S : \mathbb{R}^n \to S$ by letting $\mathsf{p}_S(x)$ stand for "the" point in S that is nearest to x. That is, we define p_S through the equation

$$d(x, \mathsf{p}_S(x)) = d(x,S).$$

The point $\mathsf{p}_S(x)$ is called the **projection of x on S**, and in a variety of contexts, it is thought of as "the best approximation of x from S." We now know that p_S is well-defined whenever S is a nonempty closed and convex subset of $\mathbb{R}^n$—in this case p_S is called the **projection operator into** S. This map is *idempotent*, that is, $\mathsf{p}_S \circ \mathsf{p}_S = \mathsf{p}_S$, and it admits x as a fixed point iff $x \in S$.

Do you think this map is continuous? If S is also known to be bounded, then the answer is easily seen to be yes. Indeed, if (x^m) is a sequence in $\mathbb{R}^n$ that converges to some point x, but $\mathsf{p}_S(x^m)$ does not converge to $\mathsf{p}_S(x)$, then,

[17] Warning. The use of d_2 is essential here. For instance, if $S := \{y \in \mathbb{R}^2_+ : y_1 + y_2 = 1\}$, then $d_1((0,0), S) = d_1((0,0), y)$ for *every* $y \in S$.

by Theorems C.1 and C.2, there exists a convergent subsequence $(\mathrm{p}_S(x^{m_k}))$ such that $y := \lim \mathrm{p}_S(x^{m_k}) \neq \mathrm{p}_S(x)$. (Why? Recall Exercise A.39.) Since S is closed, we have $y \in S$, and hence $d_2(x, \mathrm{p}_S(x)) < d_2(x, y)$. It then follows from the continuity of d_2 (Exercise 3) that $d_2(x^{m_k}, \mathrm{p}_S(x)) < d_2(x^{m_k}, \mathrm{p}_S(x^{m_k}))$ for k large enough, which is impossible. (Why?) We conclude: *In a Euclidean space, the projection operator into a nonempty convex and compact set is a continuous function.*[18] □

EXAMPLE 6
Here is a quick proof of Edelstein's Fixed Point Theorem (Exercise C.50). Let X be compact metric space and $\Phi \in X^X$ a pseudocontraction. It is plain that such a map may have at most one fixed point; the crux of the problem here is to establish the *existence* of a fixed point. Since Φ is continuous, the function $\varphi \in \mathbb{R}_+^X$ defined by $\varphi(x) := d(x, \Phi(x))$ is continuous. (Why? Recall Exercise 3 and Example 2.[1].) So, by Weierstrass' Theorem, φ attains its minimum on X, that is, $\varphi(x^*) = \min\{d(x, \Phi(x)) : x \in X\}$ for some $x^* \in X$. But since Φ is a pseudocontraction, we must have $\min\{d(x, \Phi(x)) : x \in X\} = 0$ (right?), and it follows that $d(x^*, \Phi(x^*)) = 0$, as we sought. Conclusion: *Every pseudocontraction on a compact metric space has a unique fixed point.*

Even more general results can be obtained by using this technique. For instance, you can show similarly that if X is a compact metric space and Φ is a continuous self-map on X with

$$d(\Phi(x), \Phi(y)) \leq \max\{d(\Phi(x), x), d(x, y), d(y, \Phi(y))\}$$

for each $x, y \in X$, then Φ has a fixed point. □

EXERCISE 28 Provide a single example that shows that neither compactness nor continuity is a necessary condition for the conclusion of Weierstrass' Theorem. Can either of these requirements be completely dispensed with in its statement?

[18] I have given this result here only to illustrate how nicely compactness and continuity interact with each other. Projection operators are in fact much better behaved than what this statement suggests. First, the use of the boundedness hypothesis is completely redundant here. Second, a projection operator into a closed convex set is not only continuous, it is also nonexpansive. (All this will be proved in Section G.3.3. See, in particular, Remark G.4.)

EXERCISE 29 True or false: $f([0,1]) = [\min f([0,1]), \max f([0,1])]$ for any $f \in \mathbf{C}[0,1]$.

EXERCISE 30^{H} (*The Converse of Weierstrass' Theorem in* $\mathbb{R}^n$) Let $n \in \mathbb{N}$. Show that if T is a nonempty subset of $\mathbb{R}^n$ such that *every* $\varphi \in \mathbf{C}(T)$ is bounded, then T must be compact.

EXERCISE 31^{H} Let f be a differentiable real function on $\mathbb{R}$, and $-\infty < a \leq b < \infty$. Show that if f is differentiable on (a,b) and $f'(b) \geq f'(a)$, then, for each $\alpha \in [f'(a), f'(b)]$, there exists a $t \in [a,b]$ with $f'(t) = \alpha$.[19]

4 Semicontinuity

Continuity is only a sufficient condition for a real function to assume its maximum (or minimum) over a compact set. Simple examples show that it is not at all necessary, and hence it is of interest to find other sufficient conditions to this effect. This brings us to the following continuity concepts, which were introduced in the 1899 dissertation of René Baire.

DEFINITION

Let X be any metric space, and $\varphi \in \mathbb{R}^X$. We say that φ is **upper semicontinuous at** $x \in X$ if, for any $\varepsilon > 0$, there exists a $\delta > 0$ (which may depend on both ε and x) such that

$$d(x,y) < \delta \quad \text{implies} \quad \varphi(y) \leq \varphi(x) + \varepsilon$$

for each $y \in X$. Similarly, if, for any $\varepsilon > 0$, there exists a $\delta > 0$ such that

$$d(x,y) < \delta \quad \text{implies} \quad \varphi(y) \geq \varphi(x) - \varepsilon,$$

then φ is said to be **lower semicontinuous at** x. The function φ is said to be **upper (lower) semicontinuous** if it is upper (lower) semicontinuous at each $x \in X$.

[19] A real map φ on a metric space X is called **Darboux continuous** if, for any $x, y \in X$ and $\alpha \in \mathbb{R}$ with $\varphi(x) < \alpha < \varphi(y)$, there exists a $z \in X$ such that $\varphi(z) = \alpha$. If X is connected, then continuity of φ entails its Darboux continuity—this is the Intermediate Value Theorem. (The converse is false, obviously.) The present exercise shows that the derivative of a differentiable real map on (a,b) is necessarily Darboux continuous, while, of course, it need not be continuous.

In a manner of speaking, if $\varphi \in \mathbb{R}^X$ is upper semicontinuous at x, then the images of points nearby x under φ do not exceed $\varphi(x)$ "too much," while there is no restriction about how far these images can fall below $\varphi(x)$. Similarly, if φ is lower semicontinuous at x, then the images of points nearby x under φ do not fall below $\varphi(x)$ "too much," but they can still be vastly greater than $\varphi(x)$.

It follows readily from the definitions that a real function on a metric space is continuous iff it is both upper and lower semicontinuous. (Right?) So, semicontinuity is really a weakening of the ordinary notion of continuity for real functions.

Let's search for an alternative way of looking at the notion of semicontinuity that would make the computations easier. Notice first that if $\varphi \in \mathbb{R}^X$ is upper semicontinuous at $x \in X$, then, for any $\varepsilon > 0$, there exists a $\delta > 0$ such that

$$\varphi(x) + \varepsilon \geq \sup\{\varphi(y) : y \in N_{\delta,X}(x)\}.$$

But then

$$\begin{aligned}\varphi(x) + \varepsilon &\geq \inf\{\sup\{\varphi(y) : y \in N_{\delta,X}(x)\} : \delta > 0\} \\ &= \lim_{m\to\infty} \sup\{\varphi(y) : y \in N_{\frac{1}{m},X}(x)\}\end{aligned}$$

for any $\varepsilon > 0$. (Why?) It follows that

$$\varphi(x) \geq \lim_{m\to\infty} \sup\{\varphi(y) : y \in N_{\frac{1}{m},X}(x)\}.$$

Since $x \in N_{\frac{1}{m},X}(x)$ for every $m \in \mathbb{N}$, the converse inequality holds as well, and we obtain

$$\varphi(x) = \lim_{m\to\infty} \sup\{\varphi(y) : y \in N_{\frac{1}{m},X}(x)\}. \tag{3}$$

(Go a little slow here. Understanding the above expression will make life much easier in what follows. Draw a graph of a continuous real function on $[0, 1]$, and see why it satisfies (3) everywhere. Draw next an increasing and right-continuous step function on $[0, 1]$, and observe that (3) holds at each discontinuity point of your function. How about left-continuous increasing step functions?)

We can actually reverse the reasoning that led us to (3) to show that this expression implies the upper semicontinuity of φ at x. (Check!) Thus, if we define $\varphi^\bullet : X \to \overline{\mathbb{R}}$ by

$$\varphi^\bullet(x) := \lim_{m\to\infty} \sup\{\varphi(y) : y \in N_{\frac{1}{m},X}(x)\}, \tag{4}$$

then we may conclude: φ is upper semicontinuous at x iff $\varphi(x) = \varphi^\bullet(x)$. (The map $\varphi^\bullet$ is called the **lim sup** of φ.) The analogous reasoning would show that φ is lower semicontinuous at x iff $\varphi(x) = \varphi_\bullet(x)$, where $\varphi_\bullet : X \to \overline{\mathbb{R}}$ is defined by

$$\varphi_\bullet(x) := \lim_{m\to\infty} \inf\{\varphi(y) : y \in N_{\frac{1}{m},X}(x)\}. \tag{5}$$

(The map $\varphi_\bullet$ is called the **lim inf** of φ.) These observations make it easier to check whether a real function is semicontinuous or not. (See Figure 3 for simple illustrations.) For instance, it is now quite easy to see that $\mathbf{1}_{\mathbb{R}_+}$, the indicator function of $\mathbb{R}_+$ on $\mathbb{R}$, is an upper semicontinuous function on $\mathbb{R}$ that is not lower semicontinuous at zero, whereas $\mathbf{1}_{\mathbb{R}_{++}}$ is lower semicontinuous on $\mathbb{R}$, but it is not upper semicontinuous at zero. (Yes?) A more interesting example is $\mathbf{1}_{\mathbb{Q}}$, which is upper semicontinuous at each rational, lower semicontinuous at each irrational, and discontinuous everywhere. These examples suggest that one can think of upper semicontinuity as allowing for *upward* jumps and lower semicontinuity for *downward* jumps.

The characterization of semicontinuity in terms of $\varphi^\bullet$ and $\varphi_\bullet$ also allows us to generalize the principal definition to cover the extended real-valued functions.

DEFINITION

Let X be a metric space and $\varphi : X \to \overline{\mathbb{R}}$ any function. We say that φ is **upper semicontinuous at** $x \in X$ if $\varphi(x) = \varphi^\bullet(x)$, and it is **lower semicontinuous at** x if $\varphi(x) = \varphi_\bullet(x)$. (Here $\varphi^\bullet$ and $\varphi_\bullet$ are defined by (4) and (5), respectively.) We say that φ is **upper (lower) semicontinuous** if it is upper (lower) semicontinuous at each $x \in X$.

It is a bit unusual to give two different definitions for the same concept, but there is no ambiguity here, since we showed above that our earlier definition is actually covered by this new one. Moreover, the latter definition is superior to the previous one because it allows us to talk about the

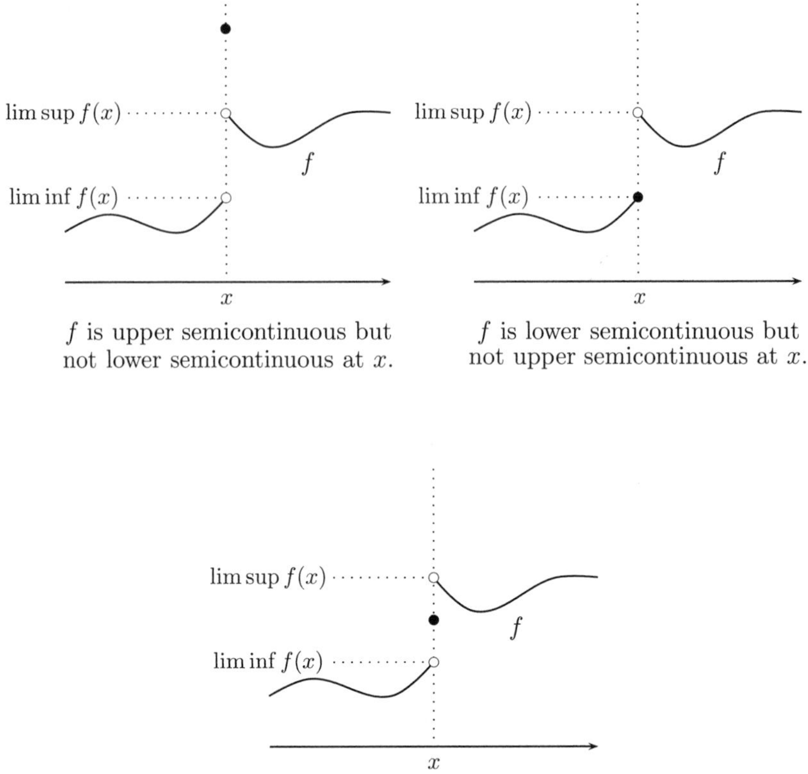

f is upper semicontinuous but not lower semicontinuous at x.

f is lower semicontinuous but not upper semicontinuous at x.

f is neither upper semicontinuous nor lower semicontinuous at x.

FIGURE 3

semicontinuity of a function such as $\varphi^\bullet$, which may be extended real-valued (even when φ is real-valued).[20]

EXERCISE 32 Let φ be any real function defined on a metric space X. Prove:

(a) $\varphi_\bullet \leq \varphi \leq \varphi^\bullet$;

(b) φ is upper semicontinuous iff $-\varphi$ is lower semicontinuous;

(c) $\varphi^\bullet$ is upper semicontinuous;

(d) $\varphi_\bullet$ is lower semicontinuous.

[20] For instance, consider the function $f : [-1, 1] \to \mathbb{R}$ defined by $f(0) := 0$ and $f(t) := \frac{1}{|t|}$ otherwise. What is $f^\bullet(0)$? Is f upper semicontinuous at 0? Is $f^\bullet$?

Here is a useful characterization of upper semicontinuous functions that parallels the characterization of continuous functions given in Proposition 1.

PROPOSITION 4

For any metric space X, and $\varphi \in \mathbb{R}^X$, the following statements are equivalent:

(a) *φ is upper semicontinuous.*

(b) *For every $\alpha \in \mathbb{R}$, the set $\{x : \varphi(x) < \alpha\}$ is open in X.*

(c) *For every $\alpha \in \mathbb{R}$, the set $\{x : \varphi(x) \geq \alpha\}$ is closed in X.*

(d) *For any $x \in X$ and $(x_m) \in X^\infty$, $x_m \to x$ implies $\varphi(x) \geq \limsup \varphi(x_m)$.*

PROOF

(a) ⇒ (b). Fix any $\alpha \in \mathbb{R}$ and assume that $\varphi^{-1}((-\infty, \alpha)) \neq \emptyset$, for otherwise the claim is trivial. Let $x \in \varphi^{-1}((-\infty, \alpha))$, and fix any $0 < \varepsilon < \alpha - \varphi(x)$. Then, by the (first definition of) upper semicontinuity at x, there exists a $\delta > 0$ such that $\alpha > \varphi(x) + \varepsilon \geq \varphi(y)$ for all $y \in N_{\delta,X}(x)$. Then $N_{\delta,X}(x) \subseteq \varphi^{-1}((-\infty, \alpha))$, so we conclude that $\varphi^{-1}((-\infty, \alpha))$ is open.

(b) ⇔ (c). This is trivial.

(b) ⇒ (d). Let (x_m) be a convergent sequence in X, and define $x := \lim x_m$. Fix an arbitrary $\varepsilon > 0$. Then, since x belongs to the open set $\{y \in X : \varphi(y) < \varphi(x) + \varepsilon\}$, there exists an $M > 0$ such that $\varphi(x^m) < \varphi(x) + \varepsilon$ for all $m \geq M$. This means that

$$\sup\{\varphi(x^m) : m \geq M\} \leq \varphi(x) + \varepsilon,$$

so $\limsup \varphi(x^m) \leq \varphi(x) + \varepsilon$. Since $\varepsilon > 0$ is arbitrary here, the claim follows.

(d) ⇒ (a). If φ was not upper semicontinuous at some $x \in X$, we could find an $\varepsilon > 0$ and a sequence (x^m) such that $d(x, x^m) < \frac{1}{m}$ and $\varphi(x^m) > \varphi(x) + \varepsilon$ for all $m \geq 1$. But in this case we would have $x^m \to x$, so, by (d), we would reach the contradiction that $\varphi(x) \geq \limsup \varphi(x^m) > \varphi(x)$. ■

Note that, by using Exercise 32.(b) and the fact that $\inf\{\varphi(x) : x \in X\} = -\sup\{-\varphi(x) : x \in X\}$ for any $\varphi \in \mathbb{R}^X$, one can easily recover from Proposition 4 the corresponding characterizations for lower semicontinuous functions. For example, φ is lower semicontinuous iff $\varphi(\lim x_m) \leq \liminf \varphi(x_m)$ for any convergent sequence (x_m) in X.

EXERCISE 33 Let $\varphi_i : \mathbb{R}^n \to \mathbb{R}$ be upper semicontinuous for each $i \in \mathbb{N}$. Show that $\varphi_1 + \varphi_2$, $\max\{\varphi_1, \varphi_2\}$ and $\inf\{\varphi_i : i \in \mathbb{N}\}$ are upper semicontinuous functions. Give an example to show that $\sup\{\varphi_i : i \in \mathbb{N}\}$ need not be upper semicontinuous.[21]

EXERCISE 34 Let S be a nonempty set in a metric space X. Show that the indicator function $\mathbf{1}_S$ is upper semicontinuous on X if S is closed, and lower semicontinuous if S is open.

After all this work, you must be wondering why one would ever need to deal with the notion of semicontinuity in practice. This concept will actually play an important role in an economic application that we will consider in the next section. Moreover, for your immediate enjoyment, we use the notion of semicontinuity to obtain a useful generalization of our beloved Weierstrass' Theorem. This is the highlight of this section.

PROPOSITION 5
(Baire) *Let X be a compact metric space, and $\varphi \in \mathbb{R}^X$. If φ is upper semicontinuous, then there exists an $x \in X$ with $\varphi(x) = \sup \varphi(X)$. If φ is lower semicontinuous, then there exists a y with $\varphi(y) = \inf \varphi(X)$.*

In words, an upper semicontinuous function always assumes its maximum (but not necessarily its minimum) over a compact set. Thus, if you are interested in the maximization of a particular function over a compact set, but if your function is not continuous (so that Weierstrass' Theorem is to no avail), upper semicontinuity should be the next thing to check. If your objective function turns out to be upper semicontinuous, then you're

[21] But the following is true: *If $\varphi \in \mathbb{R}^X$ is bounded and upper semicontinuous, then there exists a sequence (φ_m) of continuous functions such that $\varphi_m(x) \nearrow \varphi(x)$ for each $x \in X$.* I will not prove this approximation-by-continuous-functions theorem, but you might want to try it out for yourself in the case where X is a compact interval.

assured of the *existence* of a solution to your maximization problem. By contrast, lower semicontinuity is the useful property in the case of minimization problems.

EXAMPLE 7

[1] Consider the following optimization problem:

$$\text{Maximize } f(t) + \log(1+t) \quad \text{such that} \quad 0 \leq t \leq 2,$$

where $f \in \mathbb{R}^{[0,2]}$ is defined as

$$f(t) := \begin{cases} t^2 - 2t, & \text{if } 0 \leq t < 1 \\ 2t - t^2, & \text{if } 1 \leq t \leq 2 \end{cases}.$$

Does a solution to this problem exist? The answer is yes, but since f is discontinuous at $t = 1$, Weierstrass' Theorem does not deliver this answer readily. Instead, we observe that $f^{\bullet}(1) = 1 = f(1)$, so that the objective function of our problem is upper semicontinuous on $[0, 2]$, which, by Proposition 5, is all we need. On the other hand, since f is not lower semicontinuous, neither Weierstrass' Theorem nor Proposition 5 tells us if we can minimize the map $t \mapsto f(t) + \log(1+t)$ on $[0, 2]$.

[2] Let Y be the class of all piecewise linear continuous real maps f on $[0, 1]$ such that $f(0) = 0 = f(1)$ and $|f'| \leq 1$.[22] Then Y is a bounded and equicontinuous subset of $\mathbf{C}[0, 1]$, so by the "baby" Arzelà-Ascoli Theorem (Example C.9.[4]), $X := cl_{\mathbf{C}[0,1]}(Y)$ is compact in $\mathbf{C}[0, 1]$. Consider the following (calculus of variations) problem:

$$\text{Minimize } \int_0^1 (|f'(t)| - f(t))dt \quad \text{such that} \quad f \in X.$$

Does this problem have a solution? Since X is compact, one may hope to settle this question by an appeal to Weierstrass' Theorem. But this won't do! The objective function at hand is not continuous. For instance, consider the sequence (f_m) of "zig-zag" functions on $[0, 1]$ illustrated

[22] Piecewise linearity of f means that there are finitely many numbers $0 = a_0 < a_1 \ldots < a_k = 1$ such that f has a constant derivative on every $[a_i, a_{i+1}]$, $i = 0, \ldots, k$.

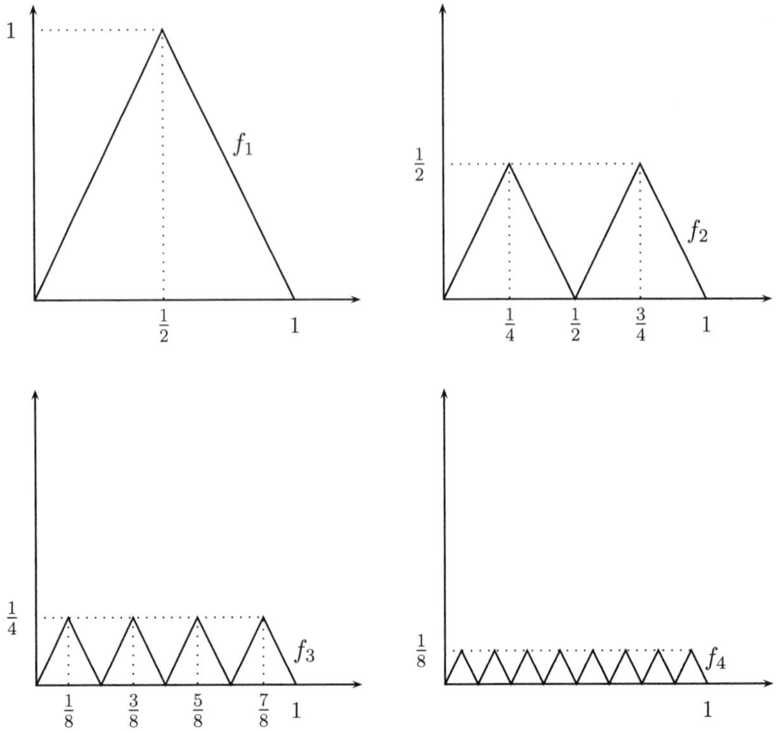

FIGURE 4

in Figure 4. It is readily checked that $\sup\{|f_m(t)| : 0 \leq t \leq 1\} \to 0$, so this sequence converges to the zero function on $[0, 1]$. And yet, $\int_0^1 (|f_m'(t)| - f_m(t))dt = 2 - \frac{1}{2^m} \to 2$. Thus the map $f \mapsto \int_0^1 (|f'(t)| - f(t))dt$ is not continuous on X, so Weierstrass' Theorem does not apply. But, as we leave for you to check, this map is lower semicontinuous, so we can apply Proposition 5 to conclude that our little problem has a solution.

[3] In view of Proposition 5, we can relax the continuity of f and u to upper semicontinuity in Examples 4.[1] and 4.[2] and reach the same conclusions. □

We next offer two alternative proofs for the first assertion of Proposition 5. The second assertion, in turn, follows from the first one in view of Exercise 32.(b).

Proof of Proposition 5
Clearly, there exists a sequence $(x^m) \in X^\infty$ such that $\varphi(x^m) \nearrow \sup \varphi(X) =: s$ (Exercise A.38). By Theorem C.2, there exists a subsequence (x^{m_k}) of this sequence that converges to some $x \in X$. But then, if φ is upper semicontinuous, we may apply Proposition 4 to find

$$s \geq \varphi(x) \geq \limsup \varphi(x^{m_k}) = \lim \varphi(x^m) = s$$

so that $\varphi(x) = s$. ∎

While elegant, this proof uses the notion of sequential compactness and hence requires us to invoke a somewhat "deep" result like Theorem C.2. It may thus be a good idea to give another proof for Proposition 5 that uses the compactness property more directly. Well, here is one.

Another Proof for Proposition 5
Assume that φ is upper semicontinuous. By Proposition 4, $\{\varphi^{-1}((-\infty, m)) : m \in \mathbb{N}\}$ is an open cover of X. Since X is compact, this class must have a finite subset that also covers X. So, there exists an $M \in \mathbb{R}$ such that $X \subseteq \varphi^{-1}((-\infty, M))$, and this means that $s := \sup \varphi(X)$ is finite. But then, by definition of s and Proposition 4, $\mathcal{A} := \{\varphi^{-1}([s - \frac{1}{m}, \infty)) : m \in \mathbb{N}\}$ is a class of closed subsets of X that has the finite intersection property. Thus, thanks to Example C.8, there exists an $x \in \bigcap \mathcal{A}$. Clearly, we must have $\varphi(x) = s$. ∎

5 Applications

In this section we allow ourselves to digress a bit to consider a few applications that draw from our development so far. We first consider a generalization of the Banach Fixed Point Theorem in which lower semicontinuity makes a major appearance. We then supplement our earlier results on utility theory by studying the problem of finding a continuous utility function that represents a given preference relation. Finally, we turn to the characterization of additive continuous functions defined on $\mathbb{R}^n$. As a corollary of this characterization, we will be able to prove here a theorem of de Finetti that is used quite often in the theory of individual and social choice.

5.1* Caristi's Fixed Point Theorem

As our first application, we wish to consider the following famed result of metric fixed point theory, which was proved by John Caristi in 1976.

> CARISTI'S FIXED POINT THEOREM
> *Let Φ be a self-map on a complete metric space X. If*
>
> $$d(x, \Phi(x)) \leq \varphi(x) - \varphi(\Phi(x)) \qquad \text{for all } x \in X$$
>
> *for some lower semicontinuous $\varphi \in \mathbb{R}^X$ that is bounded from below, then Φ has a fixed point in X.*

It is easy to see that this theorem generalizes the Banach Fixed Point Theorem. If $\Phi \in X^X$ is a contraction with the contraction coefficient $K \in (0, 1)$, then the hypothesis of Caristi's theorem is satisfied for $\varphi \in \mathbb{R}_+^X$ defined by $\varphi(x) := \frac{1}{1-K} d(x, \Phi(x))$. Indeed, φ is continuous (why?), and we have

$$\begin{aligned}
\varphi(x) - \varphi(\Phi(x)) &= \tfrac{1}{1-K}\left(d(x, \Phi(x)) - d(\Phi(x), \Phi^2(x))\right) \\
&\geq \tfrac{1}{1-K}\left(d(x, \Phi(x)) - Kd(x, \Phi(x))\right) \\
&= d(x, \Phi(x)).
\end{aligned}$$

In fact, Caristi's generalization is quite substantial. While the Banach Fixed Point Theorem requires the involved self-map to have the contraction property (which is much stronger than continuity), Caristi's Fixed Point Theorem does not even require the self-map to be continuous.

There are several ways of proving Caristi's theorem. The proof that we outline below, which is due to Bröndsted (1976), sits particularly square with the present treatment. Goebel and Kirk (1990) give a different proof, and discuss others that have appeared in the literature.

EXERCISE 35 (Bröndsted) Assume the hypotheses of Caristi's Fixed Point Theorem. Define the binary relation $\succsim$ on X by $y \succsim x$ iff $\varphi(x) - \varphi(y) \geq d(x, y)$.

(a) Show that $(X, \succsim)$ is a poset such that $U_{\succsim}(x) := \{y \in X : y \succsim x\}$ is closed for each $x \in X$.

(b) Fix any $x^0 \in X$. Use induction to construct a sequence $(x^m) \in X^\infty$ such that $\cdots \succsim x^2 \succsim x^1 \succsim x^0$ with $\varphi(x^m) \leq \frac{1}{m} + \inf \varphi(U_{\succsim}(x^{m-1}))$ for each $m \in \mathbb{N}$. By using the Cantor-Fréchet Intersection Theorem, prove that there is a unique x^* in $\bigcap_{m=0}^{\infty} U_{\succsim}(x^m)$.

(c) Show that x^* is $\succsim$-maximal in X.

(d) Show that $x^* = \Phi(x^*)$.

5.2 Continuous Representation of a Preference Relation

We now go back to the decision theoretic setting described in Sections B.4 and C.2.3, and see how one may improve the utility representation results we have obtained there by using the real analysis technology introduced thus far. Indeed, while fundamental, the results obtained in Section B.4 (and those in Section C.2.3) do need improvement. They are of limited use even in the simple setting of classic consumer theory, where one usually takes $\mathbb{R}^n_+$ as the grand commodity space. For instance, they are of no immediate help in dealing with the following apparently natural question: *What sort of preference relations defined on $\mathbb{R}^n_+$ can be represented by a continuous utility function?*

It turns out that this is not a very easy question to answer. There are, however, special cases of the basic problem that can be solved relatively easily. First of all, if all we need is the upper semicontinuity of the utility function, then we are in good shape. In particular, we have the following improvement of Rader's Utility Representation Theorem 1.

Rader's Utility Representation Theorem 2
Let X be a separable metric space, and $\succsim$ a complete preference relation on X. If $\succsim$ is upper semicontinuous, then it can be represented by an upper semicontinuous utility function $u \in \mathbb{R}^X$.

We give the proof in the form of an exercise.

Exercise 36 Assume the hypotheses of the theorem above, and use Rader's Utility Representation Theorem 1 to find a function $v \in [0,1]^X$ with $x \succsim y$ iff $v(x) \geq v(y)$ for any $x, y \in X$. Let $v^\bullet \in [0,1]^X$ be the lim sup of v. By Exercise 32, $v^\bullet$ is upper semicontinuous. Show that $v^\bullet$ represents $\succsim$.

Another interesting special case of the continuous utility representation problem obtains when $X = \mathbb{R}^n_+$ and when the preference relation to be represented is monotonic. This is explored in the next exercise.

Exercise 37[H] (*Wold's Theorem*) Let $n \in \mathbb{N}$. A preference relation $\succsim$ on $\mathbb{R}^n_+$ is called **strictly increasing** if $x > y$ implies $x \succ y$ for any $x, y \in \mathbb{R}^n_+$.

(a) Let $\succsim$ be a complete, continuous and strictly increasing preference relation on $\mathbb{R}^n_+$. Define $u : \mathbb{R}^n_+ \to \mathbb{R}$ by $u(x) := \max\{\alpha \geq 0 : x \succsim (\alpha, \ldots, \alpha)\}$. Show that u is well-defined, strictly increasing, and continuous, and that it represents $\succsim$.

*(b) How would the conclusions of part (a) change if $\succsim$ was a complete, lower semicontinuous, and strictly increasing preference relation on $\mathbb{R}^n_+$?

Let us now briefly sketch how one would approach the general problem.[23] The starting point is the following important result, which we state without proof.

The Open Gap Lemma[24]
(Debreu) *For any nonempty subset S of $\mathbb{R}$, there exists a strictly increasing function $f \in \mathbb{R}^S$ such that every $\supseteq$-maximal connected set in $\mathbb{R} \backslash f(S)$ is either a singleton or an open interval.*

The Open Gap Lemma is a technical result whose significance may not be self-evident, but it has far-reaching implications for utility theory. Informally put, it allows us to find a *continuous* utility function for a preference relation that we somehow already know to be representable. In other words, often it

[23] For a mathematically more sophisticated but still a very readable account, I recommend Richter (1980) on this matter.

[24] This result was first stated in Debreu (1954), but the proof given there had a flaw. A complete proof first appeared in Debreu (1964). Since then a number of relatively simple proofs have been obtained; see, for instance, Jaffray (1975) and Beardon (1992).

reduces the problem of finding a continuous representation to the simpler problem of finding a utility representation. The following result formalizes this point.

Lemma 1

(Debreu) *Let X be a metric space, and let $u \in \mathbb{R}^X$ represent a preference relation $\succsim$ on X. If $\succsim$ is continuous, then it is representable by a continuous utility function.*

Proof

Assume that $\succsim$ is continuous. We wish to apply the Open Gap Lemma to find a strictly increasing $f \in \mathbb{R}^{u(X)}$ such that every $\supseteq$-maximal connected set in $\mathbb{R}\backslash f(u(X))$ is either a singleton or an open interval. Define $v := f \circ u$ and observe that v represents $\succsim$. We will now prove that v is upper semicontinuous. Its lower semicontinuity is established similarly.

We wish to show that $v^{-1}([\alpha, \infty))$ is a closed subset of X for every $\alpha \in \mathbb{R}$ (Proposition 4). Fix an arbitrary real number α. If $\alpha = v(x)$ for some $x \in X$, then $v^{-1}([\alpha, \infty)) = \{y \in X : y \succsim x\}$ (since v represents $\succsim$), and we are done, by the upper semicontinuity of $\succsim$. We then consider the case where $\alpha \in \mathbb{R}\backslash v(X)$. Clearly, if $\alpha \leq \inf v(X)$, then we have $v^{-1}([\alpha, \infty)) = X$, and if $\alpha \geq \sup v(X)$, then $v^{-1}([\alpha, \infty)) = \emptyset$, so our claim is trivial in these cases. Assume then that $\inf v(X) < \alpha < \sup v(X)$, and let I be the $\supseteq$-maximal connected set in $\mathbb{R}\backslash v(X)$ that contains α. (Clearly, I is the union of all intervals in $\mathbb{R}\backslash v(X)$ that contain α.) By definition of v, either $I = \{\alpha\}$ or $I = (\alpha_*, \alpha^*)$ for some α_* and α^* in $v(X)$ with $\alpha_* < \alpha^*$. In the latter case, we have $v^{-1}([\alpha, \infty)) = \{y \in X : v(y) \geq \alpha^*\}$, which is a closed set, thanks to the upper semicontinuity of $\succsim$. In the former case, we let $\mathcal{A}_\alpha := \{\beta \in v(X) : \alpha \geq \beta\}$ and observe that

$$v^{-1}([\alpha, \infty)) = v^{-1}\left(\bigcap \{[\beta, \infty) : \beta \in \mathcal{A}_\alpha\}\right) = \bigcap \{v^{-1}([\beta, \infty)) : \beta \in \mathcal{A}_\alpha\}$$

(Exercise A.21). Since $v^{-1}([\beta, \infty))$ is closed for each $\beta \in v(X)$ (why?), and the intersection of any collection of closed sets is closed, we find again that $v^{-1}([\alpha, \infty))$ is a closed set. ■

EXERCISE 38 Consider the functions $u_i \in \mathbb{R}^{[0,1]}$, $i = 1, 2, 3$, defined as:

$$u_1(t) := \begin{cases} t, & \text{if } 0 \le t \le \frac{1}{2} \\ 2 - 2t & \text{otherwise} \end{cases}, \qquad u_2(t) := \begin{cases} t, & \text{if } 0 \le t < \frac{1}{2} \\ 1, & \text{if } t = \frac{1}{2} \\ t + 1, & \text{otherwise} \end{cases},$$

and $u_3 := \mathbf{1}_{[\frac{1}{2},1]}$. Define the preference relation $\succsim_i$ on $[0, 1]$ as $x \succsim_i y$ iff $u_i(x) \ge u_i(y)$ for each i. Is there a continuous utility function that represents $\succsim_i$, $i = 1, 2, 3$? An upper semicontinuous one?

Combining Lemma 1 and Rader's Utility Representation Theorem 1, we obtain the following fundamental theorem of utility theory.

DEBREU'S UTILITY REPRESENTATION THEOREM
Let X be a separable metric space and $\succsim$ a complete preference relation on X. If $\succsim$ is continuous, then it can be represented by a continuous utility function $u \in \mathbb{R}^X$.

This phenomenal result is the backbone of utility theory. It still makes frequent appearance in current research on individual and social choice. We didn't prove this result here, for this would take us too far afield. But at least you now know that Debreu's theorem can be reduced from Rader's Utility Representation Theorem 1 (which is a relatively simple result) by means of the Open Gap Lemma (which is not at all a simple result).

COROLLARY 3
Let X be a nonempty subset of $\mathbb{R}^n$ and $\succsim$ a complete preference relation on X. There exists a continuous (upper semicontinuous) utility representation for $\succsim$ if, and only if, $\succsim$ is continuous (upper semicontinuous).

EXERCISE 39 Prove Corollary 3 by using Debreu's Utility Representation Theorem.

5.3* Cauchy's Functional Equations: Additivity on $\mathbb{R}^n$

In this application, our objective is to understand the nature of *additive* functions defined on an arbitrary Euclidean space, and to see how continuity interacts with the property of additivity. Let's begin with clarifying what we mean by the latter property.

> DEFINITION
> For any given $n \in \mathbb{N}$, let S be a nonempty subset of $\mathbb{R}^n$, and $\varphi \in \mathbb{R}^S$. If S is closed under scalar multiplication (i.e. $\lambda x \in S$ for all $(\lambda, x) \in \mathbb{R} \times S$), and
>
> $$\varphi(\lambda x) = \lambda\varphi(x) \qquad \text{for all } x \in S \text{ and } \lambda \in \mathbb{R},$$
>
> we say that φ is **linearly homogeneous**. If, on the other hand, S is closed under vector addition (i.e. $x + y \in S$ for all $x, y \in S$), and
>
> $$\varphi(x + y) = \varphi(x) + \varphi(y) \qquad \text{for all } x, y \in S, \tag{6}$$
>
> then we say that φ is **additive**. Finally, φ is said to be **linear** if it is both linearly homogeneous and additive.

Equation (6) can be viewed as a functional equation in which the unknown variable is a function. It is often called **Cauchy's (first) functional equation.**[25] In general, there are many real functions that satisfy this equation on $\mathbb{R}^n$; that is, additivity alone does not tell us all that much about the structure of a function. (This is not entirely true; see Remark 1 below.) If, however, we combine additivity with linear homogeneity, then we can say quite a bit. For instance, if φ is a linear function on $\mathbb{R}^n$, by letting $\mathbf{e}^i$ stand for the ith unit vector (that is, $\mathbf{e}^1 := (1, 0, \ldots, 0)$, $\mathbf{e}^2 := (0, 1, 0, \ldots)$, and so on), we find

$$\varphi(x) = \varphi\left(\sum_{i=1}^{n} x_i\mathbf{e}^i\right) = \sum_{i=1}^{n} \varphi(x_i\mathbf{e}^i) = \sum_{i=1}^{n} \varphi(\mathbf{e}^i)x_i \qquad \text{for all } x \in \mathbb{R}^n$$

where we used (6) repetitively—$n - 1$ many times, to be exact—to get the second equality. Thus: A function $\varphi : \mathbb{R}^n \to \mathbb{R}$ *is linear iff there exist real numbers* $\alpha_1, \ldots, \alpha_n$ *such that* $\varphi(x) = \sum^n \alpha_i x_i$ *for all* $x \in \mathbb{R}^n$. This pins down the structure of linear functions defined on $\mathbb{R}^n$ in precise terms.

[25] Aczel (1966) is a classic reference that investigates thoroughly all four of Cauchy's functional equations and their applications.

However, linear homogeneity may not be readily available in a given application. A question of interest is whether we can replace it in our finding above with some other regularity properties, such as continuity and/or boundedness. The answer is yes, and the demonstration of this is our main task here. (*Note.* In what follows, we focus exclusively on functions that are defined on the entire $\mathbb{R}^n$. But this is only for convenience. The results of this section hold for functions defined on any subset of a Euclidean space that is closed under vector addition, such as $\mathbb{R}^n_+$.)

The following elementary fact is crucial for the subsequent analysis. It says that additivity on $\mathbb{R}$ entails the linearity of a real function on the set of rational numbers, and "solves" Cauchy's functional equation in terms of semicontinuous and/or bounded functions on $\mathbb{R}$.

LEMMA 2
Let f be an additive self-map on $\mathbb{R}$, that is,

$$f(x+y) = f(x) + f(y) \quad \text{for all } x, y \in \mathbb{R}. \tag{7}$$

Then there exists a real number α such that $f(q) = \alpha q$ for all $q \in \mathbb{Q}$. Moreover, f is linear if (a) it is continuous, or (b) it is upper (or lower) semicontinuous, or (c) it is bounded on some open interval.

PROOF
Take any $k, l \in \mathbb{N}$ and $a \in \mathbb{R}$. By (7), we have $f(lx) = lf(x)$ for any $x \in \mathbb{R}$. Choosing $x = \frac{k}{l}a$, therefore, we find $f(ka) = lf\left(\frac{k}{l}a\right)$. But by (7), $f(ka) = kf(a)$, so that $f\left(\frac{k}{l}a\right) = \frac{k}{l}f(a)$. Since k and l are arbitrary positive integers and a is any real number here, we may conclude: $f(qx) = qf(x)$ for all $q \in \mathbb{Q}_{++}$ and $x \in \mathbb{R}$.

Since $f(1) = f(1) + f(0)$ by (7), we have $f(0) = 0$. Then, for any $y \in \mathbb{R}_-$, we have $f(y) + f(-y) = f(0) = 0$ so that $f(-y) = -f(y)$. Combining this finding with the one of the previous paragraph, we may conclude that $f(qx) = qf(x)$ holds for all $q \in \mathbb{Q}$ and $x \in \mathbb{R}$. Then, by choosing $x = 1$ and letting $\alpha := f(1)$, we obtain $f(q) = \alpha q$ for all $q \in \mathbb{Q}$.

While part (a) is implied by (b), the analysis with lower semicontinuity is similar to that with upper semicontinuity, so we deal only with the latter case. Assume, then, that f is upper semicontinuous, and take any $x \in \mathbb{R}$. Since $\mathbb{Q}$ is dense in $\mathbb{R}$, there exists a $(q_m) \in \mathbb{Q}^\infty$ such that $q_m \to x$. By Proposition 4, then, we find $f(x) \geq \limsup f(q_m) = \lim \alpha q_m = \alpha x$. Thus, $f(x) \geq \alpha x$ holds for all $x \in \mathbb{R}$. But, by additivity and this finding, if $f(x) > \alpha x$ held for some $x \neq 0$, we would get $f(0) = f(-x+x) = f(-x) + f(x) > \alpha(-x) + \alpha x = 0$, a contradiction. (Yes?) Conclusion: $f(x) = \alpha x$ for all $x \in \mathbb{R}$.

To prove part (c), assume that there exists a $K > 0$ and $-\infty \leq a < b \leq \infty$ such that $|f(x)| \leq K$ for all $x \in (a, b)$. But, for each real number y, denseness of $\mathbb{Q}$ in $\mathbb{R}$ guarantees that we can find a $q(y) \in \mathbb{Q}$ such that $y + q(y) \in (a, b)$. Thus, since $f(q(y)) = \alpha q(y)$, we have, for each real y,

$$\begin{aligned} |f(y) - \alpha y| &= |f(y) + f(q(y)) - \alpha q(y) - \alpha y| \\ &= |f(y + q(y)) - \alpha(y + q(y))| \\ &\leq K \end{aligned}$$

where we used (7) to get the second equality. By using this finding and (7), we obtain

$$m\,|f(x) - \alpha x| = |f(mx) - \alpha mx| \leq K$$

for all $x \in \mathbb{R}$ and *any* $m \in \mathbb{N}$. But this is possible only if $f(x) = \alpha x$ for all $x \in \mathbb{R}$. ∎

EXERCISE 40^H Let f be a self-map on $\mathbb{R}$. Prove:

(a) If f is additive and upper semicontinuous at some $x \in \mathbb{R}$, then it must be linear;

(b) If f is additive and monotonic on some open interval, then it must be linear;

(c) Let $-\infty < a < b < \infty$ and $[a, b] \cap [2a, 2b] \neq \emptyset$. If f is continuous, and satisfies $f(x + y) = f(x) + f(y)$ for all $x, y \in [a, b]$, then there exists a real α such that $f(x) = \alpha x$ for all $x \in [a, b]$.

EXERCISE 41^H (*Jensen's Functional Equation*) Prove:

(a) If $f \in \mathbf{C}(\mathbb{R})$ satisfies $f\left(\frac{1}{2}(x+y)\right) = \frac{1}{2}(f(x) + f(y))$ for all $x, y \in \mathbb{R}$, then there exists an $(\alpha, \beta) \in \mathbb{R}^2$ such that $f(x) = \alpha x + \beta$ for all $x \in \mathbb{R}$;

*(b) If $f \in \mathbf{C}[0, 1]$ satisfies $f\left(\frac{1}{2}(x+y)\right) = \frac{1}{2}\left(f(x) + f(y)\right)$ for all $x, y \in [0, 1]$, then there exists an $(\alpha, \beta) \in \mathbb{R}^2$ such that $f(x) = \alpha x + \beta$ for all $0 \leq x \leq 1$.

EXERCISE 42 Prove: If $f \in \mathbf{C}(\mathbb{R})$ satisfies $f(x+y) = f(x) + 2xy + f(y)$ for all $x, y \in \mathbb{R}$, then there exists an $\alpha \in \mathbb{R}$ such that $f(x) = x^2 + \alpha x$ for all $x \in \mathbb{R}$. (Can continuity be relaxed to semicontinuity in this statement?)

*REMARK 1. While Exercise 40.(a) shows that continuity at a single point is enough to ensure the linearity of any solution to Cauchy's functional equation on $\mathbb{R}$, it turns out that continuity cannot be completely relaxed in this observation. Curiously, there exist nonlinear functions that satisfy (7) (see Exercise F.24). However, as you can imagine, such functions are highly irregular. A nonlinear function that satisfies (7) must actually be a geometric oddity: the graph of such a map must be dense in $\mathbb{R}^2$! Indeed, if f is a nonlinear self-map on $\mathbb{R}$, then there exist $x, y > 0$ such that $\frac{f(x)}{x} \neq \frac{f(y)}{y}$, so that the 2-vectors $(x, f(x))$ and $(y, f(y))$ span $\mathbb{R}^2$. This means that $\{q(x, f(x)) + r(y, f(y)) : q, r \in \mathbb{Q}\}$ must be dense in $\mathbb{R}^2$. (Why?) But if f satisfies (7), we have

$$\begin{aligned}\{q(x, f(x)) + r(y, f(y)) : q, r \in \mathbb{Q}\} &= \{(qx + ry, f(qx + ry)) : q, r \in \mathbb{Q}\} \\ &\subseteq Gr(f)\end{aligned}$$

by the first part of Lemma 2. Thus, the closure of $Gr(f)$ equals $\mathbb{R}^2$. (*Note.* This observation yields an alternative proof for Lemma 2.(c).) □

The following result identifies the nature of additive functions on $\mathbb{R}^n$ that satisfy very weak regularity properties.

PROPOSITION 6
For any given $n \in \mathbb{N}$, every continuous (or semicontinuous, or bounded) additive real function φ on $\mathbb{R}^n$ (or on $\mathbb{R}^n_+$) is linear.

PROOF
Take any real map φ on $\mathbb{R}^n$, and define $f_i : \mathbb{R} \to \mathbb{R}$ by $f_i(t) := \varphi(t\mathbf{e}^i)$, where $\mathbf{e}^i$ is the ith unit n-vector, $i = 1, \ldots, n$. It is easily checked that f_i satisfies (7) and inherits the continuity (or semicontinuity, or boundedness) of φ. Hence, if φ satisfies any of these properties, there exists a real number α_i such that $f_i(t) = \alpha_i t$ for all $t \in \mathbb{R}$, $i = 1, \ldots, n$ (Lemma 2). Thus, in that case, applying (6) inductively, we find $\varphi(x) = \sum^n \varphi(x_i \mathbf{e}^i) = \sum^n f_i(x_i) = \sum^n \alpha_i x_i$ for all $x \in \mathbb{R}^n$. (The proof would be analogous if φ was instead defined on $\mathbb{R}^n_+$.) ■

Exercise 43 How would Proposition 6 modify if instead of additivity we required $\varphi(x+y) = \varphi(x)\varphi(y)$ for all $x, y \in \mathbb{R}^n$?

5.4* Representation of Additive Preferences

As an immediate application of what we have accomplished in the previous subsection, we prove next a famous utility representation theorem. This result is prototypical of additive representation theorems, which play an important role in social choice theory and related fields. In later chapters, when we are equipped with more powerful methods, we will revisit this result and obtain a substantial generalization.

Theorem 1

(de Finetti) *Let $n \in \mathbb{N}$, and take any complete, continuous, and strictly increasing preorder $\succsim$ on $\mathbb{R}^n_+$. If $\succsim$ is additive, that is,*

$$x \succsim y \quad \textit{if and only if} \quad x + z \succsim y + z$$

for any $x, y \in \mathbb{R}^n_+$ and $z \in \mathbb{R}^n$ such that $x + z, y + z \in \mathbb{R}^n_+$, then it admits a positive linear representation, that is, there exist real numbers $\alpha_1, \ldots, \alpha_n > 0$ such that, for any $x, y \in \mathbb{R}^n_+$,

$$x \succsim y \quad \textit{if and only if} \quad \sum_{i=1}^n \alpha_i x_i \geq \sum_{i=1}^n \alpha_i y_i. \tag{8}$$

Proof[26]

Let us first agree to write $[t]_n$ for the n-vector $(t, \ldots, t)$, where t is any real number. For each $x \in \mathbb{R}^n_+$, define $m(x) := \max\{x_1, \ldots, x_n\}$, and let

$$\varphi(x) := \min\{\theta \in [0, m(x)] : [\theta]_n \succsim x\}.$$

This well-defines φ on $\mathbb{R}^n_+$ because, for each $x \in \mathbb{R}^n_+$, the set $\{\theta \in [0, m(x)] : [\theta]_n \succsim x\}$ is compact due to upper semicontinuity of $\succsim$. (Yes?) By definition, we have $[\varphi(x)]_n \succsim x$ for any x. Using the lower semicontinuity of $\succsim$, we

[26] The first part of the proof is essentially an argument to find a utility function that would represent a standard preference relation on the commodity space $\mathbb{R}^n_+$. (See Exercise 37.) It is only in the final part of the proof that we use the strength of additivity. (For an alternative proof, see Blackwell and Girshick (1954, pp. 118–119).)

see that $[\varphi(x)]_n \succ x$ is impossible, so by completeness of $\succsim$, we have $x \sim [\varphi(x)]_n$ for each x. (Yes? In particular, $\varphi([\theta]_n) = \theta$ for each $\theta \geq 0$. Why?) Thus, by transitivity of $\succsim$, we have $x \succsim y$ iff $[\varphi(x)]_n \succsim [\varphi(y)]_n$, and hence, by monotonicity of $\succsim$,

$$x \succsim y \quad \text{if and only if} \quad \varphi(x) \geq \varphi(y) \tag{9}$$

for any $x, y \in \mathbb{R}^n_+$. But then $\varphi^{-1}([\theta, \infty)) = \{x \in \mathbb{R}^n_+ : x \succsim [\theta]_n\}$ for each $\theta \geq 0$, and hence, the upper semicontinuity of $\succsim$ and Proposition 4 tell us that φ is upper semicontinuous on $\mathbb{R}^n_+$. Moreover, if $\succsim$ is additive, then

$$[\varphi(x+y)]_n \sim x+y \sim [\varphi(x)]_n + y \sim [\varphi(x)]_n + [\varphi(y)]_n = [\varphi(x)+\varphi(y)]_n,$$

and it follows that $\varphi(x+y) = \varphi(x) + \varphi(y)$ for any $x, y \in \mathbb{R}^n_+$. Thus additivity of $\succsim$ implies that of φ, and hence, by Proposition 6 and (9), we obtain (8) (for any $x, y \in \mathbb{R}^n_+$) for some real numbers $\alpha_1, \ldots, \alpha_n$. Positivity of these numbers is an obvious consequence of the monotonicity of $\succsim$. ■

We have seen in Exercise 37 how one may take advantage of the order structure of a Euclidean space in order to identify a class of preference relations that are representable. By contrast, Theorem 1 utilizes, in addition to its metric and order structures, the *linear* structure of $\mathbb{R}^n_+$, and hence delivers more about the nature of the representing utility function. The following exercise provides another instance to this effect. It assumes more than our previous results, but it also delivers more. Behaviorally speaking, it shows how one may be able to capture axiomatically some of those "interdependent" preferences that exhibit not only sheer selfishness, but also a spiteful (or altruistic) concern about others. These issues are discussed further in Ok and Koçkesen (2000).

EXERCISE 44 (*Negatively Interdependent Preferences*) Take any $n \in \mathbb{N}$, and let $X := \mathbb{R}^{n+1}_{++}$, which we interpret as the set of all income distributions in a given society with $n+1$ individuals. We denote the generic members of X by (a, y) and (b, x), where $x, y \in \mathbb{R}^n_{++}$. The vector (a, y) is interpreted as an income distribution in which the income of a particular individual is $a > 0$ and the income distribution of the rest of the society is y. The complete preference relation $\succsim$ of this individual on X satisfies the following two primitive properties: For all $a, b > 0$ and $y \in \mathbb{R}^n_{++}$,

(i) $(a, y) \succ (b, y)$ whenever $a > b$;

(ii) $(a, y) \sim (a, y')$ whenever y' is a permutation of y.

Moreover, the individual is *spiteful* in the following sense:

(iii) $(a, y) \succ (a, x)$ whenever $x > y$ (for all $a > 0$).

Consider the following properties for such a $\succsim$: For all $(a, y), (b, x) \in X$,

> *Additivity.* $(a, y) \succsim (b, x)$ implies $(a + c, y + z) \succsim (b + c, x + z)$ for any $(c, z) \in \mathbb{R}^{n+1}$ such that $(a + c, y + z), (b + c, x + z) \in X$.
>
> *Additivity with respect to the external effect.* $(a, y) \succsim (a, x)$ implies $(a, y + z) \succsim (a, x + z)$ for any $z \in \mathbb{R}^n$ such that $(a, y+z), (a, x+z) \in X$.
>
> *Decomposability.* $(a, y) \succsim (a, x)$ iff $(b, y) \succsim (b, x)$.

(a) Prove that $\succsim$ is continuous and additive iff the map $(a, y) \mapsto a - \theta \sum^n y_i$ on X represents $\succsim$.

(b) Prove that $\succsim$ is continuous, decomposable, and additive with respect to the external effect iff there exists a continuous $\psi : \mathbb{R}^2_{++} \to \mathbb{R}$ that is strictly increasing (decreasing) in its first (second) argument such that the map $(a, y) \mapsto \psi(a, \sum^n y_i)$ on X represents $\succsim$.

6 **CB**(T) and Uniform Convergence

6.1 The Basic Metric Structure of **CB**(T)

We now return to the theory of continuous functions defined on an arbitrary metric space. Let T be any metric space, and recall that $\mathbf{B}(T)$ stands for the set of all bounded functions defined on T, which is endowed with the sup metric d_∞ (Example C.1.[5]). An important subspace of $\mathbf{B}(T)$ consists of all continuous and bounded functions on T; we denote this subspace by $\mathbf{CB}(T)$ throughout this text. In turn, the set of all continuous functions on T is denoted as $\mathbf{C}(T)$. While the sup-metric cannot be used to metrize $\mathbf{C}(T)$ in general (because a continuous function need not be bounded), this complication disappears when T is compact. For then, by Corollary 2, we have $\mathbf{CB}(T) = \mathbf{C}(T)$. When T is compact, therefore, we can, and we will, think of $\mathbf{C}(T)$ as metrized by the sup-metric.

The following result identifies the basic structure of $\mathbf{CB}(T)$.

Proposition 7
For any metric space T, $\mathbf{CB}(T)$ is a complete metric subspace of $\mathbf{B}(T)$.

PROOF

Fix a metric space T. By Proposition C.7 and Example C.11.[5], it is enough to show that $\mathbf{CB}(T)$ is closed in $\mathbf{B}(T)$. Let (φ_m) be a sequence in $\mathbf{CB}(T)$, and let $d_\infty(\varphi_m, \varphi) \to 0$ for some $\varphi \in \mathbf{B}(T)$. We wish to show that φ is continuous. Take any $\varepsilon > 0$, and observe that there exists an integer $M \geq 1$ such that $\sup\{|\varphi(z) - \varphi_M(z)| : z \in T\} < \frac{\varepsilon}{3}$. Thus,

$$|\varphi(z) - \varphi_M(z)| < \tfrac{\varepsilon}{3} \quad \text{for all } z \in T. \tag{10}$$

(Notice that this statement is a "global" one, for it holds for *all* $z \in T$ thanks to the sup-metric.) Now take an arbitrary $x \in T$. By continuity of φ_M, there exists a $\delta > 0$ such that $\varphi_M(N_{\delta,T}(x)) \subseteq N_{\frac{\varepsilon}{3},\mathbb{R}}(\varphi_M(x))$. But then, for any $y \in N_{\delta,T}(x)$,

$$\left|\varphi(x) - \varphi(y)\right| \leq |\varphi(x) - \varphi_M(x)| + \left|\varphi_M(x) - \varphi_M(y)\right| + \left|\varphi_M(y) - \varphi(y)\right| < \varepsilon$$

where we used (10) twice. Since x was chosen arbitrarily in T, we thus conclude that φ is continuous. ■

EXERCISE 45 Let T be a compact metric space. For any given $n \in \mathbb{N}$, we denote by $\mathbf{C}(T, \mathbb{R}^n)$ the set of all continuous functions that map T into $\mathbb{R}^n$. This space is metrized by $d_\infty : \mathbf{C}(T, \mathbb{R}^n)^2 \to \mathbb{R}_+$ with $d_\infty(\Phi, \Psi) := \sup\{d_2(\Phi(x), \Psi(x)) : x \in T\}$.

(a) Show that d_∞ is a metric on $\mathbf{C}(T, \mathbb{R}^n)$.

(b) Prove that $\mathbf{C}(T, \mathbb{R}^n)$ is a complete metric space.

6.2 Uniform Convergence

Proposition 7 is the main reason why endowing the set of all continuous and bounded functions with the sup metric is more suitable than using other "natural" candidates. If we endowed this set with, say, the d_1 (integral) metric introduced in Exercise C.42, then we would not necessarily end up with a complete metric space—this was the main point of that exercise.

In general, "convergence" in $\mathbf{CB}(T)$ has very desirable properties. If we know that a sequence (φ_m) of continuous and bounded real functions on a metric space T converges to some $\varphi \in \mathbf{B}(T)$ with respect to d_∞, that is, if

$$\lim_{m\to\infty} \sup\{|\varphi_m(x) - \varphi(x)| : x \in T\} = 0, \tag{11}$$

Proposition 7 assures us that φ is a continuous function. In fact, we can strengthen this observation a bit further. If you go back and reexamine the proof of Proposition 7, you will notice that we never used there the boundedness of the functions φ_m and φ; all that mattered was (11) being satisfied by (φ_m) and φ. It follows that, for any sequence (φ_m) in $\mathbf{C}(T)$ and any $\varphi \in \mathbb{R}^T$, (11) implies that $\varphi \in \mathbf{C}(T)$. This suggests that (11) provides a useful convergence concept for arbitrary sequences of real functions.

DEFINITION

Let (φ_m) be any sequence of (not necessarily bounded) real functions defined on an arbitrary metric space T. If (11) holds for some real function φ on T, or equivalently, if, for all $\varepsilon > 0$, there exists an $M > 0$ (which may depend on ε) such that $|\varphi_m(x) - \varphi(x)| < \varepsilon$ for all $x \in T$ whenever $m \geq M$, then we refer to φ as the **uniform limit** of (φ_m). In this case, we say that (φ_m) **converges to** φ **uniformly**, and write $\varphi_m \to \varphi$ uniformly.

It is crucial to understand that uniform convergence is a global notion of convergence. If f is the uniform limit of a sequence (f_m) in $\mathbb{R}^{[0,1]}$, what this means is that, for any $\varepsilon > 0$, eventually the entire graph of each member of the sequence lies within the ε-strip around the graph of f, that is, $f - \varepsilon < f_m < f + \varepsilon$ holds for all m large enough. (See Figure 5.) This statement is global in the sense that it concerns the entire domain of f, not a particular point in $[0, 1]$.

Of course, in the case of bounded functions, uniform convergence is identical to convergence with respect to d_∞. Thus one may think of the ordinary convergence structure of $\mathbf{B}(T)$ (and hence of $\mathbf{CB}(T)$) as uniform. Moreover, as discussed above, we have the following fact: *If each φ_m is a continuous real function on T, and $\varphi_m \to \varphi$ uniformly for some $\varphi \in \mathbb{R}^T$, then φ must be continuous.* This result, which is slightly more general then saying that $\mathbf{CB}(T)$ is a closed subset of $\mathbf{B}(T)$, is extremely useful in applications, and illustrates the general idea that "uniform convergence preserves good behavior."

As another simple example, let us show that the uniform convergence of bounded real functions guarantees the boundedness of the limit function. Indeed, if (φ_m) is a sequence in $\mathbf{B}(T)$ and $\varphi_m \to \varphi$ uniformly for some

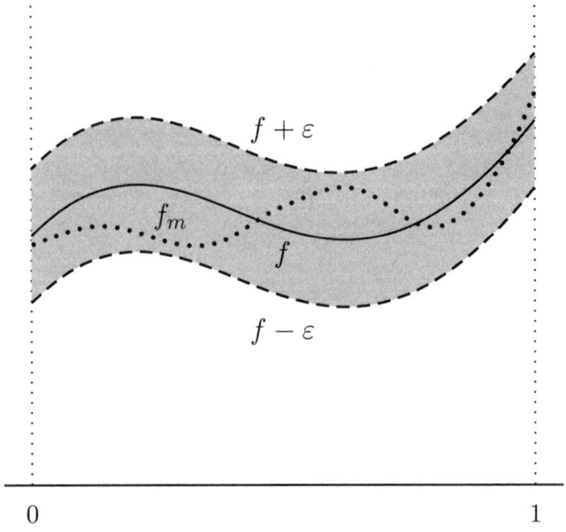

FIGURE 5

$\varphi \in \mathbb{R}^T$, then there exists an $M \in \mathbb{N}$ such that $|\varphi(x) - \varphi_M(x)| < 1$ for *all* $x \in T$. Thus, by the triangle inequality,

$$|\varphi(x)| \leq |\varphi(x) - \varphi_M(x)| + |\varphi_M(x)| \leq 1 + \sup\{|\varphi_M(y)| : y \in T\}$$

for all $x \in T$. It follows that $\varphi \in \mathbf{B}(T)$, as we sought. (*Alternative proof.* (φ_m) must be Cauchy, while $\mathbf{B}(T)$ is complete.)

For still another example, we note that uniform convergence allows one to interchange the operations of taking limits, which is an issue that arises in practice frequently. For instance, let (x^k) be a sequence in T such that $x^k \to x$, and let (φ_m) be a sequence in $\mathbf{C}(T)$ such that $\varphi_m \to \varphi$ uniformly for some $\varphi \in \mathbb{R}^T$. Suppose that we wish to compute the limit of the double sequence $(\varphi_m(x^k))$ as m and k approach infinity. Should we first let m or k go to infinity, that is, should we compute $\lim_{k\to\infty} \lim_{m\to\infty} \varphi_m(x^k)$ or $\lim_{m\to\infty} \lim_{k\to\infty} \varphi_m(x^k)$? (And yes, the order of taking limits matters in general; read on.) Since $\varphi_m \to \varphi$ uniformly, we don't need to worry about this issue. Uniform convergence guarantees the continuity of φ, so we get

$$\begin{aligned}&\lim_{k\to\infty} \lim_{m\to\infty} \varphi_m(x^k)\\ &\quad = \lim_{k\to\infty} \varphi(x^k) = \varphi(x) = \lim_{m\to\infty} \varphi_m(x) = \lim_{m\to\infty} \lim_{k\to\infty} \varphi_m(x^k),\end{aligned}$$

that is,

$$\lim_{k\to\infty} \lim_{m\to\infty} \varphi_m(x^k) = \lim_{m\to\infty} \lim_{k\to\infty} \varphi_m(x^k). \tag{12}$$

This is a good illustration of the power of uniform convergence.[27]

Uniform convergence is strictly more demanding than the perhaps more familiar notion of **pointwise convergence** of a sequence of functions (φ_m) in $\mathbb{R}^T$, which, by definition, requires only that

$$\lim_{m\to\infty} |\varphi_m(x) - \varphi(x)| = 0 \qquad \text{for all } x \in T. \tag{13}$$

Put differently, $\varphi_m \to \varphi$ **pointwise** if, for all $\varepsilon > 0$ and $x \in T$, there exists an $M > 0$ (which may depend on *both* ε *and* x) such that $|\varphi_m(x) - \varphi(x)| < \varepsilon$. (This situation is denoted simply as $\varphi_m \to \varphi$.) Clearly, equation (11), being a "global" (i.e. x-independent) statement, implies (13) which is a "local" (x-dependent) condition. The converse of this implication does not hold. For instance, consider the sequence (f_m) in $\mathbf{C}[0,1]$ defined by $f_m(t) := t^m$ for each m (Figure 6). Clearly, we have $f_m \to \mathbf{1}_{\{1\}}$ (pointwise), where $\mathbf{1}_{\{1\}}$

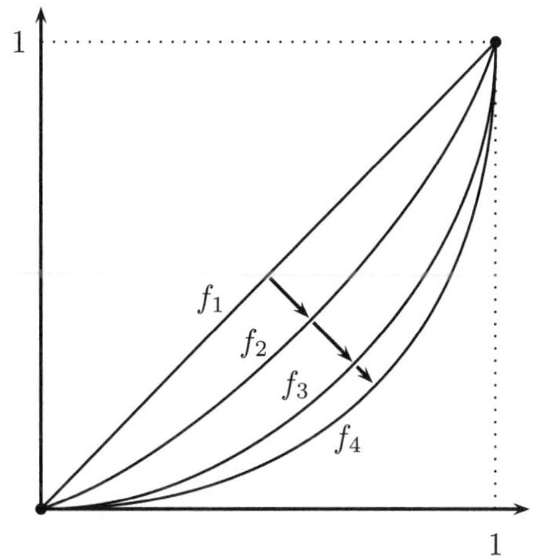

FIGURE 6

[27] Uniform convergence also preserves the Riemann integrability of self-maps on $\mathbb{R}$. While it doesn't in general preserve differentiability, it still plays a crucial role in theorems that provide sufficient conditions for the preservation of this property. A good introduction to issues of this sort is provided by Rudin (1976, pp. 147–154).

By the way, we owe the idea of uniform convergence and the fact that uniform convergence preserves continuity to Karl Weierstrass. (Who else?)

is the indicator function of $\{1\}$ in $[0,1]$.[28] Does $f_m \to \mathbf{1}_{\{1\}}$ uniformly? No, because if $d_\infty(f_m, \mathbf{1}_{\{1\}}) \to 0$ was the case, then the closedness of $\mathbf{C}[0,1]$ in $\mathbf{B}[0,1]$ would imply the continuity of $\mathbf{1}_{\{1\}}$, which is not the case.[29] Indeed, here we have

$$\lim_{m\to\infty} \sup\{|f_m(t) - \mathbf{1}_{\{1\}}(t)| : 0 \le t \le 1\} = 1.$$

In particular, (12) does not hold in this example for some convergent sequence $(t_k) \in [0,1]^\infty$. For instance, if $(t_k) := (0, \frac{1}{2}, \ldots, 1 - \frac{1}{k}, \ldots)$, we get

$$\lim_{k\to\infty} \lim_{m\to\infty} f_m(t_k) = 0 \ne 1 = \lim_{m\to\infty} \lim_{k\to\infty} f_m(t_k).$$

Here are some more examples.

Exercise 46 Consider the sequences (f_m) and (g_m) in $\mathbb{R}^{[0,1]}$ defined by

$$f_m(t) := mt(1-t^2)^m \qquad \text{and} \qquad g_m(t) := \begin{cases} m, & \text{if } 0 < t < \frac{1}{m} \\ 0, & \text{otherwise} \end{cases},$$

$m = 1, 2, \ldots$. Show that both of these sequences converge to the zero function pointwise, but neither does so uniformly. Check that $\lim \int_0^1 f_m(t)dt \ne 0$. What do you conclude?

Exercise 47 Find two uniformly convergent sequences (f_m) and (g_m) in $\mathbf{C}(\mathbb{R})$ such that $(f_m g_m)$ is not uniformly convergent.

Exercise 48^{H} Show that $f \in \mathbb{R}^{[-1,1]}$, defined by $f(t) := \sum^\infty \frac{t^i}{i^2}$, is continuous.

Exercise 49 Let S be a dense subset of a metric space T. Prove: If (φ_m) is a sequence in $\mathbf{C}(T)$ such that $(\varphi_m|_S)$ is uniformly convergent, then (φ_m) is uniformly convergent.

Exercise 50^{H} Let T be any metric space, $\varphi \in \mathbb{R}^T$, and (φ_m) a sequence in $\mathbf{B}(T)$. Prove: If $\varphi_m \to \varphi$ uniformly, then there exists a $K > 0$ such that

$$\max\{|\varphi(x)|, \sup\{|\varphi_m(x)| : m \in \mathbb{N}\}\} \le K \qquad \text{for all } x \in T.$$

[28] This example shows that pointwise convergence does not preserve continuity. However, if T is a complete metric space and (φ_m) is a sequence in $\mathbf{C}(T)$ such that $\varphi_m \to \varphi$ pointwise, it is at least true that the continuity points of φ are dense in T. This fact may be useful in those gloomy days when all one has is pointwise (and not uniform) convergence. But this is no time to digress, so we will not prove this result here.

[29] *Quiz.* Use this example (and Theorem C.2) to show that the closed unit ball of $\mathbf{C}[0,1]$ (i.e., $\{f \in \mathbf{C}[0,1] : \sup\{|f(t)| : t \in [0,1]\} \le 1\}$) is *not* compact.

While uniform convergence has all sorts of advantages, it is often not so easy to verify, especially when one is dealing with abstract function sequences. It would therefore be nice to find conditions under which uniform convergence is implied by pointwise convergence, which is often easier to check. To this end, let us agree to write $\varphi_m \searrow \varphi$ when (φ_m) is a sequence in $\mathbb{R}^T$ that is *decreasing* (that is, $\varphi_1(t) \geq \varphi_2(t) \geq \cdots$ for all $t \in T$) and that converges to φ pointwise. The expression $\varphi_m \nearrow \varphi$ is similarly interpreted.

Here is a very nice result that teaches us yet again how compactness allows one to convert a local fact to a global one. It was obtained first by Ulisse Dini in 1878. (Of course, in Dini's day there was no concept of a metric space; he proved the result in the context of $\mathbf{C}[0, 1]$.)

DINI'S THEOREM

Let T be a compact metric space, $\varphi \in \mathbf{C}(T)$ and (φ_m) a sequence in $\mathbf{C}(T)$. If $\varphi_m \searrow \varphi$ (or $\varphi_m \nearrow \varphi$), then $\varphi_m \to \varphi$ uniformly.[30]

PROOF

The proof is by the local-to-global method (Section 3.2). Assume first that $\varphi_m \searrow 0$, and fix an arbitrary $\varepsilon > 0$. We will say that an open subset O of T satisfies the property Λ_ε if there exists an $M \in \mathbb{R}$ such that

$$|\varphi_m(z)| < \varepsilon \qquad \text{for all } z \in O \text{ and } m \geq M.$$

We wish to show that T satisfies the property Λ_ε. Clearly, if the open subsets O and U of T satisfy the property Λ_ε, then so does $O \cup U$. So, since T is compact, the local-to-global method says that all we need to show here is that for any given $x \in T$, there is a $\delta > 0$ (which may depend on x, of course) such that $N_{\delta,T}(x)$ satisfies the property Λ_ε. Fix any $x \in T$. By pointwise convergence, there exists an $M > 0$ such that $0 \leq \varphi_M(x) \leq \frac{\varepsilon}{2}$, whereas by continuity of φ_M, there exists a $\delta > 0$ with $\varphi_M(N_{\delta,T}(x)) \subseteq N_{\frac{\varepsilon}{2},\mathbb{R}}(\varphi_M(x))$. Thus $0 \leq \varphi_M(z) < \varepsilon$ for all $z \in N_{\delta,T}(x)$. Since (φ_m) is decreasing, therefore,

$$0 \leq \varphi_m(z) < \varepsilon \qquad \text{for all } z \in N_{\delta,T}(x) \text{ and } m \geq M,$$

[30] It is actually enough to take each φ_m to be upper semicontinuous when $\varphi_m \searrow \varphi$ (and lower semicontinuous when $\varphi_m \nearrow \varphi$). The proof that follows settles this generalization as well.

that is, $N_{\delta,T}(x)$ satisfies the property Λ_ε. Thus, thanks to the local-to-global method, we may conclude that T satisfies the property Λ_ε. Since $\varepsilon > 0$ is arbitrary here, this means that $\varphi_m \to 0$ uniformly.

To complete the proof, we relax the assumption that $\varphi = 0$, and define $\psi_m := \varphi_m - \varphi \in \mathbf{C}(T)$ for each m. Of course, $\psi_m \searrow 0$. By what we have just proved, therefore, $\psi_m \to 0$ uniformly, and it is trivial to check that this implies $\varphi_m \to \varphi$ uniformly. The case in which $\varphi_m \nearrow \varphi$ is similarly analyzed. ■

EXAMPLE 8

[1] One cannot drop the requirement of the continuity of the limit function φ in Dini's Theorem. As we have seen earlier, the function sequence (f_m) with $f_m(t) = t^m$, $0 \le t \le 1$, does not converge in $\mathbf{C}[0,1]$. Looking at the sequence $(f_m|_{[0,1)})$ shows that compactness cannot be omitted in the statement of the result either. On the other hand, an examination of the sequence (g_m) with $g_m(0) = 0$ and

$$g_m(t) := \begin{cases} 0, & \text{if } \frac{1}{m} \le t \le 1 \\ 1, & \text{if } 0 < t < \frac{1}{m} \end{cases}$$

shows that upper semicontinuity of the members of the converging sequence is also needed here (in the assertion concerning decreasing sequences). We leave it to you to show that the monotonicity requirement is also essential in the statement of Dini's Theorem.

[2] Here is a quick illustration of how one may "use" Dini's Theorem in practice. Define $f \in \mathbf{C}[-1,1]$ by $f(t) := |t|$, and suppose we wish to approximate f uniformly by means of differentiable functions. (You would be surprised how often one needs to solve this sort of an exercise.) The Weierstrass Approximation Theorem tells us that we can actually do this using the polynomials on $[-1,1]$. The idea is to approximate f by polynomials pointwise (this is much easier), and then to apply Dini's Theorem to ensure that the convergence is uniform. To this end, define $f_1 := 0$, and let

$$f_m(t) := f_{m-1}(t) + \tfrac{1}{2}(t^2 - (f_{m-1}(t))^2), \qquad -1 \le t \le 1, \; m = 2, 3, \ldots$$

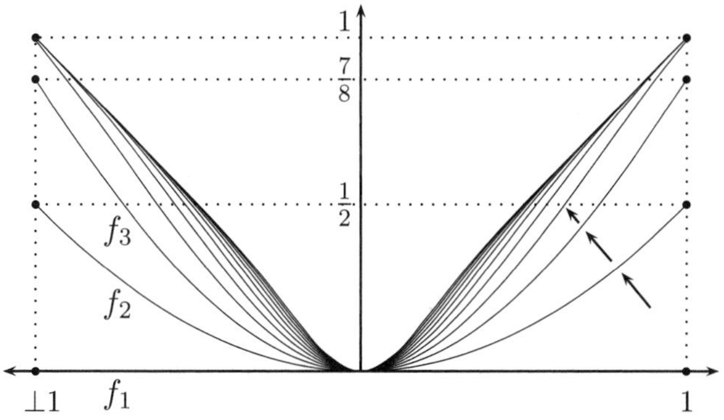

FIGURE 7

(Figure 7). Now pick any $0 < t \leq 1$, and let $y_m := f_m(t)$ for each m. Clearly,

$$0 \leq y_{m-1} \leq y_{m-1} + \tfrac{1}{2}(t^2 - y_{m-1}^2) = y_m \leq t, \qquad m = 2, 3, \ldots,$$

and a little effort shows that $y_m \to t$. Reasoning similarly for the case $-1 \leq t < 0$ lets us conclude that $f_m \to f$. Thus $f_m \nearrow f$, and hence $f_m \to f$ uniformly, by Dini's Theorem. □

EXERCISE 51 Consider the function $f \in \mathbf{C}[0,1]$ defined by $f(t) := \sqrt{t}$. Show that there exists a sequence (f_m) in $\mathbf{P}[0,1]$ such that $f_m \to f$ uniformly.

6.3* The Stone-Weierstrass Theorem and Separability of **C**(T)

We now turn to the issue of separability of $\mathbf{C}(T)$ when T is a compact metric space. We noted in Section C.2.2 that $\mathbf{C}[a,b]$ is separable precisely because the set of all polynomials on any compact interval $[a,b]$ with rational coefficients is dense in $\mathbf{C}[a,b]$. Using analogous reasoning through multivariate polynomials, one may show that $\mathbf{C}([a,b]^n)$ is separable as well. It turns out that we can extend these observations outside the realm of Euclidean spaces. To this end, we next introduce one of the most powerful results of the theory of real functions, a substantial generalization of the Weierstrass Approximation Theorem. It was obtained in 1937 by Marshall Stone and thus bears Stone's name along with that of Weierstrass.

THE STONE-WEIERSTRASS THEOREM

Take any compact metric space T, and let $\mathcal{P}$ be a subset of $\mathbf{C}(T)$ with the following properties:

(i) $\alpha\varphi + \beta\psi \in \mathcal{P}$ *for all* $\varphi, \psi \in \mathcal{P}$ *and* $\alpha, \beta \in \mathbb{R}$;

(ii) $\varphi\psi \in \mathcal{P}$ *for all* $\varphi, \psi \in \mathcal{P}$;

(iii) *all constant functions belong to* $\mathcal{P}$;

(iv) *for any distinct* $x, y \in T$, *there exists a* φ *in* $\mathcal{P}$ *such that* $\varphi(x) \neq \varphi(y)$.

Then $\mathcal{P}$ is dense in $\mathbf{C}(T)$.

PROOF

(Throughout the proof the closure operator cl is applied relative to $\mathbf{C}(T)$.) We need a few preliminary observations before we launch the main argument.

CLAIM 1

If $\varphi \in \mathcal{P}$, then $|\varphi| \in cl(\mathcal{P})$.

PROOF

If $\varphi = 0$ there is nothing to prove, so let $s := \sup\{|\varphi(x)| : x \in T\} > 0$. By Example 8.[2], there exists a sequence (p_m) in $\mathbf{P}[-1, 1]$ that converges to the absolute value function on $[-1, 1]$ uniformly. Then, $s\big(p_m \circ \big(\frac{1}{s}\varphi\big)\big) \to s\left|\frac{1}{s}\varphi\right|$ uniformly, that is,

$$d_\infty\left(s\left(p_m \circ \left(\tfrac{1}{s}\varphi\right)\right), |\varphi|\right) \to 0.$$

But, by hypothesis (i), $\frac{1}{s}\varphi \in \mathcal{P}$, and by hypotheses (i)–(iii), $s\big(p_m \circ \big(\frac{1}{s}\varphi\big)\big) \in \mathcal{P}$ for all m. (Why?) It follows that $|\varphi| \in cl(\mathcal{P})$. ‖

CLAIM 2

Let $k \in \mathbb{N}$. If $\varphi_1, \ldots, \varphi_k \in \mathcal{P}$, then both $\min\{\varphi_1, \ldots, \varphi_k\}$ and $\max\{\varphi_1, \ldots, \varphi_k\}$ belong to $cl(\mathcal{P})$.

PROOF

It is enough to prove this claim for $k = 2$. (Why?) But this case is immediate from hypothesis (i), Claim 1, and the algebraic identities

$$\min\{\varphi_1, \varphi_2\} = \tfrac{1}{2}(\varphi_1 + \varphi_2 - |\varphi_1 - \varphi_2|) \quad \text{and}$$
$$\max\{\varphi_1, \varphi_2\} = \tfrac{1}{2}(\varphi_1 + \varphi_2 + |\varphi_1 - \varphi_2|)$$

that are valid for any $\varphi_1, \varphi_2 \in \mathbb{R}^T$. ||

CLAIM 3

Given any distinct $x, y \in T$ and $a, b \in \mathbb{R}$, there exists a $\vartheta \in \mathcal{P}$ with $\vartheta(x) = a$ and $\vartheta(y) = b$.

PROOF

By hypothesis (iv), there exists a $\psi \in \mathcal{P}$ with $\psi(x) \neq \psi(y)$. Define

$$\vartheta := a + \frac{b-a}{\psi(y) - \psi(x)}(\psi - \psi(x)).$$

By hypotheses (i) and (iii), $\vartheta \in \mathcal{P}$. ||

We are now ready to prove that $cl(\mathcal{P}) = \mathbf{C}(T)$.[31] Take any $\varphi \in \mathbf{C}(T)$, and fix any $\varepsilon > 0$ and $x \in T$. By Claim 3, for any $y \in T\backslash\{x\}$, there exists a $\vartheta_y \in \mathcal{P}$ with $\vartheta_y(x) = \varphi(x)$ and $\vartheta_y(y) = \varphi(y)$. Clearly,

$$O(y) := \{z \in T : \vartheta_y(z) > \varphi(z) - \varepsilon\}$$

is open (Proposition 1), and $x, y \in O(y)$ for all $y \in T\backslash\{x\}$. Then $\{O(y) : y \in T\backslash\{x\}\}$ is an open cover of T, so by compactness, there exist a $k \in \mathbb{N}$ and $y^1, \ldots, y^k \in T\backslash\{x\}$ such that $T \subseteq \cup^k O(y^i)$. Now let $\psi_x := \max\{\vartheta_{y_1}, \ldots, \vartheta_{y_k}\} \in cl(\mathcal{P})$ (Claim 2). We have $\psi_x(x) = \varphi(x)$ and $\psi_x(z) > \varphi(z) - \varepsilon$ for all $z \in T$. (Why?)

We next work with ψ_x (instead of ϑ_y), but this time go in the other direction. That is, for any $x \in T$, we define

$$U(x) := \{z \in T : \psi_x(z) < \varphi(z) + \varepsilon\}$$

[31] The argument is beautifully simple. Just reflect on this paragraph before moving on to the next one. The idea will become transparent bit by bit.

which is an open set that contains x. By compactness of T, there exist an $l \in \mathbb{N}$ and $x^1, \ldots, x^l \in T$ with $T \subseteq \cup^l U(x^i)$. We let $\psi := \min\{\psi_{x^1}, \ldots, \psi_{x^l}\} \in cl(\mathcal{P})$ (Claim 2). Clearly, $\psi(z) > \varphi(z) - \varepsilon$ for all $z \in T$ (since each ψ_{x^i} has this property), and $\psi(z) < \varphi(z) + \varepsilon$ for all $z \in T$ (since at least one ψ_{x^i} has this property). Thus $d_\infty(\varphi, \psi) < \varepsilon$. Since $\varepsilon > 0$ was arbitrary in the entire argument, this shows that $\varphi \in cl(\mathcal{P})$, and the proof is complete. ∎

The Weierstrass Approximation Theorem is obtained as an immediate corollary of the Stone-Weierstrass Theorem by setting $\mathcal{P} := \mathbf{P}[a, b]$ in the latter result. (Check! *Hint.* To verify (iv), use $\mathrm{id}_{[a,b]}$.) The following exercise extends this fact to the case of $\mathbb{R}^n$.

EXERCISE 52 For any given $n \in \mathbb{N}$, take any nonempty compact set T in $\mathbb{R}^n$ and recall that $g \in \mathbb{R}^T$ is called a *(multivariate) polynomial* on T if there exist real numbers $\alpha_{i_1,\ldots,i_n}$ such that

$$g(t_1, \ldots, t_n) = \sum \alpha_{i_1,\ldots,i_n} \prod_{j=1}^{n} t_j^{i_j} \qquad \text{for all } (t_1, \ldots, t_n) \in T$$

where the sum runs through a finite set of n-tuples of indices $(i_1, \ldots, i_n) \in \mathbb{N}^n$. Prove that the set of all polynomials on T is dense in $\mathbf{C}(T)$. (Conclusion: *A continuous function defined on a nonempty compact subset of* $\mathbb{R}^n$ *is the uniform limit of a sequence of polynomials defined on that set.*)

EXERCISE 53 Prove: For any $f \in \mathbf{C}[0, 1]$, there exists a $K > 0$ and a sequence (f_m) in $\mathbf{P}[0, 1]$ such that $f_m \to f$ uniformly, and each f_m is Lipschitz continuous with a Lipschitz constant of at most K.

EXERCISE 54[32] Assume the hypotheses of the Stone-Weierstrass Theorem, and let $\varphi \in \mathbf{C}(T)$. Fix finitely many points $x^1, \ldots, x^k$ in T, and show that, for any $\varepsilon > 0$, there exists a $\psi \in \mathcal{P}$ with $d_\infty(\varphi, \psi) < \varepsilon$ and $\varphi(x^i) = \psi(x^i)$ for all $i = 1, \ldots, k$.

As another application of the Stone-Weierstrass Theorem, we prove that the set of all continuous real maps on a compact metric space T is a separable metric subspace of $\mathbf{B}(T)$. This fact will be of great use to us later on.

[32] See Boel, Carlsen, and Hansen (2001) for a more general version of this fact.

COROLLARY 4
Let T be a metric space. If T is compact, then $\mathbf{C}(T)$ is separable.

PROOF
(Throughout the proof the closure operator cl is applied relative to $\mathbf{C}(T)$.) Assume that T is a compact metric space. Then it must also be separable (Exercise C.28). So, by Proposition C.3, there exists a countable class $\mathcal{O}$ of open subsets of T such that $U = \cup\{O \in \mathcal{O} : O \subseteq U\}$ for any open subset U of T. Define $\varphi_O \in \mathbf{C}(T)$ by $\varphi_O(z) := d(z, T\backslash O)$. (Recall Example 1.[3].) Now let $\mathcal{P}'$ be the set of all finite products of the members of $\{\varphi_O : O \in \mathcal{O}\}$ along with the real function that equals 1 everywhere on T. It is easy to see that this set is countable (Exercise B.3 and Proposition B.2). Now let $\mathcal{P}$ stand for the set of all finite linear combinations of the members of $\mathcal{P}'$. (From linear algebra, you may recall that $\mathcal{P}$ is the subspace generated by $\mathcal{P}'$; see Section F.1.4.) $\mathcal{P}$ obviously satisfies the first three conditions of the Stone-Weierstrass Theorem. It also satisfies the fourth one, because for any distinct $x, y \in T$, there is an $O \in \mathcal{O}$ such that $x \in O$ and $y \notin O$ (why?), and hence $\varphi_O(x) \neq 0 = \varphi_O(y)$. So $\mathcal{P}$ must be dense in $\mathbf{C}(T)$. (*Note.* We are not done yet, because $\mathcal{P}$ need not be countable!)

Define

$$\mathcal{P}'' := \left\{ \sum_{i=1}^{k} q_i \psi_i : \psi_i \in \mathcal{P}', \ q_i \in \mathbb{Q}, \ i = 1, \ldots, k, \ k \in \mathbb{N} \right\}$$

which is easily checked to be countable. Observe that for any $k \in \mathbb{N}$ and any $(q_i, p_i) \in \mathbb{Q} \times \mathbb{R}$, $i = 1, \ldots, k$, we have

$$d_\infty\left(\sum_{i=1}^{k} q_i \psi_i, \sum_{i=1}^{k} p_i \psi_i \right) \leq \left(\sum_{i=1}^{k} |q_i - p_i| \right) \sup\{|\psi_i(z)| : z \in T\}.$$

Using the denseness of $\mathbb{Q}$ in $\mathbb{R}$ and this inequality, we find that $\mathcal{P} \subseteq cl(\mathcal{P}'')$. But then $\mathbf{C}(T) = cl(\mathcal{P}) \subseteq cl(cl(\mathcal{P}'')) = cl(\mathcal{P}'')$, which means $\mathbf{C}(T) = cl(\mathcal{P}'')$. ■

EXERCISE 55^{H} Give an example of a metric space T such that $\mathbf{C}(T)$ is not separable.

*EXERCISE 56 Let T be a compact metric space. Prove that the set of all Lipschitz continuous real functions on T is a separable subspace of $\mathbf{B}(T)$.

6.4* The Arzelà-Ascoli Theorem

We have just seen that $\mathbf{C}(T)$ is separable whenever T is a compact metric space. While at first glance it may seem that we could sharpen this result by deducing the compactness of $\mathbf{C}(T)$ from that of T, this is, in fact, impossible. Indeed, if $f_m \in \mathbf{C}[0,1]$ is defined by $f_m(t) := t^m$, $m = 1, 2, \ldots$, then (f_m) is a sequence in $\mathbf{C}[0,1]$ without a convergent subsequence. Thus, by Theorem C.2, $\mathbf{C}[0,1]$—in fact, the closed unit ball of $\mathbf{C}[0,1]$—is not compact.

This observation points to the fact that determining whether or not a given subset of $\mathbf{C}(T)$ is compact is not a trivial matter, even when T is compact. Nevertheless, as was hinted at in Example C.9.[4], there is a noteworthy method of tackling these sorts of problems.

DEFINITION
Let T be a metric space and $\mathcal{F} \subseteq \mathbf{C}(T)$. We say that $\mathcal{F}$ is **equicontinuous at** $x \in T$ if, for any given $\varepsilon > 0$, there is a $\delta > 0$ (which may depend on both ε and x) such that

$$\left|\varphi(x) - \varphi(y)\right| < \varepsilon \qquad \text{for all } \varphi \in \mathcal{F} \text{ and } y \in N_{\delta,T}(x).$$

If $\mathcal{F}$ is equicontinuous at each $x \in T$, then it is called **equicontinuous**. Finally, we say that $\mathcal{F}$ is **uniformly equicontinuous** if, for any $\varepsilon > 0$, there is a $\delta > 0$ (which may depend on ε) such that

$$\left|\varphi(x) - \varphi(y)\right| < \varepsilon \qquad \text{for all } \varphi \in \mathcal{F} \text{ and } x, y \in T \text{ with } d(x,y) < \delta.$$

Obviously, an equicontinuous family $\mathcal{F}$ of continuous functions need not be uniformly equicontinuous.[33] However, just as a continuous real function on a compact set is uniformly continuous (Exercise 27), an equicontinuous $\mathcal{F} \subseteq \mathbf{C}(T)$ is uniformly equicontinuous whenever T is compact. We leave proving this as an exercise, but note that the proof is analogous to the corresponding result in the case of a single function.[34]

[33] Think of the case where $\mathcal{F}$ is a singleton.

[34] You can settle this by adapting either the proof of Proposition A.11 or the lengthy hint I gave for Exercise 27. Either way is pretty easy.

EXERCISE 57 Let T be a compact metric space and $\mathcal{F}$ an equicontinuous subset of $\mathbf{C}(T)$. Show that $\mathcal{F}$ is uniformly equicontinuous.

Question: Why are we interested in the notion of equicontinuity?

Answer: Because we are interested in compact subsets of $\mathbf{CB}(T)$.

To see what we mean here, take any compact $\mathcal{F} \subseteq \mathbf{CB}(T)$. By Lemma C.1 and Theorem C.2, $\mathcal{F}$ is totally bounded, so if $\varepsilon > 0$, then there exists a finite subset $\mathcal{G}$ of $\mathcal{F}$ such that $\mathcal{F} \subseteq \cup\{N_{\frac{\varepsilon}{3},T}(\vartheta) : \vartheta \in \mathcal{G}\}$, that is, for any $\varphi \in \mathcal{F}$ there is a $\vartheta_\varphi \in \mathcal{G}$ such that $d_\infty(\varphi, \vartheta_\varphi) < \frac{\varepsilon}{3}$. Moreover, since each $\vartheta \in \mathcal{G}$ is continuous and $\mathcal{G}$ is finite, for any given $x \in T$ we can find a $\delta > 0$ such that $\left|\vartheta(y) - \vartheta(x)\right| < \frac{\varepsilon}{3}$ for all $\vartheta \in \mathcal{G}$ and $y \in N_{\delta,T}(x)$. It follows that

$$\left|\varphi(x) - \varphi(y)\right| \leq \left|\varphi(x) - \vartheta_\varphi(x)\right| + \left|\vartheta_\varphi(x) - \vartheta_\varphi(y)\right| + \left|\vartheta_\varphi(y) - \varphi(y)\right| < \varepsilon$$

for any $\varphi \in \mathcal{F}$ and $y \in N_{\delta,T}(x)$. Since $\varepsilon > 0$ and $x \in T$ are arbitrary here, we conclude: *For any metric space T, if $\mathcal{F} \subseteq \mathbf{CB}(T)$ is compact, then it is equicontinuous.*[35] (Notice that T is an arbitrary metric space here.)

This fact points to the close connection between the compactness and equicontinuity properties for subsets of $\mathbf{CB}(T)$. This connection is especially tight in the case where T is a compact metric space. In that case one can show that the closure of any bounded and equicontinuous subset of $\mathbf{C}(T)$ is necessarily compact.[36] To be able to state this result cleanly, we need one final bit of jargon.

DEFINITION

A subset S of a metric space X is said to be **relatively compact** if $cl_X(S)$ is a compact subset of X.

The equivalence of the notions of compactness and sequential compactness (Theorem C.2) allows us to readily obtain the following useful sequential characterization of the relatively compact subsets of any given metric space.

[35] In the case of $\mathbf{C}[0,1]$, this observation was made first by Césare Arzelà in 1889.

[36] This is a significantly deeper observation than the previous one. In the case of $\mathbf{C}[0,1]$, it was proved first by Giulio Ascoli in 1883. In turn, the generalization to the case of arbitrary compact T was carried out by Fréchet. (Who else?)

LEMMA 3
A subset S of a metric space X is relatively compact if, and only if, every sequence in S has a subsequence that converges in X.

PROOF
The "only if" part of the assertion is a trivial consequence of Theorem C.2. To see the "if" part, assume that every sequence in S has a subsequence that converges in X, and pick any sequence (x^m) in $cl_X(S)$. We wish to show that (x^m) has a subsequence that converges *in* $cl_X(S)$. (Thanks to Theorem C.2, this will complete the proof.) Clearly, for any term of (x^m) we can find a member of S that is as close as we want to that term. In particular, for any $m \in \mathbb{N}$ there is a $y^m \in S$ such that $d(x^m, y^m) < \frac{1}{m}$. By hypothesis, (y^m) has a subsequence, say (y^{m_k}), that converges in X. Obviously, $y := \lim y^{m_k} \in cl_X(S)$. But (x^{m_k}) must also converge to y, because

$$d(x^{m_k}, y) \leq d(x^{m_k}, y^{m_k}) + d(y^{m_k}, y) < \frac{1}{m_k} + d(y^{m_k}, y) \to 0$$

(as $k \to \infty$). We are done. ∎

Here is our main course:

THE ARZELÀ-ASCOLI THEOREM
Let T be a compact metric space, and $\mathcal{F} \subseteq \mathbf{C}(T)$. Then $\mathcal{F}$ is relatively compact if, and only if, it is bounded and equicontinuous. In particular, a subset of $\mathbf{C}(T)$ is compact if, and only if, it is closed, bounded, and equicontinuous.

PROOF
The second statement is easily deduced from the first one, so here we only need to show that a bounded and equicontinuous subset $\mathcal{F}$ of $\mathbf{C}(T)$ is relatively compact. Take any $(\varphi_m) \in \mathcal{F}^\infty$ and observe that, by Lemma 3, we will be done if we can show that (φ_m) has a subsequence that converges in $\mathbf{C}(T)$. Since $\mathbf{C}(T)$ is complete (Proposition 7), it is then enough to show that (φ_m) has a Cauchy subsequence.[37]

[37] Just to get a better handle on the forthcoming argument, first think about how you would prove this claim—by using the Bolzano-Weierstrass Theorem finitely many times—if T was finite. This will demystify the following argument, which is a showcase for Cantor's diagonal method.

Since T is totally bounded (why?), for each $n \in \mathbb{N}$ there exists a finite $T_n \subseteq T$ such that $T \subseteq \cup\{N_{\frac{1}{n},T}(x) : x \in T_n\}$. Define $S := \cup\{T_n : n \in \mathbb{N}\}$. Then S is countable (why?), so we may enumerate it as $\{x^1, x^2, \ldots\}$.

Now, since $\mathcal{F}$ is bounded, $(\varphi_m(x^1))$ is a bounded real sequence. (Yes?) So, it has a convergent subsequence, say $(\varphi_{m,1}(x^1))$. Similarly, $(\varphi_{m,1}(x^2))$ has a convergent subsequence, say $(\varphi_{m,2}(x^2))$. (Notice that $(\varphi_{m,2}(x^1))$ also converges.) Continuing this way inductively, we obtain a sequence of subsequences $((\varphi_{m,1}), (\varphi_{m,2}), \ldots)$ of (φ_m) such that

(i) $(\varphi_{m,t})$ is a subsequence of $(\varphi_{m,t-1})$ for each $t = 2, 3, \ldots$; and

(ii) $(\varphi_{m,t}(x^i))$ converges for each $i = 1, \ldots, t$.

Define $\psi_m := \varphi_{m,m}$ for each $m \in \mathbb{N}$. (Cantor's diagonal method!) Obviously, (ψ_m) is a subsequence of (φ_m). Moreover, $(\psi_m(x^i))$ converges for each $i = 1, 2, \ldots$, that is,

$$\lim \psi_m(z) \in \mathbb{R} \qquad \text{for all } z \in S. \tag{14}$$

(Why? Think about it!) We will complete the proof by showing that (ψ_m) is Cauchy.

Fix any $\varepsilon > 0$. Clearly, $\{\psi_1, \psi_2, \ldots\}$ is uniformly equicontinuous (Exercise 57), and hence there exists an $n \in \mathbb{N}$ such that

$$\left|\psi_m(x) - \psi_m(y)\right| < \tfrac{\varepsilon}{3} \qquad \text{for all } m \in \mathbb{N} \text{ and } x, y \in T \text{ with } d(x, y) < \tfrac{1}{n}. \tag{15}$$

Moreover, by (14), $(\psi_m(z))$ is Cauchy for each $z \in T_n$, and hence, since T_n is a finite set, there exists an $M \in \mathbb{R}$ such that

$$|\psi_k(z) - \psi_l(z)| < \tfrac{\varepsilon}{3} \qquad \text{for all } k, l \geq M \text{ and all } z \in T_n. \tag{16}$$

(Why?) Since $T \subseteq \cup\{N_{\frac{1}{n},T}(z) : z \in T_n\}$, for any $x \in T$ there is a $z \in T_n$ such that $x \in N_{\frac{1}{n},T}(z)$, so, by (15) and (16), we have

$$|\psi_k(x) - \psi_l(x)| \leq |\psi_k(x) - \psi_k(z)| + |\psi_k(z) - \psi_l(z)| + |\psi_l(z) - \psi_l(x)| < \varepsilon$$

for all $k, l \geq M$. That is, $|\psi_k(x) - \psi_l(x)| \leq \varepsilon$ for any $x \in T$, or put differently, $d_\infty(\psi_k, \psi_l) \leq \varepsilon$ for all $k, l \geq M$. Since $\varepsilon > 0$ is arbitrary here, we may thus conclude that (ψ_m) is a Cauchy subsequence of (φ_m), and rest the proof in peace. ■

In Chapter J we will see some pretty impressive applications of the Arzelà-Ascoli Theorem. For now, we conclude with some preliminary corollaries of this important result.

EXERCISE 58[H] Let T be a compact metric space, and $\mathcal{F}$ a closed subset of $\mathbf{C}(T)$ such that $\sup\{|\varphi(x)| : \varphi \in \mathcal{F}\} < \infty$ for each $x \in T$. Show that $\mathcal{F}$ is compact iff it is equicontinuous.

EXERCISE 59 For any given $H \in \mathbf{C}(\mathbb{R})$, define the self-map Φ on $\mathbf{C}[0, 1]$ by

$$\Phi(f)(x) := \int_0^x H(f(t))dt.$$

Show that $\Phi(\mathcal{F})$ is relatively compact in $\mathbf{C}[0, 1]$ for any bounded subset $\mathcal{F}$ of $\mathbf{C}[0, 1]$.

EXERCISE 60 For any Lipschitz continuous function $f \in \mathbf{C}[0, 1]$, let ℓ_f stand for the Lipschitz constant of f. Denote the zero function on $[0, 1]$ by $\mathbf{0}$, and prove that $\{f \in \mathbf{C}[0, 1] : d_\infty(f, \mathbf{0}) + \ell_f \leq 1\}$ is compact.

7* Extension of Continuous Functions

One is sometimes confronted with the problem of extending a given continuous function defined on some set to a larger set in such a way that the extended function is continuous on this larger set.[38] This section is devoted to the basic analysis of this problem.

We begin with the following surprisingly useful result of Pavel Urysohn.[39]

[38] This sort of a problem arises, for instance, when proving equilibrium existence theorems where a standard fixed point argument does not readily apply. But let me not get ahead of the story; I will discuss such issues later in some detail. For the moment, it is best to view the following analysis as a contribution to our understanding of the nature of continuous functions.

[39] Pavel Urysohn (1898–1924) was an extraordinary mathematician who was one of the founders of the field of point set topology; Urysohn's Lemma (the above version of which is extremely simplified) is sometimes referred to as "the first nontrivial fact of point set topology." Urysohn published his first paper when he was 17 (on a physics topic), and died when he was only 26; he apparently drowned while swimming off the coast of France.

URYSOHN'S LEMMA

Let A and B be two nonempty closed subsets of a metric space X with $A \cap B = \emptyset$. For any $-\infty < a \leq b < \infty$, there exists a continuous function $\varphi \in [a, b]^X$ such that $\varphi(x) = a$ for all $x \in A$ and $\varphi(x) = b$ for all $x \in B$.

PROOF

Given any $-\infty < a \leq b < \infty$, define $\varphi \in [a, b]^X$ by $\varphi(x) := (b - a)\psi(x) + a$ where $\psi \in [0, 1]^X$ is defined by

$$\psi(x) := \frac{d(x, A)}{d(x, A) + d(x, B)}.$$

Since A and B are closed, $d(x, A) > 0$ for all $x \in X \backslash A$ and $d(x, B) > 0$ for all $x \in X \backslash B$ (Exercise 2). Since $A \cap B = \emptyset$, we thus have $d(x, A) + d(x, B) > 0$ for all $x \in X$, so ψ is well-defined. It is also continuous (Example 1.[3]), and $\psi|_A = 0$ and $\psi|_B = 1$. That φ has the desired properties is immediate from these observations. ■

This result may not seem surprising to you in the case where $X = \mathbb{R}$. (By the way, $\mathbb{R}$ is a good space to show that none of the regularity conditions used in Urysohn's Lemma can be omitted in its statement.) The amazing thing is that the result applies with cunning generality, even in spaces where we completely lack geometric intuition. Do not be fooled by the simplicity of its proof. Urysohn's Lemma is what one might call a "deep" theorem.[40]

The following result, the highlight of this section, was first proved by Heindrich Tietze in 1915. There are a variety of ways to prove this theorem. We choose a method of proof that is based on a certain approximation procedure, for this method proves useful in other contexts as well. A more direct method of proof is sketched in Exercise 63 below.

THE TIETZE EXTENSION THEOREM

Let T be a nonempty closed subset of a metric space X. For every continuous $\varphi \in \mathbf{C}(T)$ there exists a continuous $\varphi^* \in \mathbf{C}(X)$ with $\varphi^*|_T = \varphi$. Moreover, if $a \leq \varphi \leq b$ for some real numbers a and b, then φ^* can be chosen so that $a \leq \varphi^* \leq b$.

[40] If you are wondering where on earth one would need to apply such a result, don't worry; I did that too when I first learned about it. Seeing it in action is the best way to understand how useful Urysohn's Lemma really is, and in what sort of situations it may provide the help one seeks. The main result of this section provides a good case in point. I will give other applications later.

PROOF

(Urysohn) We first prove the second claim. Let $\varphi \in \mathbf{C}(T)$ be a continuous map with $-1 \leq \varphi \leq 1$.[41] Let $\varphi_0 := \varphi$ and define $A_0 := \left\{z \in T : \varphi_0(z) \leq -\frac{1}{2}\right\}$ and $B_0 := \left\{z \in T : \varphi_0(z) \geq \frac{1}{2}\right\}$. Clearly, $A_0 \cap B_0 = \emptyset$ while $A_0 \neq \emptyset$ (since $\inf \varphi_0(T) = -1$) and $B_0 \neq \emptyset$. Moreover, by continuity of φ_0, these sets are closed in T, and hence in X (since T is closed in X). Thus, by Urysohn's Lemma, there exists a continuous $\vartheta_0 : X \to \left[-\frac{1}{2}, \frac{1}{2}\right]$ such that $\vartheta_0|_{A_0} = -\frac{1}{2}$ and $\vartheta_0|_{B_0} = \frac{1}{2}$. (Notice that ϑ_0, while defined on X, is a (very rough) approximation of φ on T. You might want to draw a picture here to follow the logic of the following construction.) We define next $\varphi_1 := \varphi - \vartheta_0$, and observe that $\inf \varphi_1(T) = -\frac{1}{2}$ and $\sup \varphi_1(T) = \frac{1}{2}$. Now define $A_1 := \left\{z \in T : \varphi_1(z) \leq -\frac{1}{4}\right\}$ and $B_1 := \left\{z \in T : \varphi_1(z) \geq \frac{1}{4}\right\}$, and apply Urysohn's Lemma again to find a continuous $\vartheta_1 : X \to [-\frac{1}{4}, \frac{1}{4}]$ such that $\vartheta_1|_{A_1} = -\frac{1}{4}$ and $\vartheta_1|_{B_1} = \frac{1}{4}$. (While defined on the entire X, $\vartheta_0 + \vartheta_1$ approximates φ on T, and it does so "better" than ϑ_0.) As the next step of induction, we define $\varphi_2 := \varphi - (\vartheta_0 + \vartheta_1)$, and use Urysohn's Lemma again to find a continuous $\vartheta_2 : X \to [-\frac{1}{8}, \frac{1}{8}]$ with $\vartheta_2|_{A_2} = -\frac{1}{8}$ and $\vartheta_2|_{B_2} = \frac{1}{8}$, where $A_2 := \left\{z \in T : \varphi_2(z) \leq -\frac{1}{8}\right\}$ and $B_2 := \left\{z \in T : \varphi_2(z) \geq \frac{1}{8}\right\}$. (The idea of the proof must be transpiring at this point.) Proceeding inductively, then, we obtain a sequence (ϑ_m) in $\mathbf{CB}(X)$ such that, for each m,

$$|\vartheta_m(x)| \leq \frac{1}{2^{m+1}} \qquad \text{for all } x \in X$$

and

$$|\varphi(z) - \phi_m(z)| \leq \frac{1}{2^{m+1}} \qquad \text{for all } z \in T, \tag{17}$$

[41] It is without loss of generality to set $a = -1$ and $b = 1$. Indeed, once we learn that we can extend a continuous map in $[-1,1]^T$ to a continuous map in $[-1,1]^X$, then we can extend any continuous map in $[a,b]^T$ to a continuous map in $[a,b]^X$ (for any $-\infty < a < b < \infty$) by a simple normalization trick. Indeed, $\varphi \in [a,b]^T$ is continuous iff $\psi \in [-1,1]^T$ is continuous, where

$$\psi(x) := \tfrac{2}{b-a}\left(\varphi(x) - \left(\tfrac{a+b}{2}\right)\right).$$

(This is not a magical formula: I tried $\psi := \alpha\varphi + \beta$, and chose the real numbers α and β to guarantee $\alpha a + \beta = -1$ and $\alpha b + \beta = 1$.) Moreover, if $\psi^* \in [-1,1]^X$ is a continuous extension of ψ, then $\varphi^* = \frac{b-a}{2}\psi^* + \frac{a+b}{2}$ is a continuous extension of φ with $a \leq \varphi^* \leq b$.

where $\phi_m := \vartheta_0 + \cdots + \vartheta_m$. It is plain that $\phi_m \in \mathbf{CB}(X)$ for each m. (Why?)[42] Moreover, (ϕ_m) is Cauchy in $\mathbf{CB}(X)$, since, for any $k > l$ and $x \in X$, we have

$$|\phi_k(x) - \phi_l(x)| = \left|\sum_{i=l+1}^{k} \vartheta_i(x)\right| \leq \sum_{i=l+1}^{k} |\vartheta_i(x)| \leq \sum_{i=l+1}^{k} \frac{1}{2^{i+1}} \leq \sum_{i=l}^{\infty} \frac{1}{2^i} \to 0$$

(as $l \to \infty$). Since $\mathbf{CB}(X)$ is complete, then, $\lim \phi_m$ exists and belongs to $\mathbf{CB}(X)$. Using (17), we also find $\varphi(z) = \lim \phi_m(z)$ for all $z \in T$, so setting $\varphi^* := \lim \phi_m$ we are done.

It remains to consider the case where φ is not bounded. To this end, define $f : \mathbb{R} \to (-1, 1)$ by $f(t) := \frac{t}{1+|t|}$, which is a homeomorphism. By what we have established above, we know that there is a continuous extension of $f \circ \varphi$ defined on X. Denoting this extension by ψ^*, it is readily checked that $\varphi^* := f^{-1} \circ \psi^*$ is a continuous extension of φ that is defined on X. ■

Exercise 61 Let Y be a closed subset of a compact metric space X, and define $\Phi : \mathbf{C}(X) \to \mathbf{C}(Y)$ by $\Phi(\varphi) := \varphi|_Y$. Prove that Φ is a continuous surjection.

Exercise 62[H] (*The Converse of Weierstrass' Theorem*) Show that if T is a nonempty subset of a metric space X such that *every* $\varphi \in \mathbf{C}(T)$ is bounded, then T must be compact in X.

The following exercise provides an alternative proof of the Tietze Extension Theorem.

Exercise 63 (Dieudonné) Let T be a nonempty closed subset of a metric space X, and let $\varphi \in [1, 2]^T$ be continuous. Define $\varphi^* \in [1, 2]^X$ by

$$\varphi^*(x) := \begin{cases} \frac{1}{d(x,T)} \inf\{\varphi(y)d(x, y) : y \in T\}, & \text{if } x \in X \backslash T \\ \varphi(x), & \text{if } x \in T \end{cases},$$

and show that φ^* is continuous.

[42] It may be tempting to conclude the proof at this point as follows. "By (17),

$$d_\infty(\varphi, \phi_m) = \sup\{|\varphi(z) - \phi_m(z)| : z \in T\} \leq \frac{1}{2^{m+1}} \to 0.$$

Thus $\lim \phi_m$ must exist and equal φ on T. Since $\mathbf{CB}(X)$ is closed, and $\phi_m \in \mathbf{CB}(X)$ for each m, we also have $\lim \phi_m \in \mathbf{CB}(X)$." This is a seriously erroneous argument. Why?

It is important to note that we can readily generalize the Tietze Extension Theorem to the case of vector-valued functions. Indeed, any continuous (and bounded) function $f : T \to \mathbb{R}^n$, where T is a closed subset of a metric space, can be extended to a continuous (and bounded) function defined on the entire space. This fact is obtained by applying the Tietze Extension Theorem component by component (Example 2.[3]).

EXERCISE 64 Take any $n \in \mathbb{N}$ and let T be a nonempty closed subset of a metric space X. Show that if $\varphi : T \to [0,1]^n$ is continuous, then there exists a continuous $\varphi^* : X \to [0,1]^n$ with $\varphi^*|_T = \varphi$.

EXERCISE 65 Let T be a nonempty closed subset of a metric space X. For any $m \in \mathbb{N}$, suppose that Y_m is a metric space with the following property: Every continuous $\varphi : T \to Y_m$ has a continuous extension defined on X. Now let $Y := \mathsf{X}^\infty Y_i$, and metrize this set by means of the product metric. Show that every continuous $\varphi : T \to Y$ has a continuous extension defined on X.

The following exercise shows that a Tietze type theorem can be proved even when the domain of the function to be extended is not closed, provided that the function is uniformly continuous and has a compact codomain. In this case the extension can be guaranteed to be uniformly continuous as well.

EXERCISE 66[H] Let T be a nonempty set in a metric space X, and let $\varphi \in [-1,1]^T$ be uniformly continuous.

(a) Define A_0 and B_0 as in the proof of the Tietze Extension Theorem, and show that $d(A_0, B_0) > 0$.

(b) Let u and v be two uniformly continuous real functions on a metric space Y. If $\inf\{|v(y)| : y \in Y\} > 0$, and $\frac{u}{v}$ is bounded, then $\frac{u}{v}$ is uniformly continuous on Y. Prove this.

(c) Recursively define the sequences (A_m), (B_m), and (φ_m) as in the proof of the Tietze Extension Theorem, but this time by means of the sequence (ϑ_m) in $\mathbf{CB}(X)$ defined by $\vartheta_m(x) := \frac{1}{2^m} k_m(x) - \frac{1}{2^{m+1}}$ where

$$\varphi_m(x) := \frac{d(x, A_m)}{d(x, A_m) + d(x, B_m)} \qquad \text{for all } x \in X,$$

$m = 1, 2, \ldots$. Use parts (a) and (b) to show that each ϑ_m is well-defined and uniformly continuous on X. (*Note.* We are basically following the strategy of the proof of the Tietze Extension Theorem, but without invoking Urysohn's Lemma.)

(d) Prove that there exists a uniformly continuous $\varphi^* \in [-1,1]^X$ that extends φ.

(e) Show that this uniformly continuous extension result would not hold if the range of φ was not contained in a compact metric space.

The final result of this section is an interesting variant of the Tietze Extension Theorem, which was proved in 1934 by Edward McShane (in the case where the ambient metric space is Euclidean). It demands more from the function to be extended, namely its Lipschitz continuity, but it ensures that the extension is Lipschitz continuous as well. Its proof is more direct than the one we gave for the Tietze Extension Theorem. In fact, it provides an explicit formula for computing the sought extension.

Theorem 2

(McShane) *Let T be a nonempty set in a metric space X. Any Lipschitz continuous $\varphi \in \mathbb{R}^T$ can be extended to a Lipschitz continuous function $\varphi^* \in \mathbb{R}^X$.*[43]

Proof

Take any Lipschitz continuous $\varphi \in \mathbb{R}^T$, and define $\varphi^* \in \mathbb{R}^X$ by

$$\varphi^*(x) := \inf\{\varphi(w) + Kd(w,x) : w \in T\},$$

where $K > 0$ is the Lipschitz constant of φ. To show that φ^* is a well-defined real function, we need to prove that $\varphi^*(x) > -\infty$ for all $x \in X$. To this end, fix any $y \in T$, and observe that, by Lipschitz continuity of φ, we have $|\varphi(w) - \varphi(y)| \leq Kd(w,y)$, and hence $\varphi(w) \geq \varphi(y) - Kd(w,y)$, for all $w \in T$. Then, by the triangle inequality,

$$\varphi(w) + Kd(x,w) \geq \varphi(y) - Kd(w,y) + Kd(x,w) \geq \varphi(y) - Kd(x,y)$$

[43] More is true: the forthcoming proof shows that the Lipschitz constants of φ and φ^* can be chosen to be identical.

for all $w \in T$, and it follows that

$$\varphi^*(x) \geq \varphi(y) - Kd(x, y) > -\infty.$$

Since $\varphi(x) \leq \varphi(w) + Kd(x, w)$ for all $x, w \in T$ by Lipschitz continuity of φ, we have $\varphi^*|_T = \varphi$. (Yes?) It remains to establish the Lipschitz continuity of φ^*. To this end, take any $x, y \in X$ and observe that, for any $\varepsilon > 0$, there exists a $z \in T$ such that $\varphi^*(x) + \varepsilon \geq \varphi(z) + Kd(x, z)$. But $\varphi^*(y) \leq \varphi(z) + Kd(y, z)$ (because $z \in T$), so by Lipschitz continuity of φ and the triangle inequality,

$$\varphi^*(y) - \varphi^*(x) \leq \varphi(z) + Kd(y, z) - (\varphi(z) + Kd(x, z)) + \varepsilon \leq Kd(x, y) + \varepsilon.$$

Interchanging the roles of x and y, and noting that $\varepsilon > 0$ is arbitrary here, we get $\left|\varphi^*(x) - \varphi^*(y)\right| \leq Kd(x, y)$. Since x and y are arbitrary points in X, we are done. ■

The following exercise shows that Hölder continuous functions too can be extended in a way similar to Lipschitz continuous functions.

Exercise 67^{H} Let T be a nonempty set in a metric space X, and let $\phi : T \to \ell^\infty$ be a function such that there exist a $K > 0$ and an $\alpha \in (0, 1]$ such that

$$d_\infty(\phi(x), \phi(y)) \leq Kd(x, y)^\alpha \qquad \text{for all } x, y \in X.$$

Prove that any $\phi : T \to \ell^\infty$ can be extended to a Hölder continuous function $\phi^* : X \to \ell^\infty$. (*Note.* This observation also shows that Theorem 2 applies to $\mathbb{R}^n$-valued functions. Why?)

8 Fixed Point Theory II

One of the subfields of mathematical analysis that is most commonly used in economics is fixed point theory. We have already seen two major results of this theory: Tarski's Fixed Point Theorem (Section B.3.1) and the Banach Fixed Point Theorem (Section C.6.2). As powerful as they are, these results are to little avail when one has to work with self-maps that are neither monotonic nor contractive. Yet as we shall see in this section, there is in fact much more to fixed point theory. It turns out that we can guarantee the existence of a fixed point of a continuous self-map, provided that the domain of this map is sufficiently well-behaved.

8.1 The Fixed Point Property

Let us begin by giving a name to those metric spaces on which continuous self-maps are guaranteed to have a fixed point.

> DEFINITION
> A metric space X is said to have the **fixed point property** if every continuous self-map on X has a fixed point.

EXAMPLE 9

[1] A discrete metric space X has the fixed point property iff it contains a single point. The "if" part of this assertion is trivial. To prove the "only if" part, observe that if x and y are two distinct points in X, then $f \in X^X$ defined by $f(x) := y$ and $f(z) := x$ for all $z \in X\backslash\{x\}$, is continuous (as any map on a discrete space is continuous). Clearly, f does not have a fixed point.

[2] $(0, 1)$ does not have the fixed point property. For instance, $x \mapsto x^2$ is a continuous self-map on $(0, 1)$ without a fixed point. Similarly, neither $\mathbb{R}$ nor $\mathbb{R}_+$ has the fixed point property. How about $[0, 1] \cup [2, 3]$?

[3] Let $n \in \mathbb{N}$. The sphere $S := \{x \in \mathbb{R}^n : \sum^n x_i^2 = 1\}$, when metrized by any metric d_p, $1 \leq p \leq \infty$, does not have the fixed point property. Indeed, $x \mapsto -x$ is a continuous self-map on S without a fixed point.

[4] $[0, 1]$ has the fixed point property (Corollary 1). □

In general, it is not easy to see whether a given metric space has the fixed point property or not. Determination of those metric spaces that have this property is one of the major areas of research in fixed point theory. A common methodology that is used in this theory is to deduce new metric spaces with the fixed point property from the ones that are already known to possess this property. The following simple fact often proves useful in this regard.

PROPOSITION 8
Let X and Y be two homeomorphic metric spaces. If X has the fixed point property, then so does Y.

PROOF
If f is a continuous self-map on Y and $g \in X^Y$ is a homeomorphism, then $g \circ f \circ g^{-1}$ is a continuous self-map on X. So, if X has the fixed point property, then this map must have a fixed point, say x, in X. Then $g^{-1}(x)$ is a fixed point of f. ∎

EXERCISE 68[H] Let X and Y be two metric spaces such that there exists a continuous bijection from X onto Y.

(a) Show that if X is compact and has the fixed point property, then Y has the fixed point property as well.

(b) Give an example to show that if Y is compact and has the fixed point property, then X need not have the fixed point property.

8.2 Retracts

A metric subspace of a metric space X need not have the fixed point property even if X has this property. For instance, $[0, 1]$ has the fixed point property, but $(0, 1)$ does not. Roughly speaking, this is because $[0, 1]$ cannot be condensed into $(0, 1)$ continuously in a way that leaves each point of $(0, 1)$ intact. Let us make this point precise.

DEFINITION
Let X be a metric space and S a metric subspace of X. A continuous map $r : X \to S$ is called a **retraction** (from X onto S) if $r(x) = x$ for all $x \in S$. If such a map exists, then we say that S is a **retract** of X.

For instance, $[0, 1]$ is a retract of $\mathbb{R}$. Indeed, $r \in [0, 1]^{\mathbb{R}}$ with $r(t) = 0$ for all $t < 0$, $r|_{[0,1]} = \mathrm{id}_{[0,1]}$, and $r(t) = 1$ for all $t > 1$, is a retraction. By contrast, $(0, 1)$ is not a retract of $[0, 1]$. (Proof?) Here are some other examples.

EXAMPLE 10

[1] Every singleton subspace of a metric space is a retract of that space.

[2] Every metric subspace of a discrete space is a retract of that space.

[3] For any $n \in \mathbb{N}$, the unit cube $[0,1]^n$ is a retract of $\mathbb{R}^n$. Indeed, since $\mathrm{id}_{[0,1]^n}$ is continuous, the Tietze Extension Theorem (or more precisely, Exercise 64) guarantees that there is a continuous extension of $\mathrm{id}_{[0,1]^n}$ that maps $\mathbb{R}^n$ onto $[0,1]^n$. Clearly, this extension is a retraction from $\mathbb{R}^n$ onto $[0,1]^n$.

[4] For any $n \in \mathbb{N}$ and $\alpha \geq 0$, define

$$B_\alpha^n := \left\{ x \in \mathbb{R}^n : \sum_{i=1}^n x_i^2 \leq \alpha \right\}.$$

We claim that any nonempty closed and convex subset S of any B_α^n is a retract B_α^n. The idea is simple. Let $r : B_\alpha^n \to S$ be the restriction (to B_α^n) of the projection operator into S, that is, define $r(x)$ through the equation $d(x, r(x)) = d(x, S)$. By what is shown in Example 5, this map is well-defined and continuous. But it is obvious that $r|_S = \mathrm{id}_S$, so r is retraction from B_α^n onto S. □

Retracts play an important role in fixed point theory. One of the main reasons for this is the fact that any retract of a metric space X inherits the fixed point property of X (provided that X has this property to begin with, of course). We will use this property in the next subsection to reduce a difficult fixed point problem to an "easier" one.

PROPOSITION 9
(Borsuk) *Let S be a retract of a metric space X. If X has the fixed point property, then so does S.*

PROOF
If f is a continuous self-map on S, and r is a retraction from X onto S, then $f \circ r$ is a continuous self-map on X. Thus, if X has the fixed point property, then this map has a fixed point, say x, in X. Clearly, $x = f(r(x)) \in S$. Since r is a retraction, then $r(x) = x$, which implies that x is a fixed point of f. ■

In what follows we will write B^n for B_1^n, that is,

$$B^n := \left\{ x \in \mathbb{R}^n : \sum_{i=1}^n x_i^2 \leq 1 \right\}$$

for any $n \in \mathbb{N}$. This set is called the **closed unit ball** of $\mathbb{R}^n$, and equals the closure of the 1-neighborhood of the origin. Its boundary $bd_{\mathbb{R}^n}(B^n)$ is called the $n-1$ **dimensional unit sphere**. (So the unit circle in $\mathbb{R}^2$ is the one-dimensional sphere.) It is customary to denote $bd_{\mathbb{R}^n}(B^n)$ by S^{n-1}, that is,

$$S^{n-1} := \left\{ x \in \mathbb{R}^n : \sum_{i=1}^{n} x_i^2 = 1 \right\}.$$

The following geometric observation, which was proved by Karol Borsuk in 1931, is extremely important: $n-1$ *dimensional unit sphere is not a retract of* B^n, $n = 1, 2, \ldots$. This is almost a triviality in the case of $n = 1$, for $\{-1, 1\}$ is clearly not a retract of $[-1, 1]$. (Why?) Unfortunately, it gets pretty difficult to prove as soon as we increase the dimension of the space to 2. But the basic claim is not an incredible one. Try to move in your mind, continuously, all points of B^2 to the boundary of this set without moving the points in S^1. Then if $x \in B^2 \backslash S^1$ is near $y \in S^1$, you should map it somewhere close to y. You will soon realize that there is a problem; you can't continue on doing this in a continuous manner, for eventually you will have to move nearby points in S^1 to points in S^1 that are quite apart.[44]

We state this fact as a lemma, but omit its difficult proof.[45]

> BORSUK'S LEMMA
> *There is no retraction from B^n onto S^{n-1}, $n = 1, 2, \ldots$.*

What does this have to do with fixed points? Hold on, we are almost there.

[44] I apologize for being so informal, but I do not want to get into details here. All I want is that you sort of "feel" that the claim I am making is true. Let us consider a sample attempt (which is doomed to fail) at constructing a retraction from B^2 onto S^1. For any $x \in B^2 \backslash \{\mathbf{0}\}$, think of the line that goes through $\mathbf{0}$ and x, and define $r(x)$ as the nearest point to x that is on both S^1 and this line. (Here $\mathbf{0}$ stands for the 2-vector $(0, 0)$.) This defines $r : B^2 \backslash \{\mathbf{0}\} \to S^1$ as a continuous function with $r(x) = x$ for all $x \in S^1$. Looks promising, no? But here is the catch: we need to extend this r to the entire B^2, that is, we need to decide which point in S^1 we should assign to $\mathbf{0}$. No matter what you choose as a value for $r(\mathbf{0})$, the resulting map will not be continuous at $\mathbf{0}$! (Why?)

[45] The friendliest proofs of this fact that I know of are the ones given by Kannai (1981) and Gamelin and Greene (1999). Both of these proofs require some backround in vector calculus, however. The former one is based on the Divergence Theorem and the latter on Stokes' Theorem.

EXERCISE 69 [H] For any $(n, \alpha) \in \mathbb{N} \times \mathbb{R}_+$, show that B^n_α is a retract of any subset of $\mathbb{R}^n$ that contains it.

EXERCISE 70 A nonempty subset S of a metric space X is said to be a **neighborhood retract** of X if there is an open subset O of X such that S is a retract of O.

(a) Show that the S^{n-1} is a neighborhood retract of $\mathbb{R}^n$ for any $n \in \mathbb{N}$.

(b) Is the boundary of $\{(x_m) \in \mathbb{R}^\infty : \sum^\infty \frac{1}{2^i} \min\{1, x_i^2\} \leq 1\}$ a neighborhood retract of $\mathbb{R}^\infty$?

EXERCISE 71 Let S be a nonempty subset of a metric space X. Show that S is a retract of X iff for every metric space Y and every continuous $f : S \to Y$, there exists a continuous $F : X \to Y$ with $F|_S = f$.

8.3 The Brouwer Fixed Point Theorem

The following theorem is the most important fixed point theorem that applies to continuous self-maps on a Euclidean space. Or, as some would say, the most important fixed point theorem, period.[46]

> BROUWER'S FIXED POINT THEOREM FOR THE UNIT BALL
> *B^n has the fixed point property, $n = 1, 2, \ldots$.*

PROOF

Suppose the claim is not true, so we can find an $n \in \mathbb{N}$ and a continuous self-map Φ on B^n without a fixed point. Define $r : B^n \to S^{n-1}$ as follows. For each $x \in B^n$, consider the line that goes through x and $\Phi(x)$, and the two points on the intersection of S^{n-1} and this line (Figure 8). Obviously, exactly

[46] I will prove this result by using Borsuk's Lemma. Since I'm able to offer this lemma only at face value in this text, I wouldn't blame you if you felt dissatisfied with this situation. In that case, you might want to look at Franklin (1980), who presents three different proofs (in an exceptionally detailed manner, I might add). One of these is combinatorial, so if you like combinatorial arguments, read the one that focuses on Sperner's Lemma (which is also covered in many other related texts, such as Border (1989)). Another type of proof considered by Franklin is based on an ingenious idea of John Milnor, which was later simplified by Rogers (1980). Finally, there are some "elementary" proofs that use advanced calculus techniques. But if you are familiar with these (such as the Divergence Theorem), you can follow the proof given by Kannai (1981) for Borsuk's Lemma anyway.

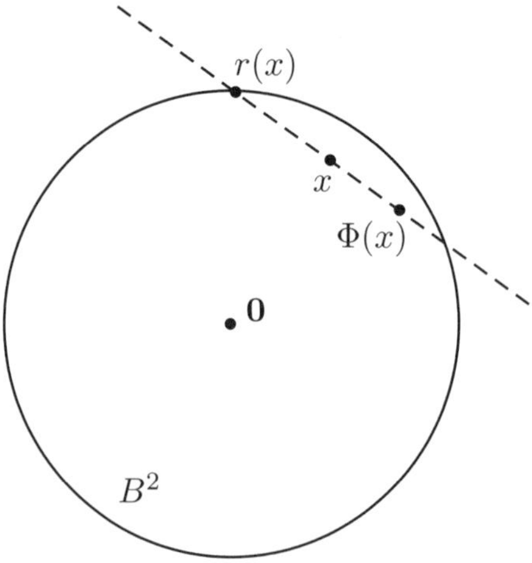

Figure 8

one of these points has the property that $\Phi(x)$ is not on the line segment x and that point. Let $r(x)$ be that point. Since $x \neq \Phi(x)$ for every $x \in B^n$, r is well-defined. (There is something to be checked here!) A little reflection also shows that r is continuous, and we have $r(x) = x$ for all $x \in S^{n-1}$. (Verify!) Thus we are forced to conclude that S^{n-1} is a retract of B^n, which cannot be true due to Borsuk's Lemma. ∎

Exercise 72 Show that Brouwer's Fixed Point Theorem for the Unit Ball implies Borsuk's Lemma, and conclude that these two theorems are equivalent.

A good number of fixed point theorems that are invoked routinely in certain parts of economic theory can be derived by using Brouwer's Fixed Point Theorem for the Unit Ball. This is somewhat surprising, for this result does not seem at first to be terribly useful. After all, it applies only to self-maps that are defined on the closed unit ball of a Euclidean space. How often, you may ask, do we get to work with such self-maps? Well, not very often indeed. But in many cases one may reduce the fixed point problem at hand to a fixed point problem on a closed unit ball, and then invoke Brouwer's Fixed Point Theorem for the Unit Ball to solve our original problem. In particular, this method allows us to generalize Brouwer's Fixed Point Theorem for the Unit Ball to the following result, which is a fixed point theorem that should be familiar to all students of economic theory.[47]

THE BROUWER FIXED POINT THEOREM
For any given $n \in \mathbb{N}$, let S be a nonempty, closed, bounded, and convex subset of $\mathbb{R}^n$. If Φ is a continuous self-map on S, then there exists an $x \in S$ such that $\Phi(x) = x$.

PROOF
Since S is bounded, there exists an $\alpha > 0$ such that $S \subseteq B^n_\alpha$. Since B^n has the fixed point property (Brouwer's Fixed Point Theorem for the Unit Ball) and B^n_α and B^n are homeomorphic (why?), B^n_α has the fixed point property (Proposition 8). But S is a retract of B^n_α (Example 10.[4]), so it must have the fixed point property as well (Proposition 9). ∎

All of the requirements of the Brouwer Fixed Point Theorem are needed in its statement, as simple examples in the case $n = 1$ would illustrate. Whether the result (or some ramification of it) is true in spaces that are more general than Euclidean spaces is, however, a delicate issue that has to wait for a later section in the text (Section J.3.3). Even at this stage, however, we can easily show that the exact statement of the result would be false, for instance, in $\mathbf{C}[0, 1]$ or in ℓ^1. This is the content of the next two exercises.

EXERCISE 73[H] Denote the zero function on $[0, 1]$ by $\mathbf{0}$, and let $S := \{f \in \mathbf{C}[0, 1] : d_\infty(f, \mathbf{0}) \leq 1, f(0) = 0 \text{ and } f(1) = 1\}$. Define the self-map Φ on S by $\Phi(f)(t) := f(t^2)$. Show that S is a nonempty, closed, bounded, and

[47] Jan Brouwer (1881–1966) is a distinguished mathematician who is considered one of the main founders of modern topology. (Suffice it to say that before him it was not known that $\mathbb{R}^m$ and $\mathbb{R}^n$ cannot be homeomorphic unless $m = n$.) It is worth noting that in later stages of his career, he became the most forceful proponent of the so-called intuitionist philosophy of mathematics, which not only forbids the use of the Axiom of Choice but also rejects the axiom that a proposition is either true or false (thereby disallowing the method of proof by contradiction). The consequences of taking this position are dire. For instance, an intuitionist would not accept the existence of an irrational number! In fact, in his later years, Brouwer did not view the Brouwer Fixed Point Theorem as a theorem. (He had proved this result in 1912, when he was functioning as a "standard" mathematician.)

If you want to learn about intuitionism in mathematics, I suggest reading—in your spare time, please—the four articles by Heyting and Brouwer in Benacerraf and Putnam (1983). Especially Heyting's second article is a beautiful read.

convex subset of $\mathbf{C}[0,1]$, and Φ is a nonexpansive self-map on S without a fixed point.

EXERCISE 74 (*Kakutani's Example*) Define $S := \{(x_m) \in \ell^1 : \sum^\infty |x_i| \leq 1\}$, and consider the self-map Φ on S defined by $\Phi((x_m)) := (1 - \sum^\infty |x_i|, x_1, x_2, \ldots)$. Show that S is a nonempty, closed, bounded, and convex subset of ℓ^1, and Φ is a Lipschitz continuous map without a fixed point.

8.4 Applications

Economic theorists often use the Brouwer Fixed Point Theorem to establish the existence of equilibrium in models of competitive or strategic interaction. You can check your favorite graduate microeconomics text to see a few such examples. Here we present some other kinds of applications. An illustration of how fixed point arguments are used in equilibrium analysis is given in the next chapter.

Our first application illustrates how new fixed point theorems may be obtained from the Brouwer Fixed Point Theorem.

PROPOSITION 10
For any $n \in \mathbb{N}$, any continuous $\Phi : B^n \to \mathbb{R}^n$ with $\Phi(S^{n-1}) \subseteq B^n$, has a fixed point.

The Brouwer Fixed Point Theorem does not apply readily to settle the claim here, because Φ is not a self-map. We will instead transform Φ into a self-map, find a fixed point of this transformed map by using the Brouwer Fixed Point Theorem, and check the behavior of Φ at that point.

PROOF OF PROPOSITION 10
Take any continuous $\Phi : B^n \to \mathbb{R}^n$, and define the retraction $r : \mathbb{R}^n \to B^n$ by

$$r(x) := \begin{cases} x, & \text{if } x \in B^n \\ \frac{x}{d_2(x,\mathbf{0})}, & \text{otherwise} \end{cases},$$

where $\mathbf{0}$ is the n-vector $(0, \ldots, 0)$. Clearly, $r \circ \Phi$ is a continuous self-map on B^n, and hence it has a fixed point by the Brouwer Fixed Point Theorem. That is, $r(\Phi(x)) = x$ for some $x \in B^n$.

Now if $x \in S^{n-1}$, then $\Phi(x) \in B^n$ by hypothesis. If, on the other hand, $x \in B^n \backslash S^{n-1}$, then $r(\Phi(x)) \in B^n \backslash S^{n-1}$, which is possible only if $\Phi(x) \in B^n$ by definition of r. Therefore, we have $\Phi(x) \in B^n$ for sure. But then $x = r(\Phi(x)) = \Phi(x)$ by definition of r. ■

Exercise 75

(a) Prove the *Antipodal Theorem*: For any $n \in \mathbb{N}$, if $\Phi : B^n \to \mathbb{R}^n$ is a nonexpansive map such that $\Phi(-x) = -\Phi(x)$ for all $x \in S^{n-1}$, then Φ has a fixed point.

(b) Can the word "nonexpansive" be replaced with "continuous" in this statement?

Our second application comes from matrix theory.

Example 11

Let $n \in \mathbb{N}$, and take any $n \times n$ matrix $\mathbf{A} := [a_{ij}]_{n\times n}$. For any $x \in \mathbb{R}^n$, we write $\mathbf{A}x$ for the n-vector $\left(\sum^n a_{1j}x_j, \ldots, \sum^n a_{nj}x_j\right)$. You might recall from linear algebra that a nonzero $\lambda \in \mathbb{R}$ is said to be an **eigenvalue** of $\mathbf{A}$ if $\mathbf{A}x = \lambda x$ holds for some $x \in \mathbb{R}^n \backslash \{\mathbf{0}\}$ (in which case x is called an **eigenvector** of $\mathbf{A}$). If we agree to say that A is **strictly positive** when $a_{ij} > 0$ for each $i, j = 1, \ldots, n$, then we are ready for the following famed theorem of matrix theory.

> Perron's Theorem
>
> *Every strictly positive $n \times n$ matrix has a positive eigenvalue and a positive eigenvector.*

The algebraic proof of this result is not all that straightforward. But we can give an almost immediate proof by using the Brouwer Fixed Point Theorem. Here it goes. Let $S := \left\{x \in \mathbb{R}^n_+ : \sum^n x_i = 1\right\}$, and take any strictly positive $\mathbf{A} = [a_{ij}]_{n\times n}$. Since $a_{ij} > 0$ for all i and j, and since there exists a j with $x_j > 0$ for any $x \in S$, it is obvious that $\sum^n (\mathbf{A}x)_i > 0$ for all $x \in S$, where $(\mathbf{A}x)_i$ denotes the ith term of n-vector $\mathbf{A}x$. We may then define the map $\Phi : S \to \mathbb{R}^n_+$ by

$$\Phi(x) := \frac{\mathbf{A}x}{\sum^n (\mathbf{A}x)_i}.$$

Clearly, Φ is continuous and $\Phi(S) \subseteq S$, whereas S is a nonempty, closed, bounded, and convex subset of $\mathbb{R}^n$. Thus, by the Brouwer Fixed Point

Theorem, there exists an $x^* \in S$ such that $\Phi(x^*) = x^*$. But then, letting $\lambda := \sum^n (\mathbf{A}x^*)_i$, we find $\mathbf{A}x^* = \lambda x^*$, and Perron's Theorem is proved. □

EXERCISE 76 For any $n \in \mathbb{N}$, an $n \times n$ matrix $\mathbf{A} = [a_{ij}]_{n \times n}$ is said to be **stochastic** if $a_{ij} \geq 0$ for all i and j, and $\sum_{j=1}^n a_{ij} = 1$ for all i. Prove that, for every strictly positive $n \times n$ stochastic matrix $\mathbf{A}$, there is an $x \in \mathbb{R}^n_{++}$ such that $\mathbf{A}x = x$ and $\sum^n x_i = 1$. (*Note.* In fact, this x is unique! It's a bit tricky, but this fact can be proved by the Banach Fixed Point Theorem.)

Our final application, which we present here in the form of an exercise, concerns the theory of nonlinear equations, and concludes the present chapter.

EXERCISE 77[H] (Zeidler) Let $n \in \mathbb{N}$, $\alpha > 0$, and $f_i \in \mathbf{C}(\mathbb{R}^n)$, $i = 1, \ldots, n$. Prove: If the boundary condition

$$\sum_{i=1}^n f_i(y) y_i \geq 0 \qquad \text{for all } y \in bd_{\mathbb{R}^n}(B^n_\alpha)$$

holds, then there is at least one $x \in B^n_\alpha$ with $f_i(x) = 0$ for all $i = 1, \ldots, n$.

Chapter E

Continuity II

A function that maps every element of a given set to a nonempty *subset* of another set is called a *correspondence* (or a *multifunction*). Such maps arise quite frequently in optimization theory and theoretical economics. Although this is not really a standard topic in real analysis, this chapter is devoted to the analysis of correspondences, because of their importance for economists.

Our first task is to understand in what sense a correspondence can be viewed as "continuous." After a brief set-theoretical overview of correspondences, therefore, we spend some time examining various continuity concepts for correspondences. These concepts are needed to state and prove Berge's Maximum Theorem, which tells one when the solution to an optimization problem depends on the parameters of the problem in a "continuous" way. Along with a few relatively straightforward applications that attest to the importance of this result, the chapter also touches on a major topic of optimization theory, the theory of stationary dynamic programming. In particular, we discuss at some length the issue of existence, uniqueness, and monotonicity of solutions of a dynamic programming problem, and illustrate our findings with a standard topic in macroeconomics, the one-sector optimal growth model.

Delving into the theory of continuous correspondences more deeply, the rest of the chapter is a bit more advanced than the earlier sections. In this part we introduce the partition of unity and prove the famous Michael Selection Theorem, which provides sufficient conditions to "find" a continuous function within a given correspondence (in the sense that the graph of the function is contained in that of the correspondence). In turn, using this result and the Brouwer Fixed Point Theorem, we derive the celebrated Kakutani Fixed Point Theorem, which is frequently used in equilibrium analysis. Nadler's Contraction Correspondence Theorem, although it is of a different flavor, also gets some attention in this section, because this result ties in closely with our earlier work on contractions. Some of these fixed point theorems are put to good use in the final section of the chapter, where we elaborate on the notion of Nash equilibrium.[1]

1 Correspondences

By a **correspondence** Γ from a nonempty set X into another nonempty set Y, we mean a map from X into $2^Y \backslash \{\emptyset\}$. Thus, for each $x \in X$, $\Gamma(x)$ is a nonempty subset of Y.[2] We write $\Gamma : X \rightrightarrows Y$ to denote that Γ is a correspondence from X into Y. Here X is called the **domain** of Γ and Y the **codomain** of Γ. For any $S \subseteq X$, we let

$$\Gamma(S) := \bigcup \{\Gamma(x) : x \in S\}.$$

(*Note.* $\Gamma(\emptyset) = \emptyset$.) The set $\Gamma(X)$ is called the **range** of Γ. If $\Gamma(X) = Y$, we say that Γ is a **surjective correspondence**, and if $\Gamma(X) \subseteq X$, we refer to Γ as a **self-correspondence on** X.

Of course, every function $f \in Y^X$ can be viewed as a particular correspondence from X into Y. Indeed, there is no difference between f and the correspondence $\Gamma : X \rightrightarrows Y$, defined by $\Gamma(x) := \{f(x)\}$.[3] Conversely, if Γ is **single-valued**, that is, $|\Gamma(x)| = 1$ for all $x \in X$, then it can be thought of as a function mapping X into Y. We will thus identify the terms "single-valued correspondence" and "function" in the following discussion.

[1] As for general references, I should note that Berge (1963) and Border (1989) provide more comprehensive treatments of continuous correspondences than is given here, Berge in topological spaces, Border in Euclidean spaces. (But be warned that Berge includes the property of compact-valuedness in the definition of upper hemicontinuity, so, for instance, an upper hemicontinuous correspondence always satisfies the closed graph property in his definition.) These two books also provide quite comprehensive analyses of topics like parametric continuity (in optimization problems), continuous selections, and fixed point theory for correspondences. These issues are also studied in Sundaram (1998) at an introductory level, and in Klein and Thompson (1984) and Aliprantis and Border (1999) at a more advanced level. I will mention more specialized references as we go along.

[2] Alternatively, one may view a correspondence simply as a (binary) relation from X to Y, thereby identifying a correspondence with its graph. A moment's reflection shows that the situation is analogous to the way one may think of a function as a particular relation (Section A.1.5).

By the way, the math literature is not unified in the way it refers to correspondences. Some mathematicians call them *multifunctions*, some *many-valued maps*, and still others refer to them as *set-valued maps*. In the economics literature, however, the term correspondence seems widely agreed upon.

[3] It is not true that $f(x) = \Gamma(x)$ here, so from a purely formal point of view we cannot quite say that "f is Γ." But this is splitting hairs, really.

Before we move on to the formal investigation of continuous correspondences, we present some basic examples from economics, which may serve as a motivation for much of what follows. Some of these examples will be further developed in due course.

EXAMPLE 1

[1] For any $n \in \mathbb{N}$, $p \in \mathbb{R}^n_{++}$ and $\iota > 0$, define

$$B(p,\iota) := \{x \in \mathbb{R}^n_+ : \sum_{i=1}^n p_i x_i \leq \iota\},$$

which is called, as you surely recall, the **budget set** of a consumer with income ι at prices p (Example D.4.[2]). If we treated p and ι as variables, then it would be necessary to view B as a correspondence. We have $B : \mathbb{R}^{n+1}_{++} \rightrightarrows \mathbb{R}^n_+$.

[2] Let X be a nonempty set and $\succsim$ a preference relation on X. Recall that the *upper contour* set of an alternative $x \in X$ with respect to $\succsim$ is defined as

$$U_{\succsim}(x) := \{y \in X : y \succsim x\}.$$

Since $\succsim$ is reflexive, $x \in U_{\succsim}(x)$, so $U_{\succsim}(x) \neq \emptyset$ for any $x \in X$. Thus we may view $U_{\succsim}$ as a well-defined self-correspondence on X. This correspondence contains all the information that $\succsim$ has: $y \succsim x$ iff $y \in U_{\succsim}(x)$ for any $x, y \in X$. In practice, it is simply a matter of convenience to work with $\succsim$ or with $U_{\succsim}$.

[3] Let X be a metric space and $\succsim$ an upper semicontinuous preference relation on X. Let $\mathbf{c}(X)$ denote the set of all nonempty compact subsets of X, and define $C_{\succsim} : \mathbf{c}(X) \rightrightarrows X$ by

$$C_{\succsim}(S) := \{x \in S : y \succ x \text{ for no } y \in S\}.$$

By what is proved in Example C.7, $C_{\succsim}$ is a (compact-valued) correspondence. It is called the **choice correspondence induced by** $\succsim$.

[4] Let T be any nonempty set, $\emptyset \neq S \subseteq T$ and $\varphi \in \mathbb{R}^T$. The canonical optimization problem is to find the maximum value that φ attains on S. (We say that φ is the **objective function** of the problem, and S is its **constraint set**.) Put more precisely, to "solve" this problem means

to identify all $y \in S$ such that $\varphi(y) \geq \varphi(x)$ for all $x \in S$, that is, to compute

$$\arg\max\{\varphi(x) : x \in S\} := \{y \in S : \varphi(y) \geq \varphi(x) \text{ for all } x \in S\}.$$

This set is referred to as the **solution set** of the problem.

Now take any nonempty set Θ, call it the **parameter space**, and suppose that the constraint set of our canonical optimization problem depends on the parameter $\theta \in \Theta$. The constraint of the problem would then be modeled by means of a correspondence of the form $\Gamma : \Theta \rightrightarrows T$, which is called the **constraint correspondence** of the problem. This leads to the following formulation, which indeed captures many optimization problems that arise in economics: *For each $\theta \in \Theta$, maximize $\varphi(x)$ subject to $x \in \Gamma(\theta)$.* Clearly, the solution set $\arg\max\{\varphi(x) : x \in \Gamma(\theta)\}$ of the problem depends in this case on the value of θ.

Assume next that T is a metric space, and $\varphi \in \mathbf{C}(T)$. If $\Gamma : \Theta \rightrightarrows T$ is compact-valued (that is, $\Gamma(\theta)$ is a compact subset of T for each $\theta \in \Theta$), we can then think of the solution to our problem as a correspondence from Θ into T. For, in this case, Weierstrass' Theorem makes sure that $\sigma : \Theta \rightrightarrows X$ is well-defined by

$$\sigma(\theta) := \arg\max\{\varphi(x) : x \in \Gamma(\theta)\}. \tag{1}$$

Naturally enough, σ is called the **solution correspondence** of the problem.

Understanding how the continuity (and other) properties of σ depend on those of φ and Γ is one of the main issues studied in optimization theory. We will turn to this matter in Section 3. □

EXERCISE 1^H Define $\Gamma : \mathbb{R}_+ \rightrightarrows \mathbb{R}_+$ by $\Gamma(\theta) := [0, \theta]$, and consider the function $f \in \mathbf{C}(\mathbb{R})$ defined by $f(t) := \sin t$. Now define σ as in (1), and give an explicit formula for it. What is $\sigma(\mathbb{R}_+)$?

EXERCISE 2^H Let (X, d) be a discrete metric space, and let $\emptyset \neq S \subseteq X$. Define the self-correspondence Γ on X by $\Gamma(x) := \{y \in S : d(x, y) = d(x, S)\}$. Give an explicit formula for Γ. What is $\Gamma(X)$?

EXERCISE 3 Let Γ be a self-correspondence on a nonempty set X. Show that $\{\Gamma(x) : x \in X\}$ is a partition of X iff there exists an equivalence relation $\sim$ on X such that $\Gamma(x) = [x]_\sim$ for all $x \in X$ (Section A.1.3).

Exercise 4[H] Let X and Y be two nonempty sets and $\Gamma : X \rightrightarrows Y$ a correspondence. We say that Γ is **injective** if $\Gamma(x) \cap \Gamma(x') = \emptyset$ for any distinct $x, x' \in X$, and that it is **bijective** if it is both injective and surjective. Prove that Γ is bijective iff $\Gamma = f^{-1}$ for some $f \in X^Y$.

Exercise 5 Let Γ be a self-correspondence on a nonempty set X. An element x of X is called a **fixed point** of Γ if $x \in \Gamma(x)$, and a nonempty subset S of X is called a **fixed set** of Γ if $S = \Gamma(S)$.

(a) Give an example of a self-correspondence that does not have a fixed point but that does have a fixed set.

(b) Prove that if a self-correspondence has a fixed point, then it has a fixed set.

2 Continuity of Correspondences

Since correspondences are generalizations of functions, it seems reasonable to ask if we can extend the notion of continuity, which we have originally defined for functions, to the realm of correspondences. Of course, we should be consistent with our original definition in the sense that the notion of continuity that we might define for a correspondence should reduce to ordinary continuity when that correspondence is single-valued. There are at least two reasonable continuity notions for correspondences that satisfy this requirement. We will take each of these in turn.

2.1 Upper Hemicontinuity

Definition

For any two metric spaces X and Y, a correspondence $\Gamma : X \rightrightarrows Y$ is said to be **upper hemicontinuous at** $x \in X$ if, for every open subset O of Y with $\Gamma(x) \subseteq O$, there exists a $\delta > 0$ such that

$$\Gamma(N_{\delta,X}(x)) \subseteq O.$$

Γ is called **upper hemicontinuous on** $S \subseteq X$ if it is upper hemicontinuous at each $x \in S$, and **upper hemicontinuous** if it is upper hemicontinuous on the entire X.

Clearly, this definition mimics that of continuity of a function, and it reduces to the definition of the latter when Γ is single-valued. That is to say, every upper hemicontinuous single-valued correspondence "is" a continuous function, and conversely.

Intuitively speaking, upper hemicontinuity at x says that a small perturbation of x does not cause the image set $\Gamma(x)$ to "suddenly" get large. What we mean by this is illustrated in Figure 1, which depicts three correspondences mapping $\mathbb{R}_+$ into $\mathbb{R}_+$. You should not pass this point in the lecture before becoming absolutely certain that Γ_1 and Γ_3 are upper

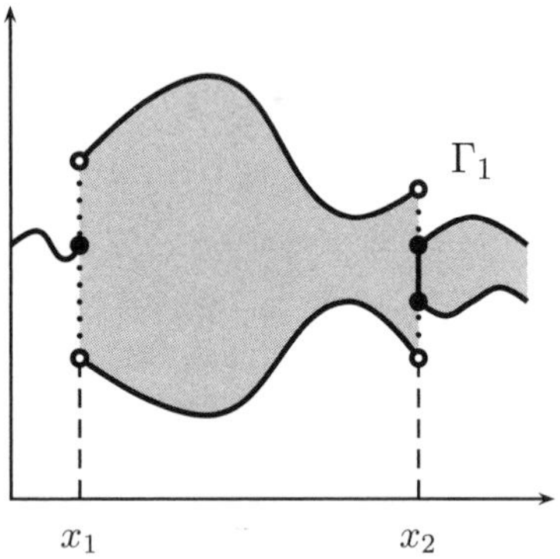

Not upper hemicontinuous at x_1
Not upper hemicontinuous at x_2
Lower hemicontinuous

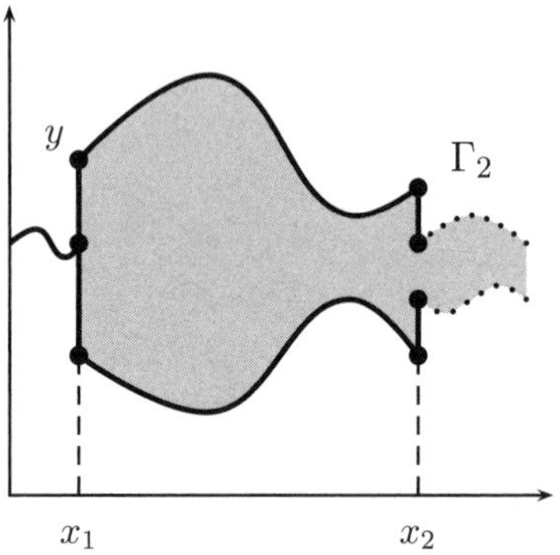

Not lower hemicontinuous at x_1
Not lower hemicontinuous at x_2
Upper hemicontinuous

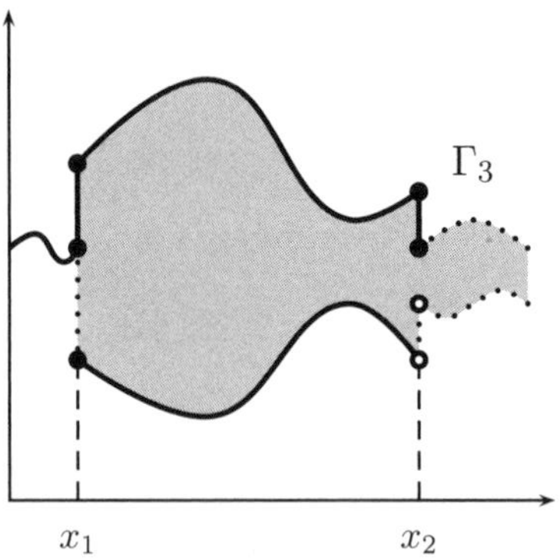

Not upper hemicontinuous at x_1 and x_2
Not lower hemicontinuous at x_1 and x_2

FIGURE 1

hemicontinuous everywhere but x_1 and x_2, respectively, while Γ_2 is upper hemicontinuous.

EXERCISE 6 Let X and Y be two metric spaces and $\Gamma : X \rightrightarrows Y$ a correspondence. Define

$$\Gamma^{-1}(O) := \{x \in X : \Gamma(x) \subseteq O\} \qquad \text{for all } O \subseteq Y.$$

($\Gamma^{-1}(O)$ is called the **upper inverse image** of O under Γ.) Show that Γ is upper hemicontinuous iff $\Gamma^{-1}(O)$ is open in X for every open subset O of Y.

While this exercise highlights the analogy between the notions of continuity and upper hemicontinuity (recall Proposition D.1), there are some important differences regarding the implications of these properties. Most notably, although a continuous function maps compact sets to compact sets (Proposition D.3), this is not the case with upper hemicontinuous correspondences. For instance, while $\Gamma : [0,1] \rightrightarrows \mathbb{R}_+$ defined by $\Gamma(x) := \mathbb{R}_+$ is obviously upper hemicontinuous, it doesn't map even a singleton set to a compact set. (Suppose you wanted to imitate the proof of Proposition D.3 to show that an upper hemicontinuous image of a compact set is compact. Where would the argument fail?) However, in most applications, the correspondences that one deals with have some additional structure that might circumvent this problem. In particular, if every value of the correspondence was a compact set, then we would be okay here.

DEFINITION

For any two metric spaces X and Y, a correspondence $\Gamma : X \rightrightarrows Y$ is said to be **compact-valued** if $\Gamma(x)$ is a compact subset of Y for each $x \in X$. Similarly, Γ is said to be **closed-valued** if the image of every x under Γ is a closed subset of Y. Finally, if Y is a subset of a Euclidean space and $\Gamma(x)$ is convex for each $x \in X$, then we say that Γ is **convex-valued**.

Under the hypothesis of compact-valuedness, we can prove a result for upper hemicontinuous correspondences that parallels Proposition D.3.

Proposition 1

Let X and Y be two metric spaces. If $\Gamma : X \rightrightarrows Y$ is a compact-valued and upper hemicontinuous correspondence, then $\Gamma(S)$ is compact in Y for any compact subset S of X.

Proof

Take any compact-valued and upper hemicontinuous correspondence $\Gamma : X \rightrightarrows Y$. Let $\mathcal{O}$ be an open cover of $\Gamma(S)$, where $S \subseteq X$ is compact. We wish to find a finite subset of $\mathcal{O}$ that would also cover $\Gamma(S)$. Note that for each $x \in S$, $\mathcal{O}$ is also an open cover of $\Gamma(x)$, so, since $\Gamma(x)$ is compact, there exist finitely many open sets $O_1(x), \ldots, O_{m_x}(x)$ in $\mathcal{O}$ such that $\Gamma(x) \subseteq \bigcup^{m_x} O_i(x) =: O(x)$. By Exercise 6, $\Gamma^{-1}(O(x))$ is open in X for each $x \in X$. Moreover, $\Gamma(S) \subseteq \bigcup\{O(x) : x \in S\}$, so that $S \subseteq \bigcup\{\Gamma^{-1}(O(x)) : x \in S\}$, that is, $\{\Gamma^{-1}(O(x)) : x \in S\}$ is an open cover of S. By compactness of S, therefore, there exist finitely many points $x^1, \ldots, x^m \in S$ such that $\{\Gamma^{-1}(O(x^i)) : i = 1, \ldots, m\}$ covers S. But then $\{O(x^1), \ldots, O(x^m)\}$ must cover $\Gamma(S)$. (Why?) Therefore, $\{O_j(x^i) : j = 1, \ldots, m_{x^i}, i = 1, \ldots, m\}$ is a finite subset of $\mathcal{O}$ that covers $\Gamma(S)$. ■

Recalling how useful the sequential characterization of continuity of a function is, we now ask if it is possible to give such a characterization for upper hemicontinuity of a correspondence. The answer is yes, at least in the case of compact-valued correspondences. (Try the following characterization in Figure 1.)

Proposition 2

Let X and Y be two metric spaces and $\Gamma : X \rightrightarrows Y$ a correspondence. Γ is upper hemicontinuous at $x \in X$ if, for any $(x^m) \in X^\infty$ and $(y^m) \in Y^\infty$ with $x^m \to x$ and $y^m \in \Gamma(x^m)$ for each m, there exists a subsequence of (y^m) that converges to a point in $\Gamma(x)$. If Γ is compact-valued, then the converse is also true.

Proof

If Γ is not upper hemicontinuous at x, then there exists an open subset O of Y with $\Gamma(x) \subseteq O$ and $\Gamma(N_{\frac{1}{m},X}(x)) \backslash O \neq \emptyset$ for any $m \in \mathbb{N}$. But then there exist $(x^m) \in X^\infty$ and $(y^m) \in Y^\infty$ such that $x^m \to x$ and $y^m \in \Gamma(x^m) \backslash O$

for each m. (Formally, we invoke the Axiom of Choice here.) Since $\Gamma(x) \subseteq O$ and each y^m belongs to the closed set $Y \backslash O$, no subsequence of (y^m) can converge to a point in $\Gamma(x)$. (Recall Proposition C.1.)

Conversely, assume that Γ is a compact-valued correspondence that is upper hemicontinuous at $x \in X$. If (x^m) is a sequence in X with $x^m \to x$, then $S := \{x, x^1, x^2, \ldots\}$ is sequentially compact in X. (Why?) By Theorem C.2 and Proposition 1, then, $\Gamma(S)$ is sequentially compact in Y. So, if $y^m \in \Gamma(x^m)$ for each m, then (y^m), being a sequence in $\Gamma(S)$, must possess a subsequence that converges to some $y \in \Gamma(S)$. With an innocent abuse of notation, we denote this subsequence also by (y^m), and write $y^m \to y$.

We wish to show that $y \in \Gamma(x)$. Let $y \notin \Gamma(x)$, and observe that this implies $\varepsilon := d_Y(y, \Gamma(x)) > 0$ because $\Gamma(x)$ is a compact (hence closed) set (Exercise D.2). Now define

$$T := \left\{z \in Y : d_Y(z, \Gamma(x)) \leq \tfrac{\varepsilon}{2}\right\}.$$

Since $d_Y(\cdot, \Gamma(x))$ is a continuous map (Example D.1.[3]), T is a closed set. Moreover, by definition, we have $\Gamma(x) \subseteq int_Y(T)$ and $y \notin T$ (Figure 2). But, since Γ is upper hemicontinuous at x, there exists a $\delta > 0$ such that

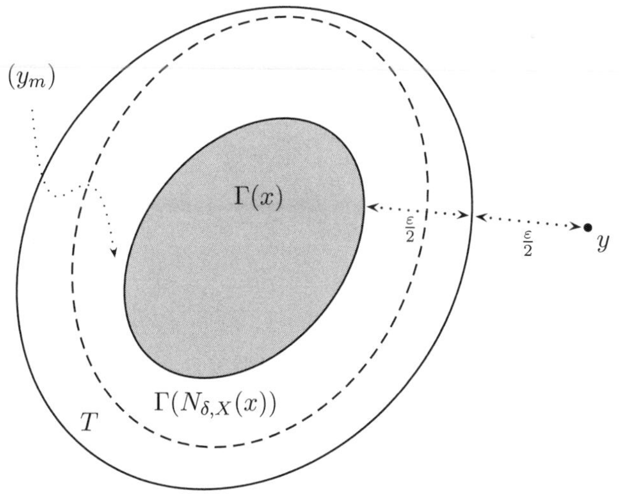

Figure 2

$\Gamma(N_{\delta,X}(x)) \subseteq int_Y(T)$. Thus there is an $M \in \mathbb{R}$ such that $y^m \in \Gamma(x^m) \subseteq int_Y(T)$ for all $m \geq M$. But then, since T is a closed set and $y^m \to y$, we must have $y \in T$, a contradiction. ■

Here is a nice application of this result.

EXAMPLE 2
(*The Budget Correspondence*) Fix any $n \in \mathbb{N}$, and define $B : \mathbb{R}^{n+1}_{++} \rightrightarrows \mathbb{R}^n_+$ by

$$B(p, \iota) := \{x \in \mathbb{R}^n_+ : px \leq \iota\}, \tag{2}$$

where px stands for the inner product of the n-vectors p and x, that is, $px := \sum^n p_i x_i$. We claim that B, which is aptly called the **budget correspondence**, is upper hemicontinuous. The proof is by means of Proposition 2. Take an arbitrary $(p, \iota) \in \mathbb{R}^{n+1}_{++}$, and sequences (p^m, ι_m) and (x^m) (in $\mathbb{R}^{n+1}_{++}$ and $\mathbb{R}^n_+$, respectively) such that $\lim(p^m, \iota_m) = (p, \iota)$ and $x^m \in B(p^m, \iota_m)$ for each m. Since $(p_i^m) \in \mathbb{R}^\infty_{++}$ converges to a strictly positive real number, we have $p_i^* := \inf\{p_i^m : m \in \mathbb{N}\} > 0$ for each $i = 1, \ldots, n$. Similarly, $\iota^* := \sup\{\iota_m : m \in \mathbb{N}\} < \infty$. But it is plain that $x^m \in B(p^*, \iota^*)$ for each m, while $B(p^*, \iota^*)$ is obviously a closed and bounded subset of $\mathbb{R}^n_+$. By the Heine-Borel Theorem and Theorem C.2, therefore, there exists a subsequence (x^{m_k}) that converges to some $x \in \mathbb{R}^n_+$. But then, by a straightforward continuity argument, $px = \lim p^{m_k} x^{m_k} \leq \lim \iota_{m_k} = \iota$, that is, $x \in B(p, \iota)$. Since (p, ι) was arbitrarily chosen in $\mathbb{R}^{n+1}_{++}$, we may invoke Proposition 2 to conclude that B is upper hemicontinuous. □

EXERCISE 7 Consider the self-correspondence Γ on $[0, 1]$ defined as: $\Gamma(0) := (0, 1]$ and $\Gamma(t) := (0, t)$ for all $0 < t \leq 1$. Is Γ upper hemicontinuous? Does Γsatisfy the sequential property considered in Proposition 2?

EXERCISE 8[H] Give an example of a compact-valued and upper hemicontinuous correspondence $\Gamma : [0, 1] \rightrightarrows \mathbb{R}$ such that $\Upsilon : [0, 1] \rightrightarrows \mathbb{R}$, defined by $\Upsilon(t) := bd_{\mathbb{R}}(\Gamma(t))$, is not upper hemicontinuous.

EXERCISE 9 For any given $n \in \mathbb{N}$, define $\Gamma : \mathbb{R}^n \rightrightarrows S^{n-1}$ by

$$\Gamma(x) := \{y \in S^{n-1} : d(x, y) = d(x, S^{n-1})\}.$$

(Recall that S^{n-1} is the $n-1$ dimensional unit sphere (Section D.8.2).) Is Γ well-defined? Is it compact-valued? Is it upper hemicontinuous?

EXERCISE 10^H Define $B : \mathbb{R}^{n+1}_{+} \rightrightarrows \mathbb{R}^n_{+}$ by (2). Is B upper hemicontinuous?

EXERCISE 11^H Give an alternative proof for Proposition 1 using Theorem C.2 and Proposition 2.

EXERCISE 12^H Let X and Y be two metric spaces and $f \in Y^X$. We say that f is a **closed map** if $f(S)$ is closed in Y for every closed subset S of X.

(a) Show that f is a closed surjection iff f^{-1} is an upper hemicontinuous correspondence.

(b) If f is a continuous surjection, does f^{-1} need to be upper hemicontinuous?

EXERCISE 13 Let X and Y be two metric spaces and $\Gamma : X \rightrightarrows Y$ a correspondence. Define $\overline{\Gamma} : X \rightrightarrows Y$ by $\overline{\Gamma}(x) := cl_Y(\Gamma(x))$. Show that if Γ is upper hemicontinuous, then so is $\overline{\Gamma}$. Is the converse necessarily true?

EXERCISE 14 Let X and Y be two metric spaces and Γ_1 and Γ_2 correspondences from X into Y.

(a) Define $\Phi : X \rightrightarrows Y$ by $\Phi(x) := \Gamma_1(x) \cup \Gamma_2(x)$. Show that if Γ_1 and Γ_2 are upper hemicontinuous, then so is Φ.

(b) Assume that $\Gamma_1(x) \cap \Gamma_2(x) \neq \emptyset$ for any $x \in X$, and define $\Psi : X \rightrightarrows Y$ by $\Psi(x) := \Gamma_1(x) \cap \Gamma_2(x)$. Show that if Γ_1 and Γ_2 are compact-valued and upper hemicontinuous, then so is Ψ.

EXERCISE 15 Let Γ_1 and Γ_2 be two correspondences that map a metric space X into $\mathbb{R}^n$. Define $\Phi : X \rightrightarrows \mathbb{R}^{2n}$ and $\Psi : X \rightrightarrows \mathbb{R}^n$ by

$$\Phi(x) := \Gamma_1(x) \times \Gamma_2(x) \quad \text{and}$$

$$\Psi(x) := \{y_1 + y_2 : (y_1, y_2) \in \Gamma_1(x) \times \Gamma_2(x)\},$$

respectively. Show that if Γ_1 and Γ_2 are compact-valued and upper hemicontinuous, then so are Φ and Ψ.

EXERCISE 16 A metric space is said to have the **fixed set property** if for every upper hemicontinuous self-correspondence Γ on X, there is a nonempty closed subset S of X such that $\Gamma(S) = S$.

(a) Is the fixed set property a topological property?

(b) Is it true that the fixed set property of a metric space is inherited by all of its retracts?

EXERCISE 17 Let T be a metric space and $\mathcal{F}$ a nonempty subset of $\mathbf{C}(T)$. Define $\Gamma : T \rightrightarrows \mathbb{R}$ by $\Gamma(t) := \cup\{f(t) : f \in \mathcal{F}\}$. Show that Γ is compact-valued and upper hemicontinuous if (*a*) $\mathcal{F}$ is finite, or more generally, (*b*) $\mathcal{F}$ is compact in $\mathbf{C}(T)$.

2.2 The Closed Graph Property

By analogy with the graph of a function, the **graph** of a correspondence $\Gamma : X \rightrightarrows Y$, denoted by $Gr(\Gamma)$, is defined as

$$Gr(\Gamma) := \{(x, y) \in X \times Y : y \in \Gamma(x)\}.$$

In turn, we say that Γ has a *closed graph* if $Gr(\Gamma)$ is closed in the product metric space $X \times Y$. The following definition expresses this in slightly different words.

> DEFINITION
> Let X and Y be two metric spaces. A correspondence $\Gamma : X \rightrightarrows Y$ is said to be **closed at** $x \in X$ if, for any convergent sequences $(x^m) \in X^\infty$ and $(y^m) \in Y^\infty$ with $x^m \to x$ and $y^m \to y$, we have $y \in \Gamma(x)$ whenever $y^m \in \Gamma(x^m)$ for each $m = 1, 2, \ldots$. Γ is said to have a **closed graph** (or to satisfy **the closed graph property**) if it is closed at every $x \in X$.

EXERCISE 18 Let X and Y be two metric spaces. Prove that a correspondence $\Gamma : X \rightrightarrows Y$ has a closed graph iff $Gr(\Gamma)$ is closed in $X \times Y$.

Although it is somewhat standard, the choice of terminology here is rather unfortunate. In particular, if $\Gamma(x)$ is a closed set, this does not mean that Γ is closed at x. Indeed, even if $\Gamma : X \rightrightarrows Y$ is closed-valued, Γ need not have a closed graph. After all, any self-map f on $\mathbb{R}$ is closed-valued, but of course, a discontinuous self-map on $\mathbb{R}$ does not have a closed graph. But the converse is true, that is, if Γ is closed at x, then $\Gamma(x)$ must be a closed subset of Y. (Why?) In short, the closed graph property is (much!) more demanding than being closed-valued.

It is worth noting that having a closed graph cannot really be considered a continuity property. Indeed, this property does not reduce to our ordinary notion of continuity in the case of single-valued correspondences. For instance, the graph of the function $f : \mathbb{R}_+ \to \mathbb{R}_+$, where $f(0) := 0$ and $f(t) := \frac{1}{t}$ for all $t > 0$, is closed in $\mathbb{R}^2$, but this function exhibits a serious discontinuity at the origin. Thus, for single-valued correspondences, having a closed graph is in general a *weaker* property than continuity—the latter implies the former, but not conversely. (As we shall see shortly, however, the converse would hold if the codomain was compact.)

Nevertheless, the closed graph property is still an interesting property to impose on a correspondence. After all, closedness at x simply says that "if some points in the images of points nearby x concentrate around a particular point in the codomain, that point must be contained in the image of x." This statement, at least intuitively, brings to mind the notion of continuity. What is more, there is in fact a tight connection between upper hemicontinuity and the closed graph property.

PROPOSITION 3

Let X and Y be two metric spaces and $\Gamma : X \rightrightarrows Y$ a correspondence.

(a) *If Γ has a closed graph, then it need not be upper hemicontinuous. But if Γ has a closed graph and Y is compact, then it is upper hemicontinuous.*

(b) *If Γ is upper hemicontinuous, then it need not have a closed graph. But if Γ is upper hemicontinuous and closed-valued, then it has a closed graph.*

PROOF

(a) It is observed above that the closed graph property does not imply upper hemicontinuity even for single-valued correspondences. To see the second claim, assume that Γ is closed at some $x \in X$ and Y is compact. Take any $(x^m) \in X^\infty$ and $(y^m) \in Y^\infty$ with $x^m \to x$ and

$y^m \in \Gamma(x^m)$ for each m. By Theorem C.2, there exists a strictly increasing sequence $(m_k) \in \mathbb{N}^\infty$ and a point $y \in Y$ such that $y^{m_k} \to y$ (as $k \to \infty$). Thus, in view of Proposition 2, it is enough to show that $y \in \Gamma(x)$. But since $x^{m_k} \to x$ and $y^{m_k} \in \Gamma(x^{m_k})$ for each k, this is an obvious consequence of the closedness of Γ at x.

(b) The self-correspondence Γ on $\mathbb{R}_+$, defined by $\Gamma(t) := (0, 1)$, is upper hemicontinuous, but it is closed nowhere. The proof of the second claim, on the other hand, is identical to the argument given in the third paragraph of the proof of Proposition 2. (Use, again, the idea depicted in Figure 2.) ■

The closed graph property is often easier to verify than upper hemicontinuity. For this reason, Proposition 3.(a) is used in practice quite frequently. In fact, even when we don't have the compactness of its codomain, we may be able to make use of the closed graph property to verify that a given correspondence is upper hemicontinuous. A case in point is illustrated in the following exercise.

EXERCISE 19 Let X and Y be two metric spaces and Γ_1 and Γ_2 two correspondences from X into Y, with $\Gamma_1(x) \cap \Gamma_2(x) \neq \emptyset$ for any $x \in X$. Define $\Psi : X \rightrightarrows Y$ by $\Psi(x) := \Gamma_1(x) \cap \Gamma_2(x)$. Prove: If Γ_1 is compact-valued and upper hemicontinuous at $x \in X$, and if Γ_2 is closed at x, then Ψ is upper hemicontinuous at x.

EXERCISE 20 Let X be a compact metric space and Γ a self-correspondence on X. Prove: If Γ has a closed graph, then $\Gamma(S)$ is a closed set whenever S is closed. Give an example to show that the converse claim is false.

EXERCISE 21^H Let X be any metric space, Y a compact metric space, and $f \in Y^X$ a continuous function. For any fixed $\varepsilon > 0$, define $\Gamma : X \rightrightarrows Y$ by

$$\Gamma(x) := \{y \in Y : d_Y(y, f(x)) \leq \varepsilon\}.$$

Show that Γ is upper hemicontinuous.[4]

[4] Here is an example (which was suggested to me by Kim Border) that shows that compactness of Y is essential for this result. Pick any $(q, x) \in \mathbb{Q}\times\mathbb{R}\backslash\mathbb{Q}$ with $q > x$, and let

EXERCISE 22[H] (*A Fixed Set Theorem*) Let X be a compact metric space and Γ an upper hemicontinuous self-correspondence on X. Prove: There exists a nonempty compact subset S of X with $S = \Gamma(S)$.[5]

2.3 Lower Hemicontinuity

We now turn to the second continuity concept that we will consider for correspondences. While upper hemicontinuity of a correspondence $\Gamma : X \rightrightarrows Y$ guarantees that the image set $\Gamma(x)$ of a point $x \in X$ does not "explode" consequent on a small perturbation of x, in some sense it allows for it to "implode." For instance, Γ_2 in Figure 1 is upper hemicontinuous at x_1, even though there is an intuitive sense in which Γ_2 is discontinuous at this point, since the "value" of Γ_2 changes dramatically when we perturb x_1 marginally. Let us look at this situation a bit more closely. The feeling that Γ_2 behaves in some sense discontinuously at x_1 stems from the fact that the image sets of *some* of the points that are very close to x_1 seem "far away" from *some* of the points in the image of x_1. To be more precise, let us fix a small $\varepsilon > 0$. A reasonable notion of continuity would demand that if $x' \in X$ is close enough to x, then its image $\Gamma_2(x')$ should not be "far away" from any point in $\Gamma(x)$, say y, in the sense that it should at least intersect $N_{\varepsilon,\mathbb{R}}(y)$. It is this property that Γ_2 lacks. No matter how close is x' to x_1, we have $\Gamma_2(x') \cap N_{\varepsilon,\mathbb{R}}(y) = \emptyset$ so long as $x' < x_1$ (for ε small enough). The next continuity concept that we introduce is a property that rules out precisely this sort of a thing.

$\varepsilon := q - x$. The self-correspondence Γ on $\mathbb{R}\backslash\mathbb{Q}$, defined by $\Gamma(t) := \{t \in \mathbb{R}\backslash\mathbb{Q} : |t - x| \leq \varepsilon\}$, is not upper hemicontinuous at x. To see this, note first that $\Gamma(x) := [2x - q, q) \cap \mathbb{R}\backslash\mathbb{Q}$. (Why?) Now let $O := (2x - q - 1, q) \cap \mathbb{R}\backslash\mathbb{Q}$, and observe that O is an open subset of $\mathbb{R}\backslash\mathbb{Q}$ with $\Gamma(x) \subseteq O$. But it is clear that $\Gamma(N_\delta(x))\backslash O \neq \emptyset$ for any $\delta > 0$.

5 *Digression*. There is more to the story. For any metric space X, a nonempty subset S of X is said to be an almost-fixed set of $\Gamma : X \rightrightarrows X$, if $\Gamma(S) \subseteq S \subseteq cl_X(\Gamma(S))$. The following identifies exactly when Γ has a compact almost-fixed set.

Theorem. Γ has a compact almost-fixed set iff X is compact.

To prove the "if" part, we let $\mathcal{A} := \{A \in 2^X\backslash\{\emptyset\} : A$ is closed and $A \supseteq cl_X(\Gamma(A))\}$, and apply Zorn's Lemma to the poset $(\mathcal{A}, \supseteq)$. (The "only if" part is a bit trickier.) I have actually written on this topic elsewhere; see Ok (2004) for a detailed treatment of fixed set theory.

DEFINITION

For any two metric spaces X and Y, a correspondence $\Gamma : X \rightrightarrows Y$ is said to be **lower hemicontinuous at** $x \in X$ if, for every open set O in Y with $\Gamma(x) \cap O \neq \emptyset$, there exists a $\delta > 0$ such that

$$\Gamma(x') \cap O \neq \emptyset \qquad \text{for all } x' \in N_{\delta,X}(x).$$

Γ is called **lower hemicontinuous on** $S \subseteq X$ if it is lower hemicontinuous at each $x \in S$, and **lower hemicontinuous** if it is lower hemicontinuous on the entire X.

Here is an alternative way of saying this (compare with Exercise 6).

EXERCISE 23 Let X and Y be two metric spaces and $\Gamma : X \rightrightarrows Y$ a correspondence. Define

$$\Gamma_{-1}(O) := \{x \in X : \Gamma(x) \cap O \neq \emptyset\} \qquad \text{for all } O \subseteq Y.$$

($\Gamma_{-1}(O)$ is called the **lower inverse image** of O under Γ.) Show that Γ is lower hemicontinuous iff $\Gamma_{-1}(O)$ is open in X for every open subset O of Y.

This observation provides us with a different perspective about the nature of the upper and lower hemicontinuity properties. Recall that a function f that maps a metric space X into another metric space Y is continuous iff $f^{-1}(O)$ is open in X for every open subset O of Y (Proposition D.1). Suppose we wished to extend the notion of continuity to the case of a correspondence $\Gamma : X \rightrightarrows Y$ by using this way of looking at things. Then the issue is to decide how to define the "inverse image" of a set under Γ. Of course, we should make sure that this definition reduces to the usual one when Γ is single-valued. Since $f^{-1}(O) = \{x \in X : \{f(x)\} \subseteq O\}$ and $f^{-1}(O) = \{x \in X : \{f(x)\} \cap O \neq \emptyset\}$, there are at least two ways of doing this. The first one leads us to the notion of the *upper inverse image* of O under Γ, and hence to *upper* hemicontinuity (Exercise 6); the second way leads to the notion of the *lower inverse image* of O under Γ, and hence to *lower* hemicontinuity (Exercise 23).

Among other things, this discussion shows that lower hemicontinuity is a genuine continuity condition in the sense that it reduces to our ordinary notion of continuity in the case of single-valued correspondences. It is also logically independent of upper hemicontinuity (see Figure 1), and also of the closed graph property. But of course, some nice correspondences satisfy all of these properties.

EXAMPLE 3

(*The Budget Correspondence, Again*) Let us show that the budget correspondence $B : \mathbb{R}^{n+1}_{++} \rightrightarrows \mathbb{R}^n_+$ defined by (2) is lower hemicontinuous (recall Example 2). Take an arbitrary $(p, \iota) \in \mathbb{R}^{n+1}_{++}$ and any open subset O of $\mathbb{R}^n_+$ with $B(p, \iota) \cap O \neq \emptyset$. To derive a contradiction, suppose that for every $m \in \mathbb{N}$ (however large), there exists an $(n+1)$-vector (p^m, ι_m) within $\frac{1}{m}$-neighborhood of (p, ι) such that $B(p^m, \iota_m) \cap O = \emptyset$. Now pick any $x \in B(p, \iota) \cap O$. Since O is open in $\mathbb{R}^n_+$, we have $\lambda x \in B(p, \iota) \cap O$ for $\lambda \in (0, 1)$ close enough to 1. (Why?) But, since $p^m \to p$ and $\iota_m \to \iota$, a straightforward continuity argument yields $\lambda p^m x < \iota_m$ for m large enough. Then, for any such m, we have $\lambda x \in B(p^m, \iota_m)$, that is, $B(p^m, \iota_m) \cap O \neq \emptyset$, a contradiction. □

How about a sequential characterization of lower hemicontinuity? Let's try to see first what sort of a conjecture we may come up with by examining the behavior of Γ_2 in Figure 1 again. Recall that Γ_2 is not lower hemicontinuous at x_1, because the images of *some* of the points that are nearby x_1 seem "far away" from *some* of the points in the image of x_1. For instance, take a point like y in $\Gamma_2(x_1)$. While the sequence $(x_1 - \frac{1}{m})$ obviously converges to x_1, the sequence of image sets $(\Gamma_2(x_1 - \frac{1}{m}))$ does not get "close" to $\Gamma_2(x_1)$. Although this statement is ambiguous in the sense that we do not know at present how to measure the "distance" between two sets, it is intuitive that if $(\Gamma_2(x_1 - \frac{1}{m}))$ is to be viewed as getting "close" to $\Gamma_2(x_1)$, then there must be at least one sequence (y_m) with $y_m \in \Gamma_2(x_1 - \frac{1}{m})$ for each m, and $y_m \to y$. Such a sequence, however, does not exist in this example, hence the lack of lower hemicontinuity. This motivates the following characterization result (which is "cleaner" than Proposition 2, because it is free of the compact-valuedness requirement).

PROPOSITION 4

Let X and Y be two metric spaces and $\Gamma : X \rightrightarrows Y$ a correspondence. Γ is lower hemicontinuous at $x \in X$ if, and only if, for any $(x^m) \in X^\infty$ with $x^m \to x$ and any $y \in \Gamma(x)$, there exists a $(y^m) \in Y^\infty$ such that $y^m \to y$ and $y^m \in \Gamma(x^m)$ for each m.

PROOF

Suppose Γ is lower hemicontinuous at some $x \in X$, and take any $((x^m), y) \in X^\infty \times \Gamma(x)$ with $x^m \to x$. By lower hemicontinuity, for every $k \in \mathbb{N}$, there exists a $\delta(k) > 0$ such that $\Gamma(x') \cap N_{\frac{1}{k},Y}(y) \neq \emptyset$ for each $x' \in N_{\delta(k),X}(x)$. Since $x^m \to x$, there exists an $m_1 \in \mathbb{N}$ such that $d(x^m, x) < \delta(1)$ for each $m \geq m_1$, and for any $k = 2, 3, \ldots$, there exists an $m_k \in \{m_{k-1}+1, m_{k-1}+2, \ldots\}$ such that $d(x^{m_k}, x) < \delta(k)$ for each $m \geq m_k$. This gives us a subsequence (x^{m_k}) such that $\Gamma(x^m) \cap N_{\frac{1}{k},Y}(y) \neq \emptyset$ for each $k \geq 1$ and $m \geq m_k$. Now pick any $(y^m) \in Y^\infty$ such that

$$y^m \in \Gamma(x^m) \cap N_{\frac{1}{k},Y}(y) \qquad \text{for all } m \in \{m_k, \ldots, m_{k+1} - 1\},\ k = 1, 2, \ldots$$

It is readily checked that $y^m \to y$ and $y^m \in \Gamma(x^m)$ for each m.

Conversely, suppose Γ is not lower hemicontinuous at some $x \in X$. Then there exists an open subset O of Y such that $\Gamma(x) \cap O \neq \emptyset$ and, for every $m \in \mathbb{N}$, there exists an $x^m \in N_{\frac{1}{m},X}(x)$ with $\Gamma(x^m) \cap O = \emptyset$. Note that $x^m \to x$, and pick any $y \in \Gamma(x) \cap O$. By hypothesis, there must exist a sequence $(y^m) \in Y^\infty$ such that $y^m \to y$ and $y^m \in \Gamma(x^m)$ for each m. But since $y \in O$ and O is open, $y^m \in O$ for m large enough, contradicting that $\Gamma(x^m) \cap O = \emptyset$ for all m. ■

EXERCISE 24[H] Show that the word "upper" can be replaced with "lower" in Exercise 14.(a) and Exercise 15, but not in Exercise 14.(b) and Proposition 3. Can the word "upper" be replaced with "lower" in Proposition 1?

EXERCISE 25 Let X and Y be two metric spaces and $\Gamma : X \rightrightarrows Y$ a correspondence. Define $\overline{\Gamma} : X \rightrightarrows Y$ by $\overline{\Gamma}(x) := cl_Y(\Gamma(x))$. Prove or disprove: Γ is lower hemicontinuous iff $\overline{\Gamma}$ is lower hemicontinuous.

2.4 Continuous Correspondences

A correspondence is said to be *continuous* when it satisfies both of the continuity notions we considered above. The behavior of such a correspondence is nicely regular in that small perturbations in its statement do not cause the image sets of nearby points to show drastic (upward or downward) alterations.

DEFINITION
Let X and Y be two metric spaces. A correspondence $\Gamma : X \rightrightarrows Y$ is said to be **continuous at** $x \in X$ if it is both upper and lower hemicontinuous at x. It is called **continuous on** $S \subseteq X$ if it is continuous at each $x \in S$, and **continuous** if it is continuous on the entire X.

For example, combining Examples 2 and 3, we see that the budget correspondence $B : \mathbb{R}^{n+1}_{++} \rightrightarrows \mathbb{R}^n_+$ defined by (2) is a continuous and compact-valued correspondence.[6] We consider a few other examples below.

EXERCISE 26 Define the correspondence $\Gamma : [0,1] \rightrightarrows [-1,1]$ as

$$\Gamma(0) := [-1,1] \quad \text{and} \quad \Gamma(t) := \left\{\sin \tfrac{1}{t}\right\} \text{ for any } 0 < t \leq 1.$$

Is Γ continuous?

EXERCISE 27[H] Let X be a metric space and $\varphi \in \mathbf{C}(X)$. Show that the correspondence $\Gamma : X \rightrightarrows \mathbb{R}$, defined by $\Gamma(x) := [0, \varphi(x)]$, is continuous.

EXERCISE 28[H] Define the correspondence $\Gamma : [0,1] \rightrightarrows [0,1]$ as

$$\Gamma(t) := \begin{cases} [0,1] \cap \mathbb{Q}, & \text{if } t \in [0,1]\backslash\mathbb{Q} \\ [0,1]\backslash\mathbb{Q}, & \text{if } t \in [0,1] \cap \mathbb{Q} \end{cases}.$$

Show that Γ is not continuous, but it is lower hemicontinuous. Is Γ upper hemicontinuous at any rational? (How about at 0?) At any irrational? Does this correspondence have the closed graph property?

EXERCISE 29[H] Let $a > 0$, $n \in \mathbb{N}$, and $T := [0,a]^n$. For any $u \in \mathbb{R}^T$, define the correspondence $\Gamma_u : u(T) \to \mathbb{R}^n_+$ by

$$\Gamma_u(v) := \{x \in T : u(x) \geq v\}.$$

Prove that Γ_u is continuous for any strictly increasing $u \in \mathbf{C}(T)$.

EXERCISE 30 Let X, Y, and Z be metric spaces, and let $\Gamma : X \rightrightarrows Y$ and $\Upsilon : Y \rightrightarrows Z$ be any two correspondences. We define $\Upsilon \circ \Gamma : X \rightrightarrows Z$ by $(\Upsilon \circ \Gamma)(x) := \Upsilon(\Gamma(x))$. If Γ and Υ are continuous, does $\Upsilon \circ \Gamma$ have to be continuous?

[6] B would remain continuous if we allowed some of the prices of the commodities to be zero. But it would then cease being compact-valued.

2.5* The Hausdorff Metric and Continuity

An alternative way of thinking about the continuity of a correspondence stems from viewing that correspondence as a *function* that maps a set to a "point" in a power set, and then to impose the usual continuity property on this function. Of course, this approach necessitates that we have a sensible metric on the associated codomain of sets, and the formulation of this is not really a trivial matter. However, in the case of compact-valued correspondences (where the range of the correspondence is contained in the set of all *compact* subsets of its codomain), this approach becomes quite useful. We explore this issue next.

Let Y be a metric space, and let us denote by $\mathbf{c}(Y)$ the class of all nonempty compact subsets of Y.[7] For any two nonempty subsets A and B of Y, we define

$$\omega(A, B) := \sup\{d_Y(z, B) : z \in A\}$$

and

$$d_{\mathrm{H}}(A, B) := \max\{\omega(A, B), \omega(B, A)\},$$

which are well-defined as real numbers, provided that A and B are bounded. (See Figure 3.) When d_{H} is viewed as a map on $\mathbf{c}(Y) \times \mathbf{c}(Y)$, it is called the **Hausdorff metric**. (*Note.* In that case sup can be replaced with max in the definition of the map ω, thanks to Example D.1.[3] and Weierstrass' Theorem.)

You are asked below to verify that $(\mathbf{c}(Y), d_{\mathrm{H}})$ is indeed a metric space. We also provide several other exercises here to help you get better acquainted with the Hausdorff metric. (For a more detailed analysis, see the first chapter of Nadler (1978).)

EXERCISE 31 Let Y be a metric space.
(a) True or false: If Y is bounded, $(2^Y\{\emptyset\}, d_{\mathrm{H}})$ is a semimetric space.
(b) Give an example of Y such that $(2^Y\backslash\{\emptyset\}, d_{\mathrm{H}})$ is not a metric space.
(c) Show that $(\mathbf{c}(Y), d_{\mathrm{H}})$ is a metric space.

7 Why not metrize $2^Y\backslash\{\emptyset\}$ by the map $d : 2^Y\backslash\{\emptyset\} \times 2^Y\backslash\{\emptyset\} \to \mathbb{R}_+$ defined by $d(A, B) := \inf\{d(x, y) : (x, y) \in A \times B\}$? This won't do! After all, if $A \cap B \neq \emptyset$, we have $d(A, B) = 0$ even when $A \neq B$. (In fact, d is not even a semimetric on $2^Y\backslash\{\emptyset\}$; why?) The metrization problem at hand is more subtle than this.

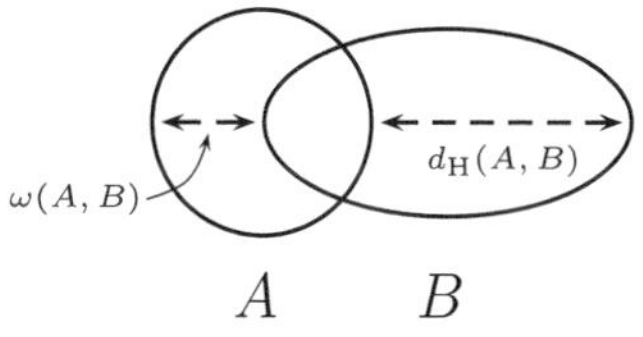

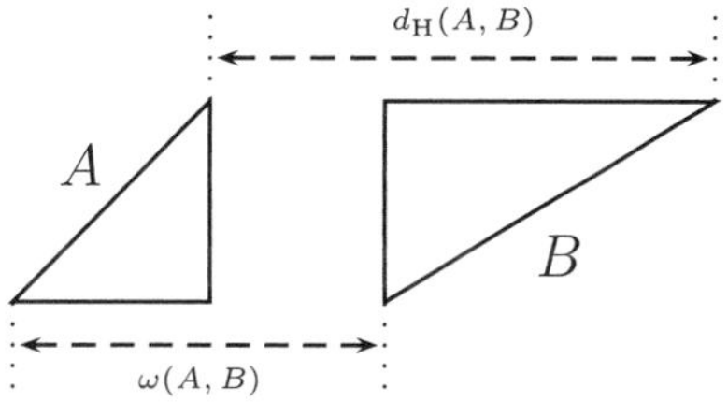

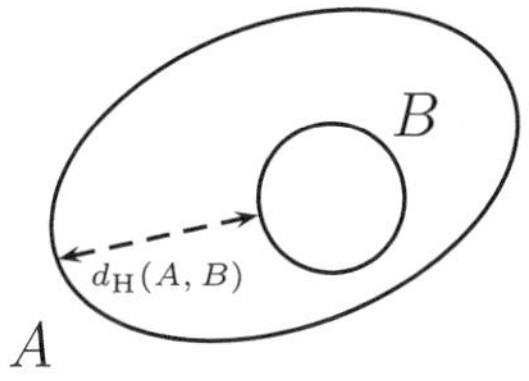

FIGURE 3

EXERCISE 32[H]

(a) Compute $d_H([0,1],[a,b])$ for any $-\infty < a < b < \infty$.

(b) If Y is a discrete space, does $(\mathbf{c}(Y), d_H)$ have to be discrete as well?

EXERCISE 33 Let Y be a metric space.

(a) Define $N_{\varepsilon,Y}(S) := \cup\{N_{\varepsilon,Y}(z) : z \in S\}$ for any $S \in \mathbf{c}(Y)$ and $\varepsilon > 0$. Show that

$$d_H(A, B) = \inf\{\varepsilon > 0 : A \subseteq N_{\varepsilon,Y}(B) \text{ and } B \subseteq N_{\varepsilon,Y}(A)\}$$

for any $A, B \in \mathbf{c}(Y)$.

(b) Take any $S \in \mathbf{c}(Y)$. Show that, for any $y \in Y$ and $(y_m) \in Y^\infty$ with $y_m \to y$, we have $S \cup \{y_m\} \to S \cup \{y\}$ in $(\mathbf{c}(Y), d_H)$.

*EXERCISE 34 Prove: If Y is a compact metric space, then so is $(\mathbf{c}(Y), d_H)$.

EXERCISE 35[H] Prove: If Y is a complete metric space, then so is $(\mathbf{c}(Y), d_H)$.

EXERCISE 36[H] Let Y be a complete metric space and $\mathcal{F}$ a nonempty class of self-maps on Y. A nonempty subset S of Y is called **self-similar** with respect to $\mathcal{F}$ if $S = \cup\{f(S) : f \in \mathcal{F}\}$. Prove: If $\mathcal{F}$ is a nonempty finite set of contractions in Y^Y, then there exists a unique compact self-similar set S with respect to $\mathcal{F}$.[8]

Now consider a compact-valued $\Gamma : X \rightrightarrows Y$ where X and Y are arbitrary metric spaces. Since $\Gamma(x) \in \mathbf{c}(Y)$ for each $x \in X$, we can actually think of Γ as a function of the form $\Gamma : X \to \mathbf{c}(Y)$. Since we now have a way of thinking about $\mathbf{c}(Y)$ as a metric space, it seems reasonable to declare Γ as "continuous" whenever, for any $x \in X$ and $\varepsilon > 0$, we can find a $\delta > 0$ such that

$$d(x, y) < \delta \quad \text{implies} \quad d_H(\Gamma(x), \Gamma(y)) < \varepsilon.$$

We refer to Γ as **Hausdorff continuous** when it satisfies this property. Given that this continuity notion is quite natural—if you like the Hausdorff metric, that is—it would be a shame if it did not link well with the continuity of correspondences as defined in the previous section. Fortunately, these two concepts turn out to be identical.

PROPOSITION 5
Let X and Y be two metric spaces and $\Gamma : X \rightrightarrows Y$ a compact-valued correspondence. Γ is Hausdorff continuous if, and only if, it is continuous.

PROOF
Fix an arbitrary $x \in X$.

CLAIM 1. Γ is lower hemicontinuous at x iff, for any $(x^m) \in X^\infty$ with $x^m \to x$,

$$\max\left\{d_Y(y, \Gamma(x^m)) : y \in \Gamma(x)\right\} \to 0. \tag{3}$$

CLAIM 2. Γ is upper hemicontinuous at x iff, for any $(x^m) \in X^\infty$ with $x^m \to x$,

$$\max\left\{d_Y(y, \Gamma(x)) : y \in \Gamma(x^m)\right\} \to 0.$$

[8] Here is a little follow-up for this problem. Let Y be a compact metric space, and let $\mathbf{C}(Y, Y)$ denote the set of all continuous self-maps on Y. We metrize this space by the metric $d_\infty : (f, g) \mapsto \sup\{d'(f(y), g(y)) : y \in Y\}$, where $d' := \frac{d_Y}{1+d_Y}$. Now prove: *If $\mathcal{F}$ is a nonempty compact subset of $\mathbf{C}(Y, Y)$, then there exists a compact self-similar set S with respect to $\mathcal{F}$.* (If you're stuck (and interested), a proof is given in Ok (2004).)

A moment's reflection shows that it is enough to prove these claims to conclude that our main assertion is correct. We will only prove Claim 1 here, the proof of Claim 2 being similar (and easier).

Let Γ be lower hemicontinuous at x, and take any $(x^m) \in X^\infty$ with $x^m \to x$. Pick any

$$y^m \in \arg\max \left\{ d_Y(y, \Gamma(x^m)) : y \in \Gamma(x) \right\}, \qquad m = 1, 2, \ldots$$

We wish to show that $s := \limsup d_Y(y^m, \Gamma(x^m)) = 0$. (Obviously, $0 \leq s \leq \infty$.) Take a subsequence of $(d_Y(y^m, \Gamma(x^m)))$ that converges to s, say $(d_Y(y^{m_k}, \Gamma(x^{m_k})))$. Since (y^{m_k}) is a sequence in the compact set $\Gamma(x)$, it must have a subsequence that converges to a point y in $\Gamma(x)$, which, without loss of generality, we may again denote by (y^{m_k}). We have

$$d_Y(y^{m_k}, \Gamma(x^{m_k})) \leq d_Y(y^{m_k}, y) + d_Y(y, \Gamma(x^{m_k})) \qquad \text{for all } k = 1, 2, \ldots \tag{4}$$

(Yes?) Given that $y \in \Gamma(x)$ and $x^{m_k} \to x$, Proposition 4 implies that there exists a sequence $(z^{m_k}) \in Y^\infty$ such that $z^{m_k} \to y$ and $z^{m_k} \in \Gamma(x^{m_k})$ for each k. But $d_Y(y, \Gamma(x^{m_k})) \leq d_Y(y, z^{m_k}) \to 0$, and combining this with (4) shows that $d_Y(y^{m_k}, \Gamma(x^{m_k})) \to 0$. Thus $s = 0$, as we sought.

To prove the converse, we will again use Proposition 4. Take any $(x^m) \in X^\infty$ with $x^m \to x$, and let $y \in \Gamma(x)$. Then (3) implies

$$d_Y(y, \Gamma(x^m)) \leq \max \left\{ d_Y(z, \Gamma(x^m)) : z \in \Gamma(x) \right\} \to 0$$

so that $d_Y(y, \Gamma(x^m)) \to 0$. By Example D.5, then, there is a sequence $(y^m) \in Y^\infty$ such that $y^m \in \Gamma(x^m)$ for each m, and $y^m \to y$. (Why?) By Proposition 4, therefore, Γ is lower hemicontinuous at x. ■

Is this result good for anything? Well, it provides an alternative method for checking whether or not a given correspondence is continuous. For instance, recall that we have shown earlier that the budget correspondence $B : \mathbb{R}^{n+1}_{++} \rightrightarrows \mathbb{R}^n_+$ defined by (2) is compact-valued and continuous (Examples 2 and 3). By using Proposition 5 we can give a quicker proof of this fact. Indeed, for any $(p, \iota), (p', \iota') \in \mathbb{R}^{n+1}_{++}$, we have

$$d_H(B(p, \iota), B(p', \iota')) = \max \left\{ \left| \frac{\iota}{p_i} - \frac{\iota'}{p'_i} \right| : i = 1, \ldots, n \right\}.$$

(Verify!) It follows readily from this observation that $B : \mathbb{R}^{n+1}_{++} \rightrightarrows \mathbb{R}^n_+$ is Hausdorff continuous. (Yes?) Thus, thanks to Proposition 5, it is continuous.

EXERCISE 37 For any positive integers m and n, let $F : \mathbb{R}^n \to \mathbb{R}^m$ be a continuous function and $\varepsilon > 0$. Prove that $\Gamma : \mathbb{R}^n \rightrightarrows \mathbb{R}^m$, defined by $\Gamma(x) := cl_{\mathbb{R}^m}(N_{\varepsilon,\mathbb{R}^m}(F(x)))$, is continuous.

3 The Maximum Theorem

The stage is now set for one of the most important theorems of optimization theory, the so-called *Maximum Theorem*.[9]

THE MAXIMUM THEOREM
Let Θ *and* X *be two metric spaces,* $\Gamma : \Theta \rightrightarrows X$ *a compact-valued correspondence, and* $\varphi \in \mathbf{C}(X \times \Theta)$. *Define*

$$\sigma(\theta) := \arg\max \{\varphi(x,\theta) : x \in \Gamma(\theta)\} \quad \textit{for all } \theta \in \Theta \tag{5}$$

and

$$\varphi^*(\theta) := \max \{\varphi(x,\theta) : x \in \Gamma(\theta)\} \quad \textit{for all } \theta \in \Theta, \tag{6}$$

and assume that Γ *is continuous at some* $\theta \in \Theta$. *Then:*

(a) $\sigma : \Theta \rightrightarrows X$ *is compact-valued, upper hemicontinuous, and closed at* θ.

(b) $\varphi^* : \Theta \to \mathbb{R}$ *is continuous at* θ.

PROOF
Thanks to Weierstrass' Theorem, $\sigma(\theta) \neq \emptyset$ for all $\theta \in \Theta$, so $\sigma : \Theta \rightrightarrows X$ is well-defined. Since $\sigma(\theta) \subseteq \Gamma(\theta)$ for each θ, and $\Gamma(\theta)$ is compact, the compact-valuedness of σ would follow from its closed-valuedness (Proposition C.4). The latter property is, on the other hand, easily verified

[9] This result was first proved by Claude Berge in 1959 in the case of nonparametric objective functions. The formulation we give here, due to Gerard Debreu, is a bit more general. Walker (1979) and Leininger (1984) provide more general formulations.

by using the closedness of $\Gamma(\theta)$ and continuity of $\varphi(\cdot, \theta)$ for each θ. (This is an easy exercise.)

We wish to show that σ is closed at θ. To this end, take any $(\theta^m) \in \Theta^\infty$ and $(x^m) \in X^\infty$ such that $\theta^m \to \theta$, $x^m \in \sigma(\theta^m)$ for each m, and $x^m \to x$. Claim: $x \in \sigma(\theta)$. Since Γ has a closed graph (Proposition 3), we have $x \in \Gamma(\theta)$. Thus, if $x \notin \sigma(\theta)$, then there must exist a $y \in \Gamma(\theta)$ such that $\varphi(y, \theta) > \varphi(x, \theta)$. By the lower hemicontinuity of Γ, then, we can find a sequence $(y^m) \in X^\infty$ such that $y^m \in \Gamma(\theta^m)$ for each m, and $y^m \to y$. But since $(y^m, \theta^m) \to (y, \theta)$ and $(x^m, \theta^m) \to (x, \theta)$, the inequality $\varphi(y, \theta) > \varphi(x, \theta)$ and continuity of φ force that $\varphi(y^m, \theta^m) > \varphi(x^m, \theta^m)$ for m large enough.[10] Since $y^m \in \Gamma(\theta^m)$ for each m, this contradicts the hypothesis that $x^m \in \sigma(\theta^m)$ for each m. Conclusion: $x \in \sigma(\theta)$.

Given its closedness at θ, proving that σ is upper hemicontinuous at θ is now easy. Simply observe that $\sigma(\theta) = \Gamma(\theta) \cap \sigma(\theta)$ for all $\theta \in \Theta$, and apply Exercise 19.

We now turn to assertion (b). Since $\varphi(y, \theta) = \varphi(z, \theta)$ must hold for any $y, z \in \sigma(\theta)$, φ^* is well-defined on Θ. To prove that it is continuous at θ, pick any $(\theta^m) \in \Theta^\infty$ with $\theta^m \to \theta$. We wish to show that $\varphi^*(\theta^m) \to \varphi^*(\theta)$. Of course, $(\varphi^*(\theta^m))$ has a subsequence, say $(\varphi^*(\theta^{m_k}))$ with $\varphi^*(\theta^{m_k}) \to \limsup \varphi^*(\theta^m)$. Now pick any $x^{m_k} \in \sigma(\theta^{m_k})$ so that $\varphi^*(\theta^{m_k}) = \varphi(x^{m_k}, \theta^{m_k})$ for each m_k. Since σ is compact-valued and upper hemicontinuous at x, we can use Proposition 2 to find a subsequence of (x^{m_k}), which we again denote by (x^{m_k}) for convenience, that converges to a point x in $\sigma(\theta)$. By continuity of φ, then

$$\varphi^*(\theta^{m_k}) = \varphi(x^{m_k}, \theta^{m_k}) \to \varphi(x, \theta) = \varphi^*(\theta),$$

which proves that $\varphi^*(\theta) = \limsup \varphi^*(\theta^m)$. But the same argument also shows that $\varphi^*(\theta) = \liminf \varphi^*(\theta^m)$. ∎

REMARK 1

[1] The lower hemicontinuity of σ does not follow from the hypotheses of the Maximum Theorem. (For instance, let $X := \Theta := [0, 1] =: \Gamma(\theta)$ for all $\theta \in \Theta$, and define $\varphi \in \mathbf{C}([0, 1]^2)$ by $\varphi(t, \theta) := t\theta$. Then the solution correspondence σ is not lower hemicontinuous at 0.) But if, in addition

[10] Although this is reasonable (well, it's true), it is not entirely obvious. So mark this point, I'll elaborate on it shortly.

to the assumptions of the Maximum Theorem, it is the case that there is a *unique* maximum of $\varphi(\cdot,\theta)$ on $\Gamma(\theta)$ for each θ, then σ must be a continuous function. For instance, if $X \subseteq \mathbb{R}^n$, $\Gamma(\theta)$ is a convex set and $\varphi(\cdot,\theta)$ is strictly quasiconcave on $\Gamma(\theta)$ for each θ, this situation ensues. (Recall Section A.4.6.)

[2] Let $X := \Theta := \mathbb{R}$, and define $\varphi \in \mathbf{C}(X \times \Theta)$ by $\varphi(t,\theta) := t$. Consider the following correspondences from X into Θ:

$$\Gamma_1(\theta) := \begin{cases} [0,1), & \theta = 0 \\ \{0\}, & \theta \neq 0 \end{cases}, \qquad \Gamma_2(\theta) := \begin{cases} [0,1], & \theta = 0 \\ \{0\}, & \theta \neq 0 \end{cases} \qquad \text{and}$$

$$\Gamma_3(\theta) := \begin{cases} \{0\}, & \theta = 0 \\ [0,1], & \theta \neq 0 \end{cases}.$$

Now define $\sigma_i(\theta)$ by (5) with Γ_i playing the role of Γ, $i = 1,2,3$. Then, while σ_1 is not a well-defined correspondence on Θ (because $\sigma_1(0) = \emptyset$), neither σ_2 nor σ_3 is upper hemicontinuous at 0. Thus we need the compact-valuedness of Γ in the Maximum Theorem, and cannot replace the continuity of Γ with either upper or lower hemicontinuity.

[3] By Proposition D.5, σ would be well-defined in the Maximum Theorem, if φ was assumed only to be upper semicontinuous on $X \times \Theta$. So it is natural to ask if continuity of φ can be replaced in this theorem with upper semicontinuity. The answer is no! For example, let $X := [0,1]$, $\Theta := (0,1]$, $\Gamma(\theta) := [0,\theta]$ for all $\theta \in \Theta$, and define $\varphi \in \mathbb{R}^{X\times\Theta}$ as

$$\varphi(t,\theta) := \begin{cases} 1-2t, & \text{if } 0 \leq t < \frac{1}{2} \\ 3-2t, & \text{if } \frac{1}{2} \leq t \leq 1 \end{cases}$$

for all $\theta \in \Theta$. In this case, we have

$$\sigma(\theta) = \begin{cases} \{0\}, & \text{if } 0 < \theta < \frac{1}{2} \\ \{\frac{1}{2}\}, & \text{if } \frac{1}{2} \leq \theta \leq 1 \end{cases},$$

which obviously is not upper hemicontinuous at $\frac{1}{2}$.[11] Conclusion: *Lower semicontinuity of φ is essential for the validity of the Maximum Theorem.*

[11] But we have $\varphi(\sigma(\theta),\theta) = \begin{cases} 1, & 0 < \theta < \frac{1}{2} \\ 2, & \frac{1}{2} \leq \theta \leq 1 \end{cases}$ which is an upper semicontinuous real function on Θ. Is this a coincidence, you think, or can you generalize this? (See Exercise 43 below.)

Given this observation, it may be instructive to find out exactly where we have used the lower semicontinuity of φ while proving the Maximum Theorem. We used it in the second paragraph of the proof when saying,

> But since $(y^m, \theta^m) \to (y, \theta)$ and $(x^m, \theta^m) \to (x, \theta)$, the inequality $\varphi(y, \theta) > \varphi(x, \theta)$ and continuity of φ force that $\varphi(y^m, \theta^m) > \varphi(x^m, \theta^m)$ for m large enough.

Here is a detailed verification of this claim. Choose an $0 < \varepsilon < \varphi(y, \theta) - \varphi(x, \theta)$, and note that upper semicontinuity of φ implies that there exists an $M_1 > 0$ such that

$$\varphi(y, \theta) > \varphi(x, \theta) + \varepsilon \geq \varphi(x^m, \theta^m) \qquad \text{for all } m \geq M_1.$$

So $\varphi(y, \theta) > \sup\{\varphi(x^m, \theta^m) : m \in \mathbb{N}\} =: s$. Now pick any $0 < \varepsilon' < \varphi(y, \theta) - s$, and note that, *by lower semicontinuity of* φ, there exists an $M_2 > 0$ such that

$$\varphi(y^m, \theta^m) \geq \varphi(y, \theta) - \varepsilon' > s \qquad \text{for all } m \geq M_2.$$

Thus, for any integer $m \geq \max\{M_1, M_2\}$, we have $\varphi(y^m, \theta^m) > \varphi(x^m, \theta^m)$. □

Here is a simple example that illustrates how useful the Maximum Theorem may prove in economic applications. (We will later encounter more substantial applications of this theorem.)

EXAMPLE 4

(*The Demand Correspondence*) Consider an agent whose income is $\iota > 0$ and whose utility function over n-vectors of commodity bundles is $u : \mathbb{R}^n_+ \to \mathbb{R}$. The standard choice problem of this consumer is:

$$\text{Maximize } u(x) \qquad \text{such that} \qquad x \in B(p, \iota),$$

where $p \in \mathbb{R}^n_{++}$ stands for the price vector in the economy and $B : \mathbb{R}^{n+1}_{++} \rightrightarrows \mathbb{R}$, defined by (2), is the budget correspondence of the consumer. Clearly, the optimum choice of the individual is conditional on the parameter (p, ι), and so is modeled by the correspondence $\mathbf{d} : \mathbb{R}^{n+1}_{++} \rightrightarrows \mathbb{R}$, defined by

$$\mathbf{d}(p, \iota) := \arg\max\left\{u(x) : x \in B(p, \iota)\right\}.$$

As you may recall, the correspondence **d** is called the **demand correspondence** of the individual. By Weierstrass' Theorem, it is well-defined. Moreover, since B is continuous (Examples 2 and 3), we can apply the Maximum Theorem to conclude that **d** is compact-valued, closed, and upper hemicontinuous. Moreover, the **indirect utility function** $u^* : \mathbb{R}^{n+1}_{++} \to \mathbb{R}$, defined by $u^*(p,\iota) := \max\{u(x) : x \in B(p,\iota)\}$, is continuous by the same theorem, and is increasing in ι and decreasing in p by Exercise 39 below. Finally, if we further knew that u is strictly quasiconcave, then we could conclude that **d** is a continuous function. □

EXERCISE 38 Let $f : \mathbb{R}_+ \to \mathbb{R}_+$ be a continuous function, and define $U : \mathbb{R}_+ \to \mathbb{R}$ by

$$U(x) := \max\{t - (f(t) - x)^2 : 0 \le t \le f(x)\}.$$

Is U a continuous function?

EXERCISE 39 Let $n \in \mathbb{N}$. Assume that X and Θ are convex subsets of $\mathbb{R}^n$, $\varphi \in \mathbf{C}(X \times \Theta)$, and $\Gamma : \Theta \rightrightarrows X$ is a compact-valued and continuous correspondence. Define φ^* as in the Maximum Theorem.

(a) Show that if $\varphi(x,\cdot)$ is increasing for any given $x \in X$, and $\Gamma(\theta) \supseteq \Gamma(\theta')$ holds for all $\theta, \theta' \in \Theta$ with $\theta \ge \theta'$, then φ^* is an increasing and continuous function.

(b) Show that if φ is a concave function, and $Gr(\Gamma)$ is convex, then φ^* is a concave and continuous function.

EXERCISE 40 (Robinson-Day) Let Θ be any metric space, $\Gamma : \Theta \rightrightarrows \mathbb{R}$ a continuous, compact- and convex-valued correspondence, $\phi \in \mathbf{C}(\mathbb{R} \times \Theta)$, and $\psi \in \mathbf{C}(\Theta)$. Define $\varphi \in \mathbf{C}(\mathbb{R} \times \Theta)$ by $\varphi(x,\theta) := \min\{\phi(x,\theta), \psi(\theta)\}$. Show that if $\phi(\cdot,\theta)$ is strictly quasiconcave for each $\theta \in \Theta$, then $\sigma : \Theta \rightrightarrows \mathbb{R}$, defined by (5), is a continuous and convex-valued correspondence.

EXERCISE 41 Let X be a separable metric space, and make the set $\mathbf{c}(X)$ of all nonempty compact sets in X a metric space by using the Hausdorff metric. Show that if $\succsim$ is a continuous and complete preference relation on X, then the choice correspondence $C_{\succsim}$ defined in Example 1.[3] is an upper hemicontinuous and compact-valued correspondence on $\mathbf{c}(X)$.

EXERCISE 42[H] Let Θ and X be two metric spaces and $\varphi : X \times \Theta \to \mathbb{R}$ an upper semicontinuous map. If X is compact and $\varphi(x,\cdot)$ is lower semicontinuous for any given $x \in X$, then $\varphi^* : \Theta \to \mathbb{R}$, defined by $\varphi^*(\theta) := \max\{\varphi(x, \theta) : x \in X\}$, is a continuous function.

EXERCISE 43^{H} (Berge) Let Θ and X be two metric spaces and $\Gamma : \Theta \rightrightarrows X$ a compact-valued and upper hemicontinuous correspondence. Show that if $\varphi : X \times \Theta \to \mathbb{R}$ is upper semicontinuous, then $\varphi^* : \Theta \to \mathbb{R}$, defined by (6), is upper semicontinuous.

EXERCISE 44^{H} Let $a > 0$, $n \in \mathbb{N}$, and $T := [0, a]^n$. For any strictly increasing $u \in \mathbf{C}(T)$, we define $\mathbf{e}_u : \mathbb{R}^n_{++} \times u(T) \to \mathbb{R}$ by

$$\mathbf{e}_u(p, \upsilon) := \min\{px : u(x) \geq \upsilon\}.$$

($\mathbf{e}_u$ is called the **expenditure function** associated with the utility function u.)

(a) Prove that $\mathbf{e}_u$ is a continuous function.

(b) Defining u^* as in Example 4, show that

$$\mathbf{e}_u(p, u^*(p, \iota)) = \iota \qquad \text{and} \qquad u^*(p, \mathbf{e}_u(p, \upsilon)) = \upsilon$$

for all $(p, \iota) \in \mathbb{R}^{n+1}_{++}$ and $\upsilon \in u(T) \setminus \max u(T)$.

4 Application: Stationary Dynamic Programming

In contrast to the one we analyzed in Example 4, the optimization problems that arise in economic theory often possess a structure that is inherently dynamic. Such problems are significantly more complex than the static ones in general, and the application of the Maximum Theorem to them may be somewhat indirect. The stationary discounted dynamic programming theory, the elements of which are introduced in this section, is a perfect case in point.[12]

[12] The dynamic programming theory was invented in the early 1950s by Richard Bellman (1920–1984) who was a very influential figure in the development of applied mathematics in the United States and whose work was central to "operations research" establishing itself as a prominent field of study at large. If you are interested to know more about Bellman and the origins of dynamic programming, Bellman (1984) is an autobiography that is fun to read. Or, if you don't have much spare time, try Dreyfus (2000), where an interesting selection of excerpts from this autobiography is presented to underline Bellman's take on the birth of dynamic programming (including the story behind the title, *Dynamic Programming*).

4.1 The Standard Dynamic Programming Problem

Put concretely, the basic problem is to find a sequence (x^m) that would

$$\text{Maximize } \varphi(x^0, x^1) + \sum_{i=1}^{\infty} \delta^i \varphi(x^i, x^{i+1}) \tag{7}$$

such that

$$x^1 \in \Gamma(x^0) \quad \text{and} \quad x^{m+1} \in \Gamma(x^m), \ m = 1, 2, \ldots \tag{8}$$

Here x^0 is any element of a given metric space X. While X is called the **state space** of the problem, x^0 is thought of as the **initial state** of the dynamical system at hand. Γ is any self-correspondence on X that lets one know which states are possible "tomorrow" given the state of the system "today." It is called the **transition correspondence** of the problem. A sequence $(x^m) \in X^\infty$ such that $x^1 \in \Gamma(x^0)$ and $x^{m+1} \in \Gamma(x^m)$ for each m is thought of as a **feasible plan** of action through time, with x^m acting as the state of the system at period $m+1$. The objective function of the optimization problem is defined through the map $\varphi : Gr(\Gamma) \to \overline{\mathbb{R}}$, which is usually referred to as the (*one-period*) **return function**. Finally, the parameter δ is any real number in $(0,1)$, and is called the **discount factor**. We think of $\sum_{i=0}^{\infty} \delta^i \varphi(x^i, x^{i+1})$ as the present value of the stream of returns that obtain every period (starting from "today") along the feasible plan (x^m).

The following postulate is basic.

> Assumption (A0)
> For any feasible plan $(x^m) \in X^\infty$,
>
> $$\lim_{k \to \infty} \sum_{i=0}^{k} \delta^i \varphi(x^i, x^{i+1}) \in \overline{\mathbb{R}}.$$

This assumption says that it is possible to compute the present value of the intertemporal stream of returns that is induced by any feasible plan, but it allows for this value to be $-\infty$ or ∞. Put differently, all that (A0) rules out is the possibility of endless oscillations along a feasible plan. In this sense, it can be viewed as a natural hypothesis for the general development of the theory.

The primitives of the model are then contained in the list $(X, x^0, \Gamma, \varphi, \delta)$. When it satisfies (A0), we call such list a **standard dynamic programming problem.** With an innocent abuse of terminology, we will also refer to the associated maximization problem (7)–(8) in the same way.

Let us now introduce two further assumptions that will play a decisive role in the general development of the theory. Note that the first one of these postulates implies (A0).

> ASSUMPTION (A1)
> φ is continuous and bounded.

> ASSUMPTION (A2)
> Γ is compact-valued and continuous.

Many economic problems can be modeled as standard dynamic programming problems that satisfy (A1) and (A2); the comprehensive treatment of Stokey and Lucas (1989) provides a good number of concrete examples.

The first question that we wish to address here is whether there exists a solution to a standard dynamic programming problem under the assumptions stated above. Second, if the answer is yes (and it is, of course), we would like to be able to say something about the *continuity* of the optimal solution(s).

For any given standard dynamic programming problem $(X, x^0, \Gamma, \varphi, \delta)$, we define the class

$$\mathcal{D}(X, \Gamma, \varphi, \delta) := \{(X, x, \Gamma, \varphi, \delta) : x \in X\}, \tag{9}$$

which is the collection of all dynamic programming problems that differ from our original problem only in their initial states. The class of all such collections of standard dynamic programming problems is denoted by $\mathcal{DP}$.

The basic objects of analysis in dynamic programming are the members of $\mathcal{DP}$, that is, collections of the form (9). The main reason why we work with such collections instead of an arbitrary standard dynamic programming problem with a fixed initial state is the following. Given that we choose, say x^1, in the first period, the optimization problem from the second period onward looks exactly like our original problem except that the initial state of the latter problem is x^1. Put differently, because of the basic recursive structure of dynamic programming problems, when we look from period m onward, the problem at hand would be none other than $(X, x^m, \Gamma, \varphi, \delta)$ for any $m \in \mathbb{N}$. Thus, even though the initial state of the original problem may be fixed, one would still need to develop an understanding of the problem for various choices of initial states. This point will become clearer as we go along.

Let us try to rewrite our optimization problem in more familiar terms. Let $\Omega_\Gamma(x)$ stand for the set of all feasible plans for the problem $(X, x, \Gamma, \varphi, \delta)$. That is, define the correspondence $\Omega_\Gamma : X \rightrightarrows X^\infty$ by

$$\Omega_\Gamma(x) := \{(x^m) \in X^\infty : x^1 \in \Gamma(x) \text{ and } x^{m+1} \in \Gamma(x^m),\ m \in \mathbb{N}\}.$$

Define next the map $F_{\Gamma,\varphi} : \{\Omega_\Gamma(x) \times \{x\} : x \in X\} \to \overline{\mathbb{R}}$ by

$$F_{\Gamma,\varphi}((x^m), x) := \varphi(x, x^1) + \sum_{i=1}^{\infty} \delta^i \varphi(x^i, x^{i+1}).$$

(Of course $F_{\Gamma,\varphi}$ depends also on the discount factor δ, but we will keep δ fixed throughout the discussion, so there is no need to make this dependence explicit in our notation.) Thanks to (A0), $F_{\Gamma,\varphi}$ is well-defined, but it may assume the value of $-\infty$ or ∞ along a feasible plan. If, however, φ satisfies (A1), then $F_{\Gamma,\varphi}$ is not only real-valued, it is also bounded. For in this case there exists a $K > 0$ with $|\varphi| \leq K$, so

$$\left|F_{\Gamma,\varphi}((x^m), x)\right| \leq \left|\varphi(x, x^1)\right| + \sum_{i=1}^{\infty} \delta^i \left|\varphi(x^i, x^{i+1})\right| \leq K + \sum_{i=1}^{\infty} \delta^i K = \frac{K}{1-\delta}$$

for all $x \in X$ and $(x^m) \in \Omega_\Gamma(x)$. (Recall Exercise A.45 and Example A.8.[3].)

We may now rewrite our optimization problem as

$$\text{Maximize } F_{\Gamma,\varphi}((x^m), x) \quad \text{such that} \quad (x^m) \in \Omega_\Gamma(x), \tag{10}$$

where the initial state x is considered as a parameter. Very good, this looks like a problem that we can attack with our conventional weapons (Weierstrass' Theorem, the Maximum Theorem, etc.), at least under the premises of (A1) and (A2). But to be able to do this, we need to first metrize X^∞ in such a way that the requirements of the Maximum Theorem are met. While something like this can actually be done, such a direct approach turns out to be a poor method for "solving" (10). Instead, there is an alternative, *recursive* method that would enable us to bring the power of the Maximum Theorem to the fore in a more striking manner.

4.2 The Principle of Optimality

Fix any $\mathcal{D} := \mathcal{D}(X, \Gamma, \varphi, \delta) \in \mathcal{DP}$. A major ingredient of the recursive method is the function $V : X \to \overline{\mathbb{R}}$, which is defined by

$$V(x) := \sup \left\{ F_{\Gamma,\varphi}((x^m), x) : (x^m) \in \Omega_\Gamma(x) \right\}. \tag{11}$$

(We don't make explicit the dependence of V on Γ and φ to simplify the notation.) Clearly, if $F_{\Gamma,\varphi}$ is bounded, then $V \in \mathbf{B}(X)$. (*Note.* We can replace the "sup" here with "max" iff a solution to (10) exists.) V is called the **value function** for the collection $\mathcal{D}$. Owing to the recursive nature of the problems in $\mathcal{D}$, this function will play a very important role in the subsequent analysis. For one thing, provided that a solution to (10) exists, we can deduce from V the optimal plan for our problem under quite general circumstances.

LEMMA 1
(Bellman) *Let $\mathcal{D}(X, \Gamma, \varphi, \delta) \in \mathcal{DP}$, take any $x^0 \in X$ and $(x_*^m) \in \Omega_\Gamma(x^0)$, and define $V : X \to \overline{\mathbb{R}}$ by* (11). *If $V(x^0) = F_{\Gamma,\varphi}((x_*^m), x^0)$, then*

$$V(x^0) = \varphi(x^0, x_*^1) + \delta V(x_*^1)$$

and

$$V(x_*^m) = \varphi(x_*^m, x_*^{m+1}) + \delta V(x_*^{m+1}) \tag{12}$$

for each $m = 1, 2, \ldots$.[13] If (A1) *holds, the converse is also true.*

PROOF
(To simplify the notation, we denote $F_{\Gamma,\varphi}$ by F, and Ω_Γ by Ω, throughout the proof.) By definition, $V(x^0) = F((x_*^m), x^0)$ means that

$$\varphi(x^0, x_*^1) + \sum_{i=1}^{\infty} \delta^i \varphi(x_*^i, x_*^{i+1}) \geq \varphi(x^0, x^1) + \sum_{i=1}^{\infty} \delta^i \varphi(x^i, x^{i+1})$$

[13] This is quite intuitive. In the words of Bellman (1957, p. 83), "an optimal policy has the property that, whatever the initial state and decision are, the remaining decisions must constitute an optimal policy with regard to the state fromresulting from the first decision."

for any $(x^m) \in \Omega(x^0)$. Since $(x^2, x^3, \ldots) \in \Omega(x_*^1)$ implies that $(x_*^1, x^2, x^3, \ldots) \in \Omega(x^0)$ (notice the use of the recursive structure here), we then find

$$\varphi(x^0, x_*^1) + \sum_{i=1}^{\infty} \delta^i \varphi(x_*^i, x_*^{i+1}) \geq \varphi(x^0, x_*^1) + \delta\varphi(x_*^1, x^2) + \sum_{i=2}^{\infty} \delta^i \varphi(x^i, x^{i+1}),$$

that is, $F((x_*^2, x_*^3, \ldots), x_*^1) \geq F((x^2, x^3, \ldots), x_*^1)$, for all $(x^2, x^3, \ldots) \in \Omega(x_*^1)$. Thus, $V(x_*^1) = F((x_*^2, x_*^3, \ldots), x_*^1)$, and hence

$$\begin{aligned} V(x^0) &= F((x_*^m), x^0) \\ &= \varphi(x^0, x_*^1) + \delta\left(\varphi(x_*^1, x_*^2) + \sum_{i=2}^{\infty} \delta^{i-1} \varphi(x_*^i, x_*^{i+1})\right) \\ &= \varphi(x^0, x_*^1) + \delta F((x_*^2, x_*^3, \ldots), x_*^1) \\ &= \varphi(x^0, x_*^1) + \delta V(x_*^1). \end{aligned}$$

The rest of the claim follows by induction. Conversely, assume that (A1) holds, and let $x^0 \in X$ and $(x_*^m) \in \Omega(x^0)$ satisfy (12). Then,

$$\begin{aligned} V(x^0) &= \varphi(x^0, x_*^1) + \delta V(x_*^1) \\ &= \varphi(x^0, x_*^1) + \delta\varphi(x_*^1, x_*^2) + \delta^2 V(x_*^2) \\ &= \cdots \\ &= \varphi(x^0, x_*^1) + \sum_{i=1}^{k} \delta^i \varphi(x_*^i, x_*^{i+1}) + \delta^{k+1} V(x_*^{k+1}) \end{aligned}$$

for any $k \in \mathbb{N}$. But, thanks to (A1), V is bounded, so there exists a $K > 0$ such that $|V| \leq K$, and this clearly entails that $\delta^k V(x^k) \to 0$. (Right?) Thus, letting $k \to \infty$, we obtain $V(x^0) = F((x_*^m), x^0)$, as we sought. ■

EXERCISE 45 Give an example to show that (A1) cannot be omitted in the statement of Lemma 1.

The second part of Lemma 1 tells us how to go from the value function V of a standard dynamic programming problem with (A1) to its optimal path, provided that the problem has a solution. Consequently, we define the **optimal policy correspondence** for $\mathcal{D}(X, \Gamma, \varphi, \delta) \in \mathcal{DP}$ as the self-correspondence P on X with

$$P(x) := \arg\max \left\{\varphi(x, y) + \delta V(y) : y \in \Gamma(x)\right\},$$

provided that a solution to (10) exists, that is, $V(x) = \max\{F_{\Gamma,\varphi}((x^m), x) : (x^m) \in \Omega_\Gamma(x)\}$ for each $x \in X$. Thanks to Lemma 1, if (A1) holds and a solution to (10) exists, then $(x^m) \in X^\infty$ is a solution to (10) iff $x^1 \in P(x^0)$, $x^2 \in P(x^1)$, and so on. (Really? Does Lemma 1 say all this? Please go slowly here.) In turn, we will deal with the existence problem by using the following important result, which is often referred to as the **Principle of Optimality**.

Lemma 2

(Bellman) *For any* $\mathcal{D}(X, \Gamma, \varphi, \delta) \in \mathcal{DP}$ *and* $W \in \mathbf{B}(X)$,

$$W(x) = \max\left\{\varphi(x,y) + \delta W(y) : y \in \Gamma(x)\right\} \quad \textit{for all } x \in X, \tag{13}$$

implies

$$W(x) = \max\left\{F_{\Gamma,\varphi}((x^m), x) : (x^m) \in \Omega_\Gamma(x)\right\} \quad \textit{for all } x \in X.$$

So, thanks to this observation, the existence problem at hand becomes one of finding a solution to the functional equation (13). If there is a solution W to (13), then this solution, according to Lemma 2, is the value function for our dynamic programming problem. All this is nice, because we have earlier learned some techniques (in Sections C.6 and C.7) that we can use to "solve" (13). But no need to rush! Let us first prove the claim at hand.

Proof of Lemma 2

(To simplify the notation, we again denote $F_{\Gamma,\varphi}$ by F and Ω_Γ by Ω throughout the proof.) Assume that $W \in \mathbf{B}(X)$ satisfies (13), and fix an arbitrary $x \in X$. By (13), for an arbitrarily chosen $(x^m) \in \Omega(x)$,

$$W(x) \geq \varphi(x, x^1) + \delta W(x^1) \geq \varphi(x, x^1) + \delta\varphi(x^1, x^2) + \delta^2 W(x^2) \geq \dots,$$

so, proceeding inductively,

$$W(x) \geq \left(\varphi(x, x^1) + \sum_{i=1}^{k} \delta^i \varphi(x^i, x^{i+1})\right) + \delta^{k+1} W(x^{k+1}) \quad \text{for all } k = 1, 2, \dots$$

Thus, if $\sum^\infty \delta^i \varphi(x^i, x^{i+1}) > -\infty$—otherwise $W(x) \geq F((x^m), x)$ obtains trivially—letting $k \to \infty$ yields

$$W(x) \geq F((x^m), x) + \lim \delta^k W(x^k).$$

But since W is bounded by hypothesis, the real sequence $(\delta^k W(x^k))$ converges; indeed, we have $\lim \delta^k W(x^k) = 0$. Conclusion: $W(x) \geq F((x^m), x)$ for any $(x^m) \in \Omega(x)$. (We are not done yet. Why?)

Now choose a sequence (x_*^m) in $\Omega(x)$ such that $W(x) = \varphi(x, x_*^1) + \delta W(x_*^1)$, and

$$W(x_*^m) = \varphi(x_*^m, x_*^{m+1}) + \delta W(x_*^{m+1}), \qquad m = 1, 2, \ldots$$

(By (13) there exists such a sequence.) Thus, for any $k \in \mathbb{N}$, we have

$$\begin{aligned} W(x) &= \varphi(x, x_*^1) + \delta\varphi(x_*^1, x_*^2) + \cdots + \delta^k \varphi(x_*^k, x_*^{k+1}) + \delta^{k+1} W(x_*^{k+1}) \\ &= \left(\varphi(x, x_*^1) + \sum_{i=1}^{k} \delta^i \varphi(x_*^i, x_*^{i+1}) \right) + \delta^{k+1} W(x_*^{k+1}). \end{aligned}$$

Since W is bounded, letting $k \to \infty$ yields $W(x) = F((x_*^m), x)$. Since $(x_*^m) \in \Omega(x)$, combining this with the finding of the previous paragraph completes the proof. ∎

EXERCISE 46 Prove the converse of Lemma 2.

EXERCISE 47 Let $\mathcal{D}(X, \Gamma, \varphi, \delta) \in \mathcal{DP}$, $\emptyset \neq Y \subseteq X$, and $W \in \mathbf{B}(Y)$. Show that if

$$W(x) = \max\left\{\varphi(x, y) + \delta W(y) : y \in \Gamma(x)\right\} \qquad \text{for all } x \in Y,$$

then $W(x) \geq \max\{F_{\Gamma,\varphi}((x^m), x) : (x^m) \in \Omega_\Gamma(x) \cap Y^\infty\}$ for all $x \in Y$.

The following two examples aim to demonstrate the importance of Lemma 2.

EXAMPLE 5

For any $0 \leq x_0 \leq 1$, consider the problem of choosing a real sequence (x_m) in order to

$$\text{Maximize } \sum_{i=0}^{\infty} \tfrac{1}{2^i} \ln(\sqrt{x_i} - x_{i+1})$$

such that

$$0 \leq x_{m+1} \leq \sqrt{x_m}, \quad m = 0, 1, \ldots$$

We may view this problem as a standard dynamic programming problem $(X, x_0, \Gamma, \varphi, \delta)$, where $X := [0,1]$, $\Gamma(x) := [0, \sqrt{x}]$, $\varphi(x,y) := \ln(\sqrt{x} - y)$, and $\delta = \frac{1}{2}$. (Check that (A0) is satisfied.) This problem satisfies neither (A1) nor (A2), but this is okay, since here we will solve it using directly the Principle of Optimality, which does not require either of these postulates.

Our first objective is to compute the value function of the problem. Lemma 2 says that it is sufficient to solve the following functional equation for this purpose:

$$W(x) = \max\left\{\ln(\sqrt{x} - y) + \tfrac{1}{2}W(y) : 0 \leq y \leq \sqrt{x}\right\}, \qquad 0 \leq x \leq 1. \quad (14)$$

But how do we solve this?[14] The key observation is that we just need to find one $W \in \mathbf{B}(X)$ that does the job. Thus if we "guessed" the W right, we would be readily in business. Now, here is a rabbit-out-of-the-hat guess that will work: For some $\alpha \geq 0$ and $\beta \in \mathbb{R}$, $W(x) = \alpha \ln x + \beta$ for all $x \in X$.[15] With this guess at hand, the problem is to find an $(\alpha, \beta) \in \mathbb{R}_+ \times \mathbb{R}$ such that

$$\alpha \ln x + \beta = \max\left\{\ln(\sqrt{x} - y) + \tfrac{\alpha}{2}\ln y + \tfrac{\beta}{2} : 0 \leq y \leq \sqrt{x}\right\}, \qquad 0 \leq x \leq 1.$$

This is nothing that good old calculus can't handle: $y = \frac{\alpha}{2+\alpha}\sqrt{x}$ solves the associated maximization problem, so

$$\begin{aligned}\alpha \ln x + \beta &= \ln\left(\sqrt{x} - \tfrac{\alpha}{2+\alpha}\sqrt{x}\right) + \tfrac{\alpha}{2}\ln \tfrac{\alpha}{2+\alpha}\sqrt{x} + \tfrac{\beta}{2} \\ &= \left(\tfrac{1}{2} + \tfrac{\alpha}{4}\right)\ln x + \ln\left(1 - \tfrac{\alpha}{2+\alpha}\right) + \tfrac{\alpha}{2}\ln\tfrac{\alpha}{2+\alpha} + \tfrac{\beta}{2}\end{aligned}$$

for each $x \in [0,1]$. This in turn gives us two equations

$$\alpha = \tfrac{1}{2} + \tfrac{\alpha}{4} \qquad \text{and} \qquad \beta = \ln\left(1 - \tfrac{\alpha}{2+\alpha}\right) + \tfrac{\alpha}{2}\ln\tfrac{\alpha}{2+\alpha} + \tfrac{\beta}{2}$$

which are readily solved to find $\alpha = \frac{2}{3}$ and $\beta = \ln 9 - \frac{8}{3}\ln 4$. Therefore, by Lemma 2, we may conclude that the value function we are after is $V(x) = \frac{2}{3}\ln x + \ln 9 - \frac{8}{3}\ln 4$ for each $0 \leq x \leq 1$. Furthermore, from strict concavity of V and Lemma 1, it follows that the optimal policy correspondence is single-valued: $P(x) = \{\frac{1}{4}\sqrt{x}\}$ for any $0 \leq x \leq 1$. □

[14] Unfortunately, there is a considerable (and quite annoying) wedge between the *theory* and *practice* of dynamic programming. For an excellent introduction for economists to the latter matter, I recommend Ljungqvist and Sargent (2004). I will have little to say on this issue here.

[15] I will tell you shortly where on earth this crazy guess came from.

EXERCISE 48 Take any $\delta \in (0,1)$ and $x_0 \in [0,1]$, and let f be a continuous self-map on $[0,1]$. Consider the following optimization problem:

$$\text{Maximize } \sum_{i=0}^{\infty} \delta^i (1 - (f(x_i) - x_{i+1})^2)$$

such that

$$0 \leq x_{m+1} \leq f(x_m), \quad m = 0, 1, \ldots.$$

Find the optimal policy correspondence and the value function for this problem.

4.3 Existence and Uniqueness of an Optimal Solution

Now that we have seen what the Principle of Optimality can do for us, let's get back to the theory of dynamic programming. For the general theory, we will make use of both (A1) and (A2). Let us then fix an arbitrary $\mathcal{D}(X, \Gamma, \varphi, \delta) \in \mathcal{DP}$ for which (A1) and (A2) hold. With the Maximum Theorem in our arsenal, we can in fact establish the existence of a solution to the functional equation (13) relatively easily. We have studied a powerful method of doing this at some length in Section C.7.1, and the same method will work here perfectly. First define the map $\Phi : \mathbf{CB}(X) \to \mathbb{R}^X$ by

$$\Phi(W)(x) = \max \left\{ \varphi(x, y) + \delta W(y) : y \in \Gamma(x) \right\} \quad \text{for all } x \in X. \tag{15}$$

By the Maximum Theorem, $\Phi(W)$ is a continuous function, and since both φ and W are bounded, so is $\Phi(W)$. Thus, Φ is a self-map on $\mathbf{CB}(X)$. Moreover, it is readily checked that $W \geq W'$ implies $\Phi(W) \geq \Phi(W')$ and that $\Phi(W + \alpha) = \Phi(W) + \delta\alpha$ for all $W, W' \in \mathbf{CB}(X)$ and $\alpha \geq 0$. Thus we can apply Lemma C.3 to conclude that Φ is a contraction. Then, by the Banach Fixed Point Theorem, there exists a unique fixed point of Φ in $\mathbf{CB}(X)$, call it $\hat{W}$. Clearly, $\hat{W}$ satisfies (13), and hence by Lemma 2 we have

$$\hat{W}(x) = \max \left\{ F_{\Gamma,\varphi}((x^m), x) : (x^m) \in \Omega_\Gamma(x) \right\} \quad \text{for all } x \in X,$$

as we sought.[16] This proves that a solution to (13) exists, so we conclude:

PROPOSITION 6
There exists a solution for any standard dynamic programming problem $(X, x^0, \Gamma, \varphi, \delta)$ *that satisfies* (A1)-(A2).

Given that (13) has a solution for each $x \in X$ under the hypotheses (A1) and (A2), we may conclude that, for each $x^0 \in X$, an optimal path (x_*^m) for the problem $(X, x^0, \Gamma, \varphi, \delta)$ exists and satisfies $x_*^1 \in P(x^0)$, $x_*^2 \in P(x_*^1)$, and so on, where, of course, P is the optimal policy correspondence given by

$$P(x) = \arg\max \left\{\varphi(x, y) + \delta V(y) : y \in \Gamma(x)\right\} \qquad \text{for all } x \in X, \tag{16}$$

with V being the value function for $\mathcal{D}(X, \Gamma, \varphi, \delta)$ (or equivalently, as we now know, it being the unique solution of (13) in $\mathbf{CB}(X)$). Thus, the solution to our problem (10) is given by means of a *stationary* decision correspondence, which tells us what to choose in a period given the choice of the previous period. Moreover, the Maximum Theorem tells us that P is quite well-behaved in that it is compact-valued and upper hemicontinuous.

We would also like to say something about the *uniqueness* of the solution to (10), that is, about when P is single-valued. Of course, we need some additional assumptions for this purpose. (Why of course?) To this end, we let X be a nonempty convex set in $\mathbb{R}^n$, $n \in \mathbb{N}$, and posit further regularity properties.[17]

[16] I can now let you in on the secret behind the guess $W(x) := \alpha \ln x + \beta$ that did wonders in Example 5. I had in mind using the *method of successive approximations* to solve the functional equation (14). (Recall the proof of Banach's Fixed Point Theorem.) This suggested that I start with an arbitrary choice of W_0, and iterate the map Φ to see to what kind of a fixed point it would converge to. I began with an easy choice: $W_0 := 0$. Then $W_1(x) := \Phi(W_0)(x) = \max\{\ln(\sqrt{x} - y) : 0 \le y \le \sqrt{x}\} = \frac{1}{2} \ln x$ for all x. The next step in the iteration is then to find W_2, where $W_2(x) := \Phi(W_1)(x) = \max\{\ln(\sqrt{x} - y) + \frac{1}{4} \ln y : 0 \le y \le \sqrt{x}\}$ for all x. Solving the associated maximization problem gives us $W_2(x) = \left(\frac{5}{8}\right) \ln x + (2 \ln 2 - \frac{5}{4} \ln 5)$ for all $x \in [0, 1]$. As you can see, a pattern is emerging here; this is how I got my "guess." (Iterate one more time if you like.)

[17] We confine our attention here to Euclidean spaces, because we are about to postulate properties that require us to think about linear combinations of the elements of X. It is meaningless to talk about such things in an arbitrary metric space; what we need is a linear space for this purpose. Thus we work with $\mathbb{R}^n$ here because it possesses both a linear and a metric structure simultaneously. There are, of course, more general spaces that are also endowed with both of these structures, but we will begin to study them only in Chapter I.

Assumption (A3)
$Gr(\Gamma)$ is convex, that is, for all $0 \leq \lambda \leq 1$ and $x, x' \in X$,

$$y \in \Gamma(x) \text{ and } y' \in \Gamma(x') \quad \text{imply} \quad \lambda y + (1-\lambda)y' \in \Gamma(\lambda x + (1-\lambda)x').$$

Moreover, φ is concave on $Gr(\Gamma)$, and

$$\varphi(\lambda(x,y) + (1-\lambda)(x',y')) > \lambda\varphi(x,y) + (1-\lambda)\varphi(x',y')$$

for all $0 < \lambda < 1$ and $(x, y), (x', y') \in Gr(\Gamma)$ with $x \neq x'$.

We note that this postulate does not ask for the *strict* concavity of φ on its entire domain. For instance, where Γ is a self-correspondence on $[0, 1]$ with $\Gamma(x) := [0, 1]$ for all $0 \leq x \leq 1$, the map $\varphi : Gr(\Gamma) \to \mathbb{R}$ defined by $\varphi(x, y) := \sqrt{x} - y$ satisfies (A3), but it is not strictly concave on $[0, 1]^2$.

Now, to prove the uniqueness of the solution to (10), we need to show that P is single-valued, and a sufficient condition for this is that the unique function in $\mathbf{CB}(X)$ that satisfies (13) (i.e., the unique fixed point of Φ) be strictly concave. (Why?) Recall that, by Lemma 2 and the proof of the Banach Fixed Point Theorem, we have $\lim \Phi^m(W) = V$ for any choice of W in $\mathbf{CB}(X)$. But then if we can prove that Φ maps concave functions to concave functions, it would follow that V must be concave. (*Proof.* Pick any concave $W \in \mathbf{CB}(X)$ and use this claim along with the fact that the uniform (indeed, pointwise) limit of concave functions is concave (Exercise A.67).) So, take any $0 < \lambda < 1$ and $x, x' \in X$, and note that, by the first part of (A3),

$$\Phi(W)(\lambda x+(1-\lambda)x') \geq \varphi(\lambda x+(1-\lambda)x', \lambda y+(1-\lambda)y')+\delta W(\lambda y+(1-\lambda)y')$$

where $y \in P(x)$ and $y' \in P(x')$ with P as defined in (16). Then, by concavity of φ and W,

$$\begin{aligned}\Phi(W)(\lambda x + (1-\lambda)x') &\geq \lambda(\varphi(x,y) + \delta W(y)) + (1-\lambda)(\varphi(x',y') + \delta W(y')) \\ &\geq \lambda\Phi(W)(x) + (1-\lambda)\Phi(W)(x'),\end{aligned}$$

which proves that $\Phi(W)$ is concave. By using the reasoning noted above, then, we may conclude that V is a concave function. But then we may repeat this analysis with V playing the role of W, and with $x \neq x'$ and $0 < \lambda < 1$, to get

$$\begin{aligned} V(\lambda x + (1-\lambda)x') &= \Phi(V)(\lambda x + (1-\lambda)x') \\ &> \lambda \Phi(V)(x) + (1-\lambda)\Phi(V)(x') \\ &= \lambda V(x) + (1-\lambda)V(x') \end{aligned}$$

by using the second part of (A3). Thus, V must be a strictly concave function. As noted above, this proves the following:

PROPOSITION 7
Let X be a nonempty convex subset of a Euclidean space. A standard dynamic programming problem $(X, x^0, \Gamma, \varphi, \delta)$ that satisfies (A1)–(A3) *has a unique solution.*

EXERCISE 49 Consider a standard dynamic programming problem $(X, x^0, \Gamma, \varphi, \delta)$ that satisfies (A1) and (A2). Assume further that, for any $y \in X$ such that $(x, y) \in Gr(\Gamma)$ for some $x \in X$, the map $\varphi(\cdot, y)$ is strictly increasing (on $\{x \in X : y \in \Gamma(x)\}$), and that $\Gamma(x) \subseteq \Gamma(x')$ whenever $x \leq x'$. Prove that the value function of the problem is strictly increasing.

The following definition introduces a useful concept that often plays an important role in problems that concern the *monotonicity* of a solution to an optimization problem.

DEFINITION
Let S be any nonempty set in $\mathbb{R}^2$. A function $f \in \mathbb{R}^S$ is said to be **supermodular** on S if

$$f(a,b) + f(a',b') \geq f(a',b) + f(a,b')$$

whenever $(a,b), (a',b'), (a',b), (a,b') \in S$ and $(a',b') \geq (a,b)$. If (a,b) and (a',b') are distinct, and each $\geq$ is replaced with $>$ here, then we say that f is **strictly supermodular.**

EXERCISE 50 Let $\emptyset \neq X \subseteq \mathbb{R}_+$, and consider a standard dynamic programming problem $(X, x_0, \Gamma, \varphi, \delta)$ that satisfies (A1) and (A2). Assume:

(i) $x' \geq x$, $y \geq y'$ and $y \in \Gamma(x)$ imply $y' \in \Gamma(x')$;

(ii) φ is strictly supermodular on $\mathrm{Gr}(\Gamma)$.

(a) Let P be the optimal policy correspondence for $\mathcal{D}(X, \Gamma, \varphi, \delta)$. Prove that if the self-map p on X is a selection from P, that is, $p(x) \in P(x)$ for all $x \in X$, then p must be an increasing function.[18]

(b) (*A Turnpike Theorem*) Show that if X is compact, then for each optimal path (x_m^*) for $(X, x_0, \Gamma, \varphi, \delta)$ there exists an $x \in X$ such that $x_m^* \to x \in P(x)$.

By positing further properties on the primitives of a standard dynamic programming problem, one may say more about the structure of the optimal policy and value functions. We shall, however, stop our discussion at this point, noting that there are three excellent references to further your knowledge about this important topic: Bertsekas (1976), Stokey and Lucas (1989), and Mitra (2000).[19]

We turn now to a somewhat detailed analysis of an economic model by using the ideas introduced above. For other important economic applications we can do no better than refer you again to Stokey and Lucas (1989) and Ljungqvist and Sargent (2004).

4.4 Application: The Optimal Growth Model

The primitives of the (one-sector) optimal growth model are the initial capital stock $x_0 > 0$, the discount factor $0 < \delta < 1$, the production function $F : \mathbb{R}_+ \to \mathbb{R}_+$, and the (social) utility function $u : \mathbb{R}_+ \to \mathbb{R}$. With these givens, the optimization problem of the model is to

$$\text{Maximize } \sum_{i=0}^{\infty} \delta^i u(F(x_i) - x_{i+1}) \tag{17}$$

such that

$$0 \leq x_{m+1} \leq F(x_m), \quad m = 0, 1, \ldots \tag{18}$$

[18] WARNING. This does not mean that any optimal path of the problem is increasing. Why?

[19] I recommend starting with Mitra (2000) to learn more about the basic theory, and then moving on to the problems given in Chapter 5 of Stokey and Lucas (1989).

where the maximization is over the sequence of investment levels (x_m). The idea is that there is a single good that may be consumed or invested at every period m. An investment of x_{m+1} at period m yields a return of $F(x_{m+1})$ at period $m+1$. Thus at period m, the level of consumption is $F(x_m) - x_{m+1}$, yielding, for the social planner, a utility level of $u(F(x_m) - x_{m+1})$. The problem of the planner is to choose a capital accumulation plan in order to maximize the discounted sum of the stream of utilities. In what follows, we will work with the following properties.

Assumption (B)

(i) u is a bounded and twice differentiable function with $u'_+(0) = \infty$, $u' > 0$ and $u'' < 0$.

(ii) F is a bounded and twice differentiable function with $F(0) = 0$, $F'_+(0) = \infty$, $F' > 0$ and $F'' < 0$.

The list (x_0, F, u, δ) (along with the maximization problem (17)–(18)) is called an **optimal growth model**, provided that u and F satisfy (B). This model induces a standard dynamic programming problem in a natural manner. If we define $\Gamma_F : \mathbb{R}_+ \rightrightarrows \mathbb{R}_+$ by $\Gamma_F(x) := [0, F(x)]$, and $\varphi_{u,F} \in \mathbb{R}^{Gr(\Gamma)}$ by $\varphi_{u,F}(x,y) := u(F(x) - y)$, then $(\mathbb{R}_+, x_0, \Gamma_F, \varphi_{u,F}, \delta)$ is a standard dynamic programming problem, and the problem (17)–(18) is:

$$\text{Maximize} \sum_{i=0}^{\infty} \delta^i \varphi_{u,F}(x_i, x_{i+1})$$

such that

$$x_{m+1} \in \Gamma_F(x_m), \quad m = 0, 1, \ldots$$

Clearly, Γ_F is compact-valued. It is also continuous, thanks to the continuity of F (Exercise 27). Moreover, $\varphi_{u,F}$ is continuous since so are u and F, and it is bounded since so is u. We may thus apply Proposition 6 to conclude that there exists a solution to the maximization problem (17)–(18) under the premises of (B). Consequently, the value function V for $\mathcal{D}(\mathbb{R}_+, \Gamma_F, \varphi_{u,F}, \delta)$ is well-defined and corresponds to the unique fixed point of the map $\Phi : \mathbf{CB}(\mathbb{R}_+) \rightarrow \mathbb{R}^{\mathbb{R}_+}$ defined by (15) (with Γ_F and $\varphi_{u,F}$ playing the roles of Γ and φ, respectively). Thus, we have

$$V(x) = \max\left\{u(F(x) - y) + \delta V(y) : 0 \leq y \leq F(x)\right\} \quad \text{for all } x \in \mathbb{R}_+. \tag{19}$$

Given that u and F are strictly increasing, we may use Exercise 49 to show that this function is strictly increasing. It is also easy to show that $\varphi_{u,F}$ and Γ_F satisfy (A3). For instance, $\varphi_{u,F}$ is concave, because, for any $(x, y), (x', y') \in Gr(\Gamma_F)$ and $0 \leq \lambda \leq 1$,

$$\begin{aligned}\varphi_{u,F}(\lambda(x, y) + (1-\lambda)(x', y')) &= u(F(\lambda x + (1-\lambda)x') - (\lambda y + (1-\lambda)y')) \\ &\geq u(\lambda(F(x) - y) + (1-\lambda)(F(x') - y')) \\ &\geq \lambda u(F(x) - y) + (1-\lambda)u(F(x') - y') \\ &= \lambda\varphi_{u,F}(x, y) + (1-\lambda)\varphi_{u,F}(x', y'),\end{aligned}$$

where the first inequality follows from concavity of F and monotonicity of u, and the second from concavity of u. In view of the discussion preceding Proposition 7, therefore, we reach the following conclusion.

OBSERVATION 1. For any optimal growth model there is a unique solution. Moreover, the value function for the associated dynamic programming problem is strictly increasing and strictly concave.

Consequently, for any optimal growth model (x_0, F, u, δ), the optimal policy correspondence for $\mathcal{D}(\mathbb{R}_+, \Gamma_F, \varphi_{u,F}, \delta)$ can be thought of as a self-map P on $\mathbb{R}_+$, with

$$P(x) \in \arg\max\left\{u(F(x) - y) + \delta V(y) : 0 \leq y \leq F(x)\right\}.$$

(This map is called the **optimal policy function** for (x_0, F, u, δ).) In words, $P(x)$ is the optimal level of savings (investment) at any period with capital stock x. By the Maximum Theorem, P is a continuous function. Moreover, given the special structure of the optimal growth model, we have the following.

OBSERVATION 2. The optimal policy function P for any optimal growth model (x_0, F, u, δ) is increasing.

PROOF
Take any $x, x' \in \mathbb{R}_+$ with $x' > x$, and assume that $y := P(x) > P(x') =: y'$. By definition of P and (19), we have

$$u(F(x) - y) + \delta V(y) \geq u(F(x) - y') + \delta V(y')$$

since $y' < y \leq F(x)$, and

$$u(F(x') - y') + \delta V(y') \geq u(F(x') - y) + \delta V(y)$$

since $y \leq F(x) \leq F(x')$. Thus,

$$u(F(x') - y') - u(F(x') - y) \geq \delta \left(V(y) - V(y') \right) \geq u(F(x) - y') - u(F(x) - y)$$

which is impossible in view of the strict concavity of u. (Why?) □

This observation tells us that the **optimal investment plan**

$$(x_1^*, x_2^*, \ldots) := (P(x_0), P(x_1^*), \ldots)$$

for any optimal growth model (x_0, F, u, δ) is monotonic. Indeed, if $x_1^* \geq x_0$, then $x_2^* = P(x_1^*) \geq P(x^0) = x_1^*$ and so on, that is, in this case the optimal plan has it that one invests more and more every period. If, on the other hand, $x_1^* < x_0$, then the analogous reasoning shows that the optimal plan must be a decreasing sequence. Thus, the optimal investment plan cannot have any cycles; it is a monotonic sequence in $\mathbb{R}_+$. In particular, it converges to some point of $\overline{\mathbb{R}}_+$ (Proposition A.8).

What can we say about the limit of (x_m^*)? Answering this question becomes relatively easy once it is observed that, for each $x \geq 0$,

$$P(x) \in \arg\max \left\{ u(F(x) - y) + \delta u(F(y) - P(P(x))) \right. \\ \left. + \delta^2 V(P(P(x))) : 0 \leq y \leq F(x) \right\},$$

which is an instance of the so-called *one-deviation property*. (This is a serious step. It is best if you struggle with this yourself, and see why P must satisfy this property. So take a deep breath, and try to see why $(P(x), P(P(x)), \ldots)$ would not be optimal if the assertion above failed. Once you see this, the rest is a walk in the park.) Since everything is differentiable here, and since the solution of the associated optimization problem must be interior—thanks to the boundary conditions $u'_+(0) = F'_+(0) = \infty$—we must have

$$u'(F(x) - P(x)) = \delta u' \left(F(P(x)) - P(P(x)) \right) F'(P(x)),$$

which is called the **Ramsey-Euler equation**. This recursive equation is really all that we need to know about the optimal plans of consumption and investment in the present model. Let us denote by (c_m^*) the **optimal consumption plan** for (x_0, F, u, δ), that is, $(c_m^*) := (F(x_0) - x_1^*, F(x_1^*) - x_2^*, \ldots)$. The Ramsey-Euler equation then yields

$$u'(c_m^*) = \delta u'(c_{m+1}^*)F'(x_m^*), \qquad m = 1, 2, \ldots \tag{20}$$

We leave it to you to show that this equation and the strict concavity of u entail that (c_m^*) is a monotonic, and hence convergent, sequence. (Don't forget that we do not know if P is differentiable when proving this claim.) Since F is bounded, $c_m^* \to \infty$ is impossible, so we have $\lim c_m^* \in \mathbb{R}_+$.

By a **steady state** of the model (x_0, F, u, δ), we mean a level of capital stock x in which the level of investment (and hence consumption) stays the same throughout all periods: $x = P(x)$. From the Ramsey-Euler equation and the strict concavity of F it follows that there is a unique steady state of the model, call it x^G, which is characterized by $F'(x^G) = \frac{1}{\delta}$. (Surprisingly, x^G depends only on F and δ.) In the literature on growth theory, x^G is referred to as the **golden-rule level of investment**. The main reason for choosing such a flashy name is given by the final result of this section.

OBSERVATION 3. (Turnpike Theorem) The optimal investment and consumption plans (x_m^*) and (c_m^*) for any given optimal growth model (x_0, F, u, δ) are monotonic sequences with

$$\lim x_m^* = x^G \quad \text{and} \quad \lim c_m^* = F(x^G) - x^G,$$

where $F'(x^G) = \frac{1}{\delta}$.

PROOF

The monotonicity (hence convergence) claims were already verified above. Letting now $m \to \infty$ in (20) and using the continuity of u' and F', we find $u'(\lim c_m^*) = \delta u'\left(\lim c_m^*\right) F'(\lim x_m^*)$, which yields $F'(\lim x_m^*) = \frac{1}{\delta} = F'(x^G)$. Thus, since F' is strictly decreasing, $\lim x_m^* = x^G$. That $\lim c_m^* = F(x^G) - x^G$ then follows readily from the continuity of F. ☐

This is a truly remarkable result. It tells us that regardless of the initial capital stock and the (per period) utility function, the optimal investment policy tends through time to a *fixed* level x^G.

We conclude with a few exercises from intertemporal economics, the analyses of which parallel closely the one sketched above for the optimal growth model.

EXERCISE 51 Consider the following optimization problem:

$$\text{Maximize} \sum_{i=0}^{\infty} \delta^i (x_i - x_{i+1})^\alpha$$

such that $0 \leq x_{m+1} \leq x_m$, $m = 0, 1, \ldots$, where x_0, α, and δ belong to $(0, 1)$. Find the unique solution (x_m^*) for this problem, and compute (and interpret) $\lim x_m^*$.

EXERCISE 52 (*The Cake-Eating Problem*) Let C be any positive real number (the size of the cake), and assume that $u : [0, C] \to \mathbb{R}$ is a strictly increasing and strictly concave function. Consider the following optimization problem: Choose $(c_m) \in \mathbb{R}_+^\infty$ in order to

$$\text{Maximize} \sum_{i=1}^{\infty} \delta^i u(c_i) \qquad \text{such that} \qquad \sum_{i=1}^{\infty} c_i \leq C,$$

where $\delta \in (0, 1)$. Show that this problem has a unique solution, and discuss the monotonicity properties of this solution.

EXERCISE 53[H] Consider a firm whose production depends on a single input and whose technology is modeled by means of a production function $F : \mathbb{R}_+ \to \mathbb{R}_+$, which is twice differentiable, bounded, and satisfies $F(0) = 0$, $F' > 0$, $F'' < 0$, $F'(0) = \infty$. The firm has at present an input level $x_0 > 0$. Assume that the price of the firm's output is $p > 0$ and the price of its input is 1 (in any time period). The firm must purchase its inputs one period in advance (think of them as a capital investment), and may buy at most $\theta > 0$ units of inputs at any given period. The inputs of the firm are reusable, but they depreciate through time at rate $\alpha \in (0, 1)$. (So, if the level of inputs owned by the firm at period m is x_m, then the level of input x_{m+1} available in period $m+1$ equals $(1-\alpha)x_m +$ purchase made at period m, $m = 0, 1, \ldots$.)

(a) Assume that this firm wishes to choose the level of input purchases through time in order to maximize the infinite sum of its discounted profits, where its discount factor is $0 < \delta < 1$.

Formulate the optimization problem of the firm as a standard dynamic programming problem. Does this problem necessarily have a unique solution?

(b) If $F'(x_0) > \frac{1-\delta(1-\alpha)}{\delta p} > F'(\theta)$, what is the optimal policy of the firm?

EXERCISE 54 (Benoît-Ok) Let (x_0, F, u, α) and (x_0, F, v, β) be two optimal growth models. Denote the optimal capital accumulation path of the first model by $(x_0, x_1, x_2, \ldots)$ and that of the latter by $(x_0, y_1, y_2, \ldots)$. Show that if

$$\frac{v'(a)}{v'(b)} \geq \frac{\beta}{\alpha}\frac{u'(a)}{u'(b)} \qquad \text{for any } a, b \geq 0,$$

then $x_m \geq y_m$ for all $m = 1, 2, \ldots$. (*Interpretation.* In the one-sector optimal growth model, the optimal capital stock of a country can never fall strictly below that of a more delay-averse country.)

*EXERCISE 55 (Mitra) Take any optimal growth model (x_0, F, u, δ) such that $\varphi_{u,F}$ is supermodular. Denote the value function for $\mathcal{D}(\mathbb{R}_+, \Gamma_F, \varphi_{u,F}, \delta)$ by $V(\cdot, \delta)$ and the optimal policy function for (x_0, F, u, δ) by $P(\cdot, \delta)$, making the dependence of these maps on the discount factor δ explicit. (You should thus think of the domain of these functions as $\mathbb{R}_+ \times (0, 1)$ in this exercise.) Prove:

(a) V is supermodular on $\mathbb{R}_+ \times (0, 1)$.

(b) $P(x, \cdot)$ is an increasing function for any $x \in \mathbb{R}_+$.

5 Fixed Point Theory III

In this section we extend our work on fixed point theory to the context of correspondences. Our main objective in this regard is to derive Kakutani's famous generalization of the Brouwer Fixed Point Theorem, which is indeed widely applied in economic analysis. However, our approach is slightly indirect in this regard, for we wish to do some sightseeing on the way, and talk about a few interesting properties of continuous correspondences that are of independent interest. Toward the end of the section we will also talk about how one may carry the basic idea behind the Banach Fixed Point Theorem into the realm of correspondences.

5.1 Kakutani's Fixed Point Theorem

The following is one of the most important theorems that are contained in this book. It is also commonly used in economic theory; chances are you have already encountered this result elsewhere. It was proved by Shizuo Kakutani in 1941.[20]

> KAKUTANI'S FIXED POINT THEOREM
> *For any given $n \in \mathbb{N}$, let X be a nonempty, closed, bounded, and convex subset of $\mathbb{R}^n$. If Γ is a convex-valued self-correspondence on X that has a closed graph, then Γ has a fixed point, that is, there exists an $x \in X$ with $x \in \Gamma(x)$.*

EXERCISE 56

(a) Give an example of a convex-valued self-correspondence Γ on $[0, 1]$ that does not have a fixed point.

(b) Give an example of a closed-valued and upper hemicontinuous self-correspondence Γ on $[0, 1]$ that does not have a fixed point.

WARNING. The requirement of the closed graph property in the statement of Kakutani's Fixed Point Theorem can be replaced with upper hemicontinuity when Γ is closed-valued. (This is an immediate consequence of Proposition 3.(b).)

Kakutani's Fixed Point Theorem generalizes the Brouwer Fixed Point Theorem in a straightforward way. It is thus not surprising that this result finds wide applicability in the theory of games and competitive equilibrium. We provide one such application in the next section. For the time being, let us offer the following example to illustrate the typical usage of Kakutani's Fixed Point Theorem in practice.

EXAMPLE 6

Let φ and ψ be continuous functions on $\mathbb{R}^2$ such that, for all $x \in \mathbb{R}^2$,

(i) $\varphi(\cdot, x_2)$ is quasiconcave, and $\psi(x_1, \cdot)$ is quasiconvex, and

(ii) $\varphi([0, 1], x_2) \cap \mathbb{R}_+ \neq \emptyset$ and $\psi(x_1, [0, 1]) \cap \mathbb{R}_- \neq \emptyset$.

[20] Sadly, Shizuo Kakutani (1911–2004) passed away while I was working on a revision of this section. He was a great mathematician who contributed to many subfields, from functional analysis to stochastic processes and topological groups. His fixed point theorem provided a stepping stone for game theory to attain the status it has today.

We claim that, under these conditions, there exists an $x \in [0,1]^2$ such that

$$\varphi(x) \geq 0 \geq \psi(x).$$

Let us first express this claim as a fixed point problem. Define the self-correspondence Γ_φ and Γ_ψ on $[0,1]$ by

$$\Gamma_\varphi(x_2) := \{a \in [0,1] : \varphi(a, x_2) \geq 0\} \quad \text{and}$$

$$\Gamma_\psi(x_1) := \{b \in [0,1] : \psi(x_1, b) \leq 0\}.$$

By (ii), these correspondences are well-defined. Finally, define the self-correspondence Γ on $[0,1]^2$ by $\Gamma(x) := \Gamma_\varphi(x_2) \times \Gamma_\psi(x_1)$. The upshot is that, for any $x \in [0,1]^2$, we have $x \in \Gamma(x)$ iff $\varphi(x) \geq 0 \geq \psi(x)$. This is the smart step in the proof, the rest is routine.

Continuity of φ and ψ is easily seen to entail that Γ has a closed graph. Moreover, Γ is convex-valued. Indeed, if $y, z \in \Gamma(x)$ for any given $x \in [0,1]^2$, then, by quasiconcavity of $\varphi(\cdot, x_2)$,

$$\varphi(\lambda y_1 + (1-\lambda)z_1, x_2) \geq \min\{\varphi(y_1, x_2), \varphi(z_1, x_2)\} \geq 0,$$

and similarly $\psi(x_1, \lambda y_2 + (1-\lambda)z_2) \leq 0$, that is, $\lambda y + (1-\lambda)z \in \Gamma(x)$, for all $0 \leq \lambda \leq 1$. Consequently, we may apply Kakutani's Fixed Point Theorem to find an $x \in [0,1]^2$ with $x \in \Gamma(x)$. By definition of Γ, we have $\varphi(x) \geq 0 \geq \psi(x)$. ☐

Some other applications of Kakutani's Fixed Point Theorem are given in the following set of exercises.

EXERCISE 57 [H] For any given $n \in \mathbb{N}$, let S be a nonempty, closed, bounded, and convex subset of $\mathbb{R}^{2n}$ such that if $(x, y) \in S$ for some $x, y \in \mathbb{R}^n$, then $(y, z) \in S$ for some $z \in \mathbb{R}^n$. Prove that $(x^*, x^*) \in S$ for some $x^* \in \mathbb{R}^n$.

EXERCISE 58 [H] Let X be a nonempty, closed, bounded, and convex subset of a Euclidean space. Prove: If $\varphi \in \mathbf{C}(X^2)$ is such that $\varphi(x, \cdot)$ is quasiconcave for any (fixed) $x \in X$, then there exists an $x^* \in X$ with

$$\varphi(x^*, x^*) = \max\{\varphi(x^*, y) : y \in X\}.$$

*EXERCISE 59 [H] (*The Minimax Theorem*) Let X and Y be nonempty, closed, bounded, and convex subsets of any two Euclidean spaces. Prove that if $\varphi \in \mathbb{R}^{X\times Y}$ is continuous, and if the sets $\{z \in X : \varphi(z, y) \geq \alpha\}$ and $\{w \in Y : \varphi(x, w) \leq \alpha\}$ are convex for each $(x, y, \alpha) \in X \times Y \times \mathbb{R}$, then

$$\max\{\{\min \varphi(x, y) : y \in Y\} : x \in X\} = \min\{\{\max \varphi(x, y) : x \in X\} : y \in Y\}.$$

Exercise 60^{H} Given a nonempty, closed, bounded, and convex subset X of a Euclidean space, let f be a continuous self-map on X, and let Γ be a convex-valued self-correspondence on X with a closed graph. Prove: If $f(\Gamma(x)) = \Gamma(f(x))$ for any $x \in X$, then there is a common fixed point of f and Γ, that is, $f(x^*) = x^* \in \Gamma(x^*)$ for some $x^* \in X$.

5.2* Michael's Selection Theorem

Our main objective now is to offer a proof of Kakutani's Fixed Point Theorem, which is based on the Brouwer Fixed Point Theorem.[21] While doing this, we will also encounter a number of important results about correspondences that are of independent interest. In particular, in this subsection we investigate when one may be able to "select" a continuous function from a given correspondence.

Let us begin by noting the following useful strengthening of Urysohn's Lemma (Section D.7).

Lemma 3
(Partition of Unity) *For any given positive integer k, let X be a metric space and $\{O_1, \ldots, O_k\}$ an open cover of X. Then, there exist continuous functions $\varphi_1, \ldots, \varphi_k$ in $[0, 1]^X$ such that*

$$\sum_{i=1}^{k} \varphi_i = 1 \quad \text{and} \quad \varphi_i|_{X\setminus O_i} = 0, \; i = 1, \ldots, k.$$

Proof
It is without loss of generality to assume that each O_i is nonempty. Moreover, if $O_i = X$ for some i, then we are done by setting $\varphi_i = 1$ and $\varphi_j = 0$

[21] The method of proof we adopt is due to Cellina (1969), and is also reported in Hildenbrand and Kirman (1988) and Border (1989). The original proof of Kakutani (1941) is also based on the Brouwer Fixed Point Theorem, but it is more combinatorial in nature. You may consult Chapter 7 of Klein (1973) or Chapter 9 of Smart (1974) for expositions of this alternative method of proof.

for all $j \neq i$. Then, assume that each O_i is a nonempty proper subset of X, and define $\vartheta_i \in \mathbb{R}^X$ by $\vartheta_i(x) := d(x, X\backslash O_i)$, $i = 1, \ldots, k$. By Example D.1. [3], each ϑ_i is continuous on X, and obviously, $\vartheta_i|_{X\backslash O_i} = 0$ for each i. Moreover, it is impossible that $\vartheta_1(x) = \cdots = \vartheta_k(x) = 0$ for some $x \in X$, for, in view of Exercise D.2, this would mean that x lies outside of each O_i, contradicting that $\{O_1, \ldots, O_k\}$ covers X. We may thus define $\varphi_j \in [0,1]^X$ by $\varphi_j := \vartheta_j / \sum^k \vartheta_i$, $j = 1, \ldots, k$, to obtain the desired set of functions. ■

Urysohn's Lemma is an immediate consequence of Lemma 3. Indeed, if A and B are nonempty disjoint closed sets in a metric space X, then $\{X\backslash A, X\backslash B\}$ is an open cover of X, so it follows from Lemma 3 that there exist $\psi, \varphi \in \mathbf{C}(X)$ with $\psi + \varphi = 1$ and $\psi|_{X\backslash A} = 0 = \varphi|_{X\backslash B}$. Thus $\varphi|_A = 0$ and $\varphi|_B = 1$, that is, φ satisfies the requirements of Urysohn's Lemma.[22]

Recall that for any given metric space T and positive integer n, $\mathbf{C}(T, \mathbb{R}^n)$ stands for the set of all continuous functions from T into $\mathbb{R}^n$. Moreover, we metrize this set by the sup-metric (Exercise D.45). Here is the main implication we wish to draw from Lemma 3.

The Approximate Selection Lemma

(Cellina) *For any given $n \in \mathbb{N}$, let X be a nonempty, closed, bounded, and convex subset of $\mathbb{R}^n$ and $\Psi : X \rightrightarrows \mathbb{R}^n$ a convex-valued and lower hemicontinuous correspondence. Then, for any $\varepsilon > 0$, there exists a $\varphi \in \mathbf{C}(X, \mathbb{R}^n)$ such that*

$$d_2(\varphi(x), \Psi(x)) < \varepsilon \quad \text{for each } x \in X.$$

Moreover, if Ψ is a self-correspondence, φ can be chosen as a self-map.

Proof

For any $y \in \Psi(X)$ and $\varepsilon > 0$, define

$$O(y) := \left\{x \in X : d_2(y, \Psi(x)) < \tfrac{\varepsilon}{2}\right\}.$$

22 If you check the statement of Urysohn's Lemma in the previous chapter, you see that here I assume $a = 0$ and $b = 1$. This is without loss of generality. I would use the function $(b - a)\varphi + a$ for the general case.

The lower hemicontinuity of Ψ ensures that this set is open. (Verify!) So $\{O(y) : y \in \mathbb{R}^n\}$ is an open cover of X, and by compactness of X (Theorem C.1), we can find finitely many $y^1, \ldots, y^k$ in $\Psi(X)$ such that $\{O(y^i) : i = 1, \ldots, k\}$ covers X. By Lemma 3, then, there exist continuous functions $\varphi_1, \ldots, \varphi_k$ in $[0,1]^X$ such that $\sum^k \varphi_i = 1$ and $\varphi_i(x) = 0$ whenever $x \in X \backslash O(y^i)$, $i = 1, \ldots, k$. We define $\varphi : X \to \mathbb{R}^n$ by $\varphi(x) := \sum^k \varphi_i(x) y^i$. This function is continuous because each φ_i is continuous. Now take an arbitrary $x \in X$, and let $I(x) := \{i : x \in O(y^i)\}$. Obviously, $\varphi(x) := \sum_{i \in I(x)} \varphi_i(x) y^i$ (and $\sum_{i \in I(x)} \varphi_i(x) = 1$). Moreover, for each $i \in I(x)$, we have $d_2(y^i, \Psi(x)) < \frac{\varepsilon}{2}$, so $d_2(y^i, z^i) < \varepsilon$ for some $z^i \in \Psi(x)$. Let $z := \sum_{i \in I(x)} \varphi_i(x) z^i$ and notice that $z \in \Psi(x)$ since $\Psi(x)$ is convex. But then,

$$
\begin{aligned}
d_2(\varphi(x), z) &= d_2\left(\sum_{i \in I(x)} \varphi_i(x) y^i, \sum_{i \in I(x)} \varphi_i(x) z^i\right) \\
&\leq \sum_{i \in I(x)} \varphi_i(x) d_2(y^i, z^i) \\
&< \varepsilon.
\end{aligned}
$$

(Why the first inequality?) Thus $d_2(\varphi(x), \Psi(x)) < \varepsilon$, and hence the first assertion of the lemma. The second assertion follows readily, because $\Psi(X) \subseteq X$ and convexity of X ensure that $\varphi(x) \in X$ for each $x \in X$. ∎

It is important to note that lower hemicontinuity is quite essential here. We cannot even replace it with upper hemicontinuity. For instance, the correspondence

$$
\Psi(t) := \begin{cases} \{0\}, & 0 \leq t < \frac{1}{2} \\ [0,1], & t = \frac{1}{2} \\ \{1\}, & \frac{1}{2} < t \leq 1 \end{cases}
$$

is convex-valued and upper hemicontinuous on $[0,1]$, while it is not lower hemicontinuous at $\frac{1}{2}$. It is easily checked that, for small enough $\varepsilon > 0$, no $f \in \mathbf{C}[0,1]$ satisfies $d_2(f(t), \Psi(t)) < \varepsilon$ for all $t \in [0,1]$.

The Approximate Selection Lemma identifies certain conditions on a correspondence Ψ that allows us to find a continuous function, the graph of which lies arbitrarily near to that of the correspondence. Of course, one way of finding such a function is to look for a continuous f the graph of which lies entirely in that of Ψ, for then we would have

$d_2(f(x), \Psi(x)) = 0$ for all x. Unfortunately, this is too much to ask from the postulates of the Approximate Selection Lemma. (Really?) As we show next, however, the situation is quite satisfactory for *closed-valued* correspondences.

> MICHAEL'S SELECTION THEOREM
> *For any given $n \in \mathbb{N}$, let X be a nonempty, closed, bounded, and convex subset of $\mathbb{R}^n$ and $\Psi : X \rightrightarrows \mathbb{R}^n$ be a convex- and closed-valued lower hemicontinuous correspondence. Then there exists an $f \in \mathbf{C}(X, \mathbb{R}^n)$ such that $f(x) \in \Psi(x)$ for each $x \in X$.*

PROOF

By the Approximate Selection Lemma, there exists an $f_1 \in \mathbf{C}(X, \mathbb{R}^n)$ with $d_2(f_1(x), \Psi(x)) < \frac{1}{2}$ for each $x \in X$. Define $\Psi_2 : X \rightrightarrows \mathbb{R}^n$ by

$$\Psi_2(x) := \Psi(x) \cap \left\{y \in \mathbb{R}^n : d_2(y, f_1(x)) < \tfrac{1}{2}\right\}.$$

(Why is Ψ_2 well-defined?) Ψ_2 is easily checked to be convex-valued and lower hemicontinuous. (Verify!) Apply the Approximate Selection Lemma again to get an $f_2 \in \mathbf{C}(X, \mathbb{R}^n)$ with $d_2(f_2(x), \Psi_2(x)) < \frac{1}{4}$ for each $x \in X$, and define

$$\Psi_3(x) := \Psi(x) \cap \left\{y \in \mathbb{R}^n : d_2(y, f_2(x)) < \tfrac{1}{4}\right\}, \qquad x \in X.$$

Proceeding by induction, we find a sequence (f_m) in $\mathbf{C}(X, \mathbb{R}^n)$ such that, for each $m = 2, 3, \ldots,$

$$d_2(f_m(x), \Psi_m(x)) < \tfrac{1}{2^m} \qquad \text{for all } x \in X \tag{21}$$

where

$$\Psi_m(x) := \Psi(x) \cap \left\{y \in \mathbb{R}^n : d_2(y, f_{m-1}(x)) < \tfrac{1}{2^{m-1}}\right\}, \qquad x \in X.$$

We claim that (f_m) is Cauchy in $\mathbf{C}(X, \mathbb{R}^n)$. Indeed, by construction, for each $x \in X$ and $m \geq 2$, there exists a $y^m \in \Psi_m(x)$ such that

$$d_2(f_m(x), f_{m-1}(x)) \leq d_2(f_m(x), y^m) + d_2(y^m, f_{m-1}(x)) < \tfrac{1}{2^m} + \tfrac{1}{2^{m-1}} = \tfrac{3}{2^m}.$$

Therefore, for any positive integers $k > l+1$ and any $x \in X$,

$$\begin{aligned} d_2(f_k(x), f_l(x)) &\leq d_2(f_k(x), f_{k-1}(x)) + \cdots + d_2(f_{l+1}(x), f_l(x)) \\ &< \tfrac{3}{2^k} + \cdots + \tfrac{3}{2^{l+1}} \\ &< 3 \sum_{i=l+1}^{\infty} \tfrac{1}{2^i}, \end{aligned}$$

so we have $\sup\{d_2(f_k(x), f_l(x)) : x \in X\} \to 0$ as $l \to \infty$. Hence, since $\mathbf{C}(X, \mathbb{R}^n)$ is complete (Exercise D.45), there exists an $f \in \mathbf{C}(X, \mathbb{R}^n)$ such that $f_m \to f$. Now fix an arbitrary $x \in X$, and note that, thanks to (21), for each $m \in \mathbb{N}$ there is a $y^m \in \Psi_m(x) \subseteq \Psi(x)$ such that $d_2(f_m(x), y^m) < \frac{1}{2^m}$. Since $f_m(x) \to f(x)$, therefore, we must have $y^m \to f(x)$. (Why?) Since $\Psi(x)$ is a closed set, then, we conclude: $f(x) \in \Psi(x)$. ■

When combined with the Brouwer Fixed Point Theorem, Michael's Selection Theorem yields the following result, which shows that in Kakutani's Fixed Point Theorem, the closed graph property can be replaced with closed-valuedness and lower hemicontinuity.

Corollary 1
(Michael) *For any given $n \in \mathbb{N}$, let X be a nonempty, closed, bounded, and convex subset of $\mathbb{R}^n$. If Ψ is a convex- and closed-valued lower hemicontinuous self-correspondence on X, then it has a fixed point.*

This does cover ground that is missed by Kakutani's Fixed Point Theorem. For instance, the convex- and closed-valued lower hemicontinuous self-correspondence Γ defined on $[0,1]$ by $\Gamma(0) := \{0\}$ and $\Gamma(t) := [0,1]$ for all $0 < t \leq 1$, is not upper hemicontinuous.

Warning. Convex-valuedness is essential for Michael's Selection Theorem and Corollary 1. The self-correspondence Γ defined on $[0,1]$ as $\Gamma(0) := \{1\}$, $\Gamma(1) := \{0\}$ and $\Gamma(t) := \{0,1\}$ for all $0 < t < 1$, is closed-valued and lower hemicontinuous, but it does not admit a continuous selection.

Exercise 61 Define $g : [-1,1] \to [0,1]$ and $\Psi : [-1,1] \rightrightarrows [0,1]$ by

$$g(x) := \max\{xy : 0 \leq y \leq 1\} \quad \text{and} \quad \Psi(x) := \{y \in [0,1] : xy = g(x)\}.$$

(a) Show that Ψ is not lower hemicontinuous at 0.
(b) Show that Ψ does not have a continuous selection (that is, there is no $f \in \mathbf{C}[-1,1]$ with $f(x) \in \Psi(x)$ for all $-1 \leq x \leq 1$).
(c) For any $0 < \varepsilon < 1$, define $\Psi_\varepsilon : [-1,1] \rightrightarrows [0,1]$ by

$$\Psi_\varepsilon(x) := \{y \in [0,1] : xy \geq g(x) - \varepsilon\}.$$

Draw $Gr(\Psi_\varepsilon)$. Is Ψ_ε upper hemicontinuous? Is it lower hemicontinuous? Does Ψ_ε have a continuous selection?

EXERCISE 62 For any given $n \in \mathbb{N}$, let X be a nonempty, closed, bounded, and convex subset of a Euclidean space. Let $\Psi : X \rightrightarrows \mathbb{R}^n$ be a convex- and closed-valued lower hemicontinuous correspondence. Show that there exists an $\mathcal{F} \subseteq \mathbf{C}(X, \mathbb{R}^n)$ such that $\Psi(x) = \{F(x) : F \in \mathcal{F}\}$ for all $x \in X$.

EXERCISE 63 (*A Sandwich Theorem*) Let X be a nonempty, closed, bounded, and convex subset of a Euclidean space, and let φ and ϕ be two real functions on X with $\varphi \leq \phi$. Show that if φ is upper semicontinuous and ϕ is lower semicontinuous, then there exists a $\vartheta \in \mathbf{C}(X)$ such that $\varphi \leq \vartheta \leq \phi$.

EXERCISE 64 Let X be a nonempty, closed, bounded, and convex subset of a Euclidean space and Y a nonempty closed set in X. Show that if $\Psi : X \rightrightarrows \mathbb{R}^n$ is a convex- and closed-valued lower hemicontinuous correspondence, and $f \in \mathbf{C}(Y, \mathbb{R}^n)$ satisfies $f(y) \in \Psi(y)$ for all $y \in Y$, then there exists an $F \in \mathbf{C}(X, \mathbb{R}^n)$ such that $f|_Y = F|_Y$ and $F(x) \in \Psi(x)$ for all $x \in X$.

*EXERCISE 65^H (*Browder's Selection Theorem*) Let X be a compact metric space and $n \in \mathbb{N}$. Prove:
(a) If $\Psi : X \rightrightarrows \mathbb{R}^n$ is a convex-valued correspondence such that $\{x \in X : y \in \Psi(x)\}$ is open for every (fixed) $y \in \mathbb{R}^n$, then there exists an $f \in \mathbf{C}(X, \mathbb{R}^n)$ with $f(x) \in \Psi(x)$ for all $x \in X$.
(b) If S is a nonempty, closed, bounded, and convex subset of a Euclidean space and Γ is a convex-valued self-correspondence on S such that $\{x \in X : y \in \Gamma(x)\}$ is open for every (fixed) $y \in S$, then Γ has a fixed point.

5.3* Proof of Kakutani's Fixed Point Theorem

Equipped with Michael's Selection Theorem, we are now on the brink of proving Kakutani's Fixed Point Theorem. The following is the final step of our preparation.

> LEMMA 4
> (Cellina) *Let X be a nonempty, closed, bounded, and convex subset of a Euclidean space and Γ a convex-valued self-correspondence on X that has a closed graph.*[23] *Then, for any $\varepsilon > 0$, there exists a convex-valued lower hemicontinuous self-correspondence Ψ on X such that $Gr(\Psi) \subseteq N_{\varepsilon,X\times X}(Gr(\Gamma))$.*[24]

To prove this result, we need to borrow the following theorem from convex analysis. Although its proof is quite simple, it is based on a linear algebraic argument, so we relegate its proof to the early part of the next chapter (Section F.1.5). In the mean time, for any nonempty subset S of a Euclidean space $\mathbb{R}^n$, let us agree to write $co(S)$ for the set of all $x \in \mathbb{R}^n$ such that $x = \sum^k \lambda_i x^i$ for some $k \in \mathbb{N}$, $x^1, \ldots, x^k \in S$ and $\lambda_1, \ldots, \lambda_k \geq 0$ with $\sum^k \lambda_i = 1$. You might recall that this set is called the **convex hull** of S; it is easily checked to be the smallest convex subset of $\mathbb{R}^n$ that contains S. (See Section G.1.1 for a general discussion.)

> CARATHÉODORY'S THEOREM
> *For any given $n \in \mathbb{N}$ and a nonempty subset S of $\mathbb{R}^n$, if $x \in co(S)$, then there exist $z^1, \ldots, z^{n+1} \in S$ and $\lambda_1, \ldots, \lambda_{n+1} \in [0, 1]$ such that $\sum^{n+1} \lambda_i = 1$ and $x = \sum^{n+1} \lambda_i z^i$.*

In words, for any $\emptyset \neq S \subseteq \mathbb{R}^n$, every point in $co(S)$ can be expressed as a convex combination of *at most* $n + 1$ elements of S. It is easy to prove Lemma 4 by using this fact. We give the proof in the form of an exercise.

23 The closed graph property can be omitted in this statement, but its presence simplifies things a bit.

24 *Reminder.* $N_{\varepsilon,X\times X}(Gr(\Gamma)) := \cup\{N_{\varepsilon,X\times X}(w) : w \in Gr(\Gamma)\}$.

EXERCISE 66 [H] Assume the hypotheses of Lemma 4. For any $m \in \mathbb{N}$, define the self-correspondence Γ_m on X by

$$\Gamma_m(x) := co\left(\Gamma(N_{\frac{1}{m},X}(x))\right).$$

(a) Show that Γ_m is lower hemicontinuous for each $m \in \mathbb{N}$.
(b) Use Carathéodory's Theorem to show that $Gr(\Gamma_m) \subseteq N_{\varepsilon,X\times X}(Gr(\Gamma))$ for m large enough.

Kakutani's Fixed Point Theorem is now at our fingertips. The basic idea is the following. Given a correspondence Γ that satisfies the requirements of this theorem, we can use Lemma 4 to find a lower hemicontinuous correspondence Ψ the graph of which approximates $Gr(\Gamma)$. But, by the Approximate Selection Lemma, we can actually choose a continuous function the graph of which is very close to $Gr(\Psi)$. Thus, it is possible to find a continuous function the graph of which is as close to that of Γ as we want. Roughly speaking, this would allow us to find a sequence of continuous functions the graphs of which converge somewhere in $Gr(\Gamma)$. But then, following the fixed points of these functions (the existence of which we owe to the Brouwer Fixed Point Theorem), we may reach a fixed point of Γ. It remains to formalize this idea.

PROOF OF KAKUTANI'S FIXED POINT THEOREM

Let Γ be a convex-valued self-correspondence on X that has a closed graph. By Lemma 4, for each $m \in \mathbb{N}$ there exists a convex-valued lower hemicontinuous correspondence $\Psi_m : X \rightrightarrows X$ such that $Gr(\Psi_m) \subseteq N_{\frac{1}{m},X\times X}(Gr(\Gamma))$. In turn, by the Approximate Selection Lemma there exists a continuous self-map φ_m on X with $d_2(\varphi_m(x), \Psi_m(x)) < \frac{1}{m}$, $m = 1, 2, \ldots$. Then $d_2((x, \varphi_m(x)), Gr(\Psi_m(x))) \leq \frac{1}{m}$, and it follows that

$$d_2((x, \varphi_m(x)), Gr(\Gamma)) \leq \tfrac{2}{m} \qquad \text{for all } x \in X, \tag{22}$$

$m = 1, 2, \ldots$. By the Brouwer Fixed Point Theorem, each φ_m must have a fixed point, say x^m. Besides, since (x^m) lies in a compact set, it must have a convergent subsequence, say (x^{m_k}). Then, if $x := \lim x^{m_k}$, we have $\varphi_{m_k}(x^{m_k}) = x^{m_k} \to x$, so, since $d_2(\cdot, Gr(\Gamma))$ is a continuous map on $X \times X$, (22) implies $d_2((x, x), Gr(\Gamma)) = 0$. Since $Gr(\Gamma)$ is closed, it follows that $(x, x) \in Gr(\Gamma)$, that is, $x \in \Gamma(x)$. ■

5.4* Contractive Correspondences

Kakutani's Fixed Point Theorem is a straightforward reflection of the Brouwer Fixed Point Theorem, one that applies to self-correspondences. A natural question is whether we can carry out a similar generalization of the Banach Fixed Point Theorem. The answer to this question was given by Sam Nadler in 1969. It constitutes the final result of this section.

Definition

Let X be a metric space and Γ a compact-valued self-correspondence on X. Γ is called **contractive** if there exists a real number $0 < K < 1$ such that

$$d_{\mathrm{H}}(\Gamma(x), \Gamma(y)) \leq Kd(x, y) \quad \text{for all } x, y \in X, \tag{23}$$

where d_{H} is the Hausdorff metric on the class of all nonempty compact subsets of X.

Nadler's Contractive Correspondence Theorem

Let X be a complete metric space and Γ a compact-valued self-correspondence on X. If Γ is contractive, then there exists an $x \in X$ such that $x \in \Gamma(x)$.[25]

We will offer two proofs of this theorem. Being based on the method of successive approximations, the first one parallels closely the proof of the Banach Fixed Point Theorem. We thus provide only a sketch.

Exercise 67 Let X and Γ be as in the statement of Nadler's Contractive Correspondence Theorem.

(a) Show that, for any $x \in X$ and $y \in \Gamma(x)$, there exists a $z \in \Gamma(y)$ such that $d(z, y) \leq d_{\mathrm{H}}(\Gamma(x), \Gamma(y))$.

(b) Use part (a) to construct a sequence $(x^0, x^1, \ldots)$ such that $x^{m+1} \in \Gamma(x^m)$ and $d(x^{m+1}, x^m) \leq d_{\mathrm{H}}(\Gamma(x^{m-1}), \Gamma(x^m))$ for all $m \in \mathbb{N}$. Show that $d(x^{m+1}, x^m) \leq K^m d(x^0, x^1)$ for each m, where $0 < K < 1$ satisfies (23).

[25] Nadler's original result is actually proved for correspondences whose values are closed and bounded, and thus it is more general than the theorem we report here.

(c) Using part (b) and proceeding as in the proof of the Banach Fixed Point Theorem, show that (x^m) is a Cauchy sequence in X. Conclude that $x^m \to x$ for some $x \in X$.

(d) Show that $d_H(\Gamma(x^m), \Gamma(x)) \to 0$. Use this, and the fact that $x^{m+1} \in \Gamma(x^m)$ for each m, to conclude that $x \in \Gamma(x)$.

Our second proof is based on the fact that one can always select a self-map from a compact-valued contractive correspondence that would satisfy the conditions of Caristi's Fixed Point Theorem (Section D.5.1). The argument makes direct use of the Axiom of Choice, and is due to Jachymski (1998).[26]

ANOTHER PROOF FOR NADLER'S CONTRACTIVE CORRESPONDENCE THEOREM
Let $0 < K < 1$ satisfy (23). Take any $K < k < 1$, and define

$$S_x := \{y \in \Gamma(x) : kd(x,y) \leq d(x, \Gamma(x))\} \qquad \text{for each } x \in X.$$

Since $k < 1$, S_x is obviously nonempty for each $x \in X$. (Why?) Choose an arbitrary s_x from S_x for each $x \in X$—this is where we use the Axiom of Choice. Now define the self-map Φ on X by $\Phi(x) := s_x$. By definition, $\Phi(x) \in \Gamma(x)$ and $kd(x, \Phi(x)) \leq d(x, \Gamma(x))$ for each $x \in X$. Moreover, by definition of the Hausdorff metric,

$$d(\Phi(x), \Gamma(\Phi(x))) \leq d_H(\Gamma(x), \Gamma(\Phi(x))) \leq Kd(x, \Phi(x)) \qquad \text{for all } x \in X.$$

(Why?) Therefore, for each $x \in X$, we have

$$\begin{aligned} d(x, \Phi(x)) &= \tfrac{1}{k-K}\left(kd(x, \Phi(x)) - Kd(x, \Phi(x))\right) \\ &\leq \tfrac{1}{k-K}\left(d(x, \Gamma(x)) - d(\Phi(x), \Gamma(\Phi(x)))\right). \end{aligned}$$

Defining $\varphi \in \mathbb{R}^X$ by $\varphi(x) := \frac{1}{k-K} d(x, \Gamma(x))$, we then get

$$d(x, \Phi(x)) \leq \varphi(x) - \varphi(\Phi(x)) \qquad \text{for all } x \in X.$$

Since φ is bounded below and continuous (Example D.1.[3]), we may apply Caristi's Fixed Point Theorem to conclude that Φ has a fixed point, say x. Then, $x = \Phi(x) \in \Gamma(x)$, and we are done. ■

[26] *Quiz.* Did we use the Axiom of Choice in proving Michael's Selection Theorem?

6 Application: The Nash Equilibrium

As noted earlier, Kakutani's Fixed Point Theorem is widely used in establishing the existence of equilibrium in economic models. In this section we consider an important application of this sort. We first outline a very concise introduction to the theory of strategic games, and then discuss the problem of existence of Nash equilibrium, the principal solution concept in game theory.

6.1 Strategic Games

Throughout this section, m stands for an integer, with $m \geq 2$. A strategic game is a means of modeling the strategic interaction of a group of m individuals. The first ingredient of the basic model is the set $\{1, \ldots, m\}$, which we interpret as the set of individuals (players). The second ingredient is a nonempty set X_i, which is thought of as containing all actions that are available to player $i = 1, \ldots, m$. The outcome of the game is obtained when all players choose an action; hence an **outcome** is any m-vector $(x_1, \ldots, x_m) \in X_1 \times \cdots \times X_m =: X$. While we refer to X_i as the **action space** of player i, X is called the **outcome space** of the game. The distinguishing feature of a strategic game is that the payoffs of a player depend not only on her own actions but also on the actions taken by other players. We account for this interdependence by defining the **payoff function** of player i, denoted π_i, as any real function that is defined on the entire outcome space: $\pi_i : X \to \mathbb{R}$. If $\pi_1(x) > \pi_1(y)$, for instance, we understand that player 1 is better off in the situation that obtains when (s)he has taken action x_1, player 2 has taken action x_2, and so on, than in the outcome $y = (y_1, \ldots, y_m)$. (Put differently, each player i has a complete preference relation on X that is represented by the utility function π_i.) Formally speaking, then, we define an **m-person strategic game** as any set

$$\mathcal{G} := \{(X_1, \pi_1), \ldots, (X_m, \pi_m)\}$$

where X_i is an arbitrary nonempty set and $\pi_i \in \mathbb{R}^X$. (It is customary to denote $\{(X_1, \pi_1), \ldots, (X_m, \pi_m)\}$ by $\{(X_i, \pi_i)_{i=1,\ldots,m}\}$ in game theory; we also adopt this notation below.) We say that a strategic game $\mathcal{G}$ is **finite** if each action space X_i is finite.

By far the most famous example of a strategic game is the finite game called **Prisoner's Dilemma**. This game is described by means of a bimatrix like

	α	β
α	1,1	6,0
β	0,6	5,5

where, taking the row (column) player as player 1 (player 2), we have $X_1 = X_2 = \{\alpha, \beta\}$, $\pi_1(\alpha, \alpha) = 1$, $\pi_2(\alpha, \beta) := 0$, and so on.[27] The importance of this game stems from the fact that both of the players have very strong incentives to play the game noncooperatively by choosing α, so the decentralized behavior dictates the outcome (α, α). However, at the cooperative outcome (β, β), both players are strictly better off. Thus, Prisoner's Dilemma is a beautifully simple illustration of the fact that decentralized individualistic behavior need not lead to a socially optimal outcome in strategic situations.

Considering the off-chance that you are not familiar with strategic games, let us look at a few standard (infinite) strategic games.

EXAMPLE 7

[1] (*Aggregative Games*) Let $\emptyset \neq X_i \subseteq \mathbb{R}$ for each i, and consider the strategic game $\mathcal{G} := \{(X_i, \pi_i)_{i=1,\ldots,m}\}$, where

$$\pi_i(x) = H_i\left(x_i, \sum_{j=1}^{m} x_j\right),$$

with $H_i : X_i \times \left\{\sum^m x_j : x \in X\right\} \to \mathbb{R}$ being an arbitrary function, $i = 1, \ldots, m$. This sort of a game is called an **aggregative game.**[28] As we shall see below, many interesting economic games are aggregative. (See Corchón (1996) for a detailed analysis of aggregative games.)

[27] The story of *Prisoner's Dilemma* is the following. Two suspects are arrested and put in different cells before the trial. The district attorney (who is pretty sure that both of the suspects are guilty but lacks evidence) offers them the following deal: If both of them confess to their crimes (i.e., play α), then each will be sentenced to a moderate amount of prison time. If one confesses and the other does not, then the "rat" goes free for his cooperation with the authorities and the nonconfessor is sentenced to a long prison time. Finally, if neither of them confesses, then both suspects spend a short while in prison.

[28] The terminology is due to Dubey, Mas-Colell, and Shubik (1980).

[2] (*Cournot Duopoly*) Let $a > 0$, and consider the strategic 2-person aggregative game $\mathcal{G} := \{([0,a], \pi_i)_{i=1,2}\}$, where for each i, $H_i : [0,a] \times [0,2a] \to \mathbb{R}$ is defined by $H_i(u,v) := uf(v) - c(u)$ for some decreasing $f : [0,2a] \to \mathbb{R}_+$ and increasing $c : [0,a] \to \mathbb{R}_+$. Thus, in this game, the payoff (profit) function of player (firm) i is given as

$$\pi_i(x_1, x_2) := x_i f(x_1 + x_2) - c(x_i), \qquad 0 \le x_1, x_2 \le a.$$

The standard interpretation is that f is the market demand function, c is the cost function (assumed to be identical across two firms), and a is the capacity constraint (also identical across the firms). Firms in this model decide how much to supply to the market. It is clear that this situation is strategic in that the profits of either firm cannot be determined independently of the production decisions of the other firm. After all, the market price of the good produced by these firms is determined (through f) as a result of the *total* production in the industry.

[3] (*Tragedy of the Commons*) Another interesting aggregative game is $\mathcal{G} := \{(\mathbb{R}_+, \pi_i)_{i=1,\ldots,m}\}$, where for each i, $H_i : \mathbb{R}^2_+ \to \mathbb{R}$ is defined by $H_i(a,b) := 0$ if $b = 0$, and $H_i(a,b) := \frac{a}{b}pf(b) - wa$ if $b > 0$, for some $p, w > 0$ and strictly increasing $f : \mathbb{R}_+ \to \mathbb{R}_+$ with $f(0) = 0$. In this game the payoff function of player i is defined on $\mathbb{R}^m_+$ by

$$\pi_i(x) := \frac{x_i}{\sum^m x_j} pf\left(\sum_{j=1}^m x_j\right) - wx_i.$$

This game models a scenario in which a population that consists of m individuals, each of whom has access to a common pool resource, interacts through each individual choosing the level of his or her extraction effort. The total output, the unit price of which is p, is obtained from the aggregate extraction effort via the production function f. (It is then natural to assume in this setting that $f(0) = 0$, so without extractive effort there is no production.) There is an opportunity cost $w > 0$ per unit extractive effort, and each member of the population receives a share of the total product that is proportional to his or her share of aggregate extractive effort.

[4] (*Linear Bertrand Duopoly*) Let $a > 0$, and consider the strategic game $\mathcal{G} := \{([0,a], \pi_i)_{i=1,2}\}$, where

$$\pi_1(p_1, p_2) := p_1 x_1(p_1, p_2) - c x_1(p_1, p_2),$$

with $x_1 \in \mathbb{R}^{[0,a]}$ being defined as follows: $x_1(p_1, p_2) := a - p_1$ if $p_1 < p_2$, and $x_1(p_1, p_2) := 0$ if $p_1 > p_2$, while $x_1(p_1, p_2) := \frac{1}{2}a - p_1$ when $p_1 = p_2$. (π_2 is defined analogously.) This game, which is not aggregative, models a situation in which two firms engage in price competition. Here a stands for the maximum possible price level in the market, c is the constant marginal cost, and $x_i(p_1, p_2)$ denotes the output sold by firm i at the price profile (p_1, p_2). The payoff functions reflect the hypotheses that consumers always buy the cheaper good (presumably because there is no qualitative difference between the products of the firms) and that the firms share the market equally in case of a tie. □

6.2 The Nash Equilibrium

To define an equilibrium notion for strategic games, we need to identify those outcomes such that, once reached, there is no tendency for them to be altered. Thus, it appears natural to define an outcome as an equilibrium if there is no incentive for any individual to change her action, *given the actions of others at this outcome*. Before we formalize this idea, let us agree on some notation.

NOTATION. For any strategic game $\mathcal{G} := \{(X_i, \pi_i)_{i=1,\ldots,m}\}$, we let

$$X_{-i} := \{(\omega_1, \ldots, \omega_{m-1}) : \omega_j \in X_j \text{ for } j < i \text{ and } \omega_{j-1} \in X_j \text{ for } j > i\}$$

for all $i = 1, \ldots, m$. (In words, X_{-i} is the collection of all action profiles of all the players but the player i.) For any i, a generic element of X_{-i} is denoted as x_{-i}, and by (a, x_{-i}) we denote the outcome $x \in X$ where the action taken by player i is a, and the action taken by player $j \neq i$ is $\begin{cases} \omega_j, & \text{if } j < i \\ \omega_{j-1}, & \text{if } j > i \end{cases}$, where $x_{-i} = (\omega_1, \ldots, \omega_{m-1})$.[29]

[29] For instance, when $m = 3$, $X_{-1} := X_2 \times X_3$, $X_{-2} := X_1 \times X_3$, and $X_{-3} := X_1 \times X_2$, and if $x_{-2} = (b, c)$—this means the action of player 1 is b and that of player 3 is c—then $(a, x_{-2}) = (b, a, c)$.

DEFINITION

Let $\mathcal{G} := \{(X_i, \pi_i)_{i=1,\ldots,m}\}$ be a strategic game. We say that an outcome $x^* \in X$ is a **Nash equilibrium** if

$$x_i^* \in \arg\max \left\{\pi_i(x_i, x_{-i}^*) : x_i \in X_i\right\} \qquad \text{for all } i = 1, \ldots, m.$$

A Nash equilibrium x^* is said to be **symmetric** if $x_1^* = \cdots = x_m^*$. We denote the set of all Nash and symmetric Nash equilibria of a game $\mathcal{G}$ by $NE(\mathcal{G})$ and $NE_{\text{sym}}(\mathcal{G})$, respectively.

In what follows we take the appeal of the Nash equilibrium at face value. Given the immense popularity of this concept, this is justifiable for our present purposes. To get an idea about the nature of hypotheses underlying the strategic behavior presumed by the Nash equilibrium, you should consult a proper textbook on game theory.[30]

EXAMPLE 8

[1] The unique Nash equilibrium of Prisoner's Dilemma is easily found to be (α, α).

[2] The Cournot duopoly game considered in Example 7.[2] may or may not have a Nash equilibrium, depending on the nature of the functions f and c.

[3] To find the Nash equilibria of the linear Bertrand game of Example 7.[4], observe that, if the price profile (p_1^*, p_2^*) is a Nash equilibrium, then $p_1^*, p_2^* \geq c$, because negative profits can always be avoided by charging exactly c. But $p_1^* > p_2^* > c$ cannot hold, for in this case firm 1 would be making zero profits, and thus it would be better for it to charge, say, p_2^*. Moreover, $p_1^* = p_2^* > c$ is also impossible, because in this case either firm can unilaterally increase its profits by undercutting the other firm; but this contradicts that (p_1^*, p_2^*) is a Nash equilibrium. By symmetry, $p_2^* \geq p_1^* > c$ is also impossible, and hence we conclude that at least one firm must charge precisely its unit cost c in equilibrium. But we can't have $p_1^* > p_2^* = c$ either, for in this case firm 2 would not be responding in the best possible way; it could increase its profits by charging, say, $\frac{1}{2}p_1^* + \frac{1}{2}c$.

[30] My personal favorite in this regard is, by a large margin, Osborne and Rubinstein (1994).

By symmetry, then, the only candidate for equilibrium is $(p_1^*, p_2^*) = (c, c)$, and this is indeed an equilibrium, as you can easily verify.[31] □

EXERCISE 68 Consider the strategic game considered in Example 7.[3], and assume that f is bounded and twice differentiable with $f' > 0$, $f'' < 0$ and $f'(0) > \frac{w}{p}$. (Interpretation?) Prove:

(a) If $\frac{d}{dt}\left(\frac{f(t)}{t}\right) < 0$ holds, then there exists a unique Nash equilibrium of this game.

(b) There is *overexploitation* in equilibrium in the sense that the maximization of the total industry profit occurs at a lower level of extraction than that realized in equilibrium. (This is the reason why this game is often referred to as the *tragedy of commons*.)

DEFINITION

If each X_i is a nonempty compact subset of a Euclidean space, then we say that the strategic game $\mathcal{G} := \{(X_i, \pi_i)_{i=1,\ldots,m}\}$ is a **compact Euclidean game**. If, in addition, $\pi_i \in \mathbf{C}(X)$ for each $i = 1, \ldots, m$, we say that $\mathcal{G}$ is a **continuous and compact Euclidean game**. If, instead, each X_i is convex and compact, and each $\pi_i(\cdot, x_{-i})$ is quasiconcave for any given $x_{-i} \in X_{-i}$, then $\mathcal{G}$ is called a **convex and compact Euclidean game**. Finally, a compact Euclidean game which is both convex and continuous is called a **regular Euclidean game**.

The following theorem, which is based on Kakutani's Fixed Point Theorem, is of fundamental importance in game theory. It is one of the main reasons why the notion of Nash equilibrium is such a widely used solution concept.[32]

NASH'S EXISTENCE THEOREM

If $\mathcal{G} := \{(X_i, \pi_i)_{i=1,\ldots,m}\}$ is a regular Euclidean game, then $NE(\mathcal{G}) \neq \emptyset$.

[31] This is a surprising result, since it envisages that all firms operate with *zero* profits in the equilibrium. In fact, the equilibrium outcome here is nothing but the competitive equilibrium outcome.

[32] While Kakutani's Fixed Point Theorem makes it very easy to prove this result, there are also proofs that utilize only the Brouwer Fixed Point Theorem. See, for instance, Geanakoplos (2003), and Becker and Chakrabarti (2005).

Proof

Take any regular Euclidean game $\mathcal{G} := \{(X_i, \pi_i)_{i=1,\ldots,m}\}$. For each $i = 1, \ldots, m$, define the correspondences $\mathbf{b}_i : X_{-i} \rightrightarrows X_i$ and $\mathbf{b} : X \rightrightarrows X$ by

$$\mathbf{b}_i(x_{-i}) := \arg\max \left\{\pi_i(x_i, x_{-i}) : x_i \in X_i\right\} \quad \text{and}$$
$$\mathbf{b}(x) := \mathbf{b}_1(x_{-1}) \times \cdots \times \mathbf{b}_m(x_{-m}),$$

respectively. ($\mathbf{b}_i$ is called the **best response correspondence** of i.) By Weierstrass' Theorem, $\mathbf{b}$ is well-defined. Notice that if $x \in \mathbf{b}(x)$, then $x_i \in \mathbf{b}_i(x_{-i})$ for all i, and hence $x \in NE(\mathcal{G})$. Consequently, if we can show that Kakutani's Fixed Point Theorem applies to $\mathbf{b}$, the proof will be complete.

X is easily checked to be compact and convex since each X_i has these two properties. (While convexity is routine, compactness follows from Theorem C.4.) To see that $\mathbf{b}$ is convex-valued, fix an arbitrary $x \in X$ and $0 \leq \lambda \leq 1$, and notice that, for any $y, z \in \mathbf{b}(x)$, we have $\pi_i(y_i, x_{-i}) = \pi_i(z_i, x_{-i})$. Thus, by using the quasiconcavity of π_i on X_i, we find

$$\pi_i(\lambda y_i + (1-\lambda)z_i, x_{-i}) \geq \pi_i(y_i, x_{-i}) \geq \pi_i(w_i, x_{-i})$$

for any $w_i \in X_i$, that is, $\lambda y_i + (1-\lambda)z_i \in \mathbf{b}_i(x_{-i})$. Since this holds for each i, we may conclude that $\lambda y + (1-\lambda)z \in \mathbf{b}(x)$. Conclusion: $\mathbf{b}$ is convex-valued. It remains to check that $\mathbf{b}$ has a closed graph. But this follows readily from the Maximum Theorem. (True?) ■

This is a typical way Kakutani's Fixed Point Theorem is used in applications. (Go back and solve Exercise 59, if you have not done so already. That problem will be a walk in the park for you now.) Other illustrations are provided in the following set of exercises.

Exercise 69 A strategic game $\mathcal{G} := \{(X_i, \pi_i)_{i=1,\ldots,m}\}$ is called **symmetric** if $X_i = X_j$ and $\pi_i(x) = \pi_j(x')$ for all $i, j = 1, \ldots, m$ and all $x, x' \in X$ such that x' is obtained from x by exchanging x_i and x_j. Show that if $\mathcal{G}$ is a symmetric regular Euclidean game, then $NE_{\text{sym}}(\mathcal{G}) \neq \emptyset$.

Exercise 70 (Bernheim) Let $\mathcal{G} := \{(X_i, \pi_i)_{i=1,\ldots,m}\}$ be a strategic game, and define $\mathbf{b}$ as in the proof of Nash's Existence Theorem. A set $S \subseteq X$ is called the **point-rationalizable** set of $\mathcal{G}$, denoted by $R(\mathcal{G})$, if it is the largest subset of X such that $S = \mathbf{b}(S)$ and $S = S_1 \times \cdots \times S_m$ for some $S_i \subseteq X_i$, $i = 1, \ldots, m$. (Interpretation?)

(a) Show that $NE(\mathcal{G}) \subseteq R(\mathcal{G})$, and give an example to show that $R(\mathcal{G}) \subseteq NE(\mathcal{G})$ need not be true.

*(b) Show that if $\mathcal{G}$ is a continuous and compact Euclidean game, then $R(\mathcal{G})$ is a nonempty compact set.

The following exercise shows that, under fairly general assumptions, the set of all equilibria of a strategic game responds marginally to small perturbations of the associated payoff functions.

EXERCISE 71 Let Θ be a compact metric space, and assume that, for any $\theta \in \Theta$,

$$\mathcal{G}(\theta) := \{(X_i, \pi_i(\cdot, \theta))_{i=1,\ldots,m}\}$$

is a convex and compact Euclidean game. Prove: If $\pi_i \in \mathbf{C}(X \times \Theta)$, then $\Gamma : \Theta \rightrightarrows X$, defined by $\Gamma(\theta) := NE(\mathcal{G}(\theta))$, is an upper hemicontinuous correspondence. (Don't forget to verify that Γ is well-defined.)

EXERCISE 72 (*Generalized Games*) Let $(n_1, \ldots, n_m) \in \mathbb{N}^m$ and $n := \sum^m n_i$. A **generalized game** is a list

$$\mathcal{G} := \{(X, n_i, \pi_i)_{i=1,\ldots,m}\}$$

where X is a nonempty subset of $\mathbb{R}^n$ and $\pi_i \in \mathbb{R}^X$, $i = 1, \ldots, m$. Such games are able to model situations in which a given player cannot choose an action independently of the rest of the players.[33] We define $S_i(\mathcal{G}) := \{x_{-i} \in \mathbb{R}^{n-n_i} : (x_i, x_{-i}) \in X \text{ for some } x_i \in \mathbb{R}^{n_i}\}$, and let

$$\Theta_i(x_{-i}) := \{x_i \in \mathbb{R}^{n_i} : (x_i, x_{-i}) \in X\}$$

for each $x_{-i} \in S_i(\mathcal{G})$. In turn, we say that $x^* \in X$ is a **Nash equilibrium** of $\mathcal{G}$ if

$$\pi_i(x^*) \geq \pi_i(x_i, x^*_{-i}) \qquad \text{for all } x_i \in \Theta_i(x^*_{-i}),\ i = 1, \ldots, m.$$

[33] Consider the tragedy of the commons game, and interpret each x_i as the amount of oil pumped out of a common oil field. Clearly, the model we summarized in Example 7.[3] doesn't square well with this interpretation. For, given that the total extraction cannot possibly exceed the oil reserves in the common field, it is not reasonable in this case to allow for x_is to be chosen independently. This sort of a scenario is better modeled as a generalized game.

Prove: If X is convex and compact, and if π_i is a continuous real function on X that is quasiconcave on $\Theta_i(x_{-i})$ for each $x_{-i} \in S_i(\mathcal{G})$, $i = 1, \ldots, m$, then $\mathcal{G}$ has a Nash equilibrium.[34]

Nash's Existence Theorem utilizes both the metric and *linear* structures of a Euclidean space. The following exercise provides an existence result that makes use of the metric and *order* structures of such a space. It is particularly useful for studying games with non-convex action spaces.

EXERCISE 73 (*Supermodular Games*) Let $\mathcal{G} := (\{X_i, \pi_i\}_{i=1,2})$ be a two-person continuous compact Euclidean game. Use Tarski's Fixed Point Theorem to show that $NE(\mathcal{G}) \neq \emptyset$, provided that each π_i is supermodular.

6.3* Remarks on the Equilibria of Discontinuous Games

While Nash's Existence Theorem is of primary importance for game theory, there are many situations in which it doesn't help settle the existence problem at hand, because in many economic games the payoff functions of the players are discontinuous, owing to the potential ties that may take place in the game. (This is so, for instance, in the Bertrand duopoly model considered in Example 7.[4].) More important, essentially all auction and voting games exhibit a fundamental discontinuity of this sort.

Here is a result that is sometimes useful for dealing with such situations. It generalizes Nash's Existence Theorem by relaxing, albeit slightly, its continuity requirement. The basic idea behind it is simply to postulate as a hypothesis what the Maximum Theorem gives us in the presence of continuity of the payoff functions.

PROPOSITION 8
(Dasgupta-Maskin) *Let* $\mathcal{G} := \{(X_i, \pi_i)_{i=1,\ldots,m}\}$ *be a convex and compact Euclidean game. Assume that, for each* i, π_i *is upper semicontinuous and* $\pi_i^* : X_{-i} \to \mathbb{R}$, *defined by*

$$\pi_i^*(x_{-i}) := \max\left\{\pi_i(a, x_{-i}) : a \in X\right\},$$

is lower semicontinuous. Then, $NE(\mathcal{G}) \neq \emptyset$.

[34] For more on generalized games, see Cubiotii (1997), and Banks and Duggan (1999).

PROOF

For each i, define the correspondences $\mathbf{b}_i$ and $\mathbf{b}$ as in the proof of Nash's Existence Theorem. Along with each π_i^*, these correspondences are well-defined because of the upper semicontinuity of π_i (Proposition D.5). Moreover, $\mathbf{b}$ is again convex-valued, so if we can show that $\mathbf{b}$ has a closed graph under the stated assumptions, then we may complete the proof by using Kakutani's Fixed Point Theorem. Thus, all we have to establish is that each $\mathbf{b}_i$ has the closed graph property. (Why?) To this end, fix any i, and take any convergent sequence $(x^k) \in X^\infty$ such that $x_i^k \in \mathbf{b}_i(x_{-i}^k)$ for each k. Let $x := \lim x^k$. We wish to show that $x_i \in \mathbf{b}_i(x_{-i})$. Indeed, using the upper semicontinuity of π_i, the fact that $x_i^k \in \mathbf{b}_i(x_{-i}^k)$ for each k, and the lower semicontinuity of π_i^*, we find

$$\begin{aligned}\pi_i(x) &\geq \limsup \pi_i(x_i^k, x_{-i}^k)\\ &= \limsup \pi_i^*(x_{-i}^k)\\ &\geq \liminf \pi_i^*(x_{-i}^k)\\ &\geq \pi_i^*(x_{-i})\end{aligned}$$

where we invoked Proposition D.4 twice. We thus conclude that $x_i \in \mathbf{b}_i(x_{-i})$, and the proof is complete. ∎

COROLLARY 2

Let $\mathcal{G} := \{(X_i, \pi_i)_{i=1,\ldots,m}\}$ be a convex and compact Euclidean game. If, for each i, π_i is upper semicontinuous, and for each i and $x_i \in X_i$, the map $\pi_i(x_i,\cdot)$ is lower semicontinuous (on X_{-i}), then $NE(\mathcal{G}) \neq \emptyset$.

PROOF

Apply Exercise 42 and Proposition 8. ∎

There is a substantial literature on the existence of Nash equilibria of discontinuous strategic games. See, for instance, Dasgupta and Maskin (1986), Simon (1987), Simon and Zame (1990), Baye, Tian, and Zhou (1993), Tan, Yu, and Yuan (1995), and Reny (1999). We conclude with two exercises that draw from this literature.

EXERCISE 74 Let $\mathcal{G} := \{(X_i, \pi_i)_{i=1,\ldots,m}\}$ be a compact Euclidean game. Assume that, for each i, the set X_i is convex and $\pi_i(\cdot, x_{-i})$ is concave for any given $x_{-i} \in X_{-i}$. If $\sum^m \pi_i$ is continuous, then $NE(\mathcal{G}) \neq \emptyset$.[35] Prove this fact by applying the observation noted in Exercise 58 to the map $\varphi : X^2 \to \mathbb{R}$ defined by

$$\varphi(x, y) := \sum_{i=1}^{m} \left(\pi_i(y_i, x_{-i}) - \pi_i(x_i, x_{-i})\right).$$

EXERCISE 75 (Reny) Let $\mathcal{G} := \{(X_i, \pi_i)_{i=1,\ldots,m}\}$ be a convex and compact Euclidean game. Let us write π for the map $x \mapsto (\pi_1(x), \ldots, \pi_m(x))$ on X, and for each i and x, define the **security level of player** i **at** x as $\sup\{\alpha \in \mathbb{R} :$ there exist an $x_i^* \in X_i$ and an open neighborhood O of x_{-i} (in X_{-i}) such that $\pi_i(x_i^*, y_{-i}) \geq \alpha$ for all $y_{-i} \in O\}$. In 1999 Phil Reny proved the following remarkable existence theorem: $NE(\mathcal{G}) \neq \emptyset$, provided that $\mathcal{G}$ satisfies the following property: If $x^* \notin NE(\mathcal{G})$ and $(x^*, \alpha_1, \ldots, \alpha_m) \in cl_{X \times \mathbb{R}^m}(Gr(\pi))$, then there is a player i whose security level at x^* is strictly larger than α_i. (You don't need to prove this!) Deduce from this result the following corollary: If $\sum^m \pi_i$ is upper semicontinuous, and for every $x \in X$ and $\varepsilon > 0$, each player i can secure a payoff of $\pi_i(x) - \varepsilon$ at x, then $NE(\mathcal{G}) \neq \emptyset$.

[35] In fact, it is enough to assume that $\sum^m \pi_i$ is upper semicontinuous, and for each i and $x_i \in X_i$, the map $\pi_i(x_i, \cdot)$ is lower semicontinuous (on X_{-i}). This is, however, more difficult to prove; see Tan, Yu, and Yuan (1995).

Part III
ANALYSIS ON LINEAR SPACES

Chapter F

Linear Spaces

The main goal of this chapter is to provide a foundation for our subsequent introduction to *linear functional analysis.* The latter is a vast subject, and there are many different ways in which one can provide a first pass at it. We mostly adopt a geometric viewpoint in this book. Indeed, we will later spend quite a bit of time covering the rudiments of (infinite-dimensional) convex analysis. The present chapter introduces the elementary theory of linear spaces with this objective in mind.

After going through a number of basic definitions and examples (where infinite-dimensional spaces are given a bit more emphasis than usual), we review the notions of basis and dimension, and talk about linear operators and functionals.[1] Keeping an eye on the convex analysis to come, we also discuss here the notion of *affinity* at some length. In addition, we conclude an unfinished business by proving Carathéodory's Theorem, characterize the finite dimensional linear spaces, and explore the connection between hyperplanes and linear functionals in some detail. On the whole, our exposition is fairly elementary, the only minor exception being the proof of the fact that every linear space has a basis—this proof is based on the Axiom of Choice. As economic applications, we prove some basic results of expected utility theory in the context of finite prize spaces, and introduce the elements of cooperative game theory. These applications illustrate well what a little linear algebra can do for you.

[1] You can consult on any one of the numerous texts on linear algebra for more detailed treatments of these topics and related matters. My favorite is Hoffman and Kunze (1971), but this may be due to the fact that I learned this stuff from that book first. Among the more recent and popular expositions are Broida and Williamson (1989) and Strang (1988).

1 Linear Spaces

Recall that $\mathbb{R}^n$ is naturally endowed with three basic mathematical structures: an order structure, a metric structure, and a linear structure. In the previous chapters we studied the generalizations of the first two of these structures, which led to the formulation of posets and metric spaces, respectively. In this chapter we study how such a generalization can be carried out in the case of the linear structure of $\mathbb{R}^n$, which, among other things, allows us to "add" any two n-vectors. The idea is that $\mathbb{R}^n$ is naturally equipped with an *addition* operation, and our immediate goal is to obtain a suitable axiomatization of this operation.

1.1 Abelian Groups

Let us first recall that a binary operation $\bullet$ on a nonempty set X is a map from $X \times X$ into X, but we write $x \bullet y$ instead of $\bullet(x, y)$ for any $x, y \in X$ (Section A.2.1).

DEFINITION

Let X be any nonempty set, and $+$ a binary operation on X. The doubleton $(X, +)$ is called a **group** if the following three properties are satisfied:

(i) (*Associativity*) $(x + y) + z = x + (y + z)$ for all $x, y, z \in X$.

(ii) (*Existence of an identity element*) There exists an element $\mathbf{0} \in X$ such that $\mathbf{0} + x = x = x + \mathbf{0}$ for all $x \in X$.

(iii) (*Existence of inverse elements*) For each $x \in X$, there exists an element $-x \in X$ such that $x + -x = \mathbf{0} = -x + x$.

If, in addition, we have

(iv) (*Commutativity*) $x + y = y + x$ for all $x, y \in X$,

then $(X, +)$ is said to be an **Abelian** (or **commutative**) **group**.

NOTATION. For any group $(X, +)$, and any nonempty subsets A and B of X, we let

$$A + B := \{x + y : (x, y) \in A \times B\}. \tag{1}$$

In this book we will work exclusively (and often implicitly) with Abelian groups.[2] In fact, even Abelian groups do not provide sufficiently rich algebraic structure for our purposes, so we will shortly introduce more discipline into the picture. But we should first consider some examples of Abelian groups to make things a bit more concrete.

EXAMPLE 1

[1] $(\mathbb{Z},+)$, $(\mathbb{Q},+)$, $(\mathbb{R},+)$, $(\mathbb{R}^n,+)$ and $(\mathbb{R}\backslash\{0\},\cdot)$ are Abelian groups where $+$ and $\cdot$ are the usual addition and multiplication operations. (*Note.* In $(\mathbb{R}\backslash\{0\},\cdot)$ the number 1 plays the role of the identity element.) On the other hand, $(\mathbb{R},\cdot)$ is not a group because it does not satisfy the requirement (iii). (What would be the inverse of 0 in $(\mathbb{R},\cdot)$?) Similarly, $(\mathbb{Z},\cdot)$ and $(\mathbb{Q},\cdot)$ are not groups.

[2] $(\mathbb{R}^{[0,1]},+)$ is an Abelian group where $+$ is defined pointwise (i.e., $f+g \in \mathbb{R}^{[0,1]}$ is defined by $(f+g)(t) := f(t)+g(t)$).

[3] Let X be any nonempty set and $\mathcal{X}$ the class of all bijective self-maps on X. Then $(\mathcal{X},\circ)$ is a group, but it is not Abelian unless $|X| \leq 2$.

[4] Let $X := \{x \in \mathbb{R}^2 : x_1^2 + x_2^2 = 1\}$ and define $x+y := (x_1y_1 - x_2y_2, x_1y_2 + x_2y_1)$ for any $x, y \in X$. It is easily checked that $x+y \in X$ for each $x, y \in X$, so this well-defines $+$ as a binary operation on X. In fact, $(X,+)$ is an Abelian group. (Verify! *Hint.* The identity element here is $(1,0)$.)

[5] Let $(X,+)$ be any group. The identity element $\mathbf{0}$ of this group is unique. For, if $y \in X$ is another candidate for the identity element, then $\mathbf{0}+y=\mathbf{0}$, but this implies $\mathbf{0} = \mathbf{0}+y = y$. Similarly, the inverse of any given element x is unique. Indeed, if y is an inverse of x, then

$$y = y+\mathbf{0} = y+(x+-x) = (y+x)+-x = \mathbf{0}+-x = -x.$$

In particular, the inverse of the identity element is itself.

[6] (*Cancellation Laws*) The usual cancellation laws apply to any Abelian group $(X,+)$. For instance, $x+y = z+x$ iff $y=z$, $-(-x)=x$ and $-(x+y) = -x+-y$ (Section A.2.1). □

[2] "Abelian" in the term Abelian group honors the name of Niels Abel (1802–1829), who made lasting contributions to group theory in his very short life span.

These examples should convince you that we are on the right track. The notion of Abelian group is a useful generalization of $(\mathbb{R},+)$. It allows us to "add" members of arbitrary sets in concert with the intuition supplied by $\mathbb{R}$.

EXERCISE 1 Let X be a nonempty set, and define $A\triangle B := (A\backslash B)\cup(B\backslash A)$ for any $A, B \subseteq X$. Show that $(2^X, \triangle)$ is an Abelian group.

EXERCISE 2^H Let $(X,+)$ be a group, and define the binary operation $+$ on $2^X\backslash\{\emptyset\}$ as in (1). What must be true for X so that $(2^X\backslash\{\emptyset\},+)$ is a group?

EXERCISE 3^H Is there a binary operation on $(0,1)$ that would make this set an Abelian group in which the inverse of any $x \in (0,1)$ is $1-x$?

EXERCISE 4^H Let $(X,+)$ be a group. If $\emptyset \neq Y \subseteq X$ and $(Y,+)$ is a group (where we use the restriction of $+$ to $Y \times Y$, of course), then $(Y,+)$ is called a **subgroup** of $(X,+)$.

(a) For any nonempty subset Y of X, show that $(Y,+)$ is a subgroup of $(X,+)$ iff $x + -y \in Y$ for all $x, y \in Y$.

(b) Give an example to show that if $(Y,+)$ and $(Z,+)$ are subgroups of $(X,+)$, then $(Y \cup Z,+)$ need not be a subgroup of $(X,+)$.

(c) Prove that if $(Y,+)$, $(Z,+)$ and $(Y\cup Z,+)$ are subgroups of $(X,+)$, then either $Y \subseteq Z$ or $Z \subseteq Y$.

EXERCISE 5 Let $(X,+)$ and $(Y,\oplus)$ be two groups. A function $f \in Y^X$ is said to be a **homomorphism** from $(X,+)$ into $(Y,\oplus)$ iff $f(x+x') = f(x)\oplus f(x')$ for all $x, x' \in X$. If there exists a bijective such map, then we say that these two groups are **homomorphic**.

(a) Show that $(\mathbb{R},+)$ and $(\mathbb{R}\backslash\{0\},\cdot)$ are homomorphic.

(b) Show that if f is a homomorphism from $(X,+)$ into $(Y,\oplus)$, then $f(\mathbf{0})$ is the identity element of $(Y,\oplus)$, and $(f(X),\oplus)$ is a subgroup of $(Y,\oplus)$.

(c) Show that if $(X,+)$ and $(Y,\oplus)$ are homomorphic, then $(X,+)$ is Abelian iff so is $(Y,\oplus)$.

1.2 Linear Spaces: Definition and Examples

So far so good, but we are only halfway through our abstraction process. We wish to have a generalization of the linear structure of $\mathbb{R}^n$ in a way that would allow us to algebraically represent some basic geometric objects. For instance, the line segment between the vectors $(1, 2)$ and $(2, 1)$ in $\mathbb{R}^2$ can be described algebraically as $\{\lambda(1, 2) + (1 - \lambda)(2, 1) : 0 \leq \lambda \leq 1\}$, thanks to the linear structure of $\mathbb{R}^2$. By contrast, the structure of an Abelian group falls short of letting us represent even such an elementary geometric object. This is because it is meaningful to "multiply" a vector in a Euclidean space with a real number (often called a *scalar* in this context), while there is no room for doing this within an arbitrary Abelian group.

The next step is then to enrich the structure of Abelian groups by defining a *scalar multiplication* operation on them. Once this is done, we will be able to describe algebraically a "line segment" in a very general sense. This description will indeed correspond to the usual geometric notion of line segment in the case of $\mathbb{R}^2$, and moreover, it will let us define and study a general notion of convexity for sets. As you have probably already guessed, the abstract model that we are after is none other than that of *linear space*.[3] Chapter G will demonstrate that this model indeed provides ample room for powerful geometric analysis.

DEFINITION

Let X be a nonempty set. The list $(X, +, \cdot)$ is called a **linear** (or **vector**) **space** if $(X, +)$ is an Abelian group, and if $\cdot$ is a mapping that assigns to each $(\lambda, x) \in \mathbb{R} \times X$ an element $\lambda \cdot x$ of X (which we denote simply as λx) such that, for all $\alpha, \lambda \in \mathbb{R}$ and $x, y \in X$, we have

(v) (*Associativity*) $\alpha(\lambda x) = (\alpha\lambda)x$.

(vi) (*Distributivity*) $(\alpha + \lambda)x = \alpha x + \lambda x$ and $\lambda(x + y) = \lambda x + \lambda y$.

(vii) (*The unit rule*) $1x = x$.

In a linear space $(X, +, \cdot)$, the mappings $+$ and $\cdot$ are called **addition** and **scalar multiplication** operations on X, respectively. When the context makes the nature of these operations clear, we may refer to X itself as a linear space. The identity element $\mathbf{0}$ is called the **origin** (or **zero**), and any member of X is referred to as a **vector**. If $x \in X \backslash \{\mathbf{0}\}$, then we say that x is a **nonzero** vector in X.

[3] We owe the first modern definition of linear space to the 1888 work of Giuseppe Peano. While initially ignored by the profession, the original treatment of Peano was amazingly modern. I recall Peter Lax telling me once that he would not be able to tell from reading certain parts of Peano's work that it was not written instead in 1988.

NOTATION. Let $(X, +, \cdot)$ be a linear space, and $A, B \subseteq X$ and $\lambda \in \mathbb{R}$. Then,

$$A + B := \{x + y : (x, y) \in A \times B\} \quad \text{and} \quad \lambda A := \{\lambda x : x \in A\}.$$

For simplicity, we write $A + y$ for $A + \{y\}$, and similarly $y + A := \{y\} + A$.

CONVENTION. We often talk as if a linear space $(X, +, .)$ were indeed a set when referring to properties that apply only to X. Phrases like "a vector of a linear space" or "a set in a linear space" are understood accordingly.

EXAMPLE 2

[1] The most trivial example of a linear space is a singleton set where the unique member of the set is designated as the origin. Naturally, this space is denoted as $\{\mathbf{0}\}$. Any linear space that contains more than one vector is called **nontrivial.**

[2] Let $n \in \mathbb{N}$. A very important example of a linear space is, of course, our beloved $\mathbb{R}^n$ (Remark A.1). On the other hand, endowed with the usual addition and scalar multiplication operations, $\mathbb{R}^n_{++}$ is not a linear space since it does not contain the origin. This is, of course, not the only problem. After all, $\mathbb{R}^n_+$ is not a linear space either (under the usual operations), for it does not contain the inverse of any nonzero vector. Is $[-1, 1]^n$ a linear space? How about $\{x \in \mathbb{R}^n : x_1 = 0\}$? $\{x \in \mathbb{R}^n : \sum^n x_i = 1\}$? $\{x \in \mathbb{R}^n : \sum^n x_i = 0\}$?

[3] In this book, we always think of the *sum* of two real functions f and g defined on a nonempty set T as the real function $f + g \in \mathbb{R}^T$ with $(f + g)(t) := f(t) + g(t)$. Similarly, the *product* of $\lambda \in \mathbb{R}$ and $f \in \mathbb{R}^T$ is the function $\lambda f \in \mathbb{R}^T$ defined by $(\lambda f)(t) := \lambda f(t)$. (Recall Section A.4.1.) In particular, we consider the real sequence space $\mathbb{R}^\infty$ (which is none other than $\mathbb{R}^\mathbb{N}$) and the function spaces $\mathbb{R}^T$ and $\mathbf{B}(T)$ (for any nonempty set T) as linear spaces under these operations. Similarly, ℓ^p (for any $1 \leq p \leq \infty$), along with the function spaces $\mathbf{CB}(T)$ and $\mathbf{C}(T)$ (for any metric space T), are linear spaces under these operations (why?), and when we talk about these spaces, it is these operations that we have in mind. The same goes as well for other function spaces, such as $\mathbf{P}(T)$ or the space of all polynomials on T of degree $m \in \mathbb{Z}_+$ (for any nonempty subset T of $\mathbb{R}$).

[4] Since the negative of an increasing function is decreasing, the set of all increasing real functions on $\mathbb{R}$ (or on any compact interval $[a, b]$

with $a < b$) is not a linear space under the usual operations. Less trivially, the set of all monotonic self-maps on $\mathbb{R}$ is not a linear space either. To see this, observe that the self-map f defined on $\mathbb{R}$ by $f(t) := \sin t + 2t$ is an increasing function. (For $f'(t) = \cos t + 2 \geq -1 + 2 \geq 0$ for all $t \in \mathbb{R}$.) But the self-map g defined on $\mathbb{R}$ by $g(t) := \sin t - 2t$ is a decreasing function, and yet $f + g$ is obviously not monotonic. For another example, we note that the set of all semicontinuous self-maps on $\mathbb{R}$ does not form a linear space under the usual operations. (Why?) □

It will become clear shortly that linear spaces provide an ideal structure for a proper investigation of convex sets. For the moment, however, all you need to do is to recognize that a linear space is an algebraic infrastructure relative to which the addition and scalar multiplication operations behave in concert with intuition, that is, the way the corresponding operations behave on a Euclidean space. To drive this point home, let us derive some preliminary facts about these operations that we shall later invoke in a routine manner.

Fix a linear space $(X, +, \cdot)$. First of all, we have $\lambda\mathbf{0} = \mathbf{0}$ for any $\lambda \in \mathbb{R}$. Indeed, $\lambda\mathbf{0} + \lambda\mathbf{0} = \lambda(\mathbf{0} + \mathbf{0}) = \lambda\mathbf{0} = \mathbf{0} + \lambda\mathbf{0}$, so that, by Example 1.[6], $\lambda\mathbf{0} = \mathbf{0}$. A similar reasoning gives us $0x = \mathbf{0}$ for any $x \in X$. More generally, for any $x \neq \mathbf{0}$,

$$\lambda x = \mathbf{0} \quad \text{if and only if} \quad \lambda = 0.$$

Indeed, if $\lambda \neq 0$ and $\lambda x = \mathbf{0}$, then $x = \frac{1}{\lambda}\lambda x = \frac{1}{\lambda}\mathbf{0} = \mathbf{0}$. From this observation we deduce easily that, for any $x \neq \mathbf{0}$,

$$\alpha x = \beta x \quad \text{if and only if} \quad \alpha = \beta,$$

which, in particular, implies that any linear space other than $\{\mathbf{0}\}$ contains uncountably many vectors. (Exactly which properties of a linear space did we use to get this conclusion?) Finally, let us show that we have

$$-(\lambda x) = (-\lambda)x \quad \text{for all } (\lambda, x) \in \mathbb{R} \times X. \tag{2}$$

(Here $-(\lambda x)$ is the inverse of λx, while $(-\lambda)x$ is the "product" of $-\lambda$ and x; so the claim is not trivial.) Indeed, $\mathbf{0} = 0x = (\lambda - \lambda)x = \lambda x + (-\lambda)x$ for any $(\lambda, x) \in \mathbb{R} \times X$, and if we add $-(\lambda x)$ to both sides of $\mathbf{0} = \lambda x + (-\lambda)x$, we find $-(\lambda x) = (-\lambda)x$.

Thanks to (2), there is no difference between $(-1)x$ and $-x$, and hence between $x + (-1)y$ and $x + -y$. In what follows, we shall write $x - y$ for either of the latter expressions since we are now confident that there is no room for confusion.

EXERCISE 6 Define the binary operations $+_1$ and $+_2$ on $\mathbb{R}^2$ by $x +_1 y := (x_1 + y_1, x_2 + y_2)$ and $x +_2 y := (x_1 + y_1, 0)$. Is $(\mathbb{R}^2, +_i, \cdot)$, where $\cdot$ maps each $(\lambda, x) \in \mathbb{R} \times X$ to $(\lambda x_1, \lambda x_2) \in \mathbb{R}^2$, a linear space, $i = 1, 2$? What if $\cdot$ maps each (λ, x) to $(\lambda x_1, x_2)$? What if $\cdot$ maps each (λ, x) to $(\lambda x_1, 0)$?

EXERCISE 7 Let $(X, +, \cdot)$ be a linear space, and $x, y \in X$. Show that $(\{\alpha x + \lambda y : \alpha, \lambda \in \mathbb{R}\}, +, \cdot)$ is a linear space.

Henceforth we use the notation X instead of $(X, +, \cdot)$ for a linear space, but you should always keep in mind that what makes a linear space "linear" is the two operations defined on it. Two different types of addition and scalar multiplication operations on a given set may well endow this set with different linear structures, and hence yield two very different linear spaces.

1.3 Linear Subspaces, Affine Manifolds, and Hyperplanes

One method of obtaining other linear spaces from a given linear space X is to consider those subsets of X that are themselves linear spaces under the inherited operations.

DEFINITION
Let X be a linear space and $\emptyset \neq Y \subseteq X$. If Y is a linear space with the same operations of addition and scalar multiplication as with X, then it is called a **linear subspace** of X.[4] If, further, $Y \neq X$, then Y is called a **proper linear subspace** of X.

The following exercise provides an alternative (and of course equivalent) definition of the notion of linear subspace. We will use this alternative formulation freely in what follows.

EXERCISE 8 Let X be a linear space and $\emptyset \neq Y \subseteq X$. Show that Y is a linear subspace of X iff $\lambda x + y \in Y$ for each $\lambda \in \mathbb{R}$ and $x, y \in Y$.

[4] Put more precisely, if the addition and scalar multiplication operations on X are $+$ and $\cdot$, respectively, then by a *linear subspace* Y of X we mean the linear space $(Y, \oplus, \odot)$, where Y is a nonempty subset of X, and $\oplus : Y^2 \to Y$ is the restriction of $+$ to Y^2 and $\odot : \mathbb{R} \times Y \to Y$ is the restriction of $\cdot$ to $\mathbb{R} \times Y$.

EXAMPLE 3

[1] $[0, 1]$ is not a linear subspace of $\mathbb{R}$ whereas $\{x \in \mathbb{R}^2 : x_1 + x_2 = 0\}$ is a proper linear subspace of $\mathbb{R}^2$.

[2] For any $n \in \mathbb{N}$, $\mathbb{R}^{n \times n}$ is a linear space under the usual matrix operations of addition and scalar multiplication (Remark A.1). The set of all symmetric $n \times n$ matrices, that is, $\{[a_{ij}]_{n \times n} : a_{ij} = a_{ji} \text{ for each } i, j\}$ is a linear subspace of this space.

[3] For any $n \in \mathbb{N}$ and linear $f : \mathbb{R}^n \to \mathbb{R}$ (Section D.5.3), $\{x \in \mathbb{R}^n : f(x) = 0\}$ is a linear subspace of $\mathbb{R}^n$. Would this conclusion be true if all we knew was that f is additive?

[4] For any $m \in \mathbb{N}$, the set of constant functions on $[0, 1]$ is a proper linear subspace of the set of all polynomials on $[0, 1]$ of degree at most m. The latter set is a proper linear subspace of $\mathbf{P}[0, 1]$, which is a proper linear subspace of $\mathbf{C}[0, 1]$, which is itself a proper linear subspace of $\mathbf{B}[0, 1]$. Finally, $\mathbf{B}[0, 1]$ is a proper linear subspace of $\mathbb{R}^{[0,1]}$. □

EXERCISE 9

(a) Is ℓ^1 a linear subspace of ℓ^∞?

(b) Is the set $\mathbf{c}$ of all convergent real sequences a linear subspace of ℓ^∞? Of ℓ^1?

(c) Let $\mathbf{c}^0$ be the set of all real sequences all but finitely many terms of which are zero. Is $\mathbf{c}^0$ a linear space (under the usual (pointwise) operations). Is it a linear subspace of $\mathbf{c}$?

(d) Is $\{(x_m) \in \mathbf{c}^0 : x_i = 1 \text{ for some } i\}$ a linear subspace of $\mathbf{c}^0$? Of $\mathbb{R}^\infty$?

EXERCISE 10 Show that the intersection of any collection of linear subspaces of a linear space is itself a linear subspace of that space.

EXERCISE 11 If Z is a linear subspace of Y and Y a linear subspace of the linear space X, is Z necessarily a linear subspace of X?

EXERCISE 12^H Let Y and Z be linear subspaces of a linear space X. Prove:

(a) $Y + Z$ is a linear subspace of X.

(b) If $Y \cup Z$ is a linear subspace of X, then either $Y \subseteq Z$ or $Z \subseteq Y$.

Clearly, $\{x \in \mathbb{R}^2 : x_1 + x_2 = 1\}$ is not a linear subspace of $\mathbb{R}^2$. (This set does not even contain the origin of $\mathbb{R}^2$. And yes, this is crucial!) On the other hand, geometrically speaking, this set is very "similar" to the linear subspace $\{x \in \mathbb{R}^2 : x_1 + x_2 = 0\}$. Indeed, the latter is nothing but a parallel shift (*translation*) of the former set. It is thus not surprising that such sets play an important role in geometric applications of linear algebra. They certainly deserve a name.

DEFINITION

A subset S of a linear space X is said to be an **affine manifold** of X if $S = Z + x^*$ for some linear subspace Z of X and some vector $x^* \in X$.[5] If Z is a $\supseteq$-maximal proper linear subspace of X, S is then called a **hyperplane** in X.

As for examples, note that there is no hyperplane in the trivial space $\{\mathbf{0}\}$. Since the only proper linear subspace of $\mathbb{R}$ is $\{0\}$, any one-point set in $\mathbb{R}$ and $\mathbb{R}$ itself are the only affine manifolds in $\mathbb{R}$. So, all hyperplanes in $\mathbb{R}$ are singleton sets. In $\mathbb{R}^2$, any one-point set, any line (with no endpoints), and the entire $\mathbb{R}^2$ are the only affine manifolds. A hyperplane in this space is necessarily of the form $\{x \in \mathbb{R}^2 : a_1x_1 + a_2x_2 = b\}$ for some real numbers a_1, a_2 with at least one of them being nonzero, and some real number b. Finally, all hyperplanes are of the form of (infinitely extending) planes in $\mathbb{R}^3$.

A good way of thinking intuitively about the notion of *affinity* in linear analysis is this:

affinity = linearity + translation.

Since we think of $\mathbf{0}$ as the origin of the linear space X (this is a geometric interpretation; don't forget that the definition of $\mathbf{0}$ is purely algebraic), it makes sense to view a linear subspace of X as *un*translated (relative to the origin of the space), for a linear subspace "passes through" $\mathbf{0}$. The following simple but important observation thus gives support to our informal equation above: *An affine manifold S of a linear space X is a linear subspace of X iff $\mathbf{0} \in S$.* (*Proof.* If $S = Z + x^*$ for some linear subspace Z of X and $x^* \in X$, then $\mathbf{0} \in S$ implies $-x^* \in Z$. (Yes?) Since Z is a linear space, we then have $x^* \in Z$, and hence $Z = Z + x^* = S$.) An immediate corollary of this is: *If S is an affine manifold of X, then $S - x$ is a linear subspace of X*

[5] *Reminder.* $Z + x^* := \{z + x^* : z \in Z\}$.

for any $x \in S$. Moreover, this subspace is determined independently of x, because *if* S *is an affine manifold, then*

$$S - x = S - y \quad \textit{for any } x, y \in S. \tag{3}$$

We leave the proof of this assertion as an easy exercise.[6]

The following result, a counterpart of Exercise 8, provides a useful characterization of affine manifolds.

PROPOSITION 1
Let X *be a linear space and* $\emptyset \neq S \subseteq X$. *Then* S *is an affine manifold of* X *if, and only if,*

$$\lambda x + (1-\lambda)y \in S \quad \textit{for any } x, y \in S \textit{ and } \lambda \in \mathbb{R}. \tag{4}$$

PROOF
If $S = Z + x^*$ for some linear subspace Z of X and $x^* \in X$, then for any $x, y \in S$ there exist z_x and z_y in Z such that $x = z_x + x^*$ and $y = z_y + x^*$. It follows that

$$\lambda x + (1-\lambda)y = (\lambda z_x + (1-\lambda)z_y) + x^* \in Z + x^*$$

for any $x, y \in S$ and $\lambda \in \mathbb{R}$. Conversely, assume that S satisfies (4). Pick any $x^* \in S$, and define $Z := S - x^*$. Then $S = Z + x^*$, so we will be done if we can show that Z is a linear subspace of X. Thanks to Exercise 8, all we need to do is, then, to establish that Z is closed under scalar multiplication and vector addition. To prove the former fact, notice that if $z \in Z$ then $z = x - x^*$ for some $x \in S$, so, by (4),

$$\lambda z = \lambda x - \lambda x^* = \left(\lambda x + (1-\lambda)x^*\right) - x^* \in S - x^* = Z$$

[6] Just to fix the intuition, you might want to verify everything I said in this paragraph in the special case of the affine manifolds $\{x \in \mathbb{R}^2 : x_1 + x_2 = 1\}$, $\{x \in \mathbb{R}^3 : x_1 + x_2 = 1\}$, and $\{x \in \mathbb{R}^3 : x_1 + x_2 + x_3 = 1\}$. (In particular, draw the pictures of these manifolds, and see how they are obtained as translations of specific linear subspaces.)

for any $\lambda \in \mathbb{R}$. To prove the latter fact, take any $z, w \in Z$, and note that $z = x - x^*$ and $w = y - x^*$ for some $x, y \in S$. By (4), $2x - x^* \in S$ and $2y - x^* \in S$. Therefore, applying (4) again,

$$z + w = \tfrac{1}{2}\left(2x - x^*\right) + \tfrac{1}{2}\left(2y - x^*\right) - x^* \in S - x^* = Z,$$

and the proof is complete. ■

The geometric nature of affine manifolds and hyperplanes makes them indispensable tools for convex analysis, and we will indeed use them extensively in what follows. For the time being, however, we leave their discussion at this primitive stage, and instead press on reviewing the fundamental concepts of linear algebra. We will revisit these notions at various points in the subsequent development.

1.4 Span and Affine Hull of a Set

Let X be a linear space and $(x^m) \in X^\infty$. We define

$$\sum_{i=1}^{m} x^i := x^1 + \cdots + x^m \quad \text{and} \quad \sum_{i=k}^{m} x^i := \sum_{i=1}^{m-k+1} x^{i+k-1}$$

for any $m \in \mathbb{N}$ and $k \in \{1, \ldots, m\}$, and note that there is no ambiguity in this definition, thanks to the associativity of the addition operation $+$ on X. (By the same token, for any nonempty *finite* subset T of X, it is without ambiguity to write $\sum_{x \in T} x$ for the sum of all members of T). As in the case of sums of real numbers, we will write $\sum^m x^i$ for $\sum_{i=1}^m x^i$ in the text.

DEFINITION
For any $m \in \mathbb{N}$, by a **linear combination** of the vectors $x^1, \ldots, x^m$ in a linear space X, we mean a vector $\sum^m \lambda_i x^i \in X$, where $\lambda_1, \ldots, \lambda_m$ are any real numbers (called the **coefficients** of the linear combination). If, in addition, we have $\sum^m \lambda_i = 1$, then $\sum^m \lambda_i x^i$ is referred to as an **affine combination** of the vectors $x^1, \ldots, x^m$. If, on the other hand, $\lambda_i \geq 0$ for each i, then $\sum^m \lambda_i x^i$ is called a **positive linear combination** of $x^1, \ldots, x^m$. Finally, if $\lambda_i \geq 0$ for each i and $\sum^m \lambda_i = 1$, then $\sum^m \lambda_i x^i$ is called a **convex combination** of $x^1, \ldots, x^m$. Equivalently, a linear (affine (convex)) combination of the elements of a nonempty finite subset T of X is $\sum_{x \in T} \lambda(x)x$, where $\lambda \in \mathbb{R}^T$ (and $\sum_{x \in T} \lambda(x) = 1$ (and $\lambda(T) \subseteq \mathbb{R}_+$)).

Definition

The set of all linear combinations of finitely many members of a nonempty subset S of a linear space X is called the **span** of S (in X), denoted by $span(S)$. That is, for any $\emptyset \neq S \subseteq X$,

$$span(S) := \left\{ \sum_{i=1}^{m} \lambda_i x^i : m \in \mathbb{N} \text{ and } (x^i, \lambda_i) \in S \times \mathbb{R},\ i = 1, \ldots, m \right\},$$

or equivalently,

$$span(S) = \left\{ \sum_{x \in T} \lambda(x) x : T \in \mathcal{P}(S) \text{ and } \lambda \in \mathbb{R}^T \right\},$$

where $\mathcal{P}(S)$ is the class of all nonempty finite subsets of S. By convention, we let $span(\emptyset) = \{\mathbf{0}\}$.

Given a linear space X, $span(\{\mathbf{0}\}) = \{\mathbf{0}\}$ and $span(X) = X$ while $span(\{x\}) = \{\lambda x : \lambda \in \mathbb{R}\}$. The span of a set S in a linear space is always a linear space (yes?), and it is thus a linear subspace of the mother space. What is more, $span(S)$ is the smallest (i.e., $\supseteq$-minimum) linear subspace of the mother space that contains S. Especially when S is finite, this linear subspace has a very concrete description in that every vector in it can be expressed as linear combination of all the vectors in S. (Why?)

Exercise 13 Let X be a linear space and $\emptyset \neq S \subseteq X$. Show that $span(S) \subseteq Y$ for any linear subspace Y of X with $S \subseteq Y$. Conclude that $span(S)$ is the smallest linear subspace of X that contains S.

Definition

The set of all affine combinations of finitely many members of a nonempty subset S of a linear space X is called the **affine hull** of S (in X), and is denoted by $aff(S)$. That is, for any $\emptyset \neq S \subseteq X$,

$$aff(S) := \left\{ \sum_{x \in T} \lambda(x) x : T \in \mathcal{P}(S) \text{ and } \lambda \in \mathbb{R}^T \text{ with } \sum_{x \in T} \lambda(x) = 1 \right\}$$

where $\mathcal{P}(S)$ is the class of all nonempty finite subsets of S. By convention, we let $aff(\emptyset) = \{\mathbf{0}\}$.

By Proposition 1, $aff(S)$ is an affine manifold of X. Moreover, again by Proposition 1 and the Principle of Mathematical Induction, any affine manifold of X that contains S also contains $aff(S)$. (Why?) Therefore, $aff(S)$ is the smallest affine manifold of X that contains S. Equivalently, this manifold equals the intersection of all affine manifolds of X that contain S.[7]

These observations help clarify the nature of the tight connection between the notions of span and affine hull of a set. Put precisely, for any nonempty subset S of a linear space X, we have

$$aff(S) = span(S - x) + x \qquad \text{for any } x \in S. \tag{5}$$

Indeed, for any $x \in X$, $span(S - x) + x$ is an affine manifold of X that contains S; thus, being the smallest such affine manifold, $aff(S)$ must be contained in $span(S - x) + x$. Conversely, if $x \in S$, then $aff(S) - x$ is a linear subspace of X that contains $S - x$ (why?), and hence, being the smallest such linear subspace of X, $span(S - x)$ must be contained in $aff(S) - x$, that is, $aff(S) \supseteq span(S - x) + x$.

1.5 Linear and Affine Independence

DEFINITION

Let X be a linear space. A subset S of X is called **linearly dependent** in X if it either equals $\{\mathbf{0}\}$ or at least one of the vectors, say, x in S can be expressed as a linear combination of finitely many vectors in $S \backslash \{x\}$. For any $m \in \mathbb{N}$, any distinct vectors $x^1, \ldots, x^m \in X$ are called **linearly dependent** if $\{x^1, \ldots, x^m\}$ is linearly dependent in X.

DEFINITION

Let X be a linear space. A subset of X is called **linearly independent** in X if no finite subset of it is linearly dependent in X. For any $m \in \mathbb{N}$, the vectors $x^1, \ldots, x^m \in X$ are called **linearly independent** if $\{x^1, \ldots, x^m\}$ is linearly independent in X.

[7] What is the affine hull of $\{x \in \mathbb{R}^2 : d_2(x, \mathbf{0}) = 1\}$ in $\mathbb{R}^2$? Of $\{x \in \mathbb{R}^2 : x_1 + x_2 = 1\}$? Of $\{\mathbf{0}, (1, 0), (0, 1)\}$?

That is, a nonempty subset S of X is linearly independent in X iff, for every nonempty finite subset T of S and $\lambda \in \mathbb{R}^T$,

$$\sum_{x \in T} \lambda(x)x = \mathbf{0} \quad \text{implies} \quad \lambda = 0.$$

(Why?) In particular, for any $m \in \mathbb{N}$, the vectors $x^1, \ldots, x^m \in X$ are linearly independent iff, for every $(\lambda_1, \ldots, \lambda_m) \in \mathbb{R}^m$,

$$\sum_{i=1}^{m} \lambda_i x^i = \mathbf{0} \quad \text{implies} \quad \lambda_1 = \cdots = \lambda_m = 0.$$

You have probably seen these definitions before, but let's reflect on them a bit more anyway. Note first that $\emptyset$ and any singleton set other than $\{\mathbf{0}\}$ is a linearly independent set in any linear space. Moreover, it follows from these definitions that any set S in a linear space X with $\mathbf{0} \in S$ is linearly dependent in X. (Why?) For any distinct $x, y \in X \backslash \{\mathbf{0}\}$, the set $\{x, y\}$ is linearly dependent in X iff $y \in span(\{x\})$, which holds iff $span(\{x\}) = span(\{y\})$. Hence, one says that two nonzero vectors x and y are linearly dependent iff they both lie on a line through the origin. More generally, a subset S of $X \backslash \{\mathbf{0}\}$ is linearly dependent in X iff there exists an $x \in S$ such that $x \in span(S \backslash \{x\})$.

A fundamental principle of linear algebra is that there cannot be more than m linearly independent vectors in a linear space spanned by m vectors. Many important findings concerning the structure of linear spaces follow from this observation.

PROPOSITION 2

Let X be a linear space, and $A, B \subseteq X$. If B is linearly independent in X and $B \subseteq span(A)$, then $|B| \leq |A|$.

PROOF

Assume that B is linearly independent in X and $B \subseteq span(A)$. If $A = \emptyset$, then the latter hypothesis implies that either $B = \emptyset$ or $B = \{\mathbf{0}\}$. By linear independence of B, then, $B = \emptyset$, so the claim follows. Similarly, if $|A| = \infty$, there is nothing to prove. So let $m := |A| \in \mathbb{N}$, and enumerate A as $\{x^1, \ldots, x^m\}$. To derive a contradiction, let us assume that $|B| > m$. Take any $y^1 \in B$. We have $y^1 \neq \mathbf{0}$, because B is linearly independent. Since $y^1 \in span(A)$, therefore, it can be written as a linear combination of $x^1, \ldots, x^m$, with at least one coefficient of the linear combination being nonzero. This implies that at least one member of A, say x^1 (relabeling if necessary), can be written as a linear combination of $y^1, x^2, \ldots, x^m$. (Why?) Thus,

$$span(A) = span(\{y^1, x^2, \ldots, x^m\}).$$

(Why?) But repeating the same reasoning successively (and keeping in mind that B is linearly independent), we would find vectors $y^2, \ldots, y^m \in B$ with

$$span(A) = span(\{y^1, \ldots, y^m\}).$$

(How, exactly?) This means that any $y \in B \backslash \{y^1, \ldots, y^m\}$ (and there is such a y because $|B| > m$) can be written as a linear combination of $y^1, \ldots, y^m$, contradicting the linear independence of B. ■

Recall that, for any $n \in \mathbb{N}$ and $i \in \{1, \ldots, n\}$, the **ith unit vector** in $\mathbb{R}^n$ – denoted by $\mathbf{e}^i$ – is the n-vector whose ith term equals 1 and whose all other terms equal 0. Obviously, $\mathbf{e}^1, \ldots, \mathbf{e}^n$ are linearly independent in $\mathbb{R}^n$. It then follows from Proposition 2 that there can be at most n linearly independent vectors in $\mathbb{R}^n$, or, equivalently, any $S \subseteq \mathbb{R}^n$ with $|S| > n$ is linearly dependent in $\mathbb{R}^n$.

EXERCISE 14

(a) If $f, g \in \mathbf{C}(\mathbb{R})$ are defined by $f(t) := e^t$ and $g(t) := te^t$, show that $\{f, g\}$ is linearly independent in $\mathbf{C}(\mathbb{R})$.

(b) If $u, v, w \in \mathbf{C}(\mathbb{R})$ are defined by $u(t) := \sin t$, $v(t) := \cos t$, and $w(t) := (\sin t)^2$, show that $\{u, v, w\}$ is linearly independent in $\mathbf{C}(\mathbb{R})$.

EXERCISE 15^H Define $f_a, g_b \in \mathbf{P}[0, 1]$ by $f_a(t) := a + t^2$ and $g_b(t) := t + bt^2$, for any real numbers a and b.

(a) Determine the set of all $(a, b) \in \mathbb{R}^2$ such that $\{f_a, g_b\}$ is linearly independent in $\mathbf{P}[0, 1]$.

(b) Determine the set of all $(a, b) \in \mathbb{R}^2$ such that $\{\mathrm{id}_{[0,1]}, f_a, g_b\}$ is linearly independent in $\mathbf{P}[0, 1]$.

EXERCISE 16 Take any $(x_m) \in \mathbb{R}^\infty$ such that $x_m \neq 0$ for each m. Define $(x_m^k) := (0, \ldots, 0, x_1, x_2, \ldots)$ where the first nonzero term is the $(k+1)$th one, $k = 1, 2, \ldots$.

(a) Show that $\{(x_m), (x_m^1), (x_m^2), \ldots\}$ is linearly independent in $\mathbb{R}^\infty$.

(b) Does $span(\{(x_m), (x_m^1), (x_m^2), \ldots\})$ equal $\mathbb{R}^\infty$?

We now turn to the *affine* version of the notion of linear independence, which again builds on the idea that "affinity = linearity + translation." Let S be a nonempty finite subset of a linear space X. It is an easy exercise to show that $\{z - x : z \in S \backslash \{x\}\}$ is linearly independent in X iff $\{z - y : z \in S \backslash \{y\}\}$ is linearly independent in X, for *any* $x, y \in S$.[8] This prompts the following formulation.

DEFINITION
Let X be a linear space. A finite subset T of X is called **affinely independent** in X if $\{z - x : z \in T \backslash \{x\}\}$ is linearly independent in X for any $x \in T$. An arbitrary nonempty subset S of X is **affinely independent** in X if every finite subset of S is affinely independent in X. S is called **affinely dependent** in X if it is not affinely independent in X.

Since at most n vectors can be linearly independent in $\mathbb{R}^n$, $n = 1, 2, \ldots$, it follows that there can be at most $n+1$ affinely independent vectors in $\mathbb{R}^n$. It is also useful to note that a nonempty subset S of a linear space X is affinely independent in X iff, for every nonempty finite subset T of S and $\alpha \in \mathbb{R}^T$,

$$\sum_{x \in T} \alpha(x)x = \mathbf{0} \quad \text{and} \quad \sum_{x \in T} \alpha(x) = 0 \quad \text{imply} \quad \alpha = 0.$$

(Prove![9])

Recall that we invoked Carathéodory's Theorem in Section E.5.3 when proving Kakutani's Fixed Point Theorem, but shamelessly omitted its proof. Equipped with the notion of affine independence, we are now ready to settle that score.

EXAMPLE 4
(*Proof of Carathéodory's Theorem*) Let us first recall what Carathéodory's Theorem is about. Fix any $n \in \mathbb{N}$ and recall that, for any subset A of $\mathbb{R}^n$, $co(A)$ corresponds to the set of all convex combinations of finitely many

[8] In view of (3), this fact is hardly surprising. But still, please provide a proof for it.

[9] Please don't take this for granted. The "only if" part of the assertion requires a bit of effort. (*Hint.* If $\sum_{x \in T} \alpha(x)x = \mathbf{0}$ and $\sum_{x \in T} \alpha(x) = 0$, but $\alpha(x^*) \neq 0$ for some $x^* \in T$, then observe that $x^* = \sum_{x \in T \backslash \{x^*\}} \beta(x)x$ for a certain $\beta : T \backslash \{x^*\} \to \mathbb{R}$ with $\beta \neq 0$, and show that $\{x - x^* : x \in T \backslash \{x^*\}\}$ is not linearly independent in X.)

members of A. *Carathéodory's Theorem* states the following:

> *Given any nonempty $S \subseteq \mathbb{R}^n$, for every $x \in co(S)$ there exists a $T_x \subseteq S$ such that $x \in co(T_x)$ and $|T_x| \leq n+1$.*

Here goes the proof. Take any $x \in co(S)$. By definition, there exists a finite $T \subseteq S$ such that $x \in co(T)$. Let T_x be a $\supseteq$-minimal such T. (Thanks to finiteness, such a T_x must exist. Right?) Then there exists a $\lambda : T_x \to \mathbb{R}_{++}$ such that

$$x = \sum_{y \in T_x} \lambda(y)y \quad \text{and} \quad \sum_{y \in T_x} \lambda(y) = 1.$$

To derive a contradiction, assume that $|T_x| > n+1$. Then T_x must be affinely dependent in $\mathbb{R}^n$, so there exists an $\alpha \in \mathbb{R}^T$ such that

$$\sum_{y \in T_x} \alpha(y)y = \mathbf{0} \quad \text{and} \quad \sum_{y \in T_x} \alpha(y) = 0,$$

and yet $\alpha \neq 0$. Let $A := \{y \in T_x : \alpha(y) > 0\}$ and $B := T_x \backslash A$. Since $\alpha \neq 0$ and $\sum_{y \in T_x} \alpha(y) = 0$, neither of these sets can be empty. Now define

$$\theta := \min\left\{\frac{\lambda(y)}{\alpha(y)} : y \in A\right\} \quad \text{and} \quad C := \{y \in T_x : \lambda(y) - \theta\alpha(y) = 0\}.$$

Clearly, $\theta > 0$ and $C \neq \emptyset$. Moreover, by definition of θ, $\lambda(y) - \theta\alpha(y) \geq 0$ for all $y \in T_x$ while

$$\sum_{y \in T_x \backslash C} \left(\lambda(y) - \theta\alpha(y)\right) = \sum_{y \in T_x} \lambda(y) - \theta \sum_{y \in T_x} \alpha(y) = 1.$$

But

$$x = \sum_{y \in T_x} \lambda(y)y - \theta \sum_{y \in T_x} \alpha(y)y = \sum_{y \in T_x \backslash C} \left(\lambda(y) - \theta\alpha(y)\right) y,$$

so it follows that $x \in co(T_x \backslash C)$. Since $C \neq \emptyset$, this contradicts the $\supseteq$-minimality of T_x, and the proof is complete.

Actually, this argument delivers a bit more than what is claimed by Carathéodory's Theorem. It shows that what matters is not the underlying linear space $\mathbb{R}^n$ but rather the affine hull of S. Put more precisely, what we have established above is this: *For every $x \in co(S)$ there exists a $T_x \subseteq S$ such that $x \in co(T_x)$ and $|T_x| \leq m(S)+1$, where $m(S)$ is the maximum number of affinely independent vectors in S.*[10] □

*EXERCISE 17^{H} (*Radon's Lemma*) Let $n \in \mathbb{N}$. Given any nonempty $S \subseteq \mathbb{R}^n$ with $|S| \geq n+2$, there exist two disjoint sets A and B in S such that $co(A) \cap co(B) \neq \emptyset$.[11]

Radon's Lemma makes it quite easy to prove Helly's Intersection Theorem, one of the most famous results of convex analysis.

*EXERCISE 18^{H} Let $n \in \mathbb{N}$. Prove:

(a) (*Helly's Intersection Theorem*) If $\mathcal{S}$ is a finite class of convex subsets of $\mathbb{R}^n$ such that $|\mathcal{S}| \geq n+1$, and $\cap\mathcal{T} \neq \emptyset$ for any $\mathcal{T} \subseteq \mathcal{S}$ with $|\mathcal{T}| = n+1$, then $\cap\mathcal{S} \neq \emptyset$.

(b) If $\mathcal{S}$ is a class of compact and convex subsets of $\mathbb{R}^n$ such that $|\mathcal{S}| \geq n+1$ and $\cap\mathcal{T} \neq \emptyset$ for any $\mathcal{T} \subseteq \mathcal{S}$ with $|\mathcal{T}| = n+1$, then $\cap\mathcal{S} \neq \emptyset$.

1.6 Bases and Dimension

Since $span(S)$ is the linear space that consists of all finite linear combinations of the vectors in S, all there is to know about this space is contained in S. Thus, if we wish to think of a linear space as a span of a certain set, it makes sense to choose that set as small as possible. This would allow us to economically "represent" the linear space that we are interested in, thereby letting us view S as some sort of "nucleus" of the space that it spans. This prompts the introduction of the following important concept.

[10] There are various refinements of Carathéodory's Theorem that posit further structure on S and provide better bounds for the number of elements of S that are needed to express any given $x \in co(S)$ as a convex combination. For instance, one can show that if $S \subseteq \mathbb{R}^n$ is connected, then every $x \in co(S)$ can be expressed as a convex combination of at most n members of S. This is (a special case of) the *Fenchel-Bunt Theorem.*

[11] Check first that the claim is true in $\mathbb{R}^2$ with a set S that consists of 4 vectors, but it is false with a set of cardinality 3!

DEFINITION

A **basis** for a linear space X is a $\supseteq$-minimal subset of X that spans X. That is, S is a basis for X if, and only if,

(i) $X = span(S)$; and

(ii) if $X = span(T)$, then $T \subset S$ is false.

DEFINITION

If a linear space X has a finite basis, then it is said to be **finite-dimensional**, and its **dimension**, $dim(X)$, is defined as the cardinality of any one of its bases. If X does not have a finite basis, then it is called **infinite-dimensional**, in which case we write $dim(X) = \infty$.

There may seem to be an ambiguity in the definition of "the" dimension of a finite-dimensional linear space X. What if two bases for X have a different number of elements? In fact, there is no ambiguity, for the cardinality of any two bases for a linear space must indeed coincide. The following characterization of a basis for a linear space will help settle this matter.

PROPOSITION 3

A subset S of a linear space X is a basis for X if, and only if, S is linearly independent and $X = span(S)$.

EXERCISE 19 Prove Proposition 3.

Combining Propositions 2 and 3, we obtain the following fact, which removes the potential ambiguity that surrounds the notion of dimension.

COROLLARY 1

Any two bases of a finite-dimensional linear space have the same number of elements.

It is important to note that a nontrival linear space admits numerous bases. In fact, in an m-dimensional linear space, $m \in \mathbb{N}$, any linearly independent set of vectors of cardinality m is itself a basis. For if $\{x^1, \ldots, x^m\}$

is a basis for a linear space X and $\{y^1, \ldots, y^m\}$ is an arbitrary linearly independent set in X, then the argument outlined in the proof of Proposition 2 shows that

$$span(\{x^1, \ldots, x^m\}) = span(\{y^1, \ldots, y^m\}),$$

so, by Proposition 3, $\{y^1, \ldots, y^m\}$ is a basis for X.

EXAMPLE 5

[1] $\emptyset$ is a basis for the trivial space $\{\mathbf{0}\}$ and hence $dim(\{\mathbf{0}\}) = 0$.

[2] For any $n \in \mathbb{N}$, a basis for $\mathbb{R}^n$ is the set of all unit vectors $S = \{\mathbf{e}^i : i = 1, \ldots, n\}$ where $\mathbf{e}^i$ is the real n-vector the ith component of which equals 1 and the rest of the components of which are 0. It is easy to check that S is linearly independent in $\mathbb{R}^n$, and, since $x = \sum^n x_i \mathbf{e}^i$ for all $x \in \mathbb{R}^n$, we have $span(S) = \mathbb{R}^n$. Thus, by Proposition 3, S is a basis for $\mathbb{R}^n$— it is called the **standard basis for** $\mathbb{R}^n$. We thus have the following comforting fact: $dim(\mathbb{R}^n) = n$.

As noted above, any linearly independent subset S of $\mathbb{R}^n$ with n elements serves as a basis for $\mathbb{R}^n$. For instance, $\{i\mathbf{e}^i : i = 1, \ldots, n\}$ is a basis for $\mathbb{R}^n$ that is distinct from the standard basis.

[3] Fix $m \in \mathbb{N}$, and let X denote the set of all polynomials on $\mathbb{R}$ that are of degree at most m. If $f_i \in \mathbb{R}^{[0,1]}$ is defined by $f_i(t) := t^i$ for each $i = 0, \ldots, m$, then $\{f_0, \ldots, f_m\}$ is a basis for X. (*Note.* Linear independence of $\{f_0, \ldots, f_m\}$ follows from the fact that any polynomial of degree $k \in \mathbb{N}$ has at most k roots.) We have $dim(X) = m + 1$.

[4] Let $\mathbf{e}^1 := (1, 0, 0, \ldots)$, $\mathbf{e}^2 := (0, 1, 0, \ldots)$, etc. It is easily checked that $\{\mathbf{e}^i : i = 1, 2, \ldots\}$ is linearly independent in ℓ^p, $1 \leq p \leq \infty$.

For any given $1 \leq p \leq \infty$, what is the dimension of ℓ^p? Suppose $dim(\ell^p) = m$ for some $m \in \mathbb{N}$. Then, as we noted above, any linearly independent subset of ℓ^p with cardinality m, say $\{\mathbf{e}^1, \ldots, \mathbf{e}^m\}$, would qualify as a basis for ℓ^p. But of course, this is impossible; $\mathbf{e}^{m+1}$ cannot be written as a linear combination of $\mathbf{e}^1, \ldots, \mathbf{e}^m$. Conclusion: *For any* $1 \leq p \leq \infty$, $dim(\ell^p) = \infty$.

[5] Let T be any nonempty set. Is the linear space $\mathbf{B}(T)$ finite-dimensional? The discussion of [2] and [4] is easily adapted here to yield

the complete answer. If T is finite, then $\{\mathbf{1}_{\{t\}} : t \in T\}$ is a basis for $\mathbf{B}(T)$, where $\mathbf{1}_{\{t\}}$ is the indicator function of $\{t\}$ in T. (Why?) Conversely, if $|T| = \infty$, then $\{\mathbf{1}_{\{t\}} : t \in T\}$ is linearly independent in $\mathbf{B}(T)$, and hence $\mathbf{B}(T)$ is an infinite-dimensional linear space. Conclusion: *For any nonempty set T, $dim(\mathbf{B}(T)) = |T|$.*[12] □

REMARK 1. The **dimension** of an affine manifold S in a linear space X, denoted by $dim(S)$, is defined as the dimension of the linear subspace of X that is "parallel" to S. Put more precisely,

$$dim(S) := dim(S - x) \tag{6}$$

for any $x \in S$. (There is no ambiguity in this definition; recall (3).) In particular, for any hyperplane H in X, we have $dim(H) = dim(X) - 1$. (Prove!)

We can extend this definition to the case of any nonempty subset S of X by defining the dimension of this set as the dimension of the affine hull of S. That is, for any $S \in 2^X \backslash \{\emptyset\}$, we define $dim(S) := dim(aff(S))$. This is, of course, a generalization of the previous definitions. Indeed, by (6) and (5), for any $x \in S$,

$$dim(S) = dim(aff(S)) = dim(aff(S) - x) = dim(span(S - x)).$$

For instance, $dim(\{x\}) = 0$ and $dim(\{\lambda x + (1-\lambda)y : 0 \leq \lambda \leq 1\}) = 1$ for any $x, y \in \mathbb{R}^n$, $n = 1, 2, \ldots$ Less trivially, $dim(O) = n$ for any open subset O of $\mathbb{R}^n$. (Prove!) □

The following simple *uniqueness* result will prove useful on many occasions.

COROLLARY 2
Let S be a basis for a linear space X. Any nonzero vector $x \in X$ can be expressed as a linear combination of finitely many members of S with nonzero coefficients in only one way.[13] *If X is finite-dimensional, then every vector in X can be uniquely written as a linear combination of all vectors in S.*

[12] We will sharpen this observation in Chapter J. It turns out that there does not exist even a countably infinite basis for $\mathbf{B}(T)$ when T is infinite.

[13] If $x = \mathbf{0}$, we obviously need to allow for zero coefficients; hence in this case the claim needs to be altered slightly (because $\mathbf{0} = 0x = 0x + 0y$ for any $x, y \in S$). This trivial problem doesn't arise in the finite-dimensional case because we can then use *all* of the basis elements in the associated linear expressions.

PROOF

We only need to prove the first assertion. (Why?) Take any $x \in X\backslash\{\mathbf{0}\}$. Since $span(S) = X$, there exists a finite subset A of S and a map $\lambda : A \to \mathbb{R}\backslash\{0\}$ such that $x = \sum_{y \in A} \lambda(y)y$. Now suppose that B is another finite subset of S and $\alpha : B \to \mathbb{R}\backslash\{0\}$ is a map such that $x = \sum_{z \in B} \alpha(z)z$. Then

$$\sum_{y \in A\backslash B} \lambda(y)y - \sum_{z \in B\backslash A} \alpha(z)z + \sum_{w \in A \cap B} (\lambda(w) - \alpha(w))w = x - x = \mathbf{0},$$

with the convention that any of the three terms on the left-hand side equals $\mathbf{0}$ if it is a sum over the empty set. Since, by Proposition 3, S is linearly independent in X, so is $(A\backslash B) \cup (B\backslash A) \cup (A \cap B)$. It follows that $\lambda(y) = 0$ for all $y \in A\backslash B$, $\alpha(z) = 0$ for all $z \in B\backslash A$, and $\lambda(w) = \alpha(w)$ for all $w \in A \cap B$. Since neither λ nor α ever takes value zero, this can happen only if $A = B$ and $\lambda = \alpha$. ■

EXERCISE 20 Show that $\{f_0, f_1, \ldots\}$ is a basis for $\mathbf{P}[0, 1]$, where $f_i \in \mathbf{P}[0, 1]$ is defined by $f_i(t) := t^i$, $i = 0, 1, \ldots$. Conclude that $dim(\mathbf{P}[0, 1]) = \infty$.

EXERCISE 21 Show that if Y is a proper linear subspace of a finite-dimensional linear space X, then Y is also finite-dimensional and $dim\, Y < dim\, X$.

EXERCISE 22 Show that every linearly independent set in a finite-dimensional linear space can be extended to a basis for the entire space.

EXERCISE 23 [H] *(Products of Linear Spaces)* For any linear spaces X and Y, we define the following addition and scalar multiplication operations on $X \times Y$:

$$(x, y) + (x', y') := (x + x', y + y') \qquad \text{and} \qquad \lambda(x, y) := (\lambda x, \lambda y)$$

for all $x, x' \in X$, $y, y' \in Y$, and $\lambda \in \mathbb{R}$.[14] Under these operations $X \times Y$ becomes a linear space; it is called the **linear product** (or **direct sum**) of X and Y. (This space is often denoted as $X \oplus Y$ in linear algebra, but we shall stick with the notation $X \times Y$ instead.) Show that

$$dim(X \times Y) = dim(X) + dim(Y).$$

[14] Of course, we use the addition operation on X when computing $x + x'$, and that on Y when writing $y + y'$, even though we do not use a notation that makes this explicit. A similar remark applies to λx and λy as well.

It is a triviality that every finite-dimensional linear space has a basis. In fact, by Exercise 22, we can pick an arbitrary vector x in such a space, and then pick another vector y such that $\{x, y\}$ is linearly independent (provided that the dimension of the space exceeds 1, of course), and then pick another, and so on, secure in the knowledge that we will arrive at a basis in finitely many steps. But what if the space at hand is not finite-dimensional? In that case it is not at all obvious if this procedure would actually get us anywhere. Well, it turns out that it would work out just fine, if one is ready to invoke the Axiom of Choice (Section A.1.7). Our final task in this section is to demonstrate how a basis can be obtained for *any* linear space by using this axiom. But be warned that "how" is actually the wrong word here, for our argument will hardly be constructive. We shall rather use the Axiom of Choice to establish that one can always find a basis for a linear space (exploiting this property as usual "to go from finite to infinite"), but you will notice that the argument will be silent about how to achieve this in actuality. At any rate, the following result is a beautiful illustration of the power of the Axiom of Choice, or its equivalent formulation, Zorn's Lemma.

THEOREM 1
Every linear space has a basis.

PROOF
Let X be a linear space, and pick any linearly independent set Y in X. We would like to show that Y can be extended to a basis for X. To this end, denote by $\mathcal{Y}$ the set of all linearly independent sets in X that contain Y. Since $Y \in \mathcal{Y}$, we have $\mathcal{Y} \neq \emptyset$. Moreover, $(\mathcal{Y}, \supseteq)$ is obviously a poset. Let $(\mathcal{Z}, \supseteq)$ be a loset such that $\mathcal{Z} \subseteq \mathcal{Y}$. We claim that $\cup\mathcal{Z} \in \mathcal{Y}$, that is, it is a linearly independent set in X. To see this, pick arbitrary $z^1, \ldots, z^m$ in $\cup\mathcal{Z}$, and consider any linear combination $\sum^m \lambda_i z^i$ that equals 0. By definition, each z^i belongs to some linearly independent set in X, say Z_i. Since $\mathcal{Z}$ is linearly ordered by $\supseteq$, it is without loss of generality to let $Z_1 \subseteq \cdots \subseteq Z_m$. But then all z^is must belong to Z_m, and since Z_m is a linearly independent set, it follows that $\lambda_1 = \cdots = \lambda_m = 0$. This proves our claim, and lets us conclude that any loset in $(\mathcal{Y}, \supseteq)$ has an $\supseteq$-upper bound in $\mathcal{Y}$. But then, by Zorn's Lemma, $(\mathcal{Y}, \supseteq)$ must possess a $\supseteq$-maximal element, say S. This set is obviously linearly independent in X. Moreover, we have $X = span(S)$,

because if there existed an $x \in X \backslash span(S)$, then $S \cup \{x\}$ would be a linearly independent set, which would contradict the $\supseteq$-maximality of the set S in $\mathcal{Y}$. Thus S is a basis for X. ■

Please note that this proof yields something even stronger than Theorem 1:

> *Any linearly independent subset Y of a linear space X can be enlarged to a basis for X.*

EXERCISE 24[15] A **Hamel basis** for $\mathbb{R}$ is a subset S of $\mathbb{R}$ such that the following two conditions are satisfied:

- For any $m \in \mathbb{N}, s^1, \ldots, s^m \in S$ and $q_1, \ldots, q_m \in \mathbb{Q}$, we have $\sum^m q_i s^i = 0$ only if $q_1 = \cdots = q_m = 0$; and
- any nonzero real number x can be uniquely expressed as a linear combination of finitely many members of S with nonzero *rational* coefficients.

(a) By an argument analogous to the one we used for proving Theorem 1, prove that there exists a Hamel basis for $\mathbb{R}$. The next part of the exercise points to a very surprising implication of this observation (and hence of Zorn's Lemma).

(b) Let S be a Hamel basis for $\mathbb{R}$. Then, for every nonzero $x \in \mathbb{R}$, there exist a unique $m_x \in \mathbb{N}$, a unique subset $\{s^1(x), \ldots, s^{m_x}(x)\}$ of S, and a unique subset $\{q^1(x), \ldots, q^{m_x}(x)\}$ of $\mathbb{Q}\backslash\{0\}$ such that $x = \sum^{m_x} q_i(x)s^i(x)$. Define the self-map f on $\mathbb{R}$ by $f(x) := \sum^{m_x} q_i(x)f(s^i(x))$, where f is defined arbitrarily on S, except that $f(s) = 0$ and $f(s') = 1$ for some $s, s' \in S$. (We have $|S| \geq 2$, right?) Show that $f(x+y) = f(x) + f(y)$ for all $x, y \in \mathbb{R}$, but there is no $\alpha \in \mathbb{R}$ such that $f(x) = \alpha x$ for all $x \in \mathbb{R}$. Conclusion: *Cauchy's Functional Equation admits nonlinear solutions.* (Compare with Lemma D.2).

*EXERCISE 25[16] Prove: If $\mathcal{A}$ and $\mathcal{B}$ are bases for a linear space, then $\mathcal{A} \sim_{\text{card}} \mathcal{B}$.

[15] This exercise will interest you only if you are familiar with Section D.5.3.

[16] This exercise, which generalizes Corollary 1, presumes familiarity with the cardinality theory sketched in Section B.3.1.

2 Linear Operators and Functionals

2.1 Definitions and Examples

Since in a linear space we can talk about linear combinations of vectors, those functions that map one linear space into another in a way that preserves linear combinations are of obvious importance. Such maps are said to be *linear* and are formally defined as follows.

DEFINITION

Let X and Y be two linear spaces. A function $L : X \to Y$ is called a **linear operator** (or a **linear transformation**) if

$$L(\alpha x + x') = \alpha L(x) + L(x') \qquad \text{for all } x, x' \in X \text{ and } \alpha \in \mathbb{R}, \tag{7}$$

or equivalently,

$$L\left(\sum_{i=1}^{m} \alpha_i x^i\right) = \sum_{i=1}^{m} \alpha_i L(x^i)$$

for all $m \in \mathbb{N}$ and $(x^i, \alpha_i) \in X \times \mathbb{R}$, $i = 1, \ldots, m$. The set $L^{-1}(\mathbf{0})$ is called the **null space** (or the **kernel**) of L, and is denoted by $null(L)$, that is,

$$null(L) := \{x \in X : L(x) = \mathbf{0}\}.$$

A *real-valued* linear operator is referred to as a **linear functional on** X.[17]

NOTATION. The set of all linear operators from a linear space X into a linear space Y is denoted as $\mathcal{L}(X, Y)$. So, L is a linear functional on X iff $L \in \mathcal{L}(X, \mathbb{R})$.

WARNING. While, as is customary, we don't adopt a notation that makes this explicit, the $+$ operation on the left of the equation in (7) is not the same as the $+$ operation on the right of this equation. Indeed, the former one is the addition operation *on* X while the latter is that *on* Y. (The same comment applies to the $\cdot$ operation as well: we use the scalar multiplication operation *on* X when writing αx, and that *on* Y when writing $\alpha L(x)$.)

[17] *Quiz.* What is the connection between the notion of a linear operator and that of a homomorphism (Exercise 5)?

EXAMPLE 6

For any $m, n \in \mathbb{N}$, $\mathbf{A} \in \mathbb{R}^{m\times n}$ and $x \in \mathbb{R}^n$, we define

$$\mathbf{A}x := \left(\sum_{j=1}^{n} a_{1j}x_j, \ldots, \sum_{j=1}^{n} a_{mj}x_j\right),$$

where $\mathbf{A} := [a_{ij}]_{m\times n}$ (Section A.1.6.). Then $L : \mathbb{R}^n \to \mathbb{R}^m$ defined by $L(x) := \mathbf{A}x$, is a linear operator. More important: *All linear operators from* $\mathbb{R}^n$ *to* $\mathbb{R}^m$ *arise in this way.* To see this, take any $L \in \mathcal{L}(\mathbb{R}^n, \mathbb{R}^m)$, and define $L_i : \mathbb{R}^n \to \mathbb{R}$ as the function that maps each $x \in \mathbb{R}^n$ to the ith component of $L(x)$, $i = 1, \ldots, m$. Since L is a linear operator, each L_i is a linear functional on $\mathbb{R}^n$, and hence, for any $x \in \mathbb{R}^n$, we have

$$L_i(x) = L_i\left(\sum_{j=1}^{n} x_j\mathbf{e}^j\right) = \sum_{j=1}^{n} L_i(\mathbf{e}^j)x_j$$

for each $i = 1, \ldots, m$. (Here $\{\mathbf{e}^1, \ldots, \mathbf{e}^n\}$ is the standard basis for $\mathbb{R}^n$ (Example 5.[2]).) Therefore, letting $\mathbf{A} := [L_i(\mathbf{e}^j)]_{m\times n}$, we find that $L(x) = \mathbf{A}x$ for all $x \in \mathbb{R}^n$.

In particular, $L : \mathbb{R}^n \to \mathbb{R}$ is a linear functional on $\mathbb{R}^n$ iff there exist real numbers $\alpha_1, \ldots, \alpha_n$ such that $L(x) = \alpha_1x_1 + \cdots + \alpha_nx_n$ for all $x \in \mathbb{R}^n$. Put slightly differently, and using the standard inner product notation, we can say that $L \in \mathcal{L}(\mathbb{R}^n, \mathbb{R})$ iff there is an n-vector ω such that $L(x) = \omega x$ for all $x \in \mathbb{R}^n$.[18] □

EXAMPLE 7

[1] Recall that $\mathbf{C}^1[0, 1]$ is the linear space of all continuously differentiable functions on $[0, 1]$, and define $D : \mathbf{C}^1[0, 1] \to \mathbf{C}[0, 1]$ by $D(f) := f'$. Then D is a linear operator, called the **differentiation operator** on $\mathbf{C}^1[0, 1]$. It is easily seen that $null(D)$ equals the set of all constant real maps on $[0, 1]$.

[18] This is not the whole story. Suppose L was a linear functional defined on a linear subspace Y of $\mathbb{R}^n$. What would it look like then? If you think about it for a moment, you will notice that the answer is not obvious. After all, Y need not have some (or any) of the unit vectors of $\mathbb{R}^n$, and this complicates things. But, with some effort, one can still show that, even in this case, we can find an $\omega \in \mathbb{R}^n$ such that $L(x) = \omega x$ for all $x \in Y$. (This vector will, however, not be uniquely determined by L, unless $dim(Y) = n$.) In the following section I will prove something even stronger than this (Example 9).

[2] Let $L_1, L_2 \in \mathbb{R}^{\mathbf{C}[0,1]}$ be defined by

$$L_1(f) := f(0) \quad \text{and} \quad L_2(f) := \int_0^1 f(t)dt.$$

Both L_1 and L_2 are linear functionals on $\mathbf{C}[0,1]$. (Linearity of L_2 follows from Exercise A.57.) Clearly, *null* $(L_1) = \{f \in \mathbf{C}[0,1] : f(0) = 0\}$. On the other hand, *null*(L_2) is the class of all continuous functions on $[0,1]$ whose integrals vanish. In particular, by Exercise A.61, *null*$(L_2) \cap \mathbb{R}_+^{[0,1]} = \{\mathbf{0}\}$. □

EXAMPLE 8

In the following examples X and Y stand for arbitrary linear spaces.

[1] Any $L \in \mathcal{L}(X,Y)$ maps the origin of X to that of Y, that is, $\mathbf{0} \in$ *null*(L). For, $L(\mathbf{0}) = L(\mathbf{0}+\mathbf{0}) = L(\mathbf{0}) + L(\mathbf{0}) = 2L(\mathbf{0})$ so that $L(\mathbf{0}) = \mathbf{0}$. (Here, of course, $\mathbf{0}$ on the left-hand side is the origin of X, and $\mathbf{0}$ on the right-hand side is the origin of Y.) Therefore, there is only one constant linear operator from X into Y, the one that maps the entire X to the origin of Y. If $Y = \mathbb{R}$, then this functional is equal to zero everywhere, and it is called the **zero functional** on X.

[2] Suppose $L \in \mathcal{L}(X,Y)$ is injective. Then, since $L(\mathbf{0}) = \mathbf{0}$, we must have *null*$(L) = \{\mathbf{0}\}$. Interestingly, the converse is also true. Indeed, if *null*$(L) = \{\mathbf{0}\}$, then, for any $x, x' \in X$, $L(x) = L(x')$ implies $L(x - x') = \mathbf{0}$ by linearity of L, so $x - x' = \mathbf{0}$. Conclusion: *An $L \in \mathcal{L}(X,Y)$ is injective iff* *null*$(L) = \{\mathbf{0}\}$.

[3] For any $L \in \mathcal{L}(X,Y)$, *null*(L) is a linear subspace of X and $L(X)$ is a linear subspace of Y. (Proofs?) □

EXERCISE 26[H] Let Y be the set of all polynomials on $\mathbb{R}$ of degree at most 2. Define the map $L : \mathbb{R}^{2\times 2} \to Y$ by $L([a_{ij}]_{2\times 2})(t) := a_{21}t + (a_{11} + a_{12})t^2$ for all $t \in \mathbb{R}$.

(a) Show that $L \in \mathcal{L}(\mathbb{R}^2, Y)$.

(b) Compute *null*(L) and find a basis for it.

(c) Compute $L(\mathbb{R}^{2\times 2})$ and find a basis for it.

EXERCISE 27 Let X denote the set of all polynomials on $\mathbb{R}$. Define the self-maps D and L on X by $D(f) := f'$ and $L(f)(t) := tf(t)$ for all $t \in \mathbb{R}$, respectively. Show that D and L are linear operators and compute $D \circ L - L \circ D$.

EXERCISE 28[H] Let K and L be two linear functionals on a linear space X, and define the function $KL \in \mathbb{R}^X$ by $KL(x) := K(x)L(x)$. Prove or disprove: KL is a linear functional on X iff it is the zero functional on X.

EXERCISE 29 Let X be the linear space of all polynomials on $\mathbb{R}$ of degree at most 1. Define the linear functionals K and L on X by $K(f) := f(0)$ and $L(f) := f(1)$, respectively. Show that $\{K, L\}$ is a basis for $\mathcal{L}(X, \mathbb{R})$. How would you express $L_1, L_2 \in \mathcal{L}(X, \mathbb{R})$ with

$$L_1(f) := \int_0^1 f(t)dt \qquad \text{and} \qquad L_2(f) := f',$$

as linear combinations of K and L?

The following is a very important result—it is a version of the *Fundamental Theorem of Linear Algebra.*

EXERCISE 30[H] Prove: Given any two linear spaces X and Y,

$$dim(null(L)) + dim(L(X)) = dim(X) \qquad \text{for any } L \in \mathcal{L}(X, Y).$$

EXERCISE 31[H] (*Linear Correspondences*) Let X and Y be two linear spaces and $\Gamma : X \rightrightarrows Y$ a correspondence. We say that Γ is a **linear correspondence** if

$$\alpha\Gamma(x) \subseteq \Gamma(\alpha x) \qquad \text{and} \qquad \Gamma(x) + \Gamma(x') \subseteq \Gamma(x + x') \tag{8}$$

for all $x, x' \in X$ and $\alpha \in \mathbb{R}$.

(a) Show that the following statements are equivalent for a linear correspondence $\Gamma : X \rightrightarrows Y$: (i) Γ is single-valued; (ii) $\Gamma(\mathbf{0}) = \{\mathbf{0}\}$; (iii) we can replace each $\subseteq$ with $=$ in (8).

(b) Show that the following are equivalent for any $\Gamma : X \rightrightarrows Y$: (i) Γ is linear; (ii) $Gr(\Gamma)$ is a linear subspace of $X \times Y$ (Exercise 23); (iii) $\alpha\Gamma(x) + \Gamma(x') = \Gamma(\alpha x + x')$ for all $x, x' \in X$ and $\alpha \in \mathbb{R}\backslash\{0\}$.

(c) Show that if $\Gamma : X \rightrightarrows Y$ is linear, then $\Gamma(\mathbf{0})$ and $\Gamma(X)$ are linear subspaces of Y, and $\Gamma(x) = y + \Gamma(\mathbf{0})$ for any $y \in \Gamma(x)$.

(d) Show that if $L \in \mathcal{L}(X, Y)$, then L^{-1} is a linear correspondence from Y into X. Also prove that, for any linear subspace Z of Y, the correspondence $\Gamma : X \rightrightarrows Y$ defined by $\Gamma(x) := L(x) + Z$ is linear.

(e) Let $\Gamma : X \rightrightarrows Y$ be linear and $P \in \mathcal{L}(\Gamma(X), \Gamma(X))$. Show that if P is *idempotent* (i.e., $P \circ P = P$) and $null(P) = \Gamma(\mathbf{0})$, then $P \circ \Gamma$ is a linear selection from Γ, that is, $P \circ \Gamma \in \mathcal{L}(X, Y)$ and $(P \circ \Gamma)(x) \in \Gamma(x)$ for all $x \in X$. (As usual, we think of the single-valued correspondence $P \circ \Gamma$ as a function here.)

2.2 Linear and Affine Functions

In Example 6 we derived a general characterization of all linear functionals defined on a Euclidean space. Although it is extremely useful, this result does not readily identify the structure of linear functions defined on an arbitrary subset of a given Euclidean space. We need a slight twist in the tale for this.

DEFINITION

Let S be a nonempty subset of a linear space X, and denote by $\mathcal{P}(S)$ the class of all nonempty finite subsets of S. A real map $\varphi \in \mathbb{R}^S$ is called **linear** if

$$\varphi\left(\sum_{x\in A} \lambda(x)x\right) = \sum_{x\in A} \lambda(x)\varphi(x)$$

for any $A \in \mathcal{P}(S)$ and $\lambda \in \mathbb{R}^A$ such that $\sum_{x\in A} \lambda(x)x \in S$.[19]

EXAMPLE 9

Let $n \in \mathbb{N}$ and $\emptyset \neq S \subseteq \mathbb{R}^n$. Under the shadow of Example 6, we have the following characterization: *$\varphi \in \mathbb{R}^S$ is linear iff there exist real numbers $\alpha_1, \ldots, \alpha_n$ such that $\varphi(x) = \sum^n \alpha_i x_i$ for all $x \in S$.*

We only need to prove the "only if" part of this fact. Let $A \subseteq S$ be a basis for $span(S)$. Choose any $B \subset \mathbb{R}^n$ such that $A \cup B$ is a basis for $\mathbb{R}^n$. Then,

[19] This definition is duly consistent with that of a linear functional. When S is a linear subspace of X, $f \in \mathcal{L}(S, \mathbb{R})$ iff $f \in \mathbb{R}^S$ is linear.

by Corollary 2, for any $x \in \mathbb{R}^n$ there exists a unique $\lambda_x \in \mathbb{R}^{A\cup B}$ such that $x = \sum_{z\in A\cup B} \lambda_x(z)z$. Define $L : \mathbb{R}^n \to \mathbb{R}$ by

$$L(x) := \sum_{z\in A} \lambda_x(z)\varphi(z).$$

L extends φ, because, for any $x \in S$, we have $x = \sum_{z\in A} \lambda_x(z)z$, so, by linearity of φ,

$$L(x) = \sum_{z\in A} \lambda_x(z)\varphi(z) = \varphi\left(\sum_{z\in A} \lambda_x(z)z\right) = \varphi(x).$$

Moreover, L is a linear functional. Indeed, for any $x, y \in \mathbb{R}^n$ and $\alpha \in \mathbb{R}$, an easy application of Corollary 2 shows that $\lambda_{\alpha x+y}(z) = \alpha\lambda_x(z) + \lambda_y(z)$ for each $z \in A \cup B$, so

$$\begin{aligned} L(\alpha x + y) &= \sum_{z\in A} \lambda_{\alpha x+y}(z)\varphi(z) \\ &= \alpha \sum_{z\in A} \lambda_x(z)\varphi(z) + \sum_{z\in A} \lambda_y(z)\varphi(z) \\ &= \alpha L(x) + L(y). \end{aligned}$$

But then, by Example 6, there exist real numbers $\alpha_1, \ldots, \alpha_n$ such that $L(x) = \sum^n \alpha_i x_i$ for all $x \in \mathbb{R}^n$. Since $L|_S = \varphi$, therefore, $\varphi(x) = \sum^n \alpha_i x_i$ for all $x \in S$. □

In what follows we will obtain a similar characterization of affine functions. First, let us define these functions properly.

DEFINITION

Let S be a nonempty subset of a linear space X, and denote by $\mathcal{P}(S)$ the class of all nonempty finite subsets of S. A real map $\varphi \in \mathbb{R}^S$ is called **affine** if

$$\varphi\left(\sum_{x\in A} \lambda(x)x\right) = \sum_{x\in A} \lambda(x)\varphi(x)$$

for any $A \in \mathcal{P}(S)$ and $\lambda \in \mathbb{R}^A$ such that $\sum_{x\in A} \lambda(x) = 1$ and $\sum_{x\in A} \lambda(x)x \in S$.

Recall that we have motivated *affinity* as the combination of linearity and translation in the previous section. Precisely the same idea applies here. A map φ on a linear space X is affine iff its graph is an affine manifold on $X \times \mathbb{R}$, that is, it is a translation of a linear subspace of $X \times \mathbb{R}$ (Exercise 23).[20] It follows that φ is affine iff $\varphi - \varphi(\mathbf{0})$ is a linear functional on X. It is important to note that this fact remains true even if the domain of the affine map is not a linear space.

LEMMA 1
Let T be a subset of a linear space with $\mathbf{0} \in T$. Then $\varphi \in \mathbb{R}^T$ is affine if, and only if, $\varphi - \varphi(\mathbf{0})$ is a linear real map on T.

PROOF
If $\varphi \in \mathbb{R}^T$ is affine, and $A \in \mathcal{P}(T)$ and $\lambda \in \mathbb{R}^A$ satisfy $\sum_{x\in A} \lambda(x)x \in T$, then

$$\begin{aligned}
\varphi\left(\sum_{x\in A}\lambda(x)x\right) - \varphi(\mathbf{0}) &= \varphi\left(\sum_{x\in A}\lambda(x)x + \left(1-\sum_{x\in A}\lambda(x)\right)\mathbf{0}\right) - \varphi(\mathbf{0})\\
&= \sum_{x\in A}\lambda(x)\varphi(x) + \left(1-\sum_{x\in A}\lambda(x)\right)\varphi(\mathbf{0}) - \varphi(\mathbf{0})\\
&= \sum_{x\in A}\lambda(x)(\varphi(x)-\varphi(\mathbf{0})),
\end{aligned}$$

that is, $\varphi - \varphi(\mathbf{0})$ is linear. Conversely, if $\varphi - \varphi(\mathbf{0}) \in \mathbb{R}^T$ is linear, and $A \in \mathcal{P}(T)$ and $\lambda \in \mathbb{R}^A$ satisfy $\sum_{x\in A}\lambda(x) = 1$ and $\sum_{x\in A}\lambda(x)x \in T$, then

$$\begin{aligned}
\varphi\left(\sum_{x\in A}\lambda(x)x\right) &= \left(\varphi\left(\sum_{x\in A}\lambda(x)x\right) - \varphi(\mathbf{0})\right) + \varphi(\mathbf{0})\\
&= \left(\sum_{x\in A}\lambda(x)(\varphi(x)-\varphi(\mathbf{0}))\right) + \varphi(\mathbf{0})\\
&= \sum_{x\in A}\lambda(x)\varphi(x) - \left(\sum_{x\in A}\lambda(x)\right)\varphi(\mathbf{0}) + \varphi(\mathbf{0})\\
&= \sum_{x\in A}\lambda(x)\varphi(x),
\end{aligned}$$

that is, φ is affine. ■

[20] No, this is not entirely obvious. But all will become clear momentarily.

Here is a useful application of this result that generalizes the characterization of linear functions that we obtained in Example 9.

PROPOSITION 4
Let $n \in \mathbb{N}$ and $\emptyset \neq S \subseteq \mathbb{R}^n$. Then $\varphi \in \mathbb{R}^S$ is affine if, and only if, there exist real numbers $\alpha_1, \ldots, \alpha_n, \beta$ such that $\varphi(x) = \sum^n \alpha_i x_i + \beta$ for all $x \in S$.

PROOF
We only need to prove the "only if" part of the assertion. Let $\varphi \in \mathbb{R}^S$ be an affine map. Pick any $x^* \in S$, let $T := S - x^*$, and define $\psi \in \mathbb{R}^T$ by $\psi(y) := \varphi(y+x^*)$. It is easily verified that ψ is affine, so Lemma 1 entails that $\psi - \psi(\mathbf{0})$ is a linear function on T. Then, by what we have shown in Example 9, there exist $\alpha_1, \ldots, \alpha_n \in \mathbb{R}$ such that $\psi(y) - \psi(\mathbf{0}) = \sum^n \alpha_i y_i$ for all $y \in T$. So, for any $x \in S$, we have $\varphi(x) - \psi(\mathbf{0}) = \psi(x-x^*) - \psi(\mathbf{0}) = \sum^n \alpha_i(x_i - x_i^*)$. Letting $\beta := \psi(\mathbf{0}) - \sum^n \alpha_i x_i^*$, we find $\varphi(x) = \sum^n \alpha_i x_i + \beta$. ■

EXERCISE 32[H] Let S be a nonempty subset of a linear space. We say that a map $\varphi \in \mathbb{R}^S$ is **pseudo-affine** if $\varphi(\lambda x + (1-\lambda)y) = \lambda\varphi(x) + (1-\lambda)\varphi(y)$ for all $0 \leq \lambda \leq 1$ and $x, y \in S$ with $\lambda x + (1-\lambda)y \in S$.

(a) Give an example of a nonempty subset $S \subseteq \mathbb{R}^2$ and a pseudo-affine $\varphi \in \mathbb{R}^S$ that is not affine.

(b) Show that if S is a convex set (that is, $\lambda x + (1-\lambda)y \in S$ for any $0 \leq \lambda \leq 1$ and $x, y \in S$), then $\varphi \in \mathbb{R}^S$ is affine iff it is pseudo-affine.

2.3 Linear Isomorphisms

Linear operators play an important role with regard to the identification of the basic algebraic relation between two linear spaces. We explore this issue next.

Let X and Y be two linear spaces and $L \in \mathcal{L}(X, Y)$. If L is a bijection, then it is called a **linear isomorphism** between X and Y, and we say that X and Y are **isomorphic**. Just as two isometric metric spaces are indistinguishable from each other insofar as their metric properties are concerned (Section D.1.6), the linear algebraic structures of two isomorphic linear spaces coincide. Put differently, from the perspective of linear algebra, one can regard two isomorphic linear spaces as differing from each other only in the labeling of their constituent vectors.[21]

It is quite intuitive that a finite-dimensional linear space can never be isomorphic to an infinite-dimensional linear space. But there is more to the story. Given the interpretation of the notion of isomorphism, you might expect that the dimensions of any two isomorphic linear spaces must in fact be identical. To see that this is indeed the case, let X and Y be two isomorphic finite-dimensional linear spaces. If $dim(X) = 0$, it is plain that $X = \{\mathbf{0}\} = Y$, so there is nothing to prove. We thus consider the case $dim(X) \in \mathbb{N}$. Let S be a basis for X, and pick any linear isomorphism $L \in Y^X$. We wish to show that $L(S)$ is a basis for Y. Let us first establish that this set is linearly independent in Y. Take an arbitrary $\lambda \in \mathbb{R}^S$ and assume that $\sum_{z\in S} \lambda(z)L(z) = \mathbf{0}$. By linearity of L,

$$L\left(\sum_{z\in S} \lambda(z)z\right) = \sum_{z\in S} \lambda(z)L(z) = \mathbf{0},$$

so $\sum_{z\in S} \lambda(z)z = \mathbf{0}$ (since $null(L) = \{\mathbf{0}\}$ – recall Example 8.[2]). Since S is linearly independent in X, it follows that $\lambda = 0$, as we sought. Thus: $L(S)$ is linearly independent in Y. To see that $Y = span(L(S))$, on the other hand, pick any $y \in Y$ and observe that, by surjectivity of L, there must exist an $x \in X$ with $L(x) = y$. But there exists a $\lambda \in \mathbb{R}^S$ such that $x = \sum_{z\in S} \lambda(z)z$ (since S spans X), so $y = L(x) = L\left(\sum_{z\in S} \lambda(z)z\right) = \sum_{z\in S} \lambda(z)L(z)$. This proves that $L(S)$ indeed spans Y. Thus, $L(S)$ is a basis for Y, and since $|S| = |L(S)|$ (because L is a bijection), we conclude that $dim\ X = dim\ Y$.

Curiously, the converse of this observation also holds, that is, if $dim\ X = dim\ Y < \infty$, then X and Y are isomorphic. To prove this, take any $m \in \mathbb{N}$, and let $\{x^1, \ldots, x^m\}$ and $\{y^1, \ldots, y^m\}$ be any two bases for X and Y, respectively. By Corollary 2, for each $x \in X$, there exists a unique $(\lambda_1(x), \ldots, \lambda_m(x)) \in \mathbb{R}^m$ such that $x = \sum^m \lambda_i(x)x^i$. Consequently, we may define a map $L : X \to Y$ by

$$L(x) := \sum_{i=1}^{m} \lambda_i(x)y^i.$$

21 *Quiz.* Define the binary relation $\sim_{\text{iso}}$ on a given class $\mathcal{X}$ of linear spaces by $X \sim_{\text{iso}} Y$ iff X is isomorphic to Y. Show that $\sim_{\text{iso}}$ is an equivalence relation on $\mathcal{X}$.

We wish to show that L is an isomorphism.[22] Take any $x, x' \in X$ and $\alpha \in \mathbb{R}$. By Corollary 2, we have $\lambda_i(\alpha x + x') = \alpha\lambda_i(x) + \lambda_i(x')$ for each i. (Why?) It follows that

$$\begin{aligned} L\left(\alpha x + x'\right) &= \sum_{i=1}^{m} \lambda_i(\alpha x + x')y^i \\ &= \alpha \sum_{i=1}^{m} \lambda_i(x)y^i + \sum_{i=1}^{m} \lambda_i(x')y^i \\ &= \alpha L(x) + L(x'). \end{aligned}$$

Therefore, $L \in \mathcal{L}(X, Y)$. On the other hand, $null(L) = \{\mathbf{0}\}$, because if $\sum^m \lambda_i(x)y^i = \mathbf{0}$ for some $x \in X$, then $\lambda_1(x) = \cdots = \lambda_m(x) = 0$ since $\{y^1, \ldots, y^m\}$ is linearly independent in Y. Thus, L is injective (Example 8.[2]). Finally, L is surjective, because if $y \in Y$, then $y = \sum^m \lambda_i y^i$ for some $(\lambda_1, \ldots, \lambda_m) \in \mathbb{R}^m$, so choosing $x = \sum^m \lambda_i x^i \in X$ and using the linearity of L, we find $L(x) = L\left(\sum^m \lambda_i x^i\right) = \sum^m \lambda_i L\left(x^i\right) = \sum^m \lambda_i y^i = y$.

Summing up, we have proved:

PROPOSITION 5
Two finite-dimensional linear spaces are isomorphic if, and only if, they have the same dimension.

As an immediate corollary of this observation, we obtain the following striking result:

COROLLARY 3
Every nontrivial finite-dimensional linear space is isomorphic to $\mathbb{R}^n$, for some $n \in \mathbb{N}$.

Thus, for most linear algebraic purposes there is only one finite-dimensional linear space to consider, and this is none other than the good old $\mathbb{R}^n$. We will make frequent use of this important fact in the remainder of the text.

[22] Notice that $\lambda_j(x^i)$ equals 1 if $i = j$, and 0 if $i \neq j$, and therefore, we have $L(x^i) = y^i$ for each i. This is the "trick" behind the definition of L. We wish to map each basis element x^i of X to a basis element y^i of Y, and realize (by Corollary 2) that there is only one linear operator that does this.

EXERCISE 33 Let $L \in \mathcal{L}(X, Y)$. Prove that $null(L) = \{\mathbf{0}\}$ iff L maps linearly independent subsets of X to linearly independent subsets of Y.

EXERCISE 34^{H} Let X be a finite-dimensional linear space and $L \in \mathcal{L}(X, X)$. Prove that the following statements are equivalent: (i) L is injective; (ii) L is surjective; (iii) L is invertible.

EXERCISE 35^{H} Given any two linear spaces X and Y, we define the addition and scalar multiplication operations on $\mathcal{L}(X, Y)$ pointwise: For any $L_1, L_2 \in \mathcal{L}(X, Y)$ and $\lambda \in \mathbb{R}$,

$$(L_1 + L_2)(x) := L_1(x) + L_2(x) \qquad \text{and} \qquad (\lambda L_1)(x) := \lambda L_1(x)$$

for all $x \in X$.

(a) Show that $\mathcal{L}(X, Y)$ is a linear space under these operations.

(b) Prove that if X and Y are finite-dimensional, then

$$dim(\mathcal{L}(X, Y)) = dim(X)dim(Y).$$

An important corollary of this fact is: For any finite-dimensional linear space X,

$$dim(\mathcal{L}(X, \mathbb{R})) = dim(X).$$

In the next exercise we revisit Exercise 23 and offer an explanation for why some authors refer to the product of two linear spaces as the *direct sum* of these spaces.

EXERCISE 36 (*More on Products of Linear Spaces*) Take any linear space X, and let Y and Z be two linear subspaces of X with $Y \cap Z = \{\mathbf{0}\}$ and $Y + Z = X$. (In this case we say that X is an **internal direct sum** of Y and Z.)

(a) Show that for every $x \in X$ there exists a unique $(y, z) \in Y \times Z$ with $x = y + z$.

(b) Show that X is isomorphic to $Y \times Z$.

2.4 Hyperplanes, Revisited

We now turn to the relation between linear functionals (which are purely algebraic objects) and hyperplanes (which are at least partly geometric). You might recall from your earlier courses that a hyperplane in $\mathbb{R}^n$ can be defined as the set of all real n-vectors x such that $\sum^n \lambda_i x_i = \alpha$, where each λ_i is a real number (at least one of which is nonzero) and $\alpha \in \mathbb{R}$. But we know that any $L : \mathbb{R}^n \to \mathbb{R}$ is a nonzero linear functional iff $L(x) = \sum^n \lambda_i x_i$ where $\lambda_i \in \mathbb{R}$ for all i and $\lambda_i \neq 0$ for some i. Therefore, every hyperplane is of the form $L^{-1}(\alpha)$ for some $\alpha \in \mathbb{R}$ and nonzero linear functional L, at least in the case of finite-dimensional spaces. In fact, precisely the same is true in any nontrivial linear space: *A set is a hyperplane iff it is a level set of a nonzero linear functional.*[23] We now make this precise.

PROPOSITION 6

Let Y be a subset of a linear space X. Then Y is a $\supseteq$-maximal proper linear subspace of X if, and only if,

$$Y = null(L)$$

for some nonzero linear functional L on X.

PROOF

Let Y be a $\supseteq$-maximal proper linear subspace of X. Pick any $z \in X \backslash Y$, and notice that $span(Y + z) = X$ by $\supseteq$-maximality of Y. So, for any $x \in X$, there exists a unique $(y^x, \lambda^x) \in Y \times \mathbb{R}$ such that $x = y^x + \lambda^x z$.[24] Now define $L : X \to \mathbb{R}$ by $L(x) := \lambda^x$, and check that L is a nonzero linear functional on X with $null(L) = Y$.[25]

Conversely, take any nonzero $L \in \mathcal{L}(X, \mathbb{R})$, and recall that $null(L)$ is a linear subspace of X. Since L is nonzero, $null(L) \neq X$, so it is a proper linear subspace of X. Now take any $y \in X \backslash null(L)$ and assume that $L(y) = 1$ (otherwise, we would work with $\frac{1}{L(y)} y$ instead of y). Notice that, for any $x \in X$, we have $L(x - L(x)y) = L(x) - L(x) = 0$, that is $x - L(x)y \in null(L)$. Thus, $x \in null(L) + L(x)y$ for all $x \in X$. This means that $null(L) + y$ spans the entire X, which is possible only if $null(L)$ is a $\supseteq$-maximal proper linear subspace of X. (Why?) ■

23 *Reminder.* There is no hyperplane in the trivial linear space.

24 Why unique? Because, if $x = y + \lambda z$ for another $(y, \lambda) \in Y \times \mathbb{R}$, then $(\lambda^x - \lambda)z = (y - y^x) \in Y$, but this is possible only if $\lambda^x = \lambda$ (so that $y = y^x$), since Y is a linear subspace with $z \notin Y$.

25 I hope I'm not skipping too many steps here. I know that L is nonzero, because $L(z) = 1$. Moreover, if $x \in Y$, then $(y^x, \lambda^x) = (x, 0)$, so $L(x) = 0$, and conversely, if $x \in null(L)$, then $x = y^x \in Y$. (Notice how I'm using the uniqueness of (y^x, λ^x).) Now you show that L is linear.

By way of "translating" the involved linear subspaces, we can translate this result into the language of hyperplanes as follows.

COROLLARY 4

A subset H of a linear space X is a hyperplane in X if, and only if,

$$H = \{x \in X : L(x) = \alpha\} \tag{9}$$

for some $\alpha \in \mathbb{R}$ and nonzero linear functional L on X.

PROOF

Let H be a hyperplane in X, that is, $H = Z + x^*$ for some $\supseteq$-maximal proper linear subspace Z of X and $x^* \in X$. By Proposition 6, $Z = null(L)$ for some nonzero $L \in \mathcal{L}(X, \mathbb{R})$. But it is readily checked that $null(L) + x^* = L^{-1}(\alpha)$ where $\alpha := L(x^*)$, and it follows that $H = L^{-1}(\alpha)$.

Conversely, take any $\alpha \in \mathbb{R}$ and nonzero $L \in \mathcal{L}(X, \mathbb{R})$. Pick any $x^* \in L^{-1}(\alpha)$, and observe that $L^{-1}(\alpha) = null(L) + x^*$.[26] Since $null(L)$ is a $\supseteq$-maximal proper linear subspace of X (Proposition 6), this means that $L^{-1}(\alpha)$ is a hyperplane in X. ■

EXERCISE 37[H] Let H be a hyperplane in a linear space X and α a nonzero real number. Prove that if $\mathbf{0} \notin H$, then there exists a *unique* linear functional L on X such that $H = L^{-1}(\alpha)$.

Given Corollary 4, we may identify a hyperplane H in a linear space X with a real number and a nonzero linear functional L on X through equation (9). This allows us to give an analytic form to the intuition that a hyperplane "divides" the entire space into two parts. In keeping with this, we refer to either one of the sets

$$\{x \in X : L(x) \geq \alpha\} \quad \text{and} \quad \{x \in X : L(x) \leq \alpha\}$$

as a **closed half-space** induced by H, and to either one of the sets

$$\{x \in X : L(x) > \alpha\} \quad \text{and} \quad \{x \in X : L(x) < \alpha\}$$

[26] Indeed, if $x \in L^{-1}(\alpha)$, then $x = (x - x^*) + x^* \in null(L) + x^*$, while if $y \in null(L)$, then $L(y + x^*) = \alpha$.

as an **open half-space** induced by H. It is in the nature of things to imagine that H divides X into the closed half-spaces that it induces. If we omit the vectors that belong to H itself, then on one side of H is one of its open half-spaces, and on the other is its other open half-space. However, for this interpretation to make geometric sense, we need to be sure that both of the open half-spaces induced by a hyperplane are nonempty. The next exercise asks you to show that there is nothing to worry about in this regard.

Exercise 38 Show that neither of the open half-spaces induced by a hyperplane H can be empty. Also show that, in $\mathbb{R}^n$, any open half-space is open and any closed half-space is closed.

Well done! This is all the linear algebra we need for a rigorous investigation of convex sets. Before getting our hands busy with convexity, however, we will go through two economic applications that will demonstrate the power of our linear algebraic analysis so far.

3 Application: Expected Utility Theory

This section presents a linear algebraic viewpoint of the classic theory of decision making under risk and uncertainty. Our discussion is facilitated, but not trivialized, by the assumption that there are only *finitely* many prizes that do not involve risk. We begin with a slightly nonstandard treatment of the Expected Utility Theorem. Our approach yields a generalization of this remarkable result, thereby allowing us to give a particularly simple derivation of the famed Anscombe-Aumann expected utility representation.[27]

3.1 The Expected Utility Theorem

Throughout this section we fix a nonempty finite set X, which is interpreted as a set of (monetary or nonmonetary) prizes/alternatives. By a **lottery** (or a **probability distribution**) on X, we mean a map $p \in \mathbb{R}^X_+$ such that

[27] For a detailed introduction to classic expected utility theory, I recommend Kreps (1988), which is an outstanding book. Reading Kreps' text in graduate school changed my research career. I bet it will also do wonders for you, whatever your main area of interest in economics may be.

$\sum_{x \in X} p(x) = 1$. Naturally, we interpret $p(x)$ as the probability of getting $x \in X$, and for any $S \subseteq X$, the number $\sum_{x \in S} p(x)$ as the probability of getting a prize in S. (Notice that our usual convention about summing over the empty set entails that the probability of getting nothing is zero.) We view these probabilities as "objective" in the sense that everyone agrees on the likelihood of getting each alternative in X.

The set of all lotteries on X is denoted by $\mathfrak{L}_X$. This is a compact and convex subset of $\mathbb{R}^X$.[28] By a **degenerate lottery** we mean a lottery that puts unit mass on one of the alternatives x in X; we denote this lottery by δ_x. Put differently, δ_x is the indicator function of $\{x\}$ in X, that is, for any $x \in X$, we define $\delta_x \in \mathfrak{L}_X$ as $\delta_x(y) := 1$ if $y = x$, and $\delta_x(y) := 0$ otherwise. A degenerate lottery is, then, none other than a unit vector in $\mathbb{R}^X$. In particular, the set of all degenerate lotteries spans $\mathbb{R}^X$. More important, every lottery in $\mathfrak{L}_X$ can be written as a *convex* combination of degenerate lotteries. Indeed,

$$p = \sum_{x \in X} p(x)\delta_x \qquad \text{for any } p \in \mathfrak{L}_X, \tag{10}$$

a fact that will prove useful shortly.

For any lottery $p \in \mathfrak{L}_X$ and any real map u on X, the **expected value of u with respect to p** is defined as the real number

$$\mathsf{E}_p(u) := \sum_{x \in X} p(x)u(x).$$

As you know, when u is interpreted as a utility function on X, $\mathsf{E}_p(u)$ corresponds to the *expected utility* of the lottery p.

While interesting real-life scenarios in which the choice problems involve only objective uncertainty are rare—save for gambling problems—lotteries are widely used in economics, so it makes sense to start our analysis by studying the preferences of a decision maker over a set of lotteries. The following definition plays a major role in this analysis.

[28] Since X is finite, $\mathbb{R}^X$ is none other than a Euclidean space. Naturally, we think of $\mathfrak{L}_X$ as a metric subspace of this Euclidean space. To clarify this further, enumerate X as $\{x_1, \ldots, x_{|X|}\}$ and view $\mathfrak{L}_X$ as the set of all $|X|$-vectors in $\mathbb{R}_+^{|X|}$ the components of which add up to 1. From this point of view, a lottery $p \in \mathfrak{L}_X$ is an $|X|$-vector $(p_1, \ldots, p_{|X|})$, where p_i is the probability of getting the prize x_i, $i = 1, \ldots, |X|$.

DEFINITION

Let S be a convex subset of a linear space.[29] A preference relation $\succsim$ on S is called **affine** if

$$p \succsim q \quad \text{if and only if} \quad \lambda p + (1-\lambda)r \succsim \lambda q + (1-\lambda)r$$

for any $p, q, r \in S$ and $0 < \lambda \leq 1$.

The classic theory of expected utility works with affine preference relations. That is, given a convex subset S of $\mathfrak{L}_X$, it imposes on a preference relation $\succsim$ on S the following property:

THE INDEPENDENCE AXIOM. $\succsim$ is affine.

The interpretation of this assumption becomes transparent when a lottery like $\lambda p + (1-\lambda)r$ is viewed as a *compound* lottery. Suppose you are given a pair of lotteries p and q on X, and you have decided that you like p better than q. Now, somebody tells you that you cannot choose between p and q directly; a possibly unfair coin toss has to take place first. That is, the actual choice problem is to choose between the following two lotteries.

Lottery 1: If the coin comes up heads, then you get to play p and otherwise you play r.

Lottery 2: If the coin comes up heads, then you get to play q and otherwise you play r.

Under our compounding interpretation, the lottery $\lambda p + (1-\lambda)r$ corresponds to Lottery 1, where $\lambda \in (0, 1]$ is the probability that the coin comes up heads. Similarly, $\lambda q + (1-\lambda)r$ corresponds to Lottery 2. Which one do you like better, Lottery 1 or 2? Since you would choose p over q if this was your choice problem, your answer *should* be Lottery 1. Contemplate that you have rather chosen Lottery 2, and the coin was tossed. If it came up tails, no problem, you would have no regrets. But if it did come up heads, then you would regret not choosing Lottery 1. Thus, it seems natural to recommend to a decision

[29] The convexity of a set S in a linear space is defined exactly how this concept is defined in a Euclidean space: S is convex iff $\lambda p + (1-\lambda)q \in S$ for any $p, q \in S$ and $0 \leq \lambda \leq 1$. (We will study convex sets in detail in the next chapter.)

maker (who is known to like p better than q) to choose $\lambda p + (1 - \lambda)r$ over $\lambda q + (1 - \lambda)r$. This is the normative gist of the Independence Axiom.[30]

Having said this, we should note that if you wish to view the Independence Axiom descriptively, life gets tougher. The preferences of an individual either satisfy this property or not; there is little we can do to settle this issue theoretically. Tests of the descriptive appeal of the Independence Axiom must be performed experimentally. In fact, quite a bit of experimental literature is devoted to this issue, and there appears to be strong evidence against the universal validity of the Independence Axiom. (See Camerer (1995) for an introductory survey.) Consequently, the following development should be taken with a grain of caution, at least when one has descriptive applications in mind. As for the mathematical structure it brings in, you should mark that the Independence Axiom is a powerful affineness postulate. This point will become amply transparent shortly.

The main objective of the expected utility theory is to find a suitable utility representation for a preference relation on $\mathfrak{L}_X$. If you studied either Section C.2.3 or D.5.2, you know that we need some sort of a continuity postulate to obtain such a representation. For a preference relation $\succsim$ defined on a convex subset S of $\mathfrak{L}_X$, we thus impose the following:

THE CONTINUITY AXIOM. $\succsim$ is continuous.[31]

We now identify the structure of all complete preference relations $\succsim$ on a convex subset S of $\mathfrak{L}_X$ that satisfy the two axioms introduced above, provided that S is **$\succsim$-bounded**, that is, there exist two lotteries $p^*, p_* \in S$ such that $p^* \succsim p \succsim p_*$ for all $p \in S$.

PROPOSITION 7

Let X be a nonempty finite set, S a nonempty convex subset of $\mathfrak{L}_X$, and $\succsim$ a complete preference relation on S such that there is a $\succsim$-maximum and a $\succsim$-minimum in S. Then, $\succsim$ satisfies the Independence and Continuity Axioms if, and only if, there exists a (utility function) $u \in \mathbb{R}^X$ such that for any $p, q \in S$,

$$p \succsim q \quad \text{if and only if} \quad \mathsf{E}_p(u) \geq \mathsf{E}_q(u). \tag{11}$$

30 This reasoning applies to the "only if" part of the axiom. We would reason backward to motivate the "if" part.

31 *Reminder.* This means that $\{q \in S : q \succ p\}$ and $\{q \in S : p \succ q\}$ are open subsets of S for any lottery $p \in S$.

When the preferences of an individual are defined on the entire $\mathfrak{L}_X$, the statement of this result simplifies nicely. In that case, since X is finite and $\succsim$ is complete, there exists an $x^* \in X$ such that $\delta_{x^*} \succsim \delta_x$ for all $x \in X$. But then, by the Independence Axiom (applied finitely many times), we find, for any $p \in \mathfrak{L}_X$,

$$\delta_{x^*} = \sum_{x \in X} p(x)\delta_{x^*} \succsim \sum_{x \in X} p(x)\delta_x = p,$$

so δ_{x^*} is a $\succsim$-maximum of $\mathfrak{L}_X$. Analogously, there is an $x_* \in X$ such that δ_{x_*} is a $\succsim$-minimum of $\mathfrak{L}_X$. Therefore, the following remarkable result, which goes back to 1944, obtains as a special case of Proposition 7. Notice how it brings into the open the combined power of the Independence and Continuity Axioms.

THE EXPECTED UTILITY THEOREM.
(von Neumann–Morgenstern) *Let X be a nonempty finite set,*[32] *and $\succsim$ a complete preference relation on $\mathfrak{L}_X$. Then, $\succsim$ satisfies the Independence and Continuity Axioms if, and only if, there exists a (utility function) $u \in \mathbb{R}^X$ such that* (11) *holds for any $p, q \in \mathfrak{L}_X$.*

We thus say that a complete preference relation $\succsim$ on $\mathfrak{L}_X$ that satisfies the Independence and Continuity Axioms admits an *expected utility representation.* An individual with such preferences can be viewed "as if" she evaluates the value of the alternatives in X by means of a utility function $u \in \mathbb{R}^X$—this is called the **von Neumann–Morgenstern utility function** for $\succsim$—and prefers one lottery to another one iff the former yields higher expectation for u than the latter.

It is important to understand that a von Neumann–Morgenstern utility function is not at all ordinal, that is, if an agent evaluates the value of the lotteries by their expected utilities in terms of $u \in \mathbb{R}^X$, and another one does the same but with respect to $f \circ u$, then the preferences of these agents need not be the same even if $f \in \mathbb{R}^{u(X)}$ is strictly increasing. (Note the sharp contrast with the utility theory covered in Section B.4.) In fact,

32 The Expected Utility Theorem is valid for any separable metric space X; I prove this more general version of the theorem in the companion volume Ok (2007). I can't do so here because the method of proof requires familiarity with some relatively advanced topics in probability theory.

a von Neumann–Morgenstern utility function is unique only up to strictly increasing affine transformations, and hence it is often said to be *cardinal*. This is the gist of the next exercise.

EXERCISE 39 Prove: Given any nonempty finite set X, two real maps u and v on X satisfy (11) for any $p, q \in \mathfrak{L}_X$ iff there exists an $(\alpha, \beta) \in \mathbb{R}_{++} \times \mathbb{R}$ such that $v = \alpha u + \beta$.

EXERCISE 40 Let $X \subseteq \mathbb{R}_{++}$ be a finite (monetary) prize set with $|X| \geq 3$. Define $V : \mathfrak{L}_X \to \mathbb{R}$ by $V(p) := \min\{x \in X : p(x) > 0\} + \max\{x \in X : p(x) > 0\}$. Let $\succsim$ be the preference relation on $\mathfrak{L}_X$ that is represented by V. Does $\succsim$ satisfy the Independence and Continuity Axioms? Does $\succsim$ admit a von Neumann-Morgenstern representation?

EXERCISE 41 (*Allais Paradox*) Let $X := \{0, 1, a\}$,where we think of the prizes as monetary outcomes, and view $a > 1$as a large number. Consider the following choice problems:

Problem 1: Do you prefer the lottery $p^1 = \delta_1$ or $p^2 = .1\delta_a + .89\delta_1 + .01\delta_0$?

Problem 2: Do you prefer the lottery $p^3 = .09\delta_a + .91\delta_0$ or $p^4 = .11\delta_1 + .89\delta_0$?

In the experiments, a lot of subjects are observed to choose p^1 in Problem 1 and p^3 in Problem 2. Show that no expected utility maximizer, whose von Neumann-Morgenstern utility function is strictly increasing, may exhibit such a behavior.[33]

It remains to prove Proposition 7. The idea is this. Since $\succsim$ is continuous, it must have a utility representation. By using the Independence Axiom we can show that at least one of the utility functions for $\succsim$ is affine. But we know exactly what an affine map defined on a subset of a finite-dimensional Euclidean space looks like; we obtained an explicit characterization of such maps in this chapter (Section 2.2). Using this characterization yields readily the representation we are after.

[33] If you want to learn more about the related experiments and modifications of the expected utility theory that account for the Allais paradox, have a look at Camerer (1995) or Starmer (2000).

To formalize this argument, we will use the following simple but very useful observation.

LEMMA 2

(Shapley-Baucells) *Let X be a nonempty set, S a nonempty convex subset of $\mathbb{R}^X$, and $\succsim$ an affine preference relation on S. Then, for any $p, q \in S$, we have $p \succsim q$ if, and only if, there exist a $\lambda > 0$ and $r, s \in S$ with $r \succsim s$ and $p - q = \lambda (r - s)$.*

PROOF

Take any $\lambda > 0$ and $r, s \in S$ such that $r \succsim s$ and $p - q = \lambda (r - s)$. By hypothesis, and affineness of $\succsim$,

$$\tfrac{1}{1+\lambda} p + \tfrac{\lambda}{1+\lambda} s = \tfrac{1}{1+\lambda} q + \tfrac{\lambda}{1+\lambda} r \succsim \tfrac{1}{1+\lambda} q + \tfrac{\lambda}{1+\lambda} s$$

so we get $p \succsim q$ by using the applying the affineness of $\succsim$ again. The converse claim is trivial. ■

PROOF OF PROPOSITION 7

The "if" part is easy and left as an exercise. To prove the "only if" part, take any complete preference relation $\succsim$ on S that satisfies the hypotheses of the assertion, denote the $\succsim$-maximum and the $\succsim$-minimum of S by p^* and p_*, respectively, and assume that $p^* \succ p_*$. (If $p^* \sim p_*$, the proof would be completed upon choosing u to be any constant function. Verify!) Then, by Lemma 2, for any $\alpha, \beta \in [0, 1]$, we have

$$\beta p^* + (1 - \beta) p_* \succ \alpha p^* + (1 - \alpha) p_* \quad \text{if and only if} \quad \beta > \alpha.$$

(Why? Think about it!) Now, by Debreu's Utility Representation Theorem, there exists a $U \in \mathbf{C}(S)$ that represents $\succsim$. By the previous observation, $\lambda \mapsto U(\lambda p^* + (1 - \lambda) p_*)$ is a continuous and strictly increasing map on $[0, 1]$. So since, $U(p^*) \geq U(p) \geq U(p_*)$ for all $p \in S$, the Intermediate Value Theorem entails that, for any $p \in S$ there exists a unique $\lambda_p \in [0, 1]$ such that $U(p) = U(\lambda_p p^* + (1 - \lambda_p) p_*)$. Define $f : U(S) \to \mathbb{R}$ by $f(U(p)) := \lambda_p$. It is plain that f is strictly increasing, so $L := f \circ U$ represents $\succsim$. We claim that L is an affine map on S. Indeed, if $p, q \in S$ and $0 \leq \theta \leq 1$, then

$$p \sim L(p) p^* + (1 - L(p)) p_* \quad \text{and} \quad q \sim L(q) p^* + (1 - L(q)) p_*,$$

so letting $r := \theta p + (1-\theta)q$ and applying the Independence Axiom twice, we find

$$L(r)p^* + (1 - L(r))p_*$$
$$\sim r \sim (\theta L(p) + (1-\theta)L(q))p^* + (1 - (\theta L(p) + (1-\theta)L(q)))p_*$$

and it follows that $L(r) = \theta L(p) + (1-\theta)L(q)$. (Yes?) Since S is convex, this is enough to conclude that L is an affine map.[34] But then, by Proposition 4, there exists a $(u, \beta) \in \mathbb{R}^X \times \mathbb{R}$ such that $L(p) = \sum_{x \in X} u(x)p(x) + \beta$ for all $x \in S$. Since L represents $\succsim$, we have (11) for any $p, q \in S$. ∎

The following exercise provides a more elementary proof for the Expected Utility Theorem, one that does not use Debreu's Utility Representation Theorem. In Chapter H, we will be able to offer still another proof by deducing the Expected Utility Theorem from a substantially more general representation result.

EXERCISE 42 (*Elementary Proof of the Expected Utility Theorem*) Let X be a nonempty finite set and $\succsim$ a complete preorder on $\mathfrak{L}_X$ that satisfies the Independence and Continuity Axioms. As shown above, there exists an $(x^*, x_*) \in X^2$ such that $\delta_{x^*} \succsim p \succsim \delta_{x_*}$ for any $p \in \mathfrak{L}_X$. Notice that if $\delta_{x^*} \sim \delta_{x_*}$, then $p \sim q$ for any $p, q \in \mathfrak{L}_X$, and everything becomes trivial (take u as any constant function). So, assume in what follows that $\delta_{x^*} \succ \delta_{x_*}$, and define $L \in [0,1]^{\mathfrak{L}_X}$ by

$$L(p) := \inf\{\alpha \in [0,1] : \alpha\delta_{x^*} + (1-\alpha)\delta_{x_*} \succsim p\}.$$

(a) Prove that, for any $p \in \mathfrak{L}_X$, we have $p \sim \lambda\delta_{x^*} + (1-\lambda)\delta_{x_*}$ iff $\lambda = L(p)$.
(b) Show that $p \succsim q$ iff $L(p) \geq L(q)$ for any $p, q \in \mathfrak{L}_X$.
(c) Use the definition of L, the Independence Axiom, and the Principle of Mathematical Induction to show that L is an affine map on $\mathfrak{L}_X$.
(d) Define $u \in \mathbb{R}^X$ by $u(x) := L(\delta_x)$. Use part (c) and (10) to conclude the proof.

[34] No, I'm not cheating. I know that I need to verify here that $\varphi(\sum^m \lambda_i p^i) = \sum^m \lambda_i \varphi(p^i)$ for all $m \in \mathbb{N}$ and $(p^i, \lambda_i) \in S \times \mathbb{R}$, $i = 1, \ldots, m$ such that $\sum^m \lambda_i = 1$ and $\sum^m \lambda_i p^i \in S$. But I also know that what I established above is enough to guarantee that this property holds, *given that S is convex*. Recall Exercise 32.

Exercise 43 (Herstein-Milnor) Show that the Expected Utility Theorem would remain valid if we replaced the Continuity Axiom with the following property: For any $p, q, r \in \mathfrak{L}_X$ with $p \succ q \succ r$, there exist $\alpha, \beta \in (0,1)$ such that $\beta p + (1-\beta)r \succ q \succ \alpha p + (1-\alpha)r$.

Exercise 44 Let X be an arbitrary nonempty set, and define the **support** of any $p \in \mathbb{R}^X$ as the set $supp(p) := \{x \in X : p(x) > 0\}$. We say that a map $p \in \mathbb{R}^X_+$ is a **simple lottery** on X if $supp(p)$ is finite and $\sum_{x \in supp(p)} p(x) = 1$. The set of all simple lotteries on X is denoted as $\mathbb{P}(X)$. Let $\succsim$ be a complete preference relation on $\mathbb{P}(X)$. Show that $\succsim$ satisfies the Independence Axiom and the property introduced in Exercise 43 (both on $\mathbb{P}(X)$) iff there exists a (utility function) $u \in \mathbb{R}^X$ such that, for any $p, q \in \mathbb{P}(X)$,

$$p \succsim q \quad \text{if and only if} \quad \sum_{x \in supp(p)} p(x)u(x) \geq \sum_{x \in supp(q)} q(x)u(x).$$

3.2 Utility Theory under Uncertainty

The expected utility theorem does not apply to a situation in which the uncertainty surrounding the possible events is *subjective*. For instance, this approach is not adequate to model the gambling behavior of individuals over, say, basketball games. In principle, everybody may have a different probability assessment of the event that "the Boston Celtics will beat the New York Knicks in their next match-up," so in such a scenario, objective lotteries cannot be the primitives of the model. Thus one says that the von Neumann–Morgenstern expected utility theory is a model of *objective uncertainty* (commonly called *risk*).

To deal with situations involving subjective uncertainty (often called simply *uncertainty*), we extend the von Neumann–Morgenstern setup by introducing to the model a nonempty finite set Ω that is interpreted as the set of **states of nature**. In the gambling example above, for instance, the state space would be $\Omega = \{\text{Celtics win, Knicks win}\}$.

Let X be a nonempty finite set, which we again view as the set of all alternatives/prizes. By an **act**, we mean a function $\mathbf{a} \in X^\Omega$ that tells us which prizes would obtain at which states of nature. For instance, suppose, in the gambling example above, an individual is contemplating whether he should bet \$10 on the Celtics or on the Knicks. Let's assume that returns are one-to-two, that is, if the team that one bets on wins, then one gets twice the size of her bet, otherwise one loses her money. Then, in this setup, we can think of "betting \$10 on the Celtics" as an act that brings \$20 in the state "Celtics win" and −\$10 in the state "Knicks win."

In the framework of decision making under uncertainty, it is the acts that are considered as the objects of choice (rather than the lotteries, as it was the case before). Although quite a far-reaching theory can indeed be developed by using only acts as such—this is called the *Savagean* theory of uncertainty—the formalities surrounding this theory are somewhat involved. Life gets much easier, fortunately, if we contemplate that people have extraneous randomization devices at their disposal so that, not a certain outcome, but rather an objective lottery obtains at each state of nature.

This trick, also used widely in game theory, leads to what is called the *Anscombe-Aumann theory* of uncertainty, in which the primitive objects of choice are the so-called horse race lotteries. By definition, a **horse race lottery** h on Ω is a map from the state space Ω to the set of all lotteries on X, that is, $h : \Omega \to \mathfrak{L}_X$. The artificial interpretation is that first a horse race is run, which in turn leads to a state of nature obtaining, thereby dissolving the subjective uncertainty. Given the state of nature, then, a lottery is played out, thereby dissolving the remaining objective uncertainty. We denote the space of all horse race lotteries by $\mathfrak{H}_{\Omega,X}$, that is, $\mathfrak{H}_{\Omega,X} := \mathfrak{L}_X^\Omega$. To simplify our notation, we write in what follows h_ω, instead of $h(\omega)$, for any $h \in \mathfrak{H}_{\Omega,X}$ and $\omega \in \Omega$. Put simply, $h_\omega(x)$ is the probability of getting prize x in state ω.

WARNING. Since Ω and X are both finite, one can think of $\mathbb{R}^{\Omega \times X}$ as a Euclidean space (which is identical to $\mathbb{R}^{|\Omega||X|}$). Moreover, we can, and will, view $\mathfrak{H}_{\Omega,X}$, that is $\mathfrak{L}_X^\Omega$, as a metric subspace of $\mathbb{R}^{\Omega \times X}$. (Just view any $h \in \mathfrak{L}_X^\Omega$ as the map $(\omega, x) \mapsto h_\omega(x)$.) A moment's reflection would show that, looked at this way, $\mathfrak{H}_{\Omega,X}$ is a closed and bounded subset of $\mathbb{R}^{\Omega \times X}$. It is thus a compact metric space in its own right.

The preference relation $\succsim$ of a decision maker in the Anscombe-Aumann setup is defined on $\mathfrak{H}_{\Omega,X}$. The two main axioms we used in the previous section extend to this framework readily.

THE INDEPENDENCE AXIOM*. For any $f, g, h \in \mathfrak{H}_{\Omega,X}$ and any $0 < \lambda \leq 1$,

$$f \succsim g \quad \text{if and only if} \quad \lambda f + (1-\lambda)h \succsim \lambda g + (1-\lambda)h.$$

The Continuity Axiom*. $\succsim$ is a continuous relation on $\mathfrak{H}_{\Omega,X}$.

The interpretation of these properties is identical to their corresponding interpretations in the von Neumann–Morgenstern setup. However, you should note that the Independence Axiom* is considerably stronger than the more desirable independence property that would condition the relation $\succsim$ to behave independently *only* with respect to acts. This is the reason why working with $\mathfrak{H}_{\Omega,X}$ instead of the set of all acts yields an easier axiomatic theory. The "bite" of the Independence Axiom is stronger on $\mathfrak{L}_X^{\Omega}$ than on X^{Ω}.

These two axioms are not strong enough to deliver us a theorem of the same caliber with the Expected Utility Theorem. However, we still get the following *additively separable representation* result.

The State-Dependent Expected Utility Theorem

Let Ω and X be two nonempty finite sets, and $\succsim$ a complete preference relation on $\mathfrak{H}_{\Omega,X}$. Then $\succsim$ satisfies the Independence and Continuity Axioms if, and only if, there exist (utility functions) $u_\omega \in \mathbb{R}^X$, $\omega \in \Omega$, such that for any $f, g \in \mathfrak{H}_{\Omega,X}$,*

$$f \succsim g \quad \text{if and only if} \quad \sum_{\omega\in\Omega} \mathsf{E}_{f_\omega}(u_\omega) \geq \sum_{\omega\in\Omega} \mathsf{E}_{g_\omega}(u_\omega). \tag{12}$$

Proof

We only need to prove the "only if" part of the assertion. The idea is to bring this claim in a format that Proposition 7 applies. To this end, define

$$S := \tfrac{1}{|\Omega|}\mathfrak{H}_{\Omega,X}.$$

Then S is a nonempty convex and compact subset of $\mathfrak{L}_{\Omega\times X}$. (Yes?) Let us define the relation $\trianglerighteq$ on S as: $\frac{1}{|\Omega|}f \trianglerighteq \frac{1}{|\Omega|}g$ iff $f \succsim g$. Obviously, $\trianglerighteq$ is a complete preference relation on S. Moreover, if $\succsim$ satisfies the Independence and Continuity Axioms*, then $\trianglerighteq$ satisfies the Independence and Continuity Axioms. By compactness of S and continuity of $\trianglerighteq$, we know that S contains a $\trianglerighteq$-maximum and a $\trianglerighteq$-minimum. (Yes?) By Proposition 7 and definition of $\trianglerighteq$, therefore, there exists a $u \in \mathbb{R}^{\Omega\times X}$ such that, for any $f, g \in \mathfrak{H}_{\Omega,X}$,

$$f \succsim g \quad \text{iff} \quad \sum_{(\omega,x)\in\Omega\times X} \tfrac{1}{|\Omega|} f_\omega(x)u(\omega,x) \geq \sum_{(\omega,x)\in\Omega\times X} \tfrac{1}{|\Omega|} g_\omega(x)u(\omega,x).$$

The proof is completed upon defining $u_\omega := \frac{1}{|\Omega|}u(\omega,\cdot)$ for each $\omega \in \Omega$. ∎

The interpretation of the preferences that can be represented as in (12) is straightforward: To compare the various horse race lotteries, an individual with such preferences computes the expected utility of these lotteries at each state according to a possibly different utility function, and then chooses the horse race lottery that yields the highest *total* of the expected utilities.[35]

Such a representation is obviously suitable only when we wish to allow the individual to possess state-dependent utility functions. For instance, if a Celtics fan is betting on the Boston Celtics winning a given basketball game, it is likely that her utility function over monetary outcomes of the bet *depends* on whether or not the Celtics win. To model this sort of a situation, state-dependent utility representations like the one reported above becomes appropriate.

How do we get a state-independent representation? There is actually a very nice answer to this question. But let us first get rid of some potential trivialities that we so far allowed in the present setting. We say that a state $\omega' \in \Omega$ is $\succsim$**-trivial** if $\succsim$ deems any two horse race lotteries f and g that differ from each other only at state ω' as indifferent (that is, $f \sim g$ whenever $f_\omega = g_\omega$ for all $\omega \in \Omega\backslash\{\omega'\}$). For instance, in our gambling example, if $\Omega :=$ {Celtics win, Knicks win, Al Pacino wins the Academy Award that year}, then it would make good sense to assume that the third state is trivial with respect to the preference relation of the gambler in question.

The upshot is that the existence of a trivial state would obviously not play a role in shaping one's decisions, so it is without loss of generality for our purposes to ignore all such states. We do so by means of the following postulate on the preference relation $\succsim$:

THE NO-TRIVIALITY AXIOM*. No state ω in Ω is $\succsim$-trivial.

Observe next that any (complete) preference relation $\succsim$ on $\mathfrak{H}_{\Omega,X}$ induces a (complete) preference relation $\succsim^*$ on $\mathfrak{L}_X$ in a natural manner:

$$p \succsim^* q \quad \text{if and only if} \quad \langle p \rangle \succsim \langle q \rangle,$$

where $\langle p \rangle$ stands for the constant horse race lottery that equals p at every state, and similarly for $\langle q \rangle$. The idea is that, in choosing between $\langle p \rangle$ and $\langle q \rangle$,

[35] The lack of uniqueness of the utility functions found in the State-Dependent Expected Utility Theorem makes this interpretation suspect. But I don't want to get into this issue at this point.

uncertainty does not matter at all, for the consequences of either of these horse race lotteries do not depend on the state to be realized. Therefore, one may argue, if $\langle p \rangle \succsim \langle q \rangle$, then this must be because the agent's risk preferences deem the lottery p better than q.[36] But then, what would the agent decide when offered to choose between two horse race lotteries f and g such that $f_\omega \succsim^* g_\omega$ for *all* $\omega \in \Omega$? Because f dominates g in terms of $\succsim^*$, if we indeed agree that $\succsim^*$ corresponds to the agent's (state-independent) preferences over lotteries, it makes good sense to presuppose that $f \succsim g$ would hold in this case. This is, at any rate, the content of our final axiom on $\succsim$:

THE MONOTONICITY AXIOM*. For any $f, g \in \mathfrak{H}_{\Omega,X}$, if $f_\omega \succsim^* g_\omega$ for all $\omega \in \Omega$, then we have $f \succsim g$. Moreover, if $f_\omega \succsim^* g_\omega$ for all $\omega \in \Omega$, and $f_\omega \succ^* g_\omega$ for some $\omega \in \Omega$, then $f \succ g$.

You should note that this is a powerful axiom that implies the *separability* of preferences across states of nature. To see what we mean by this, take any $h \in \mathfrak{H}_{\Omega,X}$, and denote by $(r, h_{-\omega})$ the horse race lottery that yields the lottery $r \in \mathfrak{L}_X$ in state ω and agrees with h in all other states, that is,

$$(r, h_{-\omega})_\tau := \begin{cases} r, & \text{if } \tau = \omega \\ h_\tau & \text{otherwise} \end{cases}.$$

Observe that if $\succsim$ is a *complete* preference relation on $\mathfrak{H}_{\Omega,X}$ that satisfies the Monotonicity Axiom*, then we must have

$$(p, h_{-\omega}) \succsim (q, h_{-\omega}) \quad \text{if and only if} \quad (p, h_{-\omega'}) \succsim (q, h_{-\omega'}) \tag{13}$$

for any lotteries $p, q \in \mathfrak{L}_X$ and any two states $\omega, \omega' \in \Omega$. (*Proof.* The completeness of $\succsim$ implies that of $\succsim^*$. Consequently, by the Monotonicity Axiom*, $(p, h_{-\omega}) \succsim (q, h_{-\omega})$ implies $p \succsim^* q$, and in turn, applying again the Monotonicity Axiom*, we find $(p, h_{-\omega'}) \succsim (q, h_{-\omega'})$.) Therefore, it should not come as a surprise that the Monotonicity Axiom* paves the way toward a state-independent expected utility theorem.

[36] This is not unexceptionable. Notice that I am sneaking in the assumption that the risk preferences of the agent is *state-independent*.

Here comes a beautiful result.

The Anscombe-Aumann Expected Utility Theorem

Let Ω and X be two nonempty finite sets and $\succsim$ a complete preference relation on $\mathfrak{H}_{\Omega,X}$. Then $\succsim$ satisfies the Independence, Continuity, and Monotonicity Axioms if, and only if, there exist a (utility function) $u \in \mathbb{R}^X$ and a (probability distribution) $\mu \in \mathfrak{L}_\Omega$ such that, for any $f, g \in \mathfrak{H}_{\Omega,X}$,*

$$f \succsim g \quad \text{if and only if} \quad \sum_{\omega \in \Omega} \mu(\omega)\mathsf{E}_{f_\omega}(u) \geq \sum_{\omega \in \Omega} \mu(\omega)\mathsf{E}_{g_\omega}(u). \tag{14}$$

If, in addition, $\succsim$ satisfies the No-Triviality Axiom, then μ is unique, and u is unique up to positive affine transformations.*

Proof

Assume that $\succsim$ is a complete preference relation on $\mathfrak{H}_{\Omega,X}$ that satisfies the Independence, Continuity, and Monotonicity Axioms*. Then, by the State-Dependent Expected Utility Theorem, there exist $u_\omega \in \mathbb{R}^X$, $\omega \in \Omega$, that satisfy (12). Then, for any ω and ω' in Ω, (13) implies that $\mathsf{E}_p(u_\omega) \geq \mathsf{E}_q(u_\omega)$ iff $\mathsf{E}_p(u_{\omega'}) \geq \mathsf{E}_q(u_{\omega'})$ for any $p, q \in \mathfrak{L}_X$, that is, u_ω and $u_{\omega'}$ are the von Neumann–Morgenstern utility functions for the same preference relation on $\mathfrak{L}_X$. Since a von Neumann–Morgenstern utility function is unique up to strictly increasing affine transformations, there must exist an $(\alpha_\omega, \beta_\omega) \in \mathbb{R}_{++} \times \mathbb{R}$ such that $u_\omega = \alpha_\omega u_{\omega'} + \beta_\omega$ (Exercise 39). Thus, defining $u := u_{\omega'}$ and using (12), we find

$$f \succsim g \quad \text{iff} \quad \sum_{\omega \in \Omega} \alpha_\omega \mathsf{E}_{f_\omega}(u) \geq \sum_{\omega \in \Omega} \alpha_\omega \mathsf{E}_{g_\omega}(u)$$

for any $f, g \in \mathfrak{H}_{\Omega,X}$. Defining $\mu \in \mathfrak{L}_\Omega$ by

$$\mu(\omega) := \frac{\alpha_\omega}{\sum_{\tau \in \Omega} \alpha_\tau},$$

completes the "only if" part of the main assertion. Establishing the "if" and uniqueness parts of the theorem is left as an exercise. ■

An individual whose preference relation over horse race lotteries satisfy the axioms used in the Anscombe-Aumann Expected Utility Theorem ranks two *acts* $\mathbf{a}$ and $\mathbf{b}$ in X^Ω as follows:

$$\mathbf{a} \succsim \mathbf{b} \quad \text{if and only if} \quad \mathsf{E}_\mu(u(\mathbf{a})) \geq \mathsf{E}_\mu(u(\mathbf{b}))$$

where $\mu \in \mathfrak{L}_\Omega$ and $u \in \mathbb{R}^X$. Moreover, the risk preferences of this individual at *any* state is the same, and is represented by means of a von Neumann–Morgenstern utility function. That is, given any state, the agent compares the lotteries at that state according to their expected utility (with respect to a fixed utility function on X). Moreover, this agent has a *subjective probability assessment* (*prior beliefs*) μ about the likelihood of each state of nature. (Clearly, μ depends on $\succsim$.) The Anscombe-Aumann Expected Utility Theorem also tells us how this individual puts these two pieces together. She ranks acts (and horse race lotteries) by computing the (subjective) expected utility of each act using her utility function u and prior beliefs μ.[37]

EXERCISE 45 Prove the two uniqueness claims made in the Anscombe-Aumann Expected Utility Theorem.

*EXERCISE 46 Obtain the counterpart of the Anscombe-Aumann Expected Utility Theorem (modifying some of the axioms if necessary) for the case in which the state space Ω is countably infinite.

4* Application: Capacities and the Shapley Value

In this section we consider an interesting application of our linear algebra review to a seemingly unrelated area, the theory of coalitional games.[38] This will also allow us to introduce the elements of the theory of capacities and Choquet integration, which is used in the theory of individual decision making under uncertainty.

[37] It doesn't get any better than this, people! By means of primitive and normatively reasonable (appealing?) assumptions on her preferences, we recover the agent's utility function over prizes, her initial beliefs about likelihood of the states of the world, and find out how she evaluates her potential acts on the basis of her beliefs and von Neumann–Morgenstern utility function. It is possible to lose perspective while working out the technicalities, but if you take a step back and examine what we have established (which is a very special case of what Leonard Savage established in 1954), I'm sure you will be impressed.

[38] Of course, what I present here is only a minor excursion that focus on certain mathematical aspects of the theory. For thorough accounts of the general theory of coalitional games, I recommend Friedman (1990, Chaps. 6–8) and Osborne and Rubinstein (1994, Chaps. 13–15).

4.1 Capacities and Coalitional Games

Let $\mathcal{N} := \{\{1, \ldots, n\} : n \in \mathbb{N}\}$. We interpret any member N of $\mathcal{N}$ as a set of individual players in a given environment. Each member of 2^N can then be thought of as a possible *coalition* within the associated society, N being the grand coalition. By definition, a **capacity** on $N \in \mathcal{N}$ is any function $\mathbf{v} : 2^N \to \mathbb{R}$ such that $\mathbf{v}(\emptyset) = 0$. We denote the class of all capacities on N as $\mathcal{V}_N$.

Fix a society $N \in \mathcal{N}$. In the classic theory of coalitional games, one assumes that any coalition in N can sign a binding agreement, and focuses on which payoff/utility profiles (for the members) can actually be achieved by each coalition. The theory thus abstracts from strategic considerations at the individual level, and concentrates rather on the behavior of coalitions. If we further assume that side payments are possible (as it would be the case where the payoffs are measured in monetary terms), then the situation simplifies further, for then all we have to specify to capture the capability of a coalition is the *total payoff* that this coalition can achieve. Provided that we envisage a situation in which the actions of the nonmembers do not affect what a coalition can actually sustain—a 'big' assumption, to be sure—capacities would serve well for this purpose. Given a capacity $\mathbf{v} \in \mathcal{V}_N$, we may then interpret $\mathbf{v}(A)$ as the "worth" of (i.e., the total payoff that can be achieved by) the coalition $A \subseteq N$. Given this interpretation, we refer to the list $(N, \mathbf{v})$, where $\mathbf{v}$ is a capacity on N, as a **coalitional game** in this section. The set of all coalitional games is denoted as $\mathcal{G}$, that is,

$$\mathcal{G} := \{(N, \mathbf{v}) : N \in \mathcal{N} \text{ and } \mathbf{v} \in \mathcal{V}_N\}.$$

DEFINITION

Given an arbitrary $N \in \mathcal{N}$, let $\mathbf{v}$ be a capacity on N. We say that $\mathbf{v}$ (and the coalitional game $(N, \mathbf{v})$) is **monotonic** if

$$\mathbf{v}(B) \geq \mathbf{v}(A)$$

for all $A, B \subseteq N$ with $A \subseteq B$, and **superadditive** if

$$\mathbf{v}(A \cup B) \geq \mathbf{v}(A) + \mathbf{v}(B)$$

for all $A, B \subseteq N$ with $A \cap B = \emptyset$. The capacity $\mathbf{v}$ is said to be **additive** if

$$\mathbf{v}(A \cup B) = \mathbf{v}(A) + \mathbf{v}(B)$$

for all $A, B \subseteq N$ with $A \cap B = \emptyset$.

In the context of coalitional games, monotonicity of a capacity $\mathbf{v}$ means that the worth of a coalition cannot decrease when a new member joins the coalition.[39] Superadditivity of $\mathbf{v}$, on the other hand, says simply that any coalition can achieve at least as much as the sum of its constituent members. Since this property is satisfied in a good number of applications, some texts actually refer to a superadditive capacity as a *coalitional game with side payments*.[40] While in general neither property implies the other, every nonnegative superadditive capacity is monotonic. (*Proof.* If $\mathbf{v}$ is superadditive and $\mathbf{v} \geq 0$, then $\mathbf{v}(B) = \mathbf{v}(A \cup (B\backslash A)) \geq \mathbf{v}(A) + \mathbf{v}(B\backslash A) \geq \mathbf{v}(A)$ for any $A \subseteq B \subseteq N$.)

EXAMPLE 10

[1] Take any $N \in \mathcal{N}$, and define the real map $\mathbf{v}$ on 2^N by $\mathbf{v}(A) := |A|$. Clearly, $\mathbf{v}$ is an additive capacity on N—it is called the **counting measure** on N.

[2] For any given $N \in \mathcal{N}$ and $\theta \geq 1$, define the real map $\mathbf{v}_\theta$ on 2^N by $\mathbf{v}_\theta(A) := |A|^\theta$ for all $A \subseteq N$. Then each $\mathbf{v}_\theta$ is a monotonic superadditive capacity, with $\mathbf{v}_1$ being the counting measure.[41]

[3] For any $N \in \mathcal{N}$, the **simple majority game** on N is defined as the capacity $\mathbf{v}_{\text{maj}} \in \mathcal{V}_N$, where

$$\mathbf{v}_{\text{maj}}(A) := \begin{cases} 1, & |A| > \frac{1}{2}|N| \\ 0, & |A| \leq \frac{1}{2}|N| \end{cases}.$$

$\mathbf{v}_{\text{maj}}$ is easily checked to be monotonic and superadditive. It can be thought of as modeling a voting situation in which an alternative for the society will be chosen only if a strict majority of the voters supports this.

[4] Let $N \in \mathcal{N}$ and fix some $A \in 2^N\backslash\{\emptyset\}$. The capacity $\mathbf{u}_A \in \mathcal{V}_N$ defined as

$$\mathbf{u}_A(B) := \begin{cases} 1, & A \subseteq B \\ 0, & A \nsubseteq B \end{cases} \tag{15}$$

39 Some authors refer to a monotonic capacity simply as a "capacity."

40 The terminology *transferable utility* (TU) *coalitional game* is also used in the literature.

41 To prove the former claim, it is enough to show that $(a+b)^\theta \geq a^\theta + b^\theta$ for all $a, b > 0$ and $\theta \geq 1$. Indeed, since $\frac{d}{d\theta}(a^\theta + b^\theta)^{\frac{1}{\theta}} \leq 0$ for any $a, b > 0$, we have $\max\{(a^\theta + b^\theta)^{\frac{1}{\theta}} : \theta \geq 1\} = a + b$.

is called a **unanimity game**. We can again think of this game as corresponding to a voting situation in which a coalition "wins" the contest only if it contains every member of A. In particular, $\mathbf{u}_{\{j\}}$ is called a **dictatorship game** (in which the dictator is the jth player). It is easy to check that any unanimity game is a monotonic superadditive capacity. These games will play an important role in what follows. □

4.2 The Linear Space of Capacities

Let us fix an arbitrary $N \in \mathcal{N}$. Since the set $\mathcal{V}_N$ of all capacities consists of real functions, we can define the vector addition and scalar multiplication operations on $\mathcal{V}_N$ in the usual way: For any $\mathbf{v}, \mathbf{w} \in \mathcal{V}_N$ and any real number λ,

$$(\mathbf{v}+\mathbf{w})(A) := \mathbf{v}(A)+\mathbf{w}(A) \quad \text{and} \quad (\lambda\mathbf{v})(A) := \lambda\mathbf{v}(A) \quad \text{for each } A \subseteq N.$$

It is easily verified that $\mathcal{V}_N$ is a linear space under these operations. The origin $\mathbf{0}$ of this space is the capacity that assigns 0 to every coalition. While the set of all additive capacities is a linear subspace of $\mathcal{V}_N$, the set of all superadditive capacities is not. (Why?)

We wish to find a basis for $\mathcal{V}_N$. To this end, let us first recall that we write $\mathcal{P}(N)$ for the set of all nonempty coalitions in N, that is, $\mathcal{P}(N) := 2^N \backslash \{\emptyset\}$.

The following is a fundamental result in the theory of capacities.

PROPOSITION 8[42]
(Shapley) *For any given $N \in \mathcal{N}$, the set of all unanimity games $\{\mathbf{u}_A : A \in \mathcal{P}(N)\}$ is a basis for $\mathcal{V}_N$.*

PROOF
Let us first show that $\{\mathbf{u}_A : A \in \mathcal{P}(N)\}$ is a linearly independent set in $\mathcal{V}_N$. For each $A \in \mathcal{P}(N)$, pick an arbitrary real number α_A such that $\sum_{A \in \mathcal{P}(N)} \alpha_A \mathbf{u}_A = \mathbf{0}$, that is,

$$\sum_{A \in \mathcal{P}(N)} \alpha_A \mathbf{u}_A(S) = 0 \quad \text{for all } S \subseteq N. \tag{16}$$

[42] That the class of unanimity games span the set of capacities seems to have been proved first by Shapley (1953). While it is difficult to believe that Shapley did not notice that this set is also linearly independent, I learned from Massimo Marinacci that the exact articulation of this fact, and its importance thereof, was given first by Rota (1964).

Claim: $\alpha_A = 0$ for each $A \in \mathcal{P}(N)$. Proof is by induction on the cardinality of A. Since, by definition, $\sum_{A\in\mathcal{P}(N)} \alpha_A \mathbf{u}_A(\{i\}) = \alpha_{\{i\}}\mathbf{u}_{\{i\}}(\{i\}) = \alpha_{\{i\}}$ for any $i \in N$, (16) yields that $\alpha_A = 0$ for all $A \in \mathcal{P}(N)$ with $|A| = 1$. Now assume that $\alpha_A = 0$ for all $A \in \mathcal{P}(N)$ with $|A| \leq t$, $1 \leq t < |N|$, and take any $B \in \mathcal{P}(N)$ with $|B| = t+1$. By the induction hypothesis and (16), we have $\sum_{A\in\mathcal{P}(N)} \alpha_A \mathbf{u}_A(B) = \alpha_B = 0$. Conclusion: $\alpha_A = 0$ for each $A \in \mathcal{P}(N)$.

It remains to prove that the set of all unanimity games spans $\mathcal{V}_N$ (Proposition 3). We shall prove this by showing that, for any $\mathbf{v} \in \mathcal{V}_N$, we have

$$\mathbf{v} = \sum_{A\in\mathcal{P}(N)} \alpha_A^{\mathbf{v}} \mathbf{u}_A \quad \text{where } \alpha_A^{\mathbf{v}} := \sum_{B\in\mathcal{P}(A)} (-1)^{|A|-|B|}\mathbf{v}(B).$$

Take any $S \subseteq N$. If $S = \emptyset$, things are trivial, so assume that $S \in \mathcal{P}(N)$, and note that

$$\begin{aligned}\sum_{A\in\mathcal{P}(N)} \alpha_A^{\mathbf{v}} \mathbf{u}_A(S) &= \sum_{A\in\mathcal{P}(S)} \alpha_A^{\mathbf{v}} \\ &= \sum_{A\in\mathcal{P}(S)} \sum_{B\in\mathcal{P}(A)} (-1)^{|A|-|B|}\mathbf{v}(B) \\ &= \sum_{B\in\mathcal{P}(S)} \left(\sum_{B\subseteq A\subseteq S} (-1)^{|A|-|B|} \right) \mathbf{v}(B).\end{aligned}$$

But a simple computation shows that

$$\sum_{B\subseteq A\subseteq S} (-1)^{|A|-|B|} = 0 \quad \text{whenever } |B| < |S|,$$

and combining this with the previous observation yields $\sum_{A\in\mathcal{P}(N)} \alpha_A^{\mathbf{v}} \mathbf{u}_A(S) = \mathbf{v}(S)$. Since S was chosen arbitrarily in $\mathcal{P}(N)$, we are done. ■

Proposition 8 shows that $\mathcal{V}_N$ is a finite-dimensional linear space with

$$dim(\mathcal{V}_N) = 2^{|N|} - 1$$

for any $N \in \mathcal{N}$. It follows that $\mathcal{V}_N$ is isomorphic to $\mathbb{R}^{2^{|N|}-1}$ (Proposition 5).

EXERCISE 47 Fix any $N \in \mathcal{N}$. The **dual** of a unanimity game $\mathbf{u}_A$, $A \in \mathcal{P}(N)$, is defined as the capacity $\mathbf{w}_A \in \mathcal{V}_N$ with

$$\mathbf{w}_A(B) := 1 - u_{N\backslash A}(N\backslash B) \qquad \text{for all } B \subseteq N.$$

Show that $\{\mathbf{w}_A : A \in 2^N\backslash\{N\}\}$ is a basis for $\mathcal{V}_N$.

The following two exercises provide a precursory introduction to the theory of Choquet integration which is widely used in the modern theory of decision making under uncertainty. A less friendlier but a far more comprehensive introduction is provided by Marinacci and Montrucchio (2004).

EXERCISE 48 (*Choquet Integration*) Fix any $N \in \mathcal{N}$, and let $n := |N|$. The **Choquet integral** of a vector $y \in \mathbb{R}^n_+$ with respect to a capacity $\mathbf{v} \in \mathcal{V}_N$ is defined as

$$\int y d\mathbf{v} := \int_0^\infty \mathbf{v}(\{i \in N : y_i \geq t\}) dt.$$

(Why is $\int yd\mathbf{v}$ well-defined? Recall Exercise A.62.) Throughout this exercise and the next, we write $y_{(i)}$ for the ith smallest component of y, $i = 1, \ldots, n$. (By convention, we let $y_{(0)} = 0$.)

(a) Show that $\int yd\mathbf{v}_1 = \sum^n y_i$, while $\int yd\mathbf{v}_{\text{maj}}$ equals the median of the set $\{y_1, \ldots, y_n\}$, where $\mathbf{v}_1$ and $\mathbf{v}_{\text{maj}}$ were defined in Example 10.

(b) Show that $\int yd\mathbf{u}_A = \min_{i \in A} y_i$ for any $A \in \mathcal{P}(N)$.

(c) Prove that, for any $(y, \mathbf{v}) \in \mathbb{R}^n_+ \times \mathcal{V}_N$,

$$\begin{aligned} \int y d\mathbf{v} &= \sum_{i=1}^n \left(y_{(i)} - y_{(i-1)}\right) \mathbf{v}(\{(i), \ldots, (n)\}) \\ &= \sum_{i=1}^n y_{(i)} \left(\mathbf{v}(\{(i), \ldots, (n)\}) - \mathbf{v}(\{(i+1), \ldots, (n)\})\right) \end{aligned}$$

where, by convention, $\{(n+1), (n)\} := \emptyset$.

(d) Use part (c) to compute $\int yd\mathbf{v}_\theta$ for each $\theta \geq 1$ (Example 10.[2]), and show that $\mathbf{v}(A \cup B) \geq \mathbf{v}(A) + \mathbf{v}(B)$ for all disjoint $A, B \subseteq N$ iff $\int yd\mathbf{v} = \sum^n \mathbf{v}(\{i\}) y_i$ for all $y \in \mathbb{R}^n_+$.

(e) Let $k \in \mathbb{N}$. Show that, for any $(y, \mathbf{v}^i, a_i) \in \mathbb{R}^n_+ \times \mathcal{V}_N \times \mathbb{R}_+$, $i = 1, \ldots, k$, we have

$$\int yd\left(\sum_{i=1}^k a_i \mathbf{v}^i\right) = \sum_{i=1}^k a_i \int yd\mathbf{v}^i.$$

(f) (Gilboa-Schmeidler) Use Proposition 8 and part (c) to show that, for any $(y, \mathbf{v}) \in \mathbb{R}^n_+ \times \mathcal{V}_N$,

$$\int y d\mathbf{v} = \sum_{A \in \mathcal{P}(N)} \alpha_A^{\mathbf{v}} \left(\min_{i \in A} y_i \right) = \sum_{A \in 2^N \setminus \{N\}} \beta_A^{\mathbf{v}} \left(\max_{i \notin A} y_i \right)$$

for some real numbers $\alpha_A^{\mathbf{v}}$ and $\beta_A^{\mathbf{v}}$, $A \in \mathcal{P}(N)$.

EXERCISE 49 (*More on Choquet Integration*) Fix any $N \in \mathcal{N}$, let $n := |N|$, and let $\mathbf{v}$ be a monotonic capacity on N such that $\mathbf{v}(N) = 1$. The **Choquet integral** of a vector $y \in \mathbb{R}^n$ with respect to $\mathbf{v}$ is defined as

$$\int y d\mathbf{v} := \int_0^\infty \mathbf{v}(\{i \in N : y_i \geq t\}) dt + \int_{-\infty}^0 \left(\mathbf{v}(\{i \in N : y_i \geq t\}) - 1\right) dt.$$

(a) Show that $x \geq y$ implies $\int x d\mathbf{v} \geq \int y d\mathbf{v}$ for any $x, y \in \mathbb{R}^n$.
(b) Show that $\int \lambda y d\mathbf{v} = \lambda \int y d\mathbf{v}$ for any $y \in \mathbb{R}^n$ and $\lambda \geq 0$.
(c) We say that two vectors x and y in $\mathbb{R}^n$ are **comonotonic** if $(x_i - y_i)(x_j - y_j) \geq 0$ for any $i, j = 1, \ldots, n$. Prove that $\int (x + y) d\mathbf{v} = \int x d\mathbf{v} + \int y d\mathbf{v}$ for any comonotonic $x, y \in \mathbb{R}^n$. (We thus say that the Choquet integral is **comonotonically additive**.)

4.3 The Shapley Value

One major normative question that the theory of coalitional games deals with is this: What sort of payoff profile should an impartial arbitrator suggest as an outcome for a given coalitional game? In what follows, we outline an axiomatic investigation that provides an interesting answer to this question.

Take any coalitional game $(N, \mathbf{v}) \in \mathcal{G}$. Observe first that if $\mathbf{v}$ is monotonic, $\sum_{i \in N} x_i \leq \mathbf{v}(N)$ is a natural feasibility condition for a payoff profile $x \in \mathbb{R}^{|N|}$, and hence, efficiency requires that $\sum_{i \in N} x_i = \mathbf{v}(N)$. One can also view this postulate as a "group rationality" requirement since, if $\sum_{i \in N} x_i < \mathbf{v}(N)$ held, then the entire coalition could form and improve the well-being of every single individual. Therefore, if $L_i(N, \mathbf{v})$ is interpreted as the payoff allocated by the arbitrator to the ith individual in the coalitional game $(N, \mathbf{v})$, then it would be natural to impose that

$$\sum_{i \in N} L_i(N, \mathbf{v}) = \mathbf{v}(N) \qquad \text{for any monotonic } \mathbf{v} \in \mathcal{V}_N. \tag{17}$$

We call any function L on $\mathcal{G}$ that maps $(N, \mathbf{v})$ to $(L_1(\mathbf{v}), \ldots, L_{|N|}(\mathbf{v}))$, where $L_1, \ldots, L_{|N|} \in \mathbb{R}^{\mathcal{V}_N}$ satisfy (17), as a **value**.

What sort of properties should a value L satisfy? A very reasonable property to impose on L is *symmetry*. To define this formally, fix any $N \in \mathcal{N}$, take any bijective self-map σ on N, and define $A_\sigma := \{i : \sigma(i) \in A\}$ for each $A \subseteq N$. For any superadditive capacity $\mathbf{v} \in \mathcal{V}_N$, we define $\mathbf{v}_\sigma(A) := \mathbf{v}(A_\sigma)$ for all $A \subseteq N$. The idea is that $\mathbf{v}_\sigma$ and $\mathbf{v}$ are really the same capacities—one is obtained from the other by relabeling the names of the players. Thus it makes good sense to require from a value that $L_i(N, \mathbf{v}) = L_{\sigma(i)}(N, \mathbf{v}_\sigma)$ for each i; this simply says that the arbitrator should act impartially. We say that a value L is **symmetric** if it satisfies this property for any $(N, \mathbf{v}) \in \mathcal{G}$ and bijective self-map σ on N.

Given any coalitional game $(N, \mathbf{v}) \in \mathcal{G}$, consider a player $i \in N$ such that $\mathbf{v}(A) - \mathbf{v}(A \backslash \{i\}) = \mathbf{v}(\{i\})$ for each $A \in \mathcal{P}(N)$ with $i \in A$. Clearly, this individual brings no extra returns to any of the coalitions she may join other than her stand-alone payoff. Thus, from the perspective of coalitional considerations, such an individual would best be ignored—she is in some sense a *dummy player*. Since i creates no increasing returns in the game, one may also argue that the stand-alone payoff $\mathbf{v}(\{i\})$ is all that i should get from an efficiency perspective. Thus, it seems appealing to posit for a value that, for any $N \in \mathcal{N}$ and superadditive capacity $\mathbf{v} \in \mathcal{V}_N$, $L_i(N, \mathbf{v}) = \mathbf{v}(\{i\})$ whenever $\mathbf{v}(A) - \mathbf{v}(A \backslash \{i\}) = \mathbf{v}(\{i\})$ for every $A \subseteq N$ that contains i. We refer to a value that satisfies this property as one with the **dummy player property**.

Here is an example of a value that satisfies these two properties. The **Shapley value** L^{S} is the value defined on $\mathcal{G}$ by

$$L_i^{\mathrm{S}}(N, \mathbf{v}) := \sum_{A \in \mathcal{P}(N)} \frac{(|N| - |A|)!(|A| - 1)!}{|N|!} \left(\mathbf{v}(A) - \mathbf{v}(A \backslash \{i\})\right),$$

$$i = 1, \ldots, |N|.$$

This value can be thought of as allocating to each individual i a weighted average of the marginal worth of i for all coalitions she might join. It is easily checked to be a symmetric value with the dummy player property.[43]

[43] *Interpretation.* Fix any $(N, \mathbf{v}) \in \mathcal{G}$. Suppose that players in N are to enter a room one by one. This way a line may form in $|N|!$ many different ways. On the other hand, for any $A \in 2^N$ that contains i, there are $(|N| - |A|)!(|A| - 1)!$ many different ways a line may form so that when i enters the room she sees precisely the coalition $A \backslash \{i\}$ already in the room. Thus, if people are to enter the room randomly, $L_i^{\mathrm{S}}(N, \mathbf{v})$ is simply the expected marginal contribution of i when she enters the room in this story.

The Shapley value has a very convenient analytic structure. Indeed, each $L^S(N,\cdot)$ is a linear operator from $\mathcal{V}_N$ into $\mathbb{R}^{|N|}$. We call any such value *linear*. That is, a value L is **linear** if

$$L(N,\cdot) \in \mathcal{L}(\mathcal{V}_N, \mathbb{R}^{|N|}) \qquad \text{for all } N \in \mathcal{N}.$$

While it is not obvious how to motivate the linearity property from a conceptual point of view (but see Example 11 below), this condition may still be viewed as an operational consistency property. What is more, we know everything there is to know about the linear values that satisfy the symmetry and dummy player properties. After all, Lloyd Shapley showed in 1953 that there is only one such thing!

THEOREM 2
(Shapley) *The Shapley value is the only symmetric linear value with the dummy player property.*

PROOF
It is routine to show that the Shapley value is a symmetric linear value with the dummy player property. The nontrivial part of the proof is to show that there is only one such value.

Fix any $(N,\mathbf{v}) \in \mathcal{G}$. By Proposition 8, for each $\mathbf{v} \in \mathcal{V}_N$ and $A \in \mathcal{P}(N)$, there exists a unique real number $\alpha_A^{\mathbf{v}}$ such that $\mathbf{v} = \sum_{A\in\mathcal{P}(N)} \alpha_A^{\mathbf{v}} \mathbf{u}_A$.[44] So, if $L(N,\cdot) \in \mathcal{L}(\mathcal{V}_N, \mathbb{R}^{|N|})$, we must have

$$L(N,\mathbf{v}) = \sum_{A\in\mathcal{P}(N)} \alpha_A^{\mathbf{v}} L(N,\mathbf{u}_A), \tag{18}$$

that is, knowing the behavior of $L(N,\cdot)$ for each unanimity game on N would determine $L(N,\cdot)$ on the entire $\mathcal{V}_N$. Now fix any $A \in \mathcal{P}(N)$ and consider the unanimity game $\mathbf{u}_A$. If $i \notin A$, then $\mathbf{u}_A(\{i\}) = 0$ and $u_A(B) = \mathbf{u}_A(B \cup \{i\})$ for all $B \in \mathcal{P}(N)$ so that, by the dummy player property, $L_i(N,\mathbf{u}_A) = 0$. On the other hand, by the symmetry of L, we must have $L_i(N,\mathbf{u}_A) = L_j(N,\mathbf{u}_A)$ for all $i,j \in A$. But we have $\sum_{i\in A} L_i(N,\mathbf{u}_A) = \sum_{i\in N} L_i(N,\mathbf{u}_A) = \mathbf{u}_A(N) = 1$ so that $L_i(N,\mathbf{u}_A) = \frac{1}{|A|}$ for each $i \in A$. Therefore, $L(N,\mathbf{u}_A)$ is uniquely determined for each $A \in \mathcal{P}(N)$. ■

44 Where did uniqueness come from? Recall Corollary 2.

Exercise 50 Show that the linearity requirement in Theorem 2 can be replaced with additivity: $L(N, \mathbf{v}+\mathbf{w}) = L(N, \mathbf{v})+L(N, \mathbf{w})$ for all $\mathbf{v}, \mathbf{w} \in \mathcal{V}_N$, $N \in \mathcal{N}$. (Would the result remain valid if L was additive only over the set of all superadditive capacities on N, $N \in \mathcal{N}$?)

Theorem 2 alone is responsible for why the Shapley value is viewed as one of the most important solution concepts in the theory of coalitional games. It also gives us a very good reason to view this value as a useful tool for normative applications. We will shortly illustrate this point by means of an example below. However, before doing this, we should note that the Shapley value may sometimes lead to coalitionally *unstable* payoff profiles. To see this, let us define

$$core(N, \mathbf{v}) := \left\{ x \in \mathbb{R}^{|N|} : \sum_{i \in N} x_i = \mathbf{v}(N) \text{ and } \sum_{i \in A} x_i \geq \mathbf{v}(A) \text{ for all } A \subseteq N \right\}$$

for any coalitional game $(N, \mathbf{v}) \in \mathcal{G}$. Any member x of $core(N, \mathbf{v})$ is coalitionally stable in a clear sense: No coalition has any incentive to deviate from x, because no coalition A can improve upon what is allocated in x to its members by means available to it (that is, by using $\mathbf{v}(A)$).[45] A natural question is if the allocation induced by the Shapley value is necessarily in the core of a coalitional game. Unless the game satisfies some further conditions, the answer is no.

Exercise 51 Let $N := \{1, 2, 3\}$, and give an example of a superadditive nonnegative capacity $\mathbf{v} \in \mathcal{V}_N$ such that $L^S(N, \mathbf{v}) \notin core(N, \mathbf{v})$.

Exercise 52^H For any $N \in \mathcal{N}$, we say that a coalitional game $(N, \mathbf{v}) \in \mathcal{G}$ is **convex** if $\mathbf{v}(A \cup B) + \mathbf{v}(A \cap B) \geq \mathbf{v}(A) + \mathbf{v}(B)$ for all $A, B \subseteq N$. Show that $L^S(N, \mathbf{v}) \in core(N, \mathbf{v})$ for any convex $(N, \mathbf{v}) \in \mathcal{G}$.

We now come to the illustration promised above.

Example 11

(*Cost Allocation Problems*) Consider a set of agents $N \in \mathcal{N}$ who face the problem of sharing the joint cost of a common facility. (For concreteness,

[45] Many other notions of coalitional stability are discussed in the literature, but it is safe to say that the core concept is the most widely used among them. See Chapters 12–15 of Aumann and Hart (1992) for a thorough introduction to the theory of the core.

you may view each of these agents as countries or cities and the common facility as a joint irrigation or hydroelectric power system. Such scenarios are played out frequently in the Middle East, with very serious financial implications.) Let $c : 2^N \to \mathbb{R}_+$ be any function, and interpret $c(A)$ as the cost of building the facility for a set A of agents in N. The function c is called a **cost allocation problem** if it is subadditive, that is, $c(A \cup B) \leq c(A) + c(B)$ for every disjoint $A, B \subseteq N$ (increasing returns to scale), and if it satisfies $c(\emptyset) = 0$ (the cost to "no one" is zero). Due to subadditivity, an efficient state of affairs is that all agents participate in building the facility, and hence the (normative) problem at hand is really about the division of the total cost $c(N)$ among the agents. Formally, the objective here is to come up with a reasonable **cost-sharing rule** ϕ that maps a cost allocation problem c to a vector $(\phi_1(c), \ldots, \phi_{|N|}(c)) \in \mathbb{R}_+^{|N|}$ such that $\sum_{i \in N} \phi_i(c) = c(N)$ with $\phi_i(c)$ being the cost allocated to agent i.

One promising approach is to formulate the problem as a coalitional game with side payments by considering the total cost savings of a group A as the "worth" of that coalition. So, given any cost allocation problem c, we define $\mathbf{v}_c : 2^N \to \mathbb{R}_+$ by

$$\mathbf{v}_c(A) := \sum_{i \in A} c(\{i\}) - c(A)$$

which is a superadditive capacity on N with $\mathbf{v}_c(\{i\}) = 0$ for all i. Clearly, for any value L, the map ϕ^L defined (on the class of all cost allocation problems) by $\phi_i^L(c) := c(\{i\}) - L_i(N, \mathbf{v}_c), i \in N$, is a cost allocation rule. If L satisfies the dummy player property, then $\phi_i^L(c) = c(\{i\})$ for each i with $c(A) - c(A \backslash \{i\}) = c(\{i\})$ for all $A \subseteq N$. This means that an agent who does not increase the returns to scale relative to *any* group is charged her stand-alone cost. Similarly, symmetry of L corresponds to an unexceptionable anonymity requirement for ϕ^L. Finally, suppose that $L(N, \cdot)$ is additive. This implies that $\phi^L(c + c') = \phi^L(c) + \phi^L(c')$ for any two cost allocation problems c and c', which is particularly appealing from the implementation point of view.[46] Then, by Exercise 50, an axiomatically based cost-sharing rule, which may be called *the Shapley rule*, is uniquely determined: $\phi^{L^S(N, \cdot)}$.

[46] One often distinguishes between different cost categories such as managerial costs, advertisement costs, capital, labor, and so on. The additivity property allows one to deal with the cost allocation problem for each such category one at a time.

For instance, if c is given by $c(\{1\}) = 3$, $c(\{2\}) = 10$, $c(\{3\}) = 13$, $c(\{1,2\}) = 13$, $c(\{2,3\}) = 19$, $c(\{1,3\}) = 14$ and $c(\{1,2,3\}) = 20$, then $L_i^S(\{1,2,3\}, \mathbf{v}_c) = i$ for each $i = 1,2,3$. Thus the cost allocation according to the Shapley rule is $(2, 8, 10)$.[47] □

We conclude by outlining an alternative approach to the characterization of the Shapley value, which is also based on Proposition 8. This approach differs from the previous one in that it is of a *variable* population nature. That is, it requires a value to behave in a consistent way across coalitional games that involve different numbers of players.

EXERCISE 53^{H} (Hart-MasCollel) For any given $P : \mathcal{G} \to \mathbb{R}$ and $(N, \mathbf{v}) \in \mathcal{G}$, we define

$$\triangle_i P(N, \mathbf{v}) := P(N, \mathbf{v}) - P(N\backslash\{i\}, \mathbf{v}|_{N\backslash\{i\}}), \qquad i \in N,$$

if $|N| \geq 2$, and by $\triangle_i P(N, \mathbf{v}) := P(N, \mathbf{v})$ if $N = \{i\}$. (You may think of this number, if vaguely, as the marginal contribution of player i to the "potential" of $(N, \mathbf{v})$.[48]) We say that $P \in \mathbb{R}^{\mathcal{G}}$ is a **potential function** if

$$\sum_{i \in N} \triangle_i P(N, \mathbf{v}) = \mathbf{v}(N) \qquad \text{for all } (N, \mathbf{v}) \in \mathcal{G}.$$

If this looks sensible to you, then you will really like what comes next.[49] There is one and only one potential function P, and we have

$$\triangle_i P(N, \mathbf{v}) = L_i^S(N, \mathbf{v}) \qquad \text{for all } (N, \mathbf{v}) \in \mathcal{G} \text{ and } i \in N.$$

Prove!

[47] This example doesn't do justice to the theory of cost allocation. If you want to learn more about this topic, have a look at Moulin (2001).

[48] If we agree that this marginal contribution is what i should get in the game, then the problem of identifying a reasonable value for a game is now transformed into finding a reasonable P that would imply that the agent i gets precisely the payoff $\triangle_i P(N, \mathbf{v})$ in the outcome of the game.

[49] The first requirement here is completely trivial. The second one, on the other hand, simply requires that the individual marginal contributions/allocations (as defined by P) sum up exactly to the total worth of the entire group: none that is available to be shared should be wasted. Given the interpretation of $\triangle_i P(N, \mathbf{v})$, and hence P, this is a sensible "efficiency" requirement, especially in contexts similar to that of Example 11. Nevertheless, note that it is quite demanding in that it does not impose the adding-up property for a single game but for *all* subgames of a given game.

Although a major power behind Theorem 2 is the additivity property, the preceding characterization is based rather on what could be called a "marginality" principle. Indeed, marginal contributions emerge in Theorem 2 as an outcome, while they appear already as an assumption here. On the other hand, this characterization has the advantage of not imposing a linearity condition on the values that may be induced by a potential function. It is remarkable that these two approaches point toward the same direction, the Shapley value. There are in fact other approaches that yield the same conclusion as well. (See, for instance, Young (1985).) But now we need to get back to our main story.

Chapter G

Convexity

One major reason why linear spaces are so important for geometric analysis is that they allow us to define the notion of "line segment" in algebraic terms. Among other things, this enables one to formulate, purely algebraically, the notion of "convex set," which figures majorly in a variety of branches of higher mathematics.[1] Immediately relevant for economic theory is the indispensable role played by convex sets in optimization theory. At least for economists, this alone is enough of a motivation for taking on a comprehensive study of convex sets and related concepts.

We begin the chapter with a fairly detailed discussion of convex sets and cones. Our emphasis is again on the infinite-dimensional side of the picture. In particular, we consider several examples that are couched within infinite-dimensional linear spaces. After all, one of our main objectives here is to provide some help for the novice to get over the sensation of shock that the strange behavior of infinite-dimensional spaces may invoke at first. We also introduce partially ordered linear spaces and discuss the important role played by convex cones thereof.

Among the topics that are likely to be new to the reader are the algebraic interior and algebraic closure of subsets of a linear space. These notions are developed relatively leisurely, for they are essential to the treatment of the high points of the chapter, namely, the fundamental extension and separation theorems in an arbitrary linear space. In particular, we prove here the linear algebraic formulations of the Hahn-Banach Theorems on the extension of a linear functional and the separation of convex sets, along with the Krein-Rutman Theorem on the extension of positive linear functionals. We then turn to Euclidean convex analysis and deduce Minkowski's Separating and Supporting Hyperplane Theorems—these are among the most widely used theorems in economic theory—as easy corollaries of our

[1] See Klee (1971) for a concise yet instructive introduction to the theory of convex sets at large.

general results. As a final order of business, we discuss the problem of best approximation from a convex set in a Euclidean space. In particular, we introduce the orthogonal projection operator and use it to obtain the Euclidean version of the Krein-Rutman Theorem. Several economic applications of the main theorems established in this chapter are considered in Chapter H.[2]

1 Convex Sets

1.1 Basic Definitions and Examples

You are familiar with the way one defines a convex set in the context of Euclidean spaces; we have already used this concept a few times in this book. This definition carries over to the case of arbitrary linear spaces without modification.

Definition

For any $0 < \lambda < 1$, a subset S of a linear space X is said to be **λ-convex** if

$$\lambda x + (1-\lambda)y \in S \qquad \text{for any } x, y \in S,$$

or equivalently, if

$$\lambda S + (1-\lambda)S = S.$$

If S is $\frac{1}{2}$-convex, we say that it is **midpoint convex**. Finally, if S is λ-convex for all $0 < \lambda < 1$, that is, if

$$\lambda S + (1-\lambda)S = S \qquad \text{for all } 0 \leq \lambda \leq 1,$$

then it is said to be a **convex set**.

[2] There are many excellent textbooks on finite-dimensional convex analysis. Rockefellar (2000) is a classic in the field, there are also more recent expositions. In particular, Hiriart-Urruty and Lemaréchal (2000) provide a very nice introduction, and Borwein and Lewis (2000) take one to the next level.

Elementary treatments of infinite-dimensional convex analysis are harder to find. Certainly the linear algebraic treatment I present here is not commonly adopted in textbooks on mathematical analysis. One major exception is the first chapter of the excellent text by Holmes (1975). However, if this is your first serious encounter with infinite-dimensional linear spaces, it would be wiser to go to this reference only after completing the present chapter.

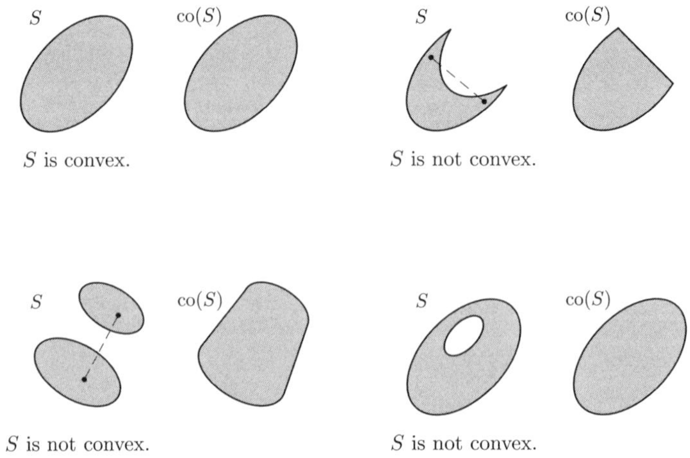

FIGURE 1

The interpretation of these concepts is straightforward. For any two distinct vectors x and y in the linear space X, we think of the set $\{\lambda x + (1-\lambda)y \in X : \lambda \in \mathbb{R}\}$ as **the line through x and y**. From this point of view, the set $\{\lambda x + (1-\lambda)y \in X : 0 \leq \lambda \leq 1\}$ corresponds to **the line segment between x and y**.[3] Consequently, a subset S of X is midpoint convex iff it contains the midpoint of the line segment between any two of its constituent vectors. It is convex iff it contains the entire line segment between any two of its elements (Figure 1).

As examples, note that $\mathbb{Q}$ is a midpoint convex subset of $\mathbb{R}$, which is not convex. Indeed, a nonempty subset of $\mathbb{R}$ is convex iff it is an interval. In any linear space X, all singleton sets, line segments, and lines, along with $\emptyset$, are convex sets. Any linear subspace, affine manifold, hyperplane, or half-space in X is also convex. For instance, for any $0 < \lambda < 1$, a hyperplane H in a linear space X is a λ-convex set, because, by Corollary F.4, $H = \{x \in X : L(x) = \alpha\}$ for some nonzero $L \in \mathcal{L}(X, \mathbb{R})$ and $\alpha \in \mathbb{R}$, and thus, for any $x, y \in H$, we have

$$L(\lambda x + (1-\lambda)y) = \lambda L(x) + (1-\lambda)L(y) = \lambda\alpha + (1-\lambda)\alpha = \alpha,$$

that is, $\lambda x + (1-\lambda)y \in H$.

[3] Observe that it is the linear structure of a linear space that lets us "talk about" these objects. We could not do so, for instance, in an arbitrary metric space.

As an immediate illustration of the appeal of convex sets—we will of course see many more examples later—let us note that the affine hull of a convex set can in general be expressed much more easily than that of a nonconvex set. Indeed, for any convex subset S of a linear space, we simply have

$$aff(S) = \{\lambda x + (1 - \lambda)y : \lambda \in \mathbb{R} \text{ and } x, y \in S\}.$$

(Geometric interpretation?) The proof requires an easy induction argument; we leave that to you. But we note that convexity is essential for this observation. For instance, the set $S := \{\mathbf{0}, \mathbf{e}^1, \mathbf{e}^2\}$ in $\mathbb{R}^2$ does not satisfy the equation above.

EXERCISE 1 Determine which of the following sets are necessarily convex (in the linear spaces in which they naturally live):
(a) a nonempty connected subset of $\mathbb{R}^2$;
(b) $\{(t, y) \in \mathbb{R}^2 : f(t) \geq y\}$, where f is a concave self-map on $\mathbb{R}$;
(c) $\{(t, y) \in \mathbb{R}^2 : \sin t \geq y\}$;
(d) an open neighborhood of a given point in $\mathbb{R}^4$;
(e) the set of all semicontinuous functions on $[0, 1]$;
(f) $\{f \in \mathbf{B}[0, 1] : f \geq 0\}$;
(g) $\{f \in \mathbf{C}[0, 1] : f(1) \neq 0\}$.

EXERCISE 2 Show that the intersection of any collection of λ-convex subsets of a linear space is λ-convex, $0 < \lambda < 1$.

EXERCISE 3 Show that if A and B are λ-convex subsets of a linear space, so is $\alpha A + B$ for any $\alpha \in \mathbb{R}$ and $0 < \lambda < 1$.

EXERCISE 4^H Prove: A convex set S is contained in one of the open half-spaces induced by a hyperplane H in a Euclidean space X iff $S \cap H = \emptyset$.

Let S be any set in a linear space X, and let $\mathcal{S}$ be the class of all convex subsets of X that contain S. We have $\mathcal{S} \neq \emptyset$—after all, $X \in \mathcal{S}$. Then, by Exercise 2, $\cap\mathcal{S}$ is a convex set in X that, obviously, contains S. Clearly, this set is the smallest (that is, $\supseteq$-minimum) convex subset of X that contains S; it is called the **convex hull** of S, and denoted by $co(S)$. (Needless to say, $S = co(S)$ iff S is convex; see Figure 1.) This notion proves useful when one needs the convexity of the set one is working with, even though that set is not known to be convex. We will encounter many situations of this sort later.

It is worth noting that the definition of $co(S)$ uses vectors that lie outside S (because the members of $\mathcal{S}$ may well contain such vectors). For this reason, this definition can be viewed as an *external* one. We may also characterize $co(S)$ *internally*, that is, by using only the vectors in S. Indeed, $co(S)$ is none other than the set of all convex combinations of finitely many members of S. That is,

$$co(S) = \left\{ \sum_{i=1}^{m} \lambda_i x^i : m \in \mathbb{N}, x^1, \ldots, x^m \in S \text{ and } (\lambda_1, \ldots, \lambda_m) \in \Delta_{m-1} \right\} \tag{1}$$

where

$$\Delta_{m-1} := \left\{ (\lambda_1, \ldots, \lambda_m) \in [0,1]^m : \sum_{i=1}^{m} \lambda_i = 1 \right\}, \qquad m = 1, 2, \ldots$$

This shows that the way we define the convex hull of a set here is in full accord with how we defined this concept in Section E.5.3 to state Carathéodory's Theorem in the context of Euclidean spaces. Moreover, for any $x, y \in X$, we see that $co(\{x, y\})$—which is denoted simply as $co\{x, y\}$ henceforth—is nothing but the line segment between x and y, and we have

$$co(S) = \bigcup \{co(T) : T \in \mathcal{P}(S)\},$$

where $\mathcal{P}(S)$ is the class of all nonempty finite subsets of S.

EXERCISE 5 Prove (1).

EXERCISE 6 Show that, for any subset S of a linear space X and $x \in X$,

$$co(S + x) = co(S) + x.$$

EXERCISE 7 Prove or disprove: If X and Y are linear spaces, and $A \subseteq X$ and $B \subseteq Y$, then $co(A \times B) = co(A) \times co(B)$, where the underlying linear space is $X \times Y$ (Exercise F.23).

The definitions of concave and convex functions (Section A.4.5) also extend in the obvious way to the case of real maps defined on a convex subset of a linear space.

DEFINITION

Let X be a linear space and T a nonempty convex subset of X. A real map φ on T is called **concave** if

$$\varphi(\lambda x + (1-\lambda)y) \geq \lambda\varphi(x) + (1-\lambda)\varphi(y) \quad \text{for all } x, y \in T \text{ and } 0 \leq \lambda \leq 1,$$

while it is called **convex** if $-\varphi$ is concave. (If both φ and $-\varphi$ are concave, then φ is an affine map; recall Exercise F.32.)

All algebraic properties of concave functions that you are familiar with are valid in this more general setup as well. For example, if T is a nonempty convex subset of a linear space, then $\varphi \in \mathbb{R}^T$ is concave iff $\{(x, a) \in T \times \mathbb{R} : \varphi(x) \geq a\}$ is a convex subset of $T \times \mathbb{R}$. Moreover, if $\psi \in \mathbb{R}^T$ is also concave, then so is $\varphi + \psi$. However, the set of all concave real functions on T is not an affine manifold of $\mathbb{R}^T$. (Why?)

The next exercise introduces the notion of convex correspondence, which plays an important role in optimization theory. In fact, we have already used convex correspondences when studying the theory of dynamic programming (recall Proposition E.7). We will come back to them in due course.

EXERCISE 8^H (*Convex Correspondences*) Let X and Y be two linear spaces and S a nonempty convex subset of X. We say that $\Gamma : S \rightrightarrows Y$ is a **convex correspondence** if $Gr(\Gamma)$ is a convex subset of $X \times Y$, that is,

$$\lambda\Gamma(x) + (1-\lambda)\Gamma(x') \subseteq \Gamma(\lambda x + (1-\lambda)x')$$
$$\text{for any } x, x' \in S \text{ and } 0 \leq \lambda \leq 1.$$

(a) Let T be a convex subset of Y, take any $L \in \mathcal{L}(X, Y)$, and define $\Gamma : S \rightrightarrows Y$ by $\Gamma(x) := L(x) + T$. Show that Γ is a convex correspondence.

(b) Let $f \in \mathbb{R}^S$ be a convex function. Show that $\Gamma : S \rightrightarrows \mathbb{R}$, defined by $\Gamma(x) := \{a \in \mathbb{R} : a \geq f(x)\}$, is a convex correspondence.

(c) Show that the budget correspondence (Example E.2) is a convex-valued correspondence that is not convex.

(d) Every linear correspondence $\Gamma : X \rightrightarrows Y$ is convex (Exercise F.31). Is the converse true?

(e) (Deutsch-Singer) Prove: If $\Gamma : X \rightrightarrows Y$ is a convex correspondence with $|\Gamma(x_0)| = 1$ for some $x_0 \in X$, then there exists an affine map $f \in Y^X$ such that $\{f(x)\} = \Gamma(x)$ for all $x \in X$.

1.2 Convex Cones

Convex sets that are closed under nonnegative scalar multiplication play an important role in convex analysis and optimization theory. Let us give them a name.

DEFINITION
A nonempty subset C of a linear space X is said to be a **cone** if it is closed under nonnegative scalar multiplication, that is, $\lambda C \subseteq C$ for all $\lambda \geq 0$, i.e.,

$$\lambda x \in C \quad \text{for any } x \in C \text{ and } \lambda \geq 0.$$

If, in addition, C is closed under addition, that is, $C + C \subseteq C$, i.e.,

$$x + y \in C \quad \text{for any } x, y \in C,$$

then it is called a **convex cone**. We say that a cone C in X is **pointed** if $C \cap -C = \{\mathbf{0}\}$,[4] **generating** if $span(C) = X$, and **nontrivial** if $C \neq \{\mathbf{0}\}$.

Geometrically speaking, a cone is a set that contains all rays that start from the origin and pass through another member of the set (Figure 2). In turn, a convex cone is none other than a cone that is also a convex set. (Why?) In a manner of speaking, the concept of convex cone lies in between that of a convex set and that of a linear subspace. This point will become clear as we develop the theory of convex sets further.

REMARK 1. Recall that the dimension of a set in a linear space X is the dimension of its affine hull (Remark F.1). But since any cone C in X contains the origin $\mathbf{0}$ (yes?), we have $aff(C) = span(C)$. (Why?) Thus $dim(C) = dim(span(C))$, so a cone C in a finite-dimensional linear space X is generating iff $dim(C) = dim(X)$. □

[4] *Reminder.* $-C := (-1)C$, that is, $-C = \{-x : x \in C\}$, and $C - C := C + -C$, that is, $C - C = \{x - y : x, y \in C\}$.

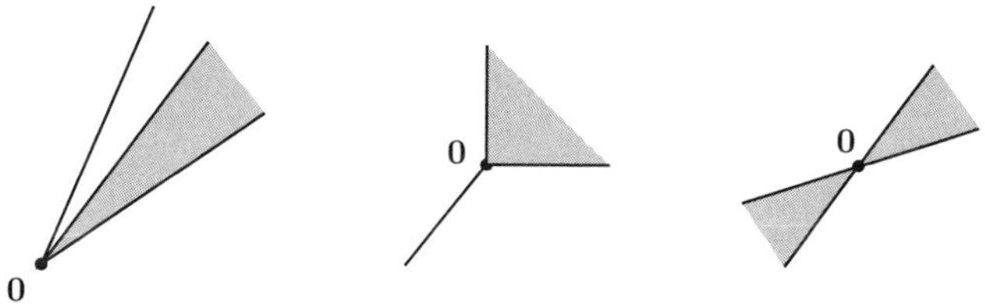

Non-convex cones in $\mathbb{R}^2$

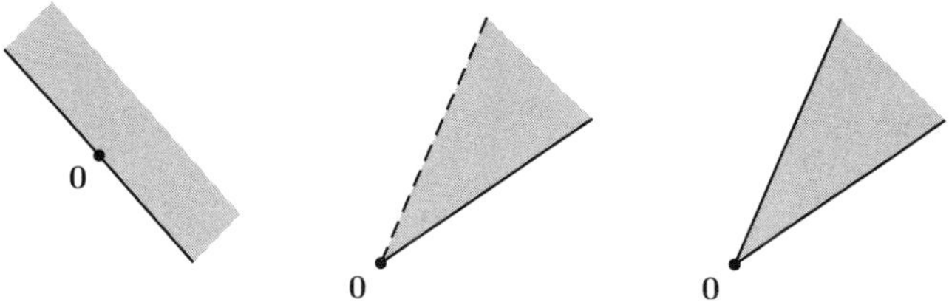

Convex cones in $\mathbb{R}^2$

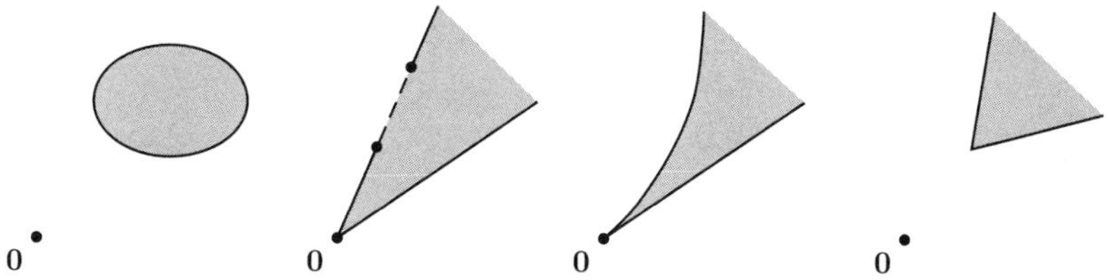

Sets in $\mathbb{R}^2$ which are not cones

Figure 2

Example 1

[1] Any linear subspace C of a linear space X is a convex cone. In this case $C \cap -C = C$, so C is a pointed cone iff $C = \{\mathbf{0}\}$. Moreover, C is generating iff $C = X$.

[2] The smallest (i.e. $\supseteq$-minimum) convex cone in a linear space X is the trivial linear subspace $\{\mathbf{0}\}$, while the largest (i.e., $\supseteq$-maximum) convex cone is X itself. Moreover, if C is a convex cone in this space, so is $-C$.

[3] For any given $n \in \mathbb{N}$, the set $\mathbb{R}^n_+$ is a pointed generating convex cone in $\mathbb{R}^n$—we have $dim(\mathbb{R}^n_+) = n$. On the other hand, $\mathbb{R}^n_{++}$ is not a

cone in $\mathbb{R}^n$ since it does not contain the origin $\mathbf{0}$. Indeed, $\mathbb{R}^n_{++} \cup \{\mathbf{0}\}$ is a pointed generating convex cone in $\mathbb{R}^n$. An example of a convex cone in $\mathbb{R}^n$, $n \geq 2$, which is not pointed is $\{x \in \mathbb{R}^n : x_1 \geq 0\}$, and an example of one that is not generating is $\{x \in \mathbb{R}^n : x_1 = 0\}$. (What are the dimensions of these cones?)

[4] If X and Y are two linear spaces and $L \in \mathcal{L}(X, Y)$, then $\{x \in X : L(x) \geq \mathbf{0}\}$ is a (pointed) convex cone in X. This observation allows us to find many convex cones. For instance, $\{(x_m) \in \mathbb{R}^\infty : x_4 \geq 0\}$, or $\{f \in \mathbf{B}[0,1] : f(\frac{1}{2}) \geq 0\}$, or $\{f \in \mathbf{C}[0,1] : \int_0^1 f(t)dt \geq 0\}$ are pointed convex cones by this very token. All of these cones are infinite-dimensional and generating.

[5] Both $\{f \in \mathbf{C}^1[0,1] : f \geq 0\}$ and $C := \{f \in \mathbf{C}^1[0,1] : f' \geq 0\}$ are convex cones in $\mathbf{C}^1[0,1]$. The former is pointed but the latter is not— all constant functions belong to $C \cap -C$. □

Let S be a nonempty set in a linear space X. It is easily checked that S is a convex cone in X iff $\sum_{x \in T} \lambda(x)x \in S$ for any nonempty finite subset T of S and $\lambda \in \mathbb{R}^T_+$. (Verify!) It follows that the smallest *convex* cone that contains S—the **conical hull** of S—exists and equals the set of all positive linear combinations of finitely many members of S. (Why?) Denoting this convex cone by $cone(S)$, therefore, we may write

$$cone(S) = \left\{ \sum_{x \in T} \lambda(x)x : T \in \mathcal{P}(S) \text{ and } \lambda \in \mathbb{R}^T_+ \right\}$$

where, as usual, $\mathcal{P}(S)$ stands for the class of all nonempty finite subsets of S. (By convention, we let $cone(\emptyset) = \{\mathbf{0}\}$.) For instance, for any $n \in \mathbb{N}$,

$$cone(\mathbb{R}^n_{++}) = \mathbb{R}^n_{++} \cup \{0\} \quad \text{and} \quad cone(\{\mathbf{e}^1, \ldots, \mathbf{e}^n\}) = \mathbb{R}^n_+,$$

where $\{\mathbf{e}^1, \ldots, \mathbf{e}^n\}$ is the standard basis for $\mathbb{R}^n$.

EXERCISE 9^H Let $\mathbf{c}^0$ be the linear space of all real sequences all but finitely many terms of which are zero. Prove:

(a) $C := \{(x_m) \in \mathbf{c}^0 : x_m > 0 \text{ for some } m\}$ is not a cone in $\mathbf{c}^0$.

(b) $C \cup \{\mathbf{0}\}$ is an infinite-dimensional cone that is not midpoint convex.

(c) $cone(C) = \mathbf{c}^0$.

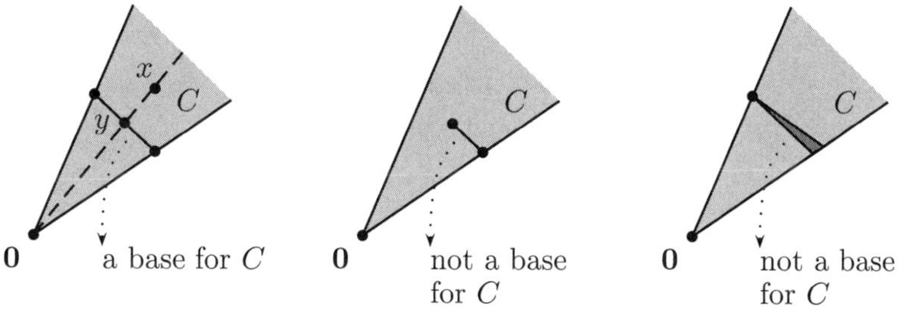

Figure 3

Exercise 10 Show that if C is a convex cone in a linear space X, then $span(C) = C - C$. Thus, C is generating iff $X = C - C$.

Exercise 11 Let S be a nonempty set in a given linear space. Show that

$$cone(S) = \cup\{\lambda co(S) : \lambda \geq 0\}.$$

Exercise 12[H] Let C be a cone in a given linear space. Prove or disprove: $cone(C) = co(C)$.

Let X be a linear space and C a nontrivial convex cone in X. A nonempty convex subset B of C is called a **base** for C if, for each $x \in C$, there exists a *unique* $(y, \lambda) \in B \times \mathbb{R}_{++}$ such that $x = \lambda y$.[5] (See Figure 3.) A lot can be learned about a convex cone with a base by studying the properties of its base alone. Since often a base for a convex cone is simpler to analyze than the cone itself, it is worth knowing if a given cone has a base. The following exercise provides two criteria for this to be the case.

*Exercise 13[H] Let C be a nontrivial convex cone in a linear space X, and B a convex subset of X. Prove that the following statements are equivalent.

(i) B is a base for C.

(ii) $C = \cup\{\lambda B : \lambda \geq 0\}$ and $\mathbf{0} \notin aff(B)$.

(iii) There exists an $L \in \mathcal{L}(X, \mathbb{R})$ such that $L(C\backslash\{\mathbf{0}\}) \subseteq \mathbb{R}_{++}$ and $B = \{x \in C : L(x) = 1\}$.

[5] The uniqueness requirement is essential here. For instance, because of this requirement, no base can contain the origin $\mathbf{0}$ of X. For, if B is a nonempty convex subset of C with $\mathbf{0} \in B \neq \{\mathbf{0}\}$, then for any nonzero $y \in B$ we also have $\frac{1}{2}y \in B$ (because B is convex), so y itself has two distinct representations: $y = 1y$ and $y = 2(\frac{1}{2}y)$. Thus in this case B cannot be a base for C. (The case $B = \{\mathbf{0}\}$ is trivially dismissed as a potential candidate for a base, of course.)

1.3 Ordered Linear Spaces

Let $(X, \succsim)$ be a preordered set (Section A.1.4). If X is a linear space and $\succsim$ is compatible with the addition and scalar multiplication operations on X in the sense that

$$x \succsim y \quad \text{if and only if} \quad \lambda x + z \succsim \lambda y + z \tag{2}$$

for any $x, y, z \in X$ and $\lambda \geq 0$, then we say that $(X, \succsim)$ is a **preordered linear space**, and refer to $\succsim$ as a **vector preorder**. More precisely, a preordered linear space is a list $(X, +, \cdot, \succsim)$ where $(X, +, \cdot)$ is a linear space, $(X, \succsim)$ is a preordered set, and these two mathematical systems are connected via the compatibility requirement (2). Of course, $(X, +, \cdot, \succsim)$ is unnecessarily mouthful. We denote this system simply as X to simplify the notation, and write $\succsim_X$ for the associated preorder on X. If X is a preordered linear space and $\succsim_X$ is a *partial* order, then we say that X is a **partially ordered linear space**, and refer to $\succsim_X$ simply as a **vector order**.

There is an intimate relation between convex cones and vector (pre)orders. First of all, by using the convex cones of a linear space X, we can obtain various vector preorders on X, and thus make X a preordered linear space in a variety of ways. To see this, take any nonempty subset C of X, and define the binary relation $\succsim^C$ on X as follows:

$$x \succsim^C y \quad \text{if and only if} \quad x \in y + C.$$

It is easy to see that $\succsim^C$ is a vector preorder on X iff C is a convex cone—in this case the preorder $\succsim^C$ is called the **preorder induced by** C. Conversely, every vector preorder on X arises in this way. Indeed, if $\succsim$ is such a preorder, then $X_+(\succsim) := \{x \in X : x \succsim \mathbf{0}\}$ is a convex cone—called the **positive cone induced by** $\succsim$—and we have $\succsim = \succsim^{X_+(\succsim)}$. Moreover, $\succsim$ is a partial order on X iff $X_+(\succsim)$ is pointed.[6]

The upshot is that there is a one-to-one correspondence between convex cones in a linear space X and vector preorders on X. The idea is rather

[6] The following two extreme situations may be worth noting here. We have $\succsim = X \times X$ iff $X_+(\succsim) = X$ (the case of complete indifference), and $\succsim = \{(x, x) : x \in X\}$ iff $X_+(\succsim) = \{\mathbf{0}\}$ (the case of full noncomparability).

transparent in $\mathbb{R}$. For any two real numbers a and b, the statement $a \geq b$ is equivalent to $a - b \geq 0$, so, since $a = b + (a - b)$, another way of saying $a \geq b$ is $a \in b + \mathbb{R}_+$. That is, the natural order $\geq$ of $\mathbb{R}$ is simply the vector order on $\mathbb{R}$ induced by the convex cone $\mathbb{R}_+$. Put differently, $\geq = \succsim^{\mathbb{R}_+}$.

For any $n \in \mathbb{N}$, exactly the same situation ensues in $\mathbb{R}^n$, that is, $x \geq y$ iff $x \in y + \mathbb{R}^n_+$. Of course, we would have obtained different ways of ordering the real n-vectors if we had used a convex cone in $\mathbb{R}^n$ distinct from $\mathbb{R}^n_+$. For instance, if we had designated $C := \{x \in \mathbb{R}^2 : x_1 > 0 \text{ or } x_1 = 0 \text{ and } x_2 \geq 0\}$ as our "positive cone" in $\mathbb{R}^2$, then we would be ordering the vectors in $\mathbb{R}^2$ *lexicographically*, that is, $\succsim^C = \succsim_{\text{lex}}$ (Example B.1).

The smoke must be clearing now. A preorder $\succsim$ on a linear space X tells us which vectors in X are "positive" in X—a vector x in X is deemed "positive" by $\succsim$ iff $x \succsim \mathbf{0}$. Thus, if $\succsim$ is a vector preorder, then

$$x \succsim y \quad \text{if and only if} \quad x = y + \text{a positive vector in } X.$$

In this case, therefore, all there is to know about $\succsim$ can be learned from the positive cone induced by $\succsim$. This simple observation brings the geometric analysis of convex cones to the fore of order-theoretic investigations.

Notation. As noted above, we denote the preorder associated with a given preordered linear space X by $\succsim_X$. In turn, the positive cone induced by $\succsim_X$ is denoted by X_+, that is,

$$X_+ := \{x \in X : x \succsim_X \mathbf{0}\}.$$

In the same vein, the **strictly positive cone** of X is defined as

$$X_{++} := \{x \in X : x \succ_X \mathbf{0}\},$$

where $\succ_X$ is the asymmetric part of $\succsim_X$ (Exercise A.7).[7] Clearly, if X is a partially ordered linear space, then $X_{++} = X_+ \backslash \{\mathbf{0}\}$.

Warning. The strictly positive cone of $\mathbb{R}^n$ (with respect to the partial order $\geq$) equals $\mathbb{R}^n_{++}$ iff $n = 1$. (Why?)

[7] Strictly speaking, I'm abusing the terminology here in that X_{++} is not a cone, for it does not contain $\mathbf{0}$. (Of course, $X_{++} \cup \{\mathbf{0}\}$ is a convex cone.)

Example 2

[1] Let X be a preordered linear space and Y a linear subspace of X. A natural way of making Y a preordered linear space is to designate $X_+ \cap Y$ as Y_+. The induced vector preorder $\succsim_Y$ on Y is then $\succsim_X \cap (Y \times Y)$. When Y is endowed with this vector preorder, we say that it is a **preordered linear subspace of** X.

[2] For any nonempty set T, recall that the natural partial order $\geq$ on $\mathbb{R}^T$ is defined pointwise, that is, $f \geq g$ iff $f(t) \geq g(t)$ for all $t \in T$. The positive cone in $\mathbb{R}^T$ is thus the class of all nonnegative-valued maps on T: $(\mathbb{R}^T)_+ := \mathbb{R}^T_+$. Unless otherwise is stated explicitly, all of the function (and hence sequence) spaces that we consider in this text (e.g., $\mathbf{C}[0,1]$, $\mathbf{C}^1[0,1]$, ℓ^p, $1 \leq p \leq \infty$, etc.) are ordered in this canonical way. That is, for any given nonempty set T and linear subspace X of $\mathbb{R}^T$, X is thought of as a preordered linear subspace of $\mathbb{R}^T$—we have $X_+ := \mathbb{R}^T_+ \cap X$. In this case, we denote the induced vector order on X simply by $\geq$ (instead of $\succsim_X$).

[3] Let X and Y be two preordered linear spaces. A natural way of making the linear product space $X \times Y$ a preordered linear space is to endow it with the product vector preorder. That is, we define $\succsim_{X \times Y}$ as

$$(x, y) \succsim_{X\times Y} (x', y') \quad \text{if and only if} \quad x \succsim_X x' \text{ and } y \succsim_Y y'$$

for all $x, x' \in X$ and $y, y' \in Y$. The natural positive cone of $X \times Y$ is thus $(X \times Y)_+ := X_+ \times Y_+$. Endowed with this order structure, $X \times Y$ is called a **preordered linear product of** X **and** Y. □

Exercise 14^{H} Let X be a preordered linear space. Show that X_+ is a generating cone in X iff, for any $x \in X$, there exists a $y \in X$ such that $y \succsim_X x$ and $y \succsim_X -x$.

Exercise 15 (Peressini) Let X be a preordered linear space.

(a) Show that if X contains a vector x_* such that, for all $x \in X$, there exists a $\lambda > 0$ with $\lambda x_* \succsim_X x$, then X_+ must be generating.

(b) Use part (a) to conclude that the positive cones of any of the following partially ordered linear spaces are generating: ℓ^p (with $1 \leq p \leq \infty$), the linear space $\mathbf{c}$ of all convergent real sequences, $\mathbf{B}(T)$ (for any nonempty set T).

(c) Show that $\mathbb{R}^\infty_+$ is a generating cone in $\mathbb{R}^\infty$, even though the result stated in part (a) does not apply in this case.

Let X and Y be two preordered linear spaces, $\emptyset \neq S \subseteq X$ and $f \in Y^S$. We say that f is **increasing** if $x \succsim_X x'$ implies $f(x) \succsim_Y f(x')$ for any $x, x' \in S$. This map is said to be **strictly increasing** if it is increasing and $x \succ_X x'$ implies $f(x) \succ_Y f(x')$ for any $x, x' \in S$. Equivalently, f is increasing iff

$$f(x) \in f(x') + Y_+ \quad \text{whenever} \quad x \in x' + X_+,$$

and strictly increasing iff it is increasing and

$$f(x) \in f(x') + Y_{++} \quad \text{whenever} \quad x \in x' + X_{++}$$

for any $x, x' \in S$. It is obvious that these definitions reduce to the usual definitions of increasing and strictly increasing functions, respectively, in the case where X is a Euclidean space and $Y = \mathbb{R}$.

EXAMPLE 3

[1] For any two preordered linear spaces X and Y, a linear operator $L \in \mathcal{L}(X, Y)$ is increasing iff $L(X_+) \subseteq Y_+$. Such an L is called a **positive linear operator**. Similarly, $L \in \mathcal{L}(X, Y)$ is strictly increasing iff $L(X_+) \subseteq Y_+$ and $L(X_{++}) \subseteq Y_{++}$. Such an L is called a **strictly positive linear operator**. Of course, when $Y = \mathbb{R}$, we instead talk of positive linear *functionals* and strictly positive linear *functionals*, respectively. For instance, the zero functional on X is a positive linear functional that is not strictly positive.

[2] Take any $m, n \in \mathbb{N}$, and let $\mathbf{A} := [a_{ij}]_{m \times n} \in \mathbb{R}^{m \times n}$. Recall that $L : \mathbb{R}^n \to \mathbb{R}^m$, defined by $L(x) := \mathbf{A}x$, is a linear operator (Example F.6). This operator is positive iff $a_{ij} \geq 0$ for all i and j, and strictly positive iff $(a_{i1}, \ldots, a_{in}) > \mathbf{0}$ for each $i = 1, \ldots, m$.

[3] Each of the maps $(x_m) \mapsto x_1$, $(x_m) \mapsto (x_2, x_3, \ldots)$, and $(x_m) \mapsto \sum^\infty \frac{1}{2^i} x_i$ are positive linear functionals on ℓ^∞. The first two of these are not strictly positive, but the third one is.

[4] The linear operator $D : \mathbf{C}^1[0, 1] \to \mathbf{C}[0, 1]$ defined by $D(f) := f'$ is not positive. (Why?) □

EXERCISE 16

(a) Define $L \in \mathbb{R}^{\mathbf{C}[0,1]}$ by $L(f) := f(0)$. Is L a positive linear functional? A strictly positive one?

(b) Define the self-map L on $\mathbf{C}[0, 1]$ by $L(f)(x) := \int_0^x f(t)dt$ for all $0 \leq x \leq 1$. Is L a positive linear operator? A strictly positive one?

EXERCISE 17 (Kantorovich) Let X be a partially ordered linear space such that X_+ is generating. Show that if $L : X_+ \to \mathbb{R}_+$ is additive, that is, $L(x + y) = L(x) + L(y)$ for all $x, y \in X_+$, then there is a unique positive linear function L^* on X such that $L^*|_{X_+} = L$.

EXERCISE 18 Let X and Y be two partially ordered linear spaces, and denote the set of all positive linear operators from X into Y by $\mathcal{L}_+(X, Y)$. Show that $\mathcal{L}_+(X, Y)$ is a convex cone in $\mathcal{L}(X, Y)$. Moreover, if X_+ is generating, then $\mathcal{L}_+(X, Y)$ is pointed.

EXERCISE 19 Let X be a preordered linear space and I an interval. We say that the correspondence $\Gamma : I \rightrightarrows X$ is **strictly increasing** if for any $a, b \in I$,

(i) no two distinct elements of $\Gamma(a)$ are comparable by $\succcurlyeq_X$, and

(ii) $a > b$ and $x \in \Gamma(a)$ imply $x \succ_X y$ for some $y \in \Gamma(b)$.

Show that, for any surjective such Γ, there exists a unique strictly increasing surjection $f : \Gamma(I) \to I$ with $\Gamma = f^{-1}$.

1.4 Algebraic and Relative Interior of a Set

The generality of the treatment set aside, the basic concepts covered so far in this chapter are probably not new to you. From this point on, however, it is likely that we will be entering a new territory. The idea is this. We are interested in talking about *open* and *closed* convex sets, the *boundary* of a convex set, and so on. We can do this in $\mathbb{R}^n$ easily by using the metric structure of $\mathbb{R}^n$. At the moment, however, all we have is a linear space, so the $\mathbb{R}^n$ example is not readily helpful. Instead, we notice that openness of convex sets in a Euclidean space can alternatively be described by using only the notion of line segments. For instance, one way of proving that $(0, 1)$ is an open interval in $\mathbb{R}$ is to observe that for any $a \in (0, 1)$, some part of the line segment between a and *any* other point in $\mathbb{R}$ (not including the endpoint a) is contained in $(0, 1)$. (That is, for any $a \in (0, 1)$ and any real number b, there is a c in $co\{a,b\}\backslash\{a\}$ such that $co\{a,c\} \subseteq (0,1)$.) In this sense all points of $(0,1)$ are in the "interior" of $(0,1)$, and hence we may think of $(0,1)$ as open. Clearly, we could not do the same for $(0,1]$. The line segment between 1 and 2, say, intersects $(0,1]$ only at the endpoint 1.

These considerations become trivial, of course, if the term "open" is understood relative to the usual metric on $\mathbb{R}$. The point is that, in this discussion, we did not use a metric at all, everything was algebraic. So, perhaps there is a way of defining an openness notion, at least *for convex sets*, that would be identical to the usual notion of openness in Euclidean spaces but would not require us to use a distance function. This would let us talk about open convex sets in an arbitrary linear space in a geometrically meaningful way.

We now develop this idea formally.

DEFINITION

Let S be a subset of a linear space X. A vector x in S is called an **algebraic interior point** of S (*in* X) if, for any $y \in X$, there exists an $\alpha_y > 0$ such that

$$(1-\alpha)x + \alpha y \in S \qquad \text{for all } 0 \leq \alpha \leq \alpha_y.$$

The set of all algebraic interior points of S in X is called the **algebraic interior** of S (*in* X) and is denoted by $\textit{al-int}_X(S)$. (*Note.* $\textit{al-int}_X(\emptyset) = \emptyset$.) If $S \subseteq \textit{al-int}_X(S)$, we say that S is **algebraically open** in X.

Geometrically speaking, $x \in \textit{al-int}_X(S)$ means that one may move linearly from x toward *any* direction in the linear space X without leaving the set S immediately. (See Figure 4.) While intuitive, this definition is completely algebraic—it is not based on a distance function—so it makes sense in the setting of an arbitrary linear space. As we shall see shortly, however, there are strong links between the algebraic interior and the usual interior of a convex set in a variety of interesting cases.

Whether or not a given set is algebraically open depends crucially on relative to which space one asks the question.[8] Indeed, if Y is a linear subspace of X and $S \subseteq Y \subseteq X$, then $\textit{al-int}_Y(S)$ and $\textit{al-int}_X(S)$ may well

[8] Go back and read the same sentence again, but this time omit the word "algebraically." Does it sound familiar?

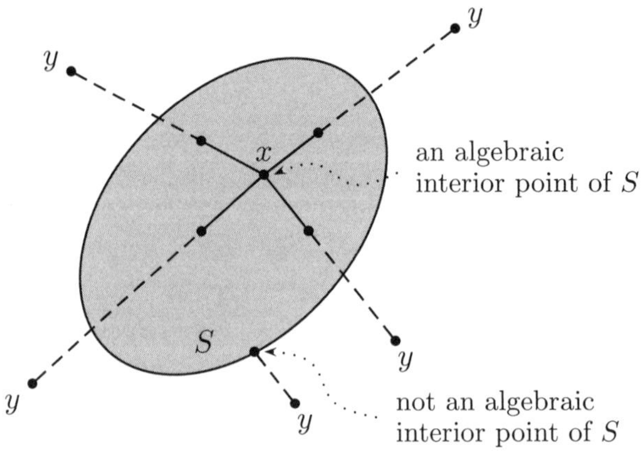

FIGURE 4

be quite different. For instance, $\{0\} \subseteq \mathbb{R}$ is algebraically open in the trivial linear space $\{0\}$, but it is not in $\mathbb{R}$. Similarly, if $S := \{(t,t) : 0 \leq t \leq 1\}$ and $Y := \{(t,t) : t \in \mathbb{R}\}$, then $\textit{al-int}_Y(S) = \{(t,t) : 0 < t < 1\}$, while $\textit{al-int}_{\mathbb{R}^2}(S) = \emptyset$.

Now, the natural habitat of a convex set is its affine hull. In particular, if S is a subset of a linear space X with $\mathbf{0} \in S$, then all algebraic properties of S can be studied *within* $\textit{span}(S)$; we do not, in effect, need the vectors in $X \backslash \textit{span}(S)$ for this purpose. Nothing changes, of course, if we drop the assumption $\mathbf{0} \in S$ here, except that the natural residence of S then becomes $\textit{aff}(S)$. Thus, the algebraic interior of a subset of a linear space relative to its affine hull is of primary importance. In convex analysis, it is referred to as the *relative interior* of that set.

DEFINITION

Let S be a subset of a linear space X. A vector x in S is called a **relative interior point** of S if for any $y \in \textit{aff}(S)$, there exists an $\alpha_y > 0$ such that

$$(1-\alpha)x + \alpha y \in S \qquad \text{for all } 0 \leq \alpha \leq \alpha_y.$$

The set of all relative interior points of S is called the **relative interior** of S and is denoted by $ri(S)$. (*Note.* $ri(\emptyset) = \emptyset$.)

Thus, a relative interior point of a set S is a vector x in S such that, when moving from x toward any direction *in* $\textit{aff}(S)$, we do not immediately leave S. For example, we have $(0,1) = ri((0,1))$, but $1 \notin ri((0,1])$. But note that $S = ri(S)$ does not imply that S is algebraically open. Indeed, $\{0\} = ri(\{0\})$, because $aff\{0\} = \{0\}$, but $\{0\}$ is not algebraically open in $\mathbb{R}$, for $al\text{-}int_{\mathbb{R}}(\{0\}) = \emptyset$. Similarly, if $S := \{(t,t) : 0 \leq t \leq 1\}$, then $ri(S) = \{(t,t) : 0 < t < 1\}$ while $al\text{-}int_{\mathbb{R}^2}(S) = \emptyset$.

Before considering less trivial examples, let us clarify the relation between the algebraic and relative interior of an arbitrary subset S of a linear space X:

$$al\text{-}int_X(S) = \begin{cases} ri(S), & \text{if } aff(S) = X \\ \emptyset, & \text{if } aff(S) \subset X \end{cases}. \tag{3}$$

(Since $S \subseteq X$, we have $aff(S) \subseteq X$, so this equation covers all contingencies.)

Here is why. If $aff(S) = X$, we trivially have $al\text{-}int_X(S) = ri(S)$, so consider instead the case $aff(S) \subset X$. Take any $x^* \in S$, and let $Y := span(S - x^*)$.[9] Since $aff(S)$ is a proper subset of X, Y is a *proper* linear subspace of X. Pick any $w \in X\backslash Y$, and notice that $(1-\alpha)z + \alpha w \notin Y$ for any $z \in Y$ and $0 < \alpha \leq 1$. (Otherwise we would contradict $w \notin Y$, wouldn't we? See Figure 5.) But then

$$(1-\alpha)(z + x^*) + \alpha(w + x^*) \notin Y + x^* = aff(S)$$

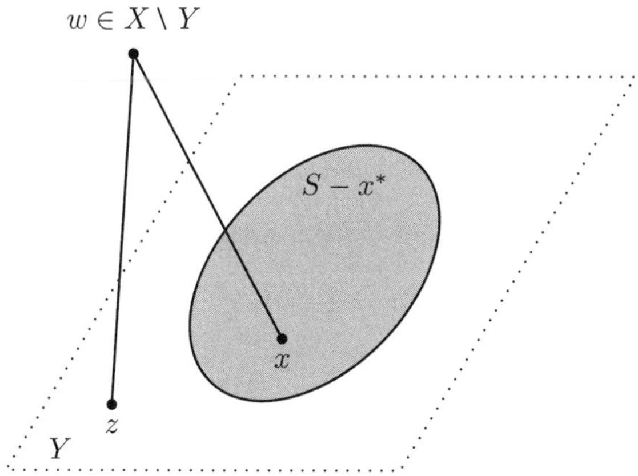

Figure 5

[9] *Reminder.* $aff(S) = span(S - x) + x$ for any $x \in S$ (Section F.1.4).

for any $z \in Y$ and $0 < \alpha \leq 1$. Letting $y := w + x^*$, therefore, we find $(1-\alpha)x + \alpha y \notin Y$ for any $x \in S$ and $0 < \alpha \leq 1$. Conclusion: $\textit{al-int}_X(S) = \emptyset$.

It is worth putting on record the following immediate consequence of (3).

OBSERVATION 1. For any subset S of a linear space X, we have $\textit{al-int}_X(S) \neq \emptyset$ if, and only if, $\textit{aff}(S) = X$ and $\textit{ri}(S) \neq \emptyset$.

Let us now move on to examples. Our first set of examples explores the connection between the notions of algebraic and metric openness in the context of Euclidean spaces. Please note that in Examples 4–6 we simplify our notation by writing $\textit{al-int}(S)$ for $\textit{al-int}_{\mathbb{R}^n}(S)$, and $\textit{int}(S)$ for $\textit{int}_{\mathbb{R}^n}(S)$.

EXAMPLE 4

Given any $n \in \mathbb{N}$, consider any nonempty subset S of $\mathbb{R}^n$.

Question: How does $\textit{al-int}(S)$ relate to $\textit{int}(S)$?

Answer: $\textit{al-int}(S)$ is surely larger than $\textit{int}(S)$, but the converse . . . well, it depends!

For, if $x \in \textit{int}(S)$, then there exists an $\varepsilon > 0$ such that $N_{\varepsilon,\mathbb{R}^n}(x) \subseteq S$, while for any (fixed) $y \in \mathbb{R}^n$, a straightforward continuity argument shows that there exists an $\alpha_y > 0$ such that $d_2((1-\alpha)x + \alpha y, x) < \varepsilon$ for all $0 \leq \alpha \leq \alpha_y$. Thus

$$(1-\alpha)x + \alpha y \in N_{\varepsilon,\mathbb{R}^n}(x) \subseteq S, \qquad 0 \leq \alpha \leq \alpha_y,$$

and we may conclude that $x \in \textit{al-int}(S)$. Conclusion:

$$\textit{int}(S) \subseteq \textit{al-int}(S) \qquad \text{for any set } S \text{ in } \mathbb{R}^n. \tag{4}$$

By the way, combining this observation with (3), we find, for any subset S of $\mathbb{R}^n$,

$$\textit{int}(S) \begin{cases} \subseteq \textit{ri}(S), & \text{if } \textit{dim}(S) = n \\ = \emptyset, & \text{if } \textit{dim}(S) < n \end{cases}.$$

This answer is only partial, however. To complete the picture, we need to find out whether we can strengthen $\subseteq$ to $=$ in (4). In fact, we can't. Consider the set

$$S := \left\{(\theta, r) : 0 \leq \theta \leq 2\pi \text{ and } 0 \leq r \leq 1 - \tfrac{\theta}{2\pi}\right\},$$

which is expressed in polar coordinates (Figure 6). It is easily checked that $\mathbf{0} \notin \textit{int}_{\mathbb{R}^2}(S)$ but $\mathbf{0} \in \textit{ri}(S) = \textit{al-int}(S)$. Thus: The interior of a subset of

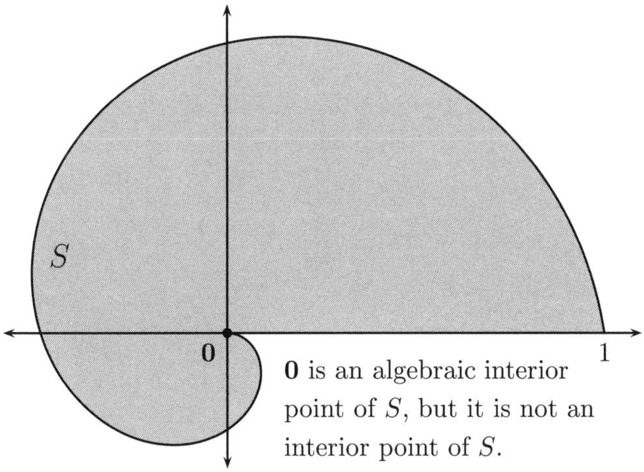
S
0
1
0 is an algebraic interior point of S, but it is not an interior point of S.

Figure 6

a Euclidean space may be a proper subset of its relative and/or algebraic interior. □

Example 5

We shall now show that the anomaly depicted in the second part of Example 4 was in fact due to the nonconvexity of the set in question. That is, we wish to show that we can in fact strengthen $\subseteq$ to $=$ in (4), *provided that S is convex.*

Take any nonempty convex subset S of $\mathbb{R}^n$, and let $x \in$ *al-int*(S). Then we can move from x in any direction without leaving S immediately. In particular, consider the vectors $y^i := x + \mathbf{e}^i$, $i = 1, \ldots, n$, and $y^i := x - \mathbf{e}^i$, $i = n+1, \ldots, 2n$, where $\mathbf{e}^i$ is the ith unit vector in $\mathbb{R}^n$. Since x is an algebraic interior point of S in $\mathbb{R}^n$, for each $i \in \{1, \ldots, n\}$ there exists an $\alpha_i > 0$ such that

$$x + \alpha\mathbf{e}^i = (1-\alpha)x + \alpha y^i \in S, \qquad 0 \leq \alpha \leq \alpha_i,$$

and similarly, for each $i \in \{n+1, \ldots, 2n\}$ there exists an $\alpha_i > 0$ such that

$$x - \alpha\mathbf{e}^i \in S, \qquad 0 \leq \alpha \leq \alpha_i.$$

Letting $\alpha^* := \min\{\alpha_1, \ldots, \alpha_{2n}\}$, therefore, we find that $x + \alpha\mathbf{e}^i \in S$ for all $i = 1, \ldots, 2n$ and $\alpha \in [-\alpha^*, \alpha^*]$. But since S is convex, we have

$$T := co\left(\left\{x + \alpha\mathbf{e}^i : -\alpha^* \leq \alpha \leq \alpha^*,\ i = 1, \ldots, n\right\}\right) \subseteq S$$

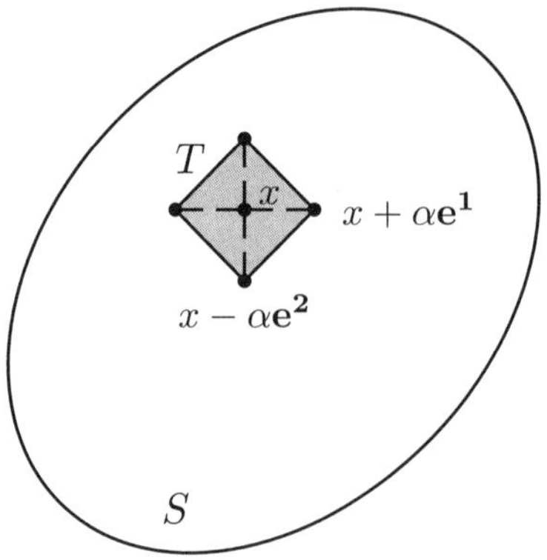

FIGURE 7

(Figure 7). Of course, $x \in int(T)$ and $int(T) \subseteq int(S)$, so $x \in int(S)$, as we sought. Conclusion:

$$int(S) = al\text{-}int(S) \qquad \text{for any convex set } S \text{ in } \mathbb{R}^n. \tag{5}$$

This is what we meant by "well, it depends!" in our answer to the question posed at the beginning of Example 4. □

The following is an immediate consequence of (3) and (5).

OBSERVATION 2. Given any $n \in \mathbb{N}$ and a convex subset S of $\mathbb{R}^n$, we have

$$int(S) = \begin{cases} ri(S), & \text{if } dim(S) = n \\ \emptyset, & \text{if } dim(S) < n \end{cases}.$$

Moreover, S is open in $\mathbb{R}^n$ if, and only if, it is algebraically open in $\mathbb{R}^n$.

EXAMPLE 6

Take any $n \in \mathbb{N}$ and let $\{\mathbf{e}^1, \ldots, \mathbf{e}^n\}$ stand for the standard basis for $\mathbb{R}^n$, as usual. Consider the set

$$\begin{aligned} S &:= co\left(\{\mathbf{0}, \mathbf{e}^1, \ldots, \mathbf{e}^n\}\right) \\ &= \left\{ \sum_{i=1}^n \lambda_i \mathbf{e}^i : \lambda_i \geq 0,\ i = 1, \ldots, n \text{ and } \sum_{i=1}^n \lambda_i \leq 1 \right\}. \end{aligned}$$

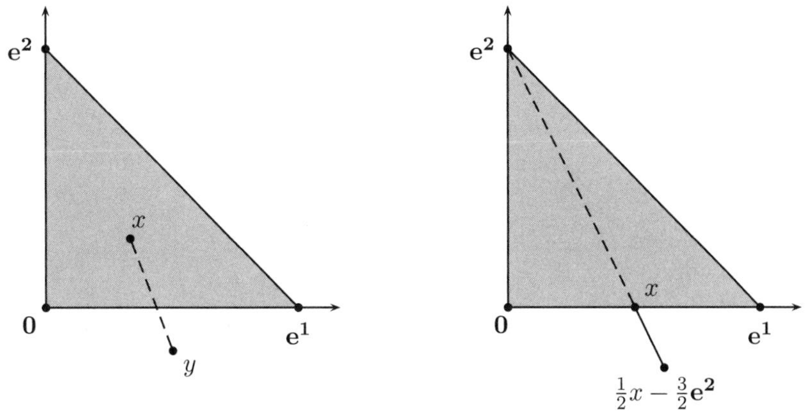

Figure 8

It is easy to see that the algebraic interior of S in $\mathbb{R}^n$ is not empty, and thus equals $ri(S)$. (See Figure 8.) For instance, the "center" of S is in the algebraic interior of S in $\mathbb{R}^n$, that is,

$$x := \sum_{i=1}^{n} \frac{1}{n+1}\mathbf{e}^i \in al\text{-}int(S).$$

Indeed, for any $y \in \mathbb{R}^n$ and $0 \leq \alpha \leq 1$, we have

$$(1-\alpha)x + \alpha y = \sum_{i=1}^{n} \theta_i(\alpha)\mathbf{e}^i \quad \text{where} \quad \theta_i(\alpha) := \tfrac{1-\alpha}{n+1} + \alpha y_i,\ i = 1, \ldots, n.$$

It is readily verified that there exists an $\alpha^* > 0$ small enough to guarantee that $\theta_i(\alpha) \geq 0$ for each i and $\sum^n \theta_i(\alpha) \leq 1$ whenever $0 < \alpha \leq \alpha^*$. This means that $(1-\alpha)x+\alpha y \in S$.[10] Keeping in mind that any finite-dimensional linear space is isomorphic to $\mathbb{R}^n$ for some n (Corollary F.3), we may then conclude: *In a finite-dimensional linear space, the convex hull of any basis and the origin has nonempty algebraic interior.* (Why exactly?) □

As we have observed earlier, if a subset S of a linear space X has a nonempty algebraic interior in X, then the affine hull of S must equal the entire space (Observation 1). We shall see shortly (in Examples 7 and 8) that

[10] Thanks to (5), it is enough to observe that $\{x \in \mathbb{R}^n_{++} : \sum^n x_i < 1\}$ is open in $\mathbb{R}^n$ to conclude that $al\text{-}int(S) \neq \emptyset$.

the converse of this statement is not true, even for convex sets. However, as you are asked to prove in the next exercise, all goes well in the finite-dimensional case.

EXERCISE 20[H] Let S be a nonempty convex subset of a finite-dimensional linear space X. Prove:
(a) $ri(S) \neq \emptyset$;
(b) $al\text{-}int_X(S) \neq \emptyset$ iff $aff(S) = X$.

The following exercise notes that the relative interior of a convex set in a Euclidean space $\mathbb{R}^n$ is none other than its interior *in* its affine hull (when the latter is viewed as a metric subspace of $\mathbb{R}^n$).

EXERCISE 21 Given any $n \in \mathbb{N}$, show that $ri(S) = int_{aff(S)}(S)$ for any nonempty convex subset S of $\mathbb{R}^n$.

EXERCISE 22[H] Do we necessarily have $al\text{-}int_X(al\text{-}int_X(S)) = al\text{-}int_X(S)$ for any subset S of a Euclidean space X?

We now turn to the infinite-dimensional case, where a few surprises are waiting for us.

EXAMPLE 7
It is obvious that $\mathbb{R}^n_{++}$ is algebraically open in $\mathbb{R}^n$. This might tempt one to think that the set $\mathbb{R}^\infty_{++}$ of all positive real sequences is also algebraically open in $\mathbb{R}^\infty$. This is, however, false. Take any $(x_m) \in \mathbb{R}^\infty_{++}$, and consider the real sequence $(y_m) := (-x_1, -2x_2, -3x_3, \ldots)$. Clearly, for any real number α,

$$(1-\alpha)(x_m) + \alpha(y_m) = ((1-2\alpha)x_1, (1-3\alpha)x_2, (1-4\alpha)x_3, \ldots).$$

But it is obvious that, for any fixed $\alpha > 0$, we cannot have $(1-(m+1)\alpha) > 0$ for *all* $m \in \mathbb{N}$. It follows that $(1-\alpha)(x_m) + \alpha(y_m) \notin \mathbb{R}^\infty_{++}$ for *any* $\alpha > 0$. Conclusion:

$$al\text{-}int_{\mathbb{R}^\infty}(\mathbb{R}^\infty_{++}) = \emptyset.$$

How about the relative interior of $\mathbb{R}^\infty_{++}$? Surely, that must be nonempty. (After all, we have seen in Exercise 20 that the relative interior of a nonempty convex set in $\mathbb{R}^n$ always has a nonempty relative interior.) But no, this is not the case. Indeed, we have $aff(\mathbb{R}^\infty_{++}) = \mathbb{R}^\infty$, so Observation 1 tells us that $ri(\mathbb{R}^\infty_{++}) = \emptyset$ as well![11] Conclusion: The relative interior of a nonempty convex subset of an infinite-dimensional linear space may well be empty. □

EXAMPLE 8

In Example 6 we saw that, in a finite-dimensional linear space, the convex hull of any basis and the origin has a nonempty algebraic interior. We now wish to show that this observation is not true in *any* infinite-dimensional linear space! So pick any such space X, and take any basis A for X (Theorem F.1). Now consider the set

$$S := co(A \cup \{\mathbf{0}\}) = \left\{ \sum_{i=1}^{k} \lambda_i x^i : k \in \mathbb{N},\ (x^i, \lambda_i) \in A \times \mathbb{R}_+, \right.$$

$$\left. i = 1, \ldots, k \text{ and } \sum_{i=1}^{k} \lambda_i \leq 1 \right\}.$$

We claim that $al\text{-}int_X(S) = \emptyset$. (You should be somewhat intrigued at this point. After all, we seem to know next to nothing about X.) To see this, take any $k \in \mathbb{N}$ and $(x^1, \lambda_1), \ldots, (x^k, \lambda_k) \in A \times [0, 1]$ with $\sum^k \lambda_i \leq 1$, and let $x := \sum^k \lambda_i x^i \in S$. Now here is what is funny about infinite-dimensional spaces; they possess infinitely many basis vectors. So, no matter how large k is, we can find an $x^{k+1} \in A \backslash \{x^1, \ldots, x^k\}$. (Of course, we could do no such thing (for $k > n$) in Example 6.) Choose, then, $y := \frac{1}{2}x - \frac{3}{2}x^{k+1}$, and observe that, for any $\alpha > 0$,

$$(1-\alpha)x + \alpha y = \sum_{i=1}^{k} \left(1 - \tfrac{\alpha}{2}\right) \lambda_i x^i + \left(\tfrac{-3\alpha}{2}\right) x^{k+1} \notin S.$$

[11] How do I know that $aff(\mathbb{R}^\infty_{++}) = \mathbb{R}^\infty$? Because I see that $span(\mathbb{R}^\infty_{++} - \mathbf{1}) = \mathbb{R}^\infty$, where $\mathbf{1} := (1, 1, \ldots)$, so $aff(\mathbb{R}^\infty_{++}) = \mathbb{R}^\infty + \mathbf{1} = \mathbb{R}^\infty$. Okay, why is $span(\mathbb{R}^\infty_{++} - \mathbf{1}) = \mathbb{R}^\infty$ true? Well, take any $(x_m) \in \mathbb{R}^\infty$, and define $I := \{i : x_i > 0\}$ and $J := \mathbb{N} \backslash I$. If (y_m) and (z_m) are the sequences defined by

$$y_i := \begin{cases} x_i + 1, & \text{if } i \in I \\ 1, & \text{if } i \in J \end{cases} \quad \text{and} \quad z_i := \begin{cases} 1, & \text{if } i \in I \\ -x_i + 1, & \text{if } i \in J \end{cases},$$

then $(x_m) = ((y_m) - \mathbf{1}) - ((z_m) - \mathbf{1})$. Since both (y_m) and (z_m) belong to $\mathbb{R}^\infty_{++}$, this shows that $(x_m) \in span(\mathbb{R}^\infty_{++} - \mathbf{1})$.

(Why is the last assertion true? If you don't see this in ten seconds, you should go back and have a good look at Corollary F.2.) Thus, we conclude, $\textit{al-int}_X(S) = \emptyset$.

Here is a nice corollary of this. Since we obviously have $\textit{aff}(S) = X$ in this example, we have $ri(S) = \emptyset$ as well. Thus: *In any infinite-dimensional linear space, there exists a nonempty convex set S with $ri(S) = \emptyset$.*

The second diagram of Figure 8 may help clarify what is going on here. Consider a vector like x in that figure. This point is expressed by using $\mathbf{0}$ and $\mathbf{e}^1$, which do not constitute a basis for $\mathbb{R}^2$. But then we can find a vector, say $\mathbf{e}^2$, that is linearly independent of $\mathbf{e}^1$, and we use the line segment through $\mathbf{e}^2$ and x to show that x cannot be in the algebraic interior of $co(\{\mathbf{0}, \mathbf{e}^1, \mathbf{e}^2\})$. (We couldn't do such a thing for the vector x in the first diagram, since x was expressed there by using both $\mathbf{e}^1$ and $\mathbf{e}^2$.) The upshot is that all points in $co(A \cup \{\mathbf{0}\})$ in the present example behave just like the vector x in the second diagram, precisely because in an infinite-dimensional space every vector is expressed as a linear combination of *finitely* many (hence not all) basis vectors with nonzero coefficients. This is the gist of infinite dimensionality. □

An important insight that stems from Example 8 is that, in an infinite-dimensional linear space, algebraically open sets, which have nonempty algebraic interiors by necessity, are "really, really large." You should carry this intuition with you at all times. After all, Example 8 shows that even if we have a convex set S so large that $\textit{aff}(S)$ is actually the entire mother space, we are still not guaranteed the existence of a single algebraic interior point. This is, of course, in stark contrast to the finite-dimensional case.

EXERCISE 23 Show that $\textit{al-int}_{\ell^p}(\ell^p_+) = \emptyset$ for all $1 \le p < \infty$.

EXERCISE 24 Prove:

$$\textit{al-int}_{\ell^\infty}(\ell^\infty_+) = \{(x_m) \in \ell^\infty_+ : \inf\{x_m : m \in \mathbb{N}\} > 0\}.$$

EXERCISE 25[H] Compute the algebraic interior of the positive cone of $\mathbf{C}[0, 1]$.

EXAMPLE 9

For any nonzero linear functional L on a linear space X, consider the open half-space $S := \{x \in X : L(x) > 0\}$. We claim that S is algebraically open in X. Indeed, let $x \in S$ and pick any $y \in X$. Since $L((1-\alpha)x+\alpha y) = (1-\alpha)L(x)+\alpha L(y)$, by choosing

$$\alpha_y = \begin{cases} \frac{1}{2}, & \text{if } L(y) \geq 0 \\ \frac{L(x)}{L(x)-L(y)}, & \text{if } L(y) < 0 \end{cases},$$

we find $(1-\alpha)x+\alpha y \in S$ for all $0 \leq \alpha < \alpha_y$. Thus: *Any open half-space in X is algebraically open in X.* Since a half-space cannot be empty, therefore: *The algebraic interior of a half-space is not empty, and equals its relative interior.* □

Exercise 26 [H] For any subset S of a linear space X, show that $x \in$ *al-int*$_X(S)$ iff for each $y \in X$, there exists an $\varepsilon_y > 0$ such that $x + \varepsilon y \in S$ for all $\varepsilon \in \mathbb{R}$ with $|\varepsilon| \leq \varepsilon_y$.

Exercise 27 [H] Let S be a convex subset of a linear space X. Prove:

(a) If $x \in$ *al-int*$_X(S)$, then $(1-\alpha)x+\alpha y \in$ *al-int*$_X(S)$ for any $(\alpha, y) \in [0,1) \times S$.

(b) If *al-int*$_X(S) \neq \emptyset$, then *aff*$(S) = X =$ *aff*(*al-int*$_X(S)$).

Exercise 28 For any algebraically open subset A of a linear space X, show that $A + \lambda B$ is algebraically open in X for any $(\lambda, B) \in \mathbb{R} \times 2^X$.

The following exercise shows how algebraic interior points of a convex set may be relevant in optimization theory. It says that if a concave real map attains its minimum on the algebraic interior of its domain, then it must be constant.

Exercise 29 Let S be a convex subset of a linear space X. Where $x^* \in$ *al-int*$_X(S)$ and $f \in \mathbb{R}^S$ is a concave function, prove:

(a) For every $x \in S$, there exists an $(\alpha, y) \in (0,1) \times S$ with $x^* = (1-\alpha)x + \alpha y$.

(b) If $x^* \in \arg\min\{f(x) : x \in S\}$, then f must be a constant function.

1.5 Algebraic Closure of a Set

We now turn to the algebraic formulation of a notion of "closedness" for subsets of a linear space.

Definition
Let S be a subset of a linear space X. A vector x in X is called an **algebraic closure point** of S (*in* X) if, for some $y \in S$, we have

$$(1-\alpha)x + \alpha y \in S \qquad \text{for all } 0 < \alpha \leq 1.$$

The set of all algebraic closure points of S (in X) is called the **algebraic closure** of S (*in* X), and is denoted by $\textit{al-cl}_X(S)$. (*Note.* $\textit{al-cl}_X(\emptyset) = \emptyset$.) If $\textit{al-cl}_X(S) \subseteq S$, we say that S is **algebraically closed in** X. Finally, the set $\textit{al-cl}_X(S) \backslash \textit{al-int}_X(S)$ is called the **algebraic boundary** of S (*in* X), and is denoted by $\textit{al-bd}_X(S)$.

Geometrically speaking, if x is an algebraic closure point of a set S (in a linear space), we understand that this vector is "reachable" from S in the sense that for at least one vector y in S, the entire line segment between x and y, possibly excluding the endpoint x, stays within S.[12] The algebraic boundary points of S are similarly interpreted.

Warning. By contrast to the notion of algebraic interior, it is not important relative to which linear space we consider the algebraic closure operator. For any subset S of a linear space X, all that matters for the determination of $\textit{al-cl}_X(S)$ is the affine hull of S.[13] Consequently, from now on, we will denote $\textit{al-cl}_X(S)$ simply by $\textit{al-cl}(S)$, and if S is algebraically closed in some linear space, then we will simply say that it is *algebraically closed*. Since the algebraic boundary of a set depends on its algebraic interior, however, "X" in the notation $\textit{al-bd}_X(S)$ is not superfluous. For instance, if $S = \{(t,0) \in \mathbb{R}^2 : 0 < t < 1\}$, then $\textit{al-bd}_{\mathbb{R}\times\{0\}}(S) = \{(0,0),(1,0)\}$ while $\textit{al-bd}_{\mathbb{R}^2}(S) = [0,1] \times \{0\}$. (Notice that the algebraic closure of S is $[0,1] \times \{0\}$ whether we consider S as a subset of $\mathbb{R} \times \{0\}$ or $\mathbb{R}^2$.)

The following is a reflection of Examples 4 and 5. (Throughout the following example, we denote the closure of S in $\mathbb{R}^n$ simply by $cl(S)$.)

[12] For this reason, some authors refer to the algebraic closure points of S as the points that are *linearly accessible* from S, and denote the algebraic closure of S by $lin(S)$.

[13] That is, for any linear spaces X and Y that contain $\textit{aff}(S)$, we have $x \in \textit{al-cl}_X(S)$ iff $x \in \textit{al-cl}_Y(S)$. (Here, of course, we implicitly assume that the vector operations of X and Y are the same with those used to define $\textit{aff}(S)$.)

EXAMPLE 10
Given any $n \in \mathbb{N}$, consider any nonempty subset S of $\mathbb{R}^n$.

Question: How does $al\text{-}cl(S)$ relate to $cl(S)$?

Answer: $cl(S)$ is surely larger than $al\text{-}cl(S)$, but the converse . . . well, it depends!

(Sounds familiar?) Indeed, if $x \in al\text{-}cl(S)$, then there exists a $y \in S$ such that $\left(1 - \frac{1}{m}\right)x + \frac{1}{m}y \in S$ for each $m \in \mathbb{N}$, so letting $m \to \infty$ we obtain $x \in cl(S)$. Conclusion:

$$al\text{-}cl(S) \subseteq cl(S) \qquad \text{for any } S \text{ in } \mathbb{R}^n. \tag{6}$$

The converse is false, however. For instance, $\mathbb{Q}$ is an algebraically closed set in $\mathbb{R}$ that is not closed. (Think about it!) Yet, the converse holds, *provided that S is convex*. Given the analysis of the previous subsection, this should not be terribly surprising to you. The proof is a little subtle, however. It is based on the following result of Euclidean analysis.

CLAIM. If S is convex and $X := aff(S)$, then

$$(1 - \lambda)cl(S) + \lambda int_X(S) \subseteq S \qquad \text{for any } 0 < \lambda \leq 1.$$ [14]

PROOF OF CLAIM
Take any $x \in cl(S)$ and $y \in int_X(S)$, and fix an arbitrary $0 < \lambda < 1$. Clearly,

$$(1 - \lambda)x + \lambda y \in S$$
$$\text{if and only if} \quad x \in \tfrac{1}{1-\lambda}S - \left(\tfrac{\lambda}{1-\lambda}\right)y = S + \left(\tfrac{\lambda}{1-\lambda}\right)(S - y).$$

Since $y \in int_X(S)$, the set $\left(\frac{\lambda}{1-\lambda}\right)(int_X(S) - y)$ is an open subset of X that contains $\mathbf{0}$. (Why?) Thus $S + \left(\frac{\lambda}{1-\lambda}\right)(int_X(S) - y)$ is an open subset of X that contains $cl_X(S) = cl(S)$. (Why?) So,

$$x \in S + \left(\tfrac{\lambda}{1-\lambda}\right)(int_X(S) - y) \subseteq S + \left(\tfrac{\lambda}{1-\lambda}\right)(S - y),$$

and it follows that $(1 - \lambda)x + \lambda y \in S$. ‖

Let us now go back to the task at hand. Assume that S is convex and $x \in cl(S)$. Of course, $int_X(S) \neq \emptyset$ (because $int_X(S) = ri(S)$ and the relative

[14] In fact, more is true. Since $(1 - \lambda)cl(S) + \lambda int_X(S)$ is an open subset of X (why?), this result entails that $(1 - \lambda)cl(S) + \lambda int_X(S) \subseteq int_X(S)$ for any $0 < \lambda \leq 1$.

interior of a nonempty convex set in $\mathbb{R}^n$ has always a nonempty relative interior; recall Exercise 20). Then pick any $y \in int_X(S)$, and use the Claim above to conclude that $(1-\lambda)x + \lambda y \in S$ for any $0 < \lambda \leq 1$. This shows that $x \in al\text{-}cl(S)$. Conclusion:

$$cl(S) = al\text{-}cl(S) \quad \text{for any convex set } S \text{ in } \mathbb{R}^n.$$

(Compare with (5).) □

We have seen earlier that the notions of openness and algebraic openness coincide for convex subsets of any Euclidean space. By Example 10, we now know that the same is true for the notions of closedness and algebraic closedness as well.

Observation 3. Given any $n \in \mathbb{N}$, a convex subset S of $\mathbb{R}^n$ is closed in $\mathbb{R}^n$ if, and only if, it is algebraically closed.

Exercise 30^H Let S be a subset of a linear space X.

(a) Show that if S is algebraically open in X, then $X \backslash S$ is algebraically closed in X, but not conversely.

(b) Prove: Provided that S is a convex set, it is algebraically open in X iff $X \backslash S$ is algebraically closed in X.

Exercise 31 Let X be a linear space and $L \in \mathcal{L}(X, \mathbb{R})$. Prove or disprove: For any $\alpha \in \mathbb{R}$, either of the closed halfspaces induced by the hyperplane $L^{-1}(\alpha)$ is algebraically closed.

Exercise 32 Show that every linear subspace of a linear space is algebraically closed. Conclude that any linear subspace of a Euclidean space is closed.

1.6 Finitely Generated Cones

In convex analysis one is frequently confronted with the problem of determining whether or not a given conical (or convex) hull of a set is algebraically closed. This issue is usually settled within the specific context under study; there are only few general results that ascertain the algebraic closedness of such sets. Indeed, the conical hulls of even very well-behaved sets may fail to be algebraically closed. For instance, in $\mathbb{R}^2$, consider the closed ball around $(1, 0)$ with radius 1, that is, the set S of all $x \in \mathbb{R}^2$ such that $0 \leq x_1 \leq 2$ and $x_2^2 + (x_1 - 1)^2 \leq 1$. This set is a compact and convex subset of $\mathbb{R}^2$, but its conical hull is not closed; it equals $\mathbb{R}^2_{++} \cup \{\mathbf{0}\}$.

The situation is quite satisfactory for conical hulls of *finite* sets, however. Indeed, it turns out that while compactness of a set in a Euclidean space may not be enough to guarantee that its conical hull is closed, its finiteness is. In fact, we will prove here the stronger result that the conical hull of a finite set in any linear space is algebraically closed.

Let us first agree on some terminology.

Definition
Let X be a linear space and C a convex cone in X. We say that C is **generated by** $S \subseteq X$ if $C = cone(S)$. If C is generated by a finite subset of X, then it is called **finitely generated**. Finally, we say that C is **basic** if it is generated by a linearly independent subset of X.

The main result we wish to prove here is the following.

Theorem 1
Every finitely generated convex cone in a linear space X is algebraically closed.

Since a convex subset of a Euclidean space is closed iff it is algebraically closed (Observation 3), the following fact is an immediate consequence of Theorem 1.

Corollary 1
For any $n \in \mathbb{N}$, every finitely generated convex cone in $\mathbb{R}^n$ is a closed set.

The rest of this section is devoted to proving Theorem 1. Although this is quite an innocent-looking claim, we will have do some work to prove it. It is actually a good idea to stick around for this proof, because we shall divide the argument into three steps, and each of these steps contains a result that is of interest in its own right. For instance, how about the following?

Lemma 1
Every basic convex cone in a linear space X is algebraically closed.

PROOF

Let S be a linearly independent subset of X, and $C := cone(S)$. If $S = \emptyset$, then $C = \{\mathbf{0}\}$, so the claim is trivial. Assume then $S \neq \emptyset$, and take any $x \in al\text{-}cl(C)$. We wish to show that $x \in C$. If $x = \mathbf{0}$, there is nothing to prove, so let $x \neq \mathbf{0}$. Of course, $x \in span(S) =: Y$.[15] So, since S is a basis for Y, there exist a unique nonempty finite subset A of S and a $\lambda : A \to \mathbb{R}\backslash\{0\}$ such that $x = \sum_{y \in A} \lambda(y)y$ (Corollary F.2). We wish to complete the proof by showing that $\lambda(A) \subseteq \mathbb{R}_+$—this would mean $x \in C$. Since $x \in al\text{-}cl(C)$, there exists a $z \in C$ such that $x^m := (1 - \frac{1}{m})x + \frac{1}{m}z \in C$ for each $m = 1, 2, \ldots$. If $z = \mathbf{0}$, all is trivial, so we let $z \neq \mathbf{0}$. Then there exist a unique nonempty finite subset B of S and a $\gamma : B \to \mathbb{R}\backslash\{0\}$ such that $z = \sum_{y \in B} \gamma(y)y$, and since $z \in C$, we have $\gamma(B) \subseteq \mathbb{R}_+$. Now, each x^m can be expressed as a linear combination of finitely many vectors in S with nonzero coefficients in only one way (Corollary F.2). Moreover, we have, for each m,

$$\begin{aligned} x^m &= \left(1 - \tfrac{1}{m}\right) \sum_{y \in A} \lambda(y)y + \tfrac{1}{m} \sum_{y \in B} \gamma(y)y \\ &= \left(1 - \tfrac{1}{m}\right) \sum_{y \in A\backslash B} \lambda(y)y + \sum_{y \in A \cap B} \left(\left(1 - \tfrac{1}{m}\right)\lambda(y) + \tfrac{1}{m}\gamma(y)\right) y \\ &\quad + \tfrac{1}{m} \sum_{y \in B\backslash A} \gamma(y)y \end{aligned}$$

with the convention that any of the terms on the right hand side equals $\mathbf{0}$ if it is a sum over the empty set. So, since each $x^m \in C$, we must have $\lambda(y) \geq 0$ for all $y \in A\backslash B$ and $\left(1 - \frac{1}{m}\right)\lambda(y) + \frac{1}{m}\gamma(y) \geq 0$ for all $y \in A \cap B$ and $m \in \mathbb{N}$. Thus, letting $m \to \infty$ shows that $\lambda(y) \geq 0$ for all $y \in A$, so we conclude that $\lambda(A) \subseteq \mathbb{R}_+$. ■

LEMMA 2

Let S be a subset of a linear space X and $\mathcal{I}(S)$ the class of all linearly independent subsets of S in X. Then,

$$cone(S) = \cup\{cone(T) : T \in \mathcal{I}(S)\}.$$

[15] First verify the claim by using the definition of the algebraic closure operator. Then reflect on why I started the sentence by saying, "Of course . . . ." (*Hint.* The span of a set = the span of the algebraic closure of that set.)

PROOF[16]

Since $cone(\emptyset) = \{\mathbf{0}\}$, the claim is trivial if $S = \emptyset$; we thus assume $S \neq \emptyset$. Moreover, we only need to prove the $\subseteq$ part of the claimed equation. Take any $x \in cone(S)$, and observe that if $x = \mathbf{0}$, we are done trivially. We thus assume $x \neq \mathbf{0}$. Clearly, there exists a nonempty finite subset T of S such that $x \in cone(T)$. (Right?) Let T_x be a $\supseteq$-minimal such T. Then there exists a $\lambda \in \mathbb{R}^T_{++}$ such that $x = \sum_{y \in T_x} \lambda(y)y$. We wish to complete the proof by showing that $T_x \in \mathcal{I}(S)$. Suppose this is not the case. Then there exists a nonzero $\alpha \in \mathbb{R}^{T_x}$ such that $\sum_{y \in T_x} \alpha(y)y = \mathbf{0}$. We may assume that $\alpha(y) > 0$ for some $y \in T_x$ (for otherwise we would work with $-\alpha$). Then $A := \{y \in T_x : \alpha(y) > 0\} \neq \emptyset$, and we may define

$$\theta := \min\left\{\frac{\lambda(y)}{\alpha(y)} : y \in A\right\} \quad \text{and} \quad C := \{y \in T_x : \lambda(y) - \theta\alpha(y) = 0\}.$$

Clearly, $\theta > 0$ and $C \neq \emptyset$. Moreover,

$$x = \sum_{y \in T_x} \lambda(y)y - \theta \sum_{y \in T_x} \alpha(y)y = \sum_{y \in T_x \setminus C} \big(\lambda(y) - \theta\alpha(y)\big)\, y.$$

But by the choice of θ, we have $\lambda(y) - \theta\alpha(y) \geq 0$ for all $y \in T_x$, so $x \in cone(T_x \setminus C)$. Since $C \neq \emptyset$, this contradicts the $\supseteq$-minimality of T_x, and the proof is complete. ∎

Combining Lemmas 1 and 2, we see that every finitely generated convex cone in a linear space equals the union of finitely many algebraically closed convex cones. Theorem 1 thus readily follows from the following fact.

LEMMA 3

Let $\mathcal{A}$ be a nonempty finite collection of convex subsets of a linear space X. Then,

$$al\text{-}cl(\cup\mathcal{A}) = \bigcup\{al\text{-}cl(A) : A \in \mathcal{A}\}.$$

Thus, if all members of $\mathcal{A}$ are algebraically closed in X, then so is $\cup\mathcal{A}$.

[16] The proof is so similar to that of Carathéodory's Theorem that I have contemplated quite a bit whether to leave it as an exercise. I decided at the end not to do so, but you might want to have a quick look at Example F.4 first, and then write down your own proof.

Proof

The $\supseteq$ part of the assertion follows from the definition of the algebraic closure operator. To prove the converse containment, take any $x \in al\text{-}cl(\cup\mathcal{A})$, and notice that, by definition, there exists a $y^1 \in \cup\mathcal{A}$ such that $(1-\alpha)x + \alpha y^1 \in \cup\mathcal{A}$ for all $0 < \alpha \leq 1$. Suppose $y^1 \in A_1$, where $A_1 \in \mathcal{A}$. If $(1-\alpha)x + \alpha y^1 \in A_1$ for all $0 < \alpha \leq 1$, then $x \in al\text{-}cl(A_1)$ and we are done. So suppose instead that there is a vector $y^2 \notin A_1$ on the line segment between x and y^1 that belongs to some other member, say A_2, of $\mathcal{A}$. Clearly, no vector on the line segment between x and y^2 may belong to A_1. (Since A_1 is convex, and y^2 is on the line segment between any such vector and y^1, we would otherwise have $y^2 \in A$.) Moreover, if $(1-\alpha)x + \alpha y^2 \in A_2$ for all $0 < \alpha \leq 1$, then $x \in al\text{-}cl(A_2)$, and we are done. Suppose this is not the case. Then there is a vector $y^3 \notin A_1 \cup A_2$ on the line segment between x and y^2. (Again, no vector on the line segment between x and y^3 may belong to $A_1 \cup A_2$.) If $(1-\alpha)x + \alpha y^3 \in A_3$ for all $0 < \alpha \leq 1$, then $x \in al\text{-}cl(A_3)$, and we are done. Otherwise there exists a vector $y^4 \notin A_1 \cup A_2 \cup A_3$ on the line segment between x and y^3. Since $\mathcal{A}$ is finite, by continuing this way we are bound to find a member B of $\mathcal{A}$ that contains a vector y such that $(1-\alpha)x + \alpha y \in B$ for every $0 < \alpha \leq 1$. ■

Exercise 33[H] Prove: If $\mathcal{A}$ is a nonempty collection of convex subsets of a linear space X such that $\cap\{ri(A) : A \in \mathcal{A}\} \neq \emptyset$, then

$$al\text{-}cl(\cap\mathcal{A}) = \bigcap\{al\text{-}cl(A) : A \in \mathcal{A}\}.$$

Exercise 34 Show that the convex hull of a finite subset of a linear space X is algebraically closed.

2 Separation and Extension in Linear Spaces

This section contains the main results of this chapter. We first study the problem of extending a given linear functional, which is defined on a subspace of a linear space and which satisfies a certain property, to the entire space in a way that preserves that property. The spirit of this problem is analogous to the continuous extension problem we studied in Section D.7. However, unlike that problem, it is intimately linked to the geometric problem of separating a convex set from a point in its exterior by means of a hyperplane. This connection is crucial, as the latter problem arises in many economic and optimization-theoretic applications.

In what follows, we first study these linear extension and separation problems in the context of an arbitrary linear space. The important, but substantially more special, setting of Euclidean spaces is taken up in Section 3.

2.1 Extension of Linear Functionals

To get a sense of the basic linear extension problem that we wish to study here, take a Euclidean space $\mathbb{R}^n$ with $n > 1$, and observe that $Y := \mathbb{R}^{n-1} \times \{0\}$ is a linear subspace of $\mathbb{R}^n$ (which is obviously isomorphic (and homeomorphic) to $\mathbb{R}^{n-1}$). Suppose that we are given two linear functionals L_1 and L_2 on $\mathbb{R}^n$ and a linear functional L on Y such that

$$L_1(x) \leq L_2(x) \qquad \text{for all } x \in \mathbb{R}^n_+$$

and

$$L_1(x) \leq L(x) \leq L_2(x) \qquad \text{for all } x \in Y \cap \mathbb{R}^n_+.$$

The question is this: Can we find a linear functional $L^* : \mathbb{R}^n \to \mathbb{R}$ such that $L^*(x) = L(x)$ for all $x \in Y$ (that is, $L^*|_Y = L$) and $L_1 \leq L^* \leq L_2$ on $\mathbb{R}^n_+$. This is a simple instance of the basic linear extension problem that we tackle below. Before moving on to the general discussion, however, let us observe that our little problem here can be answered very easily, thanks to the finite dimensionality of $\mathbb{R}^n$. We know that there exists a vector $\alpha^j \in \mathbb{R}^n$ such that $L_j(x) = \sum^n \alpha_i^j x_i$ for all $x \in \mathbb{R}^n$ and $j = 1, 2$ (Example F.6). Since $L_1 \leq L_2$ on $\mathbb{R}^n_+$, it is plain that we have $\alpha_n^1 \leq \alpha_n^2$. On the other hand, there also exists a vector $\alpha \in \mathbb{R}^{n-1}$ such that $L(x) = \sum^{n-1} \alpha_i x_i$ for all $x \in Y$. Therefore, for any $\alpha_n^1 \leq \alpha_n \leq \alpha_n^2$, the map $L^* : \mathbb{R}^n \to \mathbb{R}$ defined by $L^*(x) := \sum^n \alpha_i x_i$ satisfies all of the properties we seek—any such map is a linear functional on $\mathbb{R}^n$ that extends L and satisfies $L_1 \leq L^* \leq L_2$ on $\mathbb{R}^n_+$.

This is a typical example that shows that linear extension problems are often settled relatively easily in finite-dimensional linear spaces, because we have a complete characterization of the linear functionals defined on any such space. While the situation is quite different in the infinite-dimensional case, there are still powerful extension theorems that apply to arbitrary linear spaces. The first result of this section will provide an excellent case in point.

We begin by introducing some preliminary concepts.

DEFINITION

Let X be a linear space. We say that $\varphi \in \mathbb{R}^X$ is **positively homogeneous** if

$$\varphi(\alpha x) = \alpha\varphi(x) \qquad \text{for all } x \in X \text{ and } \alpha > 0,$$

and **subadditive** if

$$\varphi(x+y) \leq \varphi(x) + \varphi(y) \qquad \text{for all } x, y \in X.$$

If φ is subadditive and

$$\varphi(\alpha x) = |\alpha|\, \varphi(x) \qquad \text{for all } x \in X \text{ and } \alpha \in \mathbb{R},$$

then it is said to be a **seminorm** on X.

EXERCISE 35 Let X be a linear space and φ a positively homogeneous and subadditive real map on X. Show that φ is convex and

$$-\varphi(x-y) \leq \varphi(x) - \varphi(y) \leq \varphi(y-x) \qquad \text{for all } x, y \in X.$$

The following is a fundamental principle of linear analysis. It is difficult to overemphasize its importance.

THE HAHN-BANACH EXTENSION THEOREM 1[17]

Let X be a linear space and φ a positively homogeneous and subadditive real map on X. If L is a linear functional on a linear subspace Y of X such that $L(y) \leq \varphi(y)$ for all $y \in Y$, then there exists an $L^* \in \mathcal{L}(X, \mathbb{R})$ such that

$$L^*(y) = L(y) \text{ for all } y \in Y \quad \text{and} \quad L^*(x) \leq \varphi(x) \text{ for all } x \in X.$$

[17] There are a variety of results that go with the name of the Hahn-Banach Extension Theorem; I present two such results in this book. We owe these results to the 1927 contribution of Hans Hahn and the monumental 1932 work of Stefan Banach, who is justly viewed by many as the father of functional analysis. Indeed, Banach's book (which is essentially his 1920 dissertation) brought this topic to the fore of mathematical analysis, and subsequent work established it as one of the major fields in mathematics at large.

PROOF

Take a linear subspace Y of X, and fix an $L \in \mathcal{L}(Y, \mathbb{R})$ with $L \leq \varphi|_Y$. If $Y = X$, there is nothing to prove, so assume that there exists a vector z in $X \backslash Y$. We will first show that L can be extended to a linear functional K defined on the linear subspace $Z := span(Y \cup \{z\})$ such that $K(x) \leq \varphi(x)$ for all $x \in Z$. For any $x \in Z$, there exist a unique $y_x \in Y$ and $\lambda_x \in \mathbb{R}$ such that $x = y_x + \lambda_x z$. (Why unique?) So K is a linear extension of L to Z iff

$$K(x) = L(y_x) + \lambda_x \alpha \tag{7}$$

where $\alpha := K(z)$. All we need to do is to choose a value for α that would ensure that $K(x) \leq \varphi(x)$ for all $x \in Z$. Clearly, such an α exists iff

$$L(y) + \lambda\alpha \leq \varphi(y + \lambda z) \qquad \text{for all } y \in Y \text{ and } \lambda \in \mathbb{R}\backslash\{0\}. \tag{8}$$

On the other hand, given the positive homogeneity of φ and linearity of L, this is equivalent to say that, for any $y \in Y$,

$$L\left(\tfrac{1}{\lambda}y\right) + \alpha \leq \varphi\left(\tfrac{1}{\lambda}y + z\right) \quad \text{if } \lambda > 0,$$

and

$$L\left(-\tfrac{1}{\lambda}y\right) - \alpha \leq \varphi\left(-\tfrac{1}{\lambda}y - z\right) \quad \text{if } \lambda < 0.$$

Put differently, (8) holds iff $L(u) + \alpha \leq \varphi(u + z)$ and $L(v) - \alpha \leq \varphi(v - z)$ for all $u, v \in Y$. (Yes?) Our objective is thus to pick an $\alpha \in \mathbb{R}$ such that

$$L(v) - \varphi(v - z) \leq \alpha \leq \varphi(u + z) - L(u) \qquad \text{for all } u, v \in Y. \tag{9}$$

Clearly, we can do this iff the sup of the left-hand side of this expression (over all $v \in Y$) is smaller than the inf of its right-hand side. That is, (9) holds for some $\alpha \in \mathbb{R}$ iff $L(u) + L(v) \leq \varphi(u + z) + \varphi(v - z)$ for all $u, v \in Y$. But then we're just fine, because, given that L is linear, φ is subadditive, and $L \leq \varphi|_Y$, we have

$$L(u) + L(v) = L(u + v) \leq \varphi(u + v) \leq \varphi(u + z) + \varphi(v - z) \quad \text{for all } u, v \in Y.$$

So, tracing our steps back, we see that there exists a $K \in \mathcal{L}(Z, \mathbb{R})$ with $K|_Y = L$ and $K \leq \varphi|_Z$.

If you recall how we proved Theorem F.1 by Zorn's Lemma, you know what's coming next.[18] Let $\mathcal{L}$ stand for the set of all $K \in \mathcal{L}(W, \mathbb{R})$ with $K|_Y = L$ and $K \leq \varphi|_W$, for some linear subspace W of X with $Y \subset W$. By what we just proved, $\mathcal{L} \neq \emptyset$. Define next the binary relation $\succcurlyeq$ on $\mathcal{L}$ as

$$K_1 \succcurlyeq K_2 \qquad \text{if and only if} \qquad K_1 \text{ is an extension of } K_2.$$

It is easily verified that $(\mathcal{L}, \succcurlyeq)$ is a poset and that any loset in $(\mathcal{L}, \succcurlyeq)$ has an upper bound (which is a linear functional defined on the union of all the domains of the members of the family). Then, by Zorn's Lemma, there must exist a maximal element L^* in $(\mathcal{L}, \succcurlyeq)$. L^* must be defined on the entire X, otherwise we could extend it further by using the argument outlined in the previous paragraph, which would contradict the maximality of L^*. Since $L^* \in \mathcal{L}$, we are done. ■

COROLLARY 2
Every linear functional defined on a linear subspace of a linear space X can be extended to a linear functional on X.

EXERCISE 36 Prove Corollary 2.

Even though this may not be readily evident to you, the Hahn-Banach Extension Theorem 1 is of great importance for linear analysis. Rest assured that this point will become abundantly clear as we develop the theory further. As for an immediate remedy, we offer the following example which illustrates how one may need this result in answering even some of the very basic questions about linear functionals.

EXAMPLE 11
(*Existence of Linear Functionals*) We have seen plenty of examples of linear functionals in the previous chapter, but did not worry about the following fundamental question:

Is it true that in any nontrivial linear space, there exists a nonzero linear functional?

[18] The use of Zorn's Lemma (and thus the Axiom of Choice) cannot be avoided in the proof of the Hahn-Banach Extension Theorem 1. The same goes for all major theorems that are proved in this section. However, given my "applied" slant here, I will not worry about proving these claims here.

Think about it. This is not all that trivial—the arbitrariness of the linear space makes it impossible to write down a formula for the sought linear functional. Fortunately, this does not hamper our ability to prove the *existence* of such a functional, and the Hahn-Banach Extension Theorem 1 provides a very easy way of doing this.

Let X be a nontrivial linear space, and pick any $x \in X\backslash\{\mathbf{0}\}$. Let $Y := \{\lambda x : \lambda \in \mathbb{R}\}$ and define $L \in \mathbb{R}^Y$ by $L(\lambda x) := \lambda$. It is easy to see that $L \in \mathcal{L}(Y, \mathbb{R})$. (Yes?) That's it! Since Y is a linear subspace of X, we now simply apply Corollary 2 to extend L to a linear functional defined on X, thereby establishing the existence of a nonzero linear functional on X.

Observe that this is not a constructive finding in the sense that it doesn't provide us with a formula for the linear functional we seek. All is rather based on the Hahn-Banach Extension Theorem 1 (and hence, indirectly, on the Axiom of Choice). This is a prototypical, if trivial, example of what this theorem can do for you.[19] □

EXERCISE 37 Let X be a linear space and φ a positively homogeneous and subadditive real map on X. Prove that, for any $x_0 \in X$, there exists an $L \in \mathcal{L}(X, \mathbb{R})$ such that $L(x_0) = \varphi(x_0)$ and $L \leq \varphi$.

EXERCISE 38 Show that if X is an infinite dimensional linear space, then so is $\mathcal{L}(X, \mathbb{R})$.

EXERCISE 39[H] Take any $1 \leq p \leq \infty$, and let B_{ℓ^p} be the closed unit ball of ℓ^p (i.e., $B_{\ell^p} := \{(x_m) \in \ell^p : \sum^\infty |x_i|^p \leq 1\}$). Show that any linear map L on B_{ℓ^p} can be extended to a linear functional defined on ℓ^p.

[19] The use of the Hahn-Banach Extension Theorem 1 is actually an overkill here—I only wished to give a general feeling for what sorts of things can be done with this result. Indeed, the existence of nonzero linear functionals can be proved without invoking Corollary 2 (whereas the use of the Axiom of Choice is essential). Here is another proof. By Theorem F.1, there is a basis S for X. By Corollary F.2, for any $x \in X\backslash\{\mathbf{0}\}$, there exists a unique finite subset A_x of S and a unique map $\lambda_x : A_x \to \mathbb{R}\backslash\{0\}$ such that $x = \sum_{y \in A_x} \lambda_x(y)y$. Now let $A_{\mathbf{0}} := \emptyset$, fix any $z \in S$, and define $L \in \mathbb{R}^X$ by $L(x) := \lambda_x(z)$ if $z \in A_x$, and $L(x) := 0$ otherwise. Then, as you should verify, L is a nonzero linear functional on X.

A similar argument can also be used to prove Corollary 2 without invoking the Hahn-Banach Extension Theorem 1. But please do not doubt the power of this theorem: choosing a suitable φ in this result will give us much more than the trivial applications we have looked at so far. Read on!

2.2 Extension of Positive Linear Functionals

Another kind of linear extension problem that sometimes arises in applications (such as in general equilibrium theory) concerns extending a given positive linear functional on a subspace of an ordered linear space in a way that preserves both the linearity and the *positivity* of this functional. Put more precisely, the question is this: If X is a partially ordered linear space and L a positive linear functional on a partially ordered linear subspace Y of X, can we find an $L^* \in \mathcal{L}(X, \mathbb{R})$ such that $L^*|_Y = L$ and $L^*(X_+) \subseteq \mathbb{R}_+$? Unfortunately, the answer is no in general, as the next exercise illustrates.

Exercise 40 Recall that $\mathbf{c}^0$ is the linear space of all real sequences only finitely many terms of which are nonzero. This space is a partially ordered linear subspace of ℓ^∞, and $L : \mathbf{c}^0 \to \mathbb{R}$ defined by $L((x_m)) := \sum^\infty x_i$ is a positive linear functional on $\mathbf{c}^0$. Show that there is no positive linear functional on ℓ^∞ that extends L.

Roughly speaking, the reason why things don't work out well in this example is that $\mathbf{c}^0$ is too small relative to ℓ^∞, so the knowledge of the behavior of L on $\mathbf{c}^0_+$ does not give us enough ammunition to settle the extension problem in the affirmative. Indeed, there is no member of $\mathbf{c}^0$ in the algebraic interior of ℓ^∞_+ (Exercise 24). The following important result, which goes back to 1948, shows that in cases where the subspace at hand is large enough to intersect the algebraic interior of the positive cone of the mother space, the associated positive extension problem can be settled satisfactorily. We will derive it here as a corollary of the Hahn-Banach Extension Theorem 1.

The Krein-Rutman Theorem

Let X be a preordered linear space, and Y a preordered linear subspace of X such that

$$Y \cap al\text{-}int_X(X_+) \neq \emptyset. \tag{10}$$

If L is a positive linear functional on Y, then there exists a positive linear functional L^* on X with $L^*|_Y = L$.

PROOF

Let L be a positive linear functional on Y. Pick an arbitrary $x \in X$, and define

$$\varphi(x) := \inf\{L(y) : Y \ni y \succsim_X x\}. \tag{11}$$

Let us first show that $\varphi(x)$ is a real number.[20]

CLAIM. $\{y \in Y : y \succsim_X x\} \neq \emptyset$ and $\{L(y) : Y \ni y \succsim_X x\}$ is bounded from below. Therefore, $\varphi(x) \in \mathbb{R}$.[21]

PROOF OF CLAIM

Take any $y^* \in Y \cap \textit{al-int}_X(X_+)$. Clearly, there exists an $\alpha > 0$ such that $(1-\alpha)y^* + \alpha(-x) \succsim_X \mathbf{0}$, and since $\succsim_X$ is a vector preorder, this implies $Y \ni \left(\frac{1}{\alpha} - 1\right) y^* \succsim_X x$, establishing the first part of the claim. But there also exists a $\beta > 0$ such $(1-\beta)y^* + \beta x \succsim_X \mathbf{0}$. Since $\succsim_X$ is a vector preorder, then, $(1-\beta)y^* + \beta y \succsim_X \mathbf{0}$ for any $y \in Y$ with $y \succsim_X x$. So, by positivity of L, we get $L((1-\beta)y^* + \beta y) \geq 0$, and hence $L(y) \geq \left(1 - \frac{1}{\beta}\right)L(y^*) > -\infty$ for any $y \in Y$ with $y \succsim_X x$. Since β is independent of y, this proves the second part of the claim. ||

In view of this claim, it is clear that $\varphi \in \mathbb{R}^X$ is well-defined by (11). Moreover, φ is subadditive, positively homogeneous and $L = \varphi|_Y$. (Check!) Hence, by the Hahn-Banach Extension Theorem 1, there exists a $L^* \in \mathcal{L}(X,\mathbb{R})$ with $L^*|_Y = L$ and $L^* \leq \varphi$. We claim next that $L^*(X_+) \subseteq \mathbb{R}_+$. To see this, take any $x \in X_+$, and note that $L^*(-x) \leq \varphi(-x)$, since $-x \in X$. But since $x \succsim_X \mathbf{0}$, we have $\mathbf{0} \succsim_X -x$, so by definition of φ we get $\varphi(-x) \leq L(\mathbf{0}) = 0$. Therefore, $L^*(-x) \leq 0$, that is, $L^*(x) \geq 0$, as we sought. ■

EXERCISE 41 Prove: If X is a preordered linear space such that $\textit{al-int}_X(X_+) \neq \emptyset$, then there exists a nonzero positive functional on X.

The following result, due to Segal (1947), gives an alternative sufficient condition for establishing the existence of a positive linear extension of a given positive linear functional.

[20] As smart as it is, the idea is not new; you have seen this sort of an extension technique before. In fact, doesn't this remind you how we showed that $\sqrt{2} \in \mathbb{R}$ in Section A.2.3? Or how about the way we proved Theorem D.2 earlier? Here, things are a bit harder, for φ will not turn out to be linear (otherwise we would be done). But it will allow us to use the Hahn-Banach Extension Theorem 1, nevertheless. Read on.

[21] It is crucial to prove the first claim, because if that didn't hold, we would have $\varphi(x) = \inf \emptyset = \infty$.

*EXERCISE 42[H] (Segal) Let X be a preordered linear space such that there exists an $\mathbf{e} \in X$ with the following property: For every $x \in X$, there exists an $m \in \mathbb{N}$ with $-m\mathbf{e} \succsim_X x \succsim_X m\mathbf{e}$. Show that in this case every positive linear functional defined on a preordered linear subspace Y of X with $\mathbf{e} \in Y$ can be extended to a positive linear functional on X.

2.3 Separation of Convex Sets by Hyperplanes

We now turn to the geometric implications of the Hahn-Banach Extension Theorem 1. Before getting to the core of the matter, however, we need to introduce the following auxiliary concept, which will play an important role in what follows.

DEFINITION
Let S be a convex set in a linear space X such that $\mathbf{0} \in al\text{-}int_X(S)$. The **Minkowski functional** of S (*relative to* X) is the real function φ_S on X defined by

$$\varphi_S(x) := \inf \left\{\lambda > 0 : \tfrac{1}{\lambda}x \in S\right\}.$$

Intuitively, you can think of $\frac{1}{\varphi_S(x)}x$ as the first algebraic boundary point of S (in X) on the ray that originates from $\mathbf{0}$ and passes through $x \in X$. Figure 9 illustrates the definition geometrically.

The Minkowski functional of a convex set S that contains the origin in its algebraic interior is well-defined as a real function since, given that

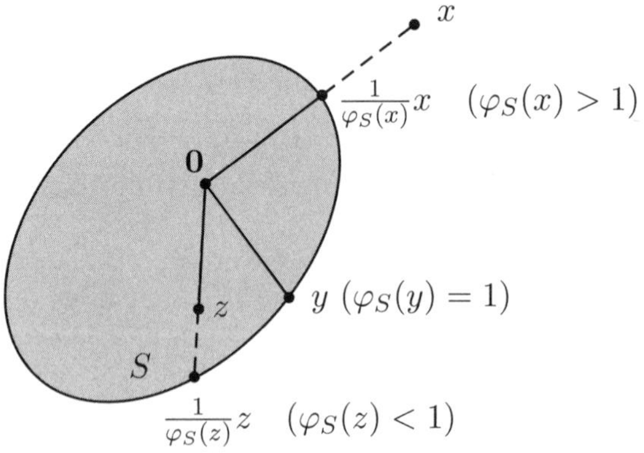

FIGURE 9

$\mathbf{0} \in \textit{al-int}_X(S)$, we have $\{\lambda > 0 : \frac{1}{\lambda}x \in S\} \neq \emptyset$ for all $x \in X$. (Why?) It readily follows from its definition that φ_S is positively homogeneous and $\varphi_S(\mathbf{0}) = 0$. Less obvious is the fact that φ_S is subadditive. To prove this, pick any $x, y \in X$, and fix any $\varepsilon > 0$. By definition of φ_S, there exists a real number $\alpha > 0$ such that $\varphi_S(x) > \alpha - \frac{\varepsilon}{2}$ and $\frac{1}{\alpha}x \in S$. Similarly, $\varphi_S(y) > \beta - \frac{\varepsilon}{2}$ must hold for some $\beta > 0$ with $\frac{1}{\beta}y \in S$. But

$$\tfrac{1}{\alpha+\beta}(x+y) = \tfrac{\alpha}{\alpha+\beta}\left(\tfrac{1}{\alpha}x\right) + \tfrac{\beta}{\alpha+\beta}\left(\tfrac{1}{\beta}x\right) \in S$$

since S is convex. Thus

$$\varphi_S(x+y) \leq \alpha + \beta < \varphi_S(x) + \varphi_S(y) + \varepsilon.$$

Since we obtained this expression for any $x, y \in X$ and $\varepsilon > 0$, it follows that φ_S is subadditive.

So far, so good. But, you may ask, why on earth should we care about Minkowski functionals? The main reason is that these functionals give us an analytic characterization of an arbitrary algebraically closed convex set with nonempty algebraic interior. As you will see shortly, it is sometimes easier to use this characterization instead of the principal definition.

LEMMA 4

Let S be a convex subset of a linear space X such that $\mathbf{0} \in \textit{al-int}_X(S)$. The Minkowski functional φ_S of S is a positively homogeneous and subadditive real function on X such that $\varphi_S(\mathbf{0}) = 0$, and

(i) $\varphi_S(x) < 1$ *if and only if* $x \in \textit{al-int}_X(S)$;

(ii) $\varphi_S(x) \leq 1$ *if and only if* $x \in \textit{al-cl}(S)$.

PROOF

We prove (i) here, and leave the proof of (ii) to you. If $\varphi_S(x) < 1$, then $\frac{1}{\lambda}x \in S$ for some $0 < \lambda < 1$. But $\mathbf{0} \in \textit{al-int}_X(S)$ and $x = \lambda\left(\frac{1}{\lambda}x\right) + (1-\lambda)\mathbf{0}$, so $x \in \textit{al-int}_X(S)$ by Exercise 27. Conversely, if $x \in \textit{al-int}_X(S)$, then there exists an $\alpha > 0$ such that

$$(1+\alpha)x = (1-\alpha)x + \alpha(2x) \in S$$

which means that $\varphi_S(x) \leq \frac{1}{1+\alpha} < 1$. ■

EXERCISE 43 Complete the proof of Lemma 4.

Now comes a very important application of the Hahn-Banach Extension Theorem 1 that also highlights the importance of the Minkowski functionals for convex analysis. In the statement of this result, and its many applications given later in Chapter H, we adopt the following notational convention.

NOTATION. For any nonempty sets A and B in a linear space X, and any $L \in \mathcal{L}(X, \mathbb{R})$, we write $L(B) \geq L(A)$ to denote

$$L(x) \geq L(y) \qquad \text{for all } (x, y) \in A \times B,$$

$L(A) > L(B)$ to denote

$$L(A) \geq L(B) \qquad \text{and} \qquad L(x) > L(y) \text{ for some } (x, y) \in A \times B,$$

and $L(A) \gg L(B)$ to denote

$$L(x) > L(y) \qquad \text{for all } (x, y) \in A \times B.$$

For any real number α, the expressions $\alpha \geq L(B)$, $\alpha > L(B)$, and $\alpha \gg L(B)$ are similarly interpreted.

PROPOSITION 1

Let S be a convex subset of a linear space X such that $\text{al-int}_X(S) \neq \emptyset$. If $z \in X \backslash \text{al-int}_X(S)$, then there exists an $(\alpha, L) \in \mathbb{R} \times \mathcal{L}(X, \mathbb{R})$ such that

$$L(z) \geq \alpha > L(S) \qquad \text{and} \qquad \alpha \gg L(\text{al-int}_X(S)).$$[22]

PROOF

Take any $z \in X \backslash \text{al-int}_X(S)$, and note that it is without loss of generality to assume that $\mathbf{0} \in \text{al-int}_X(S)$.[23] Choose $\alpha = 1$ and define L on $Y := \{\lambda z : \lambda \in \mathbb{R}\}$ by $L(\lambda z) := \lambda$ (so that $L(z) = 1$). Since $z \notin \text{al-int}_X(S)$, we have $\varphi_S(z) \geq 1$ by Lemma 4, and it follows from positive homogeneity of φ_S that $\varphi_S(\lambda z) \geq \lambda = L(\lambda z)$ for all $\lambda > 0$. Since $\varphi_S \geq 0$ by definition, this inequality also holds for any $\lambda \leq 0$. So, we have $\varphi_S|_Y \geq L$. But then by the Hahn-Banach Extension Theorem 1, L can be linearly extended to the entire X in

[22] As the following proof makes it clear, α can be taken as 1 in this statement if $\mathbf{0} \in \text{al-int}_X(S)$.

[23] Otherwise I would pick any $x \in \text{al-int}_X(S)$, and play the game with $z - x$ and $S - x$.

such a way that $\varphi_S \geq L$ holds on X. (Here we denote the extension of L to X also by L.) Since $1 \geq \varphi_S(x)$ for all $x \in S$ by Lemma 4; however, this implies that $1 \geq L(S)$. Moreover, again by Lemma 4, $1 > \varphi_S(x) \geq L(x)$ for all $x \in \textit{al-int}_X(S)$. ■

Please note that behind the simplicity of the proof of Proposition 1 lies the full power of the Hahn-Banach Extension Theorem 1. The bridge between that theorem and the present separation result is in turn established via the Minkowski functionals. This is the prototypical method of "going from a linear extension theorem to an affine separation theorem."

COROLLARY 3
Let S be a nonempty, algebraically open, and convex set in a linear space X. If $z \in X \backslash S$, then there exists an $(\alpha, L) \in \mathbb{R} \times \mathcal{L}(X, \mathbb{R})$ such that $L(z) \geq \alpha \gg L(S)$.

EXERCISE 44[H] Let S be a nonempty convex subset of a linear space X such that $ri(S) \neq \emptyset$. Show that if $z \in \textit{aff}(S)\backslash S$, then there exists an $(\alpha, L) \in \mathbb{R} \times \mathcal{L}(X, \mathbb{R})$ such that $L(z) \geq \alpha > L(S)$ and $\alpha \gg L(ri(S))$.

These are highly geometric findings that you have probably seen earlier in the special context of Euclidean spaces. We say that a hyperplane H in a linear space X **separates** the vector $z \in X$ from the set $S \subseteq X$ if z and S are contained in different closed half-spaces induced by H. This does not exclude the possibility that both z and S are contained in the hyperplane itself. When this is not the case, that is, when H separates z and S, and either z or S is not contained in H, we say that H **properly separates** z and S.[24]

Notice that the linear functional we found in Proposition 1 has to be nonzero. Therefore, by Corollary F.4 and the discussion that follows it, the statement $L(z) \geq \alpha > L(S)$ means that z is contained in the closed half-space on one side of the hyperplane $L^{-1}(\alpha)$ and S is contained on the other side of $L^{-1}(\alpha)$. (Which side is which is not important, for the direction of inequalities in Proposition 1 is inconsequential. We could as well use $-L$ in the associated statement.) Moreover, since $\alpha > L(S)$, S is not contained in

[24] For instance, in $\mathbb{R}^2$, the hyperplane $H := \mathbb{R} \times \{0\}$ separates the vector $z := (\frac{1}{2}, 0)$ and $S := [0, 1] \times \{0\}$ (even though S includes z). But notice that both z and S are contained in H, that is, H separates z and S only *improperly*.

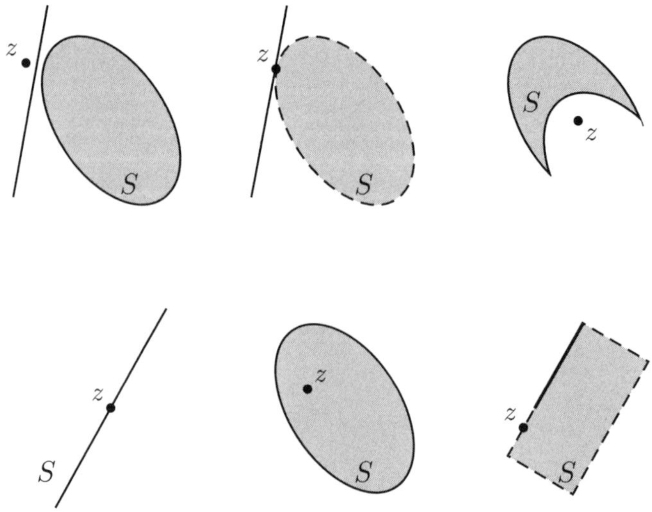

FIGURE 10

$L^{-1}(\alpha)$. It follows that the hyperplane $L^{-1}(\alpha)$ properly separates z from S. (See the first two illustrations in Figure 10.) Another (more geometric) way of putting Proposition 1 is thus the following:

> *Any convex set with a nonempty algebraic interior can be properly separated from any vector outside its algebraic interior by a hyperplane.*

This simple observation, and its many variants, play an important role in geometric functional analysis and its economic applications. It is important to note that none of the hypotheses behind it can be relaxed completely. The third, fourth, and fifth illustrations in Figure 10 show that this is the case even in $\mathbb{R}^2$. On the other hand, the last depiction in this figure illustrates that the first of the inequalities in Proposition 1 cannot be made strict, nor can the second one be strengthened to $\gg$. However, as noted in Corollary 3, if S is algebraically open, then we can strengthen the second (but not the first) inequality. By contrast, when S is algebraically closed, the first inequality can be made strict. We prove this next.

COROLLARY 4

Let S be an algebraically closed and convex subset of a linear space X such that al-int$_X(S) \neq \emptyset$. If $z \in X \backslash S$, then there exists an $(\alpha, L) \in \mathbb{R} \times \mathcal{L}(X, \mathbb{R})$ such that $L(z) > \alpha > L(S)$ and $\alpha \gg L(\text{al-int}_X(S))$.

Proof

Pick any $x \in al\text{-}int_X(S)$ and $z \in X\backslash S$. Since $z \notin S = al\text{-}cl(S)$, at least one point on the line segment between z and x does not belong to S, that is, $y := (1-\beta)z + \beta x \notin S$ for some $0 < \beta < 1$. By Proposition 1, therefore, there exists an $(\alpha, L) \in \mathbb{R} \times \mathcal{L}(X, \mathbb{R})$ such that $L(y) \geq \alpha > L(S)$ and $\alpha \gg L(al\text{-}int_X(S))$. But then $\alpha > L(x)$, so

$$L(z) = L\left(\tfrac{1}{1-\beta}(y - \beta x)\right) = \tfrac{1}{1-\beta}L(y) - \tfrac{\beta}{1-\beta}L(x) > \tfrac{1}{1-\beta}\alpha - \tfrac{\beta}{1-\beta}\alpha = \alpha.$$

Conclusion: $L(z) > \alpha > L(S)$. ∎

The following is an important generalization of Proposition 1. It is proved here as a consequence of that result, and hence of the Hahn-Banach Extension Theorem 1.

The Dieudonné Separation Theorem[25]

Let A and B be two nonempty convex subsets of a linear space X such that $al\text{-}int_X(A) \neq \emptyset$. Then, $al\text{-}int_X(A) \cap B = \emptyset$ if, and only if, there exists an $(\alpha, L) \in \mathbb{R} \times \mathcal{L}(X, \mathbb{R})$ such that $L(B) \geq \alpha > L(A)$ and $\alpha \gg L(al\text{-}int_X(A))$.

Proof

The "if" part of the assertion is trivial. To prove the "only if" part, suppose that $al\text{-}int_X(A) \cap B = \emptyset$. Note that, as was the case for Proposition 1, it is without loss of generality to assume $\mathbf{0} \in al\text{-}int_X(A)$, so we do so in what follows.

Define $S := A - B$ and observe that S is convex. Moreover, $al\text{-}int_X(S) \neq \emptyset$ since $-x \in al\text{-}int_X(S)$ for any $x \in B$. (Why?) We next claim that $\mathbf{0} \notin al\text{-}int_X(S)$. If this were not the case, for any (fixed) $z \in B$, we could find an $0 < \alpha < 1$ such that $\alpha z \in A - B$. In turn, this would mean that $\alpha z = x - y$ for some $(x, y) \in A \times B$ so that

$$\tfrac{\alpha}{\alpha+1}z + \tfrac{1}{\alpha+1}y \in B$$

because B is convex, while

$$\tfrac{\alpha}{\alpha+1}z + \tfrac{1}{\alpha+1}y = \tfrac{1}{\alpha+1}x = \tfrac{\alpha}{\alpha+1}\mathbf{0} + \tfrac{1}{\alpha+1}x \in al\text{-}int_X(A)$$

[25] This theorem is sometimes called *the geometric version of the Hahn-Banach Theorem*, a terminology due to the famous Bourbaki school. The present statement of the result was obtained first by Jean Dieudonné in 1941.

because of Exercise 27. Obviously, this contradicts the hypothesis $al\text{-}int_X(A) \cap B = \emptyset$.

Given these observations, we may apply Proposition 1 to find a nonzero $L \in \mathcal{L}(X,\mathbb{R})$ such that $L(\mathbf{0}) = 0 > L(S)$, which, by linearity, entails $L(B) > L(A)$. Choosing $\alpha := \inf L(B)$, then, we have $L(B) \geq \alpha \geq L(A)$. Moreover, we must have $\alpha > L(al\text{-}int_X(A))$, for otherwise $al\text{-}int_X(A)$ would be contained in the hyperplane $L^{-1}(\alpha)$, whereas $aff(al\text{-}int_X(A)) = X$ (Exercise 27), so in this case L would necessarily be the zero functional. (Clarify!) Thus there exists a $y \in A$ such that $\alpha > L(y)$. Now take any $x \in al\text{-}int_X(A)$, and assume that $L(x) = \alpha$, to derive a contradiction. Let $z := 2x - y$, and note that $L(z) = 2\alpha - L(y) > \alpha$. But since $x \in al\text{-}int_X(A)$, there exists a $0 < \lambda < 1$ small enough to guarantee that $(1-\lambda)x + \lambda z \in A$, and hence

$$\alpha \geq L((1-\lambda)x + \lambda z) = (1-\lambda)\alpha + \lambda L(z) > \alpha,$$

a contradiction. Thus $\alpha \gg L(al\text{-}int_X(A))$, and the proof is complete. ■

EXERCISE 45[H] Prove or disprove: If A and B are two nonempty, algebraically closed and convex sets in a linear space X such that $A \cap B \neq \emptyset$ and $al\text{-}int_X(A) \neq \emptyset$, then there exists an $(\alpha, L) \in \mathbb{R} \times \mathcal{L}(X,\mathbb{R})$ with $L(B) > \alpha > L(A)$.

EXERCISE 46[H] (*The Hahn-Banach Sandwich Theorem*) Let φ be a positively homogeneous and subadditive real function on a linear space X. Let S be a nonempty convex subset of X and $\psi \in \mathbb{R}^S$ a concave function. Prove that if $\psi \leq \varphi|_S$, then there exists a linear functional L on X such that $\psi \leq L|_S$ and $L \leq \varphi$.

EXERCISE 47 Let A and B be two nonempty convex subsets of a linear space X such that $ri(A) \neq \emptyset$ and $ri(A) \cap B = \emptyset$. Show that there exists an $(\alpha, L) \in \mathbb{R} \times \mathcal{L}(X,\mathbb{R})$ such that $L(B) \geq \alpha > L(A)$ and $\alpha \gg L(ri(A))$.

EXERCISE 48 Prove the Hahn-Banach Extension Theorem 1 assuming the validity of the Dieudonné Separation Theorem, thereby proving that these two facts are equivalent.

The geometric interpretation of the Dieudonné Separation Theorem parallels that of Proposition 1. As a matter of definition, we say that a hyperplane H in a linear space X **separates** the subsets A and B of X if A and B are contained in different closed half-spaces induced by H, and **properly separates** them if it separates them and at least one of these sets is not contained in H. Consequently, the Dieudonné Separation Theorem can be restated as follows:

Any two convex sets such that one has nonempty algebraic interior, and the other is disjoint from this interior, can be properly separated by a hyperplane.

It is important to note that the nonempty algebraic interior requirement cannot be dropped in this statement even in $\mathbb{R}^2$. (*Proof.* Envision two straight lines crossing at a single point.) But if the sets are known to be disjoint, then the basic separation principle simplifies:

A sufficient condition for two disjoint convex sets to be properly separable by a hyperplane is one of them having a nonempty algebraic interior.

In general, one cannot do away with this sufficient condition, that is, it is possible that two disjoint convex sets are nonseparable by a hyperplane. This may seem counterintuitive to you—it does to the author. This is because the claim translates in $\mathbb{R}^n$ to the following: *Two disjoint convex sets in $\mathbb{R}^n$ can be properly separated provided that one of them has a nonempty interior.* (Recall that there is no difference between the algebraic interior and the interior of a convex set in $\mathbb{R}^n$ (Observation 2).) And indeed, this is a bit of a silly statement in that two disjoint convex sets in $\mathbb{R}^n$ can be properly separated even if the interiors of both of these sets are empty. (We will prove this in Section 3.1.) So what is going on? It is again those infinite-dimensional linear spaces that are acting funny. Let's look into this matter more closely.

LEMMA 5
Let S be a subset of a linear space X, and assume that $\alpha \geq L(S)$ for some $(\alpha, L) \in \mathbb{R} \times \mathcal{L}(X, \mathbb{R})$. Then, $\alpha \geq L(\text{al-cl}(S))$.

PROOF
If $x \in \text{al-cl}(S)$, then there exists a $y \in S$ such that $\left(1 - \frac{1}{m}\right)x + \frac{1}{m}y \in S$ for all $m \in \mathbb{N}$, so $\alpha \geq L(S)$ implies

$$\alpha \geq L\left(\left(1 - \tfrac{1}{m}\right)x + \tfrac{1}{m}y\right) = \left(1 - \tfrac{1}{m}\right)L(x) + \tfrac{1}{m}L(y),$$

and letting $m \to \infty$ we find $\alpha \geq L(x)$. ∎

LEMMA 6

Let S be a subset of a linear space X. If al-cl$(S) = X$, then there does not exist an $\alpha \in \mathbb{R}$ and a nonzero $L \in \mathcal{L}(X, \mathbb{R})$ such that $\alpha \geq L(S)$.

PROOF

If this were not true, then Lemma 5 would imply that $\alpha \geq L(X)$ for some $\alpha \in \mathbb{R}$ and a nonzero $L \in \mathcal{L}(X, \mathbb{R})$. But this is impossible, because, for any $x \in X$ with $L(x) > 0$, we have $L(mx) = mL(x) \to \infty$. ■

We are now prepared to give a concrete example of two nonseparable disjoint convex sets. This will show formally that the nonempty algebraic interior requirement cannot be omitted even in the statement of Proposition 1, let alone in the Dieudonné Separation Theorem. We will subsequently argue that this example is hardly pathological.

EXAMPLE 12

We work within the linear space $\mathbf{c}^0$ of real sequences all but finitely many terms of which are zero. Define

$$S := \{(x_m) \in \mathbf{c}^0 : x_M > 0 \text{ and } x_{M+1} = x_{M+2} = \cdots = 0 \text{ for some } M \in \mathbb{N}\},$$

that is, S is the set of all sequences in $\mathbf{c}^0$ the last nonzero terms of which are positive. While S is a convex set, we have *al-int*$_{\mathbf{c}^0}(S) = \emptyset$. (Verify!) We next claim that *al-cl*$(S) = \mathbf{c}^0$. To this end, take any $(x_m) \in \mathbf{c}^0$, and let x_M be the last nonzero term of this sequence. Consider the real sequence

$$(y_m) := (x_1, \ldots, x_M, 1, 0, 0, \ldots).$$

Clearly, $(y_m) \in S$, and $(1 - \alpha)(x_m) + \alpha(y_m) \in S$ for each $0 < \alpha \leq 1$. This means that $(x_m) \in$ *al-cl*(S), and proves *al-cl*$(S) = \mathbf{c}^0$. But then, by Lemma 6, no vector in $\mathbf{c}^0 \backslash S$ (not even the origin $(0, 0, \ldots)$) can be separated from S by a hyperplane. □

*REMARK 2.[26] (*Characterization of Finite Dimensionality by Separation*) Example 12 shows that there is at least one linear space in which two nonempty disjoint convex sets are not separable. As we noted before and

[26] This remark presumes familiarity with the Well-Ordering Principle (Section B.3.2).

will prove formally in Section 3, such a linear space must necessarily be infinite-dimensional. But the question remains: In what sort of infinite-dimensional linear spaces can one actually find two disjoint but nonseparable convex sets? Answer: In all of them! That is, this particular separation property actually "characterizes" the infinite- (and hence finite-) dimensional linear spaces. Put more formally:

> FACT
> *A linear space X is infinite-dimensional if, and only if, it contains a nonempty convex set S and a vector $z \notin S$ such that z and S cannot be separated by a hyperplane.*

The "if" part of the claim follows from the Minkowski Separating Hyperplane Theorem (which is proved in Section 3.1 below) and the fact that every finite-dimensional linear space is isomorphic to some $\mathbb{R}^n$ (Corollary F.3). To prove the "only if" part, we will show that, in any infinite-dimensional linear space X, there exists a convex set $S \neq X$ with $al\text{-}cl(S) = X$. In view of Lemma 6, this is enough to prove our claim. (You see, the idea is exactly the one we used in Example 12.) Let Y be a basis for X (Theorem F.1). By the Well-Ordering Principle (Section B.3.2), there exists a well-ordering $\succsim$ on Y. So, for each $x \in X$, we may write $x = \sum^{m_x} \lambda_i(x)y^i(x)$, where $y^{m_x}(x) \succ \cdots \succ y^1(x)$ for unique $m_x \in \mathbb{N}$, $\lambda_i(x) \in \mathbb{R}\backslash\{0\}$ and $y^i(x) \in Y$, $i = 1, \ldots, m_x$. (Why?) Define

$$S := \{x \in X : \lambda_{m_x}(x) > 0\}.$$

It is obvious that S is convex and $\emptyset \neq S \neq X$. To see that $al\text{-}cl(S) = X$, let $x \in X$ and choose any $y \in Y$ such that $y \succ y^{m_x}(x)$. We must have $y \in S$. (Why?) Moreover, $(1-\alpha)x + \alpha y \in S$ for each $0 < \alpha \leq 1$, and hence $x \in al\text{-}cl(S)$. Since x was arbitrarily chosen in X, this completes the proof. □

2.4 The External Characterization of Algebraically Closed and Convex Sets

As an immediate application of our separation-by-hyperplane results, we derive here the so-called "external" characterization of algebraically closed and convex sets. Put more precisely, we provide a sufficient condition—none

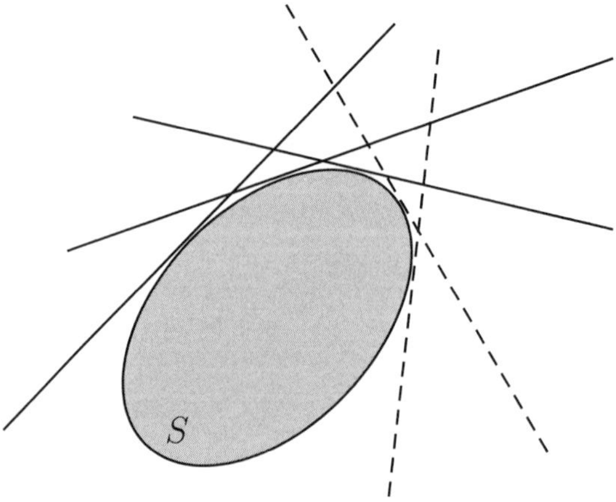

Figure 11

other than the now familiar "nonempty interior" condition—that allows one to express any such set as the intersection of a certain class of closed hyperplanes (Figure 11).

Proposition 2
Let S be an algebraically closed and convex set in a linear space X. If $\text{al-int}_X(S) \neq \emptyset$*, then either* $S = X$ *or S can be written as the intersection of all closed half-spaces that contain it.*

Proof
Assume that $\text{al-int}_X(S) \neq \emptyset$ and $S \neq X$. Define

$$\mathcal{A} := \{(\alpha, L) \in \mathbb{R} \times \mathcal{L}(X, \mathbb{R}) : \alpha \geq L(S)\}.$$

By Proposition 1, $\mathcal{A} \neq \emptyset$. We wish to show that

$$S = \bigcap \{L^{-1}((-\infty, \alpha]) : (\alpha, L) \in \mathcal{A}\}.$$

Of course, $\subseteq$ part of the claim is trivial. Conversely, by Corollary 4, $z \in X \backslash S$ implies that $\alpha \geq L(z)$ cannot hold for all $(\alpha, L) \in \mathcal{A}$. Therefore, if $\alpha \geq L(x)$ for all $(\alpha, L) \in \mathcal{A}$, then we must have $x \in S$. ■

EXERCISE 49

(a) Can one replace the statement "closed half-spaces" with "open half-spaces" in Proposition 2?

(b) Can one omit the condition $al\text{-}int_X(S) \neq \emptyset$ in Proposition 2?

EXERCISE 50 Let C be an algebraically closed convex cone in a linear space X with $al\text{-}int_X(C) \neq \emptyset$. Show that either $C = X$ or C can be written as the intersection of a class of closed half-spaces the corresponding hyperplanes of which pass through the origin.

EXERCISE 51 (Shapley-Baucells) Let X be a preordered linear space such that $al\text{-}int_X(X_+) \neq \emptyset$ and $\{\alpha \in \mathbb{R} : (1-\alpha)x + \alpha y \succsim_X \mathbf{0}\}$ is closed in $\mathbb{R}$. Show that there exists a nonempty $\mathcal{L} \subseteq \mathcal{L}(X, \mathbb{R})$ such that

$$x \succsim_X y \quad \text{if and only if} \quad L(x) \geq L(y) \ \text{ for all } L \in \mathcal{L}$$

for each $x, y \in X$. How about the converse claim?

2.5 Supporting Hyperplanes

The following definition is fundamental.

> DEFINITION
> Let S be a nonempty set in a linear space X and $x \in al\text{-}cl(S)$. A hyperplane H is said to **support** S **at** x if $x \in H$ and S is contained in one of the closed half-spaces induced by H. If, in addition, S is not contained in H, then we say that H **properly supports** S **at** x.

Thus a hyperplane H separates a given set S at $x \in al\text{-}cl(S)$ if it separates $\{x\}$ and S while "touching" S at x. (See Figure 12.) Recalling Corollary F.4, therefore, we see that H supports S at x iff we have $L(x) \geq L(S)$, where L is the nonzero linear functional associated with H. Analogously, H properly supports S at x iff $L(x) > L(S)$.[27]

Quite intuitively, the points at which a set is supportable by a hyperplane must lie in the algebraic boundary of that set. That is, if X is any linear space, $S \subseteq X$, and $x \in al\text{-}cl(S)$, then $L(x) \geq L(S)$ may hold for

[27] *Quiz.* Give an example of a set S in $\mathbb{R}^2$ that is supported by a hyperplane at some $x \in ri(S)$. Does this hyperplane properly support S at x?

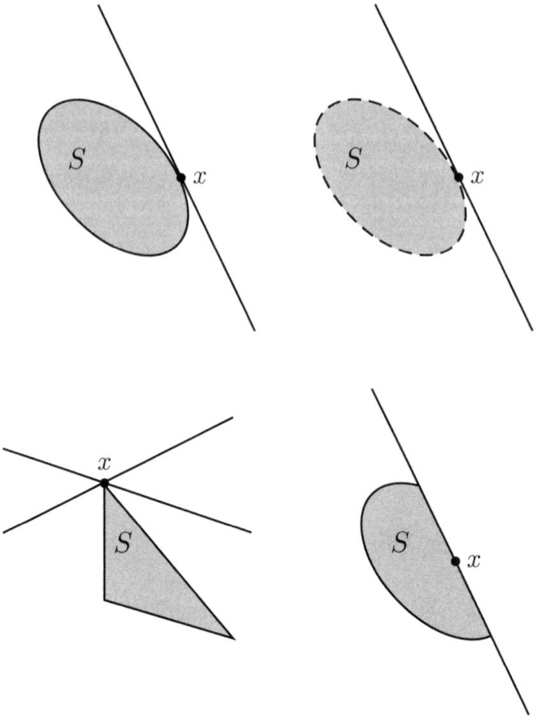

FIGURE 12

some nonzero $L \in \mathcal{L}(X, \mathbb{R})$ only if $x \in al\text{-}bd_X(S)$. (*Proof.* For any nonzero $L \in \mathcal{L}(X, \mathbb{R})$, there exists a $y \in X$ with $L(y) > L(x)$. But if $x \in al\text{-}int_X(S)$, then $z := (1-\alpha)x + \alpha y \in S$ for some $0 < \alpha < 1$, and hence $L(z) > L(x)$.) The more interesting question is the converse: When is it the case that a vector on the algebraic boundary of a given set can be supported by a hyperplane? A moment's reflection shows that we cannot expect to get a satisfactory answer to this question unless S is convex. The question then becomes: Is it true that a convex set can be supported at any of its algebraic boundary points by at least one hyperplane? The answer is no, in general.

EXERCISE 52 Show that no hyperplane in ℓ^1 can support ℓ^1_+ at $\left(\frac{1}{2}, \frac{1}{4}, \frac{1}{8}, \ldots\right)$.

You probably sense the source of the problem here. The algebraic interior of ℓ^1_+ in ℓ^1 is empty (Exercise 23), and we know that one cannot in general settle separation-by-hyperplane-type problems affirmatively unless one has some form of a "nonempty interior" condition. Our next result shows that with this sort of a condition, we would indeed be just fine.

THEOREM 2

Let S be a convex subset of a linear space X and $x \in$ al-bd$_X(S)$. If al-int$_X(S) \neq \emptyset$, then there exists a hyperplane that properly supports S at x, that is, we have $L(x) > L(S)$ for some $L \in \mathcal{L}(X, \mathbb{R})$.

PROOF

Assume that al-int$_X(S) \neq \emptyset$. Define $T := S \backslash \{x\}$, and note that $x \in$ al-cl(T).[28] Suppose T is convex, and apply Proposition 1 to find an $(\alpha, L) \in \mathbb{R} \times \mathcal{L}(X, \mathbb{R})$ such that $L(x) \geq \alpha > L(T)$. We claim that $L(x) = \alpha$. Indeed, since $x \in$ al-cl(T), there is a $y \in T$ such that $z_\lambda := (1-\lambda)x + \lambda y \in T$ for all $0 < \lambda \leq 1$. So if $L(x) > \alpha$ were the case, then we would have $(1-\lambda)L(x) + \lambda L(y) > \alpha$ for $\lambda > 0$ small enough, which would yield $L(z_\lambda) > \alpha$ (for small $\lambda > 0$), contradicting $\alpha > L(T)$. Thus we have $L(x) = \alpha > L(T)$, and this implies $L(x) > L(S)$, as we sought.

It remains to consider the case where T is not convex. In this case there must exist distinct vectors y and z in T such that $x = \lambda y + (1-\lambda)z$ for some $0 < \lambda < 1$. (Yes?) Let $A := S \backslash co\{y, z\}$ and $B := co\{y, z\}$. By Exercise 27, if any vector in B belongs to al-int$_X(S)$, then $B \backslash \{y, z\} \subseteq$ al-int$_X(S)$. (Why?) But this would imply that $x \in$ al-int$_X(S)$, a contradiction. Thus, al-int$_X(S) \cap B = \emptyset$. In turn, this means that al-int$_X(A) \neq \emptyset$ and al-int$_X(A) \cap B = \emptyset$. (Yes?) By the Dieudonné Separation Theorem, therefore, there exists an $(\alpha, L) \in \mathbb{R} \times \mathcal{L}(X, \mathbb{R})$ such that $L(co\{y, z\}) \geq \alpha > L(A)$. Since $co\{y, z\} \subseteq$ al-bd$_X(S)$, arguing exactly as we did in the previous paragraph leads us to conclude that $L(co\{y, z\}) = \alpha$. It then follows that $L(x) > L(S)$, and the proof is complete. ■

*REMARK 3 How detrimental is the "nonempty interior" hypothesis to Theorem 1? Very! If you have read Remark 2, you know that it cannot be

[28] If $x \notin S$, this claim is trivial. If $x \in S$, then pick any $y \in S \backslash \{x\}$—which we can do because al-int$_X(S) \neq \emptyset$ implies $S \neq \{x\}$—and observe that, by convexity, the entire line segment between x and y must be contained in S. Thus $(1-\lambda)x + \lambda y \in T$ for all $0 < \lambda \leq 1$.

dropped in the statement of this result, *unless X is finite-dimensional.* Put more formally:

> FACT
> *A linear space X is infinite-dimensional if, and only if, it contains a nonempty convex set S and a vector $x \in \text{al-bd}_X(S)$ such that there is no hyperplane in X that supports S at x.*

The "if" part of the claim follows from the Minkowski Supporting Hyperplane Theorem (which is proved in Section 3.2 below) and the fact that every finite-dimensional linear space is isomorphic to some $\mathbb{R}^n$ (Corollary F.3). The proof of the "only if" part is hidden in the argument given in Remark 2. □

2.6* Superlinear Maps

In convex analysis and certain branches of decision theory, real maps that are concave *and* positively homogeneous play an important role. In this section we briefly review the properties of such functions, and then provide a representation for them by using our findings on separating and supporting convex sets by hyperplanes. We will put this representation to good use later in the course.

> DEFINITION
> Let C be a convex cone in a given linear space. A real map φ on C is said to be **superlinear** if it is concave and positively homogeneous.[29] It is called **sublinear** if $-\varphi$ is superlinear.

The following exercises recount some basic properties of superlinear maps.

EXERCISE 53 Let C be a convex cone in a linear space and $\varphi \in \mathbb{R}^C$ a superlinear map. Prove:
(a) $\varphi(\mathbf{0}) = 0$.
(b) $-\varphi$ is subadditive.
(c) If C is a linear space, then $\varphi(x) \leq -\varphi(-x)$ for all $x \in C$.

[29] *Reminder.* φ is positively homogeneous iff $\varphi(\alpha x) = \alpha\varphi(x)$ for all $(\alpha, x) \in \mathbb{R}_{++} \times C$.

EXERCISE 54 Let C be a convex cone in a linear space and φ a positively homogeneous real function on C. Show that the following statements are equivalent:

(a) $-\varphi$ is subadditive.

(b) $\{(x,t) \in C \times \mathbb{R} : \varphi(x) \geq t\}$ is a convex cone in the product linear space $span(C) \times \mathbb{R}$.

(c) $\varphi(\alpha x + \beta y) \geq \alpha\varphi(x) + \beta\varphi(y)$ for all $x, y \in C$ and $\alpha, \beta \geq 0$.

EXERCISE 55 Let X be a linear space and $\varphi \in \mathbb{R}^X$ a superlinear map. Prove:

(a) φ is linear iff $\varphi(x) = -\varphi(-x)$ for all $x \in X$.

(b) $\{x \in X : \varphi(x) = -\varphi(-x)\}$ is the $\supseteq$-maximum linear subspace of X on which φ is linear.

(c) If $S \subseteq X$ is linearly independent in X, then $\varphi|_{span(S)}$ is linear iff $\varphi(x) = -\varphi(-x)$ for all $x \in S$.

An important lesson of convex analysis is that, under fairly general conditions, one can represent a superlinear map as a minimum of a certain class of linear maps. Since we know a lot about linear maps, this observation often makes working with superlinear maps rather pleasant. (And why would anyone need to work with such maps? You will see. Read on.)

PROPOSITION 3

Let C be a convex cone in a linear space with $ri(C) \neq \emptyset$. If $\varphi \in \mathbb{R}^C$ is a superlinear map, then there exists a nonempty convex set $\mathcal{L}$ of linear functionals on $span(C)$ such that

$$\varphi(y) = \min\{L(y) : L \in \mathcal{L}\} \quad \textit{for any } y \in ri(C).$$

For superlinear maps defined on a linear space, this representation reads a bit cleaner.

COROLLARY 5

If φ is a superlinear map on a linear space X, then there exists a nonempty convex set $\mathcal{L}$ of linear functionals on X such that

$$\varphi(x) = \min\{L(x) : L \in \mathcal{L}\} \quad \textit{for any } x \in X.$$

Corollary 5 is immediate from Proposition 3. On the other hand, the idea of proof of Proposition 3 is simply to find hyperplanes, in the product linear space $X := span(C) \times \mathbb{R}$, that would support the convex cone

$$C_\varphi := \{(x,t) \in C \times \mathbb{R} : \varphi(x) \geq t\}$$

at its algebraic boundary points.[30] The following elementary observation about concave maps will be useful for this purpose.

LEMMA 7

Let S be a nonempty convex subset of a linear space, and define $X := span(S) \times \mathbb{R}$. Then, for any $\varphi \in \mathbb{R}^S$,

$$Gr(\varphi) \subseteq al\text{-}bd_X(C_\varphi), \tag{12}$$

and, for any concave $\varphi \in \mathbb{R}^S$,

$$\{(x,t) \in ri(S) \times \mathbb{R} : \varphi(x) > t\} \subseteq ri(C_\varphi). \tag{13}$$

PROOF

Fix an arbitrary $\varphi \in \mathbb{R}^S$. If $(x,t) \in Gr(\varphi)$, then $(1-\lambda)(x,t) + \lambda(x,t+1) \notin C_\varphi$ for any $\lambda > 0$, so $Gr(\varphi) \cap al\text{-}int_X(C_\varphi) = \emptyset$. Since $Gr(\varphi) \subseteq C_\varphi$ by definition, therefore, $Gr(\varphi) \subseteq al\text{-}bd_X(C_\varphi)$. To prove (13), assume that φ is concave, and take any $(x,t) \in ri(S) \times \mathbb{R}$ with $\varphi(x) > t$, and any $(y,s) \in aff(C_\varphi)$. Since $aff(C_\varphi) \subseteq aff(S) \times \mathbb{R}$ and $x \in ri(S)$, there is an $\alpha'_y \in (0,1]$ such that $(1-\alpha)x + \alpha y \in S$ for all $0 \leq \alpha \leq \alpha'_y$. Also, because $\varphi(x) > t$, there is an $0 < \alpha_y < \alpha'_y$ small enough so that $(1-\alpha)(\varphi(x) - t) > \alpha \left|s - \varphi(y)\right|$ for any $0 < \alpha \leq \alpha_y$. Then, by concavity,

$$\varphi\left((1-\alpha)x + \alpha y\right) \geq (1-\alpha)\varphi(x) + \alpha\varphi(y) > (1-\alpha)t + \alpha s,$$

that is, $(1-\alpha)(x,t) + \alpha(y,s) \in C_\varphi$ for all $0 \leq \alpha \leq \alpha_y$. Hence $(x,t) \in ri(C_\varphi)$, as we sought. ■

We are now ready for the proof of Proposition 3.

[30] In convex analysis C_φ is called the **hypograph** of φ. In turn, the hypograph of $-\varphi$ is called the **epigraph** of φ.

Proof of Proposition 3

Let $\varphi \in \mathbb{R}^C$ be a superlinear map. Then C_φ must be a convex cone in $X := span(C) \times \mathbb{R}$. (Yes?) By Lemma 7, the algebraic interior of this cone in X is nonempty.[31] Now fix any $y \in ri(C)$. By Lemma 7, $(y, \varphi(y)) \in al\text{-}bd_X(C_\varphi)$, so by Theorem 2, there exists a nonzero linear functional F on X such that

$$F(y, \varphi(y)) \geq F(x,t) \qquad \text{for all } (x,t) \in C_\varphi.$$

Since C_φ is a cone, we must then have $F(y,\varphi(y)) \geq 2F(y,\varphi(y))$ and $F(y,\varphi(y)) \geq \frac{1}{2}F(y,\varphi(y))$, which is possible only if $F(y,\varphi(y)) = 0$. Thus:

$$F(x,t) \leq 0 \;\; \text{for all } (x,t) \in C_\varphi \qquad \text{and} \qquad F(y,\varphi(y)) = 0.$$

Moreover, it is obvious that there exist a real number α and a linear functional K on $span(C)$ such that $F(x,t) = K(x) + \alpha t$ for all $(x,t) \in X$. (Why?) Then

$$K(x) \leq -\alpha t \;\; \text{for all } (x,t) \in C_\varphi \qquad \text{and} \qquad K(y) = -\alpha\varphi(y). \tag{14}$$

Clearly, if $\alpha = 0$, then $K(C) \leq 0 = K(y)$. But if $K(x) < 0$ for some $x \in C$, then choosing a $\lambda > 0$ with $(1-\lambda)y + \lambda(-x) \in C$—there is such a $\lambda > 0$ because $y \in ri(C)$—we find $0 \geq K((1-\lambda)y + \lambda(-x)) = -\lambda K(x) > 0$, contradiction. Thus, if $\alpha = 0$, then $K(C) = 0$, which means that K is the zero functional on $span(C)$, contradicting the fact that F is nonzero.

We now know that $\alpha \neq 0$. Since $\{x\} \times (-\infty, \varphi(x)] \subseteq C_\varphi$ for any $x \in C$, it readily follows from (14) that $\alpha > 0$. Let $L := -\frac{1}{\alpha}K$, and observe that (14) implies $\varphi \leq L|_C$ and $\varphi(y) = L(y)$. Since y was arbitrarily chosen in $ri(C)$, therefore, we conclude: For every $y \in ri(C)$ there is a linear functional L_y on $span(C)$ such that $\varphi \leq L_y|_C$ and $\varphi(y) = L_y(y)$. Letting $\mathcal{L}$ be the convex hull of $\{L_y : y \in ri(C)\}$ completes the proof. ∎

[31] Not so fast! All Lemma 7 gives us is that the algebraic interior of C_φ *in* $span(C_\varphi)$ is not empty, while we need $al\text{-}int_X(C_\varphi) \neq \emptyset$ here. But this is no cause for worry. Obviously, $span(C_\varphi) \subseteq X$. Conversely, if $(x,t) \in C \times \mathbb{R}$, then letting $s := \varphi(x)$ and $\lambda = 1 - (s - t)$, we find $(x,t) = (x, s - (1-\lambda)) = \lambda(x,s) + (1-\lambda)(x, s-1) \in span(C_\varphi)$. So $span(C_\varphi)$ is a linear space that contains $C \times \mathbb{R}$. Thus: $span(C_\varphi) = X$.

Comparing the following observation with Proposition 3 identifies the role of the positive homogeneity property in the latter result.[32]

EXERCISE 56 Let S be a convex set in a linear space X with $al\text{-}int_X(S) \neq \emptyset$. Prove: If $\varphi \in \mathbb{R}^S$ is concave, then there exists a nonempty convex set $\mathcal{L}$ of *affine* functionals on X such that $\varphi(y) = \min\{L(y) : L \in \mathcal{L}\}$ for all $y \in al\text{-}int_X(S)$.

3 Reflections on $\mathbb{R}^n$

In many economic applications one needs to make use of a separation argument only in the context of a Euclidean space. It is thus worth noting that we can actually sharpen the statement of the Dieudonné Separation Theorem in this special case. Moreover, the connection between the inner product operation and the Euclidean metric allows one to introduce the notion of "orthogonality," thereby improving the strength of convex analysis when the ambient linear space is Euclidean. These issues are explored in this section.

NOTE. Please keep in mind that we work with a fixed, but arbitrary, positive integer n throughout this section.

3.1 Separation in $\mathbb{R}^n$

The following separation theorem, which goes back to the begining of twentieth century, is among the most widely used theorems of mathematical analysis.[33] Its advantage (over, say, Proposition 1) stems from the fact that it applies even to convex sets with empty algebraic interior.

[32] You might also want to compare Exercise 50 and Proposition 2 to get a sense of the "geometry" of the situation. You see, all is based on our ability to support the hypograph C_φ of φ at its algebraic boundary points. If φ is concave, then all we know is that C_φ is a convex set, so the supporting hyperplanes that we find need not go through the origin. But if φ is superlinear, then its hypograph is a convex *cone*, so the supporting hyperplanes that we find are sure to go through the origin.

[33] The main results of this section were discovered by Hermann Minkowski (1864–1909) during his development of the so-called geometry of numbers. (Minkowski's separation theorems appeared posthumously in 1911.) Although he also provided a very elegant model for Einstein's special relativity theory and paved the way for the discovery of general relativity (also by Albert Einstein), Minkowski made his most remarkable contributions in the realm of convex analysis. Indeed, he is often referred to as the founder of convex analysis.

PROPOSITION 4

(Minkowski) *Let S be a nonempty convex subset of $\mathbb{R}^n$ and $z \in \mathbb{R}^n \backslash S$. Then there exists a nonzero linear functional L on $\mathbb{R}^n$ such that*

$$L(z) \geq \sup L(S).$$

If, in addition, S is closed, then this inequality can be taken as strict.

PROOF

(We apply the algebraic interior and closure operators relative to $\mathbb{R}^n$ throughout the proof.) If $\textit{al-int}(S) \neq \emptyset$, then Proposition 1 yields the first claim readily, so suppose $\textit{al-int}(S) = \emptyset$. Take any $z \in \mathbb{R}^n \backslash S$, and consider first the case where $z \in \textit{al-cl}(S)$. Then, $z \in \textit{aff}(S) \neq \mathbb{R}^n$. (Recall equation (3).) There must then exist a hyperplane that contains $\textit{aff}(S)$. (Why?) So, by Corollary F.4, there is an $\alpha \in \mathbb{R}$ and a nonzero $L \in \mathcal{L}(\mathbb{R}^n, \mathbb{R})$ such that $\textit{aff}(S) \subseteq L^{-1}(\alpha)$. For this L we have $L(z) = L(S)$, so the claim holds trivially.

Now consider the case where $z \notin \textit{al-cl}(S)$.[34] Recall that $\textit{al-cl}(S) = cl(S)$, since S is a convex subset of $\mathbb{R}^n$ (Observation 3), so $\varepsilon := d_2(z, cl(S)) > 0$ (Exercise D.2). Define

$$T := cl(S) + cl(N_{\frac{\varepsilon}{2}, \mathbb{R}^n}(\mathbf{0})),$$

and observe that T is a closed and convex subset of $\mathbb{R}^n$ with $\textit{al-int}(T) = \textit{int}(T) \neq \emptyset$ and $z \notin T$. (Prove this!)[35] By Observation 3, then, T is algebraically closed, and hence, we may apply Corollary 4 to find an $(\alpha, L) \in \mathbb{R} \times \mathcal{L}(\mathbb{R}^n, \mathbb{R})$ such that

$$L(z) > \alpha > L(T) = L(cl(S)) + L\left(cl\left(N_{\frac{\varepsilon}{2}, \mathbb{R}^n}(\mathbf{0})\right)\right).$$

In particular, we have $L(z) > \alpha > L(S)$, and hence $L(z) > \sup L(S)$.

[34] The idea is simple: Enlarge $\textit{al-cl}(S)$ to an algebraically closed convex set with a nonempty interior, but one that still does not contain z. Once this is done, all that remains is to strictly separate z from that set by using Corollary 4.

[35] I leave here quite a few claims for you to settle. Most notably, the fact that T is closed is not trivial. The idea is this. If $(x^m) \in T^\infty$, then there exist sequences (s^m) in $cl(S)$ and (y^m) in $cl(N_{\frac{\varepsilon}{2}, \mathbb{R}^n}(\mathbf{0}))$ such that $x^m = s^m + y^m$ for each m. Suppose $x^m \to x \in \mathbb{R}^n$. Then, since $cl(N_{\frac{\varepsilon}{2}, \mathbb{R}^n}(\mathbf{0}))$ is compact, you can find convergent subsequences of both (s^m) and (y^m). (Why?) Since $cl(S)$ is closed, the sum of the limits of these subsequences belongs to $cl(S) + cl(N_{\frac{\varepsilon}{2}, \mathbb{R}^n}(\mathbf{0}))$, but this sum is none other than x.

We are actually done here, because the argument given in the previous paragraph (which did not presume *al-int*$(S) = \emptyset$) proves the second claim of the proposition. ■

The following result is derived from Proposition 4, just as we derived Proposition 2 from Corollary 4.

COROLLARY 6

If S is a closed and convex proper subset of $\mathbb{R}^n$, then it can be written as the intersection of all closed half-spaces (in $\mathbb{R}^n$) that contain it.

EXERCISE 57 Prove Corollary 6. Also, show that here it is enough to use *only* those closed half-spaces the associated hyperplanes of which support S at some vector.

The first part of Proposition 4 does not guarantee that any real n-vector that lies outside of a nonempty convex subset of $\mathbb{R}^n$ can be *properly* separated from this set. This is nevertheless true, as we ask you to prove next.

EXERCISE 58 Show that in the first part of Proposition 4 we can also ask from L to guarantee that $L(z) > L(S)$.

In the next chapter we shall encounter quite a few applications of Proposition 4. But it may be a good idea to give right away at least one illustration of how this sort of a separation result may be used in practice. This is the objective of the following example, which provides a novel application of Proposition 4 to linear algebra.

EXAMPLE 13

(*Farkas' Lemma*) Take any $m \in \mathbb{N}$, and any $u^1, \ldots, u^m, v \in \mathbb{R}^n$. A famous result of linear algebra, *Farkas' Lemma*, says the following: *Either*

$$\text{there exists an } x \in \mathbb{R}^m_+ \text{ such that } \sum_{j=1}^{m} x_j u^j = v \tag{15}$$

or (exclusive)

$$\text{there exists a } y \in \mathbb{R}^n \text{ such that } \sum_{i=1}^{n} y_i v_i > 0$$

and

$$\sum_{i=1}^{n} y_i u_i^j \leq 0, \; j = 1, \ldots, m. \tag{16}$$

The elementary proof of this fact (as given by Gyula Farkas in 1902) is quite tedious. Yet we can prove Farkas' Lemma by using Proposition 4 very easily. The key observation is that (15) is equivalent to saying that $v \in cone\{u^1, \ldots, u^m\} =: C$. So suppose $v \notin C$. Since C is closed (Corollary 1), we may apply Proposition 4 to find a nonzero linear functional L on $\mathbb{R}^n$ such that $L(v) > \sup L(C)$. This implies $L(v) > 0$ (since $\mathbf{0} \in C$). On the other hand, if $L(w) > 0$ for some $w \in C$, then since $mw \in C$ for all $m \in \mathbb{N}$, we would get $L(v) > L(mw) = mL(w) \to \infty$, which is a contradiction. Therefore, $L(v) > 0 \geq \sup L(C)$. But then we are done. Obviously, there is a $y \in \mathbb{R}^n$ such that $L(z) = \sum^n y_i z^i$ for all $z \in \mathbb{R}^n$ (Example F.6), and combining this with $L(v) > 0 \geq \sup L(C)$ yields (16). What could be simpler? □

Exercise 59^H (*The Fredholm Alternative*) Prove: For any $m \in \mathbb{N}$, and $u^1, \ldots, u^m, v \in \mathbb{R}^n$, either $\sum^m x_j u^j = v$ for some $x \in \mathbb{R}^m$, or (exclusive), there exists a $y \in \mathbb{R}^n$ such that $\sum^n y_i v_i \neq 0$ and $\sum^n y_i u_i^j = 0$ for all $j = 1, \ldots, m$.

Exercise 60^H Recall that an $n \times n$ matrix $[a_{ij}]_{n \times n}$ is said to be **stochastic** if $a_{ij} \geq 0$ for all i and j, and $\sum_{j=1}^n a_{ij} = 1$ for all i. Prove that, for every $n \times n$ stochastic matrix $\mathbf{A}$, there is an $x \in \mathbb{R}^n_+$ such that $\mathbf{A}x = x$ and $\sum^n x_i = 1$.

Let us now return to the problem of separating convex sets by a hyperplane within a Euclidean space. The following is the reflection of the Dieudonné Separation Theorem in this context.

The Minkowski Separating Hyperplane Theorem

Let A and B be nonempty disjoint convex sets in $\mathbb{R}^n$. Then there exists a nonzero linear functional L on $\mathbb{R}^n$ such that

$$\inf L(B) \geq \sup L(A).$$

PROOF
Define $S := A - B$ and observe that S is a nonempty convex subset of $\mathbb{R}^n$ with $\mathbf{0} \notin S$. Applying Proposition 4, therefore, we find a nonzero $L \in \mathcal{L}(\mathbb{R}^n, \mathbb{R})$ such that

$$0 = L(\mathbf{0}) \geq \sup L(S) = \sup(L(A) - L(B)) = \sup L(A) - \inf L(B),$$

and we are done. ■

Once again, we are not guaranteed of proper separation of A and B by this theorem. However, using Proposition 4 *and* Exercise 58 in the proof outlined above shows that we can in fact ask from L to satisfy $L(B) > L(A)$ in the Minkowski Separating Hyperplane Theorem. Thus:

In a Euclidean space, any two disjoint convex sets can be properly separated by a hyperplane.

We cannot, however, guarantee that this separation is *strict*. (We say that a hyperplane H in a linear space X **strictly separates** the subsets A and B of X if A and B are contained in different open half spaces induced by H.) There is no surprise here; this is not true even in $\mathbb{R}$. (For example, $[0, 1)$ and $\{1\}$ cannot be strictly separated by a hyperplane in $\mathbb{R}$.) What may be slightly more surprising is that strict separation may not be possible even if the convex sets at hand are closed. That is, adding the hypothesis that A and B be closed to the Minkowski Separating Hyperplane Theorem does not allow us to strengthen the conclusion of the theorem to $L(B) > \alpha > \sup L(A)$ for some real number α. For instance, consider the sets $A := \{(a, b) \in \mathbb{R}^2_+ : ab \geq 1\}$ and $B := \mathbb{R}_+ \times \{0\}$. A moment's reflection shows that these sets cannot be *strictly* separated by a hyperplane—there is only one hyperplane in $\mathbb{R}^2$ that separates these sets, and that's B itself!

This said, it is important to note that if one of the sets to be separated were known to be bounded, then we would be in much better shape. As the terminology goes, in that case, we may not only strictly separate the sets at hand, but we may do so *strongly* in the sense of separating (strictly) certain open neighborhoods of these sets. Put differently, in this case, we may even guarantee that small parallel shifts of our separating hyperplane continue to separate these sets. This is because if, in addition to the hypotheses of the Minkowski Separating Hyperplane Theorem, we know that A is compact and B is closed, we can show that $A - B$ is a closed set that does not contain $\mathbf{0}$

(this is an exercise for you) and then strictly separate $\mathbf{0}$ from $A - B$ by using the second part of Proposition 4. This leads us to the following separation theorem.

Proposition 5
(Minkowski) *Let A and B be nonempty closed and convex subsets of $\mathbb{R}^n$. If A is bounded and $A \cap B = \emptyset$, then there exists a nonzero linear functional L on $\mathbb{R}^n$ such that* $\inf L(B) > \max L(A)$.

Exercise 61[H]
(a) Prove Proposition 5.
(b) Prove: For any two convex subsets A and B of $\mathbb{R}^n$ with $\mathbf{0} \notin cl_{\mathbb{R}^n}(A - B)$, there exists a nonzero linear functional L on $\mathbb{R}^n$ such that $\inf L(B) > \sup L(A)$.

To drop the boundedness postulate in Proposition 5, one needs to impose further structure on the sets A and B. For instance, suppose we know that A is a closed convex cone, what then? Well, that's not enough, we need to posit more structure on B as well. (Recall that $\mathbb{R}_+ \times \{0\}$ cannot be strictly separated from $\{(a, b) \in \mathbb{R}^2_+ : ab \geq 1\}$.) For instance, if B is an hyperplane, all goes well.

Exercise 62[H] Let H be a hyperplane, and C a closed convex cone, in $\mathbb{R}^n$. Prove: If $H \cap C = \emptyset$, then there exists a nonzero linear functional L on $\mathbb{R}^n$ such that $\inf L(H) > \sup L(C)$.

Curiously, this result would not remain true if we only knew that H is an affine manifold. To give an example, consider the set $S := \{(a, b, 0) \in \mathbb{R}^3_+ : ab \geq 1\}$ and let $C := cone\{S + (0, 0, 1)\}$. Then C is a closed convex cone that is disjoint from the affine manifold $H := \mathbb{R} \times \{0\} \times \{1\}$. But H and C cannot be strongly separated. (Draw a picture to see the idea behind the construction.) After all, $\inf\{d_2(x, y) : (x, y) \in C \times H\} = 0$.

If, however, C happens to be a *finitely generated* convex cone, then there is no problem, we can strongly separate any such cone from any affine manifold disjoint from it. (We shall encounter a brilliant application of this fact in Section 3.6.) The key observation in this regard is the following.

LEMMA 8
Let Y be a linear subspace of $\mathbb{R}^n$ and C a finitely generated convex cone in $\mathbb{R}^n$. Then, $C + Y$ is a closed subset of $\mathbb{R}^n$.

PROOF
Let A be a basis for Y, and pick any finite set S such that $C = cone(S)$. It is easy to check that $C + Y = cone(A \cup -A \cup S)$. (Yes?) So $C + Y$ is a finitely generated cone, and hence it is closed (Corollary 1). ■

PROPOSITION 6
Let S be an affine manifold of $\mathbb{R}^n$ and C a finitely generated convex cone in $\mathbb{R}^n$. If $C \cap S = \emptyset$, then there exists a nonzero linear functional L on $\mathbb{R}^n$ such that $\inf L(S) > 0 = \max L(C)$.

PROOF
By definition, $S = Z + x^*$ for some linear subspace Z of X and some vector $x^* \in X$. By Lemma 8, $C - Z$, and hence $C - S$, is a closed subset of $\mathbb{R}^n$. (Yes?) Furthermore, $C \cap S = \emptyset$ entails that $\mathbf{0} \notin C - S$. Applying (the second part of) Proposition 4, therefore, we find a nonzero $L \in \mathcal{L}(\mathbb{R}^n, \mathbb{R})$ such that

$$0 = L(\mathbf{0}) > \sup L(C - S) = \sup(L(C) - L(S)) = \sup L(C) - \inf L(S),$$

and hence $\inf L(S) > \sup L(C)$. Moreover, this inequality entails that $L(C) \leq 0$ (why?), so we have $0 = \max L(C)$. ■

3.2 Support in $\mathbb{R}^n$

Another important consequence of the general development presented in Section 2 is the following observation.

THE MINKOWSKI SUPPORTING HYPERPLANE THEOREM
Let S be a convex set in $\mathbb{R}^n$ and $x \in bd_{\mathbb{R}^n}(S)$. Then, there exists a hyperplane in $\mathbb{R}^n$ that supports S at x, that is, there exists a nonzero $L \in \mathcal{L}(\mathbb{R}^n, \mathbb{R})$ with $L(x) \geq L(S)$.

PROOF

By Observations 2 and 3, $bd_{\mathbb{R}^n}(S) = al\text{-}bd_{\mathbb{R}^n}(S)$, so if $al\text{-}int_{\mathbb{R}^n}(S) \neq \emptyset$, then the claim follows readily from Theorem 2. If $al\text{-}int(S) = \emptyset$, then the entire S must lie within a hyperplane (why?), so the claim is trivial. ■

WARNING The Minkowski Supporting Hyperplane Theorem does not guarantee the existence of a hyperplane that *properly* supports S at a vector on the boundary of S. One would in general need the interior of S to be nonempty (or equivalently, $dim(S) = n$) for that purpose.

Dudley (2002, p. 199) gives a nice interpretation of the Minkowski Supporting Hyperplane Theorem: "If a convex set is an island, then from each point on the coast, there is at least a 180° unobstructed view of the ocean." But we should emphasize that this is a finite-dimensional interpretation. If our island was located in an infinite-dimensional universe, then we would also need to ascertain that it is sufficiently "solid"—in the sense of possessing a nonempty algebraic interior—for its coasts to have this desirable property. (Recall Remark 3.)

The following set of exercises should give you some idea about the typical ways in which the Minkowski Supporting Hyperplane Theorem is used in practice.

EXERCISE 63 Let C be a pointed closed convex cone in $\mathbb{R}^n$. Show that C can be written as the intersection of a family of closed half-spaces the corresponding hyperplanes of which pass through the origin.

*EXERCISE 64 (*The Minkowski-Weyl Theorem*) A cone C in $\mathbb{R}^n$ is called **polyhedral** if there is a nonempty finite subset $\mathcal{L}$ of $\mathcal{L}(\mathbb{R}^n, \mathbb{R})$ such that $C = \{x \in \mathbb{R}^n : L(x) \geq 0 \text{ for all } L \in \mathcal{L}\}$. Prove that a convex cone C in $\mathbb{R}^n$ is polyhedral iff it is finitely generated.

EXERCISE 65

(a) Show that if A and B are two closed and convex subsets of $\mathbb{R}^n$ such that $int_{\mathbb{R}^n}(A) \neq \emptyset$ and $|A \cap B| = 1$, then these sets can be separated by a hyperplane (in $\mathbb{R}^n$) that supports both A and B at $x \in A \cap B$.

(b) What sort of an assumption about A would ensure that the hyperplane found in part (a) is unique?

*EXERCISE 66^{H} Let S be a nonempty convex subset of $\mathbb{R}^n$. A vector $z \in S$ is called an **extreme point** of S if z cannot be written as $z = \lambda x + (1-\lambda)y$ for any distinct $x, y \in S$ and $0 < \lambda < 1$.

(a) Show that if S is compact, and H is a hyperplane that supports S at some $x \in al\text{-}cl(S)$, then H contains at least one extreme point of S.

(b) Prove the following special case of the *Krein-Milman Theorem*:
A nonempty compact and convex subset of $\mathbb{R}^n$ equals the convex hull of its extreme points.

EXERCISE 67 Let S be a convex set in $\mathbb{R}^n$. Show that if $\varphi \in \mathbb{R}^S$ is an upper semicontinuous and concave map, then there exists a nonempty convex subset $\mathcal{L}$ of affine functionals on $\mathbb{R}^n$ such that $\varphi(y) = \min\{L(y) : L \in \mathcal{L}\}$ for all $y \in S$. (Compare with Exercise 56.)

3.3 The Cauchy-Schwarz Inequality

One of the most useful auxiliary results in Euclidean analysis is an inequality that relates the inner product of two real n-vectors to the product of their distances from the origin. This is

THE CAUCHY-SCHWARZ INEQUALITY[36]
For any $x, y \in \mathbb{R}^n$,

$$|xy| \leq \sqrt{xx}\sqrt{yy}.^{37} \tag{17}$$

PROOF
Take any real n-vectors x and y, and assume that $y \neq \mathbf{0}$ (otherwise the claim is trivial). The trick is to exploit the fact that $(x-\lambda y)(x-\lambda y)$ is a nonnegative number for any real number λ, that is,

$$xx - 2\lambda xy + \lambda^2 yy \geq 0 \quad \text{for any } \lambda \in \mathbb{R}.$$

[36] This version of the inequality was obtained by Cauchy in 1821. More general versions had then been obtained, the most influential one of which was the one given by Hermann Schwarz in 1859. The proof I present, now standard, was given first by Hermann Weyl in 1918.

[37] Here, of course, I'm using the standard inner product notation, that is, $xy := \sum^n x_i y_i$ for any $x, y \in \mathbb{R}^n$. I'll keep using this notation in the rest of the chapter.

Indeed, if we put $\frac{xy}{yy}$ for λ here, we find $xx - \frac{(xy)^2}{yy} \geq 0$, and rearranging yields (17). ■

Note that another way of stating the Cauchy-Schwarz Inequality is

$$\left|xy\right| \leq d_2(x, \mathbf{0})d_2(y, \mathbf{0}) \qquad \text{for any } x, y \in \mathbb{R}^n,$$

and still another way is

$$\left|(x-z)(y-w)\right| \leq d_2(x, z)d_2(y, w) \qquad \text{for any } x, y, z, w \in \mathbb{R}^n.$$

3.4 Best Approximation from a Convex Set in $\mathbb{R}^n$

The results of Section 3.1 have exploited the metric structure of $\mathbb{R}^n$ (in addition to its linear structure) to sharpen the separation and support theorems we derived earlier for abstract linear spaces. It is possible to pursue this line of reasoning further to deal with certain problems of convex analysis that cannot even be addressed within an arbitrary linear space. The problem of best approximation from a convex set provides a good case in point.

Recall that, given any point x in a Euclidean space $\mathbb{R}^n$, there is a unique point that is closest to x in any given nonempty closed and convex subset S of $\mathbb{R}^n$. As in Example D.5, we call this point the **projection of x on S**. By definition, this vector "approximates" x better than any other member of S (with respect to the Euclidean metric). By using the connection between the inner product operation on $\mathbb{R}^n$ and d_2—recall that $d_2(u, v) = \sqrt{(u-v)(u-v)}$ for any $u, v \in \mathbb{R}^n$—we can provide an alternative way of looking at this situation.

Proposition 7
Let S be a convex subset of $\mathbb{R}^n$ and $x \in \mathbb{R}^n$. If $y \in S$, then

$$d_2(x, y) = d_2(x, S) \quad \textit{if and only if} \quad (x-y)(z-y) \leq 0 \ \ \textit{for all } z \in S.$$

Proof
Take any $y \in S$. We prove the "only if" part by ordinary calculus. Suppose $(x-y)(z-y) > 0$ for some $z \in S$.[38] Define the real map f on $[0, 1]$ by

$$f(t) := \left(d_2(x, (1-t)y + tz)\right)^2,$$

[38] The idea is to show that moving in the direction of z "a little bit" will have to bring us closer to x.

that is, $f(t) = \sum^n \left(x_i - y_i + t(y_i - z_i)\right)^2$ for all $0 \leq t \leq 1$. Then,

$$f'(0) = 2\sum_{i=1}^{n}(x_i - y_i)(y_i - z_i) = -2(x-y)(z-y) < 0.$$

Since $(1-t)y + tz \in S$ for each $0 \leq t \leq 1$ (because S is convex), this implies that y cannot minimize $d_2(x, \cdot)$ on S, that is, $d_2(x, y) = d_2(x, S)$ is false.

Conversely, suppose we have $(x-y)(z-y) \leq 0$ for all $z \in S$. Then, for any $z \in S$,

$$\begin{aligned}\left(d_2(x,z)\right)^2 &= (x-z)(x-z)\\ &= (x-y+y-z)(x-y+y-z)\\ &= (x-y)(x-y) - 2(x-y)(z-y) + (y-z)(y-z)\\ &\geq \left(d_2(x,y)\right)^2,\end{aligned}$$

so we conclude that $d_2(x, y) = d_2(x, S)$. ■

The following special case of this result is important for linear analysis. It provides a characterization of the projection operator onto a linear subspace of a Euclidean space.

COROLLARY 7

Let Y be a linear subspace of $\mathbb{R}^n$ and $x \in \mathbb{R}^n$. If $y \in Y$, then

$$d_2(x,y) = d_2(x,Y) \quad \text{if and only if} \quad (x-y)w = 0 \quad \text{for all } w \in Y.$$

PROOF

It is readily checked that $(x-y)(z-y) \leq 0$ holds for all $z \in Y$ iff $(x-y)w = 0$ for all $w \in Y$, and hence the claim follows from Proposition 7.[39] ■

EXERCISE 68 Given any $x \in \mathbb{R}^n$, prove that if y is the projection of x on $\mathbb{R}^n_+$, then

$$y_i = \begin{cases} x_i, & \text{if } x_i \geq 0,\\ 0, & \text{otherwise}\end{cases}.$$

EXERCISE 69 Take any $x \in \mathbb{R}^n$, and a convex cone C in $\mathbb{R}^n$. Show that if $y \in C$, then $d_2(x, y) = d_2(x, C)$ iff $(x-y)w \leq 0$ for all $w \in C$.

[39] Let $(x-y)(z-y) \leq 0$ for all $z \in Y$, and pick any $w \in Y$. Since $y \pm w \in Y$, we then have $(x-y)(y \pm w - y) \leq 0$, which means $(x-y)w = 0$.

REMARK 4 (*On the Continuity of the Projection Operators*) For any given nonempty closed and convex subset S of $\mathbb{R}^n$, define the function $\mathsf{p}_S : \mathbb{R}^n \to S$ by letting $\mathsf{p}_S(x)$ stand for the projection of x on S. We have seen in Example D.5 that this map, called the **projection operator onto** S, is a well-defined surjection. We have also proved in that example that p_S is a continuous map, provided that S is compact.

Now we can say more about p_S. In particular, thanks to Proposition 7, we presently know that

$$(x - \mathsf{p}_S(x))(z - \mathsf{p}_S(x)) \leq 0 \qquad \text{for all } z \in S. \tag{18}$$

We next use this fact to improve upon what we know about the continuity of p_S. It turns out that, let alone being continuous, p_S is a nonexpansive function (even if S is not bounded).

To see this, take any $x, y \in \mathbb{R}^n$. Since $\mathsf{p}_S(y) \in S$, (18) gives

$$(x - \mathsf{p}_S(x))(\mathsf{p}_S(y) - \mathsf{p}_S(x)) \leq 0,$$

and, similarly,

$$(y - \mathsf{p}_S(y))(\mathsf{p}_S(x) - \mathsf{p}_S(y)) \leq 0.$$

Adding these two inequalities up, we get

$$(\mathsf{p}_S(x) - \mathsf{p}_S(y))\left(y - x + \mathsf{p}_S(x) - \mathsf{p}_S(y)\right) \leq 0,$$

that is, $(\mathsf{p}_S(x) - \mathsf{p}_S(y))(\mathsf{p}_S(x) - \mathsf{p}_S(y)) \leq (\mathsf{p}_S(x) - \mathsf{p}_S(y))(x - y)$. Thus

$$\begin{aligned}\left(d_2(\mathsf{p}_S(x), \mathsf{p}_S(y))\right)^2 &\leq (\mathsf{p}_S(x) - \mathsf{p}_S(y))(x - y) \\ &\leq d_2(\mathsf{p}_S(x), \mathsf{p}_S(y))d_2(x, y),\end{aligned}$$

by the Cauchy-Schwarz Inequality, and it follows that

$$d_2(\mathsf{p}_S(x), \mathsf{p}_S(y)) \leq d_2(x, y).$$

Conclusion: p_S *is a nonexpansive map.*[40] □

[40] WARNING. p_S need not be a contraction, however. For instance, . . . (Fill in the blank!)

3.5 Orthogonal Complements

The fact that we can "multiply" two vectors x and y in $\mathbb{R}^n$, in the sense of taking their inner product xy, allows us to get a good deal of mileage in terms of Euclidean convex analysis. This is nicely demonstrated by Proposition 7. We now dig a bit deeper into the geometry of things.

You might remember from an earlier linear algebra (or calculus, or whatever) course that the angle θ between two nonzero vectors x and y in $\mathbb{R}^3$ is given by the formula $\cos\theta = \frac{xy}{\sqrt{(xx)(yy)}}$.[41] This leads one to say that x and y are *orthogonal* whenever $xy = 0$. Now, although the notion of "angle" between two nonzero vectors in $\mathbb{R}^n$ loses its geometric interpretation when $n \geq 4$, we may nonetheless generalize the notion of "orthogonality" to extend a number of important facts about $\mathbb{R}^3$ to the general case. Here is how:

DEFINITION

We say that two vectors x and y in $\mathbb{R}^n$ are **orthogonal** if $xy = 0$, and denote this by writing $x \perp y$. If x is orthogonal to every vector in a nonempty subset S of $\mathbb{R}^n$, then we write $x \perp S$, and say that x is **orthogonal to** S. Finally, the set of all vectors in $\mathbb{R}^n$ that are orthogonal to S is called the **orthogonal complement** of S (in $\mathbb{R}^n$) and is denoted as $S^\perp$. That is,

$$S^\perp := \{x \in \mathbb{R}^n : x \perp S\} = \{x \in \mathbb{R}^n : xy = 0 \text{ for all } y \in S\}.$$

For example, for any $n \in \{2, 3, \ldots\}$ and $k \in \{1, \ldots, n-1\}$, if $Y := \{(y_1, \ldots, y_k, 0, \ldots, 0) \in \mathbb{R}^n : y_i \in \mathbb{R}, i = 1, \ldots, k\}$, then $Y^\perp = \{(0, \ldots, 0, y_{k+1}, \ldots, y_n) \in \mathbb{R}^n : y_i \in \mathbb{R}, i = k+1, \ldots, n\}$. This example also illustrates in which sense $Y^\perp$ can be thought of as the "complement" of Y. Of course, the orthogonal complement of $\mathbb{R}^n$ (in $\mathbb{R}^n$) is $\{\mathbf{0}\}$. (Always remember that $\mathbf{0}$ is orthogonal to every vector.) How about the orthogonal complement of $Z := \{(-a, a) : a \in \mathbb{R}\}$? It is, naturally, the subspace $\{(a, a) : a \in \mathbb{R}\}$. (See Figure 13.)

Here are a few basic facts about orthogonal complements.

EXERCISE 70 Given any $\emptyset \neq S \subseteq \mathbb{R}^n$, prove:

(a) If $x \perp S$ for some $x \in \mathbb{R}^n$, then $x \perp span(S)$.

(b) If $x \perp y$ for every distinct $x, y \in S$, then S is linearly independent.

[41] This formula is meaningful, for the Cauchy-Schwarz Inequality ensures that the right-hand side of the equation is a number between -1 and 1.

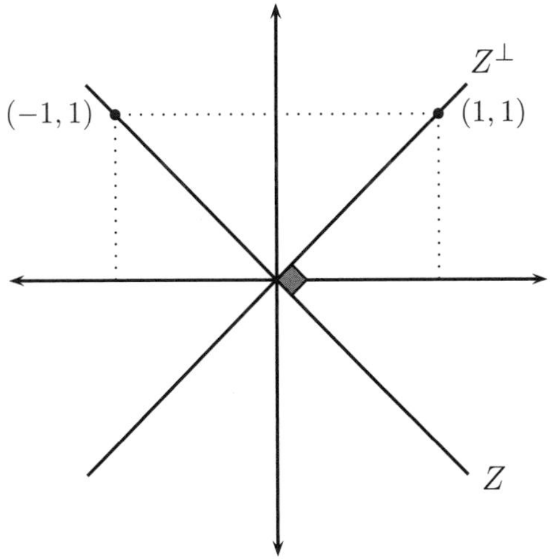

FIGURE 13

EXERCISE 71 For any linear subspace Y of $\mathbb{R}^n$, show that $Y^\perp$ is a linear subspace of $\mathbb{R}^n$ with $Y \cap Y^\perp = \{\mathbf{0}\}$.

Why should one care about the orthogonal complement of a given linear subspace Y of $\mathbb{R}^n$? Because, as Corollary 7 makes it transparent, the notion of orthogonality is essential for studying many interesting "best approximation" problems in $\mathbb{R}^n$. Indeed, for any $y \in Y$, we can think of $x - y$ as the vector that tells us the "error" we would make in approximating the vector $x \in \mathbb{R}^n$ by y. Viewed this way, Corollary 7 tells us that y is the best approximation of x in Y (that is, y is the projection of x on Y) iff the error vector $x - y$ is orthogonal to Y. (See Figure 14.)

From this discussion follows a major decomposition theorem.

THE PROJECTION THEOREM
Let Y be a linear subspace of $\mathbb{R}^n$. Then, for every $x \in \mathbb{R}^n$, there is a unique $(y, z) \in Y \times Y^\perp$ such that $x = y + z$.[42]

[42] Another way of saying this is that $\mathbb{R}^n$ is the internal direct sum of Y and $Y^\perp$ for any linear subspace Y of $\mathbb{R}^n$ (Exercise F.36). This fact gives a perspective, yet again, on the sense in which $Y^\perp$ "complements" Y.

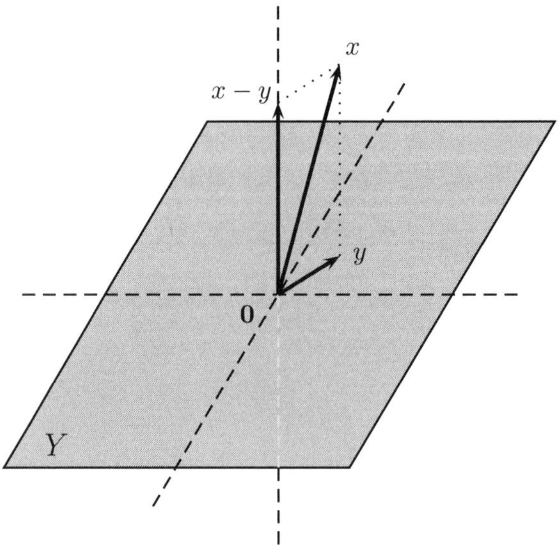

FIGURE 14

PROOF

For any given $x \in \mathbb{R}^n$, let y be the projection of x on S. Corollary 7 says that $x - y \in Y^\perp$, so letting $z := x - y$ yields the claim. (The proof of the uniqueness assertion is left as an easy exercise.) ∎

COROLLARY 8

For any linear subspace Y of $\mathbb{R}^n$, we have $Y^{\perp\perp} = Y$.

PROOF

That $Y \subseteq Y^{\perp\perp}$ follows from the definitions readily. (Check!) To prove the converse containment, take any $x \in Y^{\perp\perp}$ and use the Projection Theorem to find a $(y, z) \in Y \times Y^\perp$ such that $x = y + z$. Since $Y \subseteq Y^{\perp\perp}$, we have $y \in Y^{\perp\perp}$, and since $Y^{\perp\perp}$ is a linear subspace of $\mathbb{R}^n$, we have $z = x - y \in Y^{\perp\perp}$. So $z \in Y^\perp \cap Y^{\perp\perp}$, while, of course, $Y^\perp \cap Y^{\perp\perp} = \{\mathbf{0}\}$. Thus $z = \mathbf{0}$, and $x = y \in Y$. ∎

By way of a quick application, we use these results to derive a "representation" for the affine manifolds of a Euclidean space.

EXAMPLE 14
Take any $n \in \{2,3,\ldots\}$, and let Y be a k-dimensional linear subspace of $\mathbb{R}^n$, $k = 1,\ldots,n-1$. It follows from the Projection Theorem that $dim(Y^\perp) = n-k$. (Right?) Take any basis $\{x^1,\ldots,x^{n-k}\}$ for $Y^\perp$. Notice that, for any $y \in \mathbb{R}^n$, we have $y \perp Y^\perp$ iff $x^i y = 0$ for each $i = 1,\ldots,n-k$. Then, by Corollary 8, $Y = Y^{\perp\perp} = \{y \in \mathbb{R}^n : x^i y = 0 \text{ for each } i = 1,\ldots,n-k\}$. Defining $\mathbf{A} := \left[x^i_j\right]_{(n-k)\times n}$, therefore, we find that $y \in Y$ iff $\mathbf{A}y = \mathbf{0}$. (Notice that $\mathbf{A}$ is the matrix found by using x^is as the row vectors.) Conclusion: *For any k-dimensional proper linear subspace Y of $\mathbb{R}^n$, there exists a matrix $\mathbf{A} \in \mathbb{R}^{(n-k)\times n}$ such that*

$$Y = \{y \in \mathbb{R}^n : \mathbf{A}y = \mathbf{0}\}$$

and no row vector of A is the zero vector. By way of translation, we can deduce from this the following representation result for affine manifolds: *For any k-dimensional affine manifold S of $\mathbb{R}^n$, $k = 1,\ldots,n-1$, we have*

$$S = \{y \in \mathbb{R}^n : \mathbf{A}y = v\}$$

for some $\mathbf{A} \in \mathbb{R}^{(n-k)\times n}$ and $v \in \mathbb{R}^{n-k}$, where no row vector of A is the zero vector. (*Note.* If S is a hyperplane, this representation reduces to the one we found in Corollary F.4.) □

EXERCISE 72 Let S be a nonempty subset of $\mathbb{R}^n$ such that $x \perp y$ for every distinct $x, y \in S$. Show that if $S^\perp = \{\mathbf{0}\}$, then S is a basis for $\mathbb{R}^n$.

EXERCISE 73[H] Given any linear functional L on a linear subspace Y of $\mathbb{R}^n$, show that

$$dim\{\omega \in \mathbb{R}^n : L(y) = \omega y \text{ for all } y \in Y\} = n - dim(Y).$$

EXERCISE 74 Let Y and Z be two linear subspaces of $\mathbb{R}^n$. Prove that

$$(Y+Z)^\perp = Y^\perp \cap Z^\perp \quad \text{and} \quad (Y \cap Z)^\perp = Y^\perp + Z^\perp.$$

EXERCISE 75 Let X be a linear space and $P : X \to X$ a linear operator such that $P \circ P = P$. Show that $null(P) \cap P(X) = \{\mathbf{0}\}$ and $null(P) + P(X) = X$.

The following exercise shows that $\mathbb{R}^n$ can be orthogonally projected onto any one of its linear subspaces.

Exercise 76[H] Let Y be a linear subspace of $\mathbb{R}^n$. Show that there is a unique linear surjection $P : \mathbb{R}^n \rightarrow Y$ such that $P \circ P = P$ and $null(P) = Y^{\perp}$.

3.6 Extension of Positive Linear Functionals, Revisited

In Section 2.2 we encountered the problem of positively extending a given positive linear functional defined on a linear subspace Y of some ambient preordered linear space X. We saw there that a sufficient condition for one to be able to carry out such an extension is $Y \cap \text{al-int}_X(X_+) \neq \emptyset$ (the Krein-Rutman Theorem). If $X = \mathbb{R}^n$ (and we consider $\mathbb{R}^n$ to be endowed with its canonical partial order $\geq$), this condition reduces to $Y \cap \mathbb{R}^n_{++} \neq \emptyset$. That is, a blindfolded application of the Krein-Rutman Theorem assures us that if L is an increasing linear functional on a linear subspace Y of $\mathbb{R}^n$ and $Y \cap \mathbb{R}^n_{++} \neq \emptyset$, then there exists an increasing linear functional L^* on $\mathbb{R}^n$ with $L^*|_Y = L$. In fact, we can do better than this. Just as the "nonempty interior" condition can be relaxed in the Dieudonné Separation Theorem when the ambient space is $\mathbb{R}^n$, the condition $Y \cap \mathbb{R}^n_{++} \neq \emptyset$ can be omitted in the previous statement. As it happens, the proof of this fact draws on the entire development of the present section.

Proposition 8
Let Y be a nontrivial proper linear subspace of $\mathbb{R}^n$ and L an increasing linear functional on Y. Then there exists an increasing $L^* \in \mathcal{L}(\mathbb{R}^n, \mathbb{R})$ such that $L^*|_Y = L$, or equivalently, there exists $\alpha_1, \ldots, \alpha_n \geq 0$ such that $L(y) = \sum^n \alpha_i y_i$ for all $y \in Y$.

Proof[43]
Consider the following set:

$$S := \left\{\omega \in \mathbb{R}^n : L(y) = \omega y \text{ for all } y \in Y\right\}.$$

[43] My earlier proof of this result left a good deal to be desired. Fortunately, Ennio Stacchetti and Lin Zhou offered me two superior proofs. Ennio's proof was based on Farkas' Lemma, while Lin's proof made use of a brilliant separation argument. I present here the latter proof.

This set is obviously not empty (Example F.8). Moreover, it is an affine manifold of $\mathbb{R}^n$ with the following property:

$$S + Y^\perp \subseteq S. \tag{19}$$

(Yes?) Now, assume that $S \cap \mathbb{R}^n_+ = \emptyset$. We wish to show that, then, L cannot be increasing.

By Proposition 6, there is a nonzero $K \in \mathcal{L}(\mathbb{R}^n, \mathbb{R})$ such that $K(S) > 0 = \max K(\mathbb{R}^n_+)$. It follows that there is a $z \in \mathbb{R}^n_- \backslash \{\mathbf{0}\}$ such that $z\omega > 0$ for all $\omega \in S$.[44] By (19), therefore, $z\omega + z\omega' > 0$ for all $\omega \in S$ and $\omega' \in Y^\perp$. Since $Y^\perp$ is a nontrivial linear space (because $Y \neq \mathbb{R}^n$), this is possible only if $z\omega' = 0$ for all $\omega' \in Y^\perp$, that is, z is orthogonal to $Y^\perp$. By Corollary 8, then, $z \in Y$. So, choosing any ω in S, we may write $L(z) = \omega z > 0$. Since $z < \mathbf{0}$, it follows that L cannot be increasing. ■

EXERCISE 77 Show that if L is strictly increasing, then we can take L^* to be strictly increasing (and hence $\alpha_1, \ldots, \alpha_n > 0$) in the statement of Proposition 8.

44 Why is $z < \mathbf{0}$? Well, if, say, $z_1 > 0$, then for $x^m := \left(m, \frac{1}{m}, \ldots, \frac{1}{m}\right)$, $m = 1, 2, \ldots$, we would have $K(x^m) \to \infty$, which would contradict $0 = \max K(\mathbb{R}^n_+)$. Thus $z \leq \mathbf{0}$. Of course, $z \neq \mathbf{0}$, because K is nonzero.

Chapter H

Economic Applications

Even the limited extent of convex analysis we covered in Chapter G endows one with surprisingly powerful methods. Unfortunately, in practice, it is not always easy to recognize the situations in which these methods are applicable. To get a feeling for the sort of economic models in which convex analysis may turn out to provide the right mode of attack, one really needs a lot of practice. Our objective in this chapter is thus to present a smorgasbord of economic applications that illustrate the multifarious use of convex analysis in general, and the basic separation-by-hyperplane and linear-extension arguments in particular.

In our first application we revisit expected utility theory, but this time using preferences that are potentially incomplete. Our objective is to extend both the classic and the Anscombe-Aumann Expected Utility Theorems (Section F.3) into the realm of incomplete preferences, and to introduce the recently popular multiprior decision-making models. We then turn to welfare economics. In particular, we prove the Second Welfare Theorem, obtain a useful characterization of Pareto optima in pure distribution problems, and talk about Harsanyi's Utilitarianism Theorem. As an application to information theory, we provide a simple proof of the celebrated Blackwell's Theorem on comparing the value of information services, and, as an application to financial economics, we provide various formulations of the No-Arbitrage Theorem. Finally, in the context of cooperative game theory, we characterize the Nash bargaining solution and examine some basic applications to coalitional games without side payments. Although these applications are fairly diverse (and hence they can be read independently of each other), the methods with which they are studied here all stem from elementary convex analysis.

1 Applications to Expected Utility Theory

This section continues the investigation of expected utility theory we started in Section F.3. We adopt here the notation and definitions introduced in that section, so it may be a good idea to do a quick review of Section F.3 before commencing with the analysis provided below.

1.1 The Expected Multi-Utility Theorem

Let X be a nonempty finite set that is interpreted as a set of (monetary or nonmonetary) prizes/alternatives, and recall that a **lottery** (or a **probability distribution**) on X is a map $p \in \mathbb{R}^X_+$ such that $\sum_{x \in X} p(x) = 1$. As in Section F.3, we denote the set of all lotteries on X by $\mathfrak{L}_X$, which is a compact and convex subset of $\mathbb{R}^X$. The Expected Utility Theorem states that any complete preference relation $\succsim$ on $\mathfrak{L}_X$ that satisfies the Independence and Continuity Axioms (Section F.3.1) admits an expected utility representation. That is, for any such preference relation there is a utility function $u \in \mathbb{R}^X$ such that

$$p \succsim q \quad \text{if and only if} \quad \mathsf{E}_p(u) \geq \mathsf{E}_q(u) \tag{1}$$

for any $p, q \in \mathfrak{L}_X$.[1] Our goal here is to extend this result to the realm of *incomplete* preferences. The discussion presented in Section B.4 and a swift comparison of Propositions B.9 and B.10 suggest that our objective in this regard should be to obtain a multi-utility analogue of this theorem. And indeed, we have the following:

> THE EXPECTED MULTI-UTILITY THEOREM
> (Dubra-Maccheroni-Ok) *Let X be any nonempty finite set and $\succsim$ a preference relation on $\mathfrak{L}_X$. Then $\succsim$ satisfies the Independence and Continuity Axioms if, and only if, there exists a nonempty set $\mathcal{U} \subseteq \mathbb{R}^X$ (of utility functions) such that, for any $p, q \in \mathfrak{L}_X$,*
>
> $$p \succsim q \quad \textit{if and only if} \quad \mathsf{E}_p(u) \geq \mathsf{E}_q(u) \quad \textit{for all } u \in \mathcal{U}. \tag{2}$$

This result shows that one can think of an agent whose (possibly incomplete) preferences over lotteries in $\mathfrak{L}_X$ satisfy the Independence and

[1] *Reminder.* $\mathsf{E}_p(u) := \sum_{x \in X} p(x)u(x)$.

Continuity Axioms "as if" this agent distinguishes between the prizes in X with respect to multiple objectives (each objective being captured by one $u \in \mathcal{U}$). This agent prefers lottery p over q if the expected value of *each* of her objectives with respect to p is greater than those with respect to q. If such a domination does not take place, that is, p yields a strictly higher expectation with respect to some objective, and q with respect to some other objective, then the agent remains indecisive between p and q (and settles her choice problem by means we do not model here).

Before we move on to the proof of the Expected Multi-Utility Theorem, let us observe that the classic Expected Utility Theorem is an easy consequence of this result. All we need is the following elementary observation.

LEMMA 1
Let $n \in \mathbb{N}$, and take any $a, u \in \mathbb{R}^n$ such that $a \neq \mathbf{0}$, $\sum^n a_i = 0$, and $u_1 > \cdots > u_n$. If $\sum^n a_i u_i \geq 0$, then we must have $a_i > 0$ and $a_j < 0$ for some $i < j$.

PROOF
We must have $n \geq 2$ under the hypotheses of the assertion. Let $\alpha_i := u_i - u_{i+1}$ for each $i = 1, \ldots, n-1$. We have

$$\alpha_1 a_1 + \alpha_2(a_1 + a_2) + \cdots + \alpha_{n-1} \sum_{i=1}^{n-1} a_i + u_n \sum_{i=1}^{n} a_i = \sum_{i=1}^{n} a_i u_i.$$

So, by the hypotheses $\sum^n a_i = 0$ and $\sum^n a_i u_i \geq 0$, we have

$$\alpha_1 a_1 + \alpha_2(a_1 + a_2) + \cdots + \alpha_{n-1} \sum_{i=1}^{n-1} a_i \geq 0. \tag{3}$$

If the claim was false, there would exist a $k \in \{1, \ldots, n-1\}$ such that $a_1, \ldots, a_k \leq 0$ and $a_{k+1}, \ldots, a_n \geq 0$ with $\sum^k a_i < 0$. If $k = n-1$, this readily contradicts (3), so consider the case where $n \geq 3$ and $k \in \{1, \ldots, n-2\}$. In that case, (3) implies $\alpha_K \sum^K a_i > 0$ for some $K \in \{k+1, \ldots, n-1\}$, and we find $0 = \sum^n a_i \geq \sum^K a_i > 0$. ■

ANOTHER PROOF FOR THE EXPECTED UTILITY THEOREM
Let X be a nonempty finite set and $\succsim$ be a complete preorder on $\mathfrak{L}_X$ that satisfies the Independence and Continuity Axioms. Let us first assume that there is no indifference between the degenerate lotteries, that is, either $\delta_x \succ \delta_y$ or $\delta_y \succ \delta_x$ for any distinct $x, y \in X$.

By the Expected Multi-Utility Theorem, there exists a nonempty set $\mathcal{U} \subseteq \mathbb{R}^X$ such that (2) holds for any $p, q \in \mathfrak{L}_X$. Define $\mathcal{X} := \{(x, y) \in X^2 : \delta_x \succ \delta_y\}$, and for any $(x, y) \in \mathcal{X}$, let $u_{(x,y)}$ be an arbitrary member of $\mathcal{U}$ with $u_{(x,y)}(x) > u_{(x,y)}(y)$. (That there is such a $u_{(x,y)}$ follows from (2).) Define $u := \sum_{(x,y)\in\mathcal{X}} u_{(x,y)}$, and notice that $\delta_z \succ \delta_w$ iff $u(z) > u(w)$, for any $z, w \in X$. (Why?) In fact, u is a von Neumann–Morgenstern utility function for $\succsim$, as we show next.

Take any $p, q \in \mathfrak{L}_X$. Note first that $p \succsim q$ implies $\mathsf{E}_p(u) \geq \mathsf{E}_q(u)$ by (2), so, given that $\succsim$ is complete, if we can show that $p \succ q$ implies $\mathsf{E}_p(u) > \mathsf{E}_q(u)$, it will follow that (1) holds. To derive a contradiction, suppose $p \succ q$ but $\mathsf{E}_p(u) = \mathsf{E}_q(u)$. Then, by Lemma 1, we can find two prizes x and y in X such that $\delta_x \succ \delta_y$, $p(x) > q(x)$ and $p(y) < q(y)$. (Clearly, $q(y) > 0$.) For any $0 < \varepsilon \leq q(y)$, define the lottery $r_\varepsilon \in \mathfrak{L}_X$ as

$$r_\varepsilon(x) := q(x) + \varepsilon, \qquad r_\varepsilon(y) := q(y) - \varepsilon \qquad \text{and}$$
$$r_\varepsilon(z) := q(z) \text{ for all } z \in X\backslash\{x, y\}.$$

Since $\delta_x \succ \delta_y$, we have $u(x) > u(y)$, so $\mathsf{E}_p(u) = \mathsf{E}_q(u) < \mathsf{E}_{r_\varepsilon}(u)$ for any $0 < \varepsilon \leq q(y)$. Since $\succsim$ is complete, (2) then implies that $r_\varepsilon \succsim p$ for all $0 < \varepsilon \leq q(y)$. But $r_{\frac{1}{m}} \to q$, so this contradicts the Continuity Axiom, and we are done.

It remains to relax the assumption that there is no indifference between any two degenerate lotteries on X. Let Y be a $\supseteq$-maximal subset of X such that δ_x is not indifferent to δ_y for any distinct $x, y \in Y$. For any $p \in \mathfrak{L}_Y$, define $p' \in \mathfrak{L}_X$ with $p'|_Y = p$ and $p'(x) := 0$ for all $x \in X\backslash Y$. Define $\succsim'$ on $\mathfrak{L}_Y$ by $p \succsim' q$ iff $p' \succsim q'$. By what we have established above, there is a $u \in \mathbb{R}^Y$ such that $p \succsim' q$ iff $\mathsf{E}_p(u) \geq \mathsf{E}_q(u)$, for any $p, q \in \mathfrak{L}_Y$. We extend u to X in the obvious way by letting, for any $x \in X\backslash Y$, $u(x) := u(y_x)$, where y_x is any element of Y with $\delta_x \sim \delta_{y_x}$. It is easy to show (by using the Independence Axiom) that $u \in \mathbb{R}^X$ is a von Neumann–Morgenstern utility for $\succsim$. We leave establishing this final step as an exercise. ∎

The rest of the present subsection is devoted to proving the "only if" part of the Expected Multi-Utility Theorem. (The "if" part is straightforward.) The main argument is based on the external characterization of closed and convex sets in a Euclidean space and is contained in the following result, which is a bit more general than we need at present.

Proposition 1
Let Z be any nonempty finite set, $S \subseteq \mathbb{R}^Z$, and $Y := span(S)$. Assume that S is a compact and convex set with $\text{al-int}_Y(S) \neq \emptyset$, and $\succsim$ is a continuous affine preference relation on S. Then there exists a nonempty $\mathcal{L} \subseteq \mathcal{L}(\mathbb{R}^Z, \mathbb{R})$ such that

$$s \succsim t \quad \textit{if and only if} \quad L(s) \geq L(t) \quad \textit{for all } L \in \mathcal{L}, \tag{4}$$

for any $s, t \in S$. If $\succ \neq \emptyset$, then each $L \in \mathcal{L}$ can be taken here to be nonzero.

This is more than we need to establish the Expected Multi-Utility Theorem. Note first that if $\succsim$ in that theorem was degenerate in the sense that it declared all lotteries indifferent to each other, then we would be done by using a constant utility function. Let us then assume that $\succsim$ is not degenerate. Then, applying Proposition 1 with X and $\mathfrak{L}_X$ playing the roles of Z and S, respectively, we find a nonempty subset $\mathcal{L}$ of nonzero linear functionals on $\mathbb{R}^Z$ such that, for any $p, q \in \mathfrak{L}_X$,

$$p \succsim q \quad \text{iff} \quad L(p) \geq L(q) \text{ for all } L \in \mathcal{L}.$$

But by Example F.6, for each $L \in \mathcal{L}$ there exists a $u^L \in \mathbb{R}^X \backslash \{\mathbf{0}\}$ such that $L(\sigma) = \sum_{x \in X} \sigma(x) u^L(x) = \mathsf{E}_\sigma(u^L)$ for all $\sigma \in \mathbb{R}^X$. Letting $\mathcal{U} := \{u^L : L \in \mathcal{L}\}$, therefore, completes the proof of the Expected Multi-Utility Theorem.

It remains to prove Proposition 1. This is exactly where convex analysis turns out to be the essential tool of analysis.

Proof of Proposition 1
Define $A := \{s - t : s, t \in S \text{ and } s \succsim t\}$, and check that A is a convex subset of $\mathbb{R}^Z$. We define next

$$C := \bigcup \{\lambda A : \lambda \geq 0\}$$

which is easily verified to be a convex cone in $\mathbb{R}^Z$.[2] The following claim, which is a restatement of Lemma F.2, shows why this cone is important for us.

Claim 1. For any $s, t \in S$, we have $s \succsim t$ iff $s - t \in C$.

[2] C is the positive cone of a preordered linear space. Which space is this?

Now comes the hardest step in the proof.

CLAIM 2. C is a closed subset of $\mathbb{R}^Z$.[3]

The plan of attack is forming! Note first that if $\succ = \emptyset$, then choosing $\mathcal{L}$ as consisting only of the zero functional completes the proof. Assume then that $\succ \neq \emptyset$, which ensures that C is a proper subset of $\mathbb{R}^Z$. (Why?) Thus, since we now know that C is a closed convex cone in $\mathbb{R}^Z$, we may apply Corollary G.6 to conclude that C must be the intersection of the closed half-spaces that contain it and that are defined by its supporting hyperplanes. (Where did the second part come from?) Let $\mathcal{L}$ denote the set of all nonzero linear functionals on $\mathbb{R}^Z$ that correspond to these hyperplanes (Corollary F.4). Observe that if $L \in \mathcal{L}$, then there must exist a real number α such that $\{r \in \mathbb{R}^Z : L(r) = \alpha\}$ supports C. Since this hyperplane contains some point of C, we have $L(\sigma) \geq \alpha$ for all $\sigma \in C$ and $L(\sigma^*) = \alpha$ for some $\sigma^* \in C$. But C is a cone, so $\frac{\alpha}{2} = \frac{1}{2}L(\sigma^*) = L(\frac{1}{2}\sigma^*) \geq \alpha$ and $2\alpha = 2L(\sigma^*) = L(2\sigma^*) \geq \alpha$, which is possible only if $\alpha = 0$. Consequently, we have $\sigma \in C$ iff $L(\sigma) \geq 0$ for all $L \in \mathcal{L}$. By Claim 1, therefore, we have (4) for any $s, t \in S$.

It remains to prove Claim 2, which requires some care. Here goes the argument.[4] Take any $(\lambda_m) \in \mathbb{R}_+^\infty$, and $(s^m), (t^m) \in S^\infty$ such that

$$\sigma := \lim \lambda_m(s^m - t^m) \in Y \quad \text{and} \quad s^m \succsim t^m \quad \text{for all } m = 1, 2, \ldots.$$

We wish to show that $\sigma \in C$. Of course, if $s^m = t^m$ for infinitely many m, then we would trivially have $\sigma \in C$, so it is without loss of generality to let $s^m \neq t^m$ for each m.

Now pick any $s^* \in \textit{al-int}_Y(S)$. Since S is convex, $\textit{al-int}_Y(S)$ equals the interior of S in Y, so there exists an $\varepsilon > 0$ such that $N_{\varepsilon,Y}(s^*) \subseteq S$.[5] Take any $0 < \delta < \varepsilon$, and define

$$T := \{r \in S : r \succsim s^* \text{ and } d_2(r, s^*) \geq \delta\}.$$

[3] *Reminder.* $\mathbb{R}^Z$ is just the Euclidean space $\mathbb{R}^{|Z|}$, thanks to the finiteness of Z. For instance, for any $s, t \in \mathbb{R}^Z$ and $\varepsilon > 0$, the distance between s and t is $d_2(s,t) = \left(\sum_{z\in Z} |s(z) - t(z)|^2\right)^{\frac{1}{2}}$, and the ε-neighborhood of $s \in S$ *in* Y is $N_{\varepsilon,Y}(s) = \{r \in Y : d_2(s,r) < \varepsilon\}$.

[4] The proof I used in an earlier version of this book was somewhat clumsy. The elegant argument given below is due to Juan Dubra.

[5] *Note.* It is in this step that we invoke the finiteness of Z. (Recall Example G.5.)

Since $\succsim$ is continuous, T is closed subset of S. (Yes?) Since S is compact, therefore, T is compact as well.

Now let $d_m := d_2(s^m, t^m)$, and notice that $d_m > 0$ and

$$\lambda_m \left(s^m - t^m\right) = \frac{\lambda_m d_m}{\delta} \frac{\delta}{d_m} \left(s^m - t^m\right), \qquad m = 1, 2, \ldots$$

So, letting $\gamma_m := \frac{\lambda_m}{\delta} d_m$ and $r^m := s^* + \frac{\delta}{d_m}\left(s^m - t^m\right)$, we get

$$\lambda_m \left(s^m - t^m\right) = \gamma_m \left(r^m - s^*\right), \qquad m = 1, 2, \ldots \tag{5}$$

It is easily verified that $d_2(r^m, s^*) = \delta < \varepsilon$, so $r^m \in N_{\varepsilon,Y}(s^*)$ for each m. Since $N_{\varepsilon,Y}(s^*) \subseteq S$, therefore, we have $r^m \in S$ for each m. Moreover, $r^m - s^* \in C$, so by Claim 1, $r^m \succsim s^*$. It follows that $r^m \in T$ for each m. But since $\lambda_m(s^m - t^m) \to \sigma$, we have $\gamma_m(r^m - s^*) \to \sigma$. Since $d_2(r^m, s^*) \geq \delta$ for all m, therefore, (γ_m) must be a bounded sequence. (Why?)

We now know that there exist subsequences of (γ_m) and (r^m) that converge in $\mathbb{R}_+$ and T, say to γ and r, respectively. (Why?) Since $(\lambda_m(s^m - t^m))$ converges to σ, (5) implies that $\sigma = \gamma(r - s^*)$, so $\sigma \in C$. ■

EXERCISE 1[H] (Throughout this exercise we use the notation adopted in Proposition 1 and its proof.) We say that $\succsim$ is **weakly continuous** if $\{\alpha \in [0,1] : \alpha s + (1-\alpha)t \succsim \alpha s' + (1-\alpha)t'\}$ is a closed set for any $s, s', t, t' \in S$. In the statement of Proposition 1, we may replace the word "continuous" with "weakly continuous." To prove this, all you need is to verify that the cone C defined in the proof of Proposition 1 remains closed in $\mathbb{R}^Z$ with weak continuity. Assume that $\succ \neq \emptyset$ (otherwise Proposition 1 is vacuous), and proceed as follows.

(a) Show that $s^* - t^* \in \textit{al-int}_Y(A)$ for some $s^*, t^* \in \textit{al-int}_Y(S)$.

(b) Prove: If $\sigma \in \textit{al-cl}(C)$, then $\sigma = \lambda(s - t)$ for some $\lambda \geq 0$ and $s, t \in S$.

(c) Using the Claim proved in Example G.10, conclude that $(1-\lambda)(s^* - t^*) + \lambda(s - t) \in C$ for all $0 \leq \lambda < 1$ and $s, t \in S$. Finally, use this fact and weak continuity to show that C is algebraically closed. By Observation G.3, C is thus a closed subset of $\mathbb{R}^Z$.

EXERCISE 2 (Aumann) Let X be a nonempty finite set. Prove: If the preference relation $\succsim$ on $\mathfrak{L}_X$ satisfies the Independence and Continuity Axioms, then there exists a $u \in \mathbb{R}^X$ such that, for any $p, q \in \mathfrak{L}_X$,

$$p \succ q \quad \text{implies} \quad \mathsf{E}_p(u) > \mathsf{E}_q(u) \quad \text{and}$$

$$p \sim q \quad \text{implies} \quad \mathsf{E}_p(u) = \mathsf{E}_q(u).$$

Exercise 3 Let X be a nonempty finite set. Show that if the subsets $\mathcal{U}$ and $\mathcal{V}$ of $\mathbb{R}^X$ represent a preference relation $\succsim$ as in the Expected Multi-Utility Theorem, then $\mathcal{V}$ must belong to the closure of the convex cone generated by $\mathcal{U}$ and all constant functions on X.

1.2* Knightian Uncertainty

Let us now turn to expected utility theory under uncertainty, and recall the Anscombe-Aumann framework (Section F.3.2). Here Ω stands for a nonempty finite set of states, and X that of prizes. A horse race lottery is any map from Ω into $\mathfrak{L}_X$—we denote the set of all horse race lotteries by $\mathfrak{H}_{\Omega,X}$.[6] In the Anscombe-Aumann setup, the preference relation $\succsim$ of an individual is defined on $\mathfrak{H}_{\Omega,X}$. If this preference relation is complete and satisfies the Independence, Continuity, and Monotonicity Axioms* (Section F.3.2), then there exist a utility function $u \in \mathbb{R}^X$ and a probability distribution $\mu \in \mathfrak{L}_\Omega$ such that

$$f \succsim g \quad \text{if and only if} \quad \sum_{\omega \in \Omega} \mu(\omega)\mathsf{E}_{f_\omega}(u) \geq \sum_{\omega \in \Omega} \mu(\omega)\mathsf{E}_{g_\omega}(u) \tag{6}$$

for any $f, g \in \mathfrak{H}_{\Omega,X}$.

We now ask the following question: How would this result modify if $\succsim$ was not known to be complete? A natural conjecture in this regard is that $\succsim$ would then admit an expected multi-utility representation with multiple prior beliefs, that is, there would exist a nonempty set $\mathcal{U} \subseteq \mathbb{R}^X$ of utility functions and a nonempty set $\mathcal{M} \subseteq \mathcal{L}_\Omega$ of prior beliefs such that $f \succsim g$ iff

$$\sum_{\omega \in \Omega} \mu(\omega)\mathsf{E}_{f_\omega}(u) \geq \sum_{\omega \in \Omega} \mu(\omega)\mathsf{E}_{g_\omega}(u) \quad \text{for all } (\mu, u) \in \mathcal{M} \times \mathcal{U}$$

for any $f, g \in \mathfrak{H}_{\Omega,X}$. Unfortunately, to the best of the knowledge of this author, whether this conjecture is true or not is not known at present. What is known is that if the incompleteness of $\succsim$ stems *only* from one's inability to

[6] *Reminder.* For any $h \in \mathfrak{H}_{\Omega,X}$ and $\omega \in \Omega$, we write h_ω for $h(\omega)$, that is, $h_\omega(x)$ is the probability of getting prize x in state ω.

compare the horse race lotteries that differ across states, then the conjecture is true (with $\mathcal{U}$ being a singleton).

To make things precise, let us recall that a preference relation $\succsim$ on $\mathfrak{H}_{\Omega,X}$ induces a preference relation $\succsim^*$ on $\mathfrak{L}_X$ in the following manner:

$$p \succsim^* q \quad \text{if and only if} \quad \langle p \rangle \succsim \langle q \rangle,$$

where $\langle p \rangle$ stands for the constant horse race lottery that equals p at every state, and similarly for $\langle q \rangle$. Obviously, if $\succsim$ is complete, so is $\succsim^*$. The converse is, however, false. Indeed, the property that $\succsim$ is complete *enough* to ensure the completeness of $\succsim^*$ is much weaker than assuming outright that $\succsim$ is complete. It is this former property that the Knightian uncertainty theory is built on.

THE PARTIAL COMPLETENESS AXIOM*. The (induced) preference relation $\succsim^*$ is complete, and $\succ^* \neq \emptyset$.

When combined with the Monotonicity Axiom*, this property makes sure that, given any $h \in \mathcal{H}_{\Omega,X}$,

$$(p, h_{-\omega}) \succsim (q, h_{-\omega}) \quad \text{if and only if} \quad (p, h_{-\omega'}) \succsim (q, h_{-\omega'}) \tag{7}$$

for any lotteries $p, q \in \mathfrak{L}_X$ and any two states $\omega, \omega' \in \Omega$.[7] (Why?) Thus, if an individual cannot compare two horse race lotteries f and g, then this is because they differ in at least two states. For a theory that wishes to "blame" one's indecisiveness on uncertainty, and not on risk, the Partial Completeness Axiom* is thus quite appealing.[8]

In 1986 Truman Bewley proved the following extension of the Anscombe-Aumann Theorem for incomplete preferences on $\mathfrak{H}_{\Omega,X}$ that satisfy the Partial Completeness Axiom.[9]

[7] *Reminder.* For any $h \in \mathfrak{H}_{\Omega,X}$ and $r \in \mathfrak{L}_X$, we denote by $(r, h_{-\omega})$ the horse race lottery that yields the lottery $r \in \mathfrak{L}_X$ in state ω and agrees with h in all other states, that is, $(r, h_{-\omega})_\tau := \begin{cases} r, & \text{if } \tau = \omega \\ h_\tau, & \text{otherwise.} \end{cases}$

[8] I can't pass this point without pointing out that I fail to see why one should be expected to have complete preferences over lotteries, but not on acts. Although many authors in the field seem to take this position, it seems to me that any justification for worrying about incomplete preferences over acts would also apply to the case of lotteries.

[9] Although the importance of Bewley's related work was recognized widely, his original papers remained as working papers for a long time (mainly by Bewley's own choice). The first of the three papers that contain his seminal analysis appeared in print only in 2002; the rest of his papers remain unpublished.

Bewley's Expected Utility Theorem

Let Ω and X be any nonempty finite sets, and $\succsim$ a preference relation on $\mathfrak{H}_{\Omega,X}$. Then $\succsim$ satisfies the Independence, Continuity, Monotonicity, and Partial Completeness Axioms if, and only if, there exist a (utility function) $u \in \mathbb{R}^X$ and a nonempty set $\mathcal{M} \subseteq \mathfrak{L}_\Omega$ such that, for any $f, g \in \mathfrak{H}_{\Omega,X}$,*

$$f \succsim g \quad \textit{if and only if} \quad \sum_{\omega\in\Omega} \mu(\omega)\mathsf{E}_{f_\omega}(u) \geq \sum_{\omega\in\Omega} \mu(\omega)\mathsf{E}_{g_\omega}(u) \quad \textit{for all} \quad \mu \in \mathcal{M}.$$

An individual whose preference relation over horse race lotteries satisfies the axioms of Bewley's Expected Utility Theorem holds initial beliefs about the true state of the world, but her beliefs are *imprecise* in the sense that she does not hold one but *many* beliefs. In ranking two horse race lotteries, she computes the expected utility of each horse race lottery using each of her prior beliefs (and hence attaches to every act a multitude of expected utilities). If f yields higher expected utility than g for every prior belief that the agent holds, then she prefers f over g. If f yields strictly higher expected utility than g for some prior belief, and the opposite holds for another, then she remains indecisive as to the ranking of f and g.

This model is called the **Knightian uncertainty model**, and has recently been applied in various economic contexts ranging from financial economics to contract theory and political economy.[10] This is not the place to get into these matters at length, but let us note that all of these applications are based on certain behavioral assumptions about how an agent would make her choices when she cannot compare some (undominated) feasible acts, and hence introduces a different (behavioral) dimension to the associated decision analysis.[11]

[10] The choice of terminology is due to Bewley. But I should say that reading Frank Knight's 1921 treatise did not clarify for me why he chose this terminology. Because it is widely used in the literature, I will stick with it. As for applications of the theory, see, for instance, Billot et al. (2000), Dow and Werlang (1992), Epstein and Wang (1994), Mukerji (1998), Rigotti and Shannon (2005), and Ghirardato and Katz (2006).

[11] For instance, Bewley's original work presupposes that, when there is a status quo act in the choice problem of an agent, then she would stick to her status quo if she could not find a better feasible alternative according to her *incomplete* preferences. An axiomatic foundation for this behavioral postulate is recently provided by Masatlioglu and Ok (2005).

We now turn to the proof of Bewley's Expected Utility Theorem. The structure of this result is reminiscent of that of the Expected Multi-Utility Theorem, so you may sense that convex analysis (by way of Proposition 1) could be of use here. This is exactly the case.

Proof of Bewley's Expected Utility Theorem
We only need to prove the "only if" part of the assertion. Assume that $\succsim$ satisfies the Independence, Continuity, Monotonicity, and Partial Completeness Axioms*. By the Expected Utility Theorem, there exists a $u \in \mathbb{R}^X$ such that $p \succsim^* q$ iff $\mathsf{E}_p(u) \geq \mathsf{E}_q(u)$, for any $p, q \in \mathfrak{L}_X$. Let $T := \{\mathsf{E}_p(u) : p \in \mathfrak{L}_X\}$ and $\mathcal{S} := T^\Omega$. The proof of the following claim is relatively easy; we leave it as an exercise.

Claim 1. $\mathcal{S}$ is a compact and convex subset of $\mathbb{R}^\Omega$, and $\textit{al-int}_{\textit{span}(\mathcal{S})}(\mathcal{S}) \neq \emptyset$.

Now define the binary relation $\trianglerighteq$ on $\mathcal{S}$ by

$$F \trianglerighteq G \quad \text{if and only if} \quad f \succsim g$$

for any $f, g \in \mathfrak{H}_{\Omega,X}$ such that $F(\omega) = \mathsf{E}_{f_\omega}(u)$ and $G(\omega) = \mathsf{E}_{g_\omega}(u)$ for all $\omega \in \Omega$.

Claim 2. $\trianglerighteq$ is well-defined. Moreover, $\trianglerighteq$ is affine and continuous.

Proof of Claim 2. We will only prove the first assertion, leaving the proofs of the remaining two as easy exercises. And for this, it is clearly enough to show that, for any $f, f' \in \mathfrak{H}_{\Omega,X}$, we have $f \sim f'$ whenever $\mathsf{E}_{f_\omega}(u) = \mathsf{E}_{f'_\omega}(u)$ for all $\omega \in \Omega$. (Is it?) Take any $h \in \mathfrak{H}_{\Omega,X}$. By the Monotonicity Axiom*, if $\mathsf{E}_{f_\omega}(u) = \mathsf{E}_{f'_\omega}(u)$ for all $\omega \in \Omega$, then $(f_\omega, h_{-\omega}) \sim (f'_\omega, h_{-\omega})$ for all $\omega \in \Omega$. Thus, by the Independence Axiom* (applied $|\Omega| - 1$ many times),

$$\begin{aligned}
\tfrac{1}{|\Omega|}f + \left(1 - \tfrac{1}{|\Omega|}\right)h &= \sum_{\omega\in\Omega} \tfrac{1}{|\Omega|}(f_\omega, h_{-\omega}) \\
&\sim \sum_{\omega\in\Omega} \tfrac{1}{|\Omega|}(f'_\omega, h_{-\omega}) \\
&= \tfrac{1}{|\Omega|}f' + \left(1 - \tfrac{1}{|\Omega|}\right)h,
\end{aligned}$$

so, applying the Independence Axiom* one more time, we get $f \sim f'$. ‖

By the Partial Comparability Axiom* we have $\succ \neq \emptyset$, which implies that the strict part of $\trianglerighteq$ is nonempty as well. Then, Claims 1 and 2 show that we may apply Proposition 1 (with Ω playing the role of Z) to find a nonempty subset $\mathcal{L}$ of nonzero linear functionals on $\mathbb{R}^Z$ such that, for any $F, G \in \mathcal{S}$,

$$F \trianglerighteq G \quad \text{iff} \quad L(F) \geq L(G) \quad \text{for all} \quad L \in \mathcal{L}.$$

Clearly, there exists a $\sigma^L \in \mathbb{R}^\Omega \backslash \{\mathbf{0}\}$ such that $L(F) = \sum_{\omega \in \Omega} \sigma^L(\omega) F(\omega)$ for all $F \in \mathbb{R}^\Omega$ (Example F.6). So,

$$F \trianglerighteq G \quad \text{iff} \quad \sum_{\omega \in \Omega} \sigma^L(\omega) F(\omega) \geq \sum_{\omega \in \Omega} \sigma^L(\omega) G(\omega) \text{ for all } L \in \mathcal{L}, \tag{8}$$

for any $F, G \in \mathcal{S}$. By the Partial Comparability Axiom* there exist $p, q \in \mathfrak{L}_X$ such that $p \succ^* q$ so that $\mathsf{E}_p(u) > \mathsf{E}_q(u)$. But, by the Monotonicity Axiom*, $(p, h_{-\omega}) \succsim (q, h_{-\omega})$ holds for all $h \in \mathfrak{H}_{\Omega,X}$ and $\omega \in \Omega$. By definition of $\trianglerighteq$ and (8), therefore, we have $\sigma^L(\omega)(\mathsf{E}_p(u) - \mathsf{E}_q(u)) \geq 0$ for each $L \in \mathcal{L}$. It follows that $\sigma^L \in \mathbb{R}^\Omega_+ \backslash \{\mathbf{0}\}$ for each L, so letting $\alpha_L := \sum_{\omega \in \Omega} \sigma^L(\omega)$, and defining $\mathcal{M} := \{\frac{1}{\alpha_L} \sigma^L : L \in \mathcal{L}\}$ completes the proof. ■

Exercise 4 Derive the Anscombe-Aumann Expected Utility Theorem from Bewley's Expected Utility Theorem.

Exercise 5 State and prove a uniqueness theorem to supplement Bewley's Expected Utility Theorem (along the lines of the uniqueness part of the Anscombe-Aumann Expected Utility Theorem).

**Exercise 6 (*An open problem*) Determine how Bewley's Expected Utility Theorem would modify if we dropped the Partial Completeness Axiom* from its statement.

1.3* The Gilboa-Schmeidler Multi-Prior Model

In recent years the theory of individual decision making underwent a considerable transformation, because the descriptive power of its most foundational model, the one captured by the Anscombe-Aumann Expected Utility Theorem, was found to leave something to be desired. The most striking empirical observation that has led to this view stemmed from the 1961 experiments of Daniel Ellsberg, which we discuss next.

Consider two urns, each containing one hundred balls. The color of any one of these balls is known to be either red or black. Nothing is known about the distribution of the balls in the first urn, but it is known that exactly fifty balls in the second urn are red. One ball is drawn from each urn at random. Consider the following bets:

> Bet i_α: If the ball drawn from the ith urn is α, then you win \$10, nothing otherwise,

where $i \in \{1, 2\}$ and $\alpha \in \{\text{red, black}\}$. Most people who participate in the experiments declare that they are indifferent between the bets 1_{red} and 1_{black}, and between the bets 2_{red} and 2_{black}. Given the symmetry of the situation, this is exactly what one would expect. But how about comparing 1_{red} versus 2_{red}? The answer is surprising, although you may not think so at first. An overwhelming majority of the subjects declare that they *strictly* prefer 2_{red} over 1_{red}. (We know that their preferences are strict, because they in fact choose "2_{red} + a small fee" over "1_{red} + no fee.") It seems the fact that they do not know the ratio of black to red balls in the first urn—the so-called *ambiguity* of this urn—bothers the agents. This is a serious problem for the model envisaged by the Anscombe-Aumann Expected Utility Theorem. Indeed, no individual whose preferences can be modeled as in that result can behave this way! (For this reason, this situation is commonly called the **Ellsberg Paradox.**)

Let us translate the story at hand to the Anscombe-Aumann setup. Foremost we need to specify a state space and an outcome space. Owing to the simplicity of the situation, the choice is clear:

$$\Omega := \{0, 1, \ldots, 100\} \quad \text{and} \quad X := \{0, 10\}.$$

Here by state $\omega \in \Omega$ we mean the state in which exactly ω of the balls in the first urn are red. In turn, the outcome space X contains all possible payments an agent may receive through the bets under consideration. Given these specifications, then, the bet 1_{red} is modeled by the horse race lottery $f : \Omega \to \mathfrak{L}_X$ defined by

$$f(\omega) = \tfrac{\omega}{100}\delta_{10} + \tfrac{100-\omega}{100}\delta_0,$$

while the bet 2_{red} is modeled by the horse race lottery $f' : \Omega \to \mathfrak{L}_X$ defined by

$$f'(\omega) = \tfrac{1}{2}\delta_{10} + \tfrac{1}{2}\delta_0.$$

(Notice that f' involves risk, but not uncertainty—this is the upshot of the Ellsberg Paradox.) The horse race lotteries that correspond to the bets 1_{black} and 2_{black} are modeled similarly; we denote them by g and g', respectively.

In this formulation, the data of the Ellsberg Paradox tell us that the preference relation $\succsim$ of a decision maker on $\{f, g, f', g'\}$ may well exhibit the following structure: $f \sim g$ and $f' \succ f$. Now suppose $\succsim$ can be represented by means of a prior $\mu \in \mathcal{L}_\Omega$ and a utility function $u \in \mathbb{R}^{\{0,10\}}$, as in the Anscombe-Aumann Expected Utility Theorem. Of course, without loss of generality, we may assume that $u(0) = 0$. If $u(10) = 0$ as well, then we get $f' \sim f$ in contrast to $f' \succ f$. So, assume instead $u(10) \neq 0$. Then, $f \sim g$ (i.e., the indifference between the bets 1_{red} and 1_{black}) means

$$\sum_{\omega \in \Omega} \mu(\omega) \tfrac{\omega}{100} u(10) = \sum_{\omega \in \Omega} \mu(\omega) \tfrac{100-\omega}{100} u(10),$$

which is equivalent to $\sum_{\omega \in \Omega} \mu(\omega) \frac{\omega}{100} = \frac{1}{2}$. But the latter equation means $f' \sim f$, contradicting $f' \succ f$.

But why? What is the problem? It is the Independence Axiom*, of course. Notice that f' is nothing but a very simple mixing of f and g. Indeed, $f' = \frac{1}{2}f + \frac{1}{2}g$. Consequently, if $\succsim$ satisfies the Independence Axiom* and $f \sim g$ is true, then $f' \sim f$ must hold perforce; we cannot possibly have $f' \succ f$. The Ellsberg Paradox seems to tell us that, to increase the descriptive power of our model, we should break free from the straightjacket of the Independence Axiom* at least so as to allow for the following property in a nontrivial way.

THE UNCERTAINTY AVERSION AXIOM*[12]. For any $f, g \in \mathfrak{H}_{\Omega,X}$,

$$f \sim g \quad \text{implies} \quad \tfrac{1}{2}f + \tfrac{1}{2}g \succsim f.$$

In 1989, Itzhak Gilboa and David Schmeidler offered a brilliant weakening of the Independence Axiom* that goes along with this property very nicely. Since then this weakened version of the Independence Axiom*, called the C-Independence Axiom* ("C" for "constant"), became the industry standard.

THE C-INDEPENDENCE AXIOM*. For any $f, g \in \mathfrak{H}_{\Omega,X}$, $p \in \mathfrak{L}_X$, and any $0 < \lambda \leq 1$,

$$f \succsim g \quad \text{if and only if} \quad \lambda f + (1-\lambda)\langle p \rangle \succsim \lambda g + (1-\lambda)\langle p \rangle.$$

[12] The name of the axiom stems from the idea that the "uncertainty" involved in f and g is converted to "risk" in $\frac{1}{2}f + \frac{1}{2}g$, thereby making the latter (possibly) more valuable than both f and g.

The interpretation of this property is identical to that of the usual Independence Axiom*, but now we demand the independence property only with respect to mixing with *constant* horse race lotteries. Naturally, this is bound to increase the descriptive power of the decision-making model at hand. Shortly we shall see how it does this in exact terms.

Let us first observe that we are not sailing too far away from the Anscombe-Aumann model.

EXERCISE 7 Let Ω and X be two nonempty finite sets, and take any complete preference relation $\succsim$ on $\mathfrak{H}_{\Omega,X}$ that satisfies the C-Independence, Continuity, and Monotonicity Axioms*. Prove: If

$$f \sim g \quad \text{implies} \quad \tfrac{1}{2}f + \tfrac{1}{2}g \sim f$$

for any $f, g \in \mathfrak{H}_{\Omega,X}$, then there exists a $(\mu, u) \in \mathfrak{L}_\Omega \times \mathbb{R}^X$ such that $\succsim$ is represented by the map $h \mapsto \sum_{\omega\in\Omega} \mu(\omega)\mathsf{E}_{h_\omega}(u)$.

What next? We have weakened the axiomatic system of the Anscombe-Aumann theory to be able to cope with the Ellsberg Paradox, but so what, you may say, we could do so simply by omitting the Independence Axiom* altogether. An individual decision theory is useful only insofar as it gives us an operational representation of the decision making process of the agent. If we assume that the agent is endowed with a complete preference relation $\succsim$ and chooses from any given feasible set the horse race lottery (or act) that is a $\succsim$-maximum in that set, the "use" of such a theory lies in its entirety in the structure of representation it provides for $\succsim$. And it is exactly at this point that the Gilboa-Schmeidler theory shines bright. It turns out that replacing the Independence Axiom* with the C-Independence and Uncertainty Aversion Axioms* leads us to a beautiful utility representation for the involved preferences.

THE GILBOA-SCHMEIDLER THEOREM

Let Ω and X be any nonempty finite sets, and $\succsim$ a complete preference relation on $\mathfrak{H}_{\Omega,X}$. Then $\succsim$ satisfies the C-Independence, Continuity, Monotonicity, and Uncertainty Aversion Axioms if, and only if, there exist a (utility function) $u \in \mathbb{R}^X$ and a nonempty convex set $\mathcal{M} \subseteq \mathfrak{L}_\Omega$ such that, for any $f, g \in \mathfrak{H}_{\Omega,X}$,*

$$f \succsim g \quad \textit{if and only if} \quad \min_{\mu\in\mathcal{M}} \sum_{\omega\in\Omega} \mu(\omega)\mathsf{E}_{f_\omega}(u) \geq \min_{\mu\in\mathcal{M}} \sum_{\omega\in\Omega} \mu(\omega)\mathsf{E}_{g_\omega}(u).$$

Interpretation is easy. A decision maker whose preferences over horse race lotteries abide by the Gilboa-Schmeidler axioms may be viewed "as if" she holds imprecise beliefs about the true state of the world in that, as in Bewley's model, she possesses multiple priors. In ranking two horse race lotteries, she computes the expected utility of each horse race lottery using each of her prior beliefs, and then, *relative to her beliefs*, chooses the lottery that yields the highest of the *worst* possible expected utility. In the literature, this model, which is indeed an interesting way of *completing* Bewley's model, is often referred to as the **maxmin expected utility model with multiple priors.**[13]

To give a quick example, let us turn back to the Ellsberg Paradox that we considered at the beginning of this subsection and see how the Gilboa-Schmeidler theory fares with that. To this end, let us adopt the notation of that example, and given any nonempty $\mathcal{M} \subseteq \mathfrak{L}_\Omega$, let us define the map $U_\mathcal{M} : \mathfrak{H}_{\Omega,X} \to \mathbb{R}$ by

$$U_\mathcal{M}(h) := \min \left\{ \sum_{\omega \in \Omega} \mu(\omega)\mathsf{E}_{h_\omega}(u) : \mu \in \mathcal{M} \right\}$$

where $u \in \mathbb{R}^{\{0,10\}}$ satisfies $0 = u(0) < u(10) = 1$. Assume that $\succsim$ is the preference relation on $\mathfrak{H}_{\Omega,X}$ that is represented by $U_\mathcal{M}$. We have seen earlier that if $|\mathcal{M}| = 1$, then we cannot possibly have $f \sim g$ and $f' \succ f$. (If $|\mathcal{M}| = 1$, then $\succsim$ is represented as in the Anscombe-Aumann Theorem.) But what if $\mathcal{M} = \mathfrak{L}_\Omega$? Then we have $U_\mathcal{M}(f) = 0 = U_\mathcal{M}(g)$ and $U_\mathcal{M}(f') = \frac{1}{2} = U_\mathcal{M}(g')$, so we have $f \sim g$ and $f' \sim g'$ while $f' \succ f$ and $g' \succ g$, in full concert with the Ellsberg Paradox. In fact, the same result would obtain with much smaller sets of beliefs as well. For instance, if $\mathcal{M} := \{\mu \in \mathcal{L}_\Omega : \mu(\omega) > 0$ only if $\omega \in \{49, 50, 51\}\}$, then we have $U_\mathcal{M}(f) = \frac{49}{50} = U_\mathcal{M}(g)$ and $U_\mathcal{M}(f') = \frac{1}{2} = U_\mathcal{M}(g')$, so we remain consistent with the data of the Ellsberg Paradox.

The literature on the applications of the Gilboa-Schmeidler model is too large to be recounted here. Moreover, there are now many generalizations

[13] Many economists are critical of this model because it models a decision maker as "too pessimistic." But nothing is said about the nature of the set of beliefs in the theorem. Depending on the application, this set may be taken to comprise only optimistic beliefs so an agent who acts pessimistically relative to her (optimistic) beliefs may end up behaving not at all in a way that a pessimistic person would behave. (It should be noted that the choice of a particular set of beliefs in an application is a complicated matter that is best discussed within the specifics of that application.)

and variants of the maxmin expected utility model.[14] But going deeper into individual decision theory will get us off course here. We thus stop our present treatment here and conclude the section with the proof of the Gilboa-Schmeidler Theorem. This is a "deep" result, but one that is not all that hard to establish once one realizes that convex analysis lies at the very heart of it.

We begin with introducing the following auxiliary concept.

DEFINITION

Let Ω be a nonempty set. If φ is a real map on $\mathbb{R}^\Omega$ such that

$$\varphi(F + \alpha \mathbf{1}_\Omega) = \varphi(F) + \varphi(\alpha \mathbf{1}_\Omega) \qquad \text{for any } F \in \mathbb{R}^\Omega \text{ and } \alpha \in \mathbb{R},$$

then we say that φ is ***C*-additive**.[15]

The following result is the main step toward the proof of the Gilboa-Schmeidler Theorem.

LEMMA 2

Let Ω be a nonempty set, and $\varphi : \mathbb{R}^\Omega \to \mathbb{R}$ an increasing, superlinear, and C-additive map. Then there exists a nonempty convex subset $\mathcal{L}$ of positive linear functionals on $\mathbb{R}^\Omega$ such that

$$\varphi(F) = \min\{L(F) : L \in \mathcal{L}\} \qquad \text{for all } F \in \mathbb{R}^\Omega, \tag{9}$$

and

$$L(\mathbf{1}_\Omega) = \varphi(\mathbf{1}_\Omega) \qquad \text{for all } L \in \mathcal{L}. \tag{10}$$

PROOF

Thanks to the superlinearity of φ, Corollary G.5 ensures that, for every $F \in \mathbb{R}^\Omega$, there is a linear functional L_F on $\mathbb{R}^\Omega$ such that $\varphi \le L_F$ and $\varphi(F) = L_F(F)$. Moreover, by C-additivity of φ, we have

$$\varphi(F - \mathbf{1}_\Omega) + \varphi(\mathbf{1}_\Omega) = \varphi(F) = L_F(F) = L_F(F - \mathbf{1}_\Omega) + L_F(\mathbf{1}_\Omega)$$

[14] Particularly noteworthy, in my opinion, are Schmeidler (1989), Ghirardato and Marinacci (2001), and Maccheroni, Marinacci, and Rustichini (2005).

[15] *Reminder.* $\mathbf{1}_\Omega$ is the real function on Ω that equals 1 everywhere.

for any $F \in \mathbb{R}^\Omega$. Since $L_F(F - \mathbf{1}_\Omega) \geq \varphi(F - \mathbf{1}_\Omega)$ and $L_F(\mathbf{1}_\Omega) \geq \varphi(\mathbf{1}_\Omega)$, we must then have $L_F(F - \mathbf{1}_\Omega) = \varphi(F - \mathbf{1}_\Omega)$ and, more to the point, $L_F(\mathbf{1}_\Omega) = \varphi(\mathbf{1}_\Omega)$, for any $F \in \mathbb{R}^\Omega$. Now let $\mathcal{L}$ be the convex hull of $\{L_F : F \in \mathbb{R}^\Omega\}$. By construction, $\mathcal{L}$ satisfies (9) and (10). Furthermore, because φ is superlinear, we have $\varphi(\mathbf{0}) = 0$, so by monotonicity of φ, $F \geq \mathbf{0}$ implies $L(F) \geq \varphi(F) \geq \varphi(\mathbf{0}) = 0$ for all $L \in \mathcal{L}$. That is, each $L \in \mathcal{L}$ is positive. ■

Since we know the general representation of a linear functional on a Euclidean space (Example F.6), the following is an immediate consequence of Lemma 2.

Corollary 1
Let Ω be a nonempty finite set. If $\varphi : \mathbb{R}^\Omega \to \mathbb{R}$ is an increasing, superlinear, and C-additive map with $\varphi(\mathbf{1}_\Omega) = 1$, then there exists a nonempty convex subset $\mathcal{M}$ of $\mathfrak{L}_\Omega$ such that

$$\varphi(F) = \min\left\{\sum_{\omega\in\Omega} \mu(\omega)F(\omega) : \mu \in \mathcal{M}\right\} \qquad \text{for all } F \in \mathbb{R}^\Omega.$$

We are now ready for our main course.

Proof of the Gilboa-Schmeidler Theorem
We only need to prove the "only if" part of the assertion, and for this we may assume $\succ \neq \emptyset$, for otherwise the claim is trivial. Suppose $\succsim$ satisfies the C-Independence, Continuity, Monotonicity, and Uncertainty Aversion Axioms*. Then, the preference relation $\succsim^*$ on $\mathfrak{L}_X$ (induced by $\succsim$) is complete, and satisfies the Continuity and Independence Axioms. (Yes?) Therefore, by the Expected Utility Theorem, there exists an affine map L on $\mathfrak{L}_X$ that represents $\succsim^*$. Since X is finite, there exist (degenerate) lotteries p^* and p_* such that $p^* \succsim^* p \succsim^* p_*$ for all $p \in \mathfrak{L}_X$. (Why?) By the Monotonicity Axiom, therefore,

$$\langle p^*\rangle \succsim h \succsim \langle p_*\rangle \qquad \text{for all } h \in \mathfrak{H}_{\Omega,X}.$$

In what follows we assume that $L(p^*) = 1$ and $L(p_*) = -1$.[16]

[16] Given that $\succ \neq \emptyset$, we must have $p^* \succ^* p_*$. (Why?) So, since a von Neumann–Morgenstern utility function is unique up to strictly increasing affine transformations, the choice of these numbers is without loss of generality.

CLAIM 1. For any $h \in \mathfrak{H}_{\Omega,X}$, there is a unique $0 \leq \alpha_h \leq 1$ such that

$$h \sim \alpha_h \langle p^* \rangle + (1 - \alpha_h) \langle p_* \rangle .$$

PROOF OF CLAIM 1

Define $\alpha_h := \inf\{\alpha \in [0,1] : \alpha \langle p^* \rangle + (1-\alpha)\langle p_* \rangle \succsim h\}$, and use the C-Independence and Continuity Axioms to verify that α_h is equal to the task. ‖

CLAIM 2. There exists a unique $U : \mathfrak{H}_{\Omega,X} \to \mathbb{R}$ that represents $\succsim$ and satisfies

$$U(\langle p \rangle) = L(p) \qquad \text{for all } p \in \mathfrak{L}_X. \tag{11}$$

PROOF OF CLAIM 2

First define the real map ϕ on $\{\langle p \rangle : p \in \mathfrak{L}_X\}$ by $\phi(\langle p \rangle) = L(p)$, and then define U on $\mathfrak{H}_{\Omega,X}$ by $U(h) := \phi\left(\alpha_h \langle p^* \rangle + (1-\alpha_h)\langle p_* \rangle\right)$, where α_h is as found in Claim 1. Obviously, U represents $\succsim$ and satisfies (11). Besides, if V was another such function, we would have, for any $h \in \mathfrak{H}_{\Omega,X}$,

$$U(h) = U(\langle p_h \rangle) = L(p_h) = V(\langle p_h \rangle) = V(h),$$

where $p_h = \alpha_h p^* + (1 - \alpha_h)p_*$. ‖

CLAIM 3. There exists an increasing, superlinear, and C-additive real map φ on $\mathbb{R}^\Omega$ such that

$$\varphi(L \circ h) = U(h) \qquad \text{for all } h \in \mathfrak{H}_{\Omega,X}. \tag{12}$$

Let us suppose for the moment that Claim 3 is true. Then $U(\langle p^* \rangle) = L(p^*) = 1$, so by (12), we find $\varphi(\mathbf{1}_\Omega) = \varphi(L \circ \langle p^* \rangle) = U(\langle p^* \rangle) = 1$.[17] Therefore, combining Claims 2 and 3 with Corollary 1 completes the proof of the Gilboa-Schmeidler Theorem.[18]

[17] Try not to get confused with the notation. $L \circ \langle p^* \rangle$ is the function that maps any given $\omega \in \Omega$ to $L(p^*) = 1$, so $L \circ \langle p^* \rangle = \mathbf{1}_\Omega$.

[18] I hope you see the big picture here. We know that the Expected Utility Theorem applies to $\succsim^*$, so we may represent $\succsim^*$ with an affine function L. But any horse race lottery has an equivalent in the space of all constant horse race lotteries. This allows us to find a U on $\mathfrak{H}_{\Omega,X}$ that represents $\succsim$ and agrees with L on all constant lotteries. Now take any $h \in \mathfrak{H}_{\Omega,X}$, and replace $h(\omega)$ (which is a lottery on $\mathfrak{L}_X$) with $L(h(\omega))$ (which is a number)—you get $L \circ h$. It is easy to see that there is a map φ on the set of all $L \circ h$s such that $\varphi(L \circ h) \geq \varphi(L \circ g)$ iff $f \succsim g$; this is exactly what (12) says. The problem

It remains to prove Claim 3. This is not really difficult, but we still have to break a sweat for it. Begin by noting that $\{L \circ h : h \in \mathfrak{H}_{\Omega,X}\} = [-1,1]^\Omega$.[19] We may thus define $\psi : [-1,1]^\Omega \to \mathbb{R}$ as

$$\psi(L \circ h) := U(h) \qquad \text{for all } h \in \mathfrak{H}_{\Omega,X}.$$

Claim 4. $\psi(\lambda F) = \lambda\psi(F)$ for any $(F,\lambda) \in [-1,1]^\Omega \times \mathbb{R}_{++}$ such that $\lambda F \in [-1,1]^\Omega$.

Proof of Claim 4

Take any $F \in [-1,1]^\Omega$ and $0 < \lambda \leq 1$. Choose any $f \in \mathfrak{H}_{\Omega,X}$ such that $L \circ f = F$. We wish to show that $\psi(\lambda(L \circ f)) = \lambda\psi(L \circ f)$. To this end, take any $p \in \mathfrak{L}_X$ with $L(p) = 0$, and let $g := \lambda f + (1-\lambda)\langle p\rangle$. Let's compute $\psi(L \circ g)$ in two different ways. First, we have

$$L \circ g = \lambda\left(L \circ f\right) + (1-\lambda)\left(L \circ \langle p\rangle\right) = \lambda\left(L \circ f\right),$$

so $\psi(L \circ g) = \psi(\lambda\left(L \circ f\right)) = \psi(\lambda F)$. Second, we find a $q \in \mathfrak{L}_X$ with $\langle q\rangle \sim f$ (Claim 1) and use the C-Independence Axiom* to find $g \sim \lambda\langle q\rangle + (1-\lambda)\langle p\rangle$. Then, by Claim 2,

$$U(g) = U(\lambda\langle q\rangle + (1-\lambda)\langle p\rangle) = L(\lambda q + (1-\lambda)p) = \lambda L(q) = \lambda U(\langle q\rangle) = \lambda U(f),$$

so $\psi(L \circ g) = \lambda U(f) = \lambda\psi(L \circ f) = \lambda\psi(F)$. Conclusion: $\psi(\lambda F) = \lambda\psi(F)$ for any $F \in [-1,1]^\Omega$ and $0 < \lambda \leq 1$.

To conclude the proof of Claim 4, take any $F \in [-1,1]^\Omega$ and $\lambda > 1$ with $\lambda F \in [-1,1]^\Omega$. By what we have just established, $\psi(F) = \psi(\frac{1}{\lambda}\lambda F) = \frac{1}{\lambda}\psi(\lambda F)$, and we are done. ‖

---

reduces, then, to determining the structure of the function φ. If we had the full power of the Independence Axiom*, we would find that φ is increasing and affine — this is exactly what we did when proving the Anscombe-Aumann Theorem in Section F.3.2. Here all we got is C-Independence, so we will only be able to show that φ (actually its (unique) positive homogeneous extension to $\mathbb{R}^\Omega$) is increasing and C-additive. Adding the Uncertainty Aversion Axiom* to the picture will then give us the superlinearity of φ. But, thanks to Corollary 1, we know how to represent such functions!

19 The $\subseteq$ part of this equation is obvious. To see the $\supseteq$ part, take any $F \in [-1,1]^\Omega$, and notice that, for every $\omega \in \Omega$, there is a $p_\omega \in \mathfrak{L}_X$ such that $L(p_\omega) = F(\omega)$ by the Intermediate Value Theorem. If we define $h \in \mathfrak{H}_{\Omega,X}$ by $h(\omega) := p_\omega$, therefore, we find $L \circ h = F$.

We now extend ψ to the entire $\mathbb{R}^\Omega$ *by positive homogeneity*. That is, we define the real map φ on $\mathbb{R}^\Omega$ by

$$\varphi(F) := \tfrac{1}{\lambda}\psi(\lambda F) \qquad \text{for } any\ \lambda > 0 \text{ with } \lambda F \in [-1,1]^\Omega.$$

First, note that φ is well-defined. (Prove! *Hint.* *If* $\beta > \alpha > 0$ are such that both αF and βF belong to $[-1,1]^\Omega$, then, by Claim 4, $\frac{1}{\alpha}\psi(\alpha F) = \frac{1}{\alpha}\psi(\frac{\alpha}{\beta}\beta F) = \frac{1}{\beta}\psi(\beta F)$.) Moreover, φ is positively homogeneous. Indeed, for any $F \in \mathbb{R}^\Omega$ and $\alpha > 0$, by picking any $\lambda > 0$ such that both $\lambda\alpha F$ and λF belong to $[-1,1]^\Omega$ — since Ω is finite, such a λ obviously exists—we find

$$\varphi(\alpha F) = \tfrac{1}{\lambda}\psi(\lambda\alpha F) = \tfrac{1}{\lambda}\alpha\psi(\lambda F) = \alpha\varphi(F),$$

where we used Claim 4 to get the second equality here.

We are half way through proving Claim 3. We are now in possession of a positively homogeneous real map φ on $\mathbb{R}^\Omega$ that satisfies (12) and $\varphi(\mathbf{1}_\Omega) = 1$.[20] We next prove that φ is C-additive. Since φ is positively homogeneous, it is enough to show only that $\varphi(F+\alpha\mathbf{1}_\Omega) = \varphi(F)+\alpha\varphi(\mathbf{1}_\Omega)$ for all $F \in [-1,1]^\Omega$ and $0 < \alpha \leq 1$.[21] Of course, $L \circ f = F$ for some $f \in \mathfrak{H}_{\Omega,X}$, and $L(p) = \alpha$ for some $p \in \mathfrak{L}_X$. Again, find a $q \in \mathfrak{L}_X$ such that $f \sim \langle q\rangle$, so by the C-Independence Axiom*, $\frac{1}{2}f+\frac{1}{2}\langle p\rangle \sim \frac{1}{2}\langle q\rangle+\frac{1}{2}\langle p\rangle$. Then $U\left(\frac{1}{2}f + \frac{1}{2}\langle p\rangle\right) = U\left(\frac{1}{2}\langle q\rangle + \frac{1}{2}\langle p\rangle\right)$. But, by (12) and positive homogeneity of φ,

$$2U\left(\tfrac{1}{2}f + \tfrac{1}{2}\langle p\rangle\right) = 2\varphi(L\circ(\tfrac{1}{2}f + \tfrac{1}{2}\langle p\rangle)) = \varphi(L\circ f + \alpha\mathbf{1}_\Omega) = \varphi(F+\alpha\mathbf{1}_\Omega)$$

while, since $\langle q\rangle \sim f$ implies $L(q) = U(\langle q\rangle) = U(f)$ by Claim 2, we have

$$2U\left(\tfrac{1}{2}\langle q\rangle + \tfrac{1}{2}\langle p\rangle\right) = L(q)+L(p) = U(f)+\alpha = \varphi(L\circ f)+\alpha = \varphi(F)+\alpha.$$

It follows that

$$\varphi(F+\alpha\mathbf{1}_\Omega) = \varphi(F)+\alpha.$$

Since $\varphi(\mathbf{1}_\Omega) = 1$, we may conclude that φ is C-additive.

[20] How do I know that φ satisfies (12)? Easy! ψ is defined through (12), and φ agrees with ψ on $[-1,1]^\Omega = \{L\circ h : h \in \mathfrak{H}_{\Omega,X}\}$. (As we have seen earlier, (12), in turn, implies $\varphi(\mathbf{1}_\Omega) = 1$.)

[21] Suppose we can do this. Then for any $G \in \mathbb{R}^\Omega$ and $\beta > 0$, we take any $0 < \lambda \leq \frac{1}{\beta}$ with $\lambda G \in [-1,1]^\Omega$, and use the positive homogeneity of φ to get

$$\varphi(G+\beta\mathbf{1}_\Omega) = \tfrac{1}{\lambda}\varphi(\lambda G+\lambda\beta\mathbf{1}_\Omega) = \tfrac{1}{\lambda}\left(\varphi(\lambda G)+\lambda\beta\varphi(\mathbf{1}_\Omega)\right) = \varphi(G)+\beta\varphi(\mathbf{1}_\Omega).$$

Notice that we did not use the Uncertainty Aversion Axiom* yet. We will use it now in order to show that φ is superadditive, that is,

$$\varphi(F+G) \geq \varphi(F) + \varphi(G) \tag{13}$$

for any $F, G \in \mathbb{R}^\Omega$. (Since φ is positively homogeneous, this property entails that φ is superlinear.) Once again, thanks to positive homogeneity, it is enough to establish (12) for (arbitrarily chosen) F and G in $[-1,1]^\Omega$. Pick any $f, g \in \mathfrak{H}_{\Omega,X}$ such that $L \circ f = F$ and $L \circ g = G$.

Let us first consider the case $\varphi(F) = \varphi(G)$. In this case $U(f) = U(g)$, so we have $f \sim g$. By the Uncertainty Aversion Axiom*, then, $\frac{1}{2}f + \frac{1}{2}g \succsim f$, so, by (12),

$$\begin{aligned}
\varphi(\tfrac{1}{2}F + \tfrac{1}{2}G) &= \varphi\left(L \circ \left(\tfrac{1}{2}f + \tfrac{1}{2}g\right)\right) \\
&= U(\tfrac{1}{2}f + \tfrac{1}{2}g) \\
&\geq U(f) \\
&= \varphi(F) \\
&= \tfrac{1}{2}\varphi(F) + \tfrac{1}{2}\varphi(G),
\end{aligned}$$

and (13) follows by positive homogeneity of φ.

Finally, consider the case $\varphi(F) \neq \varphi(G)$, say $\varphi(F) > \varphi(G)$. Let $\alpha := \varphi(F) - \varphi(G)$, and define $H := G + \alpha\mathbf{1}_\Omega$. Notice that C-additivity of φ and $\varphi(\mathbf{1}_\Omega) = 1$ entail $\varphi(H) = \varphi(G) + \alpha = \varphi(F)$. So, by what we have shown in the previous paragraph,

$$\varphi(F + G + \alpha\mathbf{1}_\Omega) = \varphi(F+H) \geq \varphi(F) + \varphi(H) = \varphi(F) + \varphi(G) + \alpha.$$

By using C-additivity of φ and the fact that $\varphi(\mathbf{1}_\Omega) = 1$ we then find (13) again. The proof is now complete. ■

EXERCISE 8^{H} (*Uniqueness of the Gilboa-Schmeidler Representation*) Let Ω and X be two nonempty finite sets, and $\succsim$ a complete preference relation on $\mathfrak{H}_{\Omega,X}$ that satisfies the No-Triviality Axiom* (Section F.3.2). Assume that both of the following maps on $\mathfrak{H}_{\Omega,X}$ represent $\succsim$:

$$h \mapsto \min_{\mu \in \mathcal{M}} \sum_{\omega\in\Omega} \mu(\omega)\mathsf{E}_{h_\omega}(u) \qquad \text{and} \qquad h \mapsto \min_{\nu \in \mathcal{N}} \sum_{\omega\in\Omega} \nu(\omega)\mathsf{E}_{h_\omega}(v),$$

where $u, v \in \mathbb{R}^X$ and $\mathcal{M}$ and $\mathcal{N}$ are nonempty, closed, and convex subsets of $\mathfrak{L}_\Omega$. Prove that $\mathcal{M} = \mathcal{N}$ and v is a strictly increasing affine transformation of u.

EXERCISE 9[H] (Nehring) Let Ω and X be two nonempty finite sets, and $\succsim$ a complete preference relation on $\mathfrak{H}_{\Omega,X}$ that satisfies the C-Independence, Continuity, and Monotonicity Axioms*. Define the binary relation $\trianglerighteq$ on $\mathfrak{H}_{\Omega,X}$ by:

$$f \trianglerighteq g \quad \text{iff} \quad \lambda f + (1-\lambda)h \succsim \lambda g + (1-\lambda)h$$
$$\text{for all} \quad h \in \mathfrak{H}_{\Omega,X} \quad \text{and} \quad 0 < \lambda \leq 1.$$

(When $f \trianglerighteq g$, one says that f is **unambiguously preferred** to g.)

(a) Interpret $\trianglerighteq$.

(b) Show that, for any $f, g \in \mathfrak{H}_{\Omega,X}$ and $p, q \in \mathfrak{L}_X$,

$$f \trianglerighteq g \quad \text{implies} \quad f \succsim g, \quad \text{and} \quad \langle p \rangle \trianglerighteq \langle q \rangle \quad \text{iff} \quad \langle p \rangle \succsim \langle q \rangle .$$

(c) Show that $\trianglerighteq$ is a preorder on $\mathfrak{H}_{\Omega,X}$ that satisfies the Independence Axiom*.

(d) Show that there exists a nonempty convex subset $\mathcal{M}$ of $\mathfrak{L}_\Omega$ and a $u \in \mathbb{R}^X$ such that, for any $f, g \in \mathfrak{H}_{\Omega,X}$,

$$f \trianglerighteq g \quad \text{iff} \quad \sum_{\omega\in\Omega} \mu(\omega)\mathsf{E}_{f_\omega}(u) \geq \sum_{\omega\in\Omega} \mu(\omega)\mathsf{E}_{g_\omega}(u) \quad \text{for all} \quad \mu \in \mathcal{M}.$$

How does this result fare with the interpretation you gave in part (a)?

EXERCISE 10 (*Ambiguity Aversion*) Let Ω and X be two nonempty finite sets, and $\succsim$ a preference relation on $\mathfrak{H}_{\Omega,X}$. We say that $\succsim$ is **ambiguity averse** if there exists a preference relation $\succsim'$ on $\mathfrak{H}_{\Omega,X}$ that admits a representation as in the Anscombe-Aumann Expected Utility Theorem, and

$$\langle p \rangle \succsim' f \quad \text{implies} \quad \langle p \rangle \succsim f, \qquad \text{and} \qquad \langle p \rangle \succ' f \quad \text{implies} \quad \langle p \rangle \succ f$$

for any $p \in \mathfrak{L}_X$ and $f \in \mathfrak{H}_{\Omega,X}$. (Interpretation?) Prove: If $\succsim$ satisfies the C-Independence, Continuity, Monotonicity, and Uncertainty Aversion Axioms*, then $\succsim$ is ambiguity averse.

Exercise 11[H] (*Variational Preferences*) Let Ω and X be two nonempty finite sets, and $\succsim$ a preference relation on $\mathfrak{H}_{\Omega,X}$. We say that $\succsim$ is **variational** if it is represented by a map $V : \mathcal{A} \to \mathbb{R}$ with

$$V(f) = \min\left\{\sum_{\omega\in\Omega} \mu(\omega)L(f(\omega)) + c(\mu) : \mu \in \mathfrak{L}_\Omega\right\}$$

where $L : \mathfrak{L}_X \to \mathbb{R}$ is a nonconstant affine map, and $c : \mathfrak{L}_\Omega \to [0,\infty]$ is a lower semicontinuous convex function with $\inf c(\mathfrak{L}_\Omega) = 0$.[22]

(a) Prove or disprove: If $\succsim$ admits a representation as in the Gilboa-Schmeidler Theorem, then it must be variational.

(b) Prove or disprove: If $\succsim$ is variational, then $\succsim$ satisfies the C-Independence, Continuity, and Monotonicity Axioms*.

(c) Prove or disprove: If $\succsim$ is variational, then it is ambiguity averse.

Exercise 12 (*Pessimistic Preferences over Sets of Lotteries*) Let X be a nonempty finite set, and denote by $\mathbf{c}(\mathfrak{L}_X)$ the set of all nonempty closed subsets of $\mathfrak{L}_X$. We think of $\mathbf{c}(\mathfrak{L}_X)$ as a metric space under the Hausdorff metric (Section E.2.5). Prove that $\succsim$ is a continuous and complete preference relation on $\mathbf{c}(\mathfrak{L}_X)$ such that, for any $P, Q, R \in \mathbf{c}(\mathfrak{L}_X)$,

(i) $P \subseteq Q$ implies $P \succsim Q$;

(ii) $P \succsim R$ and $Q \succsim R$ and $P \cap Q = \emptyset$ imply $P \cup Q \succsim R$,

if, and only if, there exists a continuous $U : \mathfrak{L}_X \to \mathbb{R}$ such that

$$P \succsim Q \quad \text{iff} \quad \min\{U(p) : p \in P\} \geq \min\{U(p) : p \in Q\}$$

for any $P, Q \in \mathbf{c}(\mathfrak{L}_X)$. (Interpretation?)

2 Applications to Welfare Economics

2.1 The Second Fundamental Theorem of Welfare Economics

Throughout this application, m and l stand for fixed natural numbers with $m \geq 2$. An m-person, l-commodity **exchange economy** is formally defined as the list

$$\mathcal{E} := (\{\omega^i, u_i\}_{i=1,\ldots,m}),$$

[22] The theory of variational preferences is developed thoroughly in Maccheroni, Marinacci, and Rustichini (2005).

where $\omega^i \in \mathbb{R}^l_+$ and $u_i : \mathbb{R}^l_+ \to \mathbb{R}$ stand for the endowment vector and the utility function of agent i, respectively. The following assumptions are often imposed on these primitives:

ASSUMPTION (A1). $\omega^i \gg \mathbf{0}$ for all $i = 1, \ldots, m$.

ASSUMPTION (A2). u_i is continuous and strictly increasing for all $i = 1, \ldots, m$.

ASSUMPTION (A3). u_i is strictly quasiconcave for all $i = 1, \ldots, m$.

An **allocation** in an exchange economy $\mathcal{E}$ is defined as any vector $x = (x^1, \ldots, x^m) \in \mathbb{R}^{lm}_+$ such that $\sum^m x^i = \sum^m \omega^i$, where, $x^i \in \mathbb{R}^l_+$ denotes the commodity bundle allocated to agent i. An allocation x in $\mathcal{E}$ thus assigns a commodity bundle to all agents such that these bundles are *feasible* in the aggregate. It is said to be **weakly Pareto optimal** if, for no other allocation y in $\mathcal{E}$, we have

$$u_i(x^i) < u_i(y^i) \qquad \text{for all } i = 1, \ldots, m,$$

and **strongly Pareto optimal** if, for no other allocation y in $\mathcal{E}$, we have

$$\begin{aligned} u_i(x^i) &\le u_i(y^i) \quad \text{for all} \quad i = 1, \ldots, m \quad \text{and} \\ u_i(x^i) &< u_i(y^i) \quad \text{for some} \quad i = 1, \ldots, m. \end{aligned}$$

Clearly, these are fundamental *efficiency* properties. Once an allocation fails to satisfy either of them, one can improve upon this allocation at no welfare cost to the society.

Let us take the set of all *admissible prices* as $\mathbb{R}^l_{++}$ and denote a generic *price vector* by $p \in \mathbb{R}^l_{++}$. Given an exchange economy $\mathcal{E}$, we define the **demand correspondence** of the ith agent on $\mathbb{R}^l_{++} \times \mathbb{R}^l_+$ as

$$\mathbf{d}_i(p, \omega^i) := \arg\max\{u_i(x^i) : x^i \in \mathbb{R}^l_+ \text{ and } px^i \le p\omega^i\}$$

(Example E.4). That is, if $x^i \in \mathbf{d}_i(p, \omega^i)$, we understand that x^i is one of the most desired bundles for person i among all consumption bundles that she could afford given the price vector p. Of course, if (A2) and (A3) hold, then $\mathbf{d}_i$ can be considered as a continuous function (Example E.4), a convention that we adopt below. A **competitive equilibrium** for $\mathcal{E}$ is defined as any $(p, x) \in \mathbb{R}^l_{++} \times \mathbb{R}^{lm}_{+}$ such that

$$x^i \in \mathbf{d}_i(p, \omega^i) \quad \text{for all} \quad i = 1, \ldots, m \quad \text{and} \quad \sum_{i=1}^m x^i = \sum_{i=1}^m \omega^i.$$

If (A2) and (A3) hold, we may, and will, identify a competitive equilibrium by a price vector $p \in \mathbb{R}^l_{++}$ such that $\sum^m \mathbf{d}_i(p, \omega^i) = \sum^m \omega^i$. The idea behind the notion of competitive equilibrium is straightforward. At a price vector where a competitive equilibrium is sustained, all individuals achieve the best possible consumption plan for themselves, and therefore, there is reason to believe that an equilibrium will not be altered once it is established.

The first fundamental finding of general equilibrium theory is that an exchange economy that satisfies (A2) and (A3) has an equilibrium.[23] More relevant for the discussion here is the fact that any equilibrium is weakly Pareto optimal, and under (A2), it is strongly Pareto optimal. (It is highly likely that you know all this, but why don't you supply proofs for the last two claims, to warm up?) This fact, which is a major formal argument in favor of the free market economy, is called the *First Fundamental Theorem of Welfare Economics*. There is also a second fundamental theorem, which we shall examine here formally.

The Second Fundamental Theorem of Welfare Economics
(Arrow) *Consider an exchange economy* $(\{\omega^i, u_i\}_{i=1,\ldots,m})$ *where* (A1)–(A3) *hold. For any strongly Pareto optimal allocation* $x_* \in \mathbb{R}^{lm}_{++}$ *in this economy, there exists a competitive equilibrium* $p \in \mathbb{R}^l_{++}$ *such that* $x^i_* = \mathbf{d}_i(p, x^i_*)$, $i = 1, \ldots, m$.

So, there is a closer link between Pareto optimality and competitive equilibrium than first meets the eye. Not only is a competitive equilibrium strongly Pareto optimal (under (A2)), but conversely, every strongly Pareto optimal allocation can be sustained as a competitive equilibrium (under

[23] This is a special case of the 1954 equilibrium existence theorems of Kenneth Arrow, Gerard Debreu, and Lionel McKenzie. (All of these results are based on the Brouwer Fixed Point Theorem.) Both Arrow and Debreu have received the Nobel Prize in economics for their contributions to general equilibrium theory. (Arrow shared the prize with Sir John Hicks in 1972, and Debreu received it in 1983.)

(A1)–(A3)). This is an important finding whose significance is much debated in general equilibrium theory.[24]

What is important for us here is the fact that the proof of the second fundamental theorem provides a brilliant illustration of the power of convex analysis as we have sketched it in Chapter G. Here it is.

Fix any Pareto optimal allocation $x_* \in \mathbb{R}^{lm}_{++}$. Define $S_i := \{z \in \mathbb{R}^l_+ : u_i(z) > u_i(x^i_*)\}$ for each $i = 1, \ldots, m$, and let $S := S_1 + \cdots + S_m$. Then S is a nonempty convex subset of $\mathbb{R}^l$, thanks to (A2) and (A3). (Verify!) By continuity of u_is, moreover, it is open, and hence algebraically open (Observation G.2). Observe that, since x_* is strongly Pareto optimal, we have $\sum^m x^i_* \notin S$. By Corollary G.3, therefore, there exists a $p \in \mathbb{R}^l$ such that

$$py > p\sum_{i=1}^{m} x^i_* \qquad \text{for all } y \in S.$$

(Yes?) Moreover, since each u_i is strictly increasing, $\mathbf{e}^k + \sum^m x^i_* \in S$ (where $\mathbf{e}^k$ is the kth unit vector in $\mathbb{R}^l$, $k = 1, \ldots, l$). Therefore, the above inequality yields $p_k > 0$ for each k. To complete the proof, then, it is enough to show that $x^i_* = \mathbf{d}_i(p, x^i_*)$, $i = 1, \ldots, m$. Suppose that this is not true, that is, there exists an agent, say the individual 1, and a commodity bundle $y^1 \in \mathbb{R}^l_+$ such that $py^1 \leq px^1_*$ and $u_1(y^1) > u_1(x^1_*)$. Then, due to the continuity of u_1, we may assume without loss of generality that $py^1 < px^1_*$ and $u_1(y^1) > u_1(x^1_*)$. (Yes?) Define $\theta := \frac{1}{m-1}(px^1_* - py^1)$, let

$$y^i \in \arg\max\{u_i(x^i) : x^i \in \mathbb{R}^l_+ \text{ and } px^i \leq px^i_* + \theta\}, \qquad i = 2, \ldots, m,$$

and observe that, by (A2), $u_i(y^i) > u_i(x^i_*)$ and $py^i = px^i_* + \theta$ for all $i = 2, \ldots, m$. But then $\sum^m y^i \in S$ so, by the choice of p, we must have $\sum^m py^i > \sum^m px^i_*$ whereas

$$p\sum_{i=1}^{m} y^i = py^1 + \sum_{i=2}^{m} px^i_* + (m-1)\theta = p\sum_{i=1}^{m} x^i_*.$$

This contradiction completes the proof.

[24] The basic implication of this result is truly far reaching: If a planner has determined a particular Pareto optimal allocation as a target for an economy that satisfies (A1)–(A3), then all she has to do is provide each individual with a certain endowment, and leave the rest to the competitive market mechanism. But of course, there are many caveats. See Mas-Colell, Whinston, and Green (1995, Chap. 10), for a thorough discussion.

EXERCISE 13 Show that the l-vector p found in the above argument is unique.

EXERCISE 14 Would the Second Fundamental Theorem of Welfare Economics remain valid if we knew only that $x_* \in \mathbb{R}^{lm}_+$?

EXERCISE 15 Consider an exchange economy $\mathcal{E} = (\{\omega^i, u_i\}_{i=1,...,m})$ where (A1)–(A3) hold. We say that $(p^*, x_*) \in \mathbb{R}^l_{++} \times \mathbb{R}^{lm}_+$ is a **quasi-equilibrium** for $\mathcal{E}$ if $\sum^m p^* x^i_* = \sum^m p^* \omega^i$ and

$$u_i(x_i) \geq u_i(x^i_*) \quad \text{implies} \quad p^* x^i_* \geq p^* \omega^i$$

for each i. Prove that, for any weakly Pareto optimal allocation $x_* \in \mathbb{R}^{lm}_{++}$ in $\mathcal{E}$, there exists a $p^* \in \mathbb{R}^l_{++}$ such that (p^*, x_*) is a quasi-equilibrium for $(\{x^i_*, u_i\}_{i=1,...,m})$.

2.2 Characterization of Pareto Optima

Fix again arbitrary natural numbers m and l, with $m \geq 2$. One of the most basic models of welfare economics considers an m-person society that faces the problem of allocating an l-dimensional pie among its members. The pie corresponds to the feasible set of allocations in the economy, and is formally represented by a nonempty, closed and convex subset X of $\mathbb{R}^{lm}$. The preferences of individual i (over the commodity bundles) is represented by a utility function $u_i : \mathbb{R}^l \to \mathbb{R}$. Let E stand for the $(m+1)$-tuple $(X, u_1, \ldots, u_m)$, which we call a **distribution problem.** If each u_i is continuous, concave, and strictly increasing, then we refer to E as a **regular distribution problem.**

The **utility possibility set** in a distribution problem $E = (X, u_1, \ldots, u_m)$ is defined as

$$\mathcal{U}_E := \{(u_1(x^1), \ldots, u_m(x^m)) : x \in X\}.$$

In turn, the *Pareto optimal utility profiles* in this abstract model are contained in the Pareto boundary of $\mathcal{U}_E$:

$$\mathcal{P}_E := \{u \in \mathcal{U}_E : u \leq v \in \mathcal{U}_E \text{ implies } u = v\}.$$

Any allocation $x \in X$ is then called **strongly Pareto optimal** for E if $(u_1(x^1), \ldots, u_m(x^m)) \in \mathcal{P}_E$. For instance, if $\mathcal{E} := (\{\omega^i, u_i\}_{i=1,...,m})$ is an exchange economy that satisfies (A2), we may take $X = \{x \in \mathbb{R}^{lm}_+ : \sum^m x^i \leq$

$\sum^m \omega^i\}$, and $\mathcal{P}_E$ would then correspond to the set of all strongly Pareto optimal utility profiles in $\mathcal{E}$.

Here is a general characterization of strongly Pareto optimal outcomes for regular distribution problems. This characterization is frequently used in welfare economics.

Negishi's Theorem

Let $E = (X, u_1, \ldots, u_m)$ be a regular distribution problem. Then, for any strongly Pareto optimal $x^* \in X$ for E, there exists a $(\lambda_1, \ldots, \lambda_m) \in \mathbb{R}^m \backslash \{\mathbf{0}\}$ such that

$$x^* \in \arg\max \left\{ \sum_{i=1}^m \lambda_i u_i(x^i) : x \in X \right\}. \tag{14}$$

Conversely, if (14) holds for any $(\lambda_1, \ldots, \lambda_m) \in \mathbb{R}^m_{++}$, then x^* must be strongly Pareto optimal for E.

In welfare economics, one views an increasing function $W : \mathbb{R}^m \to \mathbb{R}$ as a *social welfare function* by interpreting the number $W(a_1, \ldots, a_m)$ as the aggregate (cardinal) welfare of the m-person society when the (cardinal) utility value of agent i is a_i. A particularly interesting class of social welfare functions are the increasing linear functionals on $\mathbb{R}^m$. In welfare economics, any such function is referred to as a **Bergson-Samuelson social welfare function**. Negishi's Theorem identifies a close connection between these linear functionals and the notion of Pareto optimality by demonstrating that the class of Bergson-Samuelson social welfare functions characterizes the set of all strongly Pareto optimal allocations in a regular distribution problem.

As for the proof of this result, we have:

Exercise 16 Prove Negishi's Theorem.

The key to the proof is the proper utilization of the Minkowski Supporting Hyperplane Theorem. Figure 1 illustrates this point, we leave the details to you.

2.3* Harsanyi's Utilitarianism Theorem

One of the classic findings of social choice theory is the characterization of the Bergson-Samuelson social welfare functions from the expected

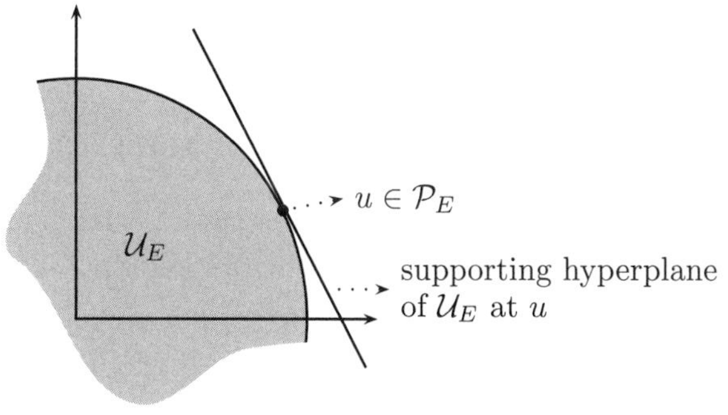

FIGURE 1

utility formulation of individual and social preferences. Informally put, the associated result shows that if all agents in a given society, along with the social planner, are expected utility maximizers, and if the preferences of the planner are linked to that of the constituent individuals by means of a Pareto condition, then the utility function of the planner must depend *linearly* on the individual utility functions.[25] There are various forms of this result which is viewed by some as a solid foundation for the Bergson-Samuelson aggregation method. We provide below a fairly general formulation.[26] Consider a nonempty set N of individuals, which may be finite or infinite. The set of (riskless) prizes is a nonempty finite set X. There is common risk in the model in that the space of alternatives on which each individual's preferences are defined is the set $\mathfrak{L}_X$ of all lotteries on X. We assume that each individual is an expected utility maximizer, that is, the preferences of individual $i \in N$ on $\mathfrak{L}_X$ are represented by a utility function $U(\cdot, i) \in \mathbb{R}^{\mathfrak{L}_X}$ with

$$U(p, i) = \sum_{x \in X} p(x) u_i(x) \qquad \text{for all } p \in \mathfrak{L}_X, \tag{15}$$

where $u_i \in \mathbb{R}^X$ is the von Neumann–Morgenstern utility function of agent i.

Now consider a social planner who wishes to aggregate the preferences of the constituent individuals. The preferences of this planner, represented

[25] This result was proved in 1955 by John Harsanyi, who is best known for his pioneering work on modeling games with incomplete information. Harsanyi (1920–2000) shared with John Nash and Reinhard Selten the 1994 Nobel Prize in Economics for his contributions to the analysis of equilibria of noncooperative games.

[26] A considerable social choice literature revolves around this result. If you are interested in this topic, Weymark (1991) is a great reference to start with. I follow Zhou (1997) here.

by a map $U \in \mathbb{R}^{\mathfrak{L}_X}$, should depend on the preferences of the individuals, at least through the following Pareto indifference relation:

$$U(p) = U(q) \quad \text{whenever} \quad U(p,i) = U(q,i) \text{ for all } i \in N. \tag{16}$$

Let us suppose now that the preferences of the planner also admit an expected utility representation, or more specifically, let

$$U(p) = \sum_{x \in X} p(x)u(x) \quad \text{for all } p \in \mathfrak{L}_X, \tag{17}$$

for some $u \in \mathbb{R}^X$. Curiously, these two apparently weak conditions actually identify the exact nature of the dependence of U on the $U(\cdot, i)$s.

HARSANYI'S UTILITARIANISM THEOREM

Let N be a nonempty set, and u and u_i, $i \in N$, real functions on a given nonempty finite set X. Then (15)–(17) *hold if, and only if, there exists an $L \in \mathcal{L}(\mathbb{R}^N, \mathbb{R})$ such that*

$$U(p) = L(U(p,\cdot)) \quad \textit{for all } p \in \mathfrak{L}_X.$$

If, in addition,

$$U(p) \geq U(q) \quad \textit{whenever} \quad U(p,i) \geq U(q,i) \textit{ for all } i \in N,$$

then $L(Y) \geq 0$ where $Y := \text{span}\{U(p,\cdot) \in \mathbb{R}^N : p \in \mathfrak{L}_X\}$.

EXERCISE 17^H Prove Harsanyi's Utilitarianism Theorem.

EXERCISE 18 Formulate Harsanyi's Utilitarianism Theorem more precisely in the special case where N is finite. (This will probably give you a better idea about the name of the result.) Improve the second part of the result in this case to the nonnegativity of L on the entire $\mathbb{R}^N$.

3 An Application to Information Theory

In this application we return to the framework of decision theory under uncertainty (Section F.3.2), and consider how an agent would compare two information services that would refine her initial beliefs about the true state of the world. Let Ω be any nonempty set, which we again think of as the set of states of nature. (For the time being we do not assume that Ω is finite.) Abusing the common terminology of probability theory a bit, we define a **simple probability distribution** on Ω as any $p \in \mathbb{R}_+^{\Omega}$ such that $supp(p) := \{\omega \in \Omega : p(\omega) > 0\}$ is a finite set and

$$\sum_{\omega \in supp(p)} p(\omega) = 1.$$

We denote the set of all simple probability distributions on Ω by $\mathbb{P}(\Omega)$. Clearly, if Ω is finite, the set $\mathfrak{L}_\Omega$ of all lotteries on Ω equals $\mathbb{P}(\Omega)$.

By an **information service** for Ω, we mean a list (I, p) where I is a nonempty finite set and $p \in \mathbb{R}_+^{\Omega \times I}$ satisfies $p(\omega, \cdot) \in \mathfrak{L}_I$ for each $\omega \in \Omega$. We interpret $p(\omega, i)$ as the probability of receiving the message $i \in I$ when the true state of the world is $\omega \in \Omega$.

DEFINITION

Let Ω be any nonempty set, and (I, p) and (J, q) two information services for Ω. We say that (I, p) is **more informative than** (J, q), if there exists a map $\Theta \in \mathbb{R}_+^{I \times J}$ such that

(i) $\sum_{j \in J} \Theta(i, j) = 1$ for each $i \in I$; and

(ii) $q(\omega, j) = \sum_{i \in I} p(\omega, i)\Theta(i, j)$ for each $(\omega, j) \in \Omega \times J$.

In words, if the information service (I, p) is more informative than (J, q), then we can think of (J, q) as sending exactly the messages of (I, p) *plus some noise*. To see this, envision a situation in which $I = \{i_1, i_2\}$ and $J = \{j_1, j_2\}$. Now suppose the true state of the world is ω. Then the probability that we will receive the message i_1 from the service (I, p) is $\alpha := p(\omega, i_1)$, while the probability that the message will be i_2 is $1 - \alpha$. How about (J, q)? This service will send us the message j_1 with probability $\Theta(i_1, j_1)\alpha + \Theta(i_2, j_1)(1 - \alpha)$ and j_2 with probability $\Theta(i_1, j_2)\alpha + \Theta(i_2, j_2)(1 - \alpha)$. Put differently, when the true state of the world is ω, it is as if the messages of (I, p) are drawn according to a lottery on the left of Figure 2, while those of (J, q) are drawn according to the *compound* lottery on the right of Figure 2. (For this interpretation to make sense, we need $\Theta(i_k, j_1)\alpha + \Theta(i_k, j_2) = 1$ for each $k = 1, 2$, which is exactly what condition (i) in our definition ensures.) It is in this sense that,

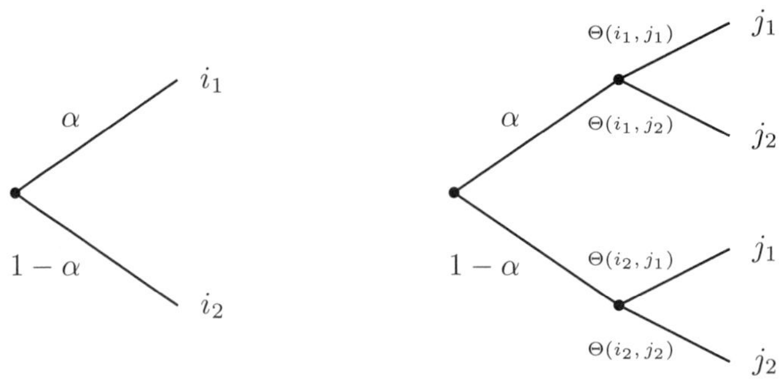

FIGURE 2

given any state of the world, we can think of (J, q) as sending us the very messages of (I, p), but after "garbling" them somewhat.

To illustrate, let $\Omega = \{\omega, \omega'\}$, and consider the case where $p(\omega, i_1) = 1$ and $p(\omega', i_2) = 1$. Thus, (I, p) is a fully informative service for Ω in the sense that once a message is received from (I, p), then the agent learns exactly what the true state of the world is.[27] Now assume that $\beta := \Theta(i_1, j_1)$ and $\gamma := \Theta(i_2, j_1)$. (That is, if (J, q) receives the message i_1 from (I, p), then it sends us j_1 with probability β and j_2 with probability $1 - \beta$, but if it receives the message i_2 from (I, p), then it sends us j_1 with probability γ and j_2 with probability $1 - \gamma$.) Clearly, unless $\beta = 1$ and $\gamma = 0$ (or $\beta = 0$ and $\gamma = 1$), the messages of (J, q) would not tell us exactly which state of the world ensues—they are less informative than those of (I, p) in an obvious sense. In the extreme situation where $\beta = \gamma$, in fact, the messages of (J, q) would convey no information whatsoever because they are drawn according to the same lottery, no matter what the true state of the world is. (In that case, there would be absolutely no reason to update our prior beliefs upon observing the messages sent to us by (J, q).)

EXERCISE 19 Let Ω be a nonempty set and $\mathcal{I}_\Omega$ the class of all information services for Ω. Define the relation $\trianglerighteq$ on $\mathcal{I}_\Omega$ by $(I, p) \trianglerighteq (J, q)$ iff (I, p) is more informative than (J, q).

(a) Show that $\trianglerighteq$ is a preorder on $\mathcal{I}_\Omega$. Is this preorder complete when $|\Omega| = 2$?

[27] With one caveat. If the agent's prior beliefs assigned probability zero to the state that (I, p) points to, life gets a bit more complicated. Let's not worry about this possibility for now, however.

(b) Characterize the strict part of $\trianglerighteq$.

(c) If they exist, find a $\trianglerighteq$-maximum and a $\trianglerighteq$-minimum of $\mathcal{I}_\Omega$. What if $|\Omega| < \infty$?

Now let us try to model the *value* of an information service for an individual. Fix a nonempty set X of prizes. Let $\mathcal{A}$ be a nonempty finite subset of X^Ω, which we view as the set of all feasible acts. We shall assume in what follows that $\mathcal{A}$ contains only *constant acts*. That is, we posit that for every $\mathbf{a} \in \mathcal{A}$ there exists an $x \in X$ such that $\mathbf{a}(\omega) = x$ for all $\omega \in \Omega$. (Put differently, one can "identify" $\mathcal{A}$ with a nonempty subset of X.) In this section we call a triplet $(\Omega, X, \mathcal{A})$ a **choice problem**, provided that this assumption is satisfied. A **decision maker** in the present model is described as a list (σ, u), where $\sigma \in \mathbb{P}(\Omega)$ corresponds to the initial (prior) beliefs about the true state of the world and $u \in \mathbb{R}^{\Omega\times X}$ is a state-dependent utility function (that is, $u(\omega, x)$ is the level of utility entailed by obtaining the prize x in state ω).[28] The expected utility of the decision maker (σ, u) who faces the choice problem $(\Omega, X, \mathcal{A})$ is

$$V_\sigma(u) := \max\left\{\mathsf{E}_\sigma(u(\cdot, \mathbf{a}(\cdot))) : \mathbf{a} \in \mathcal{A}\right\},$$

in the absence of access to any information service.

Now suppose the decision maker has access to the information service (I, p) for Ω. Given her initial beliefs σ, she would then view the probability of receiving the message i as

$$\overline{p}(i) := \sum_{\omega\in supp(\sigma)} \sigma(\omega)p(\omega, i).$$

Let us denote by $p^\sigma(\cdot, i) \in \mathbb{P}(\Omega)$ her *posterior beliefs* about the true state of the world, *conditional on receiving the message i*. The so-called *Bayesian updating* mandates that, for any $(\omega, i) \in \Omega \times I$,

$$p^\sigma(\omega, i) := \frac{\text{probability of receiving } i \text{ and the true state being } \omega}{\text{probability of receiving } i} = \frac{\sigma(\omega)p(\omega, i)}{\overline{p}(i)}$$

[28] Notice that uncertainty matters here even though $\mathcal{A}$ consists only of constant acts, precisely because the utility function of the agent depends on the state of the world. I do not know if the development outlined in this section can be carried out without assuming that $\mathcal{A}$ includes only constant acts and letting u be state independent. (This seems to me like a very good problem to go after; please come back and read this footnote again after studying the "if" part of Blackwell's Theorem.)

if $\overline{p}(i) > 0$, and $p^\sigma(\omega, i) := \sigma(\omega)$ if $\overline{p}(i) = 0$. Consequently, conditional on receiving the message $i \in I$, the expected utility of the decision maker is

$$V_\sigma^{(I,p)}(u \mid i) := \max\left\{\mathsf{E}_{p^\sigma(\cdot,i)}(u(\cdot, \mathbf{a}(\cdot))) : \mathbf{a} \in \mathcal{A}\right\}.$$

Therefore, with access to the information service (I, p) for Ω, the expected utility of (σ, u) is

$$V_\sigma^{(I,p)}(u) := \sum_{i \in I} \overline{p}(i) V_\sigma^{(I,p)}(u \mid i).$$

Using the definition of $V_\sigma^{(I,p)}(u \mid i)$ and p^σ, we can write this more explicitly as

$$V_\sigma^{(I,p)}(u) := \sum_{i \in I} \max\left\{\sum_{\omega \in supp(\sigma)} \sigma(\omega) p(\omega, i) u(\omega, \mathbf{a}(\omega)) : \mathbf{a} \in \mathcal{A}\right\}. \tag{18}$$

Using this formula it is easily checked that $V_\sigma^{(I,p)}(u) \geq V_\sigma(u)$, that is, in the present context where using the information service is costless, access to any information service for Ω is bound to yield a nonnegative utility gain. (Verify!)

The following definition introduces a natural criterion for ranking information services—one service is unambiguously better than the other if *all* decision makers agree that the former is the better service.

> DEFINITION
> Let $(\Omega, X, \mathcal{A})$ be a choice problem, and (I, p) and (J, q) two information services for Ω. We say that (I, p) is **more valuable than** (J, q) if $V_\sigma^{(I,p)}(u) \geq V_\sigma^{(J,q)}(u)$ for all decision makers $(\sigma, u) \in \mathbb{P}(\Omega) \times \mathbb{R}^{\Omega \times X}$.

It is somewhat intuitive that if the information service (I, p) is more informative than (J, q), then it must be more valuable than (J, q) for any decision maker (σ, u). This is indeed true, but there is more to the story—the converse also holds. In 1951, David Blackwell discovered that being more valuable and more informative are, in fact, equivalent ways of ordering the information services. This is a foundational result of information theory.

While Blackwell's original proof was somewhat indirect, since then easier proofs (based on convex analysis) of this important result were devised. We roughly follow here the argument given by Crémer (1982).

Blackwell's Theorem

Let Ω be any nonempty finite set, $(\Omega, X, \mathcal{A})$ a choice problem, and (I, p) and (J, q) two information services for Ω. Then (I, p) is more informative than (J, q) if, and only if, (I, p) is more valuable for than (J, q).

Proof

Suppose that (I, p) is more informative than (J, q), that is, there exists a map $\Theta \in \mathbb{R}_+^{I\times J}$ such that $\sum_{j\in J} \Theta(\cdot, j) = 1$, and $q(\omega, j) = \sum_{i\in I} p(\omega, i)\Theta(i, j)$ for each $(\omega, j) \in \Omega \times J$. Take any $(\sigma, u) \in \mathbb{P}(\Omega) \times \mathbb{R}^{\Omega\times X}$, and for each $j \in J$, pick an arbitrary $\mathbf{a}_j \in \arg\max\{\mathsf{E}_{q^\sigma(\cdot,j)}(u(\cdot, \mathbf{a}(\cdot))) : \mathbf{a} \in \mathcal{A}\}$. Then

$$
\begin{aligned}
V_\sigma^{(J,q)}(u) &= \sum_{j\in J}\sum_{\omega\in\Omega} \sigma(\omega)\left(\sum_{i\in I} p(\omega, i)\Theta(i, j)\right) u(\omega, \mathbf{a}_j(\omega)) \\
&= \sum_{j\in J}\sum_{i\in I} \Theta(i, j) \sum_{\omega\in\Omega} \sigma(\omega)p(\omega, i)u(\omega, \mathbf{a}_j(\omega)) \\
&\leq \sum_{i\in I}\sum_{j\in J} \Theta(i, j)\overline{p}(i)\, V_\sigma^{(I,p)}(u \mid i) \\
&= \sum_{i\in I} \overline{p}(i)\, V_\sigma^{(I,p)}(u \mid i) \\
&= V_\sigma^{(I,p)}(u).
\end{aligned}
$$

Conversely, let $\mathfrak{P} := \{r \in \mathbb{R}_+^{\Omega\times J} : r(\omega, \cdot) \in \mathfrak{L}_J \text{ for all } \omega \in \Omega\}$, define

$$\mathfrak{P}^* := \{r \in \mathfrak{P} : (I, p) \text{ is more informative than } (J, r)\},$$

and suppose that $(J, q) \notin \mathfrak{P}^*$. We wish to show that (I, p) is not more valuable than (J, q).

One can easily check that $\mathfrak{P}^*$ is a closed and convex subset of the Euclidean space $\mathbb{R}^{\Omega\times J}$. (Verify!) Consequently, by Proposition G.4, there exists an $(\alpha, L) \in \mathbb{R} \times \mathcal{L}(\mathbb{R}^{\Omega\times J}, \mathbb{R})$ such that

$$L(q) > \alpha > L(r) \qquad \text{for all } r \in \mathfrak{P}^*.$$

Obviously, there exists a $w \in \mathbb{R}^{\Omega\times J}$ such that $L(f) = \sum_{(\omega,j)\in\Omega\times J} w(\omega, j)f(\omega, j)$ for each $f \in \mathbb{R}^{\Omega\times J}$, so if we let $v := w - \alpha$, then

$$\sum_{(\omega,j)\in\Omega\times J} v(\omega, j)q(\omega, j) > 0 > \sum_{(\omega,j)\in\Omega\times J} v(\omega, j)r(\omega, j) \qquad \text{for all } r \in \mathfrak{P}^*.$$

By definition of $\mathfrak{P}^*$, this means that

$$\sum_{(\omega,j)\in\Omega\times J} v(\omega,j)q(\omega,j) > 0 > \sum_{(\omega,j)\in\Omega\times J} v(\omega,j)\sum_{i\in I} p(\omega,i)\Theta(i,j) \qquad (19)$$

for any $\Theta\in\mathbb{R}_+^{I\times J}$ with $\sum_{j\in J}\Theta(\cdot,j)=1$.

Now fix any $g\in J^X$ and define $u\in\mathbb{R}^{\Omega\times X}$ as

$$u(\omega,x) := \begin{cases} \frac{v(\omega,g(x))}{\sigma(\omega)}, & \text{if } \sigma(\omega)>0 \\ 0, & \text{otherwise} \end{cases}.$$

By (18) and (19), we have

$$\begin{aligned} V_\sigma^{(J,q)}(u) &= \sum_{j\in J}\max\left\{\sum_{\omega\in\Omega} q(\omega,j)v(\omega,g(\mathbf{a}(\omega))) : \mathbf{a}\in\mathcal{A}\right\} \\ &\geq \sum_{(\omega,j)\in\Omega\times J} v(\omega,j)q(\omega,j) \\ &> 0. \end{aligned}$$

Next, for any $i\in I$, pick any $\mathbf{a}_i\in\arg\max\left\{\sum_{\omega\in\Omega}\sigma(\omega)p(\omega,i)u(\omega,\mathbf{a}(\omega)) : \mathbf{a}\in\mathcal{A}\right\}$, and define $x_i\in X$ by $x_i := \mathbf{a}_i(\omega)$ for any ω. (Recall that $\mathbf{a}_i$ must be a constant act by hypothesis.) Then

$$\begin{aligned} V_\sigma^{(I,p)}(u) &= \sum_{i\in I}\max\left\{\sum_{\omega\in\Omega}\sigma(\omega)p(\omega,i)u(\omega,\mathbf{a}(\omega)) : \mathbf{a}\in\mathcal{A}\right\} \\ &= \sum_{i\in I}\sum_{\omega\in\Omega}\sigma(\omega)p(\omega,i)u(\omega,x_i) \\ &= \sum_{i\in I}\sum_{\omega\in\Omega} p(\omega,i)v(\omega,g(x_i)). \end{aligned}$$

Thus, if we define $\Theta\in\mathbb{R}_+^{I\times J}$ as $\Theta(i,j) := \begin{cases} 1, & \text{if } j=g(x_i) \\ 0, & \text{otherwise} \end{cases}$, then

$$\begin{aligned} V_\sigma^{(I,p)}(u) &= \sum_{i\in I}\sum_{\omega\in\Omega}\sum_{j\in J} p(\omega,i)\Theta(i,j)v(\omega,j) \\ &= \sum_{(\omega,j)\in\Omega\times J} v(\omega,j)\sum_{i\in I} p(\omega,i)\Theta(i,j) \\ &< 0, \end{aligned}$$

by (18). It follows that $V_\sigma^{(J,q)}(u) > V_\sigma^{(I,p)}(u)$, that is, (I,p) is not more valuable for than (J,q). ■

EXERCISE 20 Let $\Omega := \{\omega_1, \omega_2, \omega_3\}$, and consider the information services $(\{1,2\}, p)$ and $(\{1,2\}, q)$ for Ω, where $p(\omega_1, 1) = \frac{3}{4}$, $p(\omega_2, 1) = \frac{1}{4} = p(\omega_3, 1)$, and $q(\omega_1, 1) = \frac{5}{8}$, $q(\omega_2, 1) = \frac{3}{8}$ and $q(\omega_3, 1) = 1$. Is $(\{1,2\}, p)$ more valuable than $(\{1,2\}, q)$?

*EXERCISE 21 Prove Blackwell's Theorem in the case where Ω is countable.

4 Applications to Financial Economics

In this application we present a fairly general formulation of two foundational results of the theory of financial economics. We first discuss in what sense the absence of arbitrage in securities markets can be viewed as an "equilibrium" phenomenon. Then, in the context of incomplete markets, we derive the structure of arbitrage-free linear pricing.[29]

4.1 Viability and Arbitrage-Free Price Functionals

The basic framework we adopt is, again, one of uncertainty (Section F.3.2). We again let Ω stand for an arbitrary nonempty set that contains all states of nature. The interpretation is now familiar: There is uncertainty as to which particular state will obtain in the future, but we know that at least one element of Ω is bound to occur. We do not assume that Ω is finite for the moment.

We think of the prizes in terms of a consumption good, and view $\mathbb{R}$ as the outcome space. Consequently, an *act* in this framework is a map $\mathbf{a}$ in $\mathbb{R}^\Omega$. Intuitively, "choosing $\mathbf{a}$" is "choosing a particular portfolio of securities" in the sense that $\mathbf{a}$ represents the future payoff to a given portfolio of securities (in terms of the consumption good), but of course, this value depends on the state that will obtain in the future. In financial economics, such an act is

[29] I follow Harrison and Kreps (1979) here for the large part. For a comprehensive introduction to the extensions of the No-Arbitrage Theorem and other related topics in the theory of asset pricing, see Duffie (1996).

thus called a **state-contingent claim**, and we shall adhere to this terminology in this section as well. As is largely standard, we take $\mathbb{R}^\Omega$ as the space of all state-contingent claims. An important implication of this is that there is no limit to short-selling in the market for securities.

We make $\mathbb{R}^\Omega$ a preordered linear space by means of a vector preorder $\trianglerighteq$ (the strict part of which is denoted by $\triangleright$) such that $\geq \subseteq \trianglerighteq$ and $\gg \subseteq \triangleright$. We do not order $\mathbb{R}^\Omega$ simply by $\geq$, because we wish to allow for indifference in the case of zero probability events. Indeed, we would like to interpret the statement $\mathbf{a} \triangleright \mathbf{0}$ as "the claim $\mathbf{a}$ is sure to yield positive returns." The statement $\mathbf{a} > \mathbf{0}$ is in general not adequate for this purpose, because it allows for $\mathbf{a}$ to be equal to zero on a nonempty set of states. What if the probability that this set will occur in the future is 1? In that case, even though $\mathbf{a} > \mathbf{0}$, the claim $\mathbf{a}$ is sure to yield a zero return.[30] (Of course, if $\mathbf{a} \gg \mathbf{0}$, there is no such problem—in that case $\mathbf{a}$ is bound to yield positive returns.)

In this context, any $\pi \in \mathcal{L}(\mathbb{R}^\Omega, \mathbb{R})$ is referred to as a **price functional for Ω**. This formulation entails that all claims can be priced, and thus at the moment, we are working in the context of *complete markets*. Moreover, under the assumption of unlimited short-selling, the present setting maintains that the only exogenous entry of the model is the state space Ω. Thus, even though this is a bit unconventional, we shall refer to Ω as a **complete contingent claims market** in what follows. (The case of incomplete markets will be taken up below.)

Given a price functional π, we interpret $\pi(\mathbf{a})$ as the market value of the state-contingent claim $\mathbf{a}$.[31] An **arbitrage** associated with a price functional π is a state-contingent claim $\mathbf{a} \in \mathbb{R}^\Omega$ such that $\mathbf{a} \triangleright \mathbf{0}$ and $\pi(\mathbf{a}) \leq 0$, or $\mathbf{a} \trianglerighteq \mathbf{0}$ and $\pi(\mathbf{a}) < 0$. Thus, we say that the price functional π is **arbitrage-free** iff it is strictly positive relative to $\trianglerighteq$, that is,

$$\mathbf{a} \trianglerighteq \mathbf{0} \ \text{ implies } \ \pi(\mathbf{a}) \geq 0 \quad \text{and} \quad \mathbf{a} \triangleright \mathbf{0} \ \text{ implies } \ \pi(\mathbf{a}) > 0.$$

The interpretation of an arbitrage-free price functional is straightforward.

[30] No, one couldn't fix the problem by positing that all states have positive probability. What if Ω is not countable?

[31] The linearity requirement on the price functional is very sensible. For instance, we wish to have $\pi(2\mathbf{a}) = 2\pi(\mathbf{a})$, because the market value of a portfolio (say, $2\mathbf{a}$) that yields twice the payoff of another portfolio (that is, $\mathbf{a}$) should be twice the market value of that portfolio. The additivity requirement on π is similarly interpreted.

There is a sense in which one can think of the presence of an arbitrage-free price as a necessary condition for the existence of a competitive equilibrium in *some* economy, but the issue is far from trivial; it requires careful modeling. Let us first describe how the preferences of an agent should be defined in this context. Throughout this application, Z_Ω will stand for the product linear space $\mathbb{R} \times \mathbb{R}^\Omega$. We interpret $(a, \mathbf{a}) \in Z_\Omega$ as the *net trade vector* that contains a dollars and a state-contingent claim $\mathbf{a}$. By a **standard agent for** Ω in the present setting we mean a concave function $u : Z_\Omega \to \mathbb{R}$ such that

$$b > a \quad \text{implies} \quad u(b, \mathbf{a}) > u(a, \mathbf{a}) \tag{20}$$

and

$$\mathbf{b} \rhd \mathbf{a} \quad \text{implies} \quad u(a, \mathbf{b}) > u(a, \mathbf{a}) \quad \text{and}$$
$$\mathbf{a} \unrhd \mathbf{b} \quad \text{implies} \quad u(a, \mathbf{b}) \geq u(a, \mathbf{a}) \tag{21}$$

for all $(a, \mathbf{a}), (b, \mathbf{b}) \in Z_\Omega$. Thus a standard agent for Ω is identified with a utility function on Z_Ω which is concave and strictly increasing. We denote the set of all standard agents for Ω as $\mathcal{U}_\Omega$.

Given a price functional π for Ω, we define the **budget set** of a standard agent u for Ω as

$$B(\pi) := \{(a, \mathbf{a}) \in Z_\Omega : a + \pi(\mathbf{a}) \leq 0\}.$$

(Interpretation?) A price functional π for Ω is said to be **viable** if there exists at least one standard agent for Ω who is able to choose an optimal trade within her budget set at π, that is, if there exists a $u \in \mathcal{U}_\Omega$ such that

$$\arg\max\{u(a, \mathbf{a}) : (a, \mathbf{a}) \in B(\pi)\} \neq \emptyset.$$

Therefore, if π is *not* viable, then it cannot arise in an equilibrium of any securities market in which one of the traders is a standard agent. Conversely, if π is viable, then it corresponds to the equilibrium price in the case of at least one exchange economy (in which all agents are identical and no trade takes place in equilibrium; see the proof of the "only if" part of Proposition 2 below). It thus makes good sense to view "viability" as an equilibrium property. Being free of arbitrage, on the other hand, seems like a simple structural property at best—it is not defined in association with any particular economy. Remarkably, however, it turns out that there is no difference between the notions of arbitrage-free and viable price functionals.

Proposition 2
(Harrison-Kreps) *In any complete contingent claims market Ω, a price functional π for Ω is arbitrage-free if, and only if, it is viable.*

Proof
The "only if" part is easy. If π is arbitrage-free, then we define $u \in \mathcal{U}_\Omega$ by $u(a, \mathbf{a}) := a + \pi(\mathbf{a})$, and observe that $(0, \mathbf{0}) \in \arg\max\{u(a, \mathbf{a}) : (a, \mathbf{a}) \in B(\pi)\}$. To prove the "if" part of the assertion, assume that π is viable, and pick any $v \in \mathcal{U}_\Omega$ and $(a^*, \mathbf{a}^*)$ such that

$$(a^*, \mathbf{a}^*) \in \arg\max\{v(a, \mathbf{a}) : (a, \mathbf{a}) \in B(\pi)\}.$$

Mainly to simplify matters, we will work with another utility function u in $\mathcal{U}_\Omega$, which we define by

$$u(a, \mathbf{a}) := v(a + a^*, \mathbf{a} + \mathbf{a}^*).$$

Clearly, $(0, \mathbf{0}) \in \arg\max\{u(a, \mathbf{a}) : (a, \mathbf{a}) \in B(\pi)\}$. (Right?) Now define

$$A := \{(a, \mathbf{a}) \in Z_\Omega : u(a, \mathbf{a}) > u(0, \mathbf{0})\}.$$

Claim. If $(a, \mathbf{a}) \in Z_\Omega$ satisfies $a > 0$ and $\mathbf{a} \trianglerighteq \mathbf{0}$, or $a \geq 0$ and $\mathbf{a} \triangleright \mathbf{0}$, then it belongs to the algebraic interior of A in Z_Ω.

Proof of Claim
If $(a, \mathbf{a}) \in Z_\Omega$ is such that $a > 0$ and $\mathbf{a} \trianglerighteq \mathbf{0}$, or $a \geq 0$ and $\mathbf{a} \triangleright \mathbf{0}$, then $u(a, \mathbf{a}) > u(0, \mathbf{0})$ by (20) and (21). Thus, given that u is concave, for any $(b, \mathbf{b}) \in Z_\Omega$ there is a small enough $0 < \alpha^* \leq 1$ such that

$$u((1 - \alpha)(a, \mathbf{a}) + \alpha(b, \mathbf{b})) \geq (1 - \alpha)u(a, \mathbf{a}) + \alpha u(b, \mathbf{b}) > u(0, \mathbf{0})$$

for any $0 \leq \alpha \leq \alpha^*$. ‖

Since $(0, \mathbf{0})$ maximizes u on $B(\pi)$, we have $A \cap B(\pi) = \emptyset$. Moreover, concavity of u implies that A is convex, while $B(\pi)$ is clearly a convex cone. So, given that $\textit{al-int}_{Z_\Omega}(A) \neq \emptyset$ – for instance, $(1, \mathbf{0}) \in \textit{al-int}_{Z_\Omega}(A)$ by the

Claim above – we may apply the Dieudonné Separation Theorem to conclude that

$$L(A) > 0 \geq L(B(\pi)) \quad \text{and} \quad L(al\text{-}int_{Z_\Omega}(A)) \gg 0 \tag{22}$$

for some nonzero $L \in \mathcal{L}(Z_\Omega, \mathbb{R})$.[32]

Since $L(1, \mathbf{0}) > 0$, it is without loss of generality to assume that $L(1, \mathbf{0}) = 1$ (otherwise we would work with $\frac{1}{L(1,\mathbf{0})}L$ instead). In that case,

$$L(a, \mathbf{a}) = L(a, \mathbf{0}) + L(0, \mathbf{a}) = a + \phi(\mathbf{a})$$

where $\phi \in \mathcal{L}(\mathbb{R}^\Omega, \mathbb{R})$ is defined by $\phi(\mathbf{a}) := L(0, \mathbf{a})$. In fact, $\phi = \pi$, because, for any $\mathbf{a} \in \mathbb{R}^\Omega$, we have $(-\pi(\mathbf{a}), \mathbf{a}), (\pi(\mathbf{a}), -\mathbf{a}) \in B(\pi)$, and hence, by (22), $0 \geq L(-\pi(\mathbf{a}), \mathbf{a}) = -\pi(\mathbf{a}) + \phi(\mathbf{a})$ and $0 \geq \pi(\mathbf{a}) - \phi(\mathbf{a})$. Thus:

$$L(a, \mathbf{a}) = a + \pi(\mathbf{a}) \quad \text{for all } (a, \mathbf{a}) \in Z_\Omega. \tag{23}$$

We are almost done. If $\mathbf{a} \rhd \mathbf{0}$, then $(0, \mathbf{a}) \in al\text{-}int_{Z_\Omega}(A)$ by the Claim above, so by (23) and (22), $\pi(\mathbf{a}) = L(0, \mathbf{a}) > 0$. On the other hand, if $\mathbf{a} \trianglerighteq \mathbf{0}$, then $(\frac{1}{m}, \mathbf{a}) \in al\text{-}int_{Z_\Omega}(A)$ for all $m \in \mathbb{N}$ (thanks to the Claim above), so by (23) and (22), $\pi(\mathbf{a}) = L(\frac{1}{m}, \mathbf{a}) - \frac{1}{m} > -\frac{1}{m}$, so letting $m \to \infty$ yields $\pi(\mathbf{a}) \geq 0$. ∎

4.2 The No-Arbitrage Theorem

While Proposition 2 is an interesting characterization of arbitrage-free prices, its usefulness is curtailed by the hypothesis of *complete* markets. Relaxing this assumption (but retaining the feasibility of unlimited short-selling) forces us to think of a price functional π as defined on a linear subspace of $\mathbb{R}^\Omega$. If we tried to prove Proposition 2 with this formulation, in the second part of the proof we would be confronted with the problem of extending π to the entire $\mathbb{R}^\Omega$ in a strictly positive way. (If you are thinking of the Krein-Rutman Theorem, that's good, but that result gives us only positive extensions, not necessarily *strictly* positive ones. Moreover, the

[32] Wow, is this cheating or what? How did I know that I could choose the hyperplane such that it not only separates A and $B(\pi)$, it also supports $B(\pi)$ at $(0, \mathbf{0})$? Here is the full argument. The Dieudonné Separation Theorem gives us a nonzero $L \in \mathcal{L}(Z, \mathbb{R})$ and an $\alpha \in \mathbb{R}$ such that $L(A) > \alpha \geq L(B(\pi))$ and $L(al\text{-}int_{Z_\Omega}(A)) \gg \alpha$; this much is clear. Since $(0, \mathbf{0}) \in B(\pi)$ and $L(0, \mathbf{0}) = 0$, we have $\alpha \geq 0$, so $L(A) > 0$ and $L(al\text{-}int_{Z_\Omega}(A)) \gg 0$. Further, since $B(\pi)$ is a cone, we must also have $\frac{\alpha}{2^m} \geq L(B(\pi))$ for all $m = 1, 2, \ldots$. (Yes?) Letting $m \to \infty$ yields (22).

interpretation of the "nonempty interior" condition of that result would be problematic here.)

To get a handle on the problem, we will examine the incomplete markets scenario with a *finite* state space. In what follows, then, we let $0 < |\Omega| < \infty$ and take $\trianglerighteq = \geq$.[33] Moreover, we fix an arbitrary nontrivial linear subspace Y of $\mathbb{R}^\Omega$ as the set of all claims for which a market exists. The list (Ω, Y) is then referred to as a **finite incomplete contingent claims market**. In turn, a **price functional** π **for** (Ω, Y) is defined as a linear functional on Y. It is called **arbitrage-free** whenever it is strictly positive on Y, that is, for all $\mathbf{a} \in Y$,

$$\mathbf{a} \geq \mathbf{0} \quad \text{implies} \quad \pi(\mathbf{a}) \geq 0 \qquad \text{and} \qquad \mathbf{a} > \mathbf{0} \quad \text{implies} \quad \pi(\mathbf{a}) > 0.$$

What sort of price functionals are arbitrage-free? The following theorem provides a very nice answer.

THE NO-ARBITRAGE THEOREM[34]

For any finite incomplete contingent claims market (Ω, Y), the following statements are equivalent:

(a) *π is an arbitrage-free price functional for (Ω, Y).*

(b) *There exists a $q \in \mathbb{R}^\Omega_{++}$ such that*

$$\pi(\mathbf{a}) = \sum_{\omega \in \Omega} q(\omega)\mathbf{a}(\omega) \qquad \textit{for all } \mathbf{a} \in Y. \tag{24}$$

(c) *π is a linear functional on Y that can be extended to a viable price functional for Ω for the complete contingent claims market Ω.*

PROOF

To see that (b) implies (c), extend π by letting $\pi(\mathbf{a}) := \sum_{\omega\in\Omega} q(\omega)\mathbf{a}(\omega)$ for all $\mathbf{a} \in \mathbb{R}^\Omega$ and define $u \in \mathcal{U}_\Omega$ by $u(a, \mathbf{a}) = a + \pi(\mathbf{a})$, which is maximized on $B(\pi)$ at $(0, \mathbf{0})$. That (c) implies (a), on the other hand, follows from Proposition 2. It remains to prove that (a) implies (b).[35] To this end, define

$$W_\Omega := \{(\mathbf{a}, -\pi(\mathbf{a})) : \mathbf{a} \in Y\}$$

[33] We now think that each state in Ω may occur with positive probability—we can do this because $|\Omega| < \infty$—so if $\mathbf{a} > \mathbf{0}$, then the claim $\mathbf{a}$ yields a strictly positive return for sure.

[34] The equivalence of (a) and (c) is a special case of a result due to Harrison and Kreps (1979). The equivalence of (a) and (b), which is often referred to as "the" No-Arbitrage Theorem in the literature, is due to Ross (1978).

[35] While this claim would be trivial if $Y = \mathbb{R}^\Omega$, it is not so in general. Why?

which is easily checked to be a linear subspace of $\mathbb{R}^\Omega \times \mathbb{R}$. Define the linear functional $L : W_\Omega \to \mathbb{R}$ by

$$L(\mathbf{a}, -\pi(\mathbf{a})) := \pi(\mathbf{a}) - \sum_{\omega \in \Omega} \mathbf{a}(\omega) \qquad \text{for all } \mathbf{a} \in Y.$$

Observe that L is positive on W_Ω. Indeed, if $(\mathbf{a}, -\pi(\mathbf{a})) \geq \mathbf{0}$, then, by the no-arbitrage condition, $\mathbf{a} = \mathbf{0}$ must hold, and we have $L(\mathbf{a}, -\pi(\mathbf{a})) = L(\mathbf{0}, 0) = 0$. We may thus apply Proposition G.8 to positively extend L to the entire $\mathbb{R}^\Omega \times \mathbb{R}$. Denoting this extension again by L, we may then write

$$L(\mathbf{a}, a) = \sum_{\omega \in \Omega} p(\omega)\mathbf{a}(\omega) + \alpha a \qquad \text{for all } (\mathbf{a}, a) \in \mathbb{R}^\Omega \times \mathbb{R}$$

for some $p \in \mathbb{R}^\Omega_+$ and $\alpha \geq 0$. (Why?) By the extension property, then, for all $\mathbf{a} \in Y$,

$$\pi(\mathbf{a}) - \sum_{\omega \in \Omega} \mathbf{a}(\omega) = \sum_{\omega \in \Omega} p(\omega)\mathbf{a}(\omega) - \alpha\pi(\mathbf{a}).$$

But then defining $q \in \mathbb{R}^\Omega_{++}$ by $q(\omega) := \frac{1+p(\omega)}{1+\alpha}$, we obtain (24). ■

The equivalence of (a) and (b) in this theorem tells us that arbitrage-free prices in a finite-state contingent market (with unlimited short-selling) is fully characterized by "state prices." (For concreteness, you can think of the state price $q(\omega)$ as the marginal cost of obtaining an additional unit of account of the portfolio when state ω occurs.). This is thus a revealing duality result that says that the arbitrage-free price of a state contingent claim is none other than a weighted sum of the payoffs of the claim at different states, where the weights are the state prices. In turn, the equivalence of (a) and (c) tells us in what way we may think of an arbitrage-free price functional as corresponding to an equilibrium.

Exercise 22 Let (Ω, Y) be a finite incomplete contingent claims market. A price functional π for (Ω, Y) is called **weakly arbitrage-free** if, for all $\mathbf{a} \in Y$, we have $\pi(\mathbf{a}) \geq 0$ whenever $\mathbf{a} \geq \mathbf{0}$. Show that π is weakly arbitrage-free iff there exists a $q \in \mathbb{R}^\Omega_+$ such that (24) holds. How does this result relate to Farkas' Lemma?

5 Applications to Cooperative Games

5.1 The Nash Bargaining Solution

For any given integer $n \geq 2$, an **n-person Nash bargaining problem** is defined as any list (S, d) where S is a nonempty set in $\mathbb{R}^n$ and d is an element of S such that there exists at least one $x \in S$ such that $x \gg d$. We interpret S as a utility possibility set that arises from a particular (strategic or otherwise) economic situation. Each point of S is thus a payoff profile that corresponds to a particular choice of the available alternatives in the underlying economic situation. If, however, agents fail to reach an agreement, it is understood that there obtains a default outcome that gives rise to the payoff profile $d \in S$, which is called the **disagreement point** of the problem, for obvious reasons.

In what follows, we work with those bargaining problems (S, d) where S is a convex and compact set that satisfies the following *comprehensiveness* requirement: If $x \in S$ and $x \geq y \geq d$, then $y \in S$. While compactness is best viewed as a technical regularity condition, convexity of S may be justified by assuming that individuals are expected utility maximizers and that joint randomization over the set of outcomes is possible. On the other hand, our comprehensiveness postulate reflects the free disposability of the utilities relative to the disagreement point d. We denote the set of all n-person Nash bargaining problems that satisfy these properties by $\mathfrak{B}_n$.

By a **bargaining solution** on $\mathfrak{B}_n$ in this setup, we mean a map $f : \mathfrak{B}_n \to \mathbb{R}^n$ such that $f(S, d) \in S$. Normatively speaking, one may think of $f(S, d)$ as the resolution suggested by an impartial arbitrator in the case of the bargaining problem (S, d). If instead one adopts a descriptive interpretation, then $f(S, d)$ can be viewed as a prediction concerning the outcome of the underlying strategic game. A famous example is the **Nash bargaining solution** f^{Nash}, which is defined on $\mathfrak{B}_n$ by

$$f^{\text{Nash}}(S, d) \in \arg\max \left\{ \prod_{i=1}^{n} (x_i - d_i) : x \in S \right\}.$$

(*Quiz.* Why is f^{Nash} well-defined?)

What sort of bargaining solutions would we deem reasonable? Just as in Section F.4.3, it will pay plenty here to proceed axiomatically. Consider, then, the following properties imposed on an arbitrary bargaining solution f. For all $(S, d) \in \mathfrak{B}_n$:

Axiom PO. (*Pareto Optimality*) $x > f(S, d)$ implies $x \notin S$.

Axiom SYM. (*Symmetry*) If $d_1 = \cdots = d_n$ and S is a symmetric set (that is, if $\{(x_{\sigma(1)}, \ldots, x_{\sigma(n)}) : x \in S\} = S$ for any bijection σ on $\{1, \ldots, n\}$), then $f_1(S, d) = \cdots = f_n(S, d)$.

Axiom C.INV. (*Cardinal Invariance*) For any strictly increasing affine map L on $\mathbb{R}^n$, $f\left(L(S), L(d)\right) = L(f(S, d))$.

Axiom IIA. (*Independence of Irrelevant Alternatives*) If $T \subseteq S$, $(T, d) \in \mathfrak{B}_n$, and $f(S, d) \in T$, then $f(S, d) = f(T, d)$.

Especially from a normative angle, where we think of f reflecting the potential decisions of an impartial arbitrator, these properties are quite appealing. PO is once again an unexceptionable efficiency requirement.[36] Whereas SYM reflects the impartiality of the arbitrator, C.INV avoids making interpersonal utility comparisons in a cardinal way. Finally, IIA is an appealing consistency condition when viewed from the perspective of the arbitrator—it is, after all, one of the most fundamental principles of revealed preference theory. (Nevertheless, we should mention that the suitability of IIA is much debated in the literature, and many authors have axiomatically obtained other interesting bargaining solutions that violate this property. See Thomson (1994) for a comprehensive discussion.)

The following theorem, proved in 1950 by John Nash, is a cornerstone in cooperative game theory and the starting point of the extensive literature on axiomatic bargaining. Its proof is based on a brilliant application of the Minkowski Supporting Hyperplane Theorem.

Proposition 3

(Nash) *For any given integer $n \geq 2$, the only bargaining solution on $\mathfrak{B}_n$ that satisfies the Axioms PO, SYM, C.INV, and IIA is the Nash bargaining solution.*

[36] This would be hard to swallow from a descriptive angle, however, don't you think? What about Prisoner's Dilemma, for instance?

PROOF

Let f be a bargaining solution on $\mathfrak{B}_n$ that satisfies the Axioms PO, SYM, C.INV, and IIA. Fix an arbitrary problem $(S, d) \in \mathfrak{B}_n$, and define

$$B := \left\{ \left(\frac{x_1 - d_1}{f_1^{\text{Nash}}(S, d) - d_1}, \ldots, \frac{x_n - d_n}{f_n^{\text{Nash}}(S, d) - d_n} \right) : x \in S \right\}.$$

By definition of f^{Nash}, we have

$$\prod_{i=1}^{n} (f_i^{\text{Nash}}(S, d) - d_i) \geq \prod_{i=1}^{n} (x_i - d_i) \qquad \text{for all } x \in S$$

so that $1 \geq \prod^n y_i$ for all $y \in B$. Now define

$$A := \left\{ x \in \mathbb{R}^n : \prod_{i=1}^{n} x_i \geq 1 \right\}.$$

Clearly, both A and B are closed and convex subsets of $\mathbb{R}^n$, and $int_{\mathbb{R}^n}(A) \neq \emptyset$. Moreover, since $x \mapsto \prod^n x_i$ is a strictly quasiconcave map on $\mathbb{R}^n$, we have $A \cap B = \{(1, \ldots, 1)\}$. By Exercise G.65, therefore, there exists a hyperplane that separates A and B while supporting them both at $(1, \ldots, 1)$, that is, there exists a $\lambda \in \mathbb{R}^m$ such that

$$\sum_{i=1}^{n} \lambda_i x_i \geq \sum_{i=1}^{n} \lambda_i \geq \sum_{i=1}^{n} \lambda_i y_i \qquad \text{for all } (x, y) \in A \times B.$$

(See Figure 3.) But then $(1, \ldots, 1)$ minimizes the map $x \mapsto \sum^n \lambda_i x_i$ on A, which is possible only if $\lambda_1 = \cdots = \lambda_n > 0$. (Why?) Consequently, we have

$$\sum_{i=1}^{n} x_i \geq n \geq \sum_{i=1}^{n} y_i \qquad \text{for all } (x, y) \in A \times B.$$

Now define

$$C := \left\{ z \in \mathbb{R}^n : \sum_{i=1}^{n} z_i \leq n \right\}.$$

By Axioms SYM and PO, we have $f(C, \mathbf{0}) = (1, \ldots, 1)$. But $(1, \ldots, 1) \in B \subseteq C$ and hence, by Axiom IIA, $f(B, \mathbf{0}) = f(C, \mathbf{0}) = (1, \ldots, 1)$. Using the definition of B and Axiom C.INV, we then obtain $f(S, d) = f^{\text{Nash}}(S, d)$. Since it is evident that f^{Nash} satisfies the Axioms PO, SYM, C.INV and IIA, the proof is complete. ∎

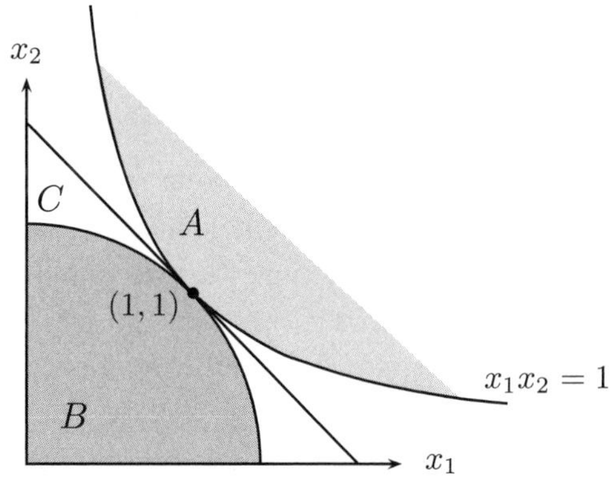

FIGURE 3

For the examination of many other axiomatic bargaining solutions, we refer the reader to the excellent survey by Thomson (1994). We stop our discussion here with a number of exercises that concern some of the variations of Proposition 3.

EXERCISE 23 Show that Proposition 3 is tight in the sense that none of the axioms used in this result is implied by the other three.

EXERCISE 24 Given any integer $n \geq 2$, show that if a bargaining solution f on $\mathfrak{B}_n$ satisfies Axioms SYM, C.INV, and IIA, then $f(S, d) \in \{d, f^{\text{Nash}}(S, d)\}$ for all $(S, d) \in \mathfrak{B}_n$.

EXERCISE 25^{H} (Roth) Given any integer $n \geq 2$, show that, in Proposition 3, one can replace Axiom PO with the axiom of *strong individual rationality* (SIR): $f(S, d) \gg d$ for all $(S, d) \in \mathfrak{B}_n$.

EXERCISE 26 (Harsanyi-Selten) Given any integer $n \geq 2$, prove that if a bargaining solution on $\mathfrak{B}_n$ satisfies the Axioms SIR, C.INV and IIA, then there exists an $\alpha \in \mathbb{R}^n_{++}$ such that $\sum^n \alpha_i = 1$ and

$$f(S, d) \in \arg\max \left\{ \prod_{i=1}^{n} (x_i - d_i)^{\alpha_i} : x \in S \right\} \qquad \text{for all } (S, d) \in \mathfrak{B}_n.$$

(Such a bargaining solution is called a **weighted Nash bargaining solution.**)

5.2* Coalitional Games without Side Payments

In Section F.4.1 we introduced the notion of capacity and used it to define coalitional games with side payments.[37] This formulation presumes that utilities/payoffs are transferable among agents, for otherwise summarizing the "worth" of a coalition by a single number would not be sensible. If we wish to desert this presumption, then we would need to specify the entire *set* of utility profiles that a coalition can sustain (by means available to it). The resulting theory is called the theory of coalitional games *without side payments*.[38]

Let $\mathcal{N} := \{\{1,\ldots,n\} : n \in \mathbb{N}\}$, and interpret any $N \in \mathcal{N}$ as a set of individual players in a given environment. For any given $N \in \mathcal{N}$, any nonempty subset S of N is said to be a **coalition** in N. The class of all coalitions in N is thus $\mathcal{P}(N) := 2^N \backslash \{\emptyset\}$.

By a **coalitional game without side payments**, we mean a tuple $(N, \mathbf{V})$, where $N \in \mathcal{N}$ and $\mathbf{V}$ is a correspondence on $\mathcal{P}(N)$ that satisfies the following properties. For any $S \in \mathcal{P}(N)$:

(i) $\mathbf{V}(S)$ is a nonempty, compact and convex subset of $\mathbb{R}^S_+$ that is comprehensive relative to $\mathbf{0}$ (that is, $x \in \mathbf{V}(S)$ implies $y \in \mathbf{V}(S)$ for any $y \in \mathbb{R}^S_+$ with $y \leq x$);[39]

(ii) (*Normalization*) $\mathbf{V}(\{i\}) = \{0\}$ for all $i \in N$;

(iii) (*Superadditivity*) If $T \in \mathcal{P}(N)$ is disjoint from S, then

$$\mathbf{V}(S \cup T) \supseteq \{x \in \mathbb{R}^{S\cup T}_+ : x|_S \in \mathbf{V}(S) \text{ and } x|_T \in \mathbf{V}(T)\}.$$

The interpretation is that $\mathbf{V}(S)$ stands for the set of all (expected) payoff allocations that the members of the coalition S can achieve through cooperation. While requirement (*i*) guarantees this set to be structurally well-behaved, (*ii*) normalizes the payoffs that any one player can get (acting alone) to be zero, thereby letting us interpret the comprehensiveness requirement in (*i*) as a *free disposability* property. Finally, condition (*iii*)

[37] The present discussion will make much more sense if you have studied (even very briefly) Section F.4.

[38] The terminology *nontransferable* (*NTU*) *coalitional game* is also used in the literature.

[39] In the language of the previous section, therefore, $(\mathbf{V}(S), \mathbf{0}) \in \mathfrak{B}_{|S|}$.

corresponds to the formalization of the idea that any coalition can achieve at least as much as what its disjoint subcoalitions can independently sustain.

Take any $N \in \mathcal{N}$. Note that a nonnegative superadditive capacity $\mathbf{v}$ on N, which is normalized to have $\mathbf{v}(\{i\}) = 0$ for each $i \in N$, induces a coalitional game $(N, \mathbf{V}_\mathbf{v})$ without side payments in a natural manner:

$$\mathbf{V}_\mathbf{v}(S) := \left\{ x \in \mathbb{R}^S_+ : \sum_{i \in S} x(i) \leq \mathbf{v}(S) \right\}, \qquad S \in \mathcal{P}(N).$$

We now pose the converse question: How can we generate a superadditive capacity from a given coalitional game without side payments? (This is important because the former concept is inherently simpler than the latter, and we have already obtained some results in Section F.4 for capacities.) Well, a quick answer is that we can't, simply because a coalitional game, as we defined it in Section F.4.1 by means of a capacity, is bound to suppress all the information concerning the *distribution* of payoffs among the members of the coalition. However, it turns out that we may actually characterize a coalitional game $(N, \mathbf{V})$ without side payments by means of a *family* of superadditive capacities. To this end, for any given $\lambda \in \mathbb{R}^N_+$, we consider the superadditive capacity $\mathbf{v}^\lambda$ on N defined by

$$\mathbf{v}^\lambda(S) := \max \left\{ \sum_{i \in S} \lambda(i) y(i) : y \in \mathbf{V}(S) \right\}.$$

(This capacity is called the **λ-transfer capacity**.) In words, $\mathbf{v}^\lambda(S)$ is the highest total λ-weighted payoffs that a coalition S in N can achieve through cooperation. Notice that, for each λ, a different kind of trade-off structure is assumed across the utilities of the agents. So a single $\mathbf{v}^\lambda$ cannot represent $\mathbf{V}$ since the latter does not make any interpersonal comparisons. But what if we consider the set of *all* $\mathbf{v}^\lambda$s? The answer is given in the next result, which is again based on a supporting hyperplane argument. (Note the close resemblance to Negishi's Theorem.)

PROPOSITION 4

Given any $N \in \mathcal{N}$, let $(N, \mathbf{V})$ be a coalitional game without side payments. For any given $S \in \mathcal{P}(N)$ and $x \in \mathbb{R}^S_+$, the following statements are equivalent:

(a) $x \in \mathbf{V}(S)$.

(b) $\sum_{i\in S} \lambda(i)x(i) \leq \mathbf{v}^\lambda(S)$, *for all* $\lambda \in \mathbb{R}^N_{++}$.

(c) $\sum_{i\in S} \lambda(i)x(i) \leq \mathbf{v}^\lambda(S)$ *for all* $\lambda \in \mathbb{R}^N_+$.

(d) $\sum_{i\in S} \lambda(i)x(i) \leq \mathbf{v}^\lambda(S)$ *for all* $\lambda \in \mathbb{R}^N_{++}$ *such that* $\sum_{i\in N} \lambda(i) = 1$.

PROOF

Evidently, (a) implies (b), and (b) and (d) are equivalent. It is thus enough to show that (b) implies (c) and (c) implies (a). To see the former claim, take any $\lambda \in \mathbb{R}^N_+$ and define $I := \{i \in S : \lambda(i) = 0\}$. By (b), we have, for any $\varepsilon > 0$,

$$\varepsilon \sum_{i\in I} x(i) + \sum_{i\in S\setminus I} \lambda(i)x(i) \leq \max\left\{\varepsilon \sum_{i\in I} y(i) + \sum_{i\in S\setminus I} \lambda(i)y(i) : y \in \mathbf{V}(S)\right\},$$

so that, by letting $\varepsilon \searrow 0$, and using the Maximum Theorem, we get

$$\sum_{i\in S} \lambda(i)x(i) = \sum_{i\in S\setminus I} \lambda(i)x(i) \leq \max\left\{\sum_{i\in S\setminus I} \lambda(i)y(i) : y \in \mathbf{V}(S)\right\} = \mathbf{v}^\lambda(S),$$

establishing (c).

To prove that (c) implies (a), assume that (a) is false, that is, $x \in \mathbb{R}^S_+ \setminus \mathbf{V}(S)$. Let $\theta^* := \inf\{\theta \geq 0 : \theta x \in \mathbf{V}(S)\}$. By compactness of $\mathbf{V}(S)$, the map $\theta^* x$ belongs to the boundary of $\mathbf{V}(S)$ in $\mathbb{R}^S$ and $\theta^* < 1$. (Verify!) Since $\mathbf{V}(S)$ is closed and convex, we may then use the Minkowski Supporting Hyperplane Theorem to find a $\lambda \in \mathbb{R}^N \setminus \{\mathbf{0}\}$ such that

$$\sum_{i\in S} \lambda(i)(\theta^* x(i)) \geq \sum_{i\in S} \lambda(i)y(i) \qquad \text{for all } y \in \mathbf{V}(S).$$

Since it is easy to verify that $\lambda \geq 0$ by using the convexity of $\mathbf{V}(S)$ (please do so), and since $\theta^* < 1$, we thus find

$$\mathbf{v}^\lambda(S) = \theta^* \sum_{i\in S} \lambda(i)x(i) < \sum_{i\in S} \lambda(i)x(i)$$

which shows that (c) is false. ∎

EXERCISE 27 Given any $N \in \mathcal{N}$, let $(N, \mathbf{V})$ be a coalitional game without side payments. For any given $\lambda \in \mathbb{R}^N_{++}$ with $\sum_{i \in N} \lambda(i) = 1$, a map $x \in \mathbf{V}(N)$ is called a λ-**transfer value** for $\mathbf{V}$ if

$$\lambda(i)x(i) = L_i^S(N, \mathbf{v}^\lambda), \qquad i = 1, \ldots, n,$$

where L^S is the Shapley value. Let $\Psi^S(N, \mathbf{V})$ denote the set of all λ-transfer values for $\mathbf{V}$.

(a) Prove that if $N = \{1, 2\}$, then $\Psi^S(N, \mathbf{V}) = \{f^{\text{Nash}}(\mathbf{V}(N), \mathbf{0})\}$.

(b) (*Roth's Example*) Let $(N, \mathbf{V})$ be the following coalitional game without side payments: $N := \{1, 2, 3\}$, $\mathbf{V}(\{i\}) = \{0\}$, $i = 1, 2, 3$, $\mathbf{V}(\{1, 2\}) = [0, \frac{1}{2}]^2$, $\mathbf{V}(\{1, 3\}) = \mathbf{V}(\{2, 3\}) = \{0\} \times [0, \frac{1}{2}]$, and $\mathbf{V}(\{1, 2, 3\}) = co\{\mathbf{0}, (\frac{1}{2}, \frac{1}{2}, 0), (0, 0, 1)\}$. Show that $\Psi^S(N, \mathbf{V}) = \{(\frac{1}{3}, \frac{1}{3}, \frac{1}{3})\}$. Do you find this normatively reasonable? How about descriptively?

*EXERCISE 28[H] Given any $N \in \mathcal{N}$, and any coalitional game $(N, \mathbf{V})$ without side payments, prove that $\Psi^S(N, \mathbf{V}) \neq \emptyset$.

Part IV

ANALYSIS ON METRIC/NORMED LINEAR SPACES

CHAPTER I

Metric Linear Spaces

In Chapters C–G we laid a foundation for studying a number of issues that arise in metric spaces (such as continuity and completeness) and others that arise in linear spaces (such as linear extensions and convexity). However, save for a few exceptions in the context of Euclidean spaces, we have so far studied such matters in isolation from each other. This situation should be remedied, for in most applications one works with a space structure that allows for the simultaneous consideration of metric and linear properties. In this chapter, therefore, we bring our earlier metric and linear analyses together and explore a framework that is general enough to encompass such situations. In particular, this setup will allow us to talk about things like continuous linear functions, complete linear spaces, closed convex sets, and so on.

We begin the chapter by discussing in what sense one may think of a metric structure to be imposed on a linear space as "compatible" with the inherent algebraic structure of that space. This leads us to the notion of *metric linear space.* After going through several examples of such linear spaces, we derive some elementary properties pertaining to them, and examine when a linear functional defined on such a space would be continuous. We discuss the basic properties of continuous linear functionals here in some detail, and highlight the significance of them from a geometric viewpoint. We then consider several characterizations of finite-dimensional metric linear spaces, and, finally, provide a primer on convex analysis on infinite-dimensional metric linear spaces. The most important results within this framework will clarify the basic connection between the notions of "openness" and "algebraic openness," and sharpen the separation theorems we obtained in Chapter G.

A common theme that keeps emerging throughout the chapter pertains to the crucial differences between the finite- and infinite-dimensional metric linear spaces. In particular, you will see that the infinite-dimensional case contains a good deal of surprises, such as the presence of discontinuous linear functionals, linear subspaces that are not closed, and open neighborhoods the closures of which are necessarily noncompact. So, hopefully, it should be fun![1]

1 Metric Linear Spaces

Take a linear space X. Given the objectives of this chapter, we wish to endow X with a metric structure that is compatible with the algebraic/geometric structure of X. We thus need to choose a distance function on X that goes well with the linear structure of X.

Let's think of the things we can do in X. We can shift around the vectors in X (by using the vector addition operation) and, for any given vector in X, we can construct the ray that originates from the origin $\mathbf{0}$ and goes through this vector (by using the scalar multiplication operation). Let's focus on shifting—that is, *translating*—vectors first. What we mean by this precisely is that, for any given $x, z \in X$, the vector $x + z$ is well-defined in X, and geometrically speaking, we think of this vector as the "translation of x by z" (or the "translation of z by x"). Now take another vector $y \in X$, and translate it by z as well to obtain the vector $y + z$. Here is a good question: How should the distance between x and y relate to that between $x + z$ and $y + z$? If we wish to model a space whose geometry is *homogeneous*—a term we will formalize later on—it makes sense to require these two distances be equal. After all, if posited globally, this sort of a property would ensure that the distance between any two given points be preserved when these vectors are translated (in the same way) anywhere else in the space.[2] This is, for instance, precisely the case for the geometry of the plane (or of any other Euclidean space).

[1] In textbooks on functional analysis, metric linear spaces are covered, at best, as special cases of either topological groups or topological linear spaces. Since I do not presume familiarity with general topology here, and since topological considerations arise in economic applications mostly through metric spaces, I chose to depart from the standard treatments. If you wish to see the results I present here in their natural (topological-algebraic) habitat, you should try outlets like Holmes (1975), Megginson (1998), or Aliprantis and Border (1999).

[2] So far I have tried to be careful in referring to the elements of a metric space as "points" and those of a linear space as "vectors." The spaces we work with from now on will be both metric and linear, so I will use these two terms interchangeably.

Definition

Let X be a linear space. A metric $d \in \mathbb{R}_+^{X \times X}$ is called **translation invariant** if

$$d(x+z, y+z) = d(x,y) \qquad \text{for all } x,y,z \in X. \tag{1}$$

Observe that (1) is a statement that ties the metric d with the group structure of the linear space X. Translation invariance is thus a concept that is partly metric and partly algebraic; it connects in a particular way the distance function on a linear space with the operation of vector addition. In fact, this connection is tighter than it may appear at first. If you endow a linear space with a translation-invariant metric, then, perforce, you make the vector addition operation of your space continuous!

Proposition 1

Let X be a linear space that is also a metric space. If the metric d of X is translation invariant, then the map $(x,y) \mapsto x+y$ is a continuous function from $X \times X$ into X.

Proof

Suppose d is translation invariant, and take any $(x^m), (y^m) \in X^\infty$ and $x, y \in X$ with $x^m \to x$ and $y^m \to y$. Then, for any $m = 1, 2, \ldots$, using (1) and the triangle inequality, we find

$$\begin{aligned} d(x^m + y^m, x+y) &= d(x^m + y^m - (x + y^m), x + y - (x + y^m)) \\ &= d(x^m - x, y - y^m) \\ &\leq d(x^m - x, \mathbf{0}) + d(\mathbf{0}, y - y^m) \\ &= d(x^m, x) + d(y^m, y), \end{aligned}$$

so $d(x^m + y^m, x + y) \to 0$, that is, $x^m + y^m \to x + y$. ∎

Exercise 1 Let X be a linear space that is also a metric space. Show that if the metric d of X is translation invariant, then, for any $k \in \mathbb{Z}$, the self-map f on X, defined by $f(x) := kx$, is continuous.

Unfortunately, translation invariance alone does not yield a sufficiently rich framework; we need a somewhat tighter link between the metric and linear structures in general. For instance, metrizing a linear space by means of a translation-invariant distance function does not guarantee the continuity of the scalar multiplication operation. As an extreme example, consider metrizing a given nontrivial metric space X by the discrete metric (which is, obviously, translation invariant). Then the map $\lambda \mapsto \lambda x$ (from $\mathbb{R}$ into X) is not continuous for any $x \in X\backslash\{\mathbf{0}\}$. (*Proof.* If $x \neq \mathbf{0}$, then $\left(\frac{1}{m}x\right)$ is a sequence that is not eventually constant, so endowing X with the discrete metric d yields $d\left(\frac{1}{m}x, \mathbf{0}\right) = 1$ for each m.)

To connect the metric and linear structures in a more satisfactory manner, therefore, we need to do better than asking for the translation invariance property.[3] In particular, we should certainly make sure that a map like $\lambda \mapsto \lambda x$ (in $X^{\mathbb{R}}$) and a map like $x \mapsto \lambda x$ (in X^X) are declared continuous by the metric we impose on X. Well, then, why don't we concentrate on the case where the metric at hand renders the map $(\lambda, x) \mapsto \lambda x$ (in $X^{\mathbb{R}\times X}$) continuous? We know that translation invariance gives us the continuity of the vector addition operation. This would, in turn, render the scalar multiplication operation on X continuous. Putting these together, we arrive at a class of spaces that are both metric and linear, and whose metric and linear structures are naturally compatible.

DEFINITION

Let X be a linear space that is also a metric space. If the metric d of X is translation invariant, and, for all convergent $(\lambda_m) \in \mathbb{R}^\infty$ and $(x^m) \in X^\infty$, we have

$$\lim \lambda_m x^m = \left(\lim \lambda_m\right)\left(\lim x^m\right),$$

then X is called a **metric linear space**. If, in addition, (X, d) is complete, then we say that X is a **Fréchet space.** A metric linear space X is said to be **nontrivial** if $X \neq \{\mathbf{0}\}$, **finite-dimensional** if $dim(X) < \infty$, and **infinite dimensional** if $dim(X) = \infty$.

Put succinctly, a metric linear space is a linear space endowed with a translation-invariant distance function that renders the scalar multiplication

[3] To be fair, I should say that a reasonably complete algebraic theory can be developed using only the translation invariance property, but this theory would lack geometric competence, so I will not pursue it here. (But see Exercises 15 and 16.)

operation continuous.[4] So, it follows from Proposition 1 that X is a metric linear space iff it is both a linear and a metric space such that

(i) the distance between any two points are preserved under the identical translations of these points,

(ii) the scalar multiplication map $(\lambda, x) \mapsto \lambda x$ is a continuous function from $\mathbb{R}\times X$ into X, and

(iii) the vector addition map $(x, y) \mapsto x + y$ is a continuous function from $X \times X$ into X.

Let's look at some examples.

Example 1

[1] $\mathbb{R}^n$ *is a Fréchet space for any* $n \in \mathbb{N}$. This is easily proved by using the fact that a sequence converges in a Euclidean space iff each of its coordinate sequences converges in $\mathbb{R}$. (Similarly, $\mathbb{R}^{n,p}$ is a Fréchet space for any $n \in \mathbb{N}$ and $1 \leq p \leq \infty$.)

[2] If we metrize $\mathbb{R}$ by the discrete metric, we do not obtain a metric linear space, even though this metric is translation invariant. By contrast, if we metrize $\mathbb{R}$ by $d \in \mathbb{R}_+^{\mathbb{R}\times\mathbb{R}}$ with $d(a, b) := \left|a^3 - b^3\right|$, then we guarantee that the scalar multiplication operation on $\mathbb{R}$ is continuous, but we do not make $\mathbb{R}$ a metric linear space because the metric d is not translation invariant.

[3] Consider the linear space $\mathbb{R}^\infty$ of all real sequences that is metrized by means of the product metric:

$$\rho((x_1, x_2, \ldots), (y_1, y_2, \ldots)) := \sum_{i=1}^{\infty} \tfrac{1}{2^i} \min\{1, \left|x_i - y_i\right|\}.$$

[4] Some authors assume that the vector addition operation is continuous, instead of positing the translation invariance of the metric when defining a metric linear space. My definition is thus a bit more demanding than the usual. Yet insofar as the topological properties are concerned, this is only a matter of convention, since, by a well-known result of Shizuo Kakutani, every metric linear space is homeomorphic to a translation-invariant metric linear space. (For a proof, see Rolewicz (1985, pp. 2–4).)

Warning. What I call here a Fréchet space is referred to as an F-space in some texts, which reserve the term "Fréchet space" for locally convex and complete metric linear spaces. (I will talk about the latter type of spaces in the next chapter.)

(Recall Section C.8.2.) This metric is obviously translation invariant. To see that it renders the scalar multiplication operation on $\mathbb{R}^\infty$ continuous, take any $(\lambda_m) \in \mathbb{R}^\infty$ and any sequence (x^m) in $\mathbb{R}^\infty$ such that $\lambda_m \to \lambda$ and $x^m \to x$ for some $(\lambda, x) \in \mathbb{R} \times \mathbb{R}^\infty$. (Of course, (x^m) is a sequence of sequences.) By Proposition C.8, $x_i^m \to x_i$, so we have $\lambda_m x_i^m \to \lambda x_i$, for each $i \in \mathbb{N}$. Applying Proposition C.8 again, we find $\lim \lambda_m x^m = \lambda x$. Combining this observation with Theorem C.4, therefore, we may conclude: *$\mathbb{R}^\infty$ is a Fréchet space.*

[4] *ℓ^p is a metric linear space for any $1 \leq p < \infty$.* Fix any such p. That d_p is translation invariant is obvious. To see the continuity of the scalar multiplication operation, take any $(\lambda_m) \in \mathbb{R}^\infty$ and any sequence (x^m) in ℓ^p such that $\lambda_m \to \lambda$ and $d_p(x^m, x) \to 0$ for some $(\lambda, x) \in \mathbb{R} \times \ell^p$. By the triangle inequality, we have

$$\begin{aligned} d_p(\lambda_m x^m, \lambda x) &\leq d_p(\lambda_m x^m, \lambda x^m) + d_p(\lambda x^m, \lambda x) \\ &= \left(\sum_{i=1}^{\infty} |\lambda_m - \lambda|^p \left|x_i^m\right|^p\right)^{\frac{1}{p}} + \left(\sum_{i=1}^{\infty} |\lambda|^p \left|x_i^m - x_i\right|^p\right)^{\frac{1}{p}} \\ &= |\lambda_m - \lambda|\, d_p(x^m, \mathbf{0}) + |\lambda|\, d_p(x_m, x). \end{aligned}$$

But since (x^m) is convergent, $(d_p(x^m, \mathbf{0}))$ is a bounded real sequence (for, $d_p(x^m, \mathbf{0}) \leq d_p(x^m, x) + d_p(x, \mathbf{0}) \to d_p(x, \mathbf{0})$). Therefore, the inequality above ensures that $d_p(\lambda_m x^m, \lambda x) \to 0$, as was to be proved.

Combining this observation with Example C.11.[4], we may conclude: *ℓ^p is a Fréchet space for any $1 \leq p < \infty$.*

[5] *ℓ^∞ is a Fréchet space.*

[6] For any metric space T, both $\mathbf{B}(T)$ and $\mathbf{CB}(T)$ are metric linear spaces. In fact, by Example C.11.[5] and Proposition D.7, both of these spaces are Fréchet spaces. □

EXERCISE 2 Supply the missing arguments in Examples 1.[5] and 1.[6].

EXERCISE 3 Show that the metric space introduced in Exercise C.42 is a metric linear space (under the pointwise defined operations) that is not Fréchet.

EXERCISE 4 Show that $\mathbf{C}^1[0, 1]$ is a Fréchet space.

EXERCISE 5 Let $X := \{(x_m) \in \mathbb{R}^\infty : \sup\{|x_m|^{\frac{1}{m}} : m \in \mathbb{N}\} < \infty\}$, and define $d \in \mathbb{R}_+^{X \times X}$ by $d((x_m), (y_m)) := \sup\{|x_m - y_m|^{\frac{1}{m}} : m \in \mathbb{N}\}$. Is X a metric linear space relative to d and the usual addition and scalar multiplication operations?

EXERCISE 6 (*Product Spaces*) Let $(X_1, X_2, \ldots)$ be a sequence of metric linear spaces, and $X := \mathsf{X}^\infty X_i$. We endow X with the product metric (Section C.8.2), and make it a linear space by defining the operations of scalar multiplication and vector addition pointwise. Show that X is a metric linear space, and it is a Fréchet space whenever each X_i is complete.

EXERCISE 7 Let X be any linear space, and φ a seminorm on X (Section G.2.1). Define $d^\varphi \in \mathbb{R}_+^{X \times X}$ by $d^\varphi(x, y) := \varphi(x - y)$. Is d^φ necessarily a distance function? Show that X would become a metric linear space when endowed with d^φ, provided that $\varphi^{-1}(0) = \{\mathbf{0}\}$.

The following proposition collects some basic facts about metric linear spaces, and provides a good illustration of the interplay between algebraic and metric considerations that is characteristic of metric linear spaces.

PROPOSITION 2
For any subsets A and B of a metric linear space X, the following are true:

(a) $cl_X(A + x) = cl_X(A) + x$ *for all* $x \in X$.

(b) *If A is open, then so is $A + B$.*

(c) *If A is compact and B is closed, then $A + B$ is closed.*

(a) *If both A and B are compact, so is $A + B$.*

PROOF
We only prove part (c) here, leaving the proofs of the remaing three claims as exercises. Let (x^m) be a sequence in $A + B$ such that $x^m \to x$ for some $x \in X$. By definition, there exist a sequence (y^m) in A and a sequence (z^m) in B such that $x^m = y^m + z^m$ for each m. Since A is compact, Theorem C.2 implies that there exists a strictly increasing sequence (m_k) in $\mathbb{N}$ such that (y^{m_k}) converges to a vector, say y, in A. But then (z^{m_k}) converges to

$x - y$ by the continuity of vector addition, so, since B is closed, it follows that $x - y \in B$. Thus, $x = y + (x - y) \in A + B$, which proves that $A + B$ is closed. ∎

Exercise 8[H] Complete the proof of Proposition 2.

Parts (b) and (c) of Proposition 2 point to a significant difference in the behavior of sums of open sets and closed sets: While the sum of any two open sets is open, the sum of two closed sets need not be closed. This difference is worth keeping in mind. It exists even in the case of our beloved real line. For instance, let $A := \mathbb{N}$ and $B := \{-2+\frac{1}{2}, -3+\frac{1}{3}, \ldots\}$. Then both A and B are closed subsets of $\mathbb{R}$, but $A + B$ is not closed in $\mathbb{R}$, for $\frac{1}{m} \in A + B$ for each $m = 1, 2, \ldots$, but $\lim \frac{1}{m} = 0 \notin A + B$. As another (more geometric example), let $A := \{(x, y) \in \mathbb{R}^2 : x \neq 0 \text{ and } y \geq \frac{1}{x}\}$ and $B := \mathbb{R} \times \{0\}$. (Draw a picture.) Then, although both A and B are closed in $\mathbb{R}^2$, $A + B$ is not a closed subset of $\mathbb{R}^2$. (*Proof.* $A + B = \mathbb{R} \times \mathbb{R}_{++}$.)

The most common method of creating a new metric linear space from a given metric linear space is to look for a subset of this space that inherits the metric linear structure of the original space. This leads us to the notion of a **metric linear subspace** of a metric linear space X, which is defined as a subset of X that is both a linear and a metric subspace of X. (For instance, $\mathbb{R}^2 \times \{0\}$ is a metric linear subspace of $\mathbb{R}^3$, and $\mathbf{C}[0, 1]$ is a metric linear subspace of $\mathbf{B}[0, 1]$.) Throughout this chapter, by a **subspace** of a metric linear space X, we mean a metric linear subspace of X.

Exercise 9[H] Find a subspace of $\mathbb{R}^\infty$ that is not closed. Is this subspace dense in $\mathbb{R}^\infty$?

Exercise 10 Show that the closure of any subspace (affine manifold) of a metric linear space X is a subspace (affine manifold) of X.

Exercise 11 For any subsets A and B of a metric linear space X, prove:

(a) $cl_X(A) + cl_X(B) \subseteq cl_X(A + B)$.

(b) $cl_X(cl_X(A) + cl_X(B)) = cl_X(A + B)$.

(c) $int_X(A) + int_X(B) \subseteq A + int_X(B) \subseteq int_X(A + B)$, provided that $int_X(B) \neq \emptyset$.

Exercise 12[H] Show that every metric linear space is connected.

EXERCISE 13[H] (Nikodem) Let X and Y be two metric linear spaces, S a nonempty convex subset of X, and $\Gamma : S \rightrightarrows Y$ a correspondence that has a closed graph. Show that if

$$\tfrac{1}{2}\Gamma(x) + \tfrac{1}{2}\Gamma(y) \subseteq \Gamma\left(\tfrac{1}{2}x + \tfrac{1}{2}y\right) \qquad \text{for any } x, y \in S,$$

then Γ is a convex correspondence (Exercise G.8).

2 Continuous Linear Operators and Functionals

The primary objects of analysis within the context of metric linear spaces are those linear operators L that map a metric linear space X (with metric d) to another metric linear space Y (with metric d_Y) such that, for any $x \in X$ and $\varepsilon > 0$, there exists a $\delta > 0$ with

$$d(x, y) < \delta \quad \text{implies} \quad d_Y(L(x), L(y)) < \varepsilon.$$

For obvious reasons, we call any such map a **continuous linear operator**, except when $Y = \mathbb{R}$, in which case we refer to it as a **continuous linear functional**.

2.1 Examples of (Dis-)Continuous Linear Operators

Let's have a look at a few concrete examples of continuous linear operators.

EXAMPLE 2

[1] Given any positive integers m and n, any *linear operator that maps* $\mathbb{R}^n$ *to* $\mathbb{R}^m$ *is a continuous linear operator.* This follows from Examples D.2.[4] and F.6. (Verify!)

[2] Let $L \in \mathbb{R}^{\mathbf{B}[0,1]}$ be defined by $L(f) := f(0)$. L is obviously a linear functional on $\mathbf{C}[0,1]$. It is also continuous—in fact, it is a nonexpansive map—because

$$|L(f) - L(g)| = |f(0) - g(0)| \le d_\infty(f, g) \qquad \text{for any } f, g \in \mathbf{B}[0,1].$$

[3] Let $L \in \mathbb{R}^{\mathbf{C}[0,1]}$ be defined by

$$L(f) := \int_0^1 f(t)dt.$$

It follows from Riemann integration theory that L is a linear functional (Proposition A.13 and Exercise A.57). Moreover, L is a nonexpansive map: For any $f, g \in \mathbf{C}[0,1]$, we have

$$\left|L(f) - L(g)\right| = \left|\int_0^1 (f(t) - g(t))dt\right| \leq d_\infty(f, g)$$

by Proposition A.12. Conclusion: L is a continuous linear functional on $\mathbf{C}[0,1]$. □

Most students are startled when they hear the term "continuous linear functional." Aren't linear functions always continuous? The answer is no, not necessarily. True, any linear real function on a Euclidean space is continuous, but this is not the case when the linear function under consideration is defined on an infinite-dimensional metric linear space. Here are a few examples that illustrate this.

EXAMPLE 3

[1] Let X be the linear space of all real sequences that are absolutely summable, that is,

$$X := \left\{(x_1, x_2, \ldots) \in \mathbb{R}^\infty : \sum_{i=1}^{\infty} |x_i| < \infty\right\}.$$

Let us make this space a metric linear space by using the sup-metric d_∞. (Notice that X is a metric linear space that differs from ℓ^1 only in its metric structure.) Define $L \in \mathbb{R}^X$ by

$$L(x_1, x_2, \ldots) := \sum_{i=1}^{\infty} x_i,$$

which is obviously a linear functional on X. Is L continuous? No! Consider the following sequence $(x^m) \in X^\infty$:

$$x^m := \left(\tfrac{1}{m}, \ldots, \tfrac{1}{m}, 0, 0, \ldots\right), \qquad m = 1, 2, \ldots,$$

where exactly m entries of x^m are nonzero. Clearly, $d_\infty(x^m, \mathbf{0}) = \frac{1}{m} \to 0$. Yet $L(x^m) = 1$ for each m, so $L(\lim x^m) = 0 \neq 1 = \lim L(x^m)$. Conclusion: L is linear but it is *not* continuous at $\mathbf{0}$.

[2] Let's play on this theme a bit more. Let

$$X := \{(x_1, x_2, \ldots) \in \mathbb{R}^\infty : \sup\{|x_i| : i \in \mathbb{N}\} < \infty\},$$

and define $L \in \mathcal{L}(X, \mathbb{R})$ by

$$L(x_1, x_2, \ldots) := \sum_{i=1}^{\infty} \delta^i x_i,$$

where $0 < \delta < 1$. It is easy to verify that if X was endowed with the sup-metric, then L would be a continuous function. (You now know that this would not be true if $\delta = 1$. Yes?) But what if we endowed X with the product metric? Then, interestingly, L would not be continuous. Indeed, if $x^1 := (\frac{1}{\delta}, 0, 0, \ldots)$, $x^2 := (0, \frac{1}{\delta^2}, 0, \ldots)$, etc., then $x^m \to (0, 0, \ldots)$ with respect to the product metric (Proposition C.8), and yet $L(x^m) = 1$ for each m while $L(0, 0, \ldots) = 0$. □

Exercise 14[H] Let X denote the linear space of all continuously differentiable real functions on $[0, 1]$, and make this space a metric linear space by using the sup-metric d_∞. (Thus X is a (dense) subspace of $\mathbf{C}[0, 1]$; it is *not* equal to $\mathbf{C}^1[0, 1]$.) Define $L \in \mathcal{L}(X, \mathbb{R})$ and $D \in \mathcal{L}(X, \mathbf{C}[0, 1])$ by $L(f) := f'(0)$ and $D(f) := f'$, respectively. Show that neither D nor L is continuous.

While it is true that linearity of a map does not necessarily guarantee its continuity, it still brings quite a bit of discipline into the picture. Indeed, linearity spreads even the tiniest bit of continuity a function may have onto the entire domain of that function. Put differently, there is no reason to distinguish between local and global continuity concepts in the presence of linearity. (This is quite reminiscent of additive real functions on $\mathbb{R}$; recall Exercise D.40.) We formalize this point next.

Proposition 3
Let X and Y be two metric linear spaces. A linear operator L from X into Y is uniformly continuous if, and only if, it is continuous at $\mathbf{0}$.

Proof
Let $L \in \mathcal{L}(X, Y)$ be continuous at $\mathbf{0}$, and take any $\varepsilon > 0$ and $x \in X$. By continuity at the origin, there exists a $\delta > 0$ such that $d_Y(L(z), \mathbf{0}) < \varepsilon$

whenever $d(z, \mathbf{0}) < \delta$. Now take any $y \in X$ with $d(x, y) < \delta$. By translation invariance of d, we have $d(x-y, \mathbf{0}) < \delta$, and hence, by translation invariance of d_Y and linearity,

$$d_Y(L(x), L(y)) = d_Y(L(x) - L(y), \mathbf{0}) = d_Y(L(x-y), \mathbf{0}) < \varepsilon.$$

Since y is arbitrary in X here, and δ does not depend on x, we may conclude that L is uniformly continuous.[5] ■

Even though its proof is easy, a newcomer to the topic may be a bit surprised by Proposition 3. Let us try to think through things more clearly here. The key observation is that any metric linear space Z is **homogeneous** in the sense that, for any given $z^0, z^1 \in Z$, we can map Z onto Z in such a way that (i) z^0 is mapped to z^1, and (ii) all topological properties of Z are left intact. More precisely, the translation map $\tau : Z \to Z$ defined by $\tau(z) := z + (z^1 - z^0)$ is a homeomorphism, thanks to the continuity of vector addition on Z. But if we know that an $L \in \mathcal{L}(X, Y)$ is continuous at a given point, say $\mathbf{0}$, to show that L must then be continuous at an arbitrary point $x^0 \in X$, all we need do is to translate X so that x^0 "becomes" $\mathbf{0}$ (of X), and translate Y so that $\mathbf{0}$ (of Y) "becomes" $y^0 := L(x^0)$. So define $\tau : X \to X$ by $\tau(x) := x - x^0$, and $\rho : Y \to Y$ by $\rho(y) := y + y^0$. Since τ and ρ are continuous everywhere, and L is continuous at $\mathbf{0}$, and since $L = \rho \circ L \circ \tau$, it follows that L is continuous at x^0. Thus linearity (in fact, additivity) spreads continuity at a single point to the entire domain precisely via translation maps that are always continuous in a metric linear space. (The situation could be drastically different if the metric and linear structures of the space were not compatible enough to yield the continuity of vector addition.)

In real analysis, one often gets a "clearer" view of things upon suitably generalizing the mathematical structure at hand. The following exercises aim to clarify the origin of Proposition 3 further by means of such a generalization.

[5] *Another proof.* If L is continuous at $\mathbf{0}$, then, for any $\varepsilon > 0$, there exists a $\delta > 0$ such that $L(N_{\delta,X}(\mathbf{0})) \subseteq N_{\varepsilon,Y}(L(\mathbf{0})) = N_{\varepsilon,Y}(\mathbf{0})$, and thus, for any $x \in X$,

$$L(N_{\delta,X}(x)) = L(x + N_{\delta,X}(\mathbf{0})) = L(x) + L(N_{\delta,X}(\mathbf{0})) \subseteq L(x) + N_{\varepsilon,Y}(\mathbf{0}) = N_{\varepsilon,Y}(L(x)),$$

and we are done. (See, all I need here is the *additivity* of L. But wait, where did I use the translation invariance of d and d_Y?)

EXERCISE 15 We say that $(X, +, d)$ is a **metric group** if (X, d) is a metric space and $(X, +)$ is a group such that the binary relation $+$ is a continuous map from $X \times X$ into X. For any $x^* \in X$, we define the self-map τ on X by $\tau(x) := x - x^*$, which is called a **left translation**. (*Right translations* are defined as maps of the form $x \mapsto -x^* + x$, and coincide with left translations when $(X, +)$ is Abelian.)

(a) Show that any left or right translation on X is a homeomorphism.

(b) Show that if O is an open set that contains the identity element $\mathbf{0}$ of X, then so is $-O$.

(c) Show that if O is an open set that contains the identity element $\mathbf{0}$ of X, then there exists an open subset U of X with $\mathbf{0} \in U$ and $U + U \subseteq O$.

EXERCISE 16 Let $(X, +_X, d_X)$ and $(Y, +_Y, d_Y)$ be two metric groups, and consider a map $h : X \to Y$ such that $h(x +_X y) = h(x) +_Y h(y)$ for all $x, y \in X$. (Recall that such a map is called a **homomorphism** from X into Y.) Prove that h is continuous iff it is continuous at $\mathbf{0}$. (How does this fact relate to Proposition 3?)

It is now time to consider a nontrivial example of a continuous linear functional defined on an infinite-dimensional metric linear space.

EXAMPLE 4

For any $n \in \mathbb{N}$, Examples 2.[1] and F.6 show that any continuous linear functional L on $\mathbb{R}^n$ is of the form $x \mapsto \sum^n \alpha_i x_i$ (for some real numbers $\alpha_1, \ldots, \alpha_n$). We now wish to determine the general structure of continuous linear functionals on $\mathbb{R}^\infty$.

Take any continuous $L \in \mathcal{L}(\mathbb{R}^\infty, \mathbb{R})$. Define $\alpha_i := L(\mathbf{e}^i)$ for each $i \in \mathbb{N}$, where $\mathbf{e}^1 := (1, 0, 0, \ldots)$, $\mathbf{e}^2 := (0, 1, 0, \ldots)$, etc. We claim that $\alpha_i = 0$ for all but finitely many i. To see this, note that

$$(x_m) = \lim_{M \to \infty} \sum_{i=1}^{M} x_i \mathbf{e}^i \quad \text{for any } (x_m) \in \mathbb{R}^\infty.$$

(*Proof.* Observe that the mth term of kth term of the sequence of sequences $(x_1\mathbf{e}^1,\ x_1\mathbf{e}^1 + x_2\mathbf{e}^2, \ldots)$ is x_m for any $k \geq m$ and $m \in \mathbb{N}$, and apply

Proposition C.8.) So, by continuity and linearity of L, for any $(x_m) \in \mathbb{R}^\infty$,

$$L((x_m)) = L\left(\lim_{M\to\infty} \sum_{i=1}^{M} x_i \mathbf{e}^i\right) = \lim_{M\to\infty} L\left(\sum_{i=1}^{M} x_i \mathbf{e}^i\right) = \lim_{M\to\infty} \sum_{i=1}^{M} \alpha_i x_i. \quad (2)$$

Since L is real-valued, this can't hold true for *all* $(x_m) \in \mathbb{R}^\infty$, unless $\alpha_i = 0$ for all but finitely many i.[6] Thus $S := \{i \in \mathbb{N} : \alpha_i \neq 0\}$ is finite, and (2) yields

$$L((x_m)) = \sum_{i\in S} \alpha_i x_i \qquad \text{for any } (x_m) \in \mathbb{R}^\infty. \quad (3)$$

Since it is easy to check that this indeed defines a continuous linear functional on $\mathbb{R}^\infty$, we conclude: *L is a continuous linear functional on $\mathbb{R}^\infty$ if, and only if, there exists a finite subset S of $\mathbb{N}$ and real numbers α_i, $i \in S$, such that* (3) *holds.*[7] □

REMARK 1. The argument given in Example 4 relies on the fact that $(x_m) = \lim_{M\to\infty} \sum^M x_i \mathbf{e}^i$ for any real sequence (x_m). This may perhaps tempt you to view the set $\{\mathbf{e}^i : i \in \mathbb{N}\}$ as a basis for $\mathbb{R}^\infty$. However, this is not the case, for we may not be able to express a real sequence as a linear combination of *finitely* many $\mathbf{e}^i$s. (Consider, for instance, the sequence $(1, 1, \ldots)$.) Instead, one says that $\{\mathbf{e}^i : i \in \mathbb{N}\}$ is a *Schauder basis* for $\mathbb{R}^\infty$. The defining feature of this concept is expressing the vectors in a metric linear space as a linear infinite series. As opposed to the standard one, this basis concept depends on the metric in question since it involves the notion of "convergence" in its definition. (More on this in Section J.2.2.) □

EXERCISE 17 Let S be a finite subset of $[0, 1]$. Show that, for any $\lambda \in \mathbb{R}^S$, the map $L \in \mathbb{R}^{\mathbf{C}[0,1]}$ defined by $L(f) := \sum_{x\in S} \lambda(x)f(x)$ is a continuous linear functional on $\mathbf{C}[0, 1]$.

EXERCISE 18 Give an example of a discontinuous linear functional on ℓ^∞.

[6] If there was a subsequence (α_{m_k}) in $\mathbb{R}\backslash\{0\}$, by defining (x_m) as $x_m := \frac{1}{\alpha_m}$ if $m = m_k$ and $x_m := 0$ otherwise, we would get $\sum^\infty \alpha_i x_i = 1 + 1 + \cdots = \infty$, contradicting (2).

[7] *Corollary*. There is no strictly positive continuous linear functional on $\mathbb{R}^\infty$. (In fact, there is no strictly positive linear functional on $\mathbb{R}^\infty$, but showing this requires more work.)

EXERCISE 19 Prove that an upper semicontinuous linear functional on a metric linear space is continuous.

EXERCISE 20[H] Show that a linear functional L on a metric linear space X is continuous iff there exists a continuous seminorm φ on X such that $|L(x)| \leq \varphi(x)$ for all $x \in X$.

EXERCISE 21[H] Let L be a linear functional on a metric linear space X. Prove:

(a) L is continuous iff it is bounded on some open neighborhood O of $\mathbf{0}$.

(b) If L is continuous, then $L(S)$ is a bounded set for any bounded subset S of X.

The following two exercises further develop the theory of linear correspondences sketched in Exercise F.31. They presume familiarity with the definitions and results of that exercise.

EXERCISE 22[H] (*Continuous Linear Correspondences*) Let X and Y be two metric linear spaces and $\Gamma : X \rightrightarrows Y$ a linear correspondence.

(a) Show that Γ is upper hemicontinuous iff it is upper hemicontinuous at $\mathbf{0}$.

(b) Show that Γ is lower hemicontinuous iff it is lower hemicontinuous at $\mathbf{0}$.

EXERCISE 23[H] (*Continuous Linear Selections*) Let X and Y be two metric linear spaces and $\Gamma : X \rightrightarrows Y$ a linear correspondence.

(a) Show that if Γ admits a continuous linear selection, then it is continuous.

(b) Prove: If $P \in \mathcal{L}(\Gamma(X), \Gamma(X))$ is continuous, idempotent (i.e., $P \circ P = P$) and $null(P) = \Gamma(\mathbf{0})$, then $P \circ \Gamma$ is a continuous linear selection from Γ.

(c) In the special case where X and Y are Euclidean spaces, prove that if Γ is upper hemicontinuous, then it admits a continuous linear selection.

2.2 Continuity of Positive Linear Functionals

Is a positive linear functional defined on a preordered metric linear space necessarily continuous? A very good question, to be sure. Monotonic real functions possess, in general, reasonably strong continuity properties. (Any such function on $\mathbb{R}$ is, for instance, continuous everywhere but countably many points.) So, while a linear functional need not be continuous in general, perhaps monotonic linear functionals are.

The bad news is that the answer is no! Indeed, all of the linear functionals considered in Example 3 are positive (with $\mathbb{R}^\infty$ being partially ordered by means of the coordinatewise order $\geq$), but as we have seen there, these functionals may well turn up discontinuous, depending on how we choose to metrize their domain. The good news is that this is not the end of the story. It is possible to pinpoint the source of the problem, and thus find out when it would not arise. Our next result, which provides us with a rich class of continuous linear functionals, does precisely this.

PROPOSITION 4

(Shaefer) *Let X be a preordered metric linear space. If $int_X(X_+) \neq \emptyset$, then any positive linear functional on X is continuous.*[8]

PROOF

Let L be a positive linear functional on X, and take any sequence $(y^m) \in X^\infty$ such that $y^m \to \mathbf{0}$. By Proposition 3, all we need to show is that $L(y^m) \to 0$. Now assume $int_X(X_+) \neq \emptyset$, and pick any $x \in int_X(X_+)$. The crux of the argument is to establish the following: For every $\alpha > 0$, we can find an $M_\alpha \in \mathbb{R}$ such that

$$\alpha x \succsim_X y^m \succsim_X -\alpha x \qquad \text{for all } m \geq M_\alpha. \tag{4}$$

Why? Because if we can prove this, then we may use the positivity of L to find, for every $k \in \mathbb{N}$, a real number M_k with

$$\tfrac{1}{k}L(x) \geq L(y^m) \geq -\tfrac{1}{k}L(x) \qquad \text{for all } m \geq M_k.$$

Letting $k \to \infty$ here yields $L(y^m) \to 0$, as we seek.[9]

[8] *Reminder.* We denote the vector preorder of X as $\succsim_X$, and the positive cone induced by $\succsim_X$ as X_+. By the way, a **preordered metric linear space** is a metric linear space endowed with a vector preorder. (In particular, no (direct) connection between the metric of the space and its vector preorder is postulated; these relate to each other only through being consistent with the operations of vector addition and scalar multiplication.)

[9] Yes, M_k depends on k here, but no matter! For any $\varepsilon > 0$, there exists a $k \in \mathbb{N}$ such that $\left|\frac{1}{k}L(x)\right| < \varepsilon$, and hence $|L(y^m)| < \varepsilon$ for all $m \geq M$, for some real number M.

Now fix any $\alpha > 0$. Clearly, $\alpha x \in int_X(X_+)$—why?—so there exists a $\delta > 0$ such that $N_{\delta,X}(\alpha x) \subseteq X_+$. Notice that, for any $y \in N_{\delta,X}(\mathbf{0})$, we have

$$y + \alpha x \in N_{\delta,X}(\mathbf{0}) + \alpha x = N_{\delta,X}(\alpha x) \subseteq X_+,$$

so $y \succsim_X -\alpha x$. But if $y \in N_{\delta,X}(\mathbf{0})$, then $-y \in N_{\delta,X}(\mathbf{0})$ by translation invariance, so the same argument yields $\alpha x \succsim_X y$. Thus, if we choose $M_\alpha \in \mathbb{R}$ such that $y^m \in N_{\delta,X}(\mathbf{0})$ for all $m \geq M_\alpha$, we obtain (4). ■

2.3 Closed versus Dense Hyperplanes

Recall that we can identify a hyperplane in a linear space with a nonzero linear functional up to an additive constant (Corollary F.4). In this subsection we show that the *closed* hyperplanes in a metric linear space have the same relationship with the *continuous* nonzero linear functionals on that space. The crux of the argument is contained in the following fact.

PROPOSITION 5

Let X be a metric linear space and $L \in \mathcal{L}(X, \mathbb{R})$. Then L is continuous if, and only if, $null(L)$ is a closed subspace of X.[10]

PROOF

The "only if" part is easy. To establish the "if" part, take any $L \in \mathcal{L}(X, \mathbb{R})$ with $Y := null(L)$ being a closed subspace of X. In view of Proposition 3, it is enough to check that L is continuous at $\mathbf{0}$. So, take any $(x^m) \in X^\infty$ with $x^m \to \mathbf{0}$. We wish to show that $L(x^m) \to 0$.

If L is the zero functional, then the claim is obviously true, so assume that it is nonzero. Then, by Proposition F.6, $null(L)$ is a $\supseteq$-maximal proper subspace of X. (Why?) So, for an arbitrarily fixed $w \in X \setminus Y$, we have $span(Y + w) = X$, and hence we may write $x^m = y^m + \lambda_m w$ for some $(\lambda_m, y^m) \in \mathbb{R} \times Y$, $m = 1, 2, \ldots$.[11] Clearly, for each m, we have $L(x^m) = L(y^m) + \lambda_m L(w) =$

[10] *Reminder.* $null(L) := L^{-1}(0)$.

[11] *False-proof.* "Given that $x^m = y^m + \lambda_m w$ for each m, and $x^m \to \mathbf{0}$, we have $\mathbf{0} = \lim y^m + (\lim \lambda_m)w$. But since Y is closed, $\lim y^m \in Y$, that is, $L(\lim y^m) = 0$, and hence applying L to both sides of this equation, we obtain $0 = (\lim \lambda_m)L(w)$. Since $w \notin Y$, we have $L(w) \neq 0$, so it follows that $\lim \lambda_m = 0$. But then, since L is linear and $y^m \in Y$ for each m, we have $L(x^m) = L(y^m) + \lambda_m L(w) = \lambda_m L(w)$ for each m, so, letting $m \to \infty$, we get $\lim L(x^m) = 0$ as we sought."

Unfortunately, things are a bit more complicated than this. Please find what's wrong with this argument before proceeding further.

$\lambda_m L(w)$ (since $y^m \in null(L)$), while $L(w) \neq 0$ (since $w \notin Y$). Thus, by the continuity of scalar multiplication, all we need to do is to show that $\lambda_m \to 0$.

Let's first verify that (λ_m) is a bounded real sequence. If this was not the case, we could find a subsequence (λ_{m_k}) of this sequence such that $\lambda_{m_k} \neq 0$ for each k, and $\frac{1}{\lambda_{m_k}} \to 0$ (as $k \to \infty$). But if we let $\theta_k := \frac{1}{\lambda_{m_k}}$, we may write $w = \theta_{m_k} x^{m_k} - \theta_{m_k} y^{m_k}$ for each k. Since the metric d of X is translation invariant and Y is a linear subspace of X, therefore,

$$\begin{aligned} d(\theta_{m_k} x^{m_k} - \theta_{m_k} y^{m_k}, Y) &= d(\theta_{m_k} x^{m_k}, Y + \theta_{m_k} y^{m_k}) \\ &= d(\theta_{m_k} x^{m_k}, Y) \\ &\leq d(\theta_{m_k} x^{m_k}, \mathbf{0}), \end{aligned}$$

$k = 1, 2, \ldots$.[12] Then, letting $k \to \infty$ and using the continuity of scalar multiplication, we get $d(w, Y) = 0$, which is impossible, given that $w \notin Y$ and Y is closed (Exercise D.2). We conclude that (λ_m) is bounded.

Now let $\lambda := \limsup \lambda_m$. Then $\lambda \in \mathbb{R}$ (because (λ_m) is bounded) and there is a subsequence (λ_{m_k}) with $\lambda_{m_k} \to \lambda$ (as $k \to \infty$). Since $y^m = x^m - \lambda_m w$ for each m, and $\lim x^m = \mathbf{0}$, we have $\lim y^{m_k} = \lambda w$ by the continuity of scalar multiplication and vector addition. But since $y^{m_k} \in Y = L^{-1}(0)$ for each k, and Y is closed, $\lim y^{m_k} \in L^{-1}(0)$, so, $\lambda L(w) = 0$. Since $w \notin Y$, we thus find $\lambda = 0$. But notice that the same argument would go through verbatim if we instead had $\lambda := \liminf \lambda_m$. We may thus conclude that $\lim \lambda_m = 0$. ■

Here is the result we promised above.

PROPOSITION 6

A subset H of a metric linear space X is a closed hyperplane in X if, and only if,

$$H = \{x \in X : L(x) = \alpha\}$$

for some $\alpha \in \mathbb{R}$ and a continuous nonzero linear functional L on X.

[12] *Reminder.* For any nonempty subset S of a metric space X, and $x \in X$, $d(x, S) := \inf\{d(x, z) : z \in S\}$.

Proof

The "if" part of the claim follows readily from Proposition D.1 and Corollary F.4. Conversely, if H is a closed hyperplane in X, then by Corollary F.4 there exist an $\alpha \in \mathbb{R}$ and a nonzero $L \in \mathcal{L}(X, \mathbb{R})$ such that $H = L^{-1}(\alpha)$. Take any $x^* \in H$ and define $Y := H - x^*$. By Proposition 2, Y is closed subspace of X. But it is evident that $Y = null(L)$ — yes? — so by Proposition 5, L is continuous. ∎

Exercise 24 Show that any open (closed) half-space induced by a closed hyperplane of a metric linear space is open (closed).

Exercise 25 Let X and Y be two metric linear spaces and $L \in \mathcal{L}(X, Y)$. Prove or disprove: If Y is finite-dimensional, then L is continuous iff $null(L)$ is a closed subspace of Y.

Proposition 6 entails that any hyperplane that is induced by a discontinuous linear functional cannot be closed. Let us inquire into the nature of such hyperplanes a bit further. In Exercise 10, we noted that the closure of a subspace Z of a metric linear space X is itself a subspace. Thus the closure of a $\supseteq$-maximal proper subspace Z of X is either Z or X. Put differently, any $\supseteq$-maximal proper linear subspace of X is either closed or (exclusive) dense. Since any hyperplane H can be written as $H = Z + x^*$ for some $\supseteq$-maximal proper linear subspace Z of X and $x^* \in X$, and since $cl_X(H) = cl_X(Z + x^*) = cl_X(Z) + x^*$ (why?), this observation gives us the following insight.

Proposition 7

A hyperplane in a metric linear space is either closed or (exclusive) dense.

How can a hyperplane be dense in the grand space that it lies in? This seemingly paradoxical situation is just another illustration of how our finite-dimensional intuition can go astray in the realm of infinite-dimensional spaces. Indeed, there cannot be a dense hyperplane in a Euclidean space. (Why?) But all such bets are off in infinite-dimensional spaces. Since a linear functional on such a space need not be continuous, a hyperplane in it is not necessarily closed.

Actually, this is not as crazy as it might first seem. After all, thanks to the Weierstrass Approximation Theorem, we know that every continuous real function on $[0,1]$ is the uniform limit of a sequence of polynomials defined on $[0,1]$. But this means that $\mathbf{P}[0,1]$ is not a closed subset of $\mathbf{C}[0,1]$, while, of course, it is a linear subspace of $\mathbf{C}[0,1]$. Thus, in an infinite-dimensional metric linear space, a proper subspace that is not even $\supseteq$-maximal can be dense!

To sum up, one important analytic lesson we learn here is that a linear functional is in general not continuous, whereas the geometric lesson is that a hyperplane is in general not closed (in which case it is dense). Although the latter finding may defy our geometric intuition (which is, unfortunately, finite-dimensional), this is just how things are in infinite-dimensional metric linear spaces.

EXAMPLE 5
Consider the metric linear space X and the discontinuous $L \in \mathcal{L}(X,\mathbb{R})$ defined in Example 3.[1]. Propositions 5 and 7 together say that $null(L)$ must be dense in X. Let us verify this fact directly. Take any $y \in X$, and define the sequence $(x^m) \in X^\infty$ by

$$x^m := \left(y_1 - \frac{1}{m}\sum_{i=1}^{\infty} y_i, \ldots, y_m - \frac{1}{m}\sum_{i=1}^{\infty} y_i, y_{m+1}, y_{m+2}, \ldots \right), \qquad m = 1, 2, \ldots$$

(Each x^m is well-defined, because, by definition of X, we have $\sum^\infty y_i \in \mathbb{R}$.) Clearly, for each $m \in \mathbb{N}$,

$$L(x^m) = \sum_{i=1}^{m} y_i - \sum_{i=1}^{\infty} y_i + \sum_{i=m+1}^{\infty} y_i = 0,$$

so that $x^m \in null(L)$. But we have

$$d_\infty(x^m, y) = \sup\left\{ \left| x_i^m - y_i \right| : i \in \mathbb{N} \right\} = \frac{1}{m}\sum_{i=1}^{\infty} y_i \to 0$$

since $\sum^\infty y_i$ is finite. This shows that, for any vector in X, we can find a sequence in $null(L)$ that converges to that vector, that is, $null(L)$ is dense in X. □

Example 6

In the previous example we inferred the denseness of a hyperplane from the discontinuity of its defining linear functional. In this example, we shall construct a discontinuous linear functional on $\mathbf{C}[0,1]$ by using the denseness of a hyperplane that this functional induces. Define $f^i \in \mathbf{P}[0,1]$ by $f^i(t) := t^i$ for each $i \in \mathbb{N}$, and recall that $B := \{f^i : i \in \mathbb{N}\}$ is a basis for $\mathbf{P}[0,1]$. Now extend B to a basis A for $\mathbf{C}[0,1]$ (which can be done as in the proof of Theorem F.1). Evidently, $A \backslash B \neq \emptyset$. Pick any $g \in A \backslash B$, let $\mathbf{1}_{\{g\}}$ stand for the indicator function of $\{g\}$ in A, and define $L : \mathbf{C}[0,1] \to \mathbb{R}$ by

$$L(f) := \lambda_1(f)\mathbf{1}_{\{g\}}(h^{1,f}) + \cdots + \lambda_{m_f}(f)\mathbf{1}_{\{g\}}(h^{m_f,f}),$$

where the numbers $m_f \in \mathbb{N}$ and $\lambda_i(f) \in \mathbb{R}\backslash\{0\}$, and the basis vectors $h^{i,f} \in A$, $i = 1, \ldots, m_f$, are uniquely defined through the equation $f = \sum^{m_f} \lambda_i(f) h^{i,f}$. (Recall Corollary F.2.) It is readily checked that L is linear, yet it is not continuous. (Why?) Because we have $\mathbf{P}[0,1] \subseteq null(L)$ (yes?), and hence by the Weierstrass Approximation Theorem, $null(L)$ must be dense in $\mathbf{C}[0,1]$. By Propositions 5 and 7, therefore, L cannot be continuous. □

2.4 Digression: On the Continuity of Concave Functions

We noted earlier that a concave real function defined on an open subset of a Euclidean space must be continuous. This need not be true for a concave function defined on an open subset of a metric linear space. After all, we now know that a linear functional (which is obviously concave) may well be discontinuous if its domain is an infinite-dimensional metric linear space. This said, we also know that linearity spreads the minimal amount of continuity a function may have onto its entire domain. Put differently, a linear functional is either continuous or it is discontinuous everywhere (Proposition 3). Remarkably, concavity matches the strength of linearity on this score. That is to say, any concave (or convex) function whose domain is an open and convex subset of a metric linear space is either continuous or discontinuous everywhere.

In fact, we can say something a bit stronger than this. Given a metric space X and a point $x \in X$, let us agree to call a function $\varphi \in \mathbb{R}^X$ **locally bounded at x from below** if there exist a real number a and an open subset U of X such that $x \in U$ and $\varphi(U) \geq a$. If φ is locally bounded at x from below for every $x \in X$, then we say that it is **locally bounded from below**. Finally, if both φ and $-\varphi$ are locally bounded from below, then we simply say that φ is **locally bounded**.

Evidently, any continuous real function on a metric space is locally bounded. (Why?) Although the converse is of course false in general,[13] it is true for concave functions. That is, if such a function is locally bounded at a given point in the interior of its domain, it must be continuous everywhere.

Proposition 8

Let O be a nonempty open and convex subset of a metric linear space X, and $\varphi : O \to \mathbb{R}$ a concave function. If φ is locally bounded at some $x_0 \in O$ from below, then it is continuous. Therefore, φ is continuous if, and only if, it is continuous at some $x_0 \in O$.

In the context of Euclidean spaces, we may sharpen this result significantly. Indeed, as we have asserted a few times earlier, any concave map on an open subset of a Euclidean space is continuous. As we show next, this is because any such map is locally bounded.

Corollary 1

Given any $n \in \mathbb{N}$, let O be a nonempty open and convex subset of $\mathbb{R}^n$. If $\varphi : O \to \mathbb{R}$ is concave, then it is continuous.

Proof

It is without loss of generality to assume that $\mathbf{0} \in O$.[14] Then, since O is open, we can find a small enough $\alpha > 0$ such that $\alpha\mathbf{e}^i \in O$ and $-\alpha\mathbf{e}^i \in O$ for each $i = 1, \ldots, n$. (Here $\{\mathbf{e}^1, \ldots, \mathbf{e}^n\}$ is the standard basis for $\mathbb{R}^n$.) Define $S := \{\alpha\mathbf{e}^1, \ldots, \alpha\mathbf{e}^n, -\alpha\mathbf{e}^1, \ldots, -\alpha\mathbf{e}^n\}$, and let $T := co(S)$. Clearly, if $x \in T$, then there exists a $\lambda \in [0,1]^S$ such that $x = \sum_{y \in S} \lambda(y)y$ and $\sum_{y \in S} \lambda(y) = 1$. So, if φ is concave,

$$\varphi(x) \geq \sum_{y \in S} \lambda(y)\varphi(y) \geq \min \varphi(S)$$

13 In a big way! An *everywhere* (uniformly) bounded function may be continuous *nowhere*! (Think of $\mathbf{1}_{\mathbb{Q}}$ on $\mathbb{R}$.)

14 Why? First follow the argument, then go back and rethink how it would modify if we didn't assume $\mathbf{0} \in O$.

for any $x \in T$, and hence $\varphi(T) \geq \min \varphi(S)$. Since $\mathbf{0} \in int_{\mathbb{R}^n}(T)$—yes?—it follows that φ must be locally bounded at $\mathbf{0}$ in this case. Applying Proposition 8 completes the proof. ∎

It remains to prove Proposition 8. The involved argument is a good illustration of how convex and metric analyses intertwine in the case of metric linear spaces. You should go through it carefully.

Of course, the second assertion of Proposition 8 is an immediate consequence of its first assertion, so we need to focus only on the latter. We divide the proof into two observations (which are stated for O and φ of Proposition 8).

OBSERVATION 1. Suppose that φ is locally bounded at some $x_0 \in O$ from below. Then φ is locally bounded at any $x \in O$ from below.

OBSERVATION 2. Suppose that φ is locally bounded at some $x \in O$ from below. Then φ is continuous at x.

Before we move to prove these facts, let's note that it is enough to establish Observation 1 under the additional hypotheses $\mathbf{0} \in O$, $x_0 = \mathbf{0}$ and $\varphi(\mathbf{0}) = 0$. To see this, suppose we were able to prove the assertion with these assumptions. Then, given any nonempty open and convex set $O \subseteq X$ with $x_0 \in O$, and concave $\varphi \in \mathbb{R}^O$, we would let $U := O - x_0$ and define $\psi \in \mathbb{R}^U$ by $\psi(x) := \varphi(x + x_0) - \varphi(x_0)$. Clearly, ψ is concave, $\mathbf{0} \in U$, and $\psi(\mathbf{0}) = 0$, while ψ is locally bounded at $y - x_0$ from below iff φ is locally bounded at y from below (for any $y \in O$). Consequently, applying what we have established to ψ would readily yield Observation 1 for φ. (Agreed?)

PROOF OF OBSERVATION 1

Let $\mathbf{0} \in O$ and $\varphi(\mathbf{0}) = 0$, and suppose that φ is locally bounded at $\mathbf{0}$ from below. Then there exists a real number a and an open subset U of O such that $\mathbf{0} \in U$ and $\varphi(U) \geq a$. Fix an arbitrary $x \in O$. Since O is open and scalar multiplication on X is continuous, we can find an $\alpha > 1$ such that $\alpha x \in O$. Let

$$V := \left(1 - \tfrac{1}{\alpha}\right) U + \tfrac{1}{\alpha}\alpha x$$

(Figure 1). Clearly, $x \in V$ and V is an open subset of O (Proposition 2). Moreover, for any $y \in V$, we have $y = \left(1 - \frac{1}{\alpha}\right) z + \frac{1}{\alpha}\alpha x$ for some $z \in U$, so,

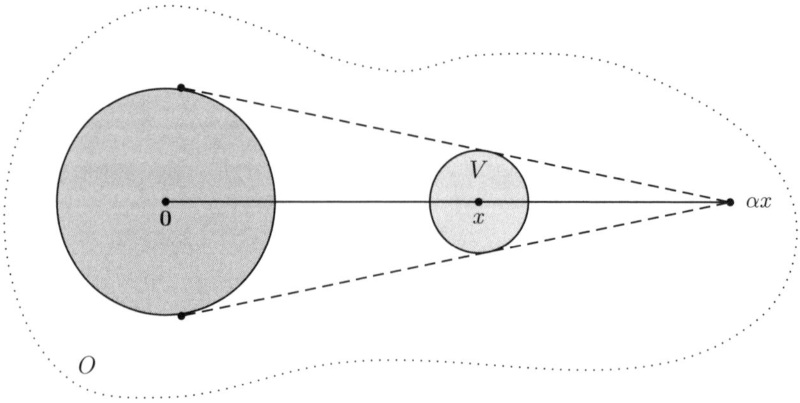

FIGURE 1

by concavity,

$$\varphi(y) \geq \left(1 - \frac{1}{\alpha}\right)\varphi(z) + \frac{1}{\alpha}\varphi(\alpha x) \geq \left(1 - \frac{1}{\alpha}\right)a + \frac{1}{\alpha}\varphi(\alpha x).$$

Thus, where $b := \left(1 - \frac{1}{\alpha}\right)a + \frac{1}{\alpha}\varphi(\alpha x)$, we have $\varphi(V) \geq b$, which proves that φ is locally bounded at x from below.[15] $\square$

It remains to prove Observation 2. Just as we could assume $x_0 = \mathbf{0}$ while proving Observation 1, we can take x in Observation 2 to be $\mathbf{0}$ without loss of generality. (Why?)

PROOF OF OBSERVATION 2

Let $\mathbf{0} \in O$ and $\varphi(\mathbf{0}) = 0$, and suppose that φ is locally bounded at $\mathbf{0}$ from below. Fix an arbitrarily small $\varepsilon > 0$. We wish to find an open subset U of O such that $\mathbf{0} \in U$ and $\varphi(U) \subseteq (-\varepsilon, \varepsilon)$. (Right?) We know that there exists a $\delta > 0$ and an $a < 0$ such that $N_{\delta,X}(\mathbf{0}) \subseteq O$ and $\varphi(N_{\delta,X}(\mathbf{0})) \geq a$. (Why?) Of course, $\lambda N_{\delta,X}(\mathbf{0})$ is an open subset of O that contains $\mathbf{0}$, for any $0 < \lambda \leq 1$. (Why?) So a good starting point is to check if we can make $\varphi(\lambda N_{\delta,X}(\mathbf{0}))$ small enough to be contained within $(-\varepsilon, \varepsilon)$ by choosing $\lambda > 0$ sufficiently small. (This is the main idea behind the proof.) Let's see. For any $y \in \lambda N_{\delta,X}(\mathbf{0})$, we have $\varphi(y) = \varphi\left((1-\lambda)\mathbf{0} + \lambda\left(\frac{1}{\lambda}y\right)\right) \geq \lambda\varphi\left(\frac{1}{\lambda}y\right) \geq \lambda a$. Moreover,

$$0 = \varphi(\mathbf{0}) = \varphi\left(\frac{1}{1+\lambda}y + \frac{\lambda}{1+\lambda}\left(-\frac{1}{\lambda}y\right)\right) \geq \frac{1}{1+\lambda}\varphi(y) + \frac{\lambda}{1+\lambda}\varphi\left(-\frac{1}{\lambda}y\right)$$

15 *Quiz.* Show that, under the givens of Proposition 8, if φ is locally bounded at x_0 from below, then it must be locally bounded. (Proving this is easier than proving Observation 2, so give it a try before studying the proof of that claim.)

and hence $-\varphi(y) \geq \lambda\varphi\left(-\frac{1}{\lambda}y\right)$. But, by translation invariance, $\frac{1}{\lambda}y \in N_{\delta,X}(\mathbf{0})$ implies $-\frac{1}{\lambda}y \in N_{\delta,X}(\mathbf{0})$, so we have $\varphi\left(-\frac{1}{\lambda}y\right) \geq a$, that is, $-\varphi(y) \geq \lambda a$. Thus, letting $b := -a > 0$, we have $\lambda b \geq \varphi(y) \geq -\lambda b$ for each $y \in \lambda N_{\delta,X}(\mathbf{0})$. So, if we take any $0 < \lambda < \frac{\varepsilon}{b}$ and set $U := \lambda N_{\delta,X}(\mathbf{0})$, we find $\varphi(U) \subseteq (-\varepsilon, \varepsilon)$. Since $\varepsilon > 0$ was arbitrarily chosen above, this completes the proof. □

We will return to the investigation of concave functions in Chapters J and K. For now, we conclude with some exercises that will extend the present analysis to the context of convex correspondences (Exercise G.8).

*EXERCISE 26^H (Borwein) Let X and Y be two metric linear spaces, O a nonempty open and convex subset of X, and $\Gamma : O \rightrightarrows Y$ a convex correspondence. Prove:

(a) If $\Gamma(x_0)$ is a bounded subset of Y for some $x_0 \in X$, then $\Gamma(x)$ is a bounded subset of Y for any $x \in X$.

(b) If Γ is lower hemicontinuous at some $x_0 \in X$, then it is lower hemicontinuous.

(c) If Γ is lower hemicontinuous at some $x_0 \in X$, and $\Gamma(x_0)$ is a bounded subset of Y, then Γ is continuous.

3 Finite-Dimensional Metric Linear Spaces

Recall that two isomorphic linear spaces cannot be distinguished from each other in terms of their linear algebraic properties (Section F.2.3). Similarly, two homeomorphic metric spaces enjoy the same structure insofar as their topological properties are concerned (Section D.1.6). It is thus plain how to identify those metric linear spaces that share the same linear algebraic *and* topological properties.

DEFINITION

Let X and Y be two metric linear spaces and $L \in \mathcal{L}(X, Y)$. If L is a continuous bijection with a continuous inverse, then it is called a **linear homeomorphism**. If such a linear operator L exists, we then say that X and Y are **linearly homeomorphic**. Similarly, if L is an isometry, then it is called a **linear isometry**, and if such an operator exists, we say that X and Y are **linearly isometric**.

We know that two linear spaces with the same *finite* dimension are necessarily isomorphic (Proposition F.5). Moreover, we can construct an explicit linear isomorphism between any two such linear spaces by mapping the basis elements of one space to the basis elements of the other. As we show next, in the case of metric linear spaces with the same *finite* dimension, any such linear isomorphism is, in fact, a linear homeomorphism. We thus have the following sharpening of Corollary F.3.

Theorem 1
(Tychonoff) *Every nontrivial finite-dimensional metric linear space is linearly homeomorphic to* $\mathbb{R}^n$, *for some* $n \in \mathbb{N}$.

We shall use the following technical observation to prove this important result.

Lemma 1
Given any $n \in \mathbb{N}$, *take any* $\lambda_i \in \mathbb{R}^{\mathbb{N}}$ *and* $\alpha_i \in \mathbb{R}$, $i = 1, \ldots, n$. *Let* Z *be a metric linear space and* $\{z^1, \ldots, z^n\}$ *a linearly independent subset of* Z. *Then,*

$$\lim_{m\to\infty} \sum_{i=1}^{n} \lambda_i(m) z^i = \sum_{i=1}^{n} \alpha_i z^i \quad \textit{if and only if}$$

$$\lim_{m\to\infty} \lambda_i(m) = \alpha_i, \; i = 1, \ldots, n.$$

This is hardly an incredible assertion, but its proof requires a small dose of razzle-dazzle. Let us then postpone its proof until the end of this section, and rather try to see what one can do with this result. Here is the main conclusion we wish to derive from it.

Proposition 9
Every linear isomorphism from one finite-dimensional linear space onto another is a homeomorphism.

It is quite easy to prove this fact by using Lemma 1. We give the proof in the form of an exercise (which replicates a chunk of the analysis presented in Section F.2.3).

EXERCISE 27 Let X and Y be two finite-dimensional metric linear spaces and $L \in \mathcal{L}(X, Y)$ a bijection.

(a) Use Lemma 1 to show that L is continuous.

(b) Let A be a basis for X. Show that $L(A)$ is a basis for Y.

(c) Define $\varphi \in X^Y$ by $\varphi(z) := \sum_{x \in A} \lambda_z(L(x))x$, where λ_z is the unique map in $\mathbb{R}^{L(A)}$ with $z = \sum_{y \in L(A)} \lambda_z(y)y$. Show that $L^{-1} = \varphi$.

(d) Use Lemma 1 to show that L^{-1} is continuous.

EXERCISE 28 Every closed and bounded subset of a finite-dimensional metric linear space is compact. True or false?

Theorem 1 is, of course, a trivial consequence of Proposition 9. Suppose X is a nontrivial metric linear space with $n := dim(X) \in \mathbb{N}$. Then, by Corollary F.3, there is a linear isomorphism from X onto $\mathbb{R}^n$. But Proposition 9 says that this linear isomorphism is a homeomorphism, so X and $\mathbb{R}^n$ are linearly homeomorphic, as asserted by Theorem 1. As an immediate corollary, we find that $\mathbb{R}^{n,p}$ and $\mathbb{R}^{n,q}$ are linearly homeomorphic for any $n \in \mathbb{N}$ and $1 \leq p, q \leq \infty$. It is in this sense that we think of these Fréchet spaces as "identical."

The main lesson here is that the linear and topological properties of any given nontrivial finite-dimensional metric linear space are exactly the same as those of a Euclidean space. The following three corollaries feast on this observation.

COROLLARY 2

Any finite-dimensional metric linear space X is a Fréchet space.

PROOF

If $X = \{\mathbf{0}\}$, there is nothing to prove, so assume that X is nontrivial, and take any Cauchy sequence $(x^m) \in X^\infty$. By Theorem 1, there exists a linear homeomorphism $L \in \mathcal{L}(X, \mathbb{R}^n)$, where $n := dim(X)$. By Proposition 3, L is uniformly continuous, so $(L(x^m))$ is a Cauchy sequence in $\mathbb{R}^n$ (Exercise D.9). Since $\mathbb{R}^n$ is complete, there is a $y \in \mathbb{R}^n$ with $L(x^m) \to y$. But then, since L^{-1} is continuous, $x^m = L^{-1}(L(x^m)) \to L^{-1}(y) \in X$, that is, (x^m) converges in X. ■

The following observation is an immediate consequence of Corollary 2 and Proposition C.7.

Corollary 3
Any finite-dimensional subspace of a metric linear space X is closed in X.

Exercise 29 Show that every affine manifold in a finite-dimensional metric linear space is closed.

Exercise 30[H] Show that $cone(S)$ is closed for any finite subset S of a metric linear space.

Another application of Theorem 1 gives us the following result, which you might have already anticipated.

Corollary 4
Any linear functional defined on a finite-dimensional metric linear space is continuous.

Again, why should you expect such a result to be true? Your argument should go along the following lines: "I know the precise structure of an arbitrary linear functional on $\mathbb{R}^n$ (Example F.6)—any such functional is continuous. Thus the claim advanced in Corollary 4 holds in $\mathbb{R}^n$ (for any n). But, insofar as continuity and linearity properties are concerned, any given nontrivial finite-dimensional metric linear space can be identified with $\mathbb{R}^n$ (for some n)—we may regard the former space as if it is obtained from $\mathbb{R}^n$ by relabelling its vectors. Thus, the claim of Corollary 4 should hold in any finite-dimensional metric linear space."

Exercise 31 Prove Corollary 4.

Finally, we note that one should be careful in identifying two linearly homeomorphic metric linear spaces. True, these spaces are indistinguishable from each other insofar as their linear algebraic and topological structures are concerned, but they may behave quite differently with regard to some other properties. The following exercise illustrates this point.

Exercise 32 A subset S of a metric space X is said to satisfy the **Unique Nearest Point Property** if for every point x outside S there is a unique point in S which is closest to x.

(a) Show that, for any positive integer n, every nonempty closed and convex subset of $\mathbb{R}^n$ has the Unique Nearest Point Property.[16]

(b) Let $p \in \{1, \infty\}$, and show that a nonempty closed and convex subset of $\mathbb{R}^{2,p}$ need not have the Unique Nearest Point Property. Conclude that having a unique nearest point is a property that need *not* be preserved under a linear homeomorphism. (Of course, this property would be preserved under any linear isometry.)

All this is good, but remember that we have left a big gap in our investigation of the finite-dimensional metric linear spaces above. Although our proof of Theorem 1 depends vitally on Lemma 1, we have not yet established the latter result. To wrap things up, then, we prove Lemma 1 now.

PROOF OF LEMMA 1

The "if" part of the claim is easy. If $\lambda_i(m) \to \alpha_i$ for each i, then, by the continuity of scalar multiplication, we have $\lambda_i(m)z^i \to \alpha_i z^i$ for each i, so, by the continuity of vector addition, $\sum^n \lambda_i(m)z^i \to \sum^n \alpha_i z^i$.

We will prove the "only if" part by an inductive argument. In what follows we assume $\sum^n \lambda_i(m)z^i \to \sum^n \alpha_i z^i$, and denote the set $\{1, \ldots, n\}$ by N.

Case 1. Assume: There is a $j_1 \in N$ such that $\lambda_i(\cdot) = 0$ for all $i \in N \backslash \{j_1\}$.[17]

Let $j_1 = 1$ for ease of reference. If $(\lambda_1(m))$ is unbounded, then there exists a subsequence $(\lambda_1(m_k))$ of this sequence with $\lambda_1(m_k) \neq 0$ for each k, and $\frac{1}{\lambda_1(m_k)} \to 0$ (as $k \to \infty$). Thus,

$$z^1 = \frac{1}{\lambda_1(m_k)} \left(\lambda_1(m_k)z^1\right) \to 0 \sum_{i=1}^{n} \alpha_i z^i = \mathbf{0}$$

by the continuity of scalar multiplication. But then $z^1 = \mathbf{0}$, which contradicts $\{z^1, \ldots, z^n\}$ being linearly independent. Therefore, $(\lambda_1(m))$ must be a bounded sequence, so if $\beta_1 \in \{\liminf \lambda_1(m), \limsup \lambda_1(m)\}$, then $\beta_1 \in \mathbb{R}$. Clearly, there exists a subsequence $(\lambda_1(m_k))$ of $(\lambda_1(m))$ with $\lambda_1(m_k) \to \beta_1$ (as $k \to \infty$), and hence $\lambda_1(m_k)z^1 \to \beta_1 z^1$ by the continuity of scalar

[16] A classic result of convex analysis maintains that the converse of this also holds: A closed subset S of $\mathbb{R}^n$ is convex if, and only if, for every vector x in $\mathbb{R}^n$, there exists a unique nearest vector in S with respect to d_2. (This is *Motzkin's Theorem*.)

[17] Observe that proving the assertion with this assumption establishes the following: For any real sequence (θ_m), we have $\theta_m z^i \to z^j$ iff $i = j$ and $\theta_m \to 1$. (Why?) We will use this observation in Case 2.

multiplication. Thus, under our case hypothesis, $\beta_1 z^1 = \sum^n \alpha_i z^i$. By linear independence, then, $\beta_1 = \alpha_1$ and $\alpha_i = 0$ for all $i \in N$ with $i > 1$.[18] Thus: $\lim \lambda_i(m) = \alpha_i$ for each $i \in N$. ∥

Case 2. Assume: There are $j_1, j_2 \in N$ such that $\lambda_i(\cdot) = 0$ for all $i \in N \backslash \{j_1, j_2\}$.

Let $j_1 = 1$ and $j_2 = 2$ for ease of reference. If $(\lambda_1(m))$ is unbounded, then there exists a subsequence $(\lambda_1(m_k))$ of this sequence with $\lambda_1(m_k) \neq 0$ for each k, and $\frac{1}{\lambda_1(m_k)} \to 0$ (as $k \to \infty$). Thus

$$z^1 + \frac{\lambda_2(m_k)}{\lambda_1(m_k)} z^2 = \frac{1}{\lambda_1(m_k)} \left(\lambda_1(m_k) z^1 + \lambda_2(m_k) z^2\right) \to \mathbf{0}$$

by the continuity of scalar multiplication. But then, since the metric of Z is translation invariant, $-\frac{\lambda_2(m_k)}{\lambda_1(m_k)} z^2 \to z^1$, which is impossible in view of what we have established in Case 1. Thus $(\lambda_1(m))$ is bounded. Besides, the analogous argument shows that $(\lambda_2(m))$ is bounded as well. Now let $\beta_i \in \{\liminf \lambda_i(m), \limsup \lambda_i(m)\}$, $i = 1, 2$. Clearly, for each $i = 1, 2$, there exists a subsequence $(\lambda_i(m_k))$ of $(\lambda_i(m))$ with $\lambda_i(m_k) \to \beta_i$ (as $k \to \infty$), and hence $\lambda_1(m_k) z^1 + \lambda_2(m_k) z^2 \to \beta_1 z^1 + \beta_2 z^2$, which implies $\beta_1 z^1 + \beta_2 z^2 = \sum^n \alpha_i z^i$. By linear independence, then, $\beta_i = \alpha_i$, $i = 1, 2$, and $\alpha_i = 0$ for all $i \in N$ with $i > 2$. Thus: $\lambda_i(m) = \alpha_i$ for all $i \in N$. ∥

Continuing this way inductively yields the proof. ■

4* Compact Sets in Metric Linear Spaces

We motivated the notion of compactness in Chapter C as some sort of a "finiteness" property that is suitable for infinite sets. So, in an intuitive sense, compact sets are "not very large." On the other hand, we argued in Chapter G that there is reason to think of algebraically open sets in infinite-dimensional linear spaces as "very large." (Recall Example G.8.) As we shall see in the next section, the same argument applies to open subsets of any infinite-dimensional metric linear space as well. But how can we reconcile these intuitive points of view? How should we think about a compact set (which is supposedly "small") that contains a nonempty open set (which is supposedly "large")?

[18] Since β_1 may equal either $\liminf \lambda_1(m)$ or $\limsup \lambda_1(m)$ here, it follows that $\lim \lambda_1(m) = \alpha_1$.

Observe first that the issue does not arise in the finite-dimensional case. For instance, there is good reason to "view" the closed unit ball $\{x \in \mathbb{R}^n : d_2(x, \mathbf{0}) \leq 1\}$ (which is obviously compact) as a "small" set. Even though this set contains an open set, the "size" interpretation of open sets applies only to *infinite*-dimensional spaces, so we are fine here. Things are more subtle in the infinite-dimensional case, however. As we show below, a compact subset of an infinite-dimensional metric linear space X can never contain an open subset of this space. So, somewhat unexpectedly, the potential conflict between our intuitive viewpoints does not arise in infinite-dimensional spaces either.

This is a pretty startling observation that you should pause and reflect on. On the one hand, it shows again that we should always keep our finite-dimensional intuitions in check when dealing with infinite-dimensional metric linear spaces. On the other hand, it tells us that our heuristic way of viewing compact sets as "small" and infinite-dimensional open sets as "large" holds water in the context of metric linear spaces:

The closure of a nonempty open subset of an infinite-dimensional metric linear space cannot be compact.

Let us now move on to the formal development. We begin with an observation that might ring a bell (if you recall Section C.4).

Lemma 2

If S is a nonempty compact subset of a metric linear space X, then, given any $\theta > 0$ and $\varepsilon > 0$, there exist finitely many vectors $x^1, \ldots, x^k$ in S such that

$$S \subseteq \bigcup_{i=1}^{k} \left(x^i + \theta N_{\varepsilon,X}(\mathbf{0}) \right).$$

Exercise 33[H] Prove Lemma 2.

Here is the main result of this section.

Theorem 2

(Riesz) *Let X be a metric linear space. If there is a compact subset S of X with $\text{int}_X(S) \neq \emptyset$, then X is finite-dimensional.*

That is, the interior of any compact set in an infinite-dimensional metric linear space must be empty. This situation is, of course, markedly different from the corresponding scenario in finite-dimensional linear spaces. For, by Theorem C.1, the closure of every bounded set (e.g., any ε-neighborhood of $\mathbf{0}$) is compact in any Euclidean space. Thus, by Theorem 1, this is the case in any finite-dimensional metric linear space. Consequently, Theorem 2 provides us with the following property that distinguishes infinite- and finite-dimensional metric linear spaces.

COROLLARY 5
(Riesz) *A metric linear space X is finite-dimensional if, and only if, the closure of every bounded subset of X is compact in X.*

A metric space X is said to be **locally compact** if every point of the space has an ε-neighborhood the closure of which is compact in X. Obviously, every compact metric space is locally compact, but not conversely. (For instance, $\mathbb{R}$ is locally compact, while, of course, it is not compact.) Thus, Corollary 5 is sometimes paraphrased as follows:

A metric linear space is locally compact iff it is finite-dimensional.[19]

EXERCISE 34 Derive Corollary 5 from Theorem 2.

An immediate application of Theorem 2 (or Corollary 5) shows that the closure of the open unit ball of an infinite-dimensional metric linear space cannot be compact. Although this fact may conflict with our "Euclidean" intuition, it is not all that mysterious. Perhaps a simple illustration may convince you. Take any $1 \leq p \leq \infty$, and let O denote the open unit ball in ℓ^p, that is, $O := \{(x_m) \in \ell^p : \sum^\infty |x_i|^p < 1\}$. The closure of this set equals the closed unit ball $B_{\ell^p} := \{(x_m) \in \ell^p : d_p((x_m), \mathbf{0}) \leq 1\}$. But an easy

[19] This important result was proved in 1918 by Frigyes Riesz (1880–1956). It is thus often referred to as *Frigyes Riesz's Theorem.* (Riesz's younger brother, Marcel Riesz, was also a well-known mathematician.) Frigyes Riesz contributed to the foundations of functional analysis and operator theory in a substantial way; his contributions accentuated the development of quantum mechanics early in the twentieth century. At least two major results in functional analysis carry his name, the Riesz-Fischer Theorem and the Riesz Representation Theorem.

computation shows that

$$d_p(\mathbf{e}^k, \mathbf{e}^l) = \begin{cases} 2^{\frac{1}{p}}, & \text{if } 1 \leq p \leq \infty \\ 1, & \text{if } p = \infty \end{cases},$$

where $\mathbf{e}^1 := (1, 0, 0, \ldots)$, $\mathbf{e}^2 := (0, 1, 0, \ldots)$, etc. It follows that $(\mathbf{e}^1, \mathbf{e}^2, \ldots)$ is a sequence in B_{ℓ^p} without a convergent subsequence. Thus, the closure of O is not (sequentially) compact.

Exercise 35 Show that $\{f \in \mathbf{C}[0,1] : d_\infty(f, \mathbf{0}) \leq 1\}$ is not a compact subset of $\mathbf{C}[0,1]$ by a direct argument, without using Theorem 2.

Without further ado, we now move on to prove Theorem 2.

Proof of Theorem 2

Assume that there is a compact subset S of X with $int_X(S) \neq \emptyset$. Take any $x \in int_X(S)$, and define $T := S - x$. Clearly, T is a compact set with $\mathbf{0} \in int_X(T)$. (Why?) Pick any $\varepsilon > 0$ such that $N_{\varepsilon,X}(\mathbf{0}) \subseteq int_X(T)$. Then $cl_X(N_{\varepsilon,X}(\mathbf{0})) \subseteq T$—yes?—and hence $cl_X(N_{\varepsilon,X}(\mathbf{0}))$ is compact, being a closed subset of a compact set. By Lemma 2, then, there exist a $k \in \mathbb{N}$ and vectors $x^1, \ldots, x^k \in cl_X(N_{\varepsilon,X}(\mathbf{0}))$ such that

$$cl_X(N_{\varepsilon,X}(\mathbf{0})) \subseteq \bigcup_{i=1}^{k} \left(x^i + \tfrac{1}{2} N_{\varepsilon,X}(\mathbf{0})\right).$$

Letting $Y := span(\{x^1, \ldots, x^k\})$, we may then write

$$cl_X(N_{\varepsilon,X}(\mathbf{0})) \subseteq Y + \tfrac{1}{2} N_{\varepsilon,X}(\mathbf{0}). \tag{5}$$

But then,

$$\begin{aligned} \tfrac{1}{2} cl_X(N_{\varepsilon,X}(\mathbf{0})) &\subseteq \tfrac{1}{2}\left(Y + \tfrac{1}{2} N_{\varepsilon,X}(\mathbf{0})\right) && \text{(by (5))} \\ &= Y + \tfrac{1}{4} N_{\varepsilon,X}(\mathbf{0}) && \text{(since } Y \text{ is a linear subspace)} \\ &\subseteq Y + \tfrac{1}{4}\left(Y + \tfrac{1}{2} N_{\varepsilon,X}(\mathbf{0})\right) && \text{(by (5))} \\ &= Y + \tfrac{1}{8} N_{\varepsilon,X}(\mathbf{0}) && \text{(since } Y \text{ is a linear subspace)} \end{aligned}$$

and hence $cl_X(N_{\varepsilon,X}(\mathbf{0})) \subseteq Y + \frac{1}{4} N_{\varepsilon,X}(\mathbf{0})$. Proceeding inductively, then, we obtain

$$cl_X(N_{\varepsilon,X}(\mathbf{0})) \subseteq Y + \tfrac{1}{2^m} N_{\varepsilon,X}(\mathbf{0}), \qquad m = 1, 2, \ldots \tag{6}$$

Let's now show that

$$Y = cl_X(Y) = \bigcap_{i=1}^{\infty} \left(Y + \tfrac{1}{2^i} N_{\varepsilon,X}(\mathbf{0}) \right). \tag{7}$$

Given that Y is a finite-dimensional subspace of X, the first equality follows readily from Corollary 3. This also shows that the $\subseteq$ part of the second equality is trivial. To see the remaining part, let $y \in Y + \frac{1}{2^m} N_{\varepsilon,X}(\mathbf{0})$ for each $m \in \mathbb{N}$. Then, there must exist a sequence (y^m) in Y and a sequence (z^m) in $N_{\varepsilon,X}(\mathbf{0})$ such that $y = y^m + \frac{1}{2^m} z^m$. Since $cl_X(N_{\varepsilon,X}(\mathbf{0}))$ is compact, there exist a subsequence (z^{m_k}) and a $z \in cl_X(N_{\varepsilon,X}(\mathbf{0}))$ with $z^{m_k} \to z$ (as $k \to \infty$). But then, by the continuity of scalar multiplication and vector addition, $y^{m_k} := y - \frac{1}{2^{m_k}} z^{m_k} \to y$ (as $k \to \infty$). Thus, there is a sequence in Y that converges to y, which means that $y \in cl_X(Y)$.

Combining (6) and (7), we get what we were after: $cl_X(N_{\varepsilon,X}(\mathbf{0})) \subseteq Y$. We are done, because this implies $Y \subseteq X = span(N_{\varepsilon,X}(\mathbf{0})) \subseteq Y$, so $dim(X) = dim(Y) \leq k < \infty$.[20] ■

EXERCISE 36 Show that the convex hull of a finite set in an infinite-dimensional metric linear space must have an empty interior.

EXERCISE 37 Let X be a metric linear space, and Y and Z two subspaces of X. Prove that if Y is closed and Z is finite-dimensional, then $Y + Z$ is a closed subspace of X.

EXERCISE 38[H] Let X and Y be two metric linear spaces and $L \in \mathcal{L}(X, Y)$. We say that L is **bounded** if $L(S)$ is a bounded subset of Y for any bounded subset S of X. It is called a **compact operator** if $cl_Y(L(S))$ is a compact subset of Y for any bounded subset S of X. Establish the following facts.

(a) If L is a compact operator, then it is bounded.
(b) If L is bounded (or continuous), it need not be a compact operator. In particular, $\mathrm{id}_X \in \mathcal{L}(X, X)$ is not compact unless $dim(X) < \infty$.
(c) If L is bounded and $dim(L(X)) < \infty$, then L is a compact operator.
(d) If L is continuous and $dim(X) < \infty$, then L is a compact operator.

[20] How do I know that $X = span(N_{\varepsilon,X}(\mathbf{0}))$? Because, for any $\omega \in X$, there exists a small enough $\lambda > 0$ with $\lambda w \in N_{\varepsilon,X}(\mathbf{0})$—thanks to the continuity of scalar multiplication—so $w = \frac{1}{\lambda}(\lambda w) \in span(N_{\varepsilon,X}(\mathbf{0}))$.

EXERCISE 39[H] Let X be a metric linear space, and $K \in \mathcal{L}(X, X)$ a compact operator.

(a) Show that if $L \in \mathcal{L}(X, X)$ is bounded, then both $K \circ L$ and $L \circ K$ are compact operators.

(b) Show that if K is an invertible compact operator, then K^{-1} is a bounded linear operator iff $dim(X) < \infty$.

5 Convex Analysis in Metric Linear Spaces

In Chapter G we investigated the basic algebraic structure of convex sets. In concert with our general program, we will now study the metric linear properties of convex sets in an arbitrary metric linear space. Our ultimate objective is to obtain a suitable counterpart of the Dieudonné Separation Theorem in the present framework. En route to this result, we will gather more information about the basic geometry of convex sets as well.

5.1 Closure and Interior of a Convex Set

You are probably familiar with the fact that the closure and interior of a convex subset of a Euclidean space are convex. This fact generalizes nicely to the case of metric linear spaces.

PROPOSITION 10
Let S be a convex subset of a metric linear space X. Then, $cl_X(S)$ and $int_X(S)$ are convex sets.

PROOF
Take any $0 \leq \lambda \leq 1$. If $x, y \in cl_X(S)$, then there exist sequences $(x^m), (y^m) \in S^\infty$ with $x^m \to x$ and $y^m \to y$. Since S is convex, $\lambda x^m + (1-\lambda)y^m \in S$ for each m, while $\lambda x^m + (1-\lambda)y^m \to \lambda x + (1-\lambda)y$ by the continuity of scalar multiplication and vector addition. Thus $\lambda x + (1-\lambda)y \in cl_X(S)$.

To prove the second assertion, take an arbitrary $0 \leq \lambda \leq 1$, and observe that, by Proposition 2, $int_X(\lambda S) + int_X((1-\lambda)S)$ is an open set that is contained in $\lambda S + (1-\lambda)S$, while $\lambda S + (1-\lambda)S = S$ (because S is convex). Since $int_X(S)$ is the $\supseteq$-maximum open subset of S, therefore,

$$\lambda int_X(S) + (1-\lambda)int_X(S) = int_X(\lambda S) + int_X((1-\lambda)S) \subseteq int_X(S).$$

(Why the first equality?) Thus: $int_X(S)$ is convex. ■

EXERCISE 40^{H} Let S be a subset of a metric linear space X. Show that if S is λ-convex for some $0 < \lambda < 1$, then both $cl_X(S)$ *and* $int_X(S)$ are convex.

The closure and interior operators thus preserve convexity. How about the converse? That is to say, does the convex hull operator preserve closedness and/or openness of a set? It turns out that it does preserve openess, but it will be easier to prove this a bit later (Exercise 45). More important at present is to see that the convex hull operator fails to preserve closedness, even in the context of Euclidean spaces. For instance, $S := \{(x, y) \in \mathbb{R}^2_+ : y \leq x^2\}$ is a closed subset of $\mathbb{R}^2$, but $co(S) = (\mathbb{R}_{++} \times \mathbb{R}_+) \cup \{\mathbf{0}\}$ is not closed in $\mathbb{R}^2$. (Draw a picture!) However, $co(\cdot)$ does preserve compactness, *provided that the underlying space is Euclidean*. We leave the proof to your able hands.

EXERCISE 41 Use Carathéodory's Theorem to prove that the convex hull of a compact set in a finite-dimensional metric linear space is compact.

It is easy to see that this observation generalizes to any metric linear space, provided that the set under consideration is finite:

The convex closure of a finite set is compact in any metric linear space.[21]

Unfortunately, the finiteness postulate cannot be relaxed here even to compactness. As we illustrate next, the compactness of a subset S of a metric linear space does not guarantee the closedness of its convex hull in general.

[21] There are various ways of proving this. For instance, if $x^1, \ldots, x^m$ are elements of a metric linear space (with $m \in \mathbb{N}$), then $(\lambda_1, \ldots, \lambda_m, x^1, \ldots, x^m) \mapsto \sum^m \lambda_i x^i$ is a continuous map from $\Delta_{m-1} \times \{(x^1, \ldots, x^m)\}$ onto $co\{x^1, \ldots, x^m\}$, where $\Delta_{m-1} := \{(\lambda_1, \ldots, \lambda_m) \in [0,1]^m : \sum^m \lambda_i = 1\}$. (Right?) Since $\Delta_{m-1} \times \{(x^1, \ldots, x^m)\}$ is compact, therefore, $co\{x^1, \ldots, x^m\}$ must be compact. Alternatively, one may give a more direct proof by using the sequential compactness of the simplex Δ_{m-1} and the continuity of scalar multiplication and vector addition.

Quiz. Show that if $S_1, \ldots, S_m$ are compact subsets of a metric linear space, then $co(S_1 \cup \cdots \cup S_m)$ is also compact.

Example 7
Let $S := \{\mathbf{0}, \frac{1}{2}\mathbf{e}^1, \frac{1}{4}\mathbf{e}^2, \frac{1}{8}\mathbf{e}^3, \ldots\}$, where $\mathbf{e}^1 := (1, 0, 0, \ldots)$, $\mathbf{e}^2 := (0, 1, 0, \ldots)$, etc. It is easy to check that S is a compact subset of ℓ^∞. (*Proof.* Any sequence in S has either a constant subsequence or a subsequence that converges to $\mathbf{0}$.) Now, for any $m \in \mathbb{N}$, consider the real sequence

$$x^m := \left(\tfrac{1}{2}, \ldots, \tfrac{1}{2^m}, 0, 0, \ldots\right), \qquad m = 1, 2, \ldots$$

Clearly, $x^m = \sum^m \frac{1}{2^i}\mathbf{e}^i \in co(S)$ for each m. (Why the second claim?) Moreover,

$$d_\infty\left(x^m, \sum_{i=1}^{\infty} \frac{1}{2^i}\mathbf{e}^i\right) = \frac{1}{2^{m+1}} \to 0.$$

Then $\lim x^m = \left(\frac{1}{2}, \frac{1}{4}, \frac{1}{8}, \ldots\right)$, but it is obvious that all but finitely many terms of any element of $co(S)$ must be zero, so we have to conclude that $\lim x^m \notin co(S)$. Thus: $co(S)$ is not closed. □

This observation suggests that to obtain the $\supseteq$-minimum closed and convex superset of a given closed (even compact) set, we may have to take the closure of the convex hull of that set.

Definition
Let S be a subset of a metric linear space X. The **closed convex hull** of S, denoted by $\overline{co}_X(S)$, is defined as the smallest (that is, $\supseteq$-minimum) closed and convex subset of X that contains S.

Let X be a metric linear space. Since the intersection of any class of closed and convex subsets of X is itself a closed and convex subset of X, the following fact is self-evident: *For any* $S \subseteq X$,

$$\overline{co}_X(S) := \bigcap \{A : A \textit{ is a closed and convex subset of } X, \textit{and } S \subseteq A\}.$$

(*Note.* $\overline{co}_X(\emptyset) = \emptyset$.)

Clearly, we can view $\overline{co}_X(\cdot)$ as a self-map on 2^X. Every closed and convex subset of X is a fixed point of this map, and $\overline{co}_X(S)$ is a closed and convex set for any $S \subseteq X$. Moreover, we have the following useful formula:

$$\overline{co}_X(S) = cl_X(co(S)) \qquad \text{for any } S \subseteq X.$$

Indeed, since $cl_X(co(S))$ is convex (Proposition 10), it is a closed and convex subset of X that contains S, so $\overline{co}_X(S) \subseteq cl_X(co(S))$. The $\supseteq$ part of the claim follows from the fact that $\overline{co}_X(S)$ is a closed set in X that includes $co(S)$. (Yes?)

EXERCISE 42 Show that if S is a subset of a metric linear space X, then $\overline{co}_X(S) \supseteq co(cl_X(S))$, but $\overline{co}_X(S) \subseteq co(cl_X(S))$ need not be true even if X is a Euclidean space.

EXERCISE 43 Prove: For any subsets A and B of a metric linear space X, if $\overline{co}_X(A)$ is compact, then $\overline{co}_X(A+B) = \overline{co}_X(A) + \overline{co}_X(B)$.

EXERCISE 44 Prove: For any subsets A and B of a metric linear space X, if both $\overline{co}_X(A)$ and $\overline{co}_X(B)$ are compact, then $\overline{co}_X(A \cup B) = co(\overline{co}_X(A) \cup \overline{co}_X(B))$.

5.2 Interior versus Algebraic Interior of a Convex Set

We now turn to the relation between the openness and algebraic openness properties of convex sets. Given the analysis of Section G.1.4, the involved arguments are mostly routine, so we will leave some of the related work to you. To fix ideas, however, we give here a complete proof of the fact that the interior and the algebraic interior of a *convex* set coincide, provided that the former is nonempty.[22]

Given a nonempty subset S of a metric linear space X, what would you "expect" the relation between $int_X(S)$ and $al\text{-}int_X(S)$ to be? Well, $x \in int_X(S)$ iff any sequence $(x^m) \in X^\infty$ that converges to x enters and stays in S eventually. Therefore, if $x \in int_X(S)$ and $y \in X$, then any sequence on the line segment between x and y that converges to x must enter and stay in S eventually. This is close to saying that $x \in al\text{-}int_X(S)$, but not quite. For this we need to be able to ascertain that a connected (and nondegenerate) portion of any such line that takes x as an endpoint should stay within S. It is at this junction that we see the importance of convexity. If S is convex, then once $z := \left(1 - \frac{1}{m}\right)x + \frac{1}{m}y \in S$ holds for some $m \in \mathbb{N}$, then the entire line segment between x and z must be contained in S. Thus, for convex S, we surely have $int_X(S) \subseteq al\text{-}int_X(S)$. How about the converse? That turns out to be a harder question to answer. We need the following lemma to get to it.

[22] In geometric functional analysis, a convex subset of a metric linear space with a nonempty interior is sometimes referred to as a **convex body**. I won't follow this practice here, however.

Lemma 3
For any subset S of a metric linear space X, if $x \in int_X(S)$ and $y \in cl_X(S)$, then

$$(1-\alpha)x + \alpha y \in int_X(S), \qquad 0 \leq \alpha < 1;$$

that is,

$$(1-\alpha)int_X(S) + \alpha cl_X(S) \subseteq int_X(S), \qquad 0 \leq \alpha < 1.$$

Proof
Take any $x \in int_X(S)$ *and* $y \in cl_X(S)$, and fix an arbitrary $0 < \alpha < 1$. Clearly,

$$(1-\alpha)x + \alpha y \in S \quad \text{iff} \quad y \in \tfrac{1}{\alpha}S - \left(\tfrac{1-\alpha}{\alpha}\right)x = S + \left(\tfrac{1-\alpha}{\alpha}\right)(S - x).$$

Since $x \in int_X(S)$, Proposition 2 implies that $\left(\frac{1-\alpha}{\alpha}\right)(int_X(S) - x)$ is an open set that contains $\mathbf{0}$. Applying Proposition 2 again, then, we find that $S + \left(\frac{1-\alpha}{\alpha}\right)(int_X(S) - x)$ is an open set that contains $cl_X(S)$. (Why?) So,

$$y \in S + \left(\tfrac{1-\alpha}{\alpha}\right)(int_X(S) - x) \subseteq S + \left(\tfrac{1-\alpha}{\alpha}\right)(S - x),$$

and it follows that $(1-\alpha)x + \alpha y \in S$.

We now know that $(1-\alpha)int_X(S) + \alpha cl_X(S) \subseteq S$ for any $0 < \alpha < 1$. But by Proposition 2, $(1-\alpha)int_X(S) + \alpha cl_X(S)$ is an open set, and since the interior of a set is the $\supseteq$-maximum open set that is contained in that set, this finding readily gives us that $(1-\alpha)int_X(S) + \alpha cl_X(S) \subseteq int_X(S)$ for any $0 < \alpha < 1$. ■

Here is the main message of the present lecture.

Proposition 11
Let S be a convex subset of a metric linear space X. Then

$$int_X(S) \subseteq al\text{-}int_X(S).$$

Moreover, if $int_X(S) \neq \emptyset$, then

$$int_X(S) = al\text{-}int_X(S).$$

Similarly, if $int_{aff(S)}(S) \neq \emptyset$, then

$$int_{aff(S)}(S) = ri(S).$$

Proof

Let $x \in int_X(S)$, and pick any $\varepsilon > 0$ with $N_{\varepsilon,X}(x) \subseteq S$. Now take any $y \in X$, and observe that $\left(1 - \frac{1}{m}\right) x + \frac{1}{m} y \to x$ (by the continuity of scalar multiplication and vector addition). So, there exists an $\alpha^* > 0$ (which may depend on y) such that $(1-\alpha^*)x + \alpha^* y \in N_{\varepsilon,X}(x) \subseteq S$. Since S is convex, all points on the line segment between $(1-\alpha^*)x + \alpha^* y$ and x belong to S. That is, by taking convex combinations of $(1-\alpha^*)x + \alpha^* y$ and x, we find

$$(1-\alpha)x + \alpha y \in S, \qquad 0 \le \alpha \le \alpha^*.$$

(Verify!) Since y is an arbitrary vector in X here, this means that $x \in$ *al-int*$_X(S)$.

To prove the second assertion, let $y \in$ *al-int*$_X(S)$, and pick any $x \in int_X(S)$. Let $x' := 2y - x$. (Note that y is the midpoint of the line segment between x and x'; see Figure 2.) By definition of the algebraic interior, we may find an $0 < \alpha^* < 1$ such that

$$(1-\alpha)y + \alpha x \in S \quad \text{and} \quad (1-\alpha)y + \alpha x' \in S, \qquad 0 \le \alpha \le \alpha^*.$$

Define $z := (1-\alpha^*)y + \alpha^* x' \in S$, and note that the line segment between z and x contains y (for, $y = \frac{1}{1+\alpha^*} z + \frac{\alpha^*}{1+\alpha^*} x$). (See Figure 2.) But, by Lemma 3, this line segment (with z possibly being excluded), and in particular y, lies in $int_X(S)$.

The proof of the third assertion is analogous and is left as an exercise. ∎

In words, the algebraic interior of a given convex subset of a metric linear space X is always larger than its interior. More precisely, there are

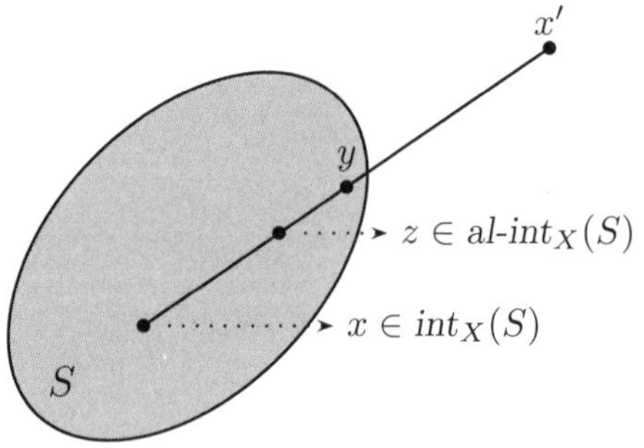

Figure 2

two possibilities in general. Either the interior of the set is empty while its algebraic interior is nonempty, or these sets coincide. By Example G.5 and Theorem 1, the former possibility can never hold if $dim(X) < \infty$. If $dim(X) = \infty$, however, there may exist a subset S of X with $int_X(S) = \emptyset \neq al\text{-}int_X(S)$. An example of this anomaly is given in Example 11 below.

Since, intuitively, we would like to think of algebraically open sets in infinite-dimensional linear spaces as "large," another lesson of Proposition 11 is that in such spaces we should also consider (again intuitively) an open convex set (or more generally, a convex set with a nonempty interior) as "large."

Here is an immediate application of Proposition 11.

EXAMPLE 8

We have seen earlier that $al\text{-}int_{\ell^p}(\ell^p_+) = \emptyset$ for any $1 \leq p < \infty$ (Exercise G.23). In view of the first part of Proposition 11, therefore, we now find that $int_{\ell^p}(\ell^p_+) = \emptyset$ for any $1 \leq p < \infty$ as well.

The situation is different in the case of ℓ^∞. Indeed, we have found in Exercise G.24 that

$$al\text{-}int_{\ell^\infty}(\ell^\infty_+) = \{(x_m) \in \ell^\infty_+ : \inf\{x_m : m \in \mathbb{N}\} > 0\}.$$

But it is easily seen that $int_{\ell^\infty}(\ell^\infty_+) \neq \emptyset$. For instance, $(1, 1, \ldots)$ belongs to $int_{\ell^\infty}(\ell^\infty_+)$. (Proof?) Thus, thanks to Proposition 11, we may conclude that

$$int_{\ell^\infty}(\ell^\infty_+) = \{(x_m) \in \ell^\infty_+ : \inf\{x_m : m \in \mathbb{N}\} > 0\},$$

nice and easy. □

EXERCISE 45[H] Let S be a subset of a metric linear space X. Prove that if S is convex, so is $al\text{-}int_X(S)$, and if S is open, so is $co(S)$.

EXERCISE 46 Let X be a metric linear space such that $N_{\varepsilon,X}(\mathbf{0})$ is convex for any $\varepsilon > 0$. Show that $int_X(S) \subseteq al\text{-}int_X(S)$ for any $S \subseteq X$.

EXERCISE 47[H] Let S be a convex set in a metric linear space X such that $int_X(S) \neq \emptyset$. Prove:

(a) $cl_X(S) = al\text{-}cl(S)$.

(b) $bd_X S = al\text{-}bd_X(S)$.

(c) $cl_X(S) = cl_X(al\text{-}int_X(S)) = cl_X(int_X(S))$.

5.3 Extension of Positive Linear Functionals, Revisited

Proposition 11 is not all that can be said about the relation between the interior and algebraic interior of a convex set, and we will return to this issue shortly. For the time being, however, we concentrate on what one can deduce from this result. Let us note first that we can readily sharpen the Krein-Rutman Theorem by using Proposition 11. Recall that the Krein-Rutman Theorem says that we can extend any positive linear functional defined on a preordered subspace Y of a preordered metric linear space X to the entire space positively, provided that $Y \cap \textit{al-int}(X_+) \neq \emptyset$. In turn, Propositions 4 and 11 combine to tell us that the extended map is in fact continuous. To sum up:

> **Proposition 12**
> *Let X be a preordered metric linear space, and Y a preordered subspace of X such that $Y \cap \mathit{int}_X(X_+) \neq \emptyset$. If L is a positive linear functional on Y, then there exists a continuous and positive linear functional L^* on X with $L^*|_Y = L$.*

Proof

Apply the Krein-Rutman Theorem first and Propositions 4 and 11 next. ■

5.4 Separation by Closed Hyperplanes

When combined with the Dieudonné Separation Theorem, Proposition 11 yields a satisfactory solution also to the problem of separating convex sets in a metric linear space by means of *closed* hyperplanes. All that remains to be established is the following simple fact.

> **Lemma 4**
> *Let X be a metric linear space and $L \in \mathcal{L}(X, \mathbb{R})$. If there exists a set $S \subseteq X$ with $\mathit{int}_X(S) \neq \emptyset$ and $\alpha > L(S)$ for some $\alpha \in \mathbb{R}$, then L must be continuous.*

Proof

Take any subset S of X, and assume $\mathit{int}_X(S) \neq \emptyset$ and $\alpha > L(S)$. Then $\alpha + 1 \gg L(S)$, so we have $S \cap L^{-1}(\alpha + 1) = \emptyset$. Thus, given that $\mathit{int}_X(S) \neq \emptyset$, the hyperplane $L^{-1}(\alpha + 1)$ cannot be dense in X. (Why?) By Propositions 6 and 7, therefore, L is continuous. ■

It follows that any linear functional (on a metric linear space) that separates two sets at least one of which has a nonempty interior, is continuous. This leads us to the following important separation theorem.[23]

The Separating Hyperplane Theorem

(Tukey) *Let A and B be two nonempty convex sets in a metric linear space X such that $\text{int}_X(A) \neq \emptyset$. Then, $\text{int}_X(A) \cap B = \emptyset$ if, and only if, there exist an $\alpha \in \mathbb{R}$ and a continuous $L \in \mathcal{L}(X, \mathbb{R})$ such that*

$$L(B) \geq \alpha > L(A) \quad \text{and} \quad \alpha \gg L(\text{int}_X(A)).$$

Proof

The "if" part of the assertion is trivial. The "only if" part follows from the Dieudonné Separation Theorem, Proposition 11 and Lemma 4. ■

The geometric interpretation of this result is, of course, identical to that of the Dieudonné Separation Theorem. But there is an important difference between the two results. While the Dieudonné Separation Theorem tells us that we can separate two disjoint convex sets (in a linear space) one of which has a nonempty algebraic interior, the Separating Hyperplane Theorem tells us not only that we can separate two disjoint convex sets (in a metric linear space) one of which has a nonempty interior, but that we can do this by means of a *closed* hyperplane (Proposition 6). This is crucial, because a hyperplane that is not closed must be dense in the mother space (Proposition 7), and thus the geometric interpretation of separating two sets with such an hyperplane is ambiguous at best. Fortunately, the Separating Hyperplane Theorem does not run into this sort of a difficulty.

Warning. The nonempty interior requirement cannot be omitted in the statement of the Separating Hyperplane Theorem. (Recall Example G.12.)

Exercise 48[H] Let A and B be two disjoint nonempty convex sets in a metric linear space X. Show that if A is open, then there exists an $\alpha \in \mathbb{R}$ and a continuous $L \in \mathcal{L}(X, \mathbb{R})$ such that $L(B) \geq \alpha > L(A)$. Can one replace $\geq$ with $>$ here?

[23] This version of the result is due to Tukey (1942).

EXAMPLE 9

(*Existence of Continuous Linear Functionals*) We have used in Example G.11 the Hahn-Banach Extension Theorem 1 to prove the existence of nonzero linear functionals in an arbitrary linear space. Similarly, we can use the Separating Hyperplane Theorem to obtain a sufficient condition for the existence of nonzero *continuous* linear functionals in an arbitrary metric linear space: *If X is a metric linear space that contains a convex set S with $int_X(S) \neq \emptyset$ and $S \neq X$, then there exists a nonzero continuous linear functional on X.* This follows readily from the Separating Hyperplane Theorem. (*Proof.* Take any $x \in X \backslash S$, and separate it from S by a closed hyperplane.) □

EXAMPLE 10

Let O be an open and convex subset of a metric linear space X. Let us prove that

$$O = int_X(cl_X(O)).$$

The method of proof we adopt will illustrate the typical use of the Separating Hyperplane Theorem in applications.[24] Observe first that the $\subseteq$ part of the claim is trivial (since $int_X(cl_X(O))$ is the largest (i.e., $\supseteq$-maximum) open set contained within $cl_X(O)$). To prove the converse containment, suppose that there is an x in $int_X(cl_X(O)) \backslash O$ to derive a contradiction. Then, by the Separating Hyperplane Theorem, there exists a continuous $L \in \mathcal{L}(X, \mathbb{R})$ such that $L(x) \gg L(O)$. (Why?) Of course, continuity of L ensures that $L(x) \geq L(cl_X(O))$. Now fix any $y \in O$, and define $z := 2x - y$. (Note that x is the midpoint of the line segment between z and y.) Clearly, $L(x) > L(y)$ implies $L(z) > L(x)$. (Yes?) But since $x \in al\text{-}int_X(cl_X(O))$ (Propositions 10 and 11), there exists a small enough $\alpha > 0$ such that $w := (1-\alpha)x + \alpha z \in cl_X(O)$. But this contradicts $L(x) \geq L(cl_X(O))$, for we have $L(w) > L(x)$. □

EXERCISE 49 Suppose that X is a metric linear space the origin of which is not contained in any proper convex subset of X. Show that the only continuous linear functional on X is the zero functional.

[24] This assertion is by no means trivial. For one thing, it is false for sets that are not open or not convex. Check if the claimed equality holds for the sets $[0, 1)$ and $(0, \frac{1}{2}) \cup (\frac{1}{2}, 1)$, for instance.

Here are two additional consequences of the Separating Hyperplane Theorem.

The Supporting Hyperplane Theorem
Let S be a closed and convex subset of a metric linear space X. If $int_X(S) \neq \emptyset$ and $x \in bd_X(S)$, then there exists a continuous $L \in \mathcal{L}(X, \mathbb{R})$ such that $L(x) \geq L(S)$ and $\alpha \gg L(int_X(S))$.

Corollary 6
Let S be a closed and convex subset of a metric linear space X. If $int_X(S) \neq \emptyset$ and $S \neq X$, then S equals the intersection of all closed halfspaces that contain it.

Exercise 50[H] Prove the Supporting Hyperplane Theorem and Corollary 6.

Exercise 51[H] Show that we cannot omit the requirement $int_X(S) \neq \emptyset$ in the statement of the Supporting Hyperplane Theorem.

Exercise 52 Let X be a preordered metric linear space. Assume that $int_X(X_+) \neq \emptyset$ and $\{\alpha : (1-\alpha)x + \alpha y \in X_+\}$ is a closed subset of $\mathbb{R}$. Show that there exists a nonempty set $\mathcal{L}$ of continuous linear functionals on X such that, for each $x, y \in X$,

$$x - y \in X_+ \quad \text{iff} \quad L(x) \geq L(y) \text{ for all } L \in \mathcal{L}.$$

(How does this relate to the Expected Multi-Utility Theorem?)

5.5* Interior versus Algebraic Interior of a Closed and Convex Set

We will revisit the problem of separating convex sets in the next chapter in a slightly more specialized setting. Before we conclude this chapter, however, we have to take care of an unfinished business about the problem of determining when the interior and the algebraic interior of a convex set coincide. We know at present that $int_X(S) \subseteq al\text{-}int_X(S)$ for any convex subset S of a metric linear space X. Moreover, if $int_X(S) \neq \emptyset$, then this containment becomes an equality (Proposition 11). As we illustrate in Example 11 below, one cannot do away with this sufficiency condition in general. However, if S is closed, and if the metric linear space that contains S is complete, then the situation is much more satisfactory.

THEOREM 3

Let S be a closed and convex subset of a Fréchet space X. Then, $int_X(S) = al\text{-}int_X(S)$.

PROOF

By Proposition 11, it is enough to show that $al\text{-}int_X(S) \neq \emptyset$ implies $int_X(S) \neq \emptyset$. Moreover, by way of translation, we may assume that $\mathbf{0} \in al\text{-}int_X(S)$.[25]

To derive a contradiction, suppose that $al\text{-}int_X(S) \neq \emptyset$ but $int_X(S) = \emptyset$. Then, for any positive integer m, mS must be a closed set with empty interior. (Why?) Hence, $X \backslash mS$ must be an open dense subset of X, $m = 1, 2, \ldots$. (Why?) Since S is closed, we may find a closed set B_1 in $X \backslash S$ with $int_X(B_1) \neq \emptyset$ and $diam(B_1) \leq 1$.[26] Since $X \backslash 2S$ is an open dense set, $int_X(B_1) \cap (X \backslash 2S)$ must be a nonempty open set. So we may find a closed set B_2 in $B_1 \backslash 2S$ with $int_X(B_2) \neq \emptyset$ and $diam(B_2) \leq \frac{1}{2}$.[27] But $X \backslash 3S$ is also open and dense, and so $int_X(B_2) \cap (X \backslash 3S)$ is a nonempty open set, allowing us to find a closed B_3 in $B_2 \backslash 3S$ with $diam(B_3) \leq \frac{1}{3}$. Continuing this way, then, we obtain a sequence (B_m) of closed sets in X such that, for each $m \in \mathbb{N}$,

(i) $diam(B_m) \leq \frac{1}{m}$; (ii) $B_m \cap mS = \emptyset$; and (iii) $B_m \supseteq B_{m+1}$. (i) and (iii) allow us to invoke the Cantor-Fréchet Intersection Theorem to find an $x \in \cap^\infty B_i$.[28] By (ii), $x \in \cap^\infty (X \backslash iS)$, and it is easy to see that this contradicts $\mathbf{0} \in al\text{-}int_X(S)$. (Verify!) ■

[25] Let $y \in al\text{-}int_X(S)$, and define $T := S - y$, which is a closed and convex set with $\mathbf{0} \in al\text{-}int_X(T)$. The subsequent argument will establish that $int_X(T) = al\text{-}int_X(T)$, which implies $int_X(S) = al\text{-}int_X(S)$.

[26] If you want to be more concrete, take any $x \in X \backslash S$, and set $B_1 := cl_X(N_{\varepsilon,X}(x))$, where $\varepsilon := \min\{\frac{1}{2}, d(x, S)\}$. By the way, you might want to follow the subsequent construction by using Figure 3. Our strategy is none other than to use the method we adopted to prove the Heine-Borel Theorem (the "butterfly hunting" method).

[27] The idea is the same. Take any $x \in int_X(B_1) \cap (X \backslash 2S)$, and set $B_2 := cl_X(N_{\varepsilon,X}(x))$ where $\varepsilon := \min\{\frac{1}{4}, d(x, 2S), d(x, bd_X(B_1))\}$.

[28] Or, if you want to prove this directly, pick any $x^m \in B_m$ for each m. By (i), (x^m) is a Cauchy sequence, and since X is complete, it converges in X. But by (iii), each B_m contains the sequence $(x^m, x^{m+1}, \ldots)$, so, being closed, it also contains $\lim x^m$.

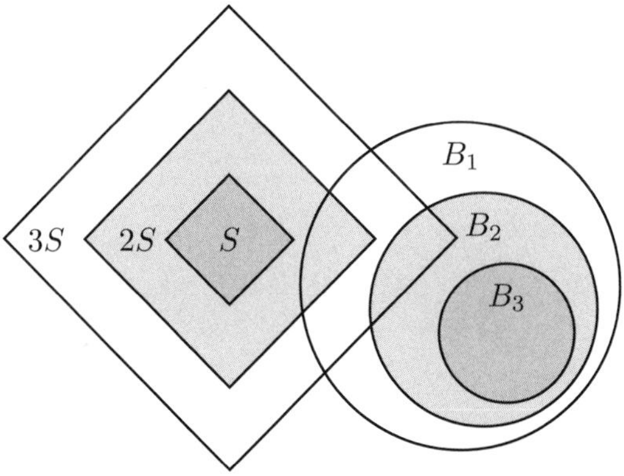

FIGURE 3

In a Fréchet space, therefore, the interior and the algebraic interior of a closed and convex set coincide. In the next chapter we will have a major opportunity to appreciate the importance of this observation. For now, we conclude by demonstrating that this result is not valid in an arbitrary metric linear space.

EXAMPLE 11
Let $X := \{(x_m) \in \mathbb{R}^\infty : \sum^\infty |x_i| < \infty\}$, and make this set a metric linear space by using the sup-metric d_∞ (Example 3.[1]). It is easily checked that this space is not complete.[29] Define

$$S := \left\{ (x_m) \in X : \sum_{i=1}^{\infty} |x_i| \leq 1 \right\},$$

which is a closed and convex subset of X. We have, $\mathbf{0} \in \textit{al-int}_X(S)$, since, for any $(y_m) \in X\backslash\{\mathbf{0}\}$, we have $\lambda(y_m) \in S$ for any $0 < \lambda < 1/\sum^\infty |y_i|$. However, $\mathbf{0} \notin \textit{int}_X(S)$. Indeed, for any positive integer k,

$$y^k := \left(\tfrac{2}{k}, \tfrac{1}{k}, \ldots, \tfrac{1}{k}, 0, 0, \ldots\right) \in X\backslash S$$

[29] Consider the sequence $(x^m) \in X^\infty$, where $x^1 := (1, 0, 0, 0, \ldots)$, $x^2 := (1, \frac{1}{2}, 0, 0, \ldots)$, $x^3 := (1, \frac{1}{2}, \frac{1}{3}, 0, \ldots)$, etc.. Then, with respect to the sup-metric, (x^m) is Cauchy, but it does not converge to any sequence that belongs to X.

(with exactly k many terms being nonzero), and yet $d_\infty(y^k, \mathbf{0}) = \frac{2}{k} \to 0$. (By Proposition 11, therefore, $int_X(S) = \emptyset$, but you may want to verify this directly.[30]) Conclusion: The nonempty interior and the completeness hypotheses cannot be omitted in Proposition 11 and Theorem 3, respectively. □

EXERCISE 53 Prove or disprove: The closedness hypothesis can be omitted in the statement of Theorem 3.

[30] The closed unit ball of a metric space can never have an empty interior, of course. Puzzled? Don't be. S is *not* the closed unit ball of X, that is, $S \neq \{(x_m) \in \ell^1 : d_\infty((x_m), \mathbf{0}) \leq 1\}$.

Chapter J

Normed Linear Spaces

This chapter introduces a very important subclass of metric linear spaces, namely, the class of *normed linear spaces.* We begin with an informal discussion that motivates the investigation of such spaces. We then formalize parts of that discussion, introduce Banach spaces, and go through a number of examples and preliminary results. The first hints of how productive mathematical analysis can be within the context of normed linear spaces are found in our final excursion to fixed point theory. Here we prove the fixed point theorems of Glicksberg, Fan, Krasnoselsky, and Schauder, and provide a few applications to game theory and functional equations. We then turn to the basic theory of continuous linear functionals defined on normed linear spaces, and sketch an introduction to classical linear functional analysis. Our treatment is guided by geometric considerations for the most part, and dovetails with that of Chapter G. In particular, we carry our earlier work on the Hahn-Banach type extension and separation theorems into the realm of normed linear spaces, and talk about a few fundamental results of infinite-dimensional convex analysis, such as the Extreme Point Theorem, the Krein-Milman Theorem, and so on. In this chapter we also bring to a conclusion our work on the classification of the differences between finite- and the infinite-dimensional linear spaces. Finally, to give at least a glimpse of the powerful Banach space methods, we establish here the famous Uniform Boundedness Principle as a corollary of our earlier geometric findings, and go through some of its applications.

The present treatment of normed linear spaces is roughly at the same level as that of the classic real analysis texts by Kolmogorov and Fomin (1970) and Royden (1994). For a more detailed introduction to Banach space theory, we recommend Kreyzig (1978), Maddox (1988), or the first chapter of Megginson (1998).[1]

[1] My coverage of linear functional analysis here is directed toward particular applications and is thus incomplete, even at an introductory level. A more leisurely introduction would

1 Normed Linear Spaces

1.1 A Geometric Motivation

The Separating Hyperplane Theorem attests to the fact that our program of marrying the metric and linear analyses of Chapters C–G has been a successful one. However, there are still a number of shortcomings we need to deal with. For instance, although we came close in Example I.9, we have so far been unable to establish the existence of a nonzero continuous linear functional on an arbitrarily given metric linear space. And this is for a good reason: there are Fréchet spaces on which the only continuous linear functional is the zero functional.[2] In view of Propositions I.6 and I.7, this means that all hyperplanes in a metric linear space may be dense, which is a truly pathologic situation. In such a space the separation of convex sets is a dubious concept at best. Moreover, by the Separating Hyperplane Theorem, the only nonempty open convex subset of such a space is the entire space itself. (Why?) Let us step back for a moment and ask, what is the source of these problems?

How would we separate two distinct points x and y in $\mathbb{R}^n$ by using the Separating Hyperplane Theorem? The answer is easy: just take an $0 < \varepsilon < d_2(x, y)$, and separate the open convex sets $N_{\varepsilon,\mathbb{R}^n}(x)$ and $N_{\varepsilon,\mathbb{R}^n}(y)$ by a closed hyperplane. So why doesn't this argument work in an arbitrary metric linear space? Because an ε-neighborhood, while necessarily nonempty and open, need *not* be convex in an arbitrary metric linear space (even if this space is finite-dimensional). This, in turn, does not let us invoke the Separating Hyperplane Theorem to find a closed hyperplane (and hence a nonzero continuous functional) at this level of generality.[3]

---

cover the open mapping and closed graph theorems, and would certainly spend some time on Hilbert spaces. I do not use these two theorems in this book, and I talk about Hilbert spaces only in passing. Moreover, my treatment completely ignores operator theory, which is an integral counterpart of linear functional analysis. In this regard, all I can do here is direct your attention to the beautiful expositions of Megginson (1998) and Schechter (2002).

[2] I don't know a simple example that would illustrate this. If you're familiar with measurable functions, you may be able to take comfort in the following example: The set of all measurable real functions on $[0, 1]$, metrized via the map $(f, g) \mapsto \int_0^1 \min\{1, |f(t) - g(t)|\}dt$, is a Fréchet space on which there is no nonzero continuous linear functional. (Even the linear functional $f \mapsto f(\mathbf{0})$ is not continuous on this space. Why?)

[3] The metric linear space discussed in the previous footnote provides a case in point. A simpler example is obtained by metrizing $\mathbb{R}^2$ with the metric $d_{1/2}(x, y) := \sqrt{|x_1 - y_1|} + \sqrt{|x_2 - y_2|}$, $x, y \in \mathbb{R}^2$. (*Proof.* Denoting this space by X, check that, for any $\varepsilon > 0$,

The problem therefore lies in the fact that the ε-neighborhoods in a metric linear space may not be convex. Or, said better, in an arbitrary metric linear space X, an ε-neighborhood of $\mathbf{0}$ may not contain an open convex set O with $\mathbf{0} \in O$. Why don't we then restrict attention to those metric linear spaces for which, for any $\varepsilon > 0$, there exists an open convex subset of $N_{\varepsilon,X}(\mathbf{0})$ that includes $\mathbf{0}$? Indeed, such a space—called a **locally convex** metric linear space—is free of the difficulties we mentioned above. After all, any two distinct points of a locally convex metric linear space can be separated by a closed hyperplane, and hence there are plenty of continuous functionals defined on such a space. Yet even locally convex metric linear spaces lack the structure that we need for a variety of problems that we will explore in this chapter. We want—and you will see why in due course—to work with metric linear spaces whose ε-neighborhoods are not only convex but also behave well with respect to dilations and contractions.

Let X be a metric linear space. Take the open unit ball $N_{1,X}(\mathbf{0})$, and stretch it in your mind to dilate it at a one-to-two ratio. What do you get, intuitively? There seem to be two obvious candidates: $2N_{1,X}(\mathbf{0})$ and $N_{2,X}(\mathbf{0})$. Our Euclidean intuition suggests that these two sets should be the same, so perhaps there is no room for choice. Let's see. Since the metric d on X is translation invariant, $x \in N_{1,X}(\mathbf{0})$ implies

$$d(2x, \mathbf{0}) = d(x, -x) \leq d(x, \mathbf{0}) + d(\mathbf{0}, -x) = 2d(x, \mathbf{0}) < 2,$$

so we have $2N_{1,X}(\mathbf{0}) \subseteq N_{2,X}(\mathbf{0})$ indeed. Conversely, take any $x \in N_{2,X}(\mathbf{0})$. Is $x \in 2N_{1,X}(\mathbf{0})$? That is to say, is $\frac{1}{2}x \in N_{1,X}(\mathbf{0})$? No, not necessarily. After all, in a metric linear space it is possible that $d(x, \mathbf{0}) = d\left(\frac{1}{2}x, \mathbf{0}\right)$. (In $\mathbb{R}^\infty$, for instance, the distance between $\mathbf{0}$ and $(2, 2, \ldots)$ and that between $\mathbf{0}$ and $(1, 1, \ldots)$ are both equal to 1.)

This is another anomaly that we wish to avoid. Since the distance between x and $\mathbf{0}$ is $d(x, \mathbf{0})$, and $\frac{1}{2}x$ is the midpoint of the line segment between $\mathbf{0}$ and

$(\frac{\varepsilon^2}{2}, 0), (0, \frac{\varepsilon^2}{2}) \in N_{\varepsilon,X}(\mathbf{0})$ but $\frac{1}{2}(\frac{\varepsilon^2}{2}, 0) + \frac{1}{2}(0, \frac{\varepsilon^2}{2}) \notin N_{\varepsilon,X}(\mathbf{0})$ with respect to $d_{1/2}$. Now draw $N_{1,X}(\mathbf{0})$ with respect to this metric.)

Warning. There is a major difference between the space $(\mathbb{R}^2, d_{1/2})$ and the one of the previous footnote. While the only linear functional on the latter space is the zero functional, every linear functional on the former is continuous. (Why?) The absence of "convex neighbor-hoods" yields a shortage of continuous linear functionals only in the case of *infinite*-dimensional spaces.

x, it makes sense to require that the distance between $\mathbf{0}$ and $\frac{1}{2}x$ be $\frac{1}{2}d(x,\mathbf{0})$. That is, we wish to have

$$d\left(\tfrac{1}{2}x,\mathbf{0}\right) = \tfrac{1}{2}d(x,\mathbf{0}) \qquad \text{for all } x \in X. \tag{1}$$

In particular, this guarantees that $2N_{1,X}(\mathbf{0}) = N_{2,X}(\mathbf{0})$.

Of course, exactly the same reasoning would also justify the requirement $d(\frac{1}{n}x,\mathbf{0}) = \frac{1}{n}d(x,\mathbf{0})$ for every $n \in \mathbb{N}$. This in turn implies $d(mx,\mathbf{0}) = md(x,\mathbf{0})$ for all $m \in \mathbb{N}$, and it follows that $d(rx,\mathbf{0}) = rd(x,\mathbf{0})$ for all $r \in \mathbb{Q}_{++}$. (Why?) Consequently, since $d(\cdot,\mathbf{0})$ is a continuous map on X, we find that $d(\lambda x,\mathbf{0}) = \lambda d(x,\mathbf{0})$ for all $\lambda > 0$. (Yes?) Finally, if $\lambda < 0$, then, by translation invariance, $d(\lambda x,\mathbf{0}) = d(\mathbf{0},-\lambda x) = d(-\lambda x,\mathbf{0}) = -\lambda d(x,\mathbf{0})$. In conclusion, the geometric anomaly that we wish to avoid points toward the following homogeneity property:

$$d(\lambda x,\mathbf{0}) = |\lambda|\, d(x,\mathbf{0}) \qquad \text{for all } x \in X. \tag{2}$$

Metric linear spaces for which (2) holds provide a remarkably rich playground. In particular, all ε-neighborhoods in such a space are convex (and hence any such space is locally convex). Thus, the difficulty that worried us at the beginning of our discussion does not arise in such metric linear spaces.[4] In fact, we gain a lot more by positing (2). It turns out that metric linear spaces with (2) provide an ideal environment to carry out a powerful convex analysis jointly with a functional analysis of linear and nonlinear operators. By the time you are done with this chapter and the next, this point will have become abundantly clear.

Exercise 1 Let X be a metric linear space. Show that every ε-neighborhood in X is convex iff $d(x,\cdot)$ is quasiconvex for each $x \in X$.

Exercise 2[H] As noted above, a metric linear space X is called **locally convex** if, for every $\varepsilon > 0$, there exists an open and convex set O_ε in $N_{\varepsilon,X}(\mathbf{0})$, with $\mathbf{0} \in O_\varepsilon$. At least one of the ε-neighborhoods in a locally convex metric linear space is convex. True or false?

Exercise 3 Let X be a metric linear space with (1). Show that there exists a nonzero continuous linear functional on X.

[4] More is true: (1) alone guarantees that $N_{\varepsilon,X}(\mathbf{0})$ (and hence $N_{\varepsilon,X}(x) = x + N_{\varepsilon,X}(\mathbf{0})$ for each $x \in X$) is a convex set for any $\varepsilon > 0$. After all, every open midpoint convex subset of a metric linear space is convex. (Proof?)

EXERCISE 4 Let X be a metric linear space with (1). Show that X is bounded iff $X = \{\mathbf{0}\}$.

EXERCISE 5 In a metric linear space X with (1), would we necessarily have (2)?

1.2 Normed Linear Spaces

Let us leave the geometric considerations we outlined above aside for a moment and instead focus on the following fundamental definition. We return to the previous discussion in about six pages.

DEFINITION

Let X be a linear space. A function $\|\cdot\| : X \to \mathbb{R}_+$ that satisfies the following properties is called a **norm** on X: For all $x, y \in X$,

(i) $\|x\| = 0$ if and only if $x = \mathbf{0}$;

(ii) (*Absolute Homogeneity*) $\|\lambda x\| = |\lambda| \, \|x\|$ for all $\lambda \in \mathbb{R}$; *and*

(iii) (*Subadditivity*) $\|x + y\| \leq \|x\| + \|y\|$.

If $\|\cdot\|$ is a norm on X, then we say that $(X, \|\cdot\|)$ is a **normed linear space**. If $\|\cdot\|$ satisfies only requirements (ii) and (iii), then $(X, \|\cdot\|)$ is called a **seminormed linear space.**[5]

Recall that the basic viewpoint of vector calculus is to regard a "vector" x in a linear space as a *directed* line segment that begins at zero and ends at x. This allows one to think of the "length" (or the "magnitude") of a vector in a natural way. For instance, we think of the magnitude of a positive real number x as the length of the interval $(0, x]$, and that of $-x$ as the length of $[-x, 0)$. Indeed, it is easily verified that the absolute value function defines

[5] The idea of normed linear space has been around since the 1906 dissertation of Fréchet, and a precursory analysis of it can be traced in the works of Eduard Helly and Hans Hahn prior to 1922. The modern definition was given first by Stefan Banach and Norbert Wiener independently in 1922. Banach then undertook a comprehensive analysis of such spaces, which culminated in his groundbreaking 1932 treatise. (See Dieudonné (1981).)

a norm on $\mathbb{R}$. Similarly, it is conventional to think of the length of a vector in $\mathbb{R}^n$ as the distance between this vector and the origin, and as you would expect, $x \mapsto d_2(x, \mathbf{0})$ defines a norm on $\mathbb{R}^n$. Just as the notion of a metric generalizes the geometric notion of "distance," therefore, the notion of a norm generalizes that of "length" or "magnitude" of a vector.

This interpretation also motivates the properties that a "norm" must satisfy. First, a norm must be nonnegative, because "length" is an inherently nonnegative notion. Second, a nonzero vector should be assigned positive length, and hence property (i). Third, the norm of a vector $-x$ should equal the norm of x, because multiplying a vector by -1 should change only the direction of the vector, not its length. Fourth, doubling a vector should double the norm of that vector, simply because the intuitive notion of "length" behaves in this manner. Property (ii) of a norm is a generalization of the latter two requirements. Finally, our Euclidean intuition about "length" suggests that the norm of a vector that corresponds to one side of a triangle should not exceed the sum of the norms of the vectors that form the other two sides of that triangle. This requirement is formalized as property (iii) in the formal definition of a norm. You may think that there may be other properties that an abstract notion of "length" should satisfy. But, as you will see in this chapter, the framework based on the properties (i)–(iii) alone turns out to be rich enough to allow for an investigation of the properties of linearity and continuity in conjunction, along with a satisfactory geometric analysis.

When the norm under consideration is apparent from the context, it is customary to dispense with the notation $(X, \|\cdot\|)$ and refer to the set X itself as a normed linear space. We frequently adopt this convention in what follows. That is, when we say that X is a normed linear space, you should understand that X is a linear space with a norm $\|\cdot\|$ lurking in the background. When we need to deal with two normed linear spaces, X and Y, the norms of these spaces will be denoted by $\|\cdot\|$ and $\|\cdot\|_Y$, respectively.

Let X be a normed linear space. For future reference, let's put on record an immediate yet important consequence of the absolute homogeneity and subadditivity properties of the norm on X. Take any $x, y \in X$. Note first that, by subadditivity, $\|x\| = \|x - y + y\| \leq \|x - y\| + \|y\|$, so that

$$\|x\| - \|y\| \leq \|x - y\| .$$

Moreover, if we change the roles of x and y in this inequality, we get $\|y\| - \|x\| \leq \|y - x\| = \|x - y\|$, where the last equality follows from the absolute homogeneity of $\|\cdot\|$. Thus,

$$\big|\|x\| - \|y\|\big| \leq \|x - y\| \qquad \text{for any } x, y \in X. \tag{3}$$

This inequality will be useful to us on several occasions, so please keep it in mind as a companion to the subadditivity property.

EXERCISE 6^{H} (*Convexity of Norms*) For any normed linear space X, show that

$$\frac{\|x + \alpha y\| - \|x\|}{\alpha} \leq \frac{\|x + \beta y\| - \|x\|}{\beta} \qquad \text{for all } x, y \in X \text{ and } \beta \geq \alpha > 0.$$

EXERCISE 7^{H} (*Rotund Spaces*) A normed linear space X is called **rotund** if $\|x + y\| < \|x\| + \|y\|$ for all linearly independent $x, y \in X$. Show that X is rotund iff $\left\|\frac{1}{2}x + \frac{1}{2}y\right\| < \frac{1}{2}\|x\| + \frac{1}{2}\|y\|$ for any distinct $x, y \in X$ with $\|x\| = \|y\| = 1$. (Neither ℓ^1 nor ℓ^∞ is rotund. Right?)

1.3 Examples of Normed Linear Spaces

For any $n \in \mathbb{N}$, we can norm $\mathbb{R}^n$ in a variety of ways to obtain a finite dimensional normed linear space. For any $1 \leq p \leq \infty$, the ***p*-norm** $\|\cdot\|_p$ on $\mathbb{R}^n$ is defined as

$$\|x\|_p := \left(\sum_{i=1}^{n} |x_i|^p\right)^{\frac{1}{p}}$$

if p is finite, and $\|x\|_p := \max\{|x_i| : i = 1, \ldots, n\}$ if $p = \infty$. Using Minkowski's Inequality 1, it is readily checked that any $\|\cdot\|_p$ is a norm on $\mathbb{R}^n$ for any $1 \leq p \leq \infty$. (Obviously, $\|x\|_p = d_p(x, \mathbf{0})$ for all $x \in \mathbb{R}^n$ and $1 \leq p \leq \infty$.)

Recall that, for any $1 \leq p, q \leq \infty$, $\mathbb{R}^{n,p}$ and $\mathbb{R}^{n,q}$ are "identical" metric linear spaces (in the sense that they are linearly homeomorphic). It would thus be reasonable to expect that $(\mathbb{R}^n, \|\cdot\|_p)$ and $(\mathbb{R}^n, \|\cdot\|_q)$ are "identical" in some formal sense as well. This is indeed the case, as we discuss later in Section 4.2.

Here are some examples of infinite-dimensional normed linear spaces.

EXAMPLE 1

[1] Let $1 \leq p < \infty$. The space ℓ^p becomes a normed linear space when endowed with the norm $\|\cdot\|_p : \ell^p \to \mathbb{R}_+$ defined by

$$\|(x_m)\|_p := \left(\sum_{i=1}^{\infty} |x_i|^p\right)^{\frac{1}{p}} = d_p\left((x_m), \mathbf{0}\right).$$

(The subadditivity of $\|\cdot\|_p$ is equivalent to Minkowski's Inequality 2.) Similarly, we make ℓ^∞ a normed linear space by endowing it with the norm $\|\cdot\|_\infty : \ell^\infty \to \mathbb{R}_+$ defined by

$$\|(x_m)\|_\infty := \sup\{|x_m| : m \in \mathbb{N}\} = d_\infty\left((x_m), \mathbf{0}\right).$$

It is easily checked that $\|\cdot\|_\infty$ is indeed a norm on ℓ^∞.

In what follows, when we consider ℓ^p as a normed linear space, the underlying norm is always $\|\cdot\|_p$, $1 \leq p \leq \infty$.

[2] For any nonempty set T, the linear space $\mathbf{B}(T)$ of all bounded real functions on T is normed by $\|\cdot\|_\infty : \mathbf{B}(T) \to \mathbb{R}_+$, where

$$\|f\|_\infty := \sup\left\{|f(t)| : t \in T\right\} = d_\infty\left(f, \mathbf{0}\right).$$

For obvious reasons, $\|\cdot\|_\infty$ is called the **sup-norm**. Of course, $\mathbf{B}(\mathbb{N})$ and ℓ^∞ are the same normed linear spaces.

[3] Let X be a normed linear space. By a **normed linear subspace** Y of X, we mean a linear subspace of X whose norm is the restriction of the norm of X to Y (that is, $\|y\|_Y := \|y\|$ for each $y \in Y$). Throughout the remainder of this book, we refer to a normed linear subspace simply as a **subspace**. If Y is a subspace of X and $Y \neq X$, then Y is called a **proper subspace** of X.

As you would surely expect, given any metric space T, we view $\mathbf{CB}(T)$ as a subspace of $\mathbf{B}(T)$. Consequently, when we talk about the norm of an $f \in \mathbf{CB}(T)$ (or of $f \in \mathbf{C}(T)$ when T is compact), what we have in mind is the sup-norm of this function, that is, $\|f\|_\infty$. Similarly, we norm $\mathbf{c}^0$, $\mathbf{c}_0$, and $\mathbf{c}$ by the sup-norm, and hence $\mathbf{c}^0$ is a subspace of $\mathbf{c}_0$, $\mathbf{c}_0$ is a subspace of $\mathbf{c}$, and $\mathbf{c}$ is a subspace of ℓ^∞.[6]

[6] *Reminder.* $\mathbf{c}^0$ is the linear space of all real sequences all but finitely many terms of which are zero; $\mathbf{c}_0$ is the linear space of all real sequences that converge to 0; and $\mathbf{c}$ is the linear space of all convergent real sequences.

[4] For any interval I, let $\mathbf{CB}^1(I)$ denote the linear space of all bounded and continuously differentiable real maps on I whose derivatives are bounded functions on I. (Obviously, $\mathbf{CB}^1([a,b]) = \mathbf{C}^1[a,b]$.) The standard norm on $\mathbf{CB}^1(I)$ is denoted by $\|\cdot\|_{\infty,\infty}$, where

$$\|f\|_{\infty,\infty} := \sup\left\{|f(t)| : t \in I\right\} + \sup\left\{|f'(t)| : t \in I\right\}.$$

(We leave it as an exercise to check that $(\mathbf{CB}^1(I), \|\cdot\|_{\infty,\infty})$ is a normed linear space.)

From now on, whenever we consider a space like $\mathbf{C}^1[a,b]$ or, more generally, $\mathbf{CB}^1(I)$, as a normed linear space, we will have the norm $\|\cdot\|_{\infty,\infty}$ in mind.

[5] Let X be the linear space of all continuous real functions on $[0,1]$. Then $\|\cdot\| \in \mathbb{R}_+^X$, defined by

$$\|f\| := \int_0^1 |f(t)|\,dt, \tag{4}$$

is a norm on X, but, as we shall see shortly, there is a major advantage of norming X via the sup-norm as opposed to this integral norm.

[6] Let X be the linear space of all bounded real functions on $[0,1]$ that are continuous everywhere but finitely many points. (So, $\mathbf{C}[0,1] \subset X \subset \mathbf{B}[0,1]$.) Then $\|\cdot\| \in \mathbb{R}_+^X$ defined by (4) is a seminorm (Exercise A.62), but not a norm, on X.

[7] (*Finite Product Spaces*) Take any $n \in \mathbb{N}$, and let $X_1, \ldots, X_n$ be normed linear spaces. We norm the product linear space $X := \mathsf{X}^n X_i$ by using the map $\|\cdot\| \in \mathbb{R}_+^X$ defined by $\|(x^1, x^2, \ldots, x^n)\| := \|x^1\|_{X_1} + \cdots + \|x^n\|_{X_n}$. (Is $\|\cdot\|$ a norm on X?)

[8] (*Countably Infinite Product Spaces*) Let X_i be a normed linear space, $i = 1, 2, \ldots$, and let $X := \mathsf{X}^\infty X_i$. We view X as a linear space under the pointwise defined addition and scalar multiplication operations. In turn, we make X a normed linear space by means of the so-called **product norm** $\|\cdot\| \in \mathbb{R}_+^X$, which is defined by

$$\|(x^1, x^2, \ldots)\| := \sum_{i=1}^{\infty} \tfrac{1}{2^i} \min\left\{1, \|x^i\|_{X_i}\right\}.$$

(Why is $\|\cdot\|$ a norm on X?) □

EXERCISE 8 Show that $\|\cdot\|_{\infty,\infty}$ is a norm on $\mathbf{CB}^1(I)$ for any interval I.

EXERCISE 9^H Define $\varphi : \ell^\infty \to \mathbb{R}_+$ by $\varphi((x_m)) := \limsup |x_m|$. Show that φ is a seminorm on ℓ^∞. Is φ a norm on ℓ^∞? Compute $\varphi^{-1}(0)$.

EXERCISE 10 Determine all p in $\overline{\mathbb{R}}_+$ such that ℓ^p is a rotund normed linear space (Exercise 7).

EXERCISE 11 Let X denote the set of all Lipschitz continuous functions on $[0,1]$, and define $\|\cdot\| \in \mathbb{R}_+^X$ by

$$\|f\| := \sup\left\{\frac{|f(a)-f(b)|}{|a-b|} : 0 \leq a < b \leq 1\right\} + |f(0)|.$$

Show that $(X, \|\cdot\|)$ is a normed linear space. Does $f \mapsto \|f\| - |f(0)|$ define a norm on X?

The next exercise introduces a special class of normed linear spaces that plays a very important role in the theory of optimization. We will talk more about them later on.

EXERCISE 12^H (*Inner Product Spaces*) Let X be a linear space, and let $\phi \in \mathbb{R}^{X\times X}$ be a function such that, for all $\lambda \in \mathbb{R}$ and $x, x', y \in X$,

(i) $\phi(x,x) \geq 0$ and $\phi(x,x) = 0$ iff $x = \mathbf{0}$;
(ii) $\phi(x,y) = \phi(y,x)$; and
(iii) $\phi(\lambda x + x', y) = \lambda\phi(x,y) + \phi(x',y)$.

(We say that ϕ is an inner product on X, and refer to (X,ϕ) as an **inner product space.**)

(a) (*Cauchy-Schwarz Inequality*) First, go back and read Section G.3.3, and second, prove that $|\phi(x,y)|^2 \leq \phi(x,x) + \phi(y,y)$ for all $x, y \in X$.
(b) Define $\|\cdot\| \in \mathbb{R}^X$ by $\|x\| := \sqrt{\phi(x,x)}$, and show that $(X, \|\cdot\|)$ is a normed linear space. (Such a space is called a **pre-Hilbert space.**)
(c) Is ℓ^2 a pre-Hilbert space?
(d) Use part (b) to show that $f \mapsto (\int_0^1 |f(t)|^2 \, dt)^{\frac{1}{2}}$ defines a norm on $\mathbf{C}[0,1]$.

1.4 Metric versus Normed Linear Spaces

Any metric linear space X that has the property (2) can be considered as a normed linear space under the norm $\|\cdot\|_d \in \mathbb{R}_+^X$ with

$$\|x\|_d := d(x, \mathbf{0}).$$

(Right?) Therefore the geometric motivation we gave in Section 1.1 to study those metric linear spaces with (2) motivates our concentration on normed linear spaces. In fact, not only does the metric of any such space arise from a norm, but there is an obvious way of "turning" a normed linear space X into a metric linear space that satisfies (2). Indeed, any given norm $\|\cdot\|$ on X readily induces a distance function $d_{\|\cdot\|}$ on X in the following manner:

$$d_{\|\cdot\|}(x, y) := \|x - y\| \qquad \text{for all } x, y \in X. \tag{5}$$

(Check that $d_{\|\cdot\|}$ is really a metric.) Endowing X with $d_{\|\cdot\|}$ makes it a metric linear space that has the property (2). To see this, take any $(\lambda_m) \in \mathbb{R}^\infty$ and $(x^m) \in X^\infty$ with $\lambda_m \to \lambda$ and $d_{\|\cdot\|}(x^m, x) \to 0$ for some $(\lambda, x) \in \mathbb{R} \times X$. By using the subadditivity and absolute homogeneity of $\|\cdot\|$, and the fact that $(\|x^m\|)$ is a bounded real sequence,[7] we obtain

$$\begin{aligned} d_{\|\cdot\|}(\lambda_m x^m, \lambda x) &= \|\lambda_m x^m - \lambda x\| \\ &\leq \|\lambda_m x^m - \lambda x^m\| + \|\lambda x^m - \lambda x\| \\ &= |\lambda_m - \lambda| \, \|x^m\| + |\lambda| \, \|x^m - x\| \\ &\to 0. \end{aligned}$$

It is obvious that $d_{\|\cdot\|}$ is translation invariant and satisfies (2).

This observation shows that there is a natural way of viewing a normed linear space as a metric linear space. For this reason, somewhat loosely speaking, one often says that *a normed linear space "is" a metric linear space.* Throughout the rest of this book, you should always keep this viewpoint in mind.

It is important to note that one cannot use (5) to derive a distance function from a norm in the absence of (2). That is, if (X, d) is a metric space, the function $R_d \in \mathbb{R}_+^X$ defined by $R_d(x) := d(x, \mathbf{0})$, is not necessarily a

[7] $(\|x^m\|)$ is bounded, because $\|x^m\| \leq \|x^m - x\| - \|x\|$ for each m, and $\|x^m - x\| \to 0$. In fact, more is true, no? $(\|x^m\|)$ is a convergent sequence. Why? (Recall (3).)

norm (even a seminorm). For instance, if d is the discrete metric, then $R_d(x) = d(\frac{1}{2}x, \mathbf{0}) = R_d(\frac{1}{2}x)$ for all $x \neq \mathbf{0}$, which shows that R_d fails to be absolutely homogeneous. For another example, notice that R_d is not a seminorm on $\mathbb{R}^\infty$ (Example I.1.[3]). In fact, since the metric of $\mathbb{R}^\infty$ fails to satisfy (2), it cannot possibly be induced by a norm on $\mathbb{R}^\infty$.[8] Therefore, we say that *a normed linear space "is" a metric linear space, but not conversely.*

Let X be a normed linear space. Now that we agree on viewing X as a metric space, namely $(X, d_{\|\cdot\|})$, let us also agree to use any notion that makes sense in the context of a metric space also for X, with that notion being defined for $(X, d_{\|\cdot\|})$. For instance, whenever we talk about the *ε-neighborhood* of a point x in X, we mean

$$N_{\varepsilon,X}(x) := \{y \in X : \|x - y\| < \varepsilon\}.$$

Similarly, the *closed unit ball* of X, which we denote henceforth as B_X, takes the form:

$$B_X := \{x \in X : \|x\| \leq 1\}.$$

Continuing in the same vein, we declare a subset of X open iff this set is open in $(X, d_{\|\cdot\|})$. By $cl_X(S)$ we mean the closure of $S \subseteq X$ *in* $(X, d_{\|\cdot\|})$, and similarly for $int_X(S)$ and $bd_X(S)$. Or, we say that $(x^m) \in X^\infty$ converges to $x \in X$ (we write this again as $x^m \to x$, of course) iff $d_{\|\cdot\|}(x^m, x) \to 0$, that is, $\|x^m - x\| \to 0$. Any topological property, along with boundedness and completeness, of a subset of X is, again, defined relative to the metric $d_{\|\cdot\|}$. By the same token, we view a real function φ on X as continuous, if, for any $x \in X$ and $\varepsilon > 0$, there exists a $\delta > 0$ such that

$$|\varphi(x) - \varphi(y)| < \varepsilon \qquad \text{for any } y \in X \text{ with } \|x - y\| < \delta.$$

Similarly, when we talk about the continuity of a function Φ that maps X into another normed linear space Y, what we mean is: For all $x \in X$ and $\varepsilon > 0$, there exists a $\delta > 0$ such that

$$\|\Phi(x) - \Phi(y)\|_Y < \varepsilon \qquad \text{for any } y \in X \text{ with } \|x - y\| < \delta.$$

By Proposition D.1, then, a map $\Phi \in Y^X$ is continuous iff, for any $x \in X$ and $(x^m) \in X^\infty$ with $\|x^m - x\| \to 0$, we have $\|\Phi(x^m) - \Phi(x)\|_Y \to 0$.

[8] This does not mean that we cannot find a norm $\|\cdot\|$ on $\mathbb{R}^\infty$ that would render $(\mathbb{R}^\infty, \rho)$ and $(\mathbb{R}^\infty, \rho_{\|\cdot\|})$ homeomorphic. Yet it is true, we cannot possibly find such a norm. (See Exercise 19.)

As an immediate example, let us ask if the norm of a normed linear space renders *itself* continuous. A moment's reflection shows that not only is this the case, but the metric induced by a norm qualify that norm as nonexpansive.

PROPOSITION 1
The norm of any normed linear space X is a nonexpansive map on X.

PROOF
Apply (3). ∎

With these definitions in mind, the findings of Chapters C–G and I apply readily to normed linear spaces. However, thanks to the absolute homogeneity axiom, there is considerably more structure in a normed linear space than in an arbitrary metric linear space. This point will become clear as we proceed.

EXERCISE 13 Let X be a normed linear space, and define the self-map Φ on $X\backslash\{\mathbf{0}\}$ by $\Phi(x) := \frac{1}{\|x\|}x$. Show that Φ is continuous.

EXERCISE 14 Let X be a normed linear space, and $(x^m) \in X^\infty$ a Cauchy sequence. Show that $(\|x^m\|)$ is a convergent real sequence.

EXERCISE 15 Prove: A subset S of a normed linear space X is bounded iff $\sup\{\|x\| : x \in S\} < \infty$.

EXERCISE 16
(a) Show that any ε-neighborhood of any point in a normed linear space is a convex set. (So, every normed linear space "is" a locally convex metric linear space.)
(b) Show that, while its metric is not induced by a norm, any ε-neighborhood of a point in $\mathbb{R}^\infty$ is convex.

EXERCISE 17 For any normed linear space X, prove that the closure of the unit open ball equals the closed unit ball, that is,

$$cl_X(\{x \in X : \|x\| < 1\}) = \{x \in X : \|x\| \leq 1\}.$$

Also show that the analogous result does not hold for an arbitrary locally convex metric linear space by establishing that

$$cl_{\mathbb{R}^\infty}(\{x \in \mathbb{R}^\infty : \rho(x,\mathbf{0}) < 1\}) \subset \{x \in \mathbb{R}^\infty : \rho(x,\mathbf{0}) \leq 1\},$$

where ρ is the product metric.

EXERCISE 18 For any normed linear space X and any $(x, \varepsilon) \in X \times \mathbb{R}_{++}$, prove:

$$cl_X(N_{\varepsilon,X}(x)) = x + cl_X(N_{\varepsilon,X}(\mathbf{0})) \quad \text{and} \quad cl_X(N_{\varepsilon,X}(\mathbf{0})) = -cl_X(N_{\varepsilon,X}(\mathbf{0})).$$

Are these equations true in an arbitrary metric linear space?

EXERCISE 19 A metric linear space X is said to be **normable** if there is a norm $\|\cdot\|$ on X such that $(X, d_{\|\cdot\|})$ and X are homeomorphic.

(a) Prove: If X is normable, then there is an $\varepsilon > 0$ such that, for *every* open convex subset O of X that contains $\mathbf{0}$, we have $N_{\varepsilon,X}(\mathbf{0}) \subseteq \lambda O$ for some $\lambda > 0$.[9]

(b) Show that $\mathbb{R}^\infty$ is a locally convex metric linear space that is not normable.

1.5 Digression: The Lipschitz Continuity of Concave Maps

In Section I.2.4 we saw that a concave (or convex) function whose domain is a subset of a metric linear space must be continuous on the interior of its domain, provided that this map is locally bounded at a point in the interior of its domain. We now revisit this result, but this time in the context of normed linear spaces. This is our first main illustration of the sorts of things one can do within the context of normed linear spaces but not within that of metric linear spaces.

Let X be a normed linear space and O a nonempty open and convex subset of X. Take any concave $\varphi : O \to \mathbb{R}$, and assume that φ is locally bounded at some $x \in X$ from below.[10] Of course, everything that we established in

[9] Provided that X is locally convex, the converse of this statement also holds. This is a special case of a classic theorem—it is called the *Kolmogorov Normability Criterion*—that was proved in 1934 by Andrei Kolmogorov, the founder of modern probability theory.

[10] Just so that all is clear, in the language of normed linear spaces, the latter requirement means that there is an $(\varepsilon, K) \in \mathbb{R}_{++} \times \mathbb{R}$ such that $f(y) \geq K$ for all $y \in O$ with $\|x - y\| < \varepsilon$.

Section I.2.4 readily applies to φ (given that X "is" a metric linear space). In particular, by Proposition I.8, φ is continuous. Thanks to the norm structure of X (that is, thanks to the fact that the metric of X is induced by a norm), we can actually say much more here. It turns out that φ is **locally Lipschitz continuous** in the sense that, for any $x \in O$, there exist a $(\delta, K) \in \mathbb{R}^2_{++}$ such that $\left|\varphi(y) - \varphi(z)\right| \leq K \left\|y - z\right\|$ for all $y, z \in N_{\delta,X}(x) \cap O$. The proof of this fact is an easy adaptation of the proof of Proposition A.14.

PROPOSITION 2

Let O be a nonempty open and convex subset of a normed linear space X and $\varphi : O \to \mathbb{R}$ a concave function. If φ is locally bounded at some $x_0 \in O$ from below, then φ is locally Lipschitz continuous.

PROOF

Assume that φ is locally bounded at some point in O from below. By Proposition I.8, then, it is continuous. Pick any $x \in X$. Since it is continuous, φ is locally bounded at x, so there exists an $\varepsilon > 0$ such that $N_{\varepsilon,X}(x) \subseteq O$, and $\alpha := \inf \varphi(N_{\varepsilon,X}(x))$ and $\beta := \sup \varphi(N_{\varepsilon,X}(x))$ are real numbers. To focus on the nontrivial case, we assume $\alpha < \beta$.

Let $0 < \delta < \varepsilon$. Fix two distinct $y, z \in N_{\delta,X}(x)$ arbitrarily, and let

$$w := y + \delta \frac{1}{\left\|y - z\right\|}(y - z) \quad \text{and} \quad \lambda := \frac{\left\|y - z\right\|}{\delta + \left\|y - z\right\|}.$$

It is readily verified that $y = \lambda w + (1 - \lambda)z$ and $\left\|w - y\right\| = \delta$.[11] We also wish to guarantee that $w \in N_{\varepsilon,X}(x)$, and for this we need to choose δ a bit more carefully. Notice that $\|w - x\| \leq \left\|w - y\right\| + \left\|y - x\right\| < 2\delta$, so if we choose, say, $\delta := \frac{\varepsilon}{2}$, we have $w \in N_{\varepsilon,X}(x)$.

[11] I chose w and λ as above in order to guarantee that (i) $y = \lambda w + (1 - \lambda)z$, and (ii) $\left\|w - y\right\| = \delta$. After all, (i) says $\lambda(y - w) = (1 - \lambda)(z - y)$, so (ii) requires $\lambda\delta = (1 - \lambda)\left\|y - z\right\|$, that is, $\lambda = \frac{\|y-z\|}{\delta+\|y-z\|}$. (Notice that I would lack the means to find such w and λ if X was known only to be a metric linear space.)

Now we are all set. The concavity of φ implies $\varphi(y) \geq \lambda(\varphi(w) - \varphi(z)) + \varphi(z)$, so

$$\begin{aligned}\varphi(z) - \varphi(y) &\leq \lambda(\varphi(z) - \varphi(w)) \\ &\leq \lambda(\beta - \alpha) \\ &= \frac{\|y - z\|}{\delta + \|y - z\|}(\beta - \alpha) \\ &< \frac{\beta - \alpha}{\delta} \|y - z\| .\end{aligned}$$

Interchanging the roles of z and y in this argument completes the proof. ■

This result is not really a generalization of Proposition A.14, because that proposition talks about the Lipschitz continuity of a concave function on any compact subset of its domain, as opposed to its *local* Lipschitz continuity. Yet it is not difficult to use Proposition 2 to obtain generalizations of Proposition A.14 proper. Two such generalizations are reported in the following exercises.

Exercise 20^H Let O be a nonempty open and convex subset of a normed linear space X, and take any concave $\varphi \in \mathbb{R}^O$. Show that if φ is locally bounded from below at a given point, then $\varphi|_S$ is Lipschitz continuous for any compact subset S of O.

Exercise 21^H For any given $n \in \mathbb{N}$, let O be a nonempty open and convex subset of $\mathbb{R}^n$. Show that if $\varphi \in \mathbb{R}^O$ is concave, then $\varphi|_S$ is Lipschitz continuous for any compact subset S of O.

2 Banach Spaces

2.1 Definition and Examples

A normed linear space X is called a **Banach space** if it is complete (that is, if $(X, d_{\|\cdot\|})$ is a complete metric space). Clearly, every Fréchet space with (2) is a Banach space. This observation supplies us with many examples of Banach spaces. For example, it follows that $\mathbb{R}^{n,p}$ and ℓ^p are Banach spaces for any $n \in \mathbb{N}$ and $1 \leq p \leq \infty$ (Examples I.1.[1] and [4]). Similarly, for any metric space T, both $\mathbf{B}(T)$ and $\mathbf{CB}(T)$ are Banach spaces (Example I.1.[6]). Here are some other examples.

EXAMPLE 2

[1] $\mathbf{CB}^1(I)$ is a Banach space for any interval I (Example 1.[4]).

[2] The metric linear space considered in Example 1.[5] is not Banach (Exercise C.42).

[3] Recall that a metric subspace of a complete metric space X is complete iff it is closed in X (Proposition C.7). Thus: *Every closed subspace of a Banach space is itself a Banach space.*

[4] The product of finitely (or countably infinitely) many Banach spaces is Banach. This follows from Theorem C.4. □

EXERCISE 22[H] Show that $\mathbf{c}_0$ is a closed subspace of ℓ^∞. Thus $\mathbf{c}_0$ is a Banach space. (How about $\mathbf{c}$?)

EXERCISE 23 Is $\mathbf{c}^0$ a Banach space?

EXERCISE 24 Let X be a Banach space, and $\ell^\infty(X) := \{(x^m) \in X^\infty : \sup\{\|x^m\| : m \in \mathbb{N}\} < \infty\}$. We make $\ell^\infty(X)$ a linear space by defining the operations of vector addition and scalar multiplication pointwise, and norm this space by $\|\cdot\|_\infty : \ell^\infty(X) \to \mathbb{R}$, where $\|(x^m)\|_\infty := \sup\{\|x^m\| : m \in \mathbb{N}\}$. Show that $\ell^\infty(X)$ is a Banach space.

EXERCISE 25 (*Quotient Spaces*) Let X be a normed linear space and Y a closed subspace of X. We define the binary relation $\sim$ on X by $x \sim y$ iff $x = y + Y$.

(a) Show that $\sim$ is an equivalence relation.

(b) Let $[x]_\sim$ be the equivalence class of x relative to $\sim$, and $X/_\sim := \{[x]_\sim : x \in X\}$ (Section A.1.3). We define the operations of vector addition and scalar multiplication on $X/_\sim$ as follows:

$$[x]_\sim + [y]_\sim := [x + y]_\sim \quad \text{and} \quad \lambda[x]_\sim := [\lambda x]_\sim.$$

Show that $X/_\sim$ is a linear space under these operations.

(c) Define $\|\cdot\| : X/_{\sim} \to \mathbb{R}_+$ by $\|[x]_{\sim}\| := d_{\|\cdot\|}(x, Y)$, and show that $(X/_{\sim}, \|\cdot\|)$ is a normed linear space. (Would this conclusion still hold if Y was not closed?)

(d) Show that if X is a Banach space, so is $(X/_{\sim}, \|\cdot\|)$.

2.2 Infinite Series in Banach Spaces

One major advantage of normed linear spaces is that they provide a suitable playground for developing a useful theory of infinite series. In particular, there is a natural way of defining the convergence and absolute convergence of an infinite series in a normed linear space.

The following definitions are obvious generalizations of the ones we have given in Section A.3.4 for infinite series of real numbers. By an **infinite series** in a normed linear space X, we mean a sequence in X of the form $(\sum^m x^i)$ for some $(x^m) \in X^\infty$. We say that this series is **convergent** (in X) if $\lim \sum^m x^i \in X$ (that is, when there exists an $x \in X$ with $\left\|\sum^m x^i - x\right\| \to 0$). In this case the vector $\lim \sum^m x^i$ is denoted as $\sum^\infty x^i$. We say that the infinite series $(\sum^m x^i)$ in X is **absolutely convergent** if $\sum^\infty \left\|x^i\right\| < \infty$.

As in the case of infinite series of real numbers, it is customary to use the notation $\sum^\infty x^i$ both for the sequence $(\sum^m x^i)$ and for the vector $\lim \sum^m x^i$ (when the latter exists, of course). Thus we often talk about the convergence (or absolute convergence) of the infinite series $\sum^\infty x^i$, but this should be understood as the convergence (or absolute convergence) of the sequence $(\sum^m x^i)$ of partial sums.

EXERCISE 26 Let (x^m) and (y^m) be two sequences in a normed linear space X, and $\lambda \in \mathbb{R}$. Prove:

(a) If $\sum^\infty x^i$ and $\sum^\infty y^i$ are convergent, so is $\sum^\infty (\lambda x^i + y^i)$.

(b) If $\sum^\infty x^i$ is convergent, then $\left\|\sum^\infty x^i\right\| \leq \sum^\infty \left\|x^i\right\|$.

EXERCISE 27 Show that if $\sum^\infty x^i$ is a convergent series in a normed linear space, then (x^m) must be convergent.

There is a tight connection between the notions of convergence and absolute convergence of infinite series in Banach spaces. In fact, we may use these concepts to obtain a useful characterization of the notion of completeness (or "Banachness") of a normed linear space. This result allows us to think about completeness without dealing with Cauchy sequences.

Proposition 3

(Banach) *Let X be a normed linear space. Then X is Banach if, and only if, every absolutely convergent series in X is convergent.*

Proof

Let X be a Banach space, and take any $(x^m) \in X^\infty$ with $\sum^\infty \|x^i\| < \infty$. For any $k, l \in \mathbb{N}$ with $k > l$,

$$d_{\|\cdot\|}\left(\sum_{i=1}^{k} x^i, \sum_{i=1}^{l} x^i\right) = \left\|\sum_{i=1}^{k} x^i - \sum_{i=1}^{l} x^i\right\| = \left\|\sum_{i=l+1}^{k} x^i\right\| \leq \sum_{i=l+1}^{k} \|x^i\|,$$

where the inequality follows from the subadditivity of $\|\cdot\|$. This inequality and a quick appeal to Exercise A.45 show that $\left(\sum^m x^i\right)$ is a Cauchy sequence in X. By completeness of X, therefore, $\sum^\infty x^i$ must be convergent.

Conversely, assume that every absolutely convergent series in X is convergent. Let (x^m) be any Cauchy sequence in X. We wish to show that (x^m) converges in X. The trick is to view the limit of this sequence as an absolutely convergent series. First find positive integers $m_1 < m_2 < \cdots$ such that

$$\left\|x^{m_{i+1}} - x^{m_i}\right\| < \tfrac{1}{2^i}, \qquad i = 1, 2, \ldots$$

(Why do such integers exist?) So $\sum^\infty \|x^{m_{i+1}} - x^{m_i}\| < \sum^\infty \frac{1}{2^i} = 1$, and hence, by hypothesis, the series $\sum^\infty (x^{m_{i+1}} - x^{m_i})$ converges. Since

$$x^{m_{k+1}} = x^{m_1} + \sum_{i=1}^{k} (x^{m_{i+1}} - x^{m_i}), \qquad k = 1, 2, \ldots,$$

this implies that the subsequence $(x^{m_1}, x^{m_2}, \ldots)$ converges in X. Thus, since a Cauchy sequence with a convergent subsequence must be convergent (Proposition C.6), we may conclude that (x^m) converges in X. ∎

Exercise 28^{H} Let (x^m) be a sequence in a normed linear space such that $\sum^\infty x^i$ is absolutely convergent. Show that $\sum^\infty x^{\sigma(i)}$ is also absolutely convergent for any bijective self-map σ on $\mathbb{N}$.

Exercise 29 Let (x^m) be a sequence in a normed linear space such that $\sum^\infty (x^{i+1} - x^i)$ is absolutely convergent. Does (x^m) have to be Cauchy? Must it be convergent?

*Exercise 30 (*Schauder Bases*) Let X be a Banach space. A set $S \subseteq X$ is said to be a **Schauder basis** for X if, for every $x \in X$, there exists a unique real sequence $(\alpha_m(x))$ and a unique $(x^m) \in S^\infty$ such that $x = \sum^\infty \alpha_i(x)x^i$. Prove:

(a) If X is finite-dimensional, a subset of X is a basis for X iff it is a Schauder basis for X.

(b) If X is infinite-dimensional, no basis for X is a Schauder basis.

(c) Each ℓ^p, $1 \leq p < \infty$, has a Schauder basis.

(d) ℓ^∞ does not have a Schauder basis.

(e) If X has a countable Schauder basis, then it is separable.

Remark 1. One major reason why an n-dimensional linear space is a well-behaved entity is that there exists a set $\{x^1, \ldots, x^n\}$ in such a space X such that, for each $x \in X$, there exists a unique $(\alpha_1, \ldots, \alpha_n)$ in $\mathbb{R}^n$ with $x = \sum^n \alpha_i x^i$ (Corollary F.2). (For instance, this is the reason why every linear functional on a finite-dimensional normed linear space is continuous.) The Banach spaces with Schauder bases are those in which this property is satisfied in terms of a *sequence* of elements in this space. Such spaces are of interest precisely because many finite-dimensional arguments have natural extensions to them through this property. Unfortunately, the analogue of Theorem F.1 is false for Schauder bases, as Exercise 30.(d) attests.[12] □

2.3* On the "Size" of Banach Spaces

The cardinality of any basis for an infinite-dimensional normed linear space is, by definition, infinite. Can this cardinality be countably infinite? Yes, of course. Take countably infinitely many linearly independent vectors in any given infinite-dimensional space, and consider the subspace spanned by these vectors. The resulting normed linear space has a countably infinite basis. Surprisingly, however, the space we have just created is sure to be incomplete. Put differently, the answer to the question above is negative in the case of Banach spaces: *A basis for an infinite-dimensional Banach space must be uncountable.*

[12] A famous problem of linear analysis (the so-called *basis problem*) was to determine if at least all separable Banach spaces have Schauder bases. After remaining open for over 40 years, this problem was settled in 1973 in the negative by Per Enflo, who constructed a closed subspace of $\mathbf{C}[0,1]$ with no Schauder basis. For a detailed account of the theory of Schauder bases, an excellent reference (for the more advanced reader) is Chapter 4 of Megginson (1998).

This curious fact gives us a good reason to think of an infinite dimensional Banach space as a "large" normed linear space. In particular, any ℓ^p is, intuitively, "much larger" than any of its subspaces that are spanned by countably infinitely many vectors. After all, the result noted above says that any basis for ℓ^p is uncountable, $1 \leq p \leq \infty$.

The rest of this section is devoted to the derivation of this interesting property of Banach spaces.[13] We first prove an elementary geometric result about finite-dimensional normed linear spaces.

LEMMA 1

Let X be a finite-dimensional normed linear space and Y a proper subspace of X. For every $(y, \alpha) \in Y \times \mathbb{R}_{++}$, there exists an $x \in X$ such that $d_{\|\cdot\|}(x, Y) = \|x - y\| = \alpha$.

PROOF

This result is proved in exactly the same way it would have been proved in $\mathbb{R}^2$ (Figure 1). Take any $z \in X \backslash Y$. Since Y is closed (Corollary I.3), there exists a $w \in Y$ such that $\beta := d_{\|\cdot\|}(z, Y) = \|z - w\| > 0$. (Why? Recall Example D.5.) Then, for any $(y, \alpha) \in Y \times \mathbb{R}_{++}$, the vector $x := \frac{\alpha}{\beta}(z - w) + y$ satisfies $d_{\|\cdot\|}(x, Y) = \|x - y\| = \alpha$. ■

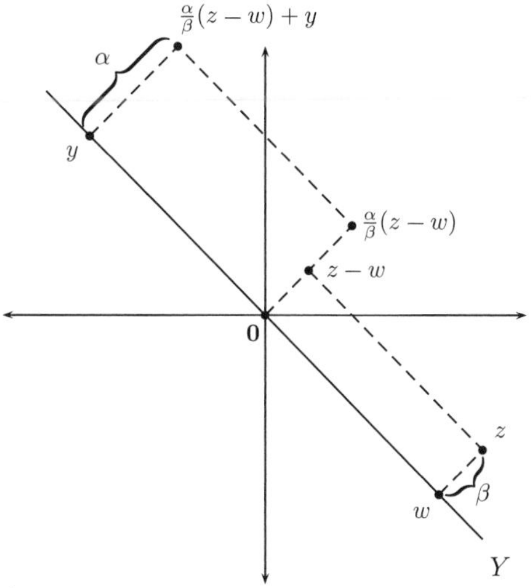

FIGURE 1

[13] For another elementary proof of this fact, see Tsing (1984).

PROPOSITION 4

No infinite-dimensional Banach space has a countable basis.

PROOF

Let X be a Banach space with $dim(X) = \infty$, and suppose $\{x^1, x^2, \ldots\}$ constitutes a basis for X. For each $m \in \mathbb{N}$, define $X_m := span\{x^1, \ldots, x^m\}$, which is a proper subspace of X. By using Lemma 1 repeatedly, we obtain a sequence $(y^m) \in X^\infty$ such that

$$y^m \in X_m \quad \text{and} \quad d_{\|\cdot\|}(y^{m+1}, X_m) = \left\|y^{m+1} - y^m\right\| = \frac{1}{4^m}, \quad m = 1, 2, \ldots$$

Obviously, $\sum^\infty (y^{i+1} - y^i)$ is absolutely convergent. Since X is Banach, this series is convergent (Proposition 3). Since $y^m = y^1 + \sum^{m-1}(y^{i+1} - y^i)$ for each $m = 2, 3, \ldots$, therefore, there is a $y \in X$ with $\left\|y^m - y\right\| \to 0$. Given that $\{x^1, x^2, \ldots\}$ is a basis for X, we have $y \in X_M$ for some $M \in \mathbb{N}$. Clearly,

$$\left\|y^{M+1} - y\right\| \geq d_{\|\cdot\|}(y^{M+1}, X_M) = \frac{1}{4^M},$$

and hence

$$\left\|y^{M+2} - y\right\| \geq \left\|y^{M+1} - y\right\| - \left\|y^{M+1} - y^{M+2}\right\| \geq \frac{1}{4^M} - \frac{1}{4^{M+1}} = \frac{3}{4^{M+1}}.$$

Proceeding inductively, we find that $\left\|y^{M+k+2} - y\right\| \geq a_k$ where $a_k := \frac{4^{k-1}(3)-4^{k-2}-\cdots-1}{4^{K+k}}$ for each $k = 1, 2, \ldots$. Since $\left\|y^m - y\right\| \to 0$, we must have $a_k \to 0$, but, in fact,

$$\lim_{k\to\infty} a_k = \lim_{k\to\infty} \frac{1}{4^{K+1}}\left(3 - \sum_{i=1}^{k-1} \frac{1}{4^i}\right) = \frac{1}{4^{K+1}}\left(3 - \sum_{i=1}^{\infty} \frac{1}{4^i}\right) = \frac{1}{4^{K+1}}\left(\frac{8}{3}\right).$$

This contradiction completes the proof. ∎

EXERCISE 31^H Let X be a normed linear space and Y a closed proper subspace of X.

(a) (*F. Riesz' Lemma*) Show that, for every $0 < \alpha < 1$, there exists an $x \in X$ such that $d_{\|\cdot\|}(x, Y) \geq \alpha$ and $\|x\| = 1$.

(b) (*A Special Case of Theorem I.2*) Use part (b) to prove that $cl_X(N_{1,X}(\mathbf{0}))$ is not a compact subset of X unless $dim(X) < \infty$.

3 Fixed Point Theory IV

Our coverage of fixed point theory so far leaves something to be desired. Tarski's Fixed Point Theorem (Section B.3.1) and the Banach Fixed Point Theorem (Section C.6.2) work in very general environments, but they require quite a bit from the self-maps they work with. By contrast, the corresponding demands of the Brouwer Fixed Point Theorem (Section D.8.3) and Kakutani's Fixed Point Theorem (Section E.5.1) are relatively modest, but these theorems apply only within Euclidean spaces. This prompts the present section, our final excursion to fixed point theory. Here we examine a few different senses in which one can extend these results at least to the context of Banach spaces. At the end of this section, you will begin to see what sort of amazing things can be accomplished within the context of normed linear spaces, as you will then possess some serious ammunition for solving existence problems in a variety of contexts.

3.1 The Glicksberg-Fan Fixed Point Theorem

The following important result is a special case of the fixed point theorems proved, independently, by Irwing Glicksberg and Ky Fan in 1952.

> THE GLICKSBERG-FAN FIXED POINT THEOREM
> *Let S be a nonempty compact and convex subset of a normed linear space X, and Γ a convex-valued self-correspondence on S that has a closed graph. Then, Γ has a fixed point, that is, there exists an $x \in S$ with $x \in \Gamma(x)$.*

PROOF[14]
Fix an arbitrary $m \in \mathbb{N}$. Since S is compact, it is totally bounded (Lemma C.1), so there exist a $k_m \in \mathbb{N}$ and $x^1(m), \ldots, x^{k_m}(m) \in S$ such that $S \subseteq \cup^{k_m} cl_X(N_{\frac{1}{m},X}(x^i(m)))$. Let

$$S(m) := co\left(\{x^1(m), \ldots, x^{k_m}(m)\}\right).$$

[14] The idea of the proof is analogous to the way we deduced Kakutani's Fixed Point Theorem from Brouwer's Fixed Point Theorem in Section E.5.3. Once again, "approximation" is the name of the game. We approximate Γ by means of a sequence of self-correspondences defined on finite-dimensional subsets of S, and use Kakutani's Fixed Point Theorem to obtain a fixed point for each term of the sequence. The limit of any convergent subsequence of the resulting sequence of fixed points is then shown to be a fixed point of Γ.

Clearly, $S(m)$ is a closed subset of S. (Why?) Since S is compact, therefore, $S(m)$ is compact. Now let

$$T_m := N_{\frac{1}{m},X}(\mathbf{0}) \qquad \text{for each } m \in \mathbb{N},$$

and define $\Gamma_m : S(m) \rightrightarrows S(m)$ by

$$\Gamma_m(x) := \big(\Gamma(x) + cl_X(T_m)\big) \cap S(m).$$

Note that, for any $x \in S$ and $y \in \Gamma(x)$, there is an $i \in \{1, \ldots, k_m\}$ such that $y \in cl_X(N_{\frac{1}{m},X}(x^i(m))) = x^i(m) + cl_X(T_m)$ so that $x^i(m) \in y + cl_X(T_m)$ (Exercise 18). Thus $(\Gamma(x) + cl_X(T_m)) \cap S(m) \neq \emptyset$, so Γ_m is well-defined. Moreover, Γ_m is convex-valued and it has a closed graph. This can be proved by using the corresponding properties of Γ. We leave working out the details as an exercise.

Let's vary m now. Since each $S(m)$ lies in a metric linear space that is linearly homeomorphic to a Euclidean space (Theorem I.1), we may apply Kakutani's Fixed Point Theorem to find an x^m in $S(m)$ such that $x^m \in \Gamma_m(x^m)$, $m = 1, 2, \ldots$ (Yes?) The sequence (x^m) lies in the compact set S, so it must have a convergent subsequence (Theorem C.2). Let's denote this subsequence again by (x^m), and write $x := \lim x^m$. We wish to show that $x \in \Gamma(x)$. Since $x^m \in \Gamma(x^m) + cl_X(T_m)$, there is a $y^m \in \Gamma(x^m)$ with $x^m - y^m \in cl_X(T_m)$, $m = 1, 2, \ldots$. The sequence (y^m) lies in S, so it has a convergent subsequence as well. Denoting this subsequence again by (y^m), we find $\lim(x^m - y^m) = \mathbf{0}$. But then

$$x = \lim x^m = \lim y^m + \lim(x^m - y^m) = \lim y^m,$$

whereas it follows from the closed graph property of Γ that $\lim y^m \in \Gamma(x)$. ∎

This result generalizes Kakutani's Fixed Point Theorem since the latter is proved in the context of Euclidean spaces, and in that context a set is closed and bounded iff it is compact. However, as the examples given in Exercises D.73 and D.74 demonstrate that, in general, the compactness requirement cannot be relaxed to being closed and bounded in the Glicksberg-Fan Fixed Point Theorem.

EXERCISE 32[H] Let S be a nonempty subset of a metric linear space X. If $\Gamma : S \rightrightarrows X$ is upper hemicontinuous and $\Gamma(x)$ is closed and convex for every $x \in S$, then does the set of all fixed points of Γ have to be closed? Convex?

Exercise 33 (Fan) Let X be a normed linear space and A and B nonempty convex and compact subsets of X. Suppose $f \in X^A$ is a continuous map with $f(A) \subseteq A + B$. Show that there is an $x \in A$ such that $f(x) \in x + B$.

*Exercise 34[H] Show that "normed linear space" can be replaced with "locally convex metric linear space" in the statement of the Glicksberg-Fan Fixed Point Theorem.

3.2 Application: Existence of the Nash Equilibrium, Revisited

In Section E.6 we introduced the theory of strategic games through games in which the action spaces of the players are subsets of a Euclidean space. Some interesting economic games, however, do not satisfy this constraint. For instance, if we wished to model a game that is repeatedly played infinitely many times, then we need to specify the action spaces of the players as a subset of an infinite-dimensional sequence space (as in dynamic programming). Or, if a player were allowed to make her decisions by randomizing between some or all of her actions, this would mean that her effective action space is in fact the set of all probability distributions defined over her original action space. Provided that the player has infinitely many (nonrandomized) action possibilities, her (effective) action space would then lie again in an infinite-dimensional linear space.

All in all, it is of interest whether we can extend the basic results we obtained in Section E.6 to the case of strategic games in which the action spaces of the players are subsets of an arbitrary normed (or metric) linear space. As you should expect, the Glicksberg-Fan Fixed Point Theorem provides immediate help in this regard.

Example 3

(*A Generalization of Nash's Existence Theorem*) If each S_i is a nonempty compact subset of a normed linear space X_i, then we say that the strategic game $\mathcal{G} := (\{S_i, \pi_i\}_{i=1,\dots,m})$ is a **compact game** (Section E.6.2). If, in addition, each $\pi_i \in \mathbb{R}^S$ is continuous (where $S := \mathsf{X}^m S_i$ lies in the product normed linear space $X := \mathsf{X}^m X_i$), we say that $\mathcal{G}$ is a **continuous compact game**. If, instead, each X_i is convex, and

$$\pi_i(\lambda x_i + (1-\lambda)y_i, x_{-i}) \geq \min\{\pi_i(x_i, x_{-i}), \pi_i(y_i, x_{-i})\}$$

for any $0 < \lambda < 1$, $x_{-i} \in X_{-i}$ and $i = 1, \ldots, n$, then $\mathcal{G}$ is called a **convex and compact game**. Finally, a compact game that is both convex and continuous is called a **regular game**.

Nash's Existence Theorem says that every Euclidean regular game has a Nash equilibrium. If you go back and check our proof of this important theorem, you will notice that the only place we really needed the hypothesis of the game being Euclidean was when we wanted to invoke Kakutani's Fixed Point Theorem. Consequently, by replacing Kakutani's Fixed Point Theorem with the Glicksberg-Fan Fixed Point Theorem in its proof, we may extend the coverage of Nash's Existence Theorem to games that are not Euclidean. Put precisely: *Every regular game has a Nash equilibrium.* □

EXERCISE 35 Prove that every regular game has a Nash equilibrium.

EXERCISE 36[H] Let $\mathcal{G} := (\{S_i, \pi_i\}_{i=1,\ldots,m})$ be a convex and compact game. Prove: If, for each i, π_i is upper semicontinuous, and, for each i and $x_i \in X_i$, the map $\pi_i(x_i, \cdot)$ is lower semicontinuous (on S_{-i}), then $\mathcal{G}$ has a Nash equilibrium.

3.3* The Schauder Fixed Point Theorems

We now turn to extending the Brouwer Fixed Point Theorem to the realm of normed linear spaces. Two major results of this sort were obtained by Julius Schauder in 1927 and 1930.[15] The first of these theorems is in fact an immediate corollary of the Glicksberg-Fan Fixed Point Theorem.

THE SCHAUDER FIXED POINT THEOREM 1
Every continuous self-map on a nonempty compact and convex subset of a normed linear space has a fixed point.[16]

[15] Julius Schauder (1899–1943) was a student of Banach and made invaluable contributions to the theory of differential equations (most of which were in collaboration with Jean Leray). It would be fair to consider him as one of the founders of modern nonlinear analysis.

[16] This result is, in fact, valid also in metric linear spaces. In fact, Schauder's original statement was in the context of an arbitrary metric linear space. His proof, however, contained a flaw, and established the result only for locally convex metric linear spaces. Whether Schauder's original claim was true remained as an open problem (known as the *Schauder Conjecture*) until Robert Cauty settled it in the affirmative in 2001.

Quiz. Why isn't the proof we gave for the Glicksberg-Fan Fixed Point Theorem valid in the case of an arbitrary metric linear space?

That is, every nonempty compact and convex subset of a normed linear space has the fixed point property. This is great news, to be sure. Unfortunately, in applications one all too often has to work with either a noncompact set or a set the compactness of which is difficult to establish. It is thus worthwhile thinking about to what extent we can extend this result to the context of continuous self-maps on a noncompact set.

It is obvious that we can't just relax the compactness requirement in the Schauder Fixed Point Theorem 1, as easy examples would demonstrate.[17] In fact, we already know that compactness cannot be relaxed in the theorem above even to being closed and bounded. (Recall Exercises D.73 and D.74.) That's the bad news. The good news is that, even when the domain of our self-map is known only to be closed, bounded and convex, we are in fact guaranteed a fixed point provided that the range of our map is suitably well-behaved, namely, it is a relatively compact subset of a Banach space.[18] This is the content of Schauder's 1930 theorem.

The Schauder Fixed Point Theorem 2
Let S be a nonempty, closed, bounded and convex subset of a Banach space X and Φ a continuous self-map on S such that $cl_X(\Phi(S))$ is compact. Then Φ has a fixed point.

This result requires a bit more work. The idea is to show that the restriction of Φ to $\overline{co}_X(\Phi(S))$ is a self-map, and then to apply the Schauder Fixed Point Theorem 1 to this map. To this end, we need to first prove the following 1930 result of Stanislaw Mazur, which is quite important in its own right.

Mazur's Compactness Theorem
Let S be a relatively compact subset of a Banach space X. Then $\overline{co}_X(S)$ is compact.

[17] Much more can be said here. In 1955 Victor Klee proved that a nonempty convex subset of a Banach space has the fixed point property if and *only if* it is compact. In fact, it is even true that on every convex and noncompact subset of a Banach space there exists a Lipschitz continuous self-map without a fixed point (Lin and Sternfeld (1985)).

[18] *Reminder.* A subset of a metric space X is called *relatively compact* if the closure of that set in X is compact (Section D.6.4).

PROOF

Since $\overline{co}_X(S) = \overline{co}_X(cl_X(S))$—yes?—we may assume that S is compact. Now recall that a metric space is compact iff it is complete and totally bounded (Theorem C.3). Since X is complete and $\overline{co}_X(S)$ is a closed subset of X, $\overline{co}_X(S)$ is a complete metric space. Therefore, all we need to do is to show that $\overline{co}_X(S)$ is also totally bounded. Take any $\varepsilon > 0$. Since S is totally bounded, we can find a $k \in \mathbb{N}$ and $x^1, \ldots, x^k \in S$ such that $S \subseteq \cup^k N_{\frac{\varepsilon}{4},X}(x^i)$, that is, for any $u \in S$, there is a $\omega(u) \in \{1, \ldots, k\}$ such that $\left\| u - x^{\omega(u)} \right\| < \frac{\varepsilon}{4}$. Now take any $z \in \overline{co}_X(S)$. Clearly, there is a $y \in co(S)$ with $\|z - y\| < \frac{\varepsilon}{4}$. Moreover, $y = \sum^l \lambda_i y^i$ for some $l \in \mathbb{N}$, $y^1, \ldots, y^l \in S$ and $(\lambda_1, \ldots, \lambda_l) \in \mathbb{R}^l_+$ with $\sum^l \lambda_i = 1$. Then,

$$\left\| y - \sum_{i=1}^{l} \lambda_i x^{\omega(y^i)} \right\| = \left\| \sum_{i=1}^{l} \lambda_i (y^i - x^{\omega(y^i)}) \right\| \leq \sum_{i=1}^{l} \lambda_i \left\| y^i - x^{\omega(y^i)} \right\| < \frac{\varepsilon}{4},$$

so

$$\left\| z - \sum_{i=1}^{l} \lambda_i x^{\omega(y^i)} \right\| \leq \|z - y\| + \left\| y - \sum_{i=1}^{l} \lambda_i x^{\omega(y^i)} \right\| < \frac{\varepsilon}{2}.$$

Since z was arbitrarily chosen in $\overline{co}_X(S)$, this proves that

$$\overline{co}_X(S) \subseteq \bigcup \{N_{\frac{\varepsilon}{2},X}(w) : w \in W\}$$

where $W := co(\{x^1, \ldots, x^k\})$. But W is compact (why?), so it is totally bounded, that is, we can find finitely many $w^1, \ldots, w^m \in W$ such that $W \subseteq \cup^m N_{\frac{\varepsilon}{2},X}(w^i)$. (Yes?) It follows that $\overline{co}_X(S) \subseteq \cup^m N_{\varepsilon,X}(w^i)$. Since $\varepsilon > 0$ was arbitrary in this discussion, we may conclude that $\overline{co}_X(S)$ is totally bounded. ∎

WARNING. The closed convex hull of a compact subset of a normed linear space may well fail to be compact in general. That is, "Banachness" is essential for the validity of Mazur's Compactness Theorem. (Curiously, this is not so for the Schauder Fixed Point Theorem 2; see Exercise 38 below.) For instance, while $S := \{\mathbf{0}, \mathbf{e}^1, \frac{1}{2}\mathbf{e}^2, \frac{1}{3}\mathbf{e}^3, \ldots\}$, where $\mathbf{e}^1 := (1, 0, 0, \ldots)$, $\mathbf{e}^2 := (0, 1, 0, \ldots)$, etc., is a compact subset of $\mathbf{c}^0$ (the subspace of ℓ^∞ that consists of all real sequences with finitely many nonzero terms), $\overline{co}_{\mathbf{c}^0}(S)$ is not compact in $\mathbf{c}^0$. Indeed, the sequence $(x^1, x^2, \ldots)$, where $x^k := \sum^k \frac{1}{2^{i+1}} \mathbf{e}^i$ for each $k = 1, 2, \ldots$, lies entirely within $\overline{co}_{\mathbf{c}^0}(S)$, but it does not have a subsequence that converges in $\mathbf{c}^0$.

PROOF OF THE SCHAUDER FIXED POINT THEOREM 2
Let $T := \overline{co}_X(\Phi(S))$, which is a nonempty and convex subset of X. Moreover, it is compact in X, by Mazur's Compactness Theorem. And, since Φ is continuous, $\Phi(S) \subseteq S$ and S is convex, we have $\Phi(T) \subseteq T$.[19] Thus $\Phi|_T$ must have a fixed point by the Schauder Fixed Point Theorem 1. ■

EXERCISE 37 (*The Leray-Schauder Fixed Point Theorem*) Let X be a Banach space and Φ a self-map on X. Prove that if $cl_X(\Phi(X))$ is compact, and $\|\Phi(x)\| \leq \|x\|$ for all $x \in X$, then Φ has a fixed point.

The following exercise derives a generalization of the Schauder Fixed Point Theorem 2. It turns out that by means of a different technique of proof, we could prove that result without assuming that X is complete and S is closed.

*EXERCISE 38[H] (*A Generalization of the Schauder Fixed Point Theorem 2*) Let S be a nonempty bounded and convex subset of a normed linear space X, and let Φ be a continuous self-map on S such that $cl_X(\Phi(S))$ is compact.

(a) Pick any $m \in \mathbb{N}$. Since $cl_X(\Phi(S))$ is totally bounded, there exist a $k_m \in \mathbb{N}$ and $y^1, \ldots, y^{k_m} \in cl_X(\Phi(S))$ such that $cl_X(\Phi(S)) \subseteq \bigcup^{k_m} N_{\frac{1}{m},X}(y^i)$. Let

$$\alpha_i(x) := \begin{cases} \frac{1}{m} - \left\|\Phi(x) - y^i\right\|, & \text{if } \Phi(x) \in N_{\frac{1}{m},X}(y^i) \\ 0, & \text{otherwise} \end{cases}, \qquad i = 1, \ldots, k_m,$$

and define $p_m \in X^S$ by

$$p_m(x) := \sum_{i=1}^{k_m} \frac{\alpha_i(x)}{\sum^{k_m} \alpha_i(x)} y^i$$

[19] What I have in mind is this:

$$\begin{aligned} \Phi(\overline{co}_X(\Phi(S))) &\subseteq cl_X(\Phi(co(\Phi(S)))) \\ &\subseteq cl_X(\Phi(S)) \\ &\subseteq cl_X(co(\Phi(S))) \\ &= \overline{co}_X(\Phi(S)). \end{aligned}$$

(p_m is called a **Schauder projection.**) Show that p_m is well-defined, and it is continuous. Moreover, prove that

$$\sup\left\{\left\|\Phi(x) - p_m(x)\right\| : x \in S\right\} \leq \tfrac{1}{m} \quad \text{and} \quad dim(span(p_m(S))) < \infty.$$

(b) Show that $p_m \circ \Phi$ is a continuous self-map on $co(\{y^1, \ldots, y^{k_m}\})$, and apply the Brouwer Fixed Point Theorem to find a $y(m) \in co(\{y^1, \ldots, y^{k_m}\})$ with $p_m(\Phi(y(m))) = y(m)$.

(c) Observe that $\left\|\Phi(y(m)) - p_m(y(m))\right\| \leq \frac{1}{m}$. Let $m \to \infty$ to conclude that Φ must have a fixed point.

(d) Describe the idea behind the proof outlined above.

3.4* Some Consequences of Schauder's Theorems

Like all fundamental fixed point theorems, Schauder's theorems yield quite a few fixed point theorems that are of interest on their own. To keep our exposition self-contained, we examine next four such offspring of these two theorems. The first of these is a generalization of Proposition D.10 that was proved by Erich Rothe in 1937. The proof of this result is analogous to how we obtained Proposition D.10 from the Brouwer Fixed Point Theorem, and is thus left as an exercise.

> ROTHE'S THEOREM
> *Let X be a Banach space and B_X the closed unit ball of X. If $\Phi : B_X \to X$ is a continuous function such that $cl_X(\Phi(B_X))$ is compact and $\Phi(bd_X(B_X)) \subseteq B_X$, then Φ has a fixed point.*

EXERCISE 39 Prove Rothe's Theorem.

EXERCISE 40 (Petryshyn) Let X be a Banach space and $\Phi : B_X \to X$ a continuous function such that $\|x - \Phi(x)\| \geq \|\Phi(x)\|$ for all $x \in bd_X(B_X)$. Show that Φ has a fixed point.

EXERCISE 41[H] In the statement of Rothe's Theorem, B_X can be replaced with any nonempty, closed, bounded and convex subset of X. Prove! (Try making use of the Minkowski functionals.)

Our next result was proved first by Andrei Markov in 1936.[20] It deals with the problem of finding a *common* fixed point of a given family $\mathcal{F}$ of continuous self-maps on a given set. Any such family $\mathcal{F}$ is called **commuting** if $f \circ g = g \circ f$ for all $f, g \in \mathcal{F}$.

MARKOV'S FIXED POINT THEOREM

Let S be a nonempty compact and convex subset of a normed linear space X. If $\mathcal{F}$ is a commuting family of affine self-maps on S, then there exists an $x \in S$ such that $x = f(x)$ for all $f \in \mathcal{F}$.

The proof is outlined in the following exercise.

EXERCISE 42 [H] Let X, S, and $\mathcal{F}$ be as in Markov's Fixed Point Theorem, and let $Fix(f)$ stand for the set of all fixed points of $f \in \mathcal{F}$.

(a) Show that $Fix(f)$ is a nonempty compact and convex subset of X for any $f \in \mathcal{F}$.

(b) Show that $\{Fix(f) : f \in \mathcal{F}\}$ has the finite intersection property (Example C.8), and hence $\cap\{Fix(f) : f \in \mathcal{F}\} \neq \emptyset$.

Our next application is of a slightly more recent vintage. It was proved in 1954 by Mark Krasnoselsky.

KRASNOSELSKY'S THEOREM

Let X be a Banach space and S a nonempty closed and convex subset of X. Take any two continuous maps g and h in X^S such that g is a contraction, $cl_X(h(S))$ is compact, and $g(S) + h(S) \subseteq S$. Then, $g + h$ has a fixed point.

This is an impressive result. It generalizes both the Banach Fixed Point Theorem and the Schauder Fixed Point Theorem 2 at one stroke. Indeed, if h equals $\mathbf{0}$ everywhere on S, and $S = X$, then Krasnoselsky's Theorem reduces to (the existence part of) the Banach Fixed Point Theorem for normed linear spaces. And, if g equals $\mathbf{0}$ everywhere, then it becomes identical to the Schauder Fixed Point Theorem 2.

[20] In 1938 Kakutani proved a generalization of this result with a direct argument. For this reason this result is often referred to as the Markov-Kakutani Fixed Point Theorem.

PROOF OF KRASNOSELSKY'S THEOREM

Let $f := \mathrm{id}_S - g$.[21] The following two observations about f are essential for the proof.

CLAIM 1. f is an embedding from S into X.

PROOF OF CLAIM 1

f is obviously a continuous map from S into X. Moreover, for any $x, y \in S$,

$$\begin{aligned} \|f(x) - f(y)\| &= \|x - y + (g(y) - g(x))\| \\ &\geq \|x - y\| - \|g(x) - g(y)\| \\ &\geq (1 + K)\|x - y\|, \end{aligned}$$

where $0 < K < 1$ is the contraction coefficient of g. This observation proves that f is injective and f^{-1} is a continuous function on $f(S)$. (Why?) ‖

CLAIM 2. $cl_X(h(S)) \subseteq f(S)$.

PROOF OF CLAIM 2

Take any $y \in cl_X(h(S))$, and define $f_y \in X^S$ by $f_y(x) := y + g(x)$. Since S is closed, we have, by hypothesis,

$$f_y(S) \subseteq cl_X(h(S)) + g(S) \subseteq cl_X(h(S) + g(S)) \subseteq S,$$

so f_y is a self-map on S. It is, moreover, a contraction. (Why?) Thus, by the Banach Fixed Point Theorem, there is an $x \in S$ with $x = f_y(x) = y + g(x)$, that is, $y \in f(S)$.‖

We are now ready to complete the proof. Define $\Phi := f^{-1} \circ h$. It follows from Claims 1 and 2 that Φ is a continuous self-map on S. Moreover, since $f : S \to f(S)$ is a homeomorphism, we have

$$cl_X(\Phi(S)) = cl_X(f^{-1}(h(S))) = cl_S(f^{-1}(h(S))) = f^{-1}(cl_{f(S)}(h(S))).$$

(Recall Exercise D.20.) But Claim 2 guarantees that $cl_{f(S)}(h(S))$ equals the closure of $h(S)$ in X, so, since the latter is compact (by hypothesis) and

[21] The idea of the proof stems from the fact that we are after a fixed point of $f^{-1} \circ h$. (Indeed, if $z = f^{-1}(h(z))$, then $z - g(z) = f(z) = h(z)$.) As you will see shortly, the hypotheses of Krasnoselsky's Theorem enables one to apply the Schauder Fixed Point Theorem 2 to this map.

f^{-1} is continuous, we may conclude that $cl_X(\Phi(S))$ is compact. (Why?) By the Schauder Fixed Point Theorem 2, therefore, there is a $z \in S$ such that $z = \Phi(z)$. Then, z is a fixed point of $g + h$. ∎

Our final example is (a slight generalization) of a fixed point theorem obtained by Boris Sadovsky in 1967 by using the Schauder Fixed Point Theorem 1. This result is one of the best illustrations of how one may improve the (existence) part of the Banach Fixed Point Theorem by using the additional (convex) structure of Banach spaces. We explore it in the form of an exercise.

Exercise 43 (*Sadovsky's Fixed Point Theorem*) Let X be a Banach space and denote by $\mathcal{B}_X$ the class of all bounded subsets of X. A map $\zeta : \mathcal{B}_X \to \mathbb{R}_+$ is called a **measure of noncompactness** on X if it satisfies the following properties: For all $A, B \in \mathcal{B}_X$,

(i) $\zeta(A) = 0$ iff A is totally bounded;
(ii) $\zeta(A) = \zeta(cl_X(A))$ and $\zeta(A) = \zeta(co(A))$; and
(iii) $\zeta(A \cup B) = \max\{\zeta(A), \zeta(B)\}$.

(For a concrete example of a measure of noncompactness, see Exercise C.46.)

Let S be a nonempty, closed, bounded, and convex subset of X, and f a continuous self-map on S. Suppose that, for some a measure of noncompactness ζ on X, f is ζ**-condensing**, that is, $\zeta(f(A)) < f(A)$ for all bounded $A \subseteq S$ that is not totally bounded. Show that f must then have a fixed point by filling in the gaps left in the following argument.

(a) Take any $x \in S$, and let $\mathcal{K}$ be the class of all closed and convex subsets A of S such that $x \in A$ and $f(A) \subseteq A$. Define $B := \bigcap\mathcal{K}$ and $C := \overline{co}_X(f(B) \cup \{x\})$. Then $B \neq \emptyset$ and $f|_B$ is a self-map on B.
(b) $B = C$.
(c) $\zeta(B) = \zeta(f(B))$. Since f is ζ-condensing, it follows that $\zeta(B) = 0$, so B is compact. By the Schauder Fixed Point Theorem 1, therefore, $f|_B$ has a fixed point.

Exercise 44 If S is bounded, then the requirement $g(S) + h(S) \subseteq S$ can be relaxed to $(g+h)(S) \subseteq S$ in the statement of Krasnoselsky's Theorem. Prove this by using Sadovsky's Fixed Point Theorem in conjunction with the Kuratowski measure of noncompactness (Exercise C.46).

3.5* Applications to Functional Equations

One of the most widely used methods of establishing that a given functional equation has a solution is to convert the problem into a fixed point problem and see if the latter can be solved using a suitable fixed point argument. You are of course familiar with this technique from our earlier work on integral and differential equations (Section C.7) and Bellman's Functional Equation (Section E.4.3). The examples considered below illustrate how Schauder's theorems commingle with this method of proof.

EXAMPLE 4

Take any $\theta \in \mathbf{C}([0,1]^2)$ and $\varphi \in \mathbf{CB}([0,1] \times \mathbb{R})$, and consider the following equation:

$$f(x) = \int_0^1 \theta(x,t)\varphi(t,f(t))dt \qquad \text{for all } 0 \leq x \leq 1. \tag{6}$$

This equation is called the **Hammerstein integral equation**. Question: Does there exist an $f \in \mathbf{C}[0,1]$ that satisfies (6)? The answer is yes, as we demonstrate next.

Define the operator $\Phi : \mathbf{C}[0,1] \to \mathbb{R}^{[0,1]}$ by

$$\Phi(f)(x) := \int_0^1 \theta(x,t)\varphi(t,f(t))dt. \tag{7}$$

It follows from the Riemann integration theory (Section A.4.3) that Φ is a self-map on $\mathbf{C}[0,1]$. (Why?) Moreover, Φ is continuous. To prove this, take any $f \in \mathbf{C}[0,1]$ and any sequence (f_m) in $\mathbf{C}[0,1]$ such that $f_m \to f$ (relative to the sup-norm, of course). It is easy to see that

$$\left\| \Phi(f_m) - \Phi(f) \right\|_\infty \leq \alpha \max\{ \left| \varphi(t, f_m(t)) - \varphi(t, f(t)) \right| : 0 \leq t \leq 1 \},$$

where $\alpha := \max\{ |\theta(x,y)| : 0 \leq x, y \leq 1 \}$. (Verify!) It will thus follow that Φ is continuous if we can show that

$$\lim_{m\to\infty} \max\{ \left| \varphi(t, f_m(t)) - \varphi(t, f(t)) \right| : 0 \leq t \leq 1 \} = 0.$$

If this equation was false, we could then find a $\gamma > 0$ and a $(t_m) \in [0,1]^\infty$ such that

$$\left| \varphi(t, f_m(t_m)) - \varphi(t, f(t_m)) \right| \geq \gamma, \qquad m = 1, 2, \ldots \tag{8}$$

Clearly, (t_m) has a subsequence that converges in $[0, 1]$. Let us denote this subsequence again by (t_m)—relabeling if necessary—and set $t^* := \lim t_m$. Since $\|f_m - f\|_\infty \to 0$, we have

$$\left|f_m(t_m) - f(t^*)\right| \leq \left|f_m(t_m) - f(t_m)\right| + \left|f(t_m) - f(t^*)\right| \to 0,$$

that is, $f_m(t_m) \to f(t^*)$. (Yes?) But then, since φ and f are continuous, $\lim \varphi(t^*, f_m(t_m)) = \varphi(t^*, f(t^*)) = \lim \varphi(t^*, f(t_m))$, contradicting (8).

Let's see what we have here. We now know that Φ is a continuous self-map on $\mathbf{C}[0, 1]$, and that the integral equation at hand has a solution iff this map has a fixed point. We wish to find a fixed point of Φ by using one of the theorems of Schauder, but since $\mathbf{C}[0, 1]$ is not bounded, we can't do this right away. We first need to find a suitable subset of $\mathbf{C}[0, 1]$ on which Φ acts still as a self-map. To get a feeling for what will do the job, let's examine the range of Φ a bit more closely. If $\beta := \sup\{|\varphi(t, z)| : (t, z) \in [0, 1] \times \mathbb{R}\}$, then it follows readily from the definition of Φ that $\Phi(f(x)) \leq \alpha\beta$ for any $0 \leq x \leq 1$ and $f \in \mathbf{C}[0, 1]$. But then $\|\Phi(f)\|_\infty \leq \alpha\beta$ for all $f \in \mathbf{C}[0, 1]$. Put differently, $\Phi(\mathbf{C}[0, 1]) \subseteq S$, where $S := \{f \in \mathbf{C}[0, 1] : \|f\|_\infty \leq \alpha\beta\}$. Aha! Then S is a nonempty, closed, bounded, and convex subset of $\mathbf{C}[0, 1]$, and $\Phi|_S$ is a self-map on S. Therefore, thanks to the Schauder Fixed Point Theorem 2, our task reduces to showing that $\Phi(S)$ is a relatively compact set. (Why?)

To show that the closure of $\Phi(S)$ is compact in $\mathbf{C}[0, 1]$, observe first that $\Phi(S)$ is obviously bounded—it is contained in S—so, by the Arzelà-Ascoli Theorem, it is enough to show that $\Phi(S)$ is equicontinuous. (Right?) To this end, fix any $(\varepsilon, x) \in \mathbb{R}_{++} \times [0, 1]$. It is easy to see that, for any $0 \leq y \leq 1$ and $f \in \mathbf{C}[0, 1]$,

$$\left|\Phi(f)(x) - \Phi(f)(y)\right| \leq \beta \int_0^1 \left|\theta(x, t) - \theta(y, t)\right| dt.$$

Since θ is continuous, there exists a $\delta > 0$ such that $\left|\theta(x, t) - \theta(y, t)\right| \leq \frac{\varepsilon}{\beta}$ for any $0 \leq y \leq 1$ with $\left|x - y\right| < \delta$. For any such δ, we then have $\left|\Phi(f)(x) - \Phi(f)(y)\right| \leq \varepsilon$ for any $0 \leq y \leq 1$ with $\left|x - y\right| < \delta$, and *any* $f \in \mathbf{C}[0, 1]$, that is, Φ is equicontinuous at x. Since x was chosen arbitrarily in $[0, 1]$ here, we are done. □

Our next application concerns a famed existence theorem that was first proved in 1890 by Giuseppe Peano (by a different and much clumsier technique). This result is a close relative of the local version of Picard's Existence Theorem (Section C.7.2) and is equally fundamental for the investigation of ordinary differential equations. The idea of its proof is analogous to that of Picard's theorem. We will have to be a bit more careful here, however, as we will use the Schauder Fixed Point Theorem 2 instead of the Banach Fixed Point Theorem.

PEANO'S EXISTENCE THEOREM

Let $-\infty < a < b < \infty$, and $(x_0, y_0) \in (a,b)^2$. If $H \in \mathbf{CB}([a,b] \times \mathbb{R})$, then there is a $\delta > 0$ such that there exists a differentiable real function f on $[x_0 - \delta, x_0 + \delta]$ with $y_0 = f(x_0)$ and

$$f'(x) = H(x, f(x)), \qquad x_0 - \delta \leq x \leq x_0 + \delta.$$

PROOF

Take any $H \in \mathbf{CB}([a,b] \times \mathbb{R})$, and, for any $\delta > 0$, let $I_\delta := [x_0 - \delta, x_0 + \delta]$. By the Fundamental Theorem of Calculus, it is enough to find a $\delta > 0$ and an $f \in \mathbf{C}(I_\delta)$ such that $I_\delta \subseteq [a,b]$ and

$$f(x) = y_0 + \int_{x_0}^{x} H(t, f(t))dt \qquad \text{for all } x \in I_\delta.$$

(Why?) The argument needed to establish this is analogous to the one given in Example 4, so we will merely sketch it. Begin by defining the self-map Φ on $\mathbf{C}(I_\delta)$ by

$$\Phi(f)(x) := y_0 + \int_{x_0}^{x} H(t, f(t))dt.$$

(Note that we did not yet choose a specific value for δ.) One can show that this operator is continuous in exactly the same way we proved the continuity of the operator defined by (7) in Example 4.

Let $\beta := \sup\{|H(x,y)| : a \leq x \leq b \text{ and } y \in \mathbb{R}\}$, and notice that $\|\Phi(f)\|_\infty \leq \beta\delta + |y_0|$ for all $f \in \mathbf{C}(I_\delta)$. (Why?) We may assume $\beta > 0$, for otherwise all is trivial. Then, if we picked any $0 < \delta < \frac{1}{\beta}$ with $I_\delta \subseteq [a,b]$, we would have $\|\Phi(f)\|_\infty \leq 1 + |y_0|$ for any $f \in \mathbf{C}(I_\delta)$. Fix any such δ, and define

$$S := \{f \in \mathbf{C}(I_\delta) : \|f\|_\infty \leq 1 + |y_0|\}.$$

We now know that $\Phi|_S$ is a continuous self-map on S. Moreover, S is clearly a nonempty, closed, bounded, and convex subset of $\mathbf{C}(I_\delta)$. By the Schauder Fixed Point Theorem 2, therefore, the proof will be complete if we can show that the closure of $\Phi(S)$ in $\mathbf{C}(I_\delta)$ is compact.

Since $\Phi(S)$ is obviously bounded—$\Phi(S)$ is contained in S—the Arzelà-Ascoli Theorem says that all we need to show is that $\Phi(S)$ is equicontinuous. To this end, fix any $\varepsilon > 0$ and $x \in I_\delta$. It is easy to see that $|\Phi(g)(x) - \Phi(g)(x')| \leq \beta|x - x'|$ for all $x' \in I_\delta$ and $g \in S$. So, if we let $\varsigma := \frac{\varepsilon}{\beta}$, we get $|\Phi(g)(x) - \Phi(g)(x')| \leq \varepsilon$ for all $x' \in I_\delta$ with $|x - x'| < \varsigma$ and all $g \in S$. Since $\varepsilon > 0$ is arbitrary here, this means that $\Phi(S)$ is equicontinuous at x, and since $x \in I_\delta$ is arbitrary, it follows that $\Phi(S)$ is equicontinuous. Thus $\Phi(S)$ is relatively compact, and we are done. ■

WARNING. In contrast to Picard's Existence Theorem, Peano's Existence Theorem is of an inherently *local* character. For instance, let us ask if there is a differentiable $f \in \mathbb{R}^{[-2,2]}$ with $f(1) = -1$ and $\sqrt{f'(x)} = f(x)$ for all $-2 \leq x \leq 2$.[22] The answer is no! If there were such an f, then it would be increasing (because $f' \geq 0$), so $f(1) = -1$ would imply $f|_{[0,1)} < 0$. By the Fundamental Theorem of Calculus, therefore, we would find

$$x = \int_0^x dt = \int_0^x \frac{f'(t)}{f(t)^2} dt = -\int_0^x \frac{d}{dt}\left(\frac{1}{f(t)}\right) dt = -\frac{1}{f(x)},$$

that is, $f(x) = -\frac{1}{x}$ for any $0 \leq x < 1$, contradicting the fact that f is real-valued.

The following exercises provide further examples of functional equations, the existence of solutions to which is established by means of a suitable fixed point argument.

EXERCISE 45 (*The Nonlinear Volterra Integral Equation*) Show that there exists an $f \in \mathbf{C}[0,1]$ such that

$$f(x) = \phi(x) + \int_0^x \theta(x, t, f(t))dt \qquad \text{for all } 0 \leq x \leq 1,$$

where $\phi \in \mathbf{C}[0,1]$ and $\theta \in \mathbf{CB}([0,1]^2 \times \mathbb{R})$.

[22] Peano's Existence Theorem tells us that there is a $0 < \delta < 1$ and a map $g : [1-\delta, 1+\delta] \to \mathbb{R}$ with $g(1) = -1$ and $\sqrt{g'(x)} = g(x)$ for all $1 - \delta \leq x \leq 1 + \delta$. The question is whether we can take $\delta = 1$ here.

Exercise 46$^{\rm H}$ (Matkowski) Let h be a Lipschitz continuous self-map on $[0, 1]$ with $h(0) = 0$; we denote the Lipschitz constant of h by K. Take any $H \in \mathbf{C}([0, 1] \times \mathbb{R})$ such that $H(0, 0) = 0$ and

$$\left|H(x, y) - H(x', y')\right| \leq p\left|x - x'\right| + q\left|y - y'\right| \qquad \text{for all } x, x', y, y' \geq 0$$

for some $p, q > 0$. Prove that if $Kq < 1$, then there exists a Lipschitz continuous real function f on $[0, 1]$ such that $f(0) = 0$ and

$$f(x) = H(x, f(h(x))) \qquad \text{for all } 0 \leq x \leq 1.$$

Exercise 47$^{\rm H}$ Show that there exists an $f \in \mathbf{C}[0, 1]$ such that

$$f(x) = \tfrac{1}{2}\cos(f(x)) + \tfrac{1}{2}\int_0^1 e^{-xt}\sin(tf(t))dt, \qquad 0 \leq x \leq 1.$$

4 Bounded Linear Operators and Functionals

In this section we revisit continuous linear operators and see what sort of things can be said about them when their domains and codomains are normed linear spaces. This section is elementary and is a basic prerequisite for a variety of topics, including optimization theory, approximation theory, and functional equations. Later we will use the concepts introduced here to substantiate the convex analysis that was sketched out in earlier chapters. These concepts also play a major role in Chapter K, where we study the differentiation (and optimization) of nonlinear functionals.

4.1 Definitions and Examples

Recall that a linear functional on a metric linear space is uniformly continuous iff it is continuous at the origin (or at any other point in the space). We can say a bit more in the case of normed linear spaces. For one thing, any continuous linear functional on a normed linear space X must be bounded on the closed unit ball B_X.[23] (*Warning*. This is not a consequence of Weierstrass' Theorem, because B_X is *not* compact if $dim(X) = \infty$; recall Theorem I.2.) To see this, note that, by continuity at $\mathbf{0}$, there exists a $0 < \delta < 1$ such that

[23] *Reminder.* $B_X := \{x \in X : \|x\| \leq 1\}$.

$|L(y)| < 1$ for all $y \in X$ with $\|y\| \leq \delta$. But then the absolute homogeneity of $\|\cdot\|$ entails that $|L(\delta x)| < 1$, that is, $|L(x)| < \frac{1}{\delta}$, for any $x \in X$ with $\|x\| \leq 1$. Thus

$$\sup\{|L(x)| : x \in B_X\} < \infty. \tag{9}$$

A moment's reflection shows that L being real-valued does not play a key role here. If L was instead a continuous linear operator from X into another normed linear space Y, (9) would simply modify to

$$\sup\{\|L(x)\|_Y : x \in B_X\} < \infty. \tag{10}$$

Linear operators that satisfy this property occupy the center stage in linear functional analysis.

Definition

Given any normed linear spaces X and Y, a linear operator $L \in \mathcal{L}(X, Y)$ is said to be **bounded** if (10) holds. If $Y = \mathbb{R}$ and (9) holds, then L is called a **bounded linear functional**.

Notation. The set of all bounded linear operators from the normed linear space X into the normed linear space Y is denoted as $\mathcal{B}(X, Y)$. So,

$$\mathcal{B}(X, Y) := \{L \in \mathcal{L}(X, Y) : \sup\{\|L(x)\|_Y : x \in B_X\} < \infty\}.$$

In functional analysis, it is customary to denote $\mathcal{B}(X, \mathbb{R})$ as X^*. We will adhere to this practice here as well.

Note that if L is a bounded linear functional, then it need not be a bounded function in the sense of having a bounded range $L(X)$.[24] Instead, in linear functional analysis, one qualifies a linear functional L as *bounded* whenever $L(B_X)$ is a bounded set.[25]

The notion of boundedness gives us a new way of thinking about the continuity of a linear operator. Indeed, these two concepts are identical for linear operators. (This should remind you of the structure of additive functionals defined on a Euclidean space; recall Lemma D.2.) We have just

[24] The latter property would be a silly requirement to impose on a linear functional. After all, the range of a nonzero linear functional cannot possibly be bounded. (Yes?)

[25] Does this entail that L maps every bounded subset of X to a bounded subset of Y? (Yes!)

seen that continuity of a linear operator implies its boundedness. To see the converse, let X and Y be two normed linear spaces, and take any $L \in \mathcal{B}(X,Y)$. Letting $M := \sup\{\|L(x)\|_Y : x \in B_X\}$, we have

$$\|L(y)\|_Y = \left\|L\left(\tfrac{1}{\|y\|}y\right)\right\|_Y \|y\| \leq M\,\|y\|$$

for all $y \in X$. Therefore, for any $(y^m) \in X^\infty$ with $y^m \to \mathbf{0}$, we have $\|L(y^m)\|_Y \leq M\,\|y^m\| \to 0$, that is, $L(y^m) \to \mathbf{0} = L(\mathbf{0})$. Thus L is continuous at $\mathbf{0}$, which implies that it is uniformly continuous everywhere (Proposition I.3). Hence follows the first part of the following basic result.

PROPOSITION 5

(Banach) *Let X and Y be two normed linear spaces and $L \in \mathcal{L}(X,Y)$. Then L is continuous if, and only if, it is bounded. Moreover, for any $L \in \mathcal{B}(X,Y)$, we have*

$$\sup\{\|L(x)\|_Y : x \in B_X\} = \sup\left\{\frac{\|L(x)\|_Y}{\|x\|} : x \in X\backslash\{\mathbf{0}\}\right\}$$
$$= \inf\{M > 0 : \|L(x)\|_Y \leq M\,\|x\| \text{ for all } x \in X\}.$$

This proposition allows us to use the phrase "bounded linear operator/functional" interchangeably with the phrase "continuous linear operator/functional" in the context of normed linear spaces. The proof of its second part is left as an exercise.

EXERCISE 48 Complete the proof of Proposition 5.

EXAMPLE 5

[1] Let X and Y be two normed linear spaces and $L \in \mathcal{L}(X,Y)$. If L is continuous, then, by Proposition 5, there exists an $M \geq 0$ such that $\|L(z)\|_Y \leq M\,\|z\|$ for all $z \in X$.[26] But then, for any $x, y \in X$, we have $\|L(x) - L(y)\|_Y = \|L(x-y)\|_Y \leq M\,\|x-y\|$. Conclusion: *A linear operator from a normed linear space into another one is continuous iff it is Lipschitz continuous.*[27]

[26] If there was no such M, then we would have, by Proposition 5, $\sup\{\|L(x)\|_Y : x \in B_X\} = \inf(\emptyset) = \infty$, contradicting the boundedness of L. (*Quiz.* Without using Proposition 5, show that (10) implies that there exists an $M \geq 0$ with $\|L(z)\|_Y \leq M\,\|z\|$ for all $z \in X$.)

[27] Did I use the absolute homogeneity property of normed linear spaces in establishing this? Where?

[2] Let $X := \{(x_m) \in \mathbb{R}^\infty : \sum^\infty |x_i| < \infty\}$, and norm this linear space by using the sup-norm. Consider the linear functional L on X defined by $L(x_1, x_2, \ldots) := \sum^\infty x_i$. We have seen earlier that this functional is not continuous (Example I.3.[1]). By Proposition 5, therefore, L must be unbounded. Indeed, for the sequence of sequences $(x^k) := ((1,0,0,\ldots),(1,1,0,\ldots),\ldots)$, we have $\left\|x^k\right\|_\infty = 1$ and $\left|L(x^k)\right| = \left|\sum^\infty x_i^k\right| = k$ for each $k \in \mathbb{N}$. Thus $\sup\{|L(x)| : x \in B_X\} = \infty$.

[3] Let X stand for the normed linear space of all continuously differentiable functions on $[0,1]$ with the sup-norm. Define $D \in \mathcal{L}(X, \mathbf{C}[0,1])$ by $D(f) := f'$. Is D continuous? No. If $i \in \mathbb{N}$, and $f_i \in \mathbb{R}^{[0,1]}$ is defined by $f_i(t) := t^i$, then $\|f_i\|_\infty = 1$ and $\|D(f_i)\|_\infty = i$. (Why?) It follows that $\sup\left\{\|D(f)\|_\infty : f \in B_X\right\} = \infty$, so by Proposition 5, D is not bounded, and hence not continuous.

[4] The differentiation operator considered in [3] would be continuous if its domain was normed suitably. For instance, as we leave for you to verify, the linear operator $D \in \mathcal{L}(\mathbf{C}^1[0,1], \mathbf{C}[0,1])$, defined by $D(f) := f'$, is continuous.[28] □

EXERCISE 49 Let X and Y be two normed linear spaces and $L \in \mathcal{B}(X,Y)$. Show that if L is surjective and there exists an $\alpha > 0$ such that $\|L(x)\|_Y \geq \alpha \|x\|$ for all $x \in X$, then $L^{-1} \in \mathcal{B}(Y,X)$.

EXERCISE 50^H Define $L \in \mathcal{B}(\mathbf{C}[0,1], \mathbf{C}[0,1])$ by $L(f)(x) := \int_0^x f(t)dt$. Let $X := \{f \in \mathbf{C}[0,1] : f(0) = 0\}$, and show that $L^{-1} \in \mathcal{L}(X, \mathbf{C}[0,1])$. Is L^{-1} bounded?

EXERCISE 51^H (Wilansky) Let X and Y be two normed linear spaces and $L : X \to Y$ an additive function (i.e., $L(x+y) = L(x) + L(y)$ for all $x, y \in X$). Show that L is continuous iff (10) holds.

EXERCISE 52 (*Bilinear Functionals*) Let X and Y be two normed linear spaces, and view $X \times Y$ as the product normed linear space (Example 1.[7]). We say that a real map f on $X \times Y$ is a **bilinear functional** if $f(\cdot, y) \in \mathcal{L}(X, \mathbb{R})$ and $f(x, \cdot) \in \mathcal{L}(Y, \mathbb{R})$ for each $(x,y) \in X \times Y$. (For instance, $f : \mathbb{R}^2 \to \mathbb{R}$ defined by $f(u,v) := uv$ is a bilinear (but not linear)

[28] Recall Example 1.[4].

functional.) Show that a bilinear functional $f \in \mathbb{R}^{X \times Y}$ is continuous iff

$$\sup\left\{\left|f(x,y)\right| : (x,y) \in B_X \times B_Y\right\} < \infty.$$

4.2 Linear Homeomorphisms, Revisited

Recall that, from the perspective of linear algebraic and topological properties, two metric linear spaces can be identified with each other if they are linearly homeomorphic. Proposition 5 provides a slightly different way of looking at this in the case of normed linear spaces. By definition, if the normed linear spaces X and Y are linearly homeomorphic, then there exists a continuous bijection $L \in \mathcal{L}(X,Y)$ such that L^{-1} is continuous. By Proposition 5, therefore, X and Y are linearly homeomorphic iff there is an invertible $L \in \mathcal{B}(X,Y)$ with $L^{-1} \in \mathcal{B}(Y,X)$. It is easy to see that the latter statement holds iff there exist $a, b \geq 0$ such that

$$a\,\|x\| \leq \|L(x)\|_Y \leq b\,\|x\| \qquad \text{for all } x \in X. \tag{11}$$

(Prove!) Thus: *Two normed linear spaces X and Y are linearly homeomorphic iff there is an invertible $L \in \mathcal{L}(X,Y)$ such that* (11) *holds.*

We say that two norms $\|\cdot\|$ and $|||\cdot|||$ on a given linear space X are **equivalent** if the resulting normed linear spaces are linearly homeomorphic, or equivalently (why?), id_X is a homeomorphism between these two spaces. By the previous observation, therefore, we understand that the norms $\|\cdot\|$ and $|||\cdot|||$ on X are equivalent iff

$$a\,\|x\| \leq |||x||| \leq b\,\|x\| \qquad \text{for all } x \in X, \tag{12}$$

for some $a, b > 0$.

Here is an immediate application of this fact. We know from the discussion of Section C.1.5 that two equivalent metrics on a linear space X need not be strongly equivalent. Evidently, at least one of these metrics cannot be induced by a norm. Put differently, the characterization of equivalence of norms given above entails that if d and D are two metrics on a linear space X that are induced by norms on X, then d and D are equivalent iff they are strongly equivalent.[29]

[29] Why? Because if d and D are equivalent, then id_X is a homeomorphism between (X,d) and (X,D), so . . .

Finally, let us note that our characterization of equivalence of norms can also be used to check if two normed linear spaces are, in fact, *not* linearly homeomorphic. Here is a simple illustration. Let X stand for the linear space of all real sequences, only finitely many terms of which are nonzero. Consider norming this space with either $\|\cdot\|_1$ or $\|\cdot\|_2$ (Example 1.[1]). The resulting normed linear spaces would not be linearly homeomorphic, because $\|\cdot\|_1$ and $\|\cdot\|_2$ are not equivalent norms on X. If they were, we could find a number $b > 0$ such that, for all $k \in \mathbb{N}$,

$$\sum_{i=1}^{k} \tfrac{1}{i} = \left\|\left(1, \tfrac{1}{2}, \ldots, \tfrac{1}{k}, 0, 0, \ldots\right)\right\|_1 \leq b \left\|\left(1, \tfrac{1}{2}, \ldots, \tfrac{1}{k}, 0, 0, \ldots\right)\right\|_2 = \left(\sum_{i=1}^{k} \tfrac{1}{i^2}\right)^{\frac{1}{2}},$$

but this is impossible, since $\sum^{\infty} \frac{1}{i}$ diverges to ∞, while $\sum^{\infty} \frac{1}{i^2}$ is convergent.[30]

Remark 2. Given any $n \in \mathbb{N}$, it follows from the discussion of Section C.1.5 that the norms $\|\cdot\|_p$ and $\|\cdot\|_q$ on $\mathbb{R}^n$ are equivalent for any $1 \leq p, q \leq \infty$. It is worth noting that, in fact, *any two norms on* $\mathbb{R}^n$ *are equivalent.*

To see this, take any norm $\|\cdot\|$ on $\mathbb{R}^n$. First, note that, for any $x \in \mathbb{R}^n$,

$$\|x\| = \left\|\sum_{i=1}^{n} x_i \mathbf{e}^i\right\| \leq \sum_{i=1}^{n} |x_i| \left\|\mathbf{e}^i\right\| \leq b \|x\|_1,$$

where $\{\mathbf{e}^1, \ldots, \mathbf{e}^n\}$ is the standard basis for $\mathbb{R}^n$ and $b := \max\{\left\|\mathbf{e}^i\right\| : i = 1, \ldots, n\}$. Conclusion: There exists a $b > 0$ such that $\|x\| \leq b \|x\|_1$ for all $x \in \mathbb{R}^n$.

It follows from this observation that $\|\cdot\|$ is a (Lipschitz) continuous map on the metric space $(\mathbb{R}^{n,1}, d_1)$. (*Proof.* $\left|\|x\| - \|y\|\right| \leq \|x - y\| \leq b \|x - y\|_1$ for any $x, y \in \mathbb{R}^n$.) Since $S := \{x \in \mathbb{R}^n : \|x\|_1 = 1\}$ is a compact subset of this space, therefore, there is an $x^* \in S$ such that $\|x^*\| \leq \|y\|$ for all $y \in S$ (Weierstrass' Theorem). Let $a := \|x^*\|$. Since $\|x^*\|_1 = 1$, we have $a > 0$. Moreover, for any $x \in \mathbb{R}^n$, $a \leq \left\|\frac{1}{\|x\|_1} x\right\|$, so $a \|x\|_1 \leq \|x\|$. We have proved: There exist $a, b > 0$ such that $a \|x\|_1 \leq \|x\| \leq b \|x\|_1$ for any $x \in \mathbb{R}^n$. Conclusion: *Any* norm on $\mathbb{R}^n$ is equivalent to $\|\cdot\|_1$. □

30 Did you know that $\sum^{\infty} \frac{1}{i^2} = \frac{\pi}{6}$, by the way?

EXERCISE 53 Find two norms on $\mathbf{C}[0,1]$ that are not equivalent.

4.3 The Operator Norm

For any given normed linear spaces X and Y, we define the real function $\|\cdot\|^* : \mathcal{B}(X,Y) \to \mathbb{R}_+$ by

$$\|L\|^* := \sup\{\|L(x)\|_Y : x \in B_X\}.$$

It is easy to check that $\|\cdot\|^*$ is a norm on the linear space $\mathcal{B}(X,Y)$. This norm is called the **operator norm**. From now on, whenever we treat $\mathcal{B}(X,Y)$ (or X^*) as a normed linear space, we will always have in mind the operator norm $\|\cdot\|^*$ as the relevant norm.

EXERCISE 54 For any normed linear spaces X and Y, show that $\mathcal{B}(X,Y)$ is a normed linear space (relative to the operator norm).

An elementary but extremely important fact of linear functional analysis is that

$$\|L(x)\|_Y \leq \|L\|^* \|x\| \qquad \text{for all } x \in X, \tag{13}$$

for any given $L \in \mathcal{B}(X,Y)$. (*Proof.* If $\|L(y)\|_Y > \|L\|^* \|y\|$ for some $y \in X$, then $y \neq \mathbf{0}$, so $x := \frac{1}{\|y\|}y \in B_X$, and we have $\|L(x)\|_Y > \|L\|^*$, which is absurd.) When combined with Proposition 5, this inequality provides an alternative way of looking at the operator norm. It says that, for any $L \in \mathcal{B}(X,Y)$, $\|L\|^*$ is the smallest number $M \geq 0$ such that $\|L(x)\|_Y \leq M\|x\|$ for all $x \in X$. This fact is used quite frequently in operator theory.[31]

EXAMPLE 6

[1] Given any $n \in \mathbb{N}$, we know that all linear functionals on $\mathbb{R}^n$ are continuous. Associated with any such linear functional L is a unique

[31] Most textbooks on functional analysis write (13) as "$\|L(x)\| \leq \|L\| \|x\|$ for all $x \in X$." Here the first norm corresponds to that of the codomain of L (namely, Y), the second is the operator norm, and finally the third one is the norm of X. The notation that I use is a bit more tedious than usual, but it makes clear which norm corresponds to which space. I thought it would be a good idea to be clear about this at this introductory stage.

vector $\alpha \in \mathbb{R}^n$ such that $L(x) = \sum^n \alpha_i x_i$ for all $x \in \mathbb{R}^n$ (why unique?), and conversely, any $\alpha \in \mathbb{R}^n$ defines a unique linear functional on $\mathbb{R}^n$ as the map $x \mapsto \sum^n \alpha_i x_i$ (Example F.6). Thus, the set of all continuous linear functionals on $\mathbb{R}^n$ is $\{L_\alpha : \alpha \in \mathbb{R}^n\}$, where we write L_α for the map $x \mapsto \sum^n \alpha_i x_i$. Question: What is the norm of an arbitrary L_α?

The answer is easy. For one thing, the Cauchy-Schwarz Inequality (Section G.3.3) gives

$$|L_\alpha(x)| = \left|\sum_{i=1}^n \alpha_i x_i\right| \leq \|\alpha\|_2 \|x\|_2 \qquad \text{for any } \alpha, x \in \mathbb{R}^n.$$

Since $\|L_\alpha\|^*$ is the smallest number $M \geq 0$ such that $|L_\alpha(x)| \leq M \|x\|_2$ for all $x \in \mathbb{R}^n$, this implies that $\|L_\alpha\|^* \leq \|\alpha\|_2$. Moreover, for any given $\alpha \in \mathbb{R}^n$, we have $|L_\alpha(\alpha)| = \sum^n |\alpha_i|^2 = \|\alpha\|_2 \|\alpha\|_2$, so it follows from (13) that $\|L_\alpha\|^* \geq \|\alpha\|_2$. Conclusion: $\|L_\alpha\|^* = \|\alpha\|_2$.

[2] Define $L \in \mathcal{L}(\mathbf{C}[0,1], \mathbb{R})$ by $L(f) := f(0)$. Clearly, we have $|L(f)| = |f(0)| \leq \|f\|_\infty$ for any $f \in \mathbf{C}[0,1]$. Thus, L is a bounded linear functional with $\|L\|^* \leq 1$. But if $f = 1$, then $|L(f)| = 1 = \|f\|_\infty$, and hence $\|L\|^* = 1$. (Why?)

[3] Define $L \in \mathcal{L}(\mathbf{C}[0,1], \mathbb{R})$ by $L(f) := \int_0^1 tf(t)dt$. For any $f \in \mathbf{C}[0,1]$,

$$|L(f)| = \left|\int_0^1 tf(t)dt\right| \leq \int_0^1 t\,|f(t)|\,dt \leq \|f\|_\infty \int_0^1 tdt = \tfrac{1}{2}\|f\|_\infty.$$

Thus, L is a bounded, and hence continuous, linear functional. This computation also shows that $\|L\|^* \leq \frac{1}{2}$. But if $f = 1$, we have $|L(f)| = \frac{1}{2} = \frac{1}{2}\|f\|_\infty$, and hence $\|L\|^* = \frac{1}{2}$. □

EXERCISE 55[H] Let $0 < \delta < 1$, and define the function $L : \ell^\infty \to \mathbb{R}$ by $L((x_m)) := \sum^\infty \delta^i x_i$. Show that L is a bounded linear functional, and compute $\|L\|^*$.

EXERCISE 56[H] Define the map $L : \mathbf{C}[0,1] \to \mathbb{R}$ by $L(f) := \int_0^1 g(t)f(t)dt$, where $g \in \mathbf{C}[0,1]$. Show that $L \in \mathbf{C}[0,1]^*$, and compute $\|L\|^*$.

EXERCISE 57
(a) Define $D \in \mathcal{B}(\mathbf{C}^1[0,1], \mathbf{C}[0,1])$ by $D(f) := f'$, and show that $\|D\|^* = 1$.
(b) Define $L \in \mathcal{B}(\mathbf{C}[0,1], \mathbf{C}[0,1])$ by $L(f)(t) := tf(t)$, and compute $\|L\|^*$.

EXERCISE 58[H] Let (X, ϕ) be a pre-Hilbert space (Exercise 12). For any given $x^* \in X$, define $L : X \to \mathbb{R}$ by $L(x) := \phi(x, x^*)$. Show that $L \in X^*$ and $\|L\|^* = \|x^*\|$.

EXERCISE 59 Take any $1 \leq p \leq \infty$, and define the **right-shift** and **left-shift operators** on ℓ^p as the self-maps R and L with $R(x_1, x_2, \ldots) := (0, x_1, x_2, \ldots)$ and $L(x_1, x_2, \ldots) := (x_2, x_3, \ldots)$, respectively. Compute $\|L\|^*$ and $\|R\|^*$.

EXERCISE 60[H] Let X, Y, and Z be normed linear spaces. Show that $\|K \circ L\|^* \leq \|K\|^* \|L\|^*$ for any $L \in \mathcal{B}(X, Y)$ and $K \in \mathcal{B}(Y, Z)$. Conclude that

$$\left(\|L^n\|^*\right)^{\frac{1}{n}} \leq \|L\|^* \qquad \text{for any } L \in \mathcal{B}(X, X) \text{ and } n \in \mathbb{N}.$$

(Here L^n is the nth iteration of L.)

EXERCISE 61 Let X and Y be two compact metric spaces and $h \in X^Y$ a continuous function. Define $L : \mathbf{C}(X) \to \mathbf{C}(Y)$ by $L(f) := f \circ h$. Show that $L \in \mathcal{B}(\mathbf{C}(X), \mathbf{C}(Y))$ and $\|L\|^* = 1$.

EXERCISE 62 Let L be a nonzero continuous linear functional on a normed linear space X. Show that

$$\|L\|^* = \frac{1}{d_{\|\cdot\|}(\mathbf{0}, L^{-1}(1))}.$$

(*Interpretation.* The norm of L equals the inverse of the distance between the origin and the hyperplane $\{x \in X : L(x) = 1\}$.)

EXERCISE 63[H] (*More on Bilinear Functionals*) Let X and Y be two normed linear spaces, and view $X \times Y$ as the product normed linear space (Example 1.[7]). Let $\mathfrak{b}_{X,Y}$ denote the set of all continuous bilinear functionals on $X \times Y$ (Exercise 52).
(a) Show that the map $|||\cdot||| : \mathfrak{b}_{X,Y} \to \mathbb{R}_+$ defined by

$$|||f||| := \sup\left\{\left|f(x, y)\right| : (x, y) \in B_X \times B_Y\right\}$$

is a norm, and we have

$$\left|f(x,y)\right| \leq |||f|||\,\|x\|\,\left\|y\right\|_Y \qquad \text{for all } f \in \mathfrak{b}_{X,Y} \text{ and } (x,y) \in X \times Y.$$

We wish to show that $\mathfrak{b}_{X,Y}$ (normed this way) "is" none other than $\mathcal{B}(X, Y^*)$. To this end, define the map $\Phi : \mathfrak{b}_{X,Y} \to \mathcal{B}(X, Y^*)$ by $\Phi(f)(x) := f(x, \cdot)$.

(b) Prove that Φ is well-defined, that is, $f(x, \cdot) \in Y^*$ for any $x \in X$.

(c) Show that

$$\Phi \in \mathcal{B}(\mathfrak{b}_{X,Y}, \mathcal{B}(X, Y^*)) \qquad \text{and} \qquad \|\Phi\|^* = 1.$$

(d) Show that Φ is a bijection and Φ^{-1} is continuous. Thus Φ is a linear isometry between $\mathfrak{b}_{X,Y}$ and $\mathcal{B}(X, Y^*)$. Conclusion: $\mathfrak{b}_{X,Y}$ is identical to $\mathcal{B}(X, Y^*)$ up to a linear isometry.

The following two exercises develop further the theory of linear correspondences sketched in Exercises F.31, I.22, and I.23. The definitions and results of those exercises are thus prerequisites for them.

EXERCISE 64 (*The Norm of a Linear Correspondence*) Let X and Y be two normed linear spaces. For any linear correspondence $\Gamma : X \rightrightarrows Y$, we define $/\Gamma(x)/ := \sup\left\{\left\|y - z\right\|_Y : (y,z) \in \Gamma(x) \times \Gamma(\mathbf{0})\right\}$ for all $x \in X$, and

$$\|\Gamma\|^* := \sup\{/\Gamma(x)/ : x \in B_X\}.$$

Show that, for any linear correspondences Γ and Υ from X into Y, and any $\lambda \in \mathbb{R}$, we have

$$\|\Gamma + \Upsilon\|^* \leq \|\Gamma\|^* + \|\Upsilon\|^* \qquad \text{and} \qquad \|\lambda\Gamma\|^* = |\lambda|\,\|\Gamma\|^*.$$

(Here $\lambda\Gamma$ is the correspondence defined by $(\lambda\Gamma)(x) := \lambda\Gamma(x)$.)

EXERCISE 65 (*Bounded Linear Correspondences*) Let X and Y be two normed linear spaces and $\Gamma : X \rightrightarrows Y$ a linear correspondence.

(a) Show that $\|\Gamma\|^* < \infty$ iff Γ is lower hemicontinuous.

(b) Prove: If Γ admits a continuous linear selection L, then $\|\Gamma\|^* \leq \|L\|^*$.

4.4 Dual Spaces

Recall that X^* is the linear space of all continuous linear functionals on the normed linear space X, and that we think of X^* itself as a normed linear space under the operator norm $\|\cdot\|^*$, where

$$\|L\|^* := \sup\{|L(x)| : x \in B_X\}.$$

The normed linear space X^* is called the **dual (space)** of X.

EXAMPLE 7

Let $n \in \mathbb{N}$. The dual of the Euclidean space $\mathbb{R}^n$ is, by definition, $\{L_\alpha : \alpha \in \mathbb{R}^n\}$, where we write L_α for the map $x \mapsto \sum^n \alpha_i x_i$ for any $\alpha \in \mathbb{R}^n$. (Why?) Define the map $\Phi : \{L_\alpha : \alpha \in \mathbb{R}^n\} \to \mathbb{R}^n$ by $\Phi(L_\alpha) := \alpha$. It is obvious that Φ is a bijection. Moreover, for any $\lambda \in \mathbb{R}$ and $\alpha, \beta \in \mathbb{R}^n$,

$$\Phi(\lambda L_\alpha + L_\beta) = \lambda\alpha + \beta = \lambda\Phi(L_\alpha) + \Phi(L_\beta)$$

so that Φ is a linear operator. Finally, by Example 6.[1],

$$\|\Phi(L_\alpha)\|_2 = \|\alpha\|_2 = \|L_\alpha\|^* \qquad \text{for any } \alpha \in \mathbb{R}^n.$$

It follows that Φ is a linear isometry from the dual space of $\mathbb{R}^n$ onto $\mathbb{R}^n$. Since linearly isometric normed linear spaces are indistinguishable from each other (in terms of both their metric and linear properties), this observation is often paraphrased as: *The dual of $\mathbb{R}^n$ "is" itself.* □

EXERCISE 66 Take any $p > 1$, and let q be the real number that satisfies $\frac{1}{p} + \frac{1}{q} = 1$. Prove that the dual of $(\mathbb{R}^n, \|\cdot\|_p)$ "is" $(\mathbb{R}^n, \|\cdot\|_q)$.

EXERCISE 67 Characterize the dual of $(\mathbb{R}^n, \|\cdot\|_\infty)$ up to linear isometry.

Determining the duals of normed linear spaces up to linear isometry is a time-honored topic in linear analysis. Unfortunately, in the case of infinite-dimensional normed linear spaces, the involved computations are rather hefty. Given the introductory nature of the present course, therefore, we will not discuss this topic here any further. It suffices for us to make note of the following fundamental fact about dual spaces.

PROPOSITION 6
For any normed linear space X, the dual space X^* is a Banach space.

We give the proof in the form of an exercise.

EXERCISE 68 [H] Let X be a normed linear space, and (L_m) a Cauchy sequence in X^*.

(a) Show that the real sequence $(L_m(x))$ is convergent for each $x \in X$.
(b) Define $L \in \mathbb{R}^X$ by $L(x) := \lim L_m(x)$, and show that $L \in X^*$.
(c) Show that $\|L_m - L\|^* \to 0$, and hence conclude that X^* is a Banach space.

Proposition 6 is very much based on the completeness of $\mathbb{R}$.[32] (Curiously, we don't need X to be complete here at all.) In fact, a straightforward extension of the argument sketched in Exercise 68 would establish readily that, for any given normed linear space X, $\mathcal{B}(X, Y)$ would be a Banach space whenever so is Y.

4.5* Discontinuous Linear Functionals, Revisited

We conclude this section by following up on a promise we made in Chapter I. Our objective is to show that there exists a discontinuous linear functional in *every* infinite-dimensional normed linear space. Therefore, the examples of discontinuous linear functionals we have encountered so far cannot be dismissed as exceptions to the rule. This fact provides yet another perspective on infinite-dimensional linear spaces.

Fix any normed linear space X. Let us compare the dual space X^* with $\mathcal{L}(X, \mathbb{R})$ (which is sometimes called the **algebraic dual** of X). In general, of course, we have $X^* \subseteq \mathcal{L}(X, \mathbb{R})$. Corollary I.4 shows us that $X^* = \mathcal{L}(X, \mathbb{R})$ when X is a finite-dimensional metric linear space. We will show next that this property actually characterizes finite dimensionality: *A metric linear space X is finite-dimensional iff $X^* = \mathcal{L}(X, \mathbb{R})$*. Remarkably, this result characterizes a purely algebraic property (namely, finite dimensionality) by using a topological property (namely, continuity). Here is the argument.

32 Where in the proof did you use the Completeness Axiom?

Assume $dim(X) = \infty$, and let T be a basis for X. Since $|T| = \infty$, we may choose countably infinitely many distinct basis vectors from T, say $x^1, x^2, \ldots$. Define

$$y^i := \frac{1}{2^i}\left(\frac{1}{\|x^i\|}x^i\right), \qquad i = 1, 2, \ldots$$

and observe that $\{y^1, y^2, \ldots\}$ is linearly independent in X. Then, by the argument given in the proof of Theorem F.1, we may extend this set to a basis S for X. By Corollary F.2, for each $x \in X$, there exist a unique finite subset $S(x)$ of S and a map $\lambda : S(x) \to \mathbb{R}\backslash\{0\}$ such that $x = \sum_{y \in S(x)} \lambda(y)y$.

Now define the function $L \in \mathbb{R}^X$ by $L(x) := \sum_{y \in S(x)} \lambda(y)$. It is readily verified that L is a linear functional and $L(y) = 1$ for any $y \in S$. (In particular, $L(y^i) = 1$ for each i.) Now consider the vector

$$z^m := \frac{1}{\|\sum^m y^i\|}\sum_{i=1}^m y^i, \qquad m = 1, 2, \ldots$$

Clearly, $z^m \in B_X$. Besides, $\|\sum^m y^i\| \leq \sum^m \|y^i\| \leq \sum^m \frac{1}{2^i} < 1$, so

$$L(z^m) = \frac{1}{\|\sum^m y^i\|}\sum_{i=1}^m L(y^i) > m, \qquad m = 1, 2, \ldots$$

It follows that $\sup\{|L(x)| : x \in B_X\} = \infty$, that is, L is an unbounded linear functional on X. By Proposition 5, therefore, L is a discontinuous linear functional on X. Conclusion: $\mathcal{L}(X, \mathbb{R}) \neq X^*$.

5 Convex Analysis in Normed Linear Spaces

In this section we revisit the separation theorems obtained in Section I.5.4 and see how one may be able to improve upon them in the context of normed linear spaces. We also investigate the basic structure of compact and convex subsets of an arbitrary normed linear space.

5.1 Separation by Closed Hyperplanes, Revisited

The fact that any ε-neighborhood in a normed linear space is necessarily convex allows us to invoke the Separating Hyperplane Theorem to separate any two distinct points of this space by means of a closed hyperplane. The first result of this section is a generalization of this observation. It turns out that a compact set in a normed linear space can be *strongly* separated from a closed set by a closed hyperplane (in the sense that small parallel shifts of the hyperplane do not cross either of these sets). More precisely, the following generalization of Proposition G.5 is true.

Theorem 1

Let A and B be two nonempty, disjoint, closed, and convex subsets of a normed linear space X. If A is compact, then there exists an $L \in X^*$ such that $\inf L(B) > \max L(A)$.

Proof

If A is compact, then $B - A$ is a closed subset of X (Proposition I.2), and $\mathbf{0} \notin B - A$ since $A \cap B = \emptyset$. Thus $\delta := d_{\|\cdot\|}(\mathbf{0}, B - A) > 0$. But then $N_{\frac{\delta}{2},X}(\mathbf{0}) \cap (B - A) = \emptyset$. So, since $N_{\frac{\delta}{2},X}(\mathbf{0})$ is an open and convex subset of O, and $B - A$ is convex, the Separating Hyperplane Theorem says that there exists an $(\alpha, L) \in \mathbb{R} \times X^*$ such that

$$L(B - A) \geq \alpha > L(N_{\frac{\delta}{2},X}(\mathbf{0})).$$

While the second inequality here entails that $\alpha > 0$ (why?), the first inequality yields $\inf L(B) \geq \alpha + \max L(A)$ by linearity of L and Weierstrass' Theorem. ■

Warning. As noted in Section G.3.1, the compactness requirement posited in Theorem 1 is essential even in $\mathbb{R}^2$. (*Example.* Consider the sets $A := \{(a, b) \in \mathbb{R}^2_+ : ab \geq 1\}$ and $B := \mathbb{R}_+ \times \{0\}$.)

One can easily deduce a number of interesting corollaries from Theorem 1. For instance, this result allows us to (strongly) separate a nonempty closed and convex subset of a normed linear space from any vector that is located outside this set. (Notice that the Separating Hyperplane Theorem does not give this to us directly.) A few other such corollaries are noted in the following exercises.

EXERCISE 69 Show that if X is a normed linear space, and $x \in null(L)$ for all $L \in X^*$, then $x = \mathbf{0}$.

EXERCISE 70 [H] Let S be a nonempty subset of a normed linear space X. Show that $\overline{co}_X(S)$ is the intersection of all closed half-spaces that contain S.

EXERCISE 71 Let A and B be two nonempty convex subsets of a normed linear space X. Show that A and B can be separated as in Theorem 1 iff $\mathbf{0} \notin cl_X(B - A)$.

EXERCISE 72 Generalize Theorem 1 to the case of locally convex metric linear spaces.

REMARK 3. Recall that the requirement $int_X(S) \neq \emptyset$ cannot be relaxed in the statement of the Supporting Hyperplane Theorem (Exercise I.51). This is also true in the case of Banach spaces, but the situation is more satisfactory in that case. After all, a famous result of infinite-dimensional convex analysis, the *Bishop-Phelps Theorem*, says: *If S is a closed and convex subset of a Banach space X, and if $E(S)$ stands for the set of all points of S at which S can be supported by a closed hyperplane, then $cl_X(E(S)) = bd_X(S)$.* The proof of this and several other related results can be found in the excellent treatises of Megginson (1988) and Aliprantis and Border (1999). For further results of this nature, and a unified summary of applications to general equilibrium theory, you might also find the paper by Aliprantis, Tourky and Yannelis (2000) useful. □

5.2* Best Approximation from a Convex Set

In Section G.3.4 we studied the best approximation from a convex subset of an arbitrary Euclidean space $\mathbb{R}^n$. We did so by utilizing the inner product operation on $\mathbb{R}^n$. Since we lack such additional structure (and hence the notion of orthogonality) in an arbitrary normed linear space, it is impossible to carry that analysis to the present realm intact. Still, this doesn't mean that we can't say anything intelligent about "projections" in a normed linear space. By way of a smart application of the Separating Hyperplane Theorem, we digress into this matter next.

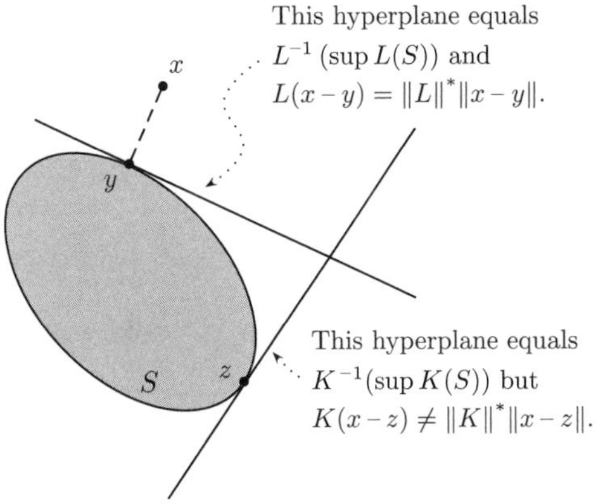
This hyperplane equals
$L^{-1}(\sup L(S))$ and
$L(x-y) = \|L\|^* \|x-y\|$.
x
y
S
z
This hyperplane equals
$K^{-1}(\sup K(S))$ but
$K(x-z) \neq \|K\|^* \|x-z\|$.

FIGURE 2

DEFINITION

Let S be a nonempty subset of a metric space X, and $x \in X$. We say that $y \in S$ is a **best approximation of** x **from** S (or a **projection of** x **on** S) if $d(x, S) = d(x, y)$.[33]

The following result provides a characterization of the best approximation of a given point from a closed and convex subset of a normed linear space. It reduces the involved approximation problem to the maximization of a continuous linear functional subject to a simple constraint. (You may find Figure 2 useful for understanding the nature of this characterization.)

PROPOSITION 7

Let S be a nonempty closed and convex subset of a normed linear space X, and $x \in X \backslash S$. Then, $y \in S$ is a best approximation of x from S if, and only if, there exists a nonzero $L \in X^*$ such that

$$L(y) = \sup L(S) \quad \text{and} \quad L(x-y) = \|L\|^* \|x-y\|. \tag{14}$$

PROOF

Let $y \in S$ be a best approximation of x from S. Clearly, $\delta := d_{\|\cdot\|}(x, y) > 0$. By the Separating Hyperplane Theorem, there exist an $\alpha \in \mathbb{R}$ and a nonzero

[33] You might want to recall Example D.5 at this point.

$L \in X^*$ such that

$$\sup L(S) \leq \inf L(N_{\delta,X}(x)) = L(x) + \inf L(N_{\delta,X}(\mathbf{0})).$$

But $\inf L(N_{\delta,X}(\mathbf{0})) = -\delta \, \|L\|^*$. (Why?) It follows that

$$L(y) \leq \sup L(S) \leq L(x) - \|L\|^* \, \|x - y\| \leq L(x) - L(x - y) = L(y),$$

and hence all inequalities in this statement hold as equalities. Thus L satisfies (14).

Conversely, assume that there is a nonzero $L \in X^*$ such that (14) holds. Then, for any $z \in X$ with $L(z) \leq L(y)$, we have

$$\|L\|^* \, \|x - y\| = L(x - y) \leq L(x - z) \leq \|L\|^* \, \|x - z\|,$$

so $\|x - y\| \leq \|x - z\|$. (We have $\|L\|^* > 0$ because L is nonzero.) Thus, y is a best approximation of x from the closed half-space $\{z \in X : L(z) \leq L(y)\}$. Since this half-space contains S, y must be a best approximation of x from S. ■

EXERCISE 73 Let S be a nonempty closed and convex subset of a normed linear space X, and $x \in X \backslash S$. Prove that

$$d_{\|\cdot\|}(x, S) = \max \left\{ \frac{L(x) - \sup L(S)}{\|L\|^*} : L \in X^* \right\}.$$

5.3 Extreme Points

You must be well aware that compact and convex sets play an important role in optimization theory. One reason for this is that such sets have a nice structure that tells us where the optimum of a linear functional defined on them may be located. To illustrate, consider the following (linear programming) problem for any $\alpha > 0$:

$$\text{Maximize } x + \alpha y \quad \text{such that} \quad x + 2y \leq 1,\ 2x + y \leq 1 \text{ and } x, y \geq 0.$$

If we plot the feasible set $S := \{(x, y) \in \mathbb{R}^2_+ : x + 2y \leq 1 \text{ and } 2x + y \leq 1\}$ of this problem (Figure 3), we immediately see that there are really only three points of this set that we should worry about: $(\frac{1}{2}, 0)$, $(0, \frac{1}{2})$ or $(\frac{1}{3}, \frac{1}{3})$. The solution to our problem must be located at one of these points. (Exactly which one depends on the value of α.) Although its feasible set contains

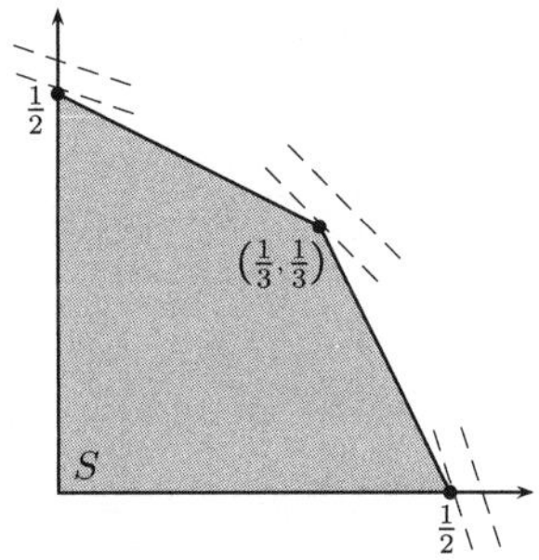

FIGURE 3

uncountably many points, the search for the solution to our problem is thus readily reduced to the comparison of the value of the objective function at only three points. What makes these points special is that they are the only points in S (other than $(0, 0)$, which is obviously not a solution) that cannot be written as the convex combination of two distinct points of S. It turns out that this is no coincidence. (Recall Exercise G.66.) As we shall see below, in a rich class of optimization problems (and certainly in all linear programming problems), it is only such points that we need to worry about.

DEFINITION
Let S be a subset of a linear space. A point $x \in S$ is said to be an **extreme point** of S provided that $x \neq \lambda y + (1 - \lambda)z$ for any distinct $y, z \in S$ and $0 < \lambda < 1$. We denote the set of all extreme points of S by $ext(S)$.

It is easy to see that, for any convex subset S of a normed linear space, we have $x \in ext(S)$ iff $S \backslash \{x\}$ is convex. (Prove!) The extreme points of any polygonal region in $\mathbb{R}^2$ are thus the vertices of that region. For instance, $ext(S) = \{(0, 0), (\frac{1}{2}, 0), (0, \frac{1}{2}), (\frac{1}{3}, \frac{1}{3})\}$ in the example considered above. Similarly, $ext([0, 1]) = \{0, 1\}$ and $ext((0, 1)) = \emptyset$. Here are some less trivial examples.

EXAMPLE 8

[1] If Y is a linear manifold in a linear space, then $ext(Y) \neq \emptyset$ iff Y is a singleton.

[2] Given any $n \in \mathbb{N}$, the set of all extreme points of $\{x \in \mathbb{R}^n : \|x\|_p = 1\}$ equals its boundary for any $1 < p < \infty$. On the other hand,

$ext(\{x \in \mathbb{R}^n : \|x\|_1 = 1\}) = \{\mathbf{e}^1, \ldots, \mathbf{e}^n\} \cup \{-\mathbf{e}^1, \ldots, -\mathbf{e}^n\}$, where $\mathbf{e}^i$ is the ith unit vector of $\mathbb{R}^n$, and $ext(\{x \in \mathbb{R}^n : \|x\|_\infty = 1\}) = \{-1, 1\}^n$.

[3] Let $S := co(\{(x, y, 0) \in \mathbb{R}^3 : x^2+y^2 = 1\} \cup \{(0, -1, 1), (0, -1, -1)\})$. (Draw a picture!) Then S is a compact and convex set, and $ext(S) = (\{(x, y, 0) \in \mathbb{R}^3 : x^2 + y^2 = 1\} \backslash \{(0, -1, 0)\}) \cup \{(0, -1, 1), (0, -1, -1)\}$. Thus the set of all extreme points of a compact and convex subset of a Euclidean space need not be closed.[34]

[4] If $S := \{f \in \mathbf{C}[0, 1] : \|f\|_\infty = 1\}$, then $ext(S) = \{-\mathbf{1}_{[0,1]}, \mathbf{1}_{[0,1]}\}$. (Proof?)

[5] If $S := \{(x_m) \in \ell^\infty : \|(x_m)\|_\infty = 1\}$, then $ext(S) = \{-1, 1\}^\infty$. (Proof?)

[6] Consider the closed unit ball $B_{\mathbf{c}_0}$ of the Banach space $\mathbf{c}_0$.[35] This set is obviously nonempty, closed, bounded, and convex. Yet it has no extreme points. Indeed, if $(x_m) \in B_{\mathbf{c}_0}$, then there must exist some $M \in \mathbb{N}$ such that $|x_M| < \frac{1}{2}$. So, if the sequences (y_m) and (z_m) are defined as

$$y_m := \begin{cases} x_m, & \text{if } m \neq M \\ x_m - \frac{1}{2}, & \text{if } m = M \end{cases} \quad \text{and} \quad z_m := \begin{cases} x_m, & \text{if } m \neq M \\ x_m + \frac{1}{2}, & \text{if } m = M \end{cases},$$

we have $(x_m) = \frac{1}{2}(y_m) + \frac{1}{2}(z_m)$, while both (y_m) and (z_m) belong to $B_{\mathbf{c}_0}$. It follows that $ext(B_{\mathbf{c}_0}) = \emptyset$. □

EXERCISE 74 Let S be a closed and convex subset of a normed linear space. Show that $x \in ext(S)$ iff $x = \frac{1}{2}y + \frac{1}{2}z$ holds for some $y, z \in S$ only when $x = y = z$. Is this true for convex sets that are not closed?

EXERCISE 75[H] Let T be a compact metric space. How many extreme points does the closed unit ball of $\mathbf{C}(T)$ have?

EXERCISE 76 Prove that $ext(S) \subseteq bd_X(S)$ for any subset S of a metric linear space X.

EXERCISE 77 As noted in Exercise 7, a normed linear space X is said to be **rotund** if $\left\|\frac{1}{2}x + \frac{1}{2}y\right\| < 1$ for any distinct $x, y \in X$ with $\|x\| = \|y\| = 1$.

[34] *Quiz.* Show that this difficulty does not arise in $\mathbb{R}^2$.

[35] *Reminder.* $\mathbf{c}_0$ is the space of all real sequences that converge to 0; it is normed by the sup-norm.

(Intuitively, this means that the boundary of the open unit ball of X does not contain a nontrivial line segment.) Prove that X is rotund iff $ext\big(N_{1,X}(\mathbf{0})\big) = bd_X(N_{1,X}(\mathbf{0}))$.

Example 8.[6] points to a rather unpleasant fact. Apparently, a nonempty subset of a Banach space need not have any extreme points even if it is closed, bounded, and convex. As an application of Theorem 1, we next show that this is nothing that compactness (and the Axiom of Choice) can't fix.

Theorem 2

For any nonempty compact subset S of a normed linear space X, we have $ext(S) \neq \emptyset$.

Proof

Consider the following property posited on the members A of 2^S:

Property $(*)$: For any $x, y \in S$, if $\lambda x + (1-\lambda)y \in A$ for some $0 < \lambda < 1$, then $x, y \in A$.

Let $\mathcal{A}$ be the class of all nonempty closed subsets of S that satisfy Property $(*)$.[36] Since $S \in \mathcal{A}$, we have $\mathcal{A} \neq \emptyset$. Moreover, $(\mathcal{A}, \subseteq)$ is obviously a poset. Let $(\mathcal{B}, \subseteq)$ be a loset such that $\mathcal{B} \subseteq \mathcal{A}$. It is easily checked that $\cap\mathcal{B} \in \mathcal{A}$.[37] Then, by Zorn's Lemma, $(\mathcal{A}, \subseteq)$ must possess a $\subseteq$-maximal element, that is, there is a set $A \in \mathcal{A}$ such that $B \subset A$ does not hold for any $B \in \mathcal{A}$.

All we need to show now is that A is a singleton (for then "the" element of A must be an extreme point of S). Suppose there exist two distinct points x and y in A. Then, by Theorem 1, there exists a nonzero $L \in X^*$ with $L(x) \neq L(y)$. Now define $B := \{z \in A : L(z) = \max L(A)\}$. Since A is compact and L is continuous, $B \neq \emptyset$, and since L is linear, B is convex. (Verify!) Moreover, thanks to the continuity of L, B is closed and satisfies Property $(*)$. (Check!) That is, $B \in \mathcal{A}$ and $B \subseteq A$. By definition of A, then, we must have $A = B$. But this means that L is constant on A, contradicting $L(x) \neq L(y)$. ■

[36] The elements of $\mathcal{A}$ are called the **extremal** (or **support**) subsets of S.

[37] Only the nonemptiness of $\cap\mathcal{B}$ is not trivial here, and that follows from the fact that $\mathcal{B}$, being a loset, satisfies the finite intersection property (Example C.8).

An immediate corollary of this result is the following theorem which plays a fundamental role in the theory of linear programming. In words, it says that a continuous linear functional achieves its maximum (or minimum) over a compact subset of a normed linear space at an extreme point of that set.

THE EXTREME POINT THEOREM

Let S be a nonempty compact subset of a normed linear space X, and L a continuous linear functional on X. Then, $L(x) = \max L(S)$ for some $x \in ext(S)$.

PROOF

Define $A := \{y \in S : L(y) = \max L(S)\}$. By Weierstrass' Theorem $A \neq \emptyset$, and by continuity of L, A is a closed set. Since S is compact, A is thus a nonempty compact subset of X. By Theorem 2, therefore, there exists an x in $ext(A)$. We wish to show that $x \in ext(S)$. Indeed, if this were not the case, we could find a $0 < \lambda < 1$ and two distinct points y and z in S such that $x = \lambda y + (1 - \lambda)z$. But, by linearity of L, this is possible only if $\lambda L(y) + (1-\lambda)L(z) = \max L(S)$, which means that $L(y) = L(z) = \max L(S)$, that is $y, z \in A$, contradicting that $x \in ext(A)$. ∎

The following is a useful generalization of the Extreme Point Theorem, which applies to some nonlinear programming problems.

THE BAUER MAXIMUM PRINCIPLE

Let S be a nonempty compact and convex subset of a normed linear space X. If φ is an upper semicontinuous and convex real function on S, then $\varphi(x) = \max \varphi(S)$ for some $x \in ext(S)$.

Here is a nice little consequence of this result.

COROLLARY 1

Let O be a nonempty open and convex subset of a normed linear space and $\varphi \in \mathbb{R}^O$ a convex function. If φ is locally bounded at some $x_0 \in O$, then

$$\emptyset \neq \arg\max \{\varphi(x) : x \in S\} \subseteq ext(S)$$

for any nonempty compact and convex subset S of O.

Proof

Apply Proposition I.8 and the Bauer Maximum Principle. ■

We outline the proof of the Bauer Maximum Principle in the following exercise.

Exercise 78

(a) Adopt the notation of the Bauer Maximum Principle, define $\mathcal{A}$ as in the proof of Theorem 2, and show that $\{x \in S : f(x) = \sup f(S)\} \in \mathcal{A}$.

(b) Define $\mathcal{B} := \{B \in \mathcal{A} : B \subseteq A\}$, and proceed with Zorn's Lemma as in the proof of Theorem 2 to show that $ext(S) \cap A \neq \emptyset$ for any $A \in \mathcal{A}$.

(c) Combine parts (a) and (b) to complete the proof of the Bauer Maximum Principle.

Playing on this theme a bit more, we can get other interesting results. The following is particularly useful.

Proposition 8

Let S be a nonempty compact subset of a normed linear space X. Then, $S \subseteq \overline{co}_X(ext(S))$.

Proof

Suppose there exists a $z \in S \backslash \overline{co}_X(ext(S))$. Since $\overline{co}_X(ext(S))$ is a nonempty closed set (Theorem 2), we can apply Theorem 1 to find an $L \in X^*$ such that $L(z) > \sup L(\overline{co}_X(ext(S)))$. (Yes?) Consider the set $A := \{y \in S : L(y) = \max L(S)\}$, which is easily checked to be a nonempty compact set. Then, by Theorem 2, there exists an extreme point x of A. As in the proof of the Extreme Point Theorem, x must also be an extreme point of S. Thus $L(x) \geq L(z) > \sup L(\overline{co}_X(ext(S)))$, a contradiction. ■

Adding convexity to the picture yields the following famous theorem, which gives a deep insight into the geometric structure of compact and convex sets. It was proved by Mark Krein and David Milman in 1940.

The Krein-Milman Theorem

Every nonempty, compact and convex subset S of a normed linear space X is the closed convex hull of its extreme points, that is, $S = \overline{co}_X(ext(S))$.

PROOF
Apply the definition of $\overline{co}_X(\cdot)$ and Proposition 8.[38] ■

It is obvious that this theorem fails for nonconvex and noncompact sets. For instance, $\overline{co}_{\mathbb{R}}(ext([0,1))) = \{0\} \neq [0,1)$ and $\overline{co}_{\mathbb{R}}(ext(\{0,1\})) = [0,1] \neq \{0,1\}$. More interesting is the fact that we really need to use the *closed* convex hull for the validity of the Krein-Milman Theorem, $co(\cdot)$ will not do. For example, consider the compact subset $S := \{\mathbf{0}, \frac{1}{2}\mathbf{e}^1, \frac{1}{4}\mathbf{e}^2, \frac{1}{8}\mathbf{e}^3, \ldots\}$ of ℓ^∞, where $\mathbf{e}^1 := (1,0,0,\ldots)$, $\mathbf{e}^2 := (0,1,0,\ldots)$, etc., and let $A := \overline{co}_{\ell^\infty}(S)$. While A is obviously convex, it is also compact by Mazur's Compactness Theorem. But $ext(A) = S$, so we cannot have $co(ext(A)) = A$. (Indeed, $co(ext(A))$ equals $co(S)$, which is not compact (not even closed; recall Example I.7), whereas A is compact.)

WARNING. If X is a Euclidean space, then we do not need to use the *closed* convex hulls in the Krein-Milman Theorem. That is, every compact and convex subset of a Euclidean space is the convex hull of its extreme points. (Why?)

We conclude this section by noting the following companion to the Krein-Milman Theorem. It often goes by the name *Milman's Converse.*

MILMAN'S CONVERSE TO THE KREIN-MILMAN THEOREM
Let S be a nonempty compact subset of a normed linear space X. If $\overline{co}_X(S)$ is compact, then $ext(\overline{co}_X(S)) \subseteq S$.

EXERCISE 79 Prove Milman's Converse to the Krein-Milman Theorem.

EXERCISE 80[H] Let X and Y be two normed linear spaces, and $L \in \mathcal{B}(X,Y)$. Show that if S is a nonempty compact and convex set in X, then $ext(L(S)) \subseteq cl_Y(L(ext(S)))$.

*EXERCISE 81[H] (Ok) Let X be a partially ordered linear space that is at the same time a Banach space. Let $L \in X^X$ be a positive linear operator such that $\|L\|^* = 1$, and denote the kth iteration of L by L^k, $k \in \mathbb{N}$. Show that if L is a compact operator (Exercise I.38) and $\liminf \|L^k(x)\| > 0$ for some $x \in X_+$, then L has a nonzero fixed point in X_+.

[38] Just in case you are curious about this, let me tell you that the Krein-Milman Theorem can*not* be proved without using the Axiom of Choice.

6 Extension in Normed Linear Spaces

6.1 Extension of Continuous Linear Functionals

We now turn to the problem of extending a continuous linear functional defined on a subspace linearly *and continuously* to a functional defined on the mother normed linear space. This problem is settled by a straightforward application of the Hahn-Banach Extension Theorem 1. In fact, not only that every continuous (equivalently, bounded) linear functional defined on a subspace of a normed linear space can be extended to a continuous linear functional defined on the entire space, but it is also true that we can do this in a way that preserves the norm of the functional.

THE HAHN-BANACH EXTENSION THEOREM 2
Let X be a normed linear space and Y a subspace of X. For any given continuous linear functional L_0 on Y, there exists a continuous linear functional L on X such that

$$L|_Y = L_0 \qquad \textit{and} \qquad \|L\|^* = \|L_0\|^*.$$

PROOF
By Proposition 5, L_0 is bounded on Y. Thus, (13) implies $\left|L_0(y)\right| \leq \|L_0\|^* \left\|y\right\|$ for each $y \in Y$. Define $\varphi \in \mathbb{R}^X$ by $\varphi(x) := \|L_0\|^* \|x\|$, and check that φ is a seminorm on X. We may then apply the Hahn-Banach Extension Theorem 1 to find a linear functional L on X with $L|_Y = L_0$ and $L \leq \varphi$. From the latter condition it follows that $L(x) \leq \|L_0\|^* \|x\|$ and $-L(x) = L(-x) \leq \|L_0\|^* \|-x\| = \|L_0\|^* \|x\|$, and hence $|L(x)| \leq \|L_0\|^* \|x\|$ for each $x \in X$. Thus L is bounded, hence continuous, and by (13), $\|L\|^* \leq \|L_0\|^*$. Moreover,

$$\begin{aligned}\|L\|^* &= \sup\{|L(x)| : x \in X \text{ and } \|x\| \leq 1\} \\ &\geq \sup\{|L(x)| : x \in Y \text{ and } \|x\| \leq 1\} \\ &= \|L_0\|^*,\end{aligned}$$

and we are done. ∎

We consider several applications of this theorem in the following set of exercises.

EXERCISE 82 [H] Show that, for any normed linear space X and $x \in X$, there exists an $L \in X^*$ such that $\|L\|^* = 1$ and $L(x) = \|x\|$.

EXERCISE 83 [H] Let X be a preordered normed linear space, Y a subspace of X, and L_0 a positive linear functional on Y. Prove: If there exists an $x \in Y$ such that $x \succsim_X y$ for all $y \in B_Y$, then L_0 can be extended to a continuous linear functional on X.

EXERCISE 84 [H] (*Duality Theorem*) Show that, for any normed linear space X and $x \in X$,

$$\|x\| = \max\{|L(x)| : L \in X^* \text{ and } \|L\|^* \leq 1\}$$

EXERCISE 85 Prove: If Y is a subspace of a normed linear space X,

$$cl_X(Y) = \bigcap\{null(L) : L \in X^* \text{ and } Y \subseteq null(L)\}.$$

EXERCISE 86 [H] Prove: If Y is a closed proper subspace of the normed linear space X, there exists an $L \in X^*$ such that $\|L\|^* = 1$, $L|_Y = 0$ and $L(x) \neq 0$ for all $x \in X \setminus Y$.

EXERCISE 87 (*Banach Limits*) Define $Z := \{(x_1, x_2 - x_1, x_3 - x_2, \ldots) : (x_m) \in \ell^\infty\}$ and $Y := cl_{\ell^\infty}(span(Z))$.

(a) Show that $(1, 1, \ldots) \notin Y$.

(b) Show that there is a continuous linear functional L on ℓ^∞ such that $\|L\|^* = 1$, $L(1, 1, \ldots) = 1$, and $L|_Y = 0$.

(c) The linear functional L found in part (b) is called a **Banach Limit** and is sometimes used to give information about the asymptotic behavior of divergent bounded sequences. Show that this functional has the following properties: For all $(x_m) \in \ell^\infty$,

   (i) $L((x_m)) = L((x_k, x_{k+1}, \ldots))$ for any $k = 1, 2, \ldots$;

   (ii) $\liminf x_m \leq L((x_m)) \leq \limsup x_m$;

   (iii) if (x_m) converges, then $L((x_m)) = \lim x_m$; and

   (iv) if $x_m \geq 0$ for each m, then $L((x_m)) \geq 0$.

EXERCISE 88 Let X and Y be two normed linear spaces with Y being Banach, and let Z be a dense subspace of X. Show that any continuous linear operator $L_0 \in \mathcal{B}(Z, Y)$ can be *uniquely* extended to a continuous linear operator $L \in \mathcal{B}(X, Y)$. Check that this extension satisfies $\|L\|^* = \|L_0\|^*$.

Exercise 89[H] (*The Taylor-Foguel Theorem*) Prove: For any normed linear space X, if X^* is rotund (Exercise 7), then every continuous linear functional on any subspace of X has a *unique* norm-preserving continuous linear extension. (*Note*. The converse is also true, but you don't have to prove that.)

6.2* Infinite-Dimensional Normed Linear Spaces

Within the body of this chapter and the previous one we have obtained several characterizations of infinite-dimensional normed (or metric) linear spaces. As an application of the Hahn-Banach Extension Theorem 2—more precisely, of Exercise 86—we now derive yet another characterization of such linear spaces.

In Example 7, we saw that the dual space of $\mathbb{R}^n$ is (linearly isometric to) itself. By Theorem I.1, therefore, we may conclude that the dual of any finite-dimensional normed linear space is finite-dimensional. The question that we ask now is whether it is possible that the dual of an infinite-dimensional normed linear space is also finite-dimensional. As you might suspect, the answer is no, that is,

$$dim(X) = dim(X^*).$$

The argument is as follows. Let X be an infinite-dimensional normed linear space so that we may find countably infinitely many linearly independent vectors $x^1, x^2, \ldots$ in X. For each $m \in \mathbb{N}$, define

$$Y_m := span(\{x^1, \ldots, x^m\}).$$

Clearly, $x^{m+1} \notin Y_m$ for each m. Moreover, each Y_m is a finite-dimensional subspace of X, and thus is closed (Corollary I.3). By Exercise 86, for each m, we can find a nonzero $L_m \in X^*$ such that $L_m|_{Y_m} = 0$ for all $y \in Y_m$, and $L_m(x) \neq 0$ for all $x \in X \setminus Y_m$. In particular, the sequence (L_m) satisfies:

$$L_m(x^i) = 0 \text{ for all } i = 1, \ldots, m,$$

and

$$L_m(x^i) \neq 0 \text{ for all } i = m+1, m+2, \ldots$$

Now pick any $k \in \mathbb{N}$, and let $\sum^k \lambda_i L_i = 0$, where $\lambda_1, \ldots, \lambda_k$ are real numbers. We have

$$\left(\sum_{i=1}^{k} \lambda_i L_i\right)(x^2) = \lambda_1 L_1(x^2) = 0$$

so that $\lambda_1 = 0$. In turn this implies that

$$\left(\sum_{i=1}^{k} \lambda_i L_i\right)(x^3) = \lambda_2 L_2(x^3) = 0$$

so that $\lambda_2 = 0$. Proceeding inductively, we then find that $\lambda_1 = \cdots = \lambda_k = 0$. Thus $\{L_1, \ldots, L_k\}$ is linearly independent in X^* for any $k \in \mathbb{N}$. But this implies that $\{L_1, L_2, \ldots\}$ is linearly independent in X^*. Conclusion: X^* must be infinite-dimensional.

This discussion culminates in our final characterization of infinite-dimensional normed linear spaces: *A normed linear space is infinite-dimensional iff its dual is infinite-dimensional.*

For bookkeeping purposes we conclude this section by summarizing all of the characterizations of infinite-dimensional normed linear spaces obtained in this book. This summary is a pretty good indication of what we were able to accomplish in our elementary introduction to linear functional analysis.

CHARACTERIZATION OF INFINITE-DIMENSIONAL NORMED LINEAR SPACES

For any normed linear space X, the following statements are equivalent:

(a) *X is infinite-dimensional.*

(b) *X is not isomorphic to any Euclidean space.*

(c) *The algebraic interior of the convex hull of any basis and the origin of X is empty.*

(d) *X contains a convex set S and a vector $x \notin S$ such that x and S cannot be separated by a hyperplane.*

(e) *X is not locally compact, that is, the closure of no open subset of X is compact.*

(f) *$\mathcal{L}(X, \mathbb{R}) \neq X^*$, that is, there exists a discontinuous linear functional on X.*

(g) *X^* is infinite-dimensional.*

Here the equivalence of (a) and (b) is trivial. We owe the equivalence of (a) and (c) to Example G.8, that of (a) and (d) to Remark G.2, and that of (a) and (e) to Theorem I.2. Finally, the equivalence of (a) and (f) was derived in Section 4.5, while we have just proved the equivalence of (a) and (g) in the present section.

It is worth mentioning that the first three of these characterizations are purely algebraic in nature (so all we need for them to be valid is that X be a linear space). The fourth one is valid in any metric linear space. The final two, on the other hand, are obtained in the context of normed linear spaces (precisely because we have defined here the notion of *dual space* only for such spaces).

7* The Uniform Boundedness Principle

We noted earlier that Banach spaces provide a rich environment in which linear analysis can be taken to the next level. Yet our treatment so far falls short of demonstrating this in a transparent way; we were able only to scratch the surface of Banach space theory in this chapter. Although we cannot offer a full remedy for this situation here, the present account would be unduly incomplete if we did not provide at least a glimpse of the sorts of things that can be accomplished using Banach space techniques. This final section of the chapter, therefore, is devoted to the investigation of a fundamental result in Banach space theory, the so-called *Uniform Boundedness Principle* (also known as the *Banach-Steinhaus Theorem*). As we shall see, our earlier work puts us in a position to provide a very simple proof for this important theorem.[39]

We have seen in Section D.6 that the pointwise limit of a sequence of continuous functions need not be continuous. It turns out that linearity would overcome this difficulty. Indeed, we show below that the pointwise limit of a sequence of continuous linear functionals is itself a continuous linear functional. This is a surprisingly far-reaching result with many interesting applications. It is also an immediate corollary of the following famous theorem, which is a special case of a result proved by Stefan Banach and Hugo Steinhaus in 1927.

[39] If you want to go deeper into matters related to Banach spaces, I recommend that you begin with Kreyzig (1978) or Maddox (1988), and then move to a more comprehensive functional analysis text such as Megginson (1998), or perhaps better, jump right to Aliprantis and Border (1999).

The Uniform Boundedness Principle

(Banach-Steinhaus) *Let X be a Banach space and Y a normed linear space. For any nonempty subset $\mathcal{L}$ of $\mathcal{B}(X,Y)$, if*

$$\sup\{\|L(x)\|_Y : L \in \mathcal{L}\} < \infty \quad \textit{for all } x \in X, \tag{15}$$

then $\sup\{\|L\|^* : L \in \mathcal{L}\} < \infty$.

Let $\mathcal{L}$ be the collection mentioned in the theorem. Since each member L of $\mathcal{L}$ is continuous, and thus bounded, we know that there exists a constant $M_L > 0$ such that $\|L(x)\|_Y \leq M_L \|x\|$ for all $x \in X$. (Of course, $M_L = \|L\|^*$ is the smallest such bound.) The question is whether we can find a *uniform* bound that will work for all $L \in \mathcal{L}$, that is, we can find a real number $M > 0$ such that $\|L(x)\|_Y \leq M \|x\|$ for all $x \in X$ *and all* $L \in \mathcal{L}$. This is not at all a trivial question. In fact, the answer may well be no if the underlying normed linear space X is not Banach. (See Exercise 90 below.) It is a remarkable fact that the answer is affirmative, provided that X is a complete normed linear space. Let us now see why this is the case.

Observe that, for each $L \in \mathcal{L}$, we can compute $\|L\|^*$ by using the values of L on *any* ε-neighborhood of the origin:

$$\begin{aligned}\|L\|^* &= \sup\{\|L(x)\|_Y : x \in B_X\}\\ &= \sup\left\{\left\|L\left(\tfrac{1}{\varepsilon}x\right)\right\|_Y : x \in N_{\varepsilon,X}(\mathbf{0})\right\}\\ &= \tfrac{1}{\varepsilon}\sup\{\|L(x)\|_Y : x \in N_{\varepsilon,X}(\mathbf{0})\}.\end{aligned}$$

So what we need is a uniform bound on $\sup\{\|L(x)\|_Y : x \in N_{\varepsilon,X}(\mathbf{0})\}$ for *some* $\varepsilon > 0$, or more generally, a uniform bound on $\sup\{|L(x)| : x \in S\}$ for *some* subset S of X that contains the origin $\mathbf{0}$ in its interior. Then why don't we try an S for which we can find such a uniform bound *by definition*, and check if $\mathbf{0} \in int_X(S)$? This may look like a long shot, but it works out beautifully here.[40]

Proof of the Uniform Boundedness Principle

Let $\mathcal{L}$ be a nonempty subset of $\mathcal{B}(X,Y)$ which satisfies (15). Define

$$S := \{x \in X : \|L(x)\|_Y \leq 1 \text{ for all } L \in \mathcal{L}\},$$

[40] There are many other ways of proving the Uniform Boundedness Principle. See, in particular, Hennefeld (1980), who provides a self-contained proof.

which is readily checked to be a closed and convex subset of X.[41] We have $\mathbf{0} \in al\text{-}int_X(S)$ since, for any $y \in X$, $\lambda y \in S$ for all $0 < \lambda < 1/\sup\{\|L(y)\|_Y : L \in \mathcal{L}\}$. Given that X is complete; therefore, Theorem I.3 yields $\mathbf{0} \in int_X(S)$, that is, there exists an $\varepsilon > 0$ such that $N_{\varepsilon,X}(\mathbf{0}) \subset S$. Thus

$$\|L\|^* = \tfrac{1}{\varepsilon}\sup\{\|L(x)\|_Y : x \in N_{\varepsilon,X}(\mathbf{0})\} \leq \tfrac{1}{\varepsilon}\sup\{\|L(x)\|_Y : x \in S\} \leq \tfrac{1}{\varepsilon}$$

for all $L \in \mathcal{L}$, and we are done. ■

Completeness of X is crucial in our argument. After all, $\mathbf{0} \in al\text{-}int_X(S)$ and $int_X(S) = \emptyset$ may well both be true for a closed convex subset S of an incomplete normed linear space (Example I.11). In fact, Example I.11 is easily modified to show that the Uniform Boundedness Principle need not hold in an arbitrary normed linear space.

Exercise 90 Let X be the set of all real sequences (x_m) with $\sum^\infty |x_m| < \infty$, and make this set a normed linear space by using the sup-norm.

(a) Find a sequence (L_k) in X^* and an unbounded linear functional L on X such that $\lim_{k\to\infty} L_k((x_m)) = L((x_m))$ for all $(x_m) \in X$.

(b) Show that the Uniform Boundedness Principle is not valid on X.

Here is the pointwise convergence result that we promised in the beginning of this section.

Corollary 2

Let X be a Banach space, Y a normed linear space, and (L_m) a sequence in $\mathcal{B}(X,Y)$ such that $\lim L_m(x) \in Y$ for each $x \in X$. Then, $\lim L_m$ is a continuous linear operator from X into Y.

Proof

Define $L \in Y^X$ by $L(x) := \lim L_m(x)$. This function is linear since, for all $x, y \in X$ and $\lambda \in \mathbb{R}$, we have

$$L(\lambda x + y) = \lim L_m(\lambda x + y) = \lambda \lim L_m(x) + \lim L_m(y) = \lambda L(x) + L(y).$$

[41] We work here with a closed and convex S, for then, by Theorem I.3, it is enough to verify that $\mathbf{0} \in al\text{-}int_X(S)$. As you know by now, this is an easier statement to check than $\mathbf{0} \in int_X(S)$.

All we need to do, then, is establish the boundedness of L. Observe that, since $\|\cdot\|_Y$ is a continuous map (Proposition 1), we have $\lim \|L_m(x)\|_Y = \|L(x)\|_Y$ for any $x \in X$. Since every convergent real sequence is bounded, it follows that $\sup\{\|L_m(x)\|_Y : m \in \mathbb{N}\} < \infty$ for each $x \in X$. So, by the Uniform Boundedness Principle, $M := \sup\{\|L_m\|^* : m \in \mathbb{N}\} < \infty$, and hence,

$$\|L(x)\|_Y \leq \sup\{\|L_m(x)\|_Y : m \in \mathbb{N}\} \leq \sup\{\|L_m\|^* \|x\| : m \in \mathbb{N}\} = M \|x\|$$

for any $x \in X$. By Proposition 5, then, L is continuous. ■

This result is actually valid in the case of metric linear spaces as well, although its proof in that case is substantially harder and thus not given here.[42]

> THE GENERALIZED UNIFORM BOUNDEDNESS PRINCIPLE
> (Mazur-Orlicz) *Let X be a Fréchet space, Y a normed linear space, and (L_m) a sequence of continuous linear operators in $\mathcal{L}(X, Y)$ such that* $\lim L_m(x) \in Y$ *for each $x \in X$. Then* $\lim L_m$ *is a continuous linear operator from X into Y.*

We next consider some concrete examples that illustrate some of the typical ways in which the Uniform Boundedness Principle and Corollary 2 are utilized in practice.

EXAMPLE 9
Let (g_m) be a convergent sequence in $\mathbf{C}[0,1]$, and define $L_m \in \mathbb{R}^{\mathbf{C}[0,1]}$ by

$$L_m(f) := \int_0^1 f(t)g_m(t)dt.$$

For any $f \in \mathbf{C}[0,1]$, one can check that $(L_m(f))$ is a Cauchy sequence in $\mathbb{R}$, and thus it converges. By Corollary 2, then, $L := \lim L_m$ is a continuous linear functional on $\mathbf{C}[0,1]$. (Indeed, we have $L(f) = \int_0^1 f(t)\left(\lim g_m\right)(t)dt$ for all $f \in \mathbf{C}[0,1]$. Why?) □

EXAMPLE 10
Let (a_m) be any real sequence. Suppose we know that the series $\sum^\infty a_i x_i$ converges *for all* $(x_m) \in \ell^1$. Can we say anything interesting about (a_m)?

[42] See Rolewicz (1985, pp. 39–41).

Yes, we can: (a_m) must be a bounded sequence! As you will see presently, this is an almost immediate implication of the Uniform Boundedness Principle.

Define $L_k : \ell^1 \to \mathbb{R}$ by

$$L_k((x_m)) := \sum_{i=1}^{k} a_i x_i.$$

Each L_k is obviously linear. Letting $M_k := \max\{|a_1|, \ldots, |a_k|\}$, we get

$$|L_k((x_m))| \leq M_k \sum_{i=1}^{k} |x_i| \leq M_k \|(x_m)\|_1$$

so that L_k is a bounded, hence continuous, linear functional on ℓ^1. By hypothesis, $L := \lim L_k$ is well-defined. Then, by Corollary 2, L is a continuous linear functional on ℓ^1, which means that $\|L\|^*$ is a real number. Thus, letting $\mathbf{e}^1 := (1, 0, 0, \ldots)$, $\mathbf{e}^2 := (0, 1, 0, \ldots)$, etc., we have $|a_i| = \left|L(\mathbf{e}^i)\right| \leq \|L\|^* \left\|\mathbf{e}^i\right\|_1 = \|L\|^*$ for each i. Conclusion: $\sup\{|a_i| : i \in \mathbb{N}\} < \infty$. □

EXERCISE 91 Let X and Y be Banach spaces. Show that $\mathcal{B}(X, Y)$ is **strongly complete** in the following sense: For any sequence (L_m) in $\mathcal{B}(X, Y)$ such that $(L_m(x))$ is Cauchy (in Y) for every $x \in X$, there is an $L \in \mathcal{B}(X, Y)$ such that $L_m(x) \to L(x)$ for every $x \in X$.

EXERCISE 92^{H} Let X be a normed linear space and $\emptyset \neq S \subseteq X$. If, for each $L \in X^*$, there exists a number M_L such that $\sup\{|L(x)| : x \in S\} \leq M_L$, then S must be a bounded set. Prove!

EXERCISE 93

(a) Let (a_m) be a real sequence. Prove that if $\sum^{\infty} a_i x_i$ converges for every convergent real sequence (x_m), then $(a_m) \in \ell^1$.

(b) Show that there is a linear isometry between the dual of $\mathbf{c}$ and ℓ^1.[43] We thus say that the dual of this space "is" ℓ^1.

[43] *Reminder.* $\mathbf{c}$ is the linear space of all convergent real sequences; it is normed by the sup-norm.

Chapter K

Differential Calculus

In the second half of this book, starting with Chapter F, we have worked on developing a thorough understanding of function spaces, be it from a geometric or an analytic viewpoint. This work allows us to move in a variety of directions. In particular, we can now extend the methods of classical differential calculus into the realm of maps defined on suitable function spaces, or more generally on normed linear spaces. In turn, this "generalized" calculus can be used to develop a theory of optimization in which the choice objects need not be real n-vectors but members of an arbitrary normed linear space (as in the calculus of variations, control theory, or dynamic programming). This task is undertaken in this chapter.

We begin with a quick review on the notion of derivative (of a real-to-real function), pointing out the advantages of viewing derivatives as linear functionals rather than numbers. Once this viewpoint is clear, it is straightforward to extend the idea of derivative to the context of functions whose domains and codomains lie within arbitrary normed linear spaces. Moreover, the resulting derivative concept, called the Fréchet derivative, inherits many properties of the derivative that you are familiar with from classical calculus. We study this concept in detail here, go through several examples, and extend to this realm some well-known results of calculus, such as the Chain Rule, the Mean Value Theorem, and so on. Keeping an eye on optimization theoretic applications, we also revisit the theory of concave functions, this time making use of Fréchet derivatives.[1]

The use of this work is demonstrated by means of a brief introduction to infinite-dimensional optimization theory. Here we see how one can easily

[1] For reasons that largely escape me, most texts on functional analysis do not cover differential calculus on normed linear spaces. A thorough treatment of this topic can be found in Dieudonné (1969), but you may find that exposition a bit heavy. Some texts on optimization theory (such as Luenberger (1969)) do contain a discussion of Fréchet differentiation, but they rarely develop the theory to the extent that we do here. The best reference I know of on differential calculus on normed linear spaces is a little book by Cartan (1971) that unfortunately is out of print.

extend first- and second-order conditions for local extremum of real functions on the line to the broader context of real maps on normed linear spaces. We also show how useful concave functions are in this context as well. As for an application, and as our final order of business here, we sketch a quick but rigorous introduction to the calculus of variations and consider a few of its economic applications.[2]

1 Fréchet Differentiation

1.1 Limits of Functions and Tangency

So far in this book we have worked almost exclusively with limits of sequences. It will be convenient to depart from this practice in this chapter and work instead with limits of functions. The basic idea is a straightforward generalization of one that you are familiar with from calculus (Section A.4.2).

Let T be a nonempty subset of a normed linear space X (whose norm is $\|\cdot\|$) and Φ a function that maps T into a normed linear space Y (whose norm is $\|\cdot\|_Y$). Let $x \in X$ be the limit of at least one sequence in $T\backslash\{x\}$.[3]

A point $y \in Y$ is said to be the **limit** of Φ at x, in which case we say "$\Phi(\omega)$ *approaches to* y *as* $\omega \to x$," provided that $\Phi(x^m) \to y$ holds for every sequence (x^m) in $T\backslash\{x\}$ with $x^m \to x$. This situation is denoted as

$$\lim_{\omega \to x} \Phi(\omega) = y.$$

Clearly, we have $\lim_{\omega \to x} \Phi(\omega) = y$ iff, for each $\varepsilon > 0$, there exists a $\delta > 0$ such that $\|y - \Phi(\omega)\|_Y < \varepsilon$ for all $\omega \in T\backslash\{x\}$ with $\|\omega - x\| < \delta$. (Yes?) Thus,

$$\lim_{\omega \to x} \Phi(\omega) = y \quad \text{iff} \quad \lim_{\omega \to x} \left\|\Phi(\omega) - y\right\|_Y = 0.$$

So, if Φ is defined on an open neighborhood of x, then it is continuous at x iff $\lim_{\omega \to x} \Phi(\omega) = \Phi(x)$.

If the limits of $\Phi_1, \Phi_2 \in Y^T$ at x exist, then we have

$$\lim_{\omega \to x} (\alpha\Phi_1(\omega) + \Phi_2(\omega)) = \alpha \lim_{\omega \to x} \Phi_1(\omega) + \lim_{\omega \to x} \Phi_2(\omega)$$

[2] Because of space constraints, I don't go into control theory here, even though this is among the standard methods of dynamic economic analysis in continuous time. However, the machinery developed here can also be used to go deep into the theory of constrained optimization over Banach spaces, from which point control theory is within a stone's throw.

[3] Of course, if $x \in int_X(T)$, this condition is automatically satisfied. (Yes?)

for any $\alpha \in \mathbb{R}$. If $Y = \mathbb{R}$, and $\lim_{\omega \to x} \Phi_1(\omega)$ and $\lim_{\omega \to x} \Phi_2(\omega)$ are real numbers, then we also have

$$\lim_{\omega \to x} \Phi_1(\omega)\Phi_2(\omega) = \lim_{\omega \to x} \Phi_1(\omega) \lim_{\omega \to x} \Phi_2(\omega). \tag{1}$$

The proofs of these results are identical to those you have seen in calculus. We use them in what follows as a matter of routine.

Now, given any two maps $\Phi_1, \Phi_2 \in Y^T$ that are continuous at x, we say that Φ_1 and Φ_2 are **tangent** at x if

$$\lim_{\omega \to x} \frac{\Phi_1(\omega) - \Phi_2(\omega)}{\|\omega - x\|} = \mathbf{0}.$$

So, if Φ_1 and Φ_2 are tangent at x, then, not only is $\Phi_1(x) = \Phi_2(x)$, but also, as $\omega \to x$, the distance between the values of these functions (i.e., $\|\Phi_1(\omega) - \Phi_2(\omega)\|_Y$) converges to 0 "faster" than ω approaches to x. Put another way, *near* x, the values of Φ_2 approximate $\Phi_1(x)$ better than ω approximates x from the same distance. In this sense, we can think of Φ_2 as a "best approximation" of Φ_1 *near* x.

1.2 What Is a Derivative?

In calculus, one is taught that the derivative of a function $f : \mathbb{R} \to \mathbb{R}$ at a point x is a number that describes the rate of instantaneous change of the value of f as x changes. This way of looking at things, while useful for certain applications, falls short of reflecting the intimate connection between the notion of the derivative of f at x and "the line that best approximates f near x." We begin our discussion by recalling this interpretation.[4]

Let O be an open subset of $\mathbb{R}$, $x \in O$, and $f \in \mathbb{R}^O$. Suppose f is differentiable at x, that is, there is a real number $f'(x)$ with

$$f'(x) = \lim_{t \to x} \frac{f(t) - f(x)}{t - x}. \tag{2}$$

[4] "In the classical teaching of calculus, this idea is immediately obscured by the accidental fact that, on a one-dimensional vector space, there is a one-to-one correspondence between linear functionals and numbers, and therefore the derivative at a point is defined as a *number* instead of a *linear functional.* This slavish subservience to the shibboleth of numerical interpretation at any cost becomes much worse when dealing with functions of several variables ...." Dieudonné (1968, p. 147).

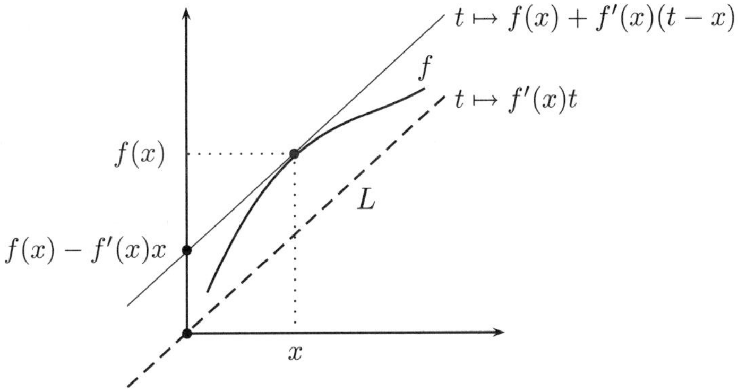

FIGURE 1

Then the affine map $t \mapsto f(x) + f'(x)(t-x)$ on $\mathbb{R}$ must be tangent to f at x, that is,

$$\lim_{t \to x} \frac{f(t) - f(x) - f'(x)(t-x)}{|t-x|} = 0.$$

(See Figure 1.) Put differently, if f is differentiable at x, then there exists a linear functional L on $\mathbb{R}$, namely, $a \mapsto f'(x)a$, such that the affine map $t \mapsto f(x) + L(t-x)$ is tangent to f at x. The converse is also true. Indeed, if $L : \mathbb{R} \to \mathbb{R}$ is a linear functional such that $t \mapsto f(x) + L(t-x)$ is tangent to f at x, then there exists an $\alpha \in \mathbb{R}$ such that $L(a) = \alpha a$ for all $a \in \mathbb{R}$, and

$$\lim_{t \to x} \left(\frac{f(t) - f(x)}{t-x} - \alpha \right) = \lim_{t \to x} \frac{f(t) - f(x) - \alpha(t-x)}{t-x} = 0.$$

It follows that f is differentiable at x, and $f'(x) = \alpha$.

This elementary argument establishes that f is differentiable at x iff there is a linear functional L on $\mathbb{R}$ such that

$$\lim_{t \to x} \frac{f(t) - f(x) - L(t-x)}{|t-x|} = 0, \tag{3}$$

that is, the affine map $t \mapsto f(x) + L(t-x)$ is tangent to f at x; it approximates f around x so well that, as $t \to x$, the error $f(t) - f(x) - L(t-x)$ of this approximation decreases to 0 faster than t tends to x.

This is the key idea behind the very concept of differentiation: The local behavior of a differentiable function is linear, just like that of an affine map. Thus, it makes sense to consider the linear functional L of (3) as central to the notion of derivative of f at x. In fact, it would be more honest to refer to L itself as the "derivative" of f at x. From this point of view, the number $f'(x)$ is simply the slope of L—it is none other than a convenient way of identifying this linear functional.

It may seem that we are arguing over semantics here. What difference does it make if we instead view the linear functional $t \mapsto f'(x)t$ as the derivative of f at x, instead of the number $f'(x)$? Well, think about it. How would you define the derivative of f at x if this function were defined on an open subset O of $\mathbb{R}^2$? The classical definition (2), which formalizes the notion of rate of change, immediately runs into difficulties in this situation. But the idea of "finding an affine map which is tangent to f at x" survives with no trouble whatsoever. All we have to do is to define the derivative of f at x as the linear functional L on $\mathbb{R}^2$ with the property that

$$\lim_{t \to x} \frac{f(t) - f(x) - L(t-x)}{\|t-x\|_2} = 0.$$

Geometrically speaking, the graph of the affine map $t \mapsto f(x) + L(t-x)$ is none other than the hyperplane tangent to the graph of f at x (Figure 2).

Looking at the derivative of a function the right way saves the day in many other circumstances. Since the notion of tangency is well-defined for functions that map a normed linear space into another, this point of view

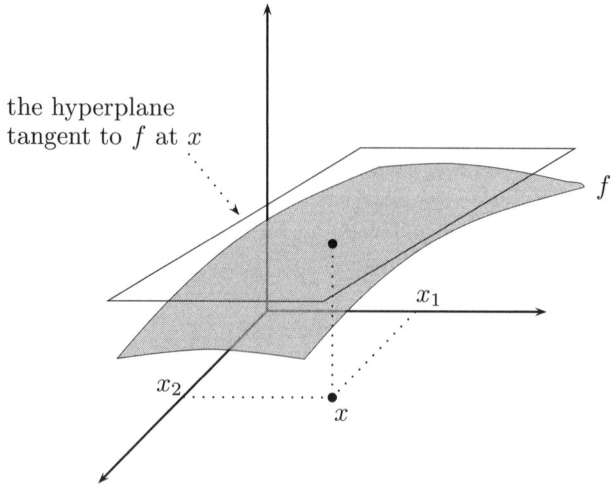

FIGURE 2

remains meaningful for any such function and paves the way toward the general theory we are about to lay out.[5]

1.3 The Fréchet Derivative

DEFINITION

Let X and Y be two normed linear spaces and T a subset of X. For any $x \in int_X(T)$, a map $\Phi : T \to Y$ is said to be **Fréchet differentiable at** x if there is a continuous linear operator $\mathsf{D}_{\Phi,x} \in \mathcal{B}(X, Y)$ such that

$$\lim_{\omega \to x} \frac{\Phi(\omega) - \Phi(x) - \mathsf{D}_{\Phi,x}(\omega - x)}{\|\omega - x\|} = \mathbf{0}. \tag{4}$$

The linear operator $\mathsf{D}_{\Phi,x}$ is called the **Fréchet derivative of Φ at** x.[6]

If O is a nonempty open subset of X with $O \subseteq T$, and if Φ is Fréchet differentiable at every $x \in O$, then we say that Φ is **Fréchet differentiable on** O. If $O = int_X(T)$ here, then Φ is said to be **Fréchet differentiable.** Finally, we say that Φ is **continuously Fréchet differentiable** if it is Fréchet differentiable and the map $\mathsf{D}_\Phi : int_X(T) \to \mathcal{B}(X, Y)$, defined by $\mathsf{D}_\Phi(x) := \mathsf{D}_{\Phi,x}$, is continuous.

The idea should be clear at this point. Just as in plain ol' calculus, we perturb $x \in int_X(T)$ infinitesimally and look at the behavior of the difference-quotient of Φ. Of course, here we can perturb x in all sorts of different ways. Indeed, since $int_X(T)$ is open in X, *any* $\omega \in X$ can be thought of as a perturbation of x, provided that $\|\omega - x\|$ is small enough. Thus, the local linear behavior of Φ at x must be captured by a linear operator defined on the entire X. The Fréchet derivative of Φ at x is, then, a linear operator $\mathsf{D}_{\Phi,x}$

[5] Part of this theory can be developed within the context of metric linear spaces as well. However, I will work exclusively with normed linear spaces in this chapter, as the applied strength of differential calculus on metric linear spaces is nowhere near that on normed linear spaces.

[6] Just to be on the safe side, let me remind you that **0** in (4) is the origin of Y. Put differently, (4) means

$$\lim_{\omega \to x} \frac{\left\|\Phi(\omega) - \Phi(x) - \mathsf{D}_{\Phi,x}(\omega - x)\right\|_Y}{\|\omega - x\|} = 0,$$

that is, for every $\varepsilon > 0$, there exists a $\delta > 0$ such that $\left\|\Phi(\omega) - \Phi(x) - \mathsf{D}_{\Phi,x}(\omega - x)\right\|_Y \leq \varepsilon \|\omega - x\|$ for each $\omega \in N_{\delta,X}(x)$.

defined on X, and one that ensures that the (globally linear) behavior of the affine map $\omega \mapsto \Phi(x) + \mathsf{D}_{\Phi,x}(\omega - x)$ approximates the (locally linear) behavior of Φ at x accurately. (See Proposition 2 below.)

Before we jump into examples, there are a few matters to clarify. First, we need to justify why we call $\mathsf{D}_{\Phi,x}$ "the" Fréchet derivative of Φ at x. How do we know that there is a *unique* $\mathsf{D}_{\Phi,x}$ in $\mathcal{B}(X, Y)$ that satisfies (4)? To settle this matter, take any two $K, L \in \mathcal{B}(X, Y)$, and suppose that (4) holds both with $\mathsf{D}_{\Phi,x} = K$ and $\mathsf{D}_{\Phi,x} = L$. We must then have

$$\lim_{\omega \to x} \frac{(K - L)(\omega - x)}{\|\omega - x\|} = \mathbf{0}.$$

(Yes?) Since $int_X(T)$ is open, this is equivalent to saying that

$$\lim_{\nu \to \mathbf{0}} \frac{(K - L)(\nu)}{\|\nu\|} = \mathbf{0}.$$

(Why?[7]) It follows that

$$\mathbf{0} = \lim_{m \to \infty} \frac{(K - L)\left(\frac{1}{m}y\right)}{\left\|\frac{1}{m}y\right\|} = \frac{(K - L)(y)}{\|y\|} \qquad \text{for all } y \in X,$$

so we have $K = L$. Conclusion: *When it exists, the Fréchet derivative of a function at any given point in the interior of its domain is unique.*

The second issue we should discuss is why we define the Fréchet derivative of a function at a point as a *continuous* linear operator. Intuitively speaking, the main reason is that the notion of tangency makes geometric sense only when the functions involved are continuous at the point of tangency. So, at least at the point that we wish to define the derivative of the function, we had better ask the linear operator we seek to be continuous. But of course, this is the same thing as asking for the continuity of that operator everywhere (why?), and hence we ask the Fréchet derivative of a function at a point to be continuous.[8] By the way, as a major side benefit of this requirement, we are able to maintain the familiar rule,

$$\text{differentiability} \implies \text{continuity}.$$

[7] Be careful here. If x did not belong to the interior of T (in X), the former equation would not imply the latter.

[8] This issue never arises in classical calculus, as any linear operator from a Euclidean space into another is, perforce, continuous.

Proposition 1
Let X and Y be two normed linear spaces, T a subset of X, and $x \in \text{int}_X(T)$. If $\Phi \in Y^T$ is Fréchet differentiable at x, then Φ is continuous at x.

Proof
Take any $\Phi \in Y^T$ such that (4) holds for some $\mathsf{D}_{\Phi,x} \in \mathcal{B}(X, Y)$. Then

$$\lim_{\omega \to x} (\Phi(\omega) - \Phi(x) - \mathsf{D}_{\Phi,x}(\omega - x)) = \mathbf{0}.$$

Since $\mathsf{D}_{\Phi,x}$ is continuous, we have $\lim_{\omega \to x} \mathsf{D}_{\Phi,x}(\omega - x) = \mathsf{D}_{\Phi,x}(\mathbf{0}) = \mathbf{0}$. It follows that $\lim_{\omega \to x} \Phi(\omega) = \Phi(x)$, that is, Φ is continuous at x.[9] ■

Finally, let us mention two alternative formulations of the definition of the Fréchet derivative at a given point x in the interior of the domain of a map $\Phi \in Y^T$. Immediate in this regard is the observation that, by changing variables, we can write (4) equivalently as

$$\lim_{\nu \to \mathbf{0}} \frac{\Phi(x + \nu) - \Phi(x) - \mathsf{D}_{\Phi,x}(\nu)}{\|\nu\|} = \mathbf{0}. \tag{5}$$

(Yes?) When convenient, this formulation can be used instead of (4).

Our second reformulation stems from the fact that, just as in the case of one-variable calculus, the derivative notion that we consider here corresponds to a *best local approximation* of a given function. This is not entirely trivial, so we state it in precise terms.

Proposition 2
Let X and Y be two normed linear spaces, T a subset of X, $x \in \text{int}_X(T)$, and $L \in \mathcal{B}(X, Y)$. For any $\Phi \in Y^T$, L is the Fréchet derivative of Φ at x if, and only if, there exists a continuous map $\mathbf{e} \in Y^T$ such that

$$\Phi(\omega) = \Phi(x) + L(\omega - x) + \mathbf{e}(\omega) \quad \textit{for all } \omega \in \text{int}_X(T),$$

$$\textit{and} \quad \lim_{\omega \to x} \frac{\mathbf{e}(\omega)}{\|\omega - x\|} = \mathbf{0}.$$

[9] Just so that we're on safe grounds here, let me note that I got $\lim_{\omega \to x} \Phi(\omega) = \Phi(x)$ from $\lim_{\omega \to x} (\Phi(\omega) - \Phi(x)) = \lim_{\omega \to x}(\Phi(\omega) - \Phi(x) - \mathsf{D}_{\Phi,x}(\omega - x)) + \lim_{\omega \to x} \mathsf{D}_{\Phi,x}(\omega - x) = \mathbf{0}$.

So, if $\Phi \in Y^T$ is Fréchet differentiable at $x \in int_X(T)$, then the affine map $\omega \mapsto \Phi(x)+\mathsf{D}_{\Phi,x}(\omega-x)$ is a best approximation of Φ at x in the sense that, as $\omega \to x$, the "size of the error" involved in this approximation, that is, $\|\mathbf{e}(\omega)\|_Y$, vanishes faster than $\|\omega - x\|$ goes to zero.

EXERCISE 1 Prove Proposition 2.

EXERCISE 2 For any two normed linear spaces X and Y, show that $\mathsf{D}_{L,x} = L$ for any $L \in \mathcal{B}(X, Y)$ and $x \in X$. (That is, the Fréchet derivative of a linear operator at any point in its domain equals the linear operator itself. Wouldn't you expect this?)

EXERCISE 3 Let X and Y be two normed linear spaces and $\varphi : X \times Y \to \mathbb{R}$ a continuous bilinear functional (Exercise J.52). Show that φ is Fréchet differentiable, and, for any $(x, y) \in X \times Y$,

$$\mathsf{D}_{\varphi,(x,y)}(z, w) = \varphi(z, y) + \varphi(x, w).$$

EXERCISE 4 Let X and Y be two normed linear spaces, O a nonempty open subset of X, and $\Phi \in Y^O$ a Fréchet differentiable map. Show that Φ would remain Fréchet differentiable if we replaced the norms of X and Y by equivalent norms, respectively. (Recall Section J.4.2.) Moreover, the Fréchet derivative of Φ would be the same in both cases.

EXERCISE 5 (*The Gateaux Derivative*) Let X and Y be two normed linear spaces and $x \in X$. The map $\Phi \in Y^X$ is said to be **Gateaux differentiable at** x if there exists an $L \in \mathcal{B}(X, Y)$ such that

$$\lim_{t\to 0} \left\| \tfrac{1}{t}(\Phi(x + ty) - \Phi(x)) - L(y)\right\|_Y = 0 \qquad \text{for all } y \in X.$$

Here L is called the **Gateaux derivative of** Φ **at** x. (The idea is the generalization of that behind the notion of directional derivatives.)

(a) Show that, when it exists, the Gateaux derivative of Φ at x is unique.
(b) Prove: If Φ is Fréchet differentiable at x, then it is Gateaux differentiable at x, and its Gateaux derivative at x equals $\mathsf{D}_{\Phi,x}$.
(c) Define $\varphi : \mathbb{R}^2 \to \mathbb{R}$ by $\varphi(\mathbf{0}) := 0$ and $\varphi(a, b) := \frac{a^3 b}{a^4+b^2}$ for all $(a, b) \neq \mathbf{0}$. Show that φ is Gateaux differentiable at $\mathbf{0}$ but it is not Fréchet differentiable there.

*EXERCISE 6^H (Vainberg) Let X and Y be two normed linear spaces and $\Phi \in Y^X$ a continuous function that maps compact subsets of X to relatively compact subsets of Y. Prove that if Φ is Fréchet differentiable at $x \in X$, then $\mathsf{D}_{\Phi,x}$ is a compact operator (Exercise I.38).

1.4 Examples

Our first set of examples aims to clarify in what sense the notion of Fréchet derivative relates to the various differentiation concepts you have seen in your earlier calculus courses.

Example 1

[1] Let I be an open interval and $x \in I$. The Fréchet derivative of a differentiable function $f \in \mathbb{R}^I$ at x is the linear map $t \mapsto f'(x)t$ on $\mathbb{R}$, that is,

$$\mathsf{D}_{f,x}(t) = f'(x)t \qquad \text{for all } t \in \mathbb{R}.$$

Thus the Fréchet derivative of f at x is exactly what we argued in Section 1.2 the "derivative of f at x" should mean. The number $f'(x)$ serves only to identify the linear map a certain shift of which gives us a best approximation of f near x.

We observe here that any differentiable function $f \in \mathbb{R}^I$ is Fréchet differentiable, and $\mathsf{D}_f : I \to \mathcal{B}(\mathbb{R}, \mathbb{R})$ satisfies

$$\mathsf{D}_f(x)(t) = f'(x)t \qquad \text{for all } x \in I \text{ and } t \in \mathbb{R}.$$

[2] Take any $m \in \mathbb{N}$ and any open interval I. If $\Phi : I \to \mathbb{R}^m$ is Fréchet differentiable at $x \in I$, then $\mathsf{D}_{\Phi,x}$ is a linear operator from $\mathbb{R}$ into $\mathbb{R}^m$ given by

$$\mathsf{D}_{\Phi,x}(t) = (\Phi_1'(x)t, \ldots, \Phi_m'(x)t) \qquad \text{for all } t \in \mathbb{R},$$

where $\Phi_i \in \mathbb{R}^I$ is the ith component map of Φ, $i = 1, \ldots, m$. This is a special case of a result we shall prove shortly.

Reminder. Given any $n \in \mathbb{N}$, let S be a subset of $\mathbb{R}^n$ with nonempty interior. Where $\mathbf{e}^j$ denotes the jth unit vector in $\mathbb{R}^n$, the **jth partial derivative of $\varphi \in \mathbb{R}^S$ at $x \in int_{\mathbb{R}^n}(S)$** is defined as

$$\partial_j\varphi(x) := \lim_{\varepsilon \to 0} \left| \frac{\varphi(x + \varepsilon\mathbf{e}^j) - \varphi(x)}{\varepsilon} \right|, \qquad j = 1, \ldots, n,$$

provided that the involved limit is well-defined. If $\partial_j\varphi(x)$ exists for each $x \in int_{\mathbb{R}^2}(S)$, then we call the map $x \mapsto \partial_j\varphi(x)$ on $int_{\mathbb{R}^2}(S)$ the **jth partial derivative** of φ.

[3] Let $n \in \mathbb{N}$, and take any open subset O of $\mathbb{R}^n$. As we show below, if $x \in O$ and $\varphi \in \mathbb{R}^O$ is Fréchet differentiable at x, then all partial derivatives of φ at x exist, and we have

$$\mathsf{D}_{\varphi,x}(t_1,\ldots,t_n) = \sum_{j=1}^{n} \partial_j\varphi(x)t_j \qquad \text{for all } (t_1,\ldots,t_n) \in \mathbb{R}^n.$$

Thus, the Fréchet derivative of φ at x is none other than the linear functional that corresponds to the n-vector $(\partial_1\varphi(x),\ldots,\partial_n\varphi(x))$, which, as you know, is called the **gradient** of φ at x.

[4] We now generalize. Given any $m, n \in \mathbb{N}$, let O be an open subset of $\mathbb{R}^n$, and fix any $x \in O$. Take any $\Phi_i \in \mathbb{R}^O$, $i = 1,\ldots,m$, and define $\Phi : O \to \mathbb{R}^m$ by

$$\Phi(t_1,\ldots,t_n) := (\Phi_1(t_1,\ldots,t_n),\ldots,\Phi_m(t_1,\ldots,t_n)).$$

(Here Φ_is are component maps of Φ.) If Φ is Fréchet differentiable at x, then the partial derivatives of each $\Phi_i \in \mathbb{R}^O$ at x exist, and we have

$$\mathsf{D}_{\Phi,x}(t_1,\ldots,t_n) = \left(\sum_{j=1}^{n} \partial_j\Phi_1(x)t_j,\ldots,\sum_{j=1}^{n} \partial_j\Phi_m(x)t_j\right) \qquad \text{for all } (t_1,\ldots,t_n) \in \mathbb{R}^n,$$

where $\partial_j\Phi_i(x)$ is the jth partial derivative of Φ_i at x, $i = 1,\ldots,m$, $j = 1,\ldots,n$.[10] Or, put differently, the linear operator $\mathsf{D}_{\Phi,x}$ satisfies

$$\mathsf{D}_{\Phi,x}(y) = \mathbf{J}_\Phi^x y \qquad \text{for all } y \in \mathbb{R}^n, \tag{6}$$

where $\mathbf{J}_\Phi^x$ is the **Jacobian matrix** of Φ at x, that is, $\mathbf{J}_\Phi^x := [\partial_j\Phi_i(x)]_{m\times n}$.

Just as $f'(x)$ in [1] turned out to be the number that identifies the Fréchet derivative of f at x, here we see that the Jacobian matrix of Φ at x

[10] Please note that I do not claim here that Φ is necessarily Fréchet differentiable at x when the partial derivatives of each $\Phi_i \in \mathbb{R}^O$ at x exist. This is, in fact, not true, simply because continuity of a map on $\mathbb{R}^n$ in each of its components does not imply the overall continuity of the map. (See Exercises 8 and 9 below.)

identifies the Fréchet derivative of Φ at x. A certain shift of this operator, namely, $(\Phi(x) - \mathsf{D}_{\Phi,x}(x)) + \mathsf{D}_{\Phi,x}$, is an affine map from $\mathbb{R}^n$ into $\mathbb{R}^m$ that best approximates Φ near x.

To prove (6), observe that, since $\mathsf{D}_{\Phi,x} \in \mathcal{L}(\mathbb{R}^n, \mathbb{R}^m)$, there must exist a matrix $\mathbf{A} := [a_{ij}]_{m\times n}$ with $\mathsf{D}_{\Phi,x}(y) = \mathbf{A}y$ for all $y \in \mathbb{R}^n$ (Example F.6). Then, by definition,

$$\lim_{\omega\to x} \frac{\Phi_i(\omega) - \Phi_i(x) - \sum_{j=1}^n a_{ij}(\omega_j - x_j)}{\|\omega - x\|_2} = 0, \qquad i = 1, \ldots, m.$$

It follows that

$$\lim_{\varepsilon\to 0} \left| \frac{\Phi_i(x + \varepsilon\mathbf{e}^j) - \Phi_i(x)}{\varepsilon} - a_{ij} \right| = 0,$$

and hence $a_{ij} = \partial_j \Phi_i(x)$ for each i and j. That is, $\mathsf{D}_{\Phi,x}(y) = \mathbf{J}_\Phi^x y$ for all $y \in \mathbb{R}^n$. □

EXERCISE 7 True or false: $f : \mathbb{R} \to \mathbb{R}$ is continuously differentiable iff it is continuously Fréchet differentiable.

EXERCISE 8 Define $\varphi : \mathbb{R}^2 \to \mathbb{R}$ by $\varphi(\mathbf{0}) := 0$ and $\varphi(x) := \frac{x_1 x_2}{\|x\|_2}$ for all $x \neq \mathbf{0}$. Show that φ is continuous and both $\partial_1\varphi(\cdot)$ and $\partial_2\varphi(\cdot)$ are well-defined on $\mathbb{R}^2$, whereas φ is not Fréchet differentiable at $\mathbf{0}$.

*EXERCISE 9[H] Given any positive integer n, let O be a nonempty open and convex subset of $\mathbb{R}^n$. Take any $\varphi \in \mathbf{C}(O)$ such that $x \mapsto \partial_i\varphi(x)$ is a continuous function on O, $i = 1, \ldots, n$. Show that φ is Fréchet differentiable.

EXERCISE 10 State and prove a generalization of the previous result that applies to continuous maps from a nonempty open and convex subset of $\mathbb{R}^n$ into $\mathbb{R}^m$, $n, m \in \mathbb{N}$.

EXERCISE 11 Given any positive integer n, let X_i be a normed linear space, $i = 1, \ldots, n$, and O a nonempty open subset of the product normed linear space $X := \mathsf{X}^n X_i$. Fix any $x \in O$, and for each i, let $O_i := \{z^i \in X_i : (z^i, x^{-i}) \in O\}$, which is an open subset of X_i. Prove: If

$\varphi : O \to \mathbb{R}$ is Fréchet differentiable at x,then $\varphi(\cdot, x^{-i}) \in \mathbb{R}^{O_i}$ is Fréchet differentiable at x^i, $i = 1, \ldots, n$, and we have

$$\mathsf{D}_{\varphi,x}(z^1, \ldots, z^n) = \sum_{i=1}^{n} \mathsf{D}_{\varphi(\cdot,x^{-i}),x^i}(z^i).$$

Compare with Example 1.[3].)

The rest of the examples considered here all work within the context of infinite-dimensional normed linear spaces. We begin with a particularly simple one and move on to more involved examples.

EXAMPLE 2
Define $\varphi : \mathbf{C}[0, 1] \to \mathbb{R}$ by

$$\varphi(f) := \int_0^1 f(t)^2 dt.$$

Let us compute D_φ, which, if it exists, is a map from $\mathbf{C}[0, 1]$ into the dual of $\mathbf{C}[0, 1]$. Fix any $f \in \mathbf{C}[0, 1]$, and note that

$$\begin{aligned}\varphi(h) - \varphi(f) &= \int_0^1 (h(t)^2 - f(t)^2)dt \\ &= 2\int_0^1 f(t)(h(t) - f(t))dt + \int_0^1 (h(t) - f(t))^2 dt\end{aligned}$$

for any $h \in \mathbf{C}[0, 1]$. Notice that, as $\|h - f\|_\infty \to 0$, the last term here would approach to 0 faster than $\|h - f\|_\infty$ vanishes. Indeed, if we define $\mathbf{e} \in \mathbb{R}^{\mathbf{C}[0,1]}$ by $\mathbf{e}(h) := \int_0^1 (h(t) - f(t))^2 dt$, then

$$\lim_{h \to f} \left| \frac{\mathbf{e}(h)}{\|h - f\|_\infty} \right| \leq \lim_{h \to f} \|h - f\|_\infty = 0.$$

So, if we define $L \in \mathbb{R}^{\mathbf{C}[0,1]}$ by $L(g) := 2\int_0^1 f(t)g(t)dt$, then $\varphi(h) = \varphi(f) + L(h - f) + \mathbf{e}(h)$ for any $h \in \mathbf{C}[0, 1]$. By Proposition 2, therefore, $\mathsf{D}_{\varphi,f} = L$. Since f was arbitrary here, we conclude that $\mathsf{D}_\varphi : \mathbf{C}[0, 1] \to \mathbf{C}[0, 1]^*$ satisfies

$$\mathsf{D}_\varphi(f)(g) = 2\int_0^1 f(t)g(t)dt$$

for any $f, g \in \mathbf{C}[0, 1]$. (Interpretation?) □

EXERCISE 12[H] Show that the map $\varphi \in map\varphi \in \mathbb{R}^{C[0,1]}$ defined by $\varphi(f) := f(0)^2$ is Fréchet differentiable, and compute D_φ.

EXERCISE 13[H] Show that the map $\varphi \in \mathbb{R}^{C[0,1]}$ defined by $\varphi(f) := \frac{1}{3}\int_0^1 f(t)^3 dt$ is Fréchet differentiable, and compute D_φ.

One-variable differential calculus is often useful in computing the Fréchet derivative of a given function, even when the domain of this function is infinite-dimensional. In particular, Taylor's Theorem is extremely useful for this purpose. Given our present purposes, all we need is the following "baby" version of that result.

> THE SECOND MEAN VALUE THEOREM
> *Let a and b be two distinct real numbers, and $I := co\{a, b\}$. If $f : I \to \mathbb{R}$ is continuously differentiable on I and twice differentiable on $int_{\mathbb{R}}(I)$, then*
>
> $$f(b) - f(a) = f'(a)(b-a) + \tfrac{1}{2}f''(c)(b-a)^2 \quad \textit{for some } c \in I\backslash\{a, b\}.$$

PROOF
The idea of the proof is reminiscent of the usual way in which one deduces the Mean Value Theorem from Rolle's Theorem (Exercise A.56). Define $g : I \to \mathbb{R}$ by

$$g(t) := f(b) - f(t) - f'(t)(b-t) - \tfrac{1}{2}M(b-t)^2,$$

where $M \in \mathbb{R}$ is chosen to guarantee that $g(a) = 0$. Clearly, g is differentiable on $int_{\mathbb{R}}(I)$, and a quick computation yields

$$g'(t) = (b-t)(M - f''(t)) \qquad \text{for any } t \in I\backslash\{a, b\}.$$

Moreover, since $g(a) = 0 = g(b)$ and $g \in \mathbf{C}^1(I)$, Rolle's Theorem guarantees that $g'(c) = 0$ for some $c \in I\backslash\{a, b\}$. But then $M = f''(c)$, and we find

$$0 = g(a) = f(b) - f(a) - f'(a)(b-a) - \tfrac{1}{2}f''(c)(b-a)^2,$$

which proves the claim. ∎

EXAMPLE 3
Let $u \in \mathbf{C}^2(\mathbb{R})$, and assume that both u' and u'' are bounded functions. Take any $0 < \delta < 1$, and define $\varphi : \ell^\infty \to \mathbb{R}$ by

$$\varphi((x_m)) := \sum_{i=1}^{\infty} \delta^i u(x_i).$$

We wish to compute D_φ, which, if it exists, is a map from ℓ^∞ into the dual of ℓ^∞. Our strategy of attack is analogous to that we followed in Example 2. Fix any bounded real sequence (x_m). By the Second Mean Value Theorem, for any $(\omega_m) \in \ell^\infty$, there exists an $(\omega_m^*) \in \ell^\infty$ such that

$$u(\omega_i) - u(x_i) = u'(x_i)(\omega_i - x_i) + \tfrac{1}{2}u''(\omega_i^*)(\omega_i - x_i)^2$$

and $\omega_i^* \in co\{\omega_i, x_i\}$ for each $i = 1, 2, \ldots$. Consequently,

$$\varphi((\omega_m)) - \varphi((x_m)) = \sum_{i=1}^{\infty} \delta^i u'(x_i)(\omega_i - x_i) + \tfrac{1}{2}\sum_{i=1}^{\infty} \delta^i u''(\omega_i^*)(\omega_i - x_i)^2 \quad (7)$$

for any $(\omega_m) \in \ell^\infty$.[11] Again, the trick here is to notice that, as $(\omega_m) \to (x_m)$, the last term of this equation would approach 0 faster than $\|(\omega_m - x_m)\|_\infty$ vanishes. Since we assume that u'' is bounded here, say by the real number $M > 0$, it is very easy to show this. Define $\mathbf{e} : \ell^\infty \to \mathbb{R}$ by $\mathbf{e}((\omega_m)) := \frac{1}{2}\sum^\infty \delta^i u''(\omega_i^*)(\omega_i - x_i)^2$, and note that

$$|\mathbf{e}((\omega_m))| \leq \left(\tfrac{M}{2}\,\|(\omega_m - x_m)\|_\infty^2\right)\sum_{i=1}^{\infty}\delta^i = \tfrac{\delta M}{2(1-\delta)}\,\|(\omega_m - x_m)\|_\infty^2$$

for any $(\omega_m) \in \ell^\infty$. It follows that $\lim_{(\omega_m)\to(x_m)} \left|\frac{\mathbf{e}((\omega_m))}{\|(\omega_m - x_m)\|_\infty}\right| = 0$, as desired. Hence, by Proposition 2, we may conclude that

$$\mathsf{D}_{\varphi,(x_m)}((y_m)) = \sum_{i=1}^{\infty} \delta^i u'(x_i) y_i \qquad \text{for all } (y_m) \in \ell^\infty.$$

What does this mean, intuitively? Well, consider the map $\psi : \ell^\infty \to \mathbb{R}$ defined by

$$\psi((y_m)) := \varphi((x_m)) + \sum_{i=1}^{\infty} \delta^i u'(x_i)(y_i - x_i).$$

This is an affine map, so its behavior is *globally* linear. And we have just found out that, near (x_m), the behavior of ψ and our original (nonlinear) map φ are "very similar," in the sense that these two maps are tangent to each other at (x_m). So, the *local* linear behavior of φ around (x_m) is best captured by the linear behavior of the affine map ψ. □

[11] For this to make sense, I need to know that the two infinite series on the right-hand side of (7) sum up to finite numbers. And I do know this. How?

Example 4
Let $H \in \mathbf{C}^2(\mathbb{R})$, and define the self-map Φ on $\mathbf{C}[0,1]$ by

$$\Phi(f)(t) := H(f(t)).$$

For an arbitrarily given $f \in \mathbf{C}[0,1]$, we wish to compute $\mathsf{D}_{\Phi,f}$. The argument is analogous to the one given above. We begin by using the Second Mean Value Theorem to find that, for any $h \in \mathbf{C}[0,1]$, there exists a map $\theta_h \in \mathbb{R}^{[0,1]}$ such that

$$\Phi(h)(t) - \Phi(f)(t) = H'(f(t))(h(t) - f(t)) + \tfrac{1}{2}H''(\theta_h(t))(h(t) - f(t))^2$$

and $\theta_h(t) \in co\{h(t), f(t)\}$ for each $0 \leq t \leq 1$. (Obviously, $\theta_h \in \mathbf{B}[0,1]$.) Define the self-map $\mathbf{e}$ on $\mathbf{C}[0,1]$ by $\mathbf{e}(h) := \frac{1}{2}(H''\circ\theta_h)(h-f)^2$. Thanks to Proposition 2, if we can show that $\frac{\|\mathbf{e}(h)\|_\infty}{\|h-f\|_\infty} \to 0$ as $h \to f$, then we may conclude that

$$\mathsf{D}_{\Phi,f}(g) := \left(H' \circ f\right) g. \tag{8}$$

But

$$\frac{\left\|\mathbf{e}(h)\right\|_\infty}{\left\|h - f\right\|_\infty} \leq \tfrac{1}{2}\left\|H'' \circ \theta_h\right\|_\infty \left\|h - f\right\|_\infty,$$

so all we need here is to show that $\left\|H'' \circ \theta_h\right\|_\infty$ is uniformly bounded for any h that is sufficiently close to f. This is quite easy. Obviously, we have $\|\theta_h\|_\infty \leq \left\|f\right\|_\infty + 1$ for any $h \in N_{1,\mathbf{C}[0,1]}(f)$. Moreover, since H'' is continuous, there is a number $M > 0$ such that $\left|H''(a)\right| \leq M$ for all $a \in \mathbb{R}$ with $|a| \leq \left\|f\right\|_\infty + 1$. Thus:

$$\frac{\left\|\mathbf{e}(h)\right\|_\infty}{\left\|h - f\right\|_\infty} \leq \tfrac{M}{2}\left\|h - f\right\|_\infty \qquad \text{for all } h \in N_{1,\mathbf{C}[0,1]}(f).$$

Conclusion: $\lim_{h\to f} \frac{\mathbf{e}(h)}{\|h-f\|_\infty} = \mathbf{0}$, and (8) is true.

Therefore, the map $g \mapsto (H' \circ f)g$ is the best linear approximation of Φ near f, or, put differently, at least on a small neighborhood of f, the behavior of the nonlinear operator Φ is "just like" that of the affine map $g \mapsto \left(H \circ f\right) + (H' \circ f)(g - f)$. □

*Exercise 14 (*The Nemyitsky Operator*) Let $H : \mathbb{R}^2 \to \mathbb{R}$ be a continuous function such that $\partial_2 H$ and $\partial_2\partial_2 H$ are continuous functions on $\mathbb{R}^2$.

Define the self-map Φ on $\mathbf{C}[0,1]$ by $\Phi(f)(t) := H(t, f(t))$. Show that Φ is Fréchet differentiable, and

$$\mathrm{D}_{\Phi}(f)(g)(t) = \partial_2 H(t, f(t))g(t) \quad \text{for all } f, g \in \mathbf{C}[0,1] \text{ and } 0 \leq t \leq 1.$$

An important topic in functional analysis concerns the determination of normed linear spaces the norms of which are Fréchet differentiable. We will have little to say on this topic in this book, but the following exercises might give you at least an idea about it.

Exercise 15 Let $(x_m) \in \ell^2 \backslash \{\mathbf{0}\}$, and show that $\|\cdot\|_2 : \ell^2 \to \mathbb{R}_+$ is Fréchet differentiable at (x_m) with

$$\mathrm{D}_{\|\cdot\|_2,(x_m)}(y_m) = \frac{1}{\|(x_m)\|_2} \sum_{i=1}^{\infty} x_i y_i \quad \text{for all } (y_m) \in \ell^2.$$

The following exercise generalizes this observation.

Exercise 16 Let X be a pre-Hilbert space (Exercise J.12) with the inner product ϕ. A famous theorem of linear analysis states that for any continuous linear functional L on X, there exists a $y \in X$ such that $L(x) = \phi(x, y)$ for all $x \in X$. (This is the *Riesz Representation Theorem.*) Assuming the validity of this fact, show that the norm $\|\cdot\|$ of X is Fréchet differentiable at each $x \in X\backslash\{\mathbf{0}\}$, and we have $\mathrm{D}_{\|\cdot\|,x}(y) = \frac{\phi(x,y)}{\|x\|}$ for all $x \in X\backslash\{\mathbf{0}\}$ and $y \in X$.

Exercise 17 It is well known that for any $L \in \mathcal{B}(\ell^1, \mathbb{R})$ there exists an $(a_m) \in \ell^\infty$ such that $L((x_m)) = \sum^\infty a_i x_i$. (You don't have to prove this result here.) Use this fact to show that the norm $\|\cdot\|_1$ on ℓ^1 is not Fréchet differentiable anywhere.

*Exercise 18[H] Determine all points at which $\|\cdot\|_\infty : \ell^\infty \to \mathbb{R}_+$ is Fréchet differentiable.

1.5 Rules of Differentiation

Most of the basic rules of differentiation of one-variable calculus have straightforward generalizations in terms of Fréchet derivatives. Just as you would suspect, for instance, the Fréchet derivative of a given linear combination of Fréchet differentiable operators equals that linear combination of the Fréchet derivatives of the involved operators.

Proposition 3
Let X and Y be two normed linear spaces, T a subset of X, and $x \in int_X(T)$. Let Φ and Ψ be two maps in Y^T that are Fréchet differentiable at x. Then, for any real number α, $\alpha\Phi + \Psi$ is Fréchet differentiable at x, and

$$\mathsf{D}_{\alpha\Phi+\Psi,x} = \alpha\mathsf{D}_{\Phi,x} + \mathsf{D}_{\Psi,x}.$$

Proof
By Proposition 2, there exist maps $\mathbf{e}_\Phi$ and $\mathbf{e}_\Psi$ in Y^T such that

$$\Phi(\omega) - \Phi(x) = \mathsf{D}_{\Phi,x}(\omega - x) + \mathbf{e}_\Phi(\omega) \quad \text{and}$$
$$\Psi(\omega) - \Psi(x) = \mathsf{D}_{\Psi,x}(\omega - x) + \mathbf{e}_\Psi(\omega)$$

for all $\omega \in int_X(T)$, and $\lim_{\omega\to x} \frac{\mathbf{e}_\Phi(\omega)}{\|\omega - x\|} = \mathbf{0} = \lim_{\omega\to x} \frac{\mathbf{e}_\Psi(\omega)}{\|\omega - x\|}$. So, given any $\alpha \in \mathbb{R}$, letting $\mathbf{e} := \alpha\mathbf{e}_\Phi + \mathbf{e}_\Psi$, we find

$$(\alpha\Phi(\omega) + \Psi(\omega)) - (\alpha\Phi(x) + \Psi(x)) = (\alpha\mathsf{D}_{\Phi,x} + \mathsf{D}_{\Psi,x})(\omega - x) + \mathbf{e}(\omega)$$

for all $\omega \in int_X(T)$, and $\lim_{\omega\to x} \frac{\mathbf{e}(\omega)}{\|\omega - x\|} = \mathbf{0}$. Applying Proposition 2 completes the proof. ■

Exercise 19 Let X be a normed linear space and O a nonempty open subset of X. Prove: If $\varphi, \psi \in \mathbb{R}^O$ are Fréchet differentiable at $x \in O$, then the product operator $\varphi\psi$ is Fréchet differentiable at x, and

$$\mathsf{D}_{\varphi\psi,x} = \psi(x)\mathsf{D}_{\varphi,x} + \varphi(x)\mathsf{D}_{\psi,x}.$$

The next result should again be familiar from ordinary calculus.

Proposition 4
(The Chain Rule) *Let X, Y, and Z be normed linear spaces, and let S and T be subsets of X and Y, respectively. Let $\Phi \in T^S$ and $\Psi \in Z^T$ be two maps such that Φ is Fréchet differentiable at $x \in int_X(S)$, and Ψ at $\Phi(x) \in int_Y(T)$. Then, $\Psi \circ \Phi$ is Fréchet differentiable at x, and*

$$\mathsf{D}_{\Psi\circ\Phi,x} = \mathsf{D}_{\Psi,\Phi(x)} \circ \mathsf{D}_{\Phi,x}.$$

PROOF

By Proposition 2, there exist maps $\mathbf{e}_\Phi \in Y^S$ and $\mathbf{e}_\Psi \in Z^T$ such that

$$\Phi(\omega) - \Phi(x) = \mathsf{D}_{\Phi,x}(\omega - x) + \mathbf{e}_\Phi(\omega) \qquad \text{for all } \omega \in int_X(S) \tag{9}$$

and

$$\Psi(w) - \Psi(\Phi(x)) = \mathsf{D}_{\Psi,\Phi(x)}(w - \Phi(x)) + \mathbf{e}_\Psi(w) \qquad \text{for all } w \in int_Y(T),$$

while $\lim_{\omega\to x} \frac{\mathbf{e}_\Phi(\omega)}{\|\omega - x\|} = \mathbf{0}$ and $\lim_{w\to\Phi(x)} \frac{\mathbf{e}_\Psi(w)}{\|w-\Phi(x)\|_Y} = \mathbf{0}$. It follows that, for any $\omega \in int_X(S)$,

$$\begin{aligned}\Psi(\Phi(\omega)) - \Psi(\Phi(x)) &= \mathsf{D}_{\Psi,\Phi(x)}(\Phi(\omega) - \Phi(x)) + \mathbf{e}_\Psi(\Phi(\omega)) \\ &= \mathsf{D}_{\Psi,\Phi(x)}(\mathsf{D}_{\Phi,x}(\omega - x) + \mathbf{e}_\Phi(\omega)) + \mathbf{e}_\Psi(\Phi(\omega)) \\ &= \mathsf{D}_{\Psi,\Phi(x)}(\mathsf{D}_{\Phi,x}(\omega - x)) + \mathsf{D}_{\Psi,\Phi(x)}(\mathbf{e}_\Phi(\omega)) \\ &\quad + \mathbf{e}_\Psi(\Phi(\omega)) \\ &= \left(\mathsf{D}_{\Psi,\Phi(x)} \circ \mathsf{D}_{\Phi,x}\right)(\omega - x) + \mathbf{e}(\omega),\end{aligned}$$

where $\mathbf{e} \in Z^S$ is defined as

$$\mathbf{e}(\omega) := \mathsf{D}_{\Psi,\Phi(x)}(\mathbf{e}_\Phi(\omega)) + \mathbf{e}_\Psi(\Phi(\omega)).$$

By Proposition 2, therefore, it remains to show that $\lim_{\omega\to x} \frac{\mathbf{e}(\omega)}{\|\omega-x\|} = \mathbf{0}$. To this end, observe first that

$$\lim_{\omega\to x} \frac{\left\|\mathsf{D}_{\Psi,\Phi(x)}(\mathbf{e}_\Phi(\omega))\right\|_Z}{\|\omega - x\|} \le \left\|\mathsf{D}_{\Psi,\Phi(x)}\right\|^* \lim_{\omega\to x} \frac{\|\mathbf{e}_\Phi(\omega)\|_Y}{\|\omega - x\|} = 0$$

(Section J.4.3). The proof will thus be complete if we can establish that $\lim_{\omega\to x} \frac{\mathbf{e}_\Psi(\Phi(\omega))}{\|\omega-x\|} = \mathbf{0}$. This requires harder work. Note first that

$$\lim_{\omega\to x} \frac{\|\mathbf{e}_\Psi(\Phi(\omega))\|_Z}{\|\omega - x\|} = \lim_{\omega\to x} \left(\frac{\|\mathbf{e}_\Psi(\Phi(\omega))\|_Z}{\|\Phi(\omega) - \Phi(x)\|_Y}\right)\left(\frac{\|\Phi(\omega) - \Phi(x)\|_Y}{\|\omega - x\|}\right). \tag{10}$$

But, since Φ is continuous, $\lim_{w\to\Phi(x)} \frac{\mathbf{e}_\Psi(w)}{\|w-\Phi(x)\|_Y} = \mathbf{0}$ implies

$$\lim_{\omega\to x} \frac{\|\mathbf{e}_\Psi(\Phi(\omega))\|_Z}{\|\Phi(\omega) - \Phi(x)\|_Y} = 0. \tag{11}$$

(Why?) Moreover, for any $\omega \in X\backslash\{x\}$, (9) gives

$$\frac{\|\Phi(\omega) - \Phi(x)\|_Y}{\|\omega - x\|} = \frac{\left\|\mathsf{D}_{\Phi,x}(\omega - x) + \mathbf{e}_\Phi(\omega)\right\|_Y}{\|\omega - x\|} \le \left\|\mathsf{D}_{\Phi,x}\right\|^* + \frac{\|\mathbf{e}_\Phi(\omega)\|_Y}{\|\omega - x\|}.$$

(Why?) Since $\lim_{\omega\to x} \frac{\mathbf{e}_\Phi(\omega)}{\|\omega-x\|} = \mathbf{0}$, therefore, there exists an $\varepsilon > 0$ such that $\frac{\|\Phi(\omega)-\Phi(x)\|_Y}{\|\omega-x\|} \leq \left\|\mathsf{D}_{\Phi,x}\right\|^* + 1$ for all $\omega \in N_{\varepsilon,X}(x)\backslash\{x\}$. Combining this observation with (11) and (10) yields $\lim_{\omega\to x} \frac{\mathbf{e}_\Psi(\Phi(\omega))}{\|\omega-x\|} = \mathbf{0}$, as we sought. ■

Here is a quick illustration of how such rules of differentiation are used in practice.

EXAMPLE 5

Let $m \in \mathbb{N}$, and for each $i = 0,\ldots,m$, define the self-map Φ_i on $\mathbf{C}[0,1]$ by

$$\Phi_i(f) := (f)^i,$$

where $(f)^i$ denotes the ith power of f, that is $(f)^i(t) = f(t)^i$ for each $0 \leq t \leq 1$. It is easily verified that each Φ_i is Fréchet differentiable, and

$$\mathsf{D}_{\Phi_0,f}(h) = \mathbf{0} \quad \text{and} \quad \mathsf{D}_{\Phi_i,f}(h) = i(f)^{i-1}h, \qquad i = 1,\ldots,m.$$

Now define $\Phi : \mathbf{C}[0,1] \to \mathbb{R}$ by

$$\Phi(f) := \sum_{i=0}^{m} a_i \int_0^1 f(t)^i dt,$$

where $a_0,\ldots,a_m \in \mathbb{R}$. Using Propositions 3 and 4 we can readily compute the Fréchet derivative of Φ. Indeed, if $L \in \mathcal{B}(\mathbf{C}[0,1],\mathbb{R})$ is defined by $L(f) := \int_0^1 f(t)dt$, then we have

$$\Phi = \sum_{i=0}^{m} a_i \left(L \circ \Phi_i\right).$$

So, since $\mathsf{D}_{L,h} = L$ for any h (why?), Propositions 3 and 4 yield

$$\mathsf{D}_{\Phi,f}(g) = \sum_{i=0}^{m} a_i \mathsf{D}_{L\circ\Phi_i,f}(g) = \sum_{i=0}^{m} a_i L\left(\mathsf{D}_{\Phi_i,f}(g)\right)$$
$$= \sum_{i=0}^{m} a_i \int_0^1 i f(t)^{i-1} g(t)dt$$

for all $f,g \in \mathbf{C}[0,1]$. (Compare with Example 2.) □

*EXERCISE 20[H] (*The Hammerstein Operator*) Define the self-map Φ on $\mathbf{C}[0,1]$ by

$$\Phi(f)(x) := \int_0^1 \theta(x,t)H(t,f(t))dt,$$

where θ is a continuous real map on $[0,1]^2$ and $H : \mathbb{R}^2 \to \mathbb{R}$ is a continuous function such that $\partial_2 H$ and $\partial_2\partial_2 H$ are well-defined and continuous functions on $\mathbb{R}^2$. Use the Chain Rule to show that Φ is Fréchet differentiable, and compute $\mathbf{D}_\Phi$.

EXERCISE 21 Take any natural numbers n, m, and k. Let O and U be nonempty open subsets of $\mathbb{R}^n$ and $\mathbb{R}^m$, respectively. For any given $x \in O$, let $\Phi : O \to U$ and $\Psi : U \to \mathbb{R}^k$ be two maps that are Fréchet differentiable at x and $\Phi(x)$, respectively. Show that $\Psi \circ \Phi$ is Fréchet differentiable at x, and

$$\mathbf{D}_{\Psi\circ\Phi,x}(y) = \mathbf{J}_\Psi^{\Phi(x)}(\mathbf{J}_\Phi^x y) \qquad \text{for all } y \in \mathbb{R}^n,$$

where $\mathbf{J}_\Phi^x$ and $\mathbf{J}_\Psi^{\Phi(x)}$ are the Jacobian matrices of Φ (at x) and Ψ (at $\Phi(x)$), respectively.

EXERCISE 22 Let X be a normed linear space, O a nonempty open and convex subset of X, and $\Phi \in \mathbb{R}^O$ a Fréchet differentiable map. Fix any distinct $x, y \in O$, and define $F : (0,1) \to \mathbb{R}$ by $F(\lambda) := \Phi(\lambda x + (1-\lambda)y)$. Show that F is differentiable, and $F'(\lambda) = \mathbf{D}_{\Phi,\lambda x+(1-\lambda)y}(x-y)$ for all $t \in \mathbb{R}$ and $0 < \lambda < 1$.

EXERCISE 23 Let X and Y be two normed linear spaces, and φ a Fréchet differentiable real map on $X \times Y$. Fix any $x \in X$, and define $\psi \in \mathbb{R}^Y$ by $\psi(y) := \varphi(x-y, y)$. Use the Chain Rule to prove that ψ is differentiable and compute $\mathbf{D}_\psi$.

1.6 The Second Fréchet Derivative of a Real Function

To outline a basic introduction to optimization theory, we also need to go through the notion of the second Fréchet derivative of real functions. Just as in ordinary calculus, the idea is to define this notion as the derivative of the derivative. Unfortunately, life gets a bit complicated here. Recall that the Fréchet derivative D_φ of a real function φ defined on an open subset O of a normed linear space X is a function that maps O into X^*. Therefore, the Fréchet derivative of D_φ at $x \in O$ is a continuous linear function that maps X into X^*. Put differently, the second Fréchet derivative of φ at x is a member of $\mathcal{B}(X, X^*)$.

In case you find this confusing, let us see how this situation compares with differentiating a function of the form $\varphi : \mathbb{R}^2 \to \mathbb{R}$ twice. In calculus, by the derivative of φ at $x \in \mathbb{R}^2$, we understand the *gradient* of φ, that is, the vector $(\partial_1\varphi(x), \partial_2\varphi(x)) \in \mathbb{R}^2$. In turn, the second derivative of φ at x is the matrix $\mathbf{H}^x := [\partial_{ij}\varphi(x)]_{2\times 2}$, where by $\partial_{ij}\varphi(x)$, we understand $\partial_i\partial_j\varphi(x)$. (You might recall that this matrix is called the *Hessian* of φ at x.) Given Example 1.[3], it is only natural that the second Fréchet derivative of φ at x is the linear operator induced by the matrix $\mathbf{H}^x$ (i.e., $y \mapsto \mathbf{H}^x y$) and hence it is a linear function that maps $\mathbb{R}^2$ into $\mathbb{R}^2$. Since the dual of $\mathbb{R}^2$ is $\mathbb{R}^2$ (Example J.7), this situation conforms perfectly with the outline of the previous paragraph.

Things will become clearer later. Let us first state the definition of the second Fréchet derivative of a real function formally.

Definition

Let T be a nonempty subset of a normed linear space X and $\varphi : T \to \mathbb{R}$ a Fréchet differentiable map. For any given $x \in int_X(T)$, if $\mathsf{D}_\varphi : int_X(T) \to X^*$ is Fréchet differentiable at x, then we say that φ is **twice Fréchet differentiable at** x. In this case, the **second Fréchet derivative of** φ **at** x, denoted by $\mathsf{D}^2_{\varphi,x}$, is a member of $\mathcal{B}(X, X^*)$; we define

$$\mathsf{D}^2_{\varphi,x} := \mathsf{D}_{\mathsf{D}_\varphi, x}.$$

If O is a nonempty open subset of X with $O \subseteq T$, and if φ is twice Fréchet differentiable at every $x \in O$, then we say that φ is **twice Fréchet differentiable on** O. If $O = int_X(T)$ here, then φ is said to be **twice Fréchet differentiable.**

The thing to get used to here is that $\mathsf{D}^2_{\varphi,x} \in \mathcal{B}(X, X^*)$, that is, $\mathsf{D}^2_{\varphi,x}(y) \in X^*$ for each $y \in X$. We should thus write $\mathsf{D}^2_{\varphi,x}(y)(z)$ for the value of the linear functional $\mathsf{D}^2_{\varphi,x}(y)$ at z. It is, however, customary to write $\mathsf{D}^2_{\varphi,x}(y, z)$ instead of $\mathsf{D}^2_{\varphi,x}(y)(z)$, thereby thinking of $\mathsf{D}^2_{\varphi,x}$ as a function that maps $X \times X$

into $\mathbb{R}$. From this viewpoint, $\mathsf{D}^2_{\varphi,x}$ is a continuous *bilinear* functional on $X \times X$ (Exercise J.52).[12]

The following is an analogue (and an easy consequence) of Proposition 2 for the second Fréchet derivative of a real function.

PROPOSITION 5

Let X be a normed linear space, T a subset of X, $x \in int_X(T)$, and $L \in \mathcal{B}(X, X^*)$. For any $\varphi \in \mathbb{R}^T$, L is the second Fréchet derivative of φ at x if, and only if, there exists a continuous map $\mathbf{e} : T \to X^*$ such that

$$\mathsf{D}_{\varphi,\omega} = \mathsf{D}_{\varphi,x} + L(\omega - x) + \mathbf{e}(\omega) \quad \text{for all } \omega \in int_X(T),$$

$$\text{and} \quad \lim_{\omega \to x} \frac{\mathbf{e}(\omega)}{\|\omega - x\|} = \mathbf{0}.$$

EXERCISE 24 Prove Proposition 5.

EXAMPLE 6

[1] Let I be an open interval, $f \in \mathbb{R}^I$ a differentiable map, and $x \in I$. If f is twice differentiable at x, then there is an error function $\mathbf{e}_1 : I \to \mathbb{R}$ such that $f'(\omega) = f'(x) + f''(x)(\omega - x) + \mathbf{e}_1(\omega)$ for all $\omega \in I$, and $\lim_{\omega \to x} \frac{\mathbf{e}_1(\omega)}{|\omega - x|} = 0$. (Why?) Hence, by Example 1.[1],

$$\mathsf{D}_{f,\omega}(t) - \mathsf{D}_{f,x}(t) = (f'(\omega) - f'(x))t = f''(x)(\omega - x)t + \mathbf{e}_1(\omega)t$$

for all $\omega \in I$ and $t \in \mathbb{R}$. We define $L : \mathbb{R} \to \mathbb{R}^*$ and $\mathbf{e} : I \to \mathbb{R}^*$ by $L(u)(v) := f''(x)uv$ and $\mathbf{e}(u)(v) := \mathbf{e}_1(u)v$, respectively. Then

$$\mathsf{D}_{f,\omega}(t) - \mathsf{D}_{f,x}(t) = L(\omega - x)(t) + \mathbf{e}(\omega)(t) \qquad \text{for all } \omega \in I \text{ and } t \in \mathbb{R},$$

and it follows from Proposition 5 that $\mathsf{D}^2_{f,x} = L$, that is,

$$\mathsf{D}^2_{f,x}(u, v) = f''(x)uv \qquad \text{for all } u, v \in \mathbb{R}.$$

REMINDER. Given any $n \in \mathbb{N}$, let S be a subset of $\mathbb{R}^n$ with nonempty interior, and $\varphi \in \mathbb{R}^S$ a map such that the jth partial derivative of φ (as

[12] This custom is fully justified, of course. After all, $\mathcal{B}(X, X^*)$ "is" the normed linear space of all continuous bilinear functionals on $X \times X$, that is, these two spaces are linearly isometric. (Recall Exercise J.63.)

a real map on $int_{\mathbb{R}^2}(S)$) exists. For any $x \in int_{\mathbb{R}^2}(S)$ and $i, j = 1, \ldots, n$, the number $\partial_{ij}\varphi(x) := \partial_i\partial_j\varphi(x)$ is referred to as a **second-order partial derivative of φ at x.** If $\partial_{ij}\varphi(x)$ exists for each $x \in int_{\mathbb{R}^2}(S)$, then we refer to the map $x \mapsto \partial_{ij}\varphi(x)$ on $int_{\mathbb{R}^2}(S)$ as a **second-order partial derivative** of φ. (*Note.* A folk theorem of advanced calculus says that if $\partial_{ij}\varphi$ and $\partial_{ji}\varphi$ are continuous maps (on $int_{\mathbb{R}^2}(S)$), then $\partial_{ij}\varphi = \partial_{ji}\varphi$.)

[2] Given any $n \in \mathbb{N}$, let O be an open subset of $\mathbb{R}^n$, and take any Fréchet differentiable map $\varphi \in \mathbb{R}^O$. If φ is twice Fréchet differentiable at $x \in O$, then all second-order partial derivatives of φ at x exist, and we have

$$\mathsf{D}^2_{\varphi,x}(u, v) = \sum_{i=1}^{n}\sum_{j=1}^{n} \partial_{ij}\varphi(x)u_i v_j \qquad \text{for all } u, v \in \mathbb{R}^n.$$

Thus, the second Fréchet derivative of φ at x is none other than the symmetric bilinear functional induced by the so-called **Hessian matrix** $[\partial_{ij}\varphi(x)]_{n\times n}$ of φ at x. □

Exercise 25 Prove the assertion made in Example 6.[2].

Example 7

Let O be a nonempty open and convex subset of a normed linear space X, and let x and y be two distinct points in O. Take any twice Fréchet differentiable map $\varphi \in \mathbb{R}^O$, and define $F \in \mathbb{R}^{(0,1)}$ by

$$F(\lambda) := \varphi(\lambda x + (1 - \lambda)y).$$

We wish to show that F is twice differentiable and compute F''. (Any guesses?)

By Exercise 22, F is differentiable, and we have

$$F'(\alpha) := \mathsf{D}_{\varphi,\alpha x+(1-\alpha)y}(x - y), \qquad 0 < \alpha < 1. \tag{12}$$

Define $G := F'$, fix any $0 < \lambda < 1$, and let us agree to write ω_α for $\alpha x + (1-\alpha)y$ for any $0 < \alpha < 1$. By Proposition 5, there exists a continuous map $\mathbf{e} : O \to X^*$ such that

$$\mathsf{D}_{\varphi,\omega} = \mathsf{D}_{\varphi,\omega_\lambda} + \mathsf{D}^2_{\varphi,\omega_\lambda}(\omega - \omega_\lambda) + \mathbf{e}(\omega) \qquad \text{for all } \omega \in O,$$

and $\lim_{\omega\to\omega_\lambda} \frac{\mathbf{e}(\omega)}{\|\omega-\omega_\lambda\|} = \mathbf{0}$. Thus, for any $\alpha \in (0,1)\backslash\{\lambda\}$, (12) gives

$$\begin{aligned} F'(\alpha) - F'(\lambda) &= \mathsf{D}_{\varphi,\omega_\alpha}(x-y) - \mathsf{D}_{\varphi,\omega_\lambda}(x-y) \\ &= \mathsf{D}^2_{\varphi,\omega_\lambda}(\omega_\alpha - \omega_\lambda, x-y) + \mathbf{e}(\omega_\alpha)(x-y). \end{aligned}$$

Since $\omega_\alpha - \omega_\lambda = (\alpha-\lambda)(x-y)$ and $\mathsf{D}^2_{\varphi,\omega_\lambda}$ is a bilinear functional (on $X \times X$), we may divide both sides of this equation by $\alpha - \lambda$ to get

$$\frac{F'(\alpha) - F'(\lambda)}{\alpha - \lambda} = \mathsf{D}^2_{\varphi,\omega_\lambda}(x-y, x-y) + \frac{\mathbf{e}(\omega_\alpha)(x-y)}{\alpha-\lambda}$$

for any $\alpha \in (0,1)\backslash\{\lambda\}$. But since $\alpha \mapsto \omega_\alpha$ is a continuous map from $(0,1)$ into X,

$$\begin{aligned} \lim_{\alpha\to\lambda} \frac{\mathbf{e}(\omega_\alpha)(x-y)}{\alpha-\lambda} &= \|x-y\| \lim_{\alpha\to\lambda} \frac{\mathbf{e}(\omega_\alpha)}{(\alpha-\lambda)\,\|x-y\|} \\ &= \|x-y\| \lim_{\alpha\to\lambda} \frac{\mathbf{e}(\omega_\alpha)}{\|\omega_\alpha - \omega_\lambda\|} \\ &= \mathbf{0}, \end{aligned}$$

so letting $\alpha \to \lambda$ in the previous equation establishes that F' is differentiable at λ, and

$$F''(\lambda) = \mathsf{D}^2_{\varphi,\lambda x+(1-\lambda)y}(x-y, x-y). \qquad \square$$

Exercise 26 Define $\varphi : \mathbf{C}[0,1] \to \mathbb{R}$ by

$$\varphi(f) := \sum_{i=0}^{m} a_i \int_0^1 f(t)^i dt,$$

where $a_0, \ldots, a_m \in \mathbb{R}$. Compute $\mathsf{D}^2_{\varphi,f}$ for any $f \in \mathbf{C}[0,1]$.

1.7 Differentiation on Relatively Open Sets

We have now at hand a potent theory of differentiation that generalizes the classic theory. There still remains one major difficulty, however. Insofar as our basic definition is concerned, we are unable to differentiate a map that is defined on a subset of a normed linear space with no interior. For instance, let $S = \{(a,1) : 0 < a < 1\}$, and define $\varphi : S \to \mathbb{R}$ by $\varphi(a,1) := a^2$. What is the Fréchet derivative of φ? Well, since S has no interior in $\mathbb{R}^2$, our basic definition does not even allow us to pose this question. That definition is based on the idea of perturbing (infinitesimally) a given point in the interior of the domain of a map in *any* direction in the space, and analyzing the behavior of the resulting difference-quotient. In this example, because $int_{\mathbb{R}^2}(S) = \emptyset$, we cannot proceed in this manner. Indeed, the variations we consider *must* be horizontal in this case. Put differently, if $x \in S$ and we wish to study the difference-quotient $\frac{\varphi(\omega)-\varphi(x)}{\omega-x}$, the variation $\omega - x$ must belong to the linear subspace $\mathbb{R} \times \{0\}$. So, in this example, the Fréchet derivative of φ needs to be viewed as a linear functional from $\mathbb{R} \times \{0\}$ into $\mathbb{R}$ (and not from $\mathbb{R}^2$ into $\mathbb{R}$).

Apparently, there is an obvious way we can generalize the definition of the Fréchet derivative and capture these sorts of examples. All we need is to take the domain of the function to be differentiated as open *in the affine manifold it generates*. Let us first give such sets a name.

DEFINITION
Let X be a normed linear space. A subset S of X is said to be **relatively open** if $|S| > 1$ and S is open in $aff(S)$.[13]

Now consider a real map φ whose domain S is relatively open in some normed linear space. Obviously, the difference-quotient $\frac{\varphi(\omega)-\varphi(x)}{\omega-x}$ makes sense iff $\omega \in S\backslash\{x\}$. Therefore, $span(S - x)$ is the linear space that contains all possible variations about x, or equivalently, $aff(S)$ is the set of all possible directions of perturbing x. (Recall the example considered above.) We are therefore led to define the Fréchet derivative of φ at x as a bounded linear functional on $span(S - x)$.

NOTATION. Let T be a subset of a normed linear space. In what follows we denote the interior of T in $aff(T)$ as $T^\diamond$, that is,

$$T^\diamond := int_{aff(T)}(T).$$

Thus, T is relatively open iff $T = T^\diamond$. Moreover, $span(T^\diamond - z) = span(T - z)$ for any $z \in T^\diamond$,[14] and if T is convex and $T^\diamond \neq \emptyset$, then $T^\diamond = ri(T)$ (Proposition I.11).

[13] In this definition I discard the cases $S = \emptyset$ and $|S| = 1$ to avoid some trivial qualifications that would otherwise accompany the subsequent propositions.
[14] *Quiz.* Prove this!

Definition

Let X and Y be two normed linear spaces and T a subset of X with $|T| > 1$ and $T^\diamond \neq \emptyset$.[15] Let

$$\mathbf{s}(T) := span(T - z) \tag{13}$$

where $z \in T^\diamond$ is arbitrary.[16] For any $x \in T^\diamond$, a map $\Phi : T \to Y$ is said to be **Fréchet differentiable at** x if there is a continuous linear operator $\mathsf{D}_{\Phi,x} \in \mathcal{B}(\mathbf{s}(T), Y)$ such that

$$\lim_{\omega \to x} \frac{\Phi(\omega) - \Phi(x) - \mathsf{D}_{\Phi,x}(\omega - x)}{\|\omega - x\|} = \mathbf{0}.$$

The linear operator $\mathsf{D}_{\Phi,x}$ is called the **Fréchet derivative of** Φ **at** x.

If Φ is Fréchet differentiable at every $x \in T^\diamond$, then we say that Φ is **Fréchet differentiable.** In this case, we say that Φ is **continuously Fréchet differentiable** if the map $\mathsf{D}_\Phi : T^\diamond \to \mathcal{B}(\mathbf{s}(T), Y)$, defined by $\mathsf{D}_\Phi(x) := \mathsf{D}_{\Phi,x}$, is continuous.

Definition

Let T be a subset of a normed linear space X with $|T| > 1$ and $T^\diamond \neq \emptyset$, and take a Fréchet differentiable map $\varphi : T \to \mathbb{R}$. For any given $x \in T^\diamond$, if $\mathsf{D}_\varphi : T^\diamond \to \mathbf{s}(T)^*$ is Fréchet differentiable at x (where $\mathbf{s}(T)$ is defined by (13)), we say that φ is **twice Fréchet differentiable at** x. In this case, the **second Fréchet derivative of** φ **at** x, denoted by $\mathsf{D}^2_{\varphi,x}$, is a member of $\mathcal{B}(\mathbf{s}(T), \mathbf{s}(T)^*)$; we define $\mathsf{D}^2_{\varphi,x} := \mathsf{D}_{\mathsf{D}_\varphi,x}$. If φ is twice Fréchet differentiable at every $x \in T^\diamond$, then we say that φ is **twice Fréchet differentiable.**

These definitions extend the ones given in Sections 1.3 and 1.6. After all, if T is a subset of X with $int_X(T) \neq \emptyset$, then $aff(T) = X = \mathbf{s}(T)$—yes?—so the two definitions become identical. Moreover, most of our findings in the previous sections apply to the Fréchet derivatives of maps defined on any

[15] If $T = \{x\}$, then T is itself an affine manifold, and hence it equals its interior in itself. But there is no sequence in $T \backslash \{x\}$ that converges to x, so we cannot possibly define the Fréchet derivative of a map on T. This is the reason why I assume $|T| > 1$ here.

[16] For any given $S \subseteq X$, we have $span(S - y) = span(S - z)$ for any $y, z \in S$. Right?

nonsingleton $T \subseteq X$ with $T^{\diamond} \neq \emptyset$. All we have to do is apply those results on $\mathbf{s}(T)$ as opposed to X.[17] For instance, Proposition 2 becomes the following in this setup.

PROPOSITION 2*
Let X and Y be two normed linear spaces, T a subset of X, $x \in T^{\diamond}$, and $L \in \mathcal{B}(\mathbf{s}(T), Y)$, where $\mathbf{s}(T)$ is defined by (13). For any $\Phi \in Y^T$, L is the Fréchet derivative of Φ at x if, and only if, there exists a continuous map $\mathbf{e} \in Y^T$ such that

$$\Phi(\omega) = \Phi(x) + L(\omega - x) + \mathbf{e}(\omega) \quad \textit{for all } \omega \in T^{\diamond},$$

$$\textit{and } \lim_{\omega \to x} \frac{\mathbf{e}(\omega)}{\|\omega - x\|} = \mathbf{0}.$$

Let us now go back to our silly little example in which the question was to find the Fréchet derivative of $\varphi : S \to \mathbb{R}$ where $S = \{(a,1) : 0 < a < 1\}$ and $\varphi(a,1) := a^2$. As S is relatively open, this question now makes sense. Besides, it is very easy to answer. For any $0 < a < 1$, $\mathsf{D}_{\varphi,(a,1)}$ is a linear functional on $\mathbb{R}\times\{0\}$, and just as one would like to see, we have $\mathsf{D}_{\varphi,(a,1)}(t,0) = 2at$ for any $t \in \mathbb{R}$. Moreover, $\mathsf{D}^2_{\varphi,(a,1)}$ is a linear operator from $\mathbb{R}\times\{0\}$ into the dual of $\mathbb{R}\times\{0\}$ (which is $\mathbb{R}\times\{0\}$ itself), or equivalently, a bilinear functional on $(\mathbb{R}\times\{0\}) \times (\mathbb{R}\times\{0\})$. We have:

$$\mathsf{D}^2_{\varphi,(a,1)}((u,0),(v,0)) = 2uv \qquad \text{for all } u, v \in \mathbb{R},$$

as you can easily check.[18]

REMARK 1. Not only that the definitions we have given in this section reduce to those of the earlier sections for maps defined on sets with nonempty interior, they are also consistent with them in the following sense. Let X and Y be two normed linear spaces, and $S \subseteq X$. Suppose that $\Phi : S \to Y$ is

[17] There is one exception, however. In the statement of the Chain Rule (Proposition 4), if we posit that S and T are relatively open, then we need the additional hypothesis that $\mathsf{D}_{\Phi,x}(\mathbf{s}(S)) \subseteq \mathsf{D}_{\Psi,\Phi(x)}(\mathbf{s}(T))$. With this modification, the proof goes through verbatim.

[18] *Quiz.* Compute $\mathsf{D}_{\varphi,(\frac{1}{2},1)}$ and $\mathsf{D}^2_{\varphi,(\frac{1}{2},1)}$, assuming this time that $S = \{(a, \frac{3}{2} - a) : 0 < a < \frac{3}{2}\}$. (*Hint.* The domain of $\mathsf{D}_{\varphi,(\frac{1}{2},1)}$ is $\{(t,-t) : t \in \mathbb{R}\}$.)

Fréchet differentiable at $x \in int_X(S)$. Then, for any $T \subseteq S$ with $|T| > 1$ and $x \in T^{\diamond}$, the map $\Phi|_T$ is Fréchet differentiable at x, and we have

$$\mathsf{D}_{\Phi|_T,x} = \mathsf{D}_{\Phi,x}|_{\mathsf{s}(T)}.$$

The proof is straightforward. □

2 Generalizations of the Mean Value Theorem

2.1 The Generalized Mean Value Theorem

One way of stating the classic Mean Value Theorem is the following: Given a differentiable real map f on an open interval I, for any $a, b \in I$ with $a < b$, there exists a $c \in (a, b)$ such that $f(b) - f(a) = f'(c)(b - a)$. This fact extends readily to real maps defined on a normed linear space.

THE GENERALIZED MEAN VALUE THEOREM
Let S be a relatively open and convex subset of a normed linear space X and $\varphi \in \mathbb{R}^S$ a Fréchet differentiable map. Then, for any distinct $x, y \in S$,

$$\varphi(x) - \varphi(y) = \mathsf{D}_{\varphi,z}(x - y)$$

for some $z \in co\{x, y\} \backslash \{x, y\}$.

Notice that the entire action takes place on the line segment $co\{x, y\}$ here. Intuitively speaking, then, we should be able to prove this result simply by applying the good ol' Mean Value Theorem on this line segment. The following elementary observation, which generalizes the result found in Exercise 22, is a means to this end.

LEMMA 1
Let S be a relatively open and convex subset of a normed linear space X, x and y distinct points in S, and $\varphi \in \mathbb{R}^S$ a Fréchet differentiable map. If $F \in \mathbb{R}^{(0,1)}$ is defined by $F(\lambda) := \varphi(\lambda x + (1 - \lambda)y)$, then F is differentiable, and

$$F'(\lambda) = \mathsf{D}_{\varphi,\lambda x+(1-\lambda)y}(x - y), \qquad 0 < \lambda < 1.$$

PROOF

Take any $0 < \lambda < 1$, and note that

$$F(\alpha) - F(\lambda) = \varphi(\alpha(x-y)+y) - \varphi(\lambda(x-y)+y)$$
$$= \mathsf{D}_{\varphi,\lambda x+(1-\lambda)y}((\alpha-\lambda)(x-y)) + \mathbf{e}(\alpha(x-y)+y)$$

for some $\mathbf{e} : S \to \mathbb{R}$ with $\lim_{\alpha\to\lambda} \frac{\mathbf{e}(\alpha(x-y)+y)}{\|(\alpha-\lambda)(x-y)\|} = 0$. Then

$$\frac{F(\alpha)-F(\lambda)}{\alpha-\lambda} = \mathsf{D}_{\varphi,\lambda x+(1-\lambda)y}(x-y) + \|x-y\| \frac{\mathbf{e}(\alpha(x-y)+y)}{\|x-y\|(\alpha-\lambda)},$$

and letting $\alpha \to \lambda$ yields the claim. ■

PROOF OF THE GENERALIZED MEAN VALUE THEOREM

Fix any distinct $x, y \in S$, and define $F : [0,1] \to \mathbb{R}$ by $F(\lambda) := \varphi(\lambda x + (1-\lambda)y)$. Being the composition of two continuous functions, F is continuous. (Yes?) Moreover, by Lemma 1, $F|_{(0,1)}$ is differentiable, and $F'(\lambda) = \mathsf{D}_{\varphi,\lambda x+(1-\lambda)y}(x-y)$ for any $0 < \lambda < 1$. So, applying the Mean Value Theorem (Exercise A.56) to F yields the proof. ■

EXERCISE 27 Let X be a preordered normed linear space. Let O be a nonempty open and convex subset of X, and $\varphi \in \mathbb{R}^O$ a Fréchet differentiable map. Show that if $\mathsf{D}_{\varphi,x}$ is a positive linear functional for any $x \in O$, then φ is increasing (that is, $\varphi(x) \geq \varphi(y)$ for any $x, y \in O$ with $x - y \in X_+$.)

EXERCISE 28 For any $n \in \mathbb{N}$, let S be a nonempty compact and convex subset of $\mathbb{R}^n$ such that $int_{\mathbb{R}^n}(S) \neq \emptyset$. Prove: If $\varphi \in \mathbf{C}(S)$ is a Fréchet differentiable function such that $\varphi(x) = 0$ for all $x \in bd_{\mathbb{R}^n}(S)$, then there is an $x^* \in int_{\mathbb{R}^n}(S)$ such that D_{φ,x^*} is the zero functional.

WARNING. The Mean Value Theorem (indeed, Rolle's Theorem) cannot be extended to the context of vector calculus without substantial modification. For instance, in the case of the map $\Phi : \mathbb{R} \to \mathbb{R}^2$ defined by $\Phi(t) := (\sin t, \cos t)$, we have $\Phi(0) = \Phi(2\pi)$ but $\mathsf{D}_{\Phi,x}(t) \neq (0,0)$ for any $0 \leq x \leq 2\pi$ and $t \in \mathbb{R}$. (Check!)

The following generalization of the Second Mean Value Theorem is also worth noting. It will prove very handy in Section 3 when we look into the properties of differentiable concave functionals.

THE GENERALIZED SECOND MEAN VALUE THEOREM
Let S be a relatively open and convex subset of a normed linear space X and $\varphi \in \mathbb{R}^S$ a continuously Fréchet differentiable map. If φ is twice Fréchet differentiable,[19] *then, for any distinct $x, y \in S$,*

$$\varphi(x) - \varphi(y) = \mathsf{D}_{\varphi,y}(x-y) + \tfrac{1}{2}\mathsf{D}^2_{\varphi,z}(x-y, x-y)$$

for some $z \in co\{x, y\}\backslash\{x, y\}$.

EXERCISE 29 [H] Prove the Generalized Second Mean Value Theorem.

EXERCISE 30 (*A Taylor's Formula with Remainder*) Let O be a nonempty open and convex subset of a normed linear space X and $\varphi \in \mathbb{R}^O$ a continuously Fréchet differentiable map that is also twice Fréchet differentiable. Show that, for each $x \in X$, there is a (remainder) function $\mathbf{r} \in \mathbb{R}^X$ such that

$$\varphi(x+z) - \varphi(x) = \mathsf{D}_{\varphi,x}(z) + \tfrac{1}{2}\mathsf{D}^2_{\varphi,x}(z,z) + \mathbf{r}(z)$$

for any $z \in X$ with $x + z \in O$, and $\lim_{z\to\mathbf{0}} \frac{\mathbf{r}(z)}{\|z\|^2} = 0$.

EXAMPLE 8
Let O be a nonempty open subset of $\mathbb{R}^2$, and take any twice continuously differentiable map $\varphi : O \to \mathbb{R}$. One can show that φ is not only continuously Fréchet differentiable, it is also twice Fréchet differentiable.[20] Moreover, as we noted earlier, a folk theorem of multivariate calculus says that $\partial_{12}\varphi(\omega) = \partial_{21}\varphi(\omega)$ for any $\omega \in O$.[21] Consequently, Example 6.[2] yields

$$\mathsf{D}^2_{\varphi,z}(u,u) = \partial_{11}\varphi(z)u_1^2 + 2\partial_{12}\varphi(z)u_1u_2 + \partial_{22}\varphi(z)u_2^2$$

[19] There may appear to be a redundancy in the hypotheses here, but in fact this is not the case. Twice Fréchet differentiability of φ does not, in general, imply its continuous Fréchet differentiability.
[20] While this may look like quite a bit to swallow, the proof is hidden in Exercises 9 and 10.
[21] You have surely seen this fact before. (A Hessian matrix is always symmetric, no?) Its proof, while a bit tedious, follows basically from the definitions (but note that the assumption of ∂_{12} (or ∂_{21}) being continuous is essential).

for any $z \in O$ and $u \in \mathbb{R}^2$. Combining this with the Generalized Second Mean Value Theorem and Example 1.[3], therefore, we obtain the following result of advanced calculus: *For every distinct $x, y \in O$, there exists a $z \in co\{x, y\}\backslash\{x, y\}$ such that*

$$\varphi(x) - \varphi(y) = \partial_1\varphi(y)(x_1 - y_1) + \partial_2\varphi(y)(x_2 - y_2) + E(z)$$

where

$$E(z) := \tfrac{1}{2}\left(\partial_{11}\varphi(z)(x_1 - y_1)^2 + 2\partial_{12}\varphi(z)(x_1 - y_1)(x_2 - y_2)\right.$$
$$\left.+ \partial_{22}\varphi(z)(x_2 - y_2)^2\right).$$

We'll make great profits out of this observation later. □

2.2* The Mean Value Inequality

We noted above that the Generalized Mean Value Theorem need not apply to maps that are not real-valued. It turns out that this is not a major difficulty. Indeed, most of the results of one-variable calculus that can be deduced from the Mean Value Theorem can also be obtained by using the so-called *Mean Value Inequality*: If O is an open subset of $\mathbb{R}$ that contains the open interval (a, b), and $f \in \mathbb{R}^O$ is differentiable, then

$$f(b) - f(a) \leq \sup\left\{\left|f'(t)\right| : t \in O\right\}(b - a).$$

Fortunately, this result extends nicely to the present framework, and this extension is all one needs for most purposes.[22]

THE MEAN VALUE INEQUALITY
Let X and Y be two normed linear spaces and O a nonempty open subset of X. Let $\Phi \in Y^O$ be Fréchet differentiable. Then, for every $x, y \in O$ with $co\{x, y\} \subseteq O$, there exists a real number $K \geq 0$ such that

$$\left\|\Phi(x) - \Phi(y)\right\|_Y \leq K\left\|x - y\right\|. \tag{14}$$

In particular, if

$$K \geq \sup\left\{\left\|\mathbf{D}_{\Phi,w}\right\|^* : w \in co\{x, y\}\right\}, \tag{15}$$

then (14) *holds.*

[22] For simplicity we state this result for maps defined on open sets, but it is straightforward to extend it to the case of maps defined on relatively open sets.

The method of proof that we will use to establish this result is one we have already used a good number of times, the method of "butterfly hunting." (Recall the proof of the Heine-Borel Theorem, for example.) We will proceed by assuming that the claim is not true, and then make a suitable nested-interval argument to derive a contradiction. The following elementary fact will be used in this argument.

Lemma 2
If x, y, and z are any points in a normed linear space with $z \in co\{x, y\}$, then

$$\|x - y\| = \|x - z\| + \|z - y\|.$$

Proof
By hypothesis, there exists a $0 \leq \lambda \leq 1$ such that $z = \lambda x + (1 - \lambda)y$. Then $x - z = (1 - \lambda)(x - y)$ and $z - y = \lambda(x - y)$. Thus

$$\|x - z\| + \|z - y\| = (1 - \lambda) \|x - y\| + \lambda \|x - y\| = \|x - y\|,$$

as we sought. ■

Proof of the Mean Value Inequality
Fix any $x, y \in O$ with $co\{x, y\} \subseteq O$, and take any $K \geq 0$ that satisfies (15).[23] Toward deriving a contradiction, suppose that there exists an $\varepsilon > 0$ such that

$$\|\Phi(x) - \Phi(y)\|_Y > (K + \varepsilon) \|x - y\|.$$

Let $x^0 := x$ and $y^0 := y$. Since, by subadditivity of $\|\cdot\|_Y$,

$$\left\|\Phi(x) - \Phi\left(\tfrac{1}{2}x + \tfrac{1}{2}y\right)\right\|_Y + \left\|\Phi\left(\tfrac{1}{2}x + \tfrac{1}{2}y\right) - \Phi(y)\right\|_Y > (K + \varepsilon) \|x - y\|,$$

either $\left\|\Phi(x) - \Phi\left(\tfrac{1}{2}x + \tfrac{1}{2}y\right)\right\|_Y$ or $\left\|\Phi\left(\tfrac{1}{2}x + \tfrac{1}{2}y\right) - \Phi(y)\right\|_Y$ must be strictly greater than $\tfrac{1}{2}(K + \varepsilon) \|x - y\|$. If the former case holds, let $x^1 := x^0$ and $y^1 := \tfrac{1}{2}x + \tfrac{1}{2}y$, and otherwise let $x^1 := \tfrac{1}{2}x + \tfrac{1}{2}y$ and $y^1 := y^0$. Either way we have $\|\Phi(x^1) - \Phi(y^1)\|_Y > (K + \varepsilon) \|x^1 - y^1\|$. Proceeding this way

[23] *Quiz.* How am I so sure that such a real number K exists?

inductively, we obtain two sequences (x^m) and (y^m) in $co\{x, y\}$ and a vector $z \in co\{x, y\}$ with the following properties: For all $m \in \mathbb{N}$,

(i) $z \in co\{x^m, y^m\}$,

(ii) $\lim x^m = z = \lim y^m$,

(iii) $\left\| \Phi(x^m) - \Phi(y^m) \right\|_Y > (K + \varepsilon) \left\| x^m - y^m \right\|$.

(Verify this carefully.[24])

Since Φ is Fréchet differentiable at z, there is a $\delta > 0$ such that $N_{\delta,X}(z) \subseteq O$ and

$$\left\| \Phi(\omega) - \Phi(z) - \mathsf{D}_{\Phi,z}(\omega - z) \right\|_Y \leq \varepsilon \left\| \omega - z \right\| \quad \text{for all } \omega \in N_{\delta,X}(z).$$

Then, by subadditivity of $\|\cdot\|_Y$, the choice of K (Section J.4.3), and (i),

$$\begin{aligned} \|\Phi(\omega) - \Phi(z)\|_Y &\leq \left\| \mathsf{D}_{\Phi,z}(\omega - z) \right\|_Y + \varepsilon \|\omega - z\| \\ &\leq \left\| \mathsf{D}_{\Phi,z} \right\|^* \|\omega - z\| + \varepsilon \|\omega - z\| \\ &\leq (K + \varepsilon) \|\omega - z\| \end{aligned}$$

for all $\omega \in N_{\delta,X}(z)$. By (ii), there exists an $M \in \mathbb{N}$ such that $x^M, y^M \in N_{\delta,X}(z)$, so by using this inequality along with (i) and Lemma 2, we find

$$\begin{aligned} \left\| \Phi(x^M) - \Phi(y^M) \right\|_Y &\leq \left\| \Phi(x^M) - \Phi(z) \right\|_Y + \left\| \Phi(z) - \Phi(y^M) \right\|_Y \\ &\leq (K + \varepsilon) \left(\left\| x^M - z \right\| + \left\| z - y^M \right\| \right) \\ &\leq (K + \varepsilon) \left\| x^M - y^M \right\| \end{aligned}$$

which contradicts (iii). Conclusion: $\left\| \Phi(x) - \Phi(y) \right\|_Y \leq (K + \varepsilon) \left\| x - y \right\|$ for all $\varepsilon > 0$. This means that (14) is true. ∎

Recall that a major corollary of the Mean Value Theorem is the fact that a real function whose derivative vanishes at an open interval must be constant on that interval. The Mean Value Inequality yields the following generalization of this fact.

[24] I am implicitly invoking Lemma 2 here along with Cantor's Nested Interval Lemma. Please make sure I'm not overlooking anything.

COROLLARY 1
Let X and Y be two normed linear spaces and O a nonempty open and convex subset of X. If $\Phi \in Y^O$ is Fréchet differentiable and $\mathbf{D}_{\Phi,x}$ is the zero operator for each $x \in O$, then there is a $y \in Y$ such that $\Phi(x) = y$ for all $x \in O$.

EXERCISE 31 Prove Corollary 1.

EXERCISE 32[H] Show that the term "convex" can be replaced with "connected" in Corollary 1.

EXERCISE 33[H] Let X and Y be two normed linear spaces, O a nonempty open and convex subset of X, and $x, y \in O$. Show that if $\Phi \in Y^O$ is Fréchet differentiable and $x_o \in O$, then

$$\left\| \Phi(x) - \Phi(y) - \mathbf{D}_{\Phi,x_o}(x - y) \right\|_Y \leq K \left\| x - y \right\|$$

for any $K \geq \sup \left\{ \left\| \mathbf{D}_{\Phi,w} - \mathbf{D}_{\Phi,x_o} \right\|^* : w \in co\{x, y\} \right\}$.

3 Fréchet Differentiation and Concave Maps

3.1 Remarks on the Differentiability of Concave Maps

You might recall that a concave function f defined on an open interval I possesses very nice differentiability properties. In particular, any such f is differentiable everywhere on I but *countably* many points. Although our main goal is to study those concave functions that are defined on convex subsets of an arbitrary normed linear space, it may still be a good idea to warm up by sketching a quick proof of this elementary fact.

EXAMPLE 9
Let I be an open interval and $f \in \mathbb{R}^I$ a concave map. It is easy to check that the right-derivative f'_+ of f is a well-defined and *decreasing* function on I. (Prove!) Thus if $(x_m) \in I^\infty$ is a decreasing sequence with $x_m \searrow x$, we have $\lim f'_+(x_m) \leq f'_+(x)$. On the other hand, concavity implies that

$$f'_+(x_m) \geq \frac{f(y) - f(x_m)}{y - x_m} \qquad \text{for all } y \in I \text{ with } y > x_m,$$

for each m, so, since f is continuous (Corollary A.2), we get

$$\lim f'_+(x_m) \geq \frac{f(y) - f(x)}{y - x} \qquad \text{for all } y \in I \text{ with } y > x.$$

In turn, this implies $\lim f'_+(x_m) \geq f'_+(x)$, so we find $\lim f'_+(x_m) = f'_+(x)$. But the analogous reasoning would yield this equation if (x_m) was an increasing sequence with $x_m \nearrow x$. (Yes?) Conclusion: f is differentiable at x iff f'_+ is continuous at x. But since f_+ is a monotonic function, it can have at most countably many points of discontinuity (Exercise B.8). Therefore, f is differentiable everywhere on I but countably many points. □

EXERCISE 34 Let I be an open interval and $f \in \mathbb{R}^I$ a concave map. Show that there exists a countable subset S of I such that $f' \in \mathbf{C}(I \backslash S)$.

Unfortunately, we are confronted with various difficulties in higher dimensions. For instance, the map $\varphi : \mathbb{R}^2 \to \mathbb{R}$ defined by $\varphi(u, v) := -|u|$ is a concave function that is not differentiable on the uncountable set $\mathbb{R} \times \{0\}$. Worse still, a concave function on an infinite-dimensional normed linear space may not possess a Fréchet derivative anywhere! For example, $\varphi : \ell^1 \to \mathbb{R}$ defined by $\varphi((x_m)) := -\|(x_m)\|_1$ is not Fréchet differentiable at any point in its domain (Exercise 17). Nevertheless, the concave functions that arise in most economic applications are in fact continuously differentiable, and hence for most practical purposes, these observations are not too problematic.

*REMARK 2. Just in case the comments above sounded overly dramatic, we mention here, without proof, two positive results about the Fréchet differentiability of concave maps.

(a) Take any $n \in \mathbb{N}$, and let us agree to say that a set S in $\mathbb{R}^n$ is **null** if, for all $\varepsilon > 0$, there exist countably many n-cubes such that (i) S is contained in the union of these cubes, and (ii) the sum of the side lengths of these cubes is at most ε. (Recall Section D.1.4.) One can show that if O is an open and convex subset of $\mathbb{R}^n$ and φ is a concave map on O, then the set of all points $x \in O$ at which φ fails to be Fréchet differentiable is null.

(b) *Asplund's Theorem.* If X is a Banach space such that X^* is separable, and φ is a continuous and concave real function defined on an open and convex subset O of X, then the set of all points at which φ is Fréchet differentiable is dense in O.[25] □

EXERCISE 35^H Let O be a nonempty open and convex subset of a normed linear space, and $\varphi \in \mathbb{R}^O$. We say that an affine map $\vartheta : X \to \mathbb{R}$ is a **support of** φ **at** $x \in O$ if $\vartheta|_O \geq \varphi$ and $\vartheta(x) = \varphi(x)$. Show that if φ is a concave map that is Fréchet differentiable at $x \in O$, then φ has a unique support at x.[26]

3.2 Fréchet Differentiable Concave Maps

Recall that there are various useful ways of characterizing the concavity of differentiable real maps defined on an open interval. It is more than likely that you are familiar with the fact that if I is an open interval and $\varphi \in \mathbb{R}^I$ is differentiable and concave, then the difference-quotient $\frac{\varphi(y)-\varphi(x)}{y-x}$ is less than $\varphi'(x)$ if $y > x$, while it exceeds $\varphi'(x)$ if $x > y$. (Draw a picture.) Put more concisely,

$$\varphi(y) - \varphi(x) \leq \varphi'(x)(y - x) \qquad \text{for all } x, y \in I.$$

A useful consequence of this observation is that a differentiable $\varphi \in \mathbb{R}^I$ is concave iff its derivative is a decreasing function on I. We now extend these facts to the context of real maps defined on open and convex subsets of an arbitrary normed linear space.

PROPOSITION 6

Let S be a relatively open and convex subset of a normed linear space and $\varphi \in \mathbb{R}^S$ a Fréchet differentiable map. Then, φ is concave if, and only if,

$$\varphi(y) - \varphi(x) \leq \mathsf{D}_{\varphi,x}(y - x) \qquad \textit{for all } x, y \in S. \tag{16}$$

[25] Asplund's Theorem is in fact more general than this. It says that the set of all points at which φ is Fréchet differentiable is a countable intersection of *open* and dense subsets of O. The easiest proof of this result that I know is the one given by Preiss and Zajicek (1984), who actually prove something even more general.

[26] WARNING. The converse of this statement is false in general, but it it is true when X is a Euclidean space.

PROOF
Suppose that φ is concave, and take any $x, y \in S$. Then,

$$\varphi(x + \lambda(y - x)) = \varphi((1-\lambda)x + \lambda y) \geq (1-\lambda)\varphi(x) + \lambda\varphi(y)$$

for all $0 \leq \lambda \leq 1$. Thus

$$\tfrac{1}{\lambda}\left(\varphi(x + \lambda(y-x)) - \varphi(x) - \mathsf{D}_{\varphi,x}(\lambda(y-x))\right) \geq \varphi(y) - \varphi(x) - \mathsf{D}_{\varphi,x}(y-x)$$

for each $0 < \lambda < 1$. We obtain (16) by letting $\lambda \to 0$ in this expression, and using the definition of $\mathsf{D}_{\varphi,x}$.[27]

Conversely, suppose (16) is true. Take any $x, y \in S$ and $0 < \lambda < 1$. If we set $z := \lambda x + (1-\lambda)y$, then $\lambda(x - z) + (1-\lambda)(y - z) = \mathbf{0}$, so (16) implies

$$\begin{aligned}\varphi(z) &= \lambda\varphi(z) + (1-\lambda)\varphi(z) + \mathsf{D}_{\varphi,z}(\lambda(x-z) + (1-\lambda)(y-z))\\ &= \lambda(\varphi(z) + \mathsf{D}_{\varphi,z}(x-z)) + (1-\lambda)(\varphi(z) + \mathsf{D}_{\varphi,z}(y-z))\\ &\geq \lambda\varphi(x) + (1-\lambda)\varphi(y),\end{aligned}$$

as we sought. ■

COROLLARY 2
Let S be a relatively open and convex subset of a normed linear space X and $\varphi \in \mathbb{R}^S$ a Fréchet differentiable map. Then, φ is concave if, and only if,

$$\left(\mathsf{D}_{\varphi,y} - \mathsf{D}_{\varphi,x}\right)(y - x) \leq 0 \quad \text{for all } x, y \in S. \tag{17}$$

EXERCISE 36 Prove Corollary 2.

EXERCISE 37 Verify that in Proposition 6 and Corollary 2 we can replace the term "concave" with "strictly concave," provided that we replace "$\leq$ " with "$<$ " in (16) and (17).

EXERCISE 38[H] Given any $n \in \mathbb{N}$, let O be a nonempty open and convex subset of $\mathbb{R}^n$, and assume that $\varphi \in \mathbb{R}^O$ is a map such that $\partial_i\varphi(x)$ exists for each $x \in O$ and $i = 1, \ldots, n$. Show that if φ is concave, then

$$\varphi(y) - \varphi(x) \leq \sum_{i=1}^{n} \partial_i\varphi(x)(y_i - x_i) \quad \text{for all } x, y \in O. \tag{18}$$

[27] Notice that I used here the Fréchet differentiability of φ only at x. Thus, $\varphi(y) - \varphi(x) \leq \mathsf{D}_{\varphi,x}(y - x)$ holds for any $y \in S$ and concave $\varphi \in \mathbb{R}^S$, provided that $\mathsf{D}_{\varphi,x}$ exists.

Conversely, if (18) holds, and each $\partial_i\varphi(\cdot)$ is a continuous function on O, then φ is concave.

If $X = \mathbb{R}$ in Corollary 2, then $\mathsf{D}_{\varphi,z}(t) = \varphi'(z)t$ for all $t \in \mathbb{R}$ and $z \in S$, so (17) reduces to $(\varphi'(y) - \varphi'(x))(y - x) \leq 0$ for all $x, y \in S$. So, a very special case of Corollary 2 is the well-known fact that a differentiable real map on an open interval is concave iff its derivative is decreasing on that interval. But a differentiable real function on a given open interval is decreasing iff its derivative is not strictly positive anywhere on that interval. It follows that a twice differentiable real map on an open interval is concave iff the second derivative of that map is less than zero everywhere. The following result generalizes this observation.

COROLLARY 3

Let S be a relatively open and convex subset of a normed linear space X and $\varphi \in \mathbb{R}^S$ a continuously Fréchet differentiable map that is also twice Fréchet differentiable. Then, φ is concave if, and only if,

$$\mathsf{D}^2_{\varphi,x}(z,z) \leq 0 \quad \textit{for all } (x,z) \in S \times X. \tag{19}$$

PROOF

If (19) is true, then, by the Generalized Second Mean Value Theorem, we have $\varphi(x) - \varphi(y) \leq \mathsf{D}_{\varphi,y}(x - y)$ for any $x, y \in S$, so the claim follows from Proposition 6. Conversely, suppose φ is concave, and take any $x \in S$ and $z \in span(S - x)$. Observe that $x + \lambda z \in x + span(S - x) = aff(S)$, for any $\lambda \in \mathbb{R}$. So, since S is relatively open and $x \in S$, we can choose a $\delta > 0$ small enough so that (i) $x + \lambda z \in S$ for all $-\delta < \lambda < \delta$; and (ii) the map $f \in \mathbb{R}^{(-\delta,\delta)}$ defined by $f(\lambda) := \varphi(x + \lambda z)$ is concave so that $f''(0) \leq 0$. (Right?) But we have $f''(\lambda) = \mathsf{D}^2_{\varphi,x+\lambda z}(z,z)$ for any $-\delta < \lambda < \delta$. (Yes?) Hence, $f''(0) \leq 0$ implies $\mathsf{D}^2_{\varphi,x}(z,z) \leq 0$, as we sought. ■

EXERCISE 39 Show that, under the conditions of Corollary 3, if $\mathsf{D}^2_{\varphi,x}(z,z) < 0$ for all $(x,z) \in S \times X$, then φ is strictly concave, but not conversely.

You of course know that the derivative of a differentiable real map need not be continuous. (That is, a differentiable real function need not be continuously differentiable.) One of the remarkable properties of concave (and hence convex) real functions defined on an open interval is that their differentiability implies their continuous differentiability. (Go back and reexamine Example 9—we proved this fact there.) As it happens, this observation generalizes to our present setting as well. Indeed, for concave real maps on a normed linear space, the notions of Fréchet differentiability and continuous Fréchet differentiability coincide.

Proposition 7
Let O be a nonempty open and convex subset of a normed linear space X and $\varphi \in \mathbb{R}^O$ a concave map. If φ is Fréchet differentiable, then it is continuously Fréchet differentiable.

Proof
Let $x \in O$, and take any $(x^m) \in O^\infty$ with $x^m \to x$. Define

$$\alpha_m := \left\| \mathsf{D}_{\varphi,x^m} - \mathsf{D}_{\varphi,x} \right\|^*, \qquad m = 1, 2, \ldots$$

We wish to show that $\alpha_m \to 0$. Note first that, by Proposition 2, for each $\varepsilon > 0$ there exists a $\delta > 0$ such that

$$\left| \varphi(\omega) - \varphi(x) - \mathsf{D}_{\varphi,x}(\omega - x) \right| \leq \varepsilon \|\omega - x\| \qquad \text{for all } \omega \in N_{\delta,X}(x). \tag{20}$$

(Yes?) On the other hand, by Proposition 6,

$$\varphi(\omega) - \varphi(x^m) \leq \mathsf{D}_{\varphi,x^m}(\omega - x^m) \qquad \text{for all } \omega \in O \text{ and } m = 1, 2, \ldots \tag{21}$$

The key is to notice that, for each m, we can find a $y^m \in X$ with $\left\| y^m - x \right\| = \delta$ and $\left(\mathsf{D}_{\varphi,x} - \mathsf{D}_{\varphi,x^m} \right)(y^m - x) \geq \frac{\delta}{2}\alpha_m$.[28] Then, setting $\omega = y^m$ in (20) and (21), respectively, we find

$$-\varphi(y^m) + \varphi(x) + \mathsf{D}_{\varphi,x}(y^m - x) \leq \varepsilon\delta$$

[28] All I'm saying here is that, for any $L \in X^*$, I can surely find a $z \in X$ with $\|z\| = \delta$ and $L(z) \geq \frac{\delta}{2} \|L\|^*$. If there was no such z, then we would have $L(w) < \frac{1}{2} \|L\|^*$ for every $w \in X$ with $\|w\| = 1$, but, since $\|L\|^* = \sup\{|L(w)| : w \in X \text{ and } \|w\| = 1\}$, this would be absurd.

and

$$\varphi(y^m) - \varphi(x^m) - \mathsf{D}_{\varphi,x^m}(y^m - x^m) \leq 0$$

for each m. It follows that

$$\begin{aligned}\varepsilon\delta &\geq \varphi(x) - \varphi(x^m) + \mathsf{D}_{\varphi,x}(y^m - x) - \mathsf{D}_{\varphi,x^m}(y^m - x^m) \\ &= \varphi(x) - \varphi(x^m) + (\mathsf{D}_{\varphi,x} - \mathsf{D}_{\varphi,x^m})(y^m - x) + \mathsf{D}_{\varphi,x^m}(x^m - x) \\ &\geq \varphi(x) - \varphi(x^m) + \tfrac{\delta}{2}\alpha_m + \mathsf{D}_{\varphi,x^m}(x^m - x)\end{aligned}$$

for each m. Since $x^m \to x$, applying the definition of D_{φ,x^m}, therefore, we find $\limsup \alpha_m \leq 2\varepsilon$. Since $\varepsilon > 0$ was arbitrary in this argument, we are done. ∎

Perhaps in an advanced calculus course you have come across the following claim: Given any $n \in \mathbb{N}$ and an open and convex subset O of $\mathbb{R}^n$, if $\varphi : O \to \mathbb{R}$ is concave and differentiable, then φ has continuous (first) partial derivatives. This is, in fact, an immediate consequence of Proposition 7 and Example 1.[3].

Here is another pleasant consequence of Proposition 7.

Corollary 4

Let O be a nonempty open and convex subset of a normed linear space X and $\varphi \in \mathbb{R}^O$ a twice Fréchet differentiable map. If φ is concave, then

$$\mathsf{D}^2_{\varphi,x}(z,z) \leq 0 \quad \textit{for all } (x,z) \in O \times X.$$

Proof

Apply Corollary 3 and Proposition 7. ∎

We conclude this section with a few preliminary remarks about the notion of the *superdifferential* of a real map. Although this concept is useful in the theory of nonsmooth optimization, we will not need it in the rest of our treatment, so our discussion proceeds only in terms of a few exercises.

DEFINITION

Let O be a nonempty open subset of a normed linear space X and $\varphi \in \mathbb{R}^O$. We say that a continuous linear functional $L \in X^*$ is a **supergradient** of φ at $x \in O$ if

$$\varphi(\omega) \leq \varphi(x) + L(\omega - x) \qquad \text{for all } \omega \in O.$$

The set of all supergradients of φ at x is called the **superdifferential** of φ at x, and is denoted by $\partial\varphi(x)$.

Geometrically speaking, L is a supergradient of φ at x iff the hyperplane $H_L := \{(\omega, t) \in X \times \mathbb{R} : L(\omega - x) - t = -\varphi(x)\}$ lies above the graph of φ. Since $(x, \varphi(x))$ is on this hyperplane, we can therefore say that L is a supergradient of φ at x iff H_L supports the set $\{(\omega, t) : \varphi(\omega) \geq t\}$ at $(x, \varphi(x))$. (Recall that the latter set is called the *hypograph* of φ.) In turn, $\partial\varphi(x)$ equals the set of all L such that H_L supports the hypograph of φ at $(x, \varphi(x))$.

EXERCISE 40^{H} Consider the real maps f, g, and h defined on $\mathbb{R}_{++}$ by

$$f(t) := \sqrt{t}, \qquad g(t) := \begin{cases} t, & \text{if } 0 < t \leq 1 \\ \frac{1}{2}(t+1), & \text{otherwise} \end{cases}, \qquad \text{and} \qquad h(t) := t^2,$$

respectively. Compute $\partial f(1)$, $\partial g(1)$, and $\partial h(1)$. (Draw pictures to illustrate the situation.)

EXERCISE 41 Let O be a nonempty open and convex subset of a normed linear space X and $\varphi \in \mathbb{R}^O$ a concave map. Show that $\partial\varphi(x) = \{F - F(\mathbf{0}) : F$ is a continuous affine map on X with $F|_O \geq \varphi$ and $F(x) = \varphi(x)\}$. (So, how does $\partial\varphi(x)$ relate to the supports of φ at x? Recall Exercise 35.)

EXERCISE 42^{H} Let O be a nonempty open and convex subset of a normed linear space X and $x \in O$. Prove: If $\varphi \in \mathbb{R}^O$ is a locally bounded concave map, then $\partial\varphi(x)$ is a nonempty, closed and convex set.

*EXERCISE 43 Let O be a nonempty open and convex subset of a normed linear space X. Prove: If $\varphi \in \mathbb{R}^O$ is a bounded and concave map, then $\partial\varphi(\cdot)$ is an upper hemicontinuous correspondence from X into X^*.

*EXERCISE 44[H] Let O be a nonempty open and convex subset of a normed linear space X and $\varphi \in \mathbb{R}^O$.

(a) Show that if φ is Fréchet differentiable, then $\partial\varphi(x) \subseteq \{\mathsf{D}_{\varphi,x}\}$.

(b) Show that if φ is Fréchet differentiable and concave, then $\partial\varphi(x) = \{\mathsf{D}_{\varphi,x}\}$.[29]

4 Optimization

Classical differential calculus provides powerful methods for optimizing a given real function of finitely many variables. Indeed, students of formal economics are bound to be big fans of the (informal) rule, "equate the first derivative of the objective function to zero and solve for the independent variables." The general theory of calculus we have developed in this chapter can be used in much the same way. Here is how.

4.1 Local Extrema of Real Maps

Let us begin by recalling the following standard terminology of optimization theory.

> DEFINITION
> Let S be a nonempty subset of a normed linear space X, and φ a real map on S. We say that a point x^* in S is a **local maximum** of φ if
>
> $$\varphi(x^*) \geq \varphi(N_{\delta,X}(x^*) \cap S) \qquad \text{for some } \delta > 0,$$
>
> and a **global maximum** of φ if
>
> $$\varphi(x^*) \geq \varphi(S).$$
>
> In turn, x^* is a **local (global) minimum** of φ if it is a local (global) maximum of $-\varphi$. By a **local (global) extremum** of φ, we mean a point in S that is either a local (global) maximum or local (global) minimum of φ.

[29] *Note.* If X is a Banach space, this fact can be sharpened nicely. In that case, if φ is upper semicontinuous and concave, then $\partial\varphi(x) = \{\mathsf{D}_{\varphi,x}\}$ holds at any $x \in O$ at which $\mathsf{D}_{\varphi,x}$ exists. The proof of this result is much harder, however. (See Brönstedt and Rockefellar (1965).)

DEFINITION

Let T be a subset of a normed linear space X and φ a real map on T.[30] We say that a point $x^* \in T^\diamond$ is a **stationary point** of φ if D_{φ,x^*} is the zero functional on $span(T - x^*)$, that is,

$$\mathsf{D}_{\varphi,x^*}(x) = 0 \qquad \text{for all } x \in span(T - x^*).$$

The rule "to maximize (or minimize) a given function, equate its first derivative to zero, and solve for the independent variable(s)" can thus be reworded as "to maximize (or minimize) a given function, find its stationary points." Or, put formally, we have the following.

PROPOSITION 8

Let T be a nonempty subset of a normed linear space X, $x^* \in T^\diamond$, and $\varphi \in \mathbb{R}^T$. If φ is Fréchet differentiable at x^* and x^* is a local extremum of φ, then x^* is a stationary point of φ.

PROOF

Assume that φ is Fréchet differentiable at x^*, and without loss of generality, let x^* be a local maximum of φ. To derive a contradiction, suppose $\mathsf{D}_{\varphi,x^*}(x) > 0$ for some $x \in span(T - x^*)$.[31] (Obviously, $x \neq \mathbf{0}$.) Now define $\omega_\lambda := x^* + \lambda x$ for any $\lambda > 0$. Then, $\omega_\lambda \in x^* + span(T - x^*) = aff(T)$ for any $\lambda > 0$, while $\lim_{\lambda \to 0} \omega_\lambda = x^*$. Since $x^* \in T^\diamond$, therefore, there exists a $\lambda_o > 0$ such that $\omega_\lambda \in T^\diamond$ for every $0 < \lambda < \lambda_o$. (Yes?) Then, by Proposition 2* (of Section 1.7),

$$\varphi(\omega_\lambda) - \varphi(x^*) = \lambda \mathsf{D}_{\varphi,x^*}(x) + \mathbf{e}(\omega_\lambda), \quad 0 < \lambda < \lambda_o, \tag{22}$$

where $\mathbf{e} \in \mathbb{R}^T$ satisfies $\lim_{\lambda \to 0} \frac{|\mathbf{e}(\omega_\lambda)|}{\lambda \|x\|} = 0$. The latter condition ensures that, for any $\alpha > 0$, there is a $0 < \lambda(\alpha) < \lambda_o$ small enough so that $|\mathbf{e}(\omega_\lambda)| < \alpha\lambda \|x\|$ for any $0 < \lambda < \lambda(\alpha)$. So, if we choose $\alpha := \frac{\mathsf{D}_{\varphi,x^*}(x)}{\|x\|}$ and denote the corresponding $\lambda(\alpha)$ by λ^*, we find $\lambda \mathsf{D}_{\varphi,x^*}(x) + \mathbf{e}(\omega_\lambda) > 0$ for all $0 < \lambda < \lambda^*$, whereas $\lambda^* < \lambda_o$. Combining this with (22), then, $\varphi(\omega_\lambda) > \varphi(x^*)$ for any $0 < \lambda < \lambda^*$, but this contradicts x^* being a local maximum of φ. Since a

[30] *Reminder.* $T^\diamond := int_{aff(T)}(T)$.

[31] The idea of the proof is duly simple: If $\mathsf{D}_{\varphi,x^*}(x) > 0$, then the value of φ must increase strictly as we slightly move from x^*, within $aff(T)$, in the direction of x.

similar contradiction would obtain under the presumption that $\mathsf{D}_{\varphi,x^*}(x) < 0$ for some $x \in T$, the assertion is proved. ∎

EXAMPLE 10

[1] Consider the map $\varphi : \mathbf{C}[0,1] \to \mathbb{R}$ defined by

$$\varphi(f) := \int_0^1 f(t)^2 dt.$$

We have seen in Example 2 that, for any $f \in \mathbf{C}[0,1]$,

$$\mathsf{D}_{\varphi,f}(g) = 2\int_0^1 f(t)g(t)dt \quad \text{for all } g \in \mathbf{C}[0,1].$$

It readily follows that $\mathsf{D}_{\varphi,f}$ is the zero functional on $\mathbf{C}[0,1]$ iff $f = 0$. Thus, by Proposition 8, there can be only one local extremum of φ, namely, the real function that equals zero everywhere on $[0,1]$. A moment's reflection would show that this function is in fact the global minimum of φ.

[2] Let us modify the previous example. Define $S := \{f \in \mathbf{C}[0,1] : f(0) = 1\}$, and consider the map $\psi := \varphi|_S$. Notice that S is an affine manifold, so, trivially, it is relatively open. Moreover, for any $f \in S$, we have

$$\mathsf{D}_{\psi,f}(g) = 2\int_0^1 f(t)g(t)dt \quad \text{for all } g \in span(S - f),$$

or equivalently,

$$\mathsf{D}_{\psi,f}(g) = 2\int_0^1 f(t)g(t)dt \quad \text{for all } g \in \mathbf{C}[0,1] \text{ with } g(0) = 0.$$

(Recall Remark 1.) Now suppose $f^* \in S$ is a local extremum of ψ. Then, by Proposition 8,

$$\int_0^1 f^*(t)g(t)dt = 0 \quad \text{for all } g \in \mathbf{C}[0,1] \text{ with } g(0) = 0.$$

But it is easy to see that this is impossible. Indeed, since $f^*(0) = 1$ and f^* is continuous, there exists an $\varepsilon > 0$ such that $f^*(t) \geq \frac{1}{2}$ for all $0 \leq t \leq \varepsilon$. So, if we define $g \in \mathbf{C}[0,1]$ by $g(t) := t(\varepsilon - t)$ for any $0 \leq t \leq \varepsilon$, and $g(t) := 0$ otherwise, then $\int_0^1 f^*(t)g(t)dt \geq \frac{1}{2}\int_0^\varepsilon t(\varepsilon - t)dt > 0$, contradicting what we have found above. Conclusion: ψ has no local extrema. □

EXERCISE 45 Consider the real functions φ_1, φ_2 and φ_3 defined on $N_{1,\ell^\infty}(\mathbf{0})$ by

$$\varphi_1((x_m)) := \sum_{i=1}^{\infty} \tfrac{1}{2^i} x_i, \quad \varphi_2((x_m)) := \sum_{i=1}^{\infty} \tfrac{1}{2^i} \sqrt{x_i}, \quad \text{and}$$

$$\varphi_3((x_m)) := \sum_{i=1}^{\infty} \tfrac{1}{2^{i+1}} x_i(x_i - 1),$$

respectively. Find all local extrema of these functions.

EXERCISE 46[H] Let O be a nonempty open and convex subset of a normed linear space X and $\varphi \in \mathbb{R}^O$ a continuously Fréchet differentiable map that is also twice Fréchet differentiable. Prove: If $x^* \in O$ is a local maximum of φ, then

$$\mathbf{D}^2_{\varphi,x^*}(z,z) \leq 0 \qquad \text{for all } z \in X.$$

Stationarity of a point in the domain of a given function is often referred to as the *first-order necessary condition for local extremum*. Indeed, Proposition 8 can be used to identify all candidates for, say, a local maximum of a Fréchet differentiable real function whose domain is a relatively open subset of a normed linear space.[32] If the function is twice Fréchet differentiable, by examining the behavior of its second Fréchet derivative in a (small) neighborhood of a stationary point, however, we may learn if that point indeed corresponds to a local extremum. This is the content of our next result.

PROPOSITION 9
Let S be a relatively open and convex subset of a normed linear space X and $\varphi \in \mathbb{R}^S$ a continuously Fréchet differentiable map that is also twice Fréchet differentiable. If $x^* \in S$ is a stationary point of φ, and there exists a $\delta > 0$ such that

$$\mathbf{D}^2_{\varphi,x}(y,y) \leq 0 \qquad \text{for all } x \in N_{\delta,X}(x^*) \cap S \text{ and } y \in span(S - x^*), \qquad (23)$$

then x^* is a local maximum of φ.

32 You don't need me to remind you that stationarity of a point does not imply that it is a local extremum. (Think of the behavior of the map $t \mapsto t^3$ at 0.)

PROOF

Suppose x^* is a stationary point of φ such that (23) holds for some $\delta > 0$, and yet x^* is not a local maximum for φ. Then there exists a $w \in N_{\delta,X}(x^*) \cap S$ such that $\varphi(x^*) < \varphi(w)$. Clearly, $w = x^* + \lambda y$ for some $\lambda > 0$ and some $y \in span(S - x^*)$ with $\|y\| = 1$. (Why?) By the Generalized Second Mean Value Theorem and stationarity of x^*, then,

$$0 < \varphi(w) - \varphi(x^*) = \tfrac{1}{2}\lambda^2 \mathbf{D}^2_{\varphi,z}(y, y)$$

for some $z \in co\{w, x^*\}$, distinct from w and x^*. Since $co\{w, x^*\} \subseteq N_{\delta,X}(x^*) \cap S$, we have $z \in N_{\delta,X}(x^*)$, so this finding contradicts (23). ■

4.2 Optimization of Concave Maps

In applications, one is frequently interested in finding the global extrema of a given real function. Unfortunately, it is in general not so easy to determine whether a local extremum is in fact a global extremum. The case where the function to be optimized is concave (or convex), is, however, a much welcome exception. In that case the involved optimization problems are simplified by the following (informal) rule: To find the global maximum (minimum) of a concave (convex) real map, it is enough to examine the stationary points of that function. You are surely familiar with this principle in the case of concave or convex functions defined on an open and convex subset of a Euclidean space. We formalize it in the general case below.

First a few preliminary observations about the optimization of concave maps.

EXERCISE 47 Let S be a convex subset of a linear space X and $\varphi \in \mathbb{R}^S$ a concave map. Show that if $x^* \in S$ is a local maximum of φ, then it is also a global maximum of φ.

EXERCISE 48^H Let S be a compact subset of a metric linear space X with $bd_X(S) \neq \emptyset$. Prove: If $\varphi \in \mathbb{R}^S$ is a locally bounded concave map, then

$$\inf\{\varphi(x) : x \in S\} = \inf\{\varphi(x) : x \in bd_X(S)\}.$$

Here is a very useful result that shows how greatly simplified the task of maximization is in the case of concave objective functions.

PROPOSITION 10
Let T be a nonempty subset of a normed linear space X, $x^* \in T^\diamond$, and $\varphi \in \mathbb{R}^T$ a concave map that is Fréchet differentiable at x^*. Then, x^* is a global maximum of φ if, and only if, x^* is a stationary point of φ.

PROOF
By Proposition 6,

$$\varphi(x) - \varphi(x^*) \leq \mathbf{D}_{\varphi,x^*}(x - x^*) \qquad \text{for all } x \in T.$$ [33]

So, if x^* is a stationary point of φ, that is, $\mathbf{D}_{\varphi,x^*}$ is the zero functional on X, we have $\varphi(x^*) \geq \varphi(x)$ for all $x \in T$. The "only if" part of the assertion is a special case of Proposition 8. ■

EXERCISE 49 Let O be a nonempty open and convex subset of a normed linear space X and $\varphi \in \mathbb{R}^O$ a concave map. For any given $x^* \in O$, prove that the zero functional on X belongs to $\partial\varphi(x^*)$ iff x^* is a global maximum of φ. (Here $\partial\varphi(x^*)$ is the superdifferential of φ at x^*.)

WARNING. The condition for a concave map to attain its global *minimum* in the interior of its domain is trivial. Indeed, if O is a convex and open subset of a normed linear space and $\varphi \in \mathbb{R}^O$ is concave, then φ has a global minimum iff it is constant on O. (Recall Exercise G.29.)

The following is a useful companion to Proposition 10. It gives sufficient conditions for a concave function to attain a *unique* global maximum.

PROPOSITION 11
Let T be a convex subset of a normed linear space X, $x^* \in T^\diamond$, and $\varphi \in \mathbb{R}^T$ a concave map that is Fréchet differentiable at x^*. If x^* is a stationary point of φ, and there exists a $\delta > 0$ such that

$$\mathbf{D}^2_{\varphi,x}(y, y) < 0 \qquad \textit{for all } x \in N_{\delta,X}(x^*) \cap T \textit{ and } y \in \mathit{span}(T - x^*),$$

then x^* is the unique global maximum of φ.

EXERCISE 50 Prove Proposition 11.

[33] Please have a look at footnote 31.

5 Calculus of Variations

The main advantage of the generalized calculus theory we have worked out in this chapter lies, at least for economists, in its applicability to optimization problems that take place in infinite-dimensional normed linear spaces. Indeed, a variety of such problems are encountered in the daily practice of macroeconomists. Such problems are often handled by some sort of a trick that enables the analyst to use finite-dimensional techniques, or by using "optimization recipes," the origins of which are left unexplored. By contrast, the present chapter tells us that the basic ideas behind finite- and infinite-dimensional optimization theory are identical; it is just that in the latter case we should work with Fréchet derivatives.

Well, that's talking the talk. In this final section of the text, we aim to walk the walk as well. As a case study, we take on the classical theory of calculus of variations and sketch a brief but rigorous introduction to it. We prove several necessary and sufficient conditions for (finite- and infinite-horizon) variational problems, all as easy consequences of the results of Section 4. We also consider a few applications, economic or otherwise, at the end of the discussion.[34]

Before we begin, let us agree on adopting the following standard notation.

NOTATION. Given any integer $n \geq 2$, let S be a convex subset of $\mathbb{R}^n$ with nonempty interior. For any $k \in \mathbb{N}$, we denote by $\mathbf{C}^k(S)$ the class of all $\varphi \in \mathbf{C}(S)$ such that each partial derivative of φ up to the kth order exists as a continuous function on $int_{\mathbb{R}^2}(S)$.[35]

5.1 Finite-Horizon Variational Problems

We begin by laying out, in the abstract, the class of variational problems that we shall be concerned with here.

[34] This section is a minicourse on the (one-dimensional) calculus of variations. There are a few texts in economics that provide such courses, mostly on the way to control theory. I find the heuristic approaches of most of these texts unsatisfactory, but let me mention that Chapter 1 of Seierstad and Sydsæter (1993) provides an honest treatment.

The math literature on this topic is extensive. Gelfand and Fomin (1968) is a classic, while Butazzo et al. (1998), Dacorogna (2003), and van Brunt (2004) provide nice up-to-date treatments. None of these references is particularly suitable for the interests of economists, however.

[35] For instance, if S is open, then we have $\mathbf{C}^2(S) := \{\varphi \in \mathbf{C}(S) : \partial_i\varphi \in \mathbf{C}(S)$ and $\partial_{ij}\varphi \in \mathbf{C}(S)$ for any $i, j = 1, \ldots, n\}$.

Let x_0 and τ be real numbers, and Ω a convex subset of $\mathbb{R}^2$. Throughout this section we assume that

$$\tau > 0 \quad \text{and} \quad int_{\mathbb{R}^2}(\Omega) \neq \emptyset.$$

To simplify our notation, for any $f \in \mathbf{C}^1[0,\tau]$, we express the statement "$(f(t),f'(t)) \in \Omega$ for all $0 \leq t \leq \tau$" by writing "$(f,f') \in \Omega$." With this convention in mind, we define

$$\mathbf{C}^1_{\Omega,x_0}[0,\tau] := \{f \in \mathbf{C}^1[0,\tau] : f(0) = x_0 \text{ and } (f,f') \in \Omega\},$$

and consider this convex set as a metric subspace of $\mathbf{C}^1[0,\tau]$. Of course, if $\Omega = \mathbb{R}^2$, then $\mathbf{C}^1_{\Omega,x_0}[0,\tau]$ is none other than the affine manifold of $\mathbf{C}^1[0,\tau]$ that contains all continuously differentiable maps on $[0,\tau]$ with $f(0) = x_0$.

Now take any nonempty set $\mathcal{F}$ such that

$$\mathcal{F} \subseteq \mathbf{C}^1_{\Omega,x_0}[0,\tau],$$

and pick any $\varphi \in \mathbf{C}^2([0,\tau] \times \Omega)$. In what follows, we refer to the list $[\Omega,\varphi,x_0,\tau,\mathcal{F}]$ as a **smooth variational problem**, and understand that this list corresponds to the following optimization problem:

$$\text{Maximize} \int_0^\tau \varphi(t,f,f')dt \quad \text{such that} \quad f \in \mathcal{F}, \tag{24}$$

where we write $\varphi(t,f,f')$ in place of the map $t \mapsto \varphi(t,f(t),f'(t))$ on $[0,\tau]$ to simplify our notation further.[36] (With an innocent abuse of terminology, we will henceforth refer to the problem (24) also as a *smooth variational problem*.) Here $\mathcal{F}$ is called the **admissible set** of the problem, and φ its **Lagrangian.**[37]

Different choices of $\mathcal{F}$ lead to different sorts of calculus of variations problems. In particular, the smooth variational problem $[\Omega,\varphi,x_0,\tau,\mathcal{F}]$,

[36] In most introductory treatments of calculus of variations, both in economics and mathematics, it is postulated at the outset that $\Omega = \mathbb{R}^2$. This is a bit troublesome, for in most applications there are basic constraints on the domain of the Lagrangians, and this pushes the involved models outside the formal coverage of the theory developed under the hypothesis $\Omega = \mathbb{R}^2$. (See Example 11, for instance.) I am exposing a slightly more complicated version of the classic theory here in order to be able to deal with such constraints properly.

[37] I use the word "smooth" here to signal that in these problems, one considers only continuously differentiable maps as admissible. The integral in (24) makes sense for a variety of other types of f, so one may in fact work in general with much larger admissible sets than I consider here. But doing this would complicate things somewhat, so I confine myself to smooth variational problems here.

where

$$\mathcal{F} = \{f \in \mathbf{C}^1_{\Omega,x_0}[0,\tau] : f(\tau) = x_\tau\} \tag{25}$$

for some $x_\tau \in \mathbb{R}$ is very important; this problem is referred to as a **simplest variational problem**. Also of interest is the so-called **free-end smooth variational problem** $[\Omega, \varphi, x_0, \tau, \mathcal{F}]$, where

$$\mathcal{F} = \mathbf{C}^1_{\Omega,x_0}[0,\tau].$$

Historically speaking, the interest in variational problems stemmed from the natural way these problems arise in Newtonian mechanics and related fields.[38] But of course, economists have their own reasons for studying calculus of variations. In fact, it is highly likely that you have already come across several dynamic optimization models in continuous time, so there is hardly any reason to motivate our interest in smooth variational problems. Just to be on the safe side, however, let us consider the following prototypical example.

EXAMPLE 11
(*The Optimal Investment Problem*) Consider a firm whose production technology enables it to make a profit of $P(x)$ by using a capital stock of x units. We assume that $P \in \mathbf{C}^2(\mathbb{R}_+)$ is a strictly increasing and twice continuously differentiable function with $P(0) = 0$. The planning horizon is finite, but time is modeled continuously. So, the "time space" is $[0, \tau]$, where 0 represents "today" and $\tau > 0$ the final date of the planning horizon. The capital accumulation of the firm through time is modeled by means of a map $f \in \mathbf{C}^1[0,\tau]$ with $f \geq 0$ and $f(0) = x_0$, where $x_0 > 0$ stands for the initial capital stock of the firm, and is exogenously given. The instantaneous change in capital stock is interpreted as gross investment, which is costly. In particular, by increasing its capital stock at a given time by $\varepsilon > 0$, the firm incurs a cost of $C(\varepsilon)$, where $C \in \mathbf{C}^2(\mathbb{R}_+)$ is strictly increasing and $C(0) = 0$. Thus, if the capital accumulation plan of the firm is f, then, at time t, the investment cost of the firm is $C(f'(t))$, and hence, its total profit is found as $P(f(t)) - C(f'(t))$. Finally, let us assume that the firm discounts the future

[38] See Goldstine (1980) for a very detailed review of the origins of calculus of variations.

according to the discount function $\eth \in \mathbf{C}^2[0,\tau]$ with $\eth' < 0$ and $\eth(0) = 1$. (That is, $\eth(t)(P(f(t)) - C(f'(t)))$ is the value of the time t profits from the "present" perspective of the firm.) Then, the associated optimization problem is this:

$$\text{Maximize} \int_0^\tau \eth(t)(P(f(t)) - C(f'(t)))dt$$

$$\text{such that, } f \in \mathbf{C}^1[0,\tau],\ f(0) = x_0 \text{ and } f, f' \geq 0.$$

This is, of course, the free-end smooth variational problem $[\Omega, \varphi, x_0, \tau, \mathcal{F}]$, where $\Omega := \mathbb{R}^2_+$, and φ is defined on $[0,\tau] \times \Omega$ by $\varphi(t,a,b) := \eth(t)(P(a) - C(b))$. □

5.2 The Euler-Lagrange Equation

We begin our investigation of the finite-horizon variational problems by putting on record a few auxiliary observations (which hold trivially for (unconstrained) problems with $\Omega = \mathbb{R}^2$). The arguments are simplified by the following elementary observation.

OBSERVATION 1. Let O be an open subset of $\mathbb{R}^2$, and take any $u \in \mathbf{C}^1[0,\tau]$ with $(u, u') \in O$. If (u_m) is a sequence in $\mathbf{C}^1[0,\tau]$ with $u_m \to u$, then there exists a real number $M > 0$ such that $(u_m, u'_m) \in O$ for all $m \geq M$.[39]

PROOF
Let us first show that there exists an $\varepsilon > 0$ such that

$$\bigcup\{N_{\varepsilon,\mathbb{R}^2}(u(t), u'(t)) : 0 \leq t \leq \tau\} \subseteq O. \tag{26}$$

If this was not the case, we could find a sequence $(t_m) \in [0,\tau]^\infty$ such that

$$N_{\frac{1}{m},\mathbb{R}^2}(u(t_m), u'(t_m)) \backslash O \neq \emptyset, \qquad m = 1, 2, \ldots$$

[39] *Reminder.* By $(u, u') \in O$, we mean $(u(t), u'(t)) \in O$ for all $0 \leq t \leq \tau$, and similarly for $(u_m, u'_m) \in O$. Recall also that $u_m \to u$ means $\|u_m - u\|_{\infty,\infty} := \|u_m - u\|_\infty + \|u'_m - u'\|_\infty \to 0$, as the entire action takes place here within $\mathbf{C}^1[0,\tau]$ (Example J.1.[4]).

By the Bolzano-Weierstrass Theorem, (t_m) has a subsequence that converges to some t^* in $[0, \tau]$. We denote this subsequence again by (t_m), relabeling if necessary. Since O is open, there exists a $\delta > 0$ such that $N_{\delta,\mathbb{R}^2}(u(t^*), u'(t^*)) \subseteq O$. But since both u and u' are continuous, we have $u(t_m) \to u(t^*)$ and $u'(t_m) \to u'(t^*)$, so, for some real number $K > 0$,

$$(u(t_m), u'(t_m)) \in N_{\frac{\delta}{2},\mathbb{R}^2}(u(t^*), u'(t^*)) \subseteq O \quad \text{for all } m \geq K.$$

But then, for any $m \geq \max\{\frac{2}{\delta}, K\}$,

$$N_{\frac{1}{m},\mathbb{R}^2}(u(t_m), u'(t_m)) \subseteq N_{\delta,\mathbb{R}^2}(u(t^*), u'(t^*)) \subseteq O,$$

a contradiction.

The rest is easy. If (u_m) is a sequence in $\mathbf{C}^1[0, \tau]$ with $\|u_m - u\|_{\infty,\infty} \to 0$, then there exists a real number $M > 0$ such that $\|u_m - u\|_\infty < \frac{\varepsilon}{\sqrt{2}}$ and $\|u'_m - u'\|_\infty < \frac{\varepsilon}{\sqrt{2}}$ for each $m \geq M$, and combining this with (26) yields the claim. □

Fix an arbitrary smooth variational problem $[\Omega, \varphi, x_0, \tau, \mathcal{F}]$. The first consequence we wish to derive from Observation 1 is that the interior of $\mathbf{C}^1_{\Omega,x_0}[0, \tau]$ in its affine hull is nonempty. We owe this fact to the hypothesis that Ω has nonempty interior.

OBSERVATION 2. If $f \in \mathbf{C}^1[0, \tau]$ satisfies $f(0) = x_0$ and $(f, f') \in int_{\mathbb{R}^2}(\Omega)$, then

$$f \in (\mathbf{C}^1_{\Omega,x_0}[0, \tau])^\diamond.$$

PROOF

Take any $f \in \mathbf{C}^1[0, \tau]$ such that $f(0) = x_0$ and $(f, f') \in int_{\mathbb{R}^2}(\Omega)$, and let (f_m) be a sequence in $aff(\mathbf{C}^1_{\Omega,x_0}[0, \tau])$ with $f_m \to f$. Obviously, $f_m(0) = x_0$ for each m. Moreover, by Observation 1, there is an $M > 0$ such that $(f_m, f'_m) \in \Omega$ for all $m \geq M$. Conclusion: Any sequence in $aff(\mathbf{C}^1_{\Omega,x_0}[0, \tau])$ that converges to f eventually enters (and stays) in $\mathbf{C}^1_{\Omega,x_0}[0, \tau]$. □

Thanks to this observation, we now see that the interior of $\mathbf{C}^1_{\Omega,x_0}[0, \tau]$ in its affine hull is none other than its relative interior (Proposition I.11). That is,

$$(\mathbf{C}^1_{\Omega,x_0}[0, \tau])^\diamond = ri(\mathbf{C}^1_{\Omega,x_0}[0, \tau]). \tag{27}$$

Here is another consequence of Observation 1.

OBSERVATION 3. $aff(\mathbf{C}^1_{\Omega,x_0}[0,\tau]) = \{f \in \mathbf{C}^1[0,\tau] : f(0) = x_0\}$.

PROOF

The $\subseteq$ part of the claim is trivial. To prove the converse containment, take any $f \in \mathbf{C}^1[0,\tau]$ such that $f(0) = x_0$. Pick any $u \in \mathbf{C}^1[0,\tau]$ such that $u(0) = x_0$ and $(u, u') \in int_{\mathbb{R}^2}(\Omega)$. Define $u_m := \left(1 - \frac{1}{m}\right) u + \frac{1}{m} f$, $m = 1, 2, \ldots$. It is easily checked that $\|u_m - f\|_{\infty,\infty} \to 0$, so, by Observation 1, there is an integer $M > 0$ such that $(u_M, u'_M) \in int_{\mathbb{R}^2}(\Omega)$. Since $u_M(0) = x_0$, then, $u_M \in \mathbf{C}^1_{\Omega,x_0}[0,\tau]$. So,

$$f = Mu_M + (1 - M)u \in aff(\mathbf{C}^1_{\Omega,x_0}[0,\tau]),$$

as we sought. □

Another way of saying what Observation 3 says is this:

$$span(\mathbf{C}^1_{\Omega,x_0}[0,\tau] - f) = \{g \in \mathbf{C}^1[0,\tau] : g(0) = 0\} \tag{28}$$

for any $f \in (\mathbf{C}^1_{\Omega,x_0}[0,\tau])^{\diamond}$. (Yes?)

You must be wondering where we are going with these boring observations. The fog is about to clear.

LEMMA 3

Let $[\Omega, \varphi, x_0, \tau, \mathcal{F}]$ be a smooth variational problem, and define $\varsigma : \mathbf{C}^1_{\Omega,x_0}[0,\tau] \to \mathbb{R}$ by

$$\varsigma(f) := \int_0^\tau \varphi(t, f, f')dt.$$

If $f \in \mathbf{C}^1[0,\tau]$ satisfies $f(0) = x_0$ and $(f, f') \in int_{\mathbb{R}^2}(\Omega)$, then ς is Fréchet differentiable at f, and

$$\mathsf{D}_{\varsigma,f}(g) = \int_0^\tau \left(\partial_2\varphi(t, f, f')g(t) + \partial_3\varphi(t, f, f')g'(t)\right) dt$$

for any $g \in \mathbf{C}^1[0,\tau]$ with $g(0) = 0$.[40]

[40] Observation 2 says that $f \in (\mathbf{C}^1_{\Omega,x_0}[0,\tau])^{\diamond}$. Besides, the domain of the linear functional $\mathsf{D}_{\varsigma,f}$ is $span(\mathbf{C}^1_{\Omega,x_0}[0,\tau] - f)$. (Right?) Thus, thanks to Observation 2 and (28), the present assertion is meaningful.

PROOF

Fix an arbitrary $f \in \mathbf{C}^1[0,\tau]$ with $f(0) = x_0$ and $(f,f') \in int_{\mathbb{R}^2}(\Omega)$. The discussion given in Example 8 allows us to state the following: For any $h \in \mathbf{C}^1_{\Omega,x_0}[0,\tau]$, there exist continuous functions α_h and β_h on $[0,\tau]$ such that $(\alpha_h, \beta_h) \in co\{(f,f'),(h,h')\}$, and

$$\varphi(\cdot,h,h') - \varphi(\cdot,f,f') = \partial_2\varphi(\cdot,f,f')(h-f) + \partial_3\varphi(\cdot,f,f')(h'-f') + E(h),$$

where E is the self-map on $\mathbf{C}[0,\tau]$ defined by

$$E(h) := \tfrac{1}{2}(\partial_{22}\varphi(\cdot,\alpha_h,\beta_h)(h-f)^2 + 2\partial_{23}\varphi(\cdot,\alpha_h,\beta_h)(h-f)(h'-f') \\ + \partial_{33}\varphi(\cdot,\alpha_h,\beta_h)(h'-f')^2).$$

Conclusion: For any $h \in \mathbf{C}^1_{\Omega,x_0}[0,\tau]$,

$$\varsigma(h) = \varsigma(f) + L(h-f) + \mathbf{e}(h),$$

where $L : \{g \in \mathbf{C}^1[0,\tau] : g(0) = 0\} \to \mathbb{R}$ is defined by

$$L(g) := \int_0^\tau \left(\partial_2\varphi(t,f,f')g(t) + \partial_3\varphi(t,f,f')g'(t)\right) dt,$$

and $\mathbf{e}(h) := \int_0^\tau E(h)(t)dt$. It is an easy matter to check that L is a bounded linear functional, so, in view of (28) and Proposition 2*, it remains to show that $\frac{\mathbf{e}(h)}{\|h-f\|_{\infty,\infty}} \to 0$ as $\|h-f\|_{\infty,\infty} \to 0$.

By Observation 1, if h is sufficiently close to f, then $\{(t,h(t),h'(t)) : 0 \le t \le \tau\} \subseteq [0,\tau] \times int_{\mathbb{R}^2}(\Omega)$, so in this case (because Ω, and hence $int_{\mathbb{R}^2}(\Omega)$, is convex), we have

$$S := \{(t,\alpha_h(t),\beta_h(t)) : 0 \le t \le \tau\} \subseteq [0,\tau] \times int_{\mathbb{R}^2}(\Omega).$$

But, since α_h and β_h are continuous, S is a compact set. (Right?) Then, given that the second-order partial derivatives of φ are continuous, there must exist a real number $K > 0$ such that $\left|\partial_{ij}\varphi(t,a,b)\right| \le K$ for any $(t,a,b) \in S$ and $i,j = 2,3$, whence

$$\left|E(h)\right| \le \tfrac{K}{2}\left(|h-f| + |h'-f'|\right)^2 \le \tfrac{K}{2}\|h-f\|^2_{\infty,\infty}.$$

Therefore,

$$\left|\mathbf{e}(h)\right| \le \int_0^\tau \left|E(h)(t)\right| dt \le \tfrac{K\tau}{2}\|h-f\|^2_{\infty,\infty},$$

provided that h is sufficiently close to f. It follows that $\frac{\mathbf{e}(h)}{\|h-f\|_{\infty,\infty}} \to 0$ as $\|h-f\|_{\infty,\infty} \to 0$, so we are done. ∎

Let's see what we got here. Suppose you are given a simplest variational problem $[\Omega, \varphi, x_0, \tau, \mathcal{F}]$, and suppose an $f \in \mathbf{C}^1[0,\tau]$, with $f(0) = x_0$ and $(f, f') \in int_{\mathbb{R}^2}(\Omega)$, solves the associated maximization problem. By Observation 1 and Proposition 8, then, f must be a stationary point of the map $f \mapsto \int_0^\tau \varphi(t, f, f')dt$. So, by Lemma 3, we find

$$\int_0^\tau \left(\partial_2\varphi(t,f,f')g(t) + \partial_3\varphi(t,f,f')g'(t)\right) dt = 0$$

for all $g \in \mathbf{C}^1[0,\tau]$ with $g(0) = 0$. This equation is called the **integral form of the Euler-Lagrange equation.**[41]

With a little more effort, we can improve on this observation substantially. Here is a key lemma we need.

Lemma 4

(DuBois-Reymond) *For any* $\tau > 0$, *if* $u \in \mathbf{C}^1[0,\tau]$ *satisfies*

$$\int_0^\tau u(t)g'(t)dt = 0 \quad \textit{for all } g \in \mathbf{C}^1[0,\tau] \textit{ with } g(0) = 0 = g(\tau),$$

then u must be a constant function.

Proof

Let $\gamma := \frac{1}{\tau}\int_0^\tau u(t)dt$, and define $g \in \mathbf{C}^1[0,\tau]$ by $g(x) := \int_0^x (u(t) - \gamma)dt$. Then, by the Fundamental Theorem of Calculus, which we apply twice, and the hypothesis of the lemma,

$$\begin{aligned}\int_0^\tau (u(t)-\gamma)^2 dt &= \int_0^\tau (u(t)-\gamma)g'(t)dt \\ &= \int_0^\tau u(t)g'(t)dt - \gamma\int_0^\tau g'(t)dt \\ &= -\gamma(g(\tau) - g(0)) \\ &= 0.\end{aligned}$$

It follows that $u = \gamma$ (Exercise A.61), and we are done. ∎

[41] Here, all I do is view a variational problem as an optimization problem over certain types of continuously differentiable functions, and then approach the problem by using the basic optimization theory sketched in Section 4. The rule "equate the first derivative of the objective function to zero and solve for the independent variable(s)," long live!

LEMMA 5

For any $\tau > 0$, *if* $v, w \in \mathbf{C}[0, \tau]$ *satisfy*

$$\int_0^\tau \big(v(t)g(t) + w(t)g'(t)\big)\, dt = 0$$

for all $g \in \mathbf{C}^1[0, \tau]$ *with* $g(0) = 0 = g(\tau)$,

then w *is differentiable and* $w' = v$.

PROOF

Define $F \in \mathbf{C}[0, \tau]$ by $F(x) := \int_0^x v(t)dt$, and note that $F' = v$ by the Fundamental Theorem of Calculus. Now, integrating by parts (Exercise A.63),

$$\int_0^\tau v(t)g(t)dt = -\int_0^\tau F(t)g'(t)dt$$

for all $g \in \mathbf{C}^1[0, \tau]$ with $g(0) = 0 = g(\tau)$. By hypothesis, therefore,

$$\int_0^\tau (-F(t) + w(t))g'(t)dt = 0$$

for all $g \in \mathbf{C}^1[0, \tau]$ with $g(0) = 0 = g(\tau)$. So, by Lemma 4, $w - F$ is a constant function, that is, $w' = F' = v$. ■

Here comes the punchline.

THEOREM 1

(Euler-Lagrange) *Let* $[\Omega, \varphi, x_0, \tau, \mathcal{F}]$ *be a simplest variational problem. If*

$$f_* \in \arg\max \left\{ \int_0^\tau \varphi(t,f,f')dt : f \in \mathcal{F} \right\}, \tag{29}$$

where $(f_*, f_*') \in int_{\mathbb{R}^2}(\Omega)$, *then*

$$\partial_2\varphi(t,f_*(t),f_*'(t)) = \tfrac{d}{dt}\partial_3\varphi(t,f_*(t),f_*'(t)), \qquad 0 < t < \tau. \tag{30}$$

PROOF

Apply Observation 2, Proposition 8, Lemma 3, and Lemma 5. ■

The functional equation (30) is called the **Euler-Lagrange equation** of the variational problem $[\Omega, \varphi, x_0, \tau, \mathcal{F}]$ and is of fundamental importance for the theory and applications of calculus of variations.[42] Unfortunately, in most cases this equation turns out to be a second-order differential equation, and it is hopeless to get explicit solutions out of it. Indeed, by differentiating we see that an $f \in \mathbf{C}^2[0,\tau]$ satisfies the Euler-Lagrange equation of $[\Omega, \varphi, x_0, \tau, \mathcal{F}]$ iff

$$\partial_2\varphi = \partial_{13}\varphi + (\partial_{23}\varphi)f' + (\partial_{33}\varphi)f''. \tag{31}$$

Despite the difficulty of solving such differential equations explicitly, one can still learn quite a bit about the nature of the solutions to the variational problem at hand by studying its Euler-Lagrange equation in implicit form. As we shall see in Sections 5.5 and 5.6, this is how the equation is used in most economic applications.[43]

Theorem 1 gives us a necessary condition for an $f \in \mathbf{C}^1[0,\tau]$ to solve a given simplest variational problem $[\Omega, \varphi, x_0, \tau, \mathcal{F}]$. If φ satisfies a suitable concavity condition, then this condition is also sufficient, as we show next.

Theorem 2

Let $[\Omega, \varphi, x_0, \tau, \mathcal{F}]$ be a simplest variational problem, and assume that $\varphi(t,\cdot,\cdot)$ is concave for each $0 \le t \le \tau$. Then, for any $f_* \in \mathcal{F}$ with $(f_*, f_*') \in \text{int}_{\mathbb{R}^2}(\Omega)$, (29) holds if, and only if, (30) holds. Moreover, if $\varphi(t,\cdot,\cdot)$ is strictly concave for each $0 \le t \le \tau$, then (29) (and hence (30)) can hold for at most one f_*.

Proof

Notice that $f \mapsto \int_0^\tau \varphi(t,f,f')dt$ is a concave map on $\mathcal{F}$, and apply Observation 2, Proposition 10, Lemma 3, and Lemma 5. The proof of the second claim is elementary. ■

42 Leonhard Euler (1707–1783) and Joseph Louis Lagrange (1736–1813) are considered to be the two greatest mathematicians of the eighteenth century. Euler contributed to virtually every branch of mathematics and in his lifetime achieved a level of productivity that is unsurpassed in the history of science, even though he went completely blind by age 59. His complete works constitute about 73 large volumes (compiled by the Swiss Society of Natural Science). Laplace is reputed to have said that the only way to learn mathematics is to read Euler. By contrast, Lagrange wrote much more concisely, but nevertheless played an irreplaceable role in the development of real function theory, optimization, differential equations, and group theory.

43 Unless $\partial_{33}\varphi = 0$, the solution of (31) depends, in general, on two constants of integration. In applications, these constants are often determined by means of the boundary conditions $f(0) = x_0$ and $f(\tau) = x_\tau$.

Theorem 2 stresses the importance of the Euler-Lagrange equation further. It shows that all solutions to a simplest variational problem are often *characterized* by the solutions to the Euler-Lagrange equation of the problem that satisfies the given boundary conditions.

WARNING. Theorem 2 gives a characterization of solutions to certain types of simplest variational problems, but it does not tell us anything about the *existence* of a solution to such a problem. Indeed, the Euler-Lagrange equation of a simplest variational problem may not have a solution, in which case, by Theorem 1, the involved optimization problem does not have a solution either. The "if" part of Theorem 2 says only that, under the said concavity hypothesis, *provided that it exists*, a solution f of the Euler-Lagrange equation (with the boundary conditions $f(0) = x_0$ and $f(\tau) = x_\tau$), is also a solution to the involved simplest variational problem.[44]

Let us now look at a few examples.

EXAMPLE 12

[1] Consider the following optimization problem:

$$\text{Minimize} \int_0^1 f'(t)^2 dt \quad \text{such that}$$

$$f \in \mathbf{C}^1[0,1],\ f(0) = 0 \text{ and } f(1) = 1.$$

(This is the simplest variational problem $[\mathbb{R}^2, \varphi, 0, 1, \mathcal{F}]$, where φ is defined on $[0,1] \times \mathbb{R}^2$ by $\varphi(t,a,b) := -b^2$, and $\mathcal{F} := \{f \in \mathbf{C}^1[0,1] : f(0) = 0 \text{ and } f(1) = 1\}$.) Theorem 2 ensures us that the solution f of the problem is characterized as: $-2f''(t) = 0$, $0 < t < 1$, with $f(0) = 0$ and $f(1) = 1$. It follows that the unique solution of our problem is $\text{id}_{[0,1]}$.[45]

[44] A major subfield of the modern theory of calculus of variations is its *existence* theory (sometimes called the *direct approach to calculus of variations*). Unfortunately, this theory, pioneered by Leonida Tonelli, necessitates some familiarity with measure theory, so I'm not going to be able to talk about it here. If you count on your measure theory, and are familiar with the notion of weak convergence, you might want to read Butazzo et al. (1998) and Dacorogna (2004) on this matter.

[45] It is readily checked that the maximization version of the problem, that is, $[\mathbb{R}^2, -\varphi, 0, 1, \mathcal{F}]$, does not have a solution.

[2] Consider the following optimization problem:

$$\text{Maximize} \int_0^1 tf'(t)dt \quad \text{such that}$$

$$f \in \mathbf{C}^1[0,1],\ f(0) = 0 \text{ and } f(1) = 1.$$

The Euler-Lagrange equation of the associated simplest variational problem is: $0 = \frac{d}{dt}\mathrm{id}_{(0,1)} = 1$. By Theorem 1, then, we conclude that there is no solution to the problem at hand.[46]

[3] Consider the following optimization problem:

$$\text{Maximize} \int_0^1 (f(t) + tf'(t))dt \quad \text{such that}$$

$$f \in \mathbf{C}^1[0,1],\ f(0) = 0 \text{ and } f(1) = 1.$$

It is easily checked that *any* $f \in \mathbf{C}^1[0,1]$ satisfies the associated Euler-Lagrange equation. By Theorem 2, then, we conclude that *any* $f \in \mathbf{C}^1[0,1]$ with $f(0) = 0$ and $f(1) = 1$ solves the problem at hand.[47]

[4] Consider the following optimization problem:

$$\text{Minimize} \int_0^1 f(t)^2 dt \quad \text{such that}$$

$$f \in \mathbf{C}^1[0,1],\ f(0) = 1 \text{ and } f(1) = 0.$$

We have seen earlier that this problem does not have a solution (Example 10.[2]). Theorem 1 confirms this. The associated Euler-Lagrange equation can be satisfied only by a function that equals zero on $(0,1)$. Then, if $f \in \mathbf{C}^1[0,1]$ solves the problem, it must equal 0 at 0, contradicting $f(0) = 1$.

[5] Given any $g \in \mathbf{C}^2[0,1]$, consider the following optimization problem:

$$\text{Minimize} \int_0^1 \left(\tfrac{1}{2}f'(t)^2 + g(t)f(t)\right) dt \quad \text{such that}$$

$$f \in \mathbf{C}^1[0,1],\ f(0) = 0 \text{ and } f(1) = 2.$$

46 This is an obvious consequence of the fact that $\int_0^1 tf'(t)dt = 1 - \int_0^1 f(t)dt$.

47 Of course! The Lagrangian of the problem equals 1 at any point on its domain.

(This is the simplest variational problem $[\mathbb{R}^2, \varphi, 0, 1, \mathcal{F}]$, where φ is defined on $[0,1] \times \mathbb{R}^2$ by $\varphi(t,a,b) := -\left(\frac{1}{2}b^2 + g(t)a\right)$, and $\mathcal{F} := \{f \in \mathbf{C}^1[0,1] : f(0) = 0 \text{ and } f(1) = 2\}$.) Theorem 2 ensures us that the solution $f \in \mathbf{C}^1[0,1]$ of this problem is characterized as: $f''(t) = g(t)$, $0 < t < 1$, with $f(0) = 0$ and $f(1) = 2$. By the Fundamental Theorem of Calculus, then, f solves our problem iff

$$f(0) = 0, \quad f(1) = 2 \quad \text{and} \quad f'(x) = f'_+(0) + \int_0^x g(t)dt, \quad 0 < x < 1.$$

For instance, if g is defined by $g(t) := 12t^2$, then the unique solution $f \in \mathbf{C}^1[0,1]$ of our problem is given by $f(t) = t + t^4$. □

EXERCISE 51^{H} Consider the following optimization problem:

$$\text{Minimize} \int_0^1 (f'(t)^2 - 1)^2 dt \quad \text{such that } f \in X \quad \text{and} \quad f(0) = 0 = f(1).$$

(a) Show that this problem has no solution if $X = \mathbf{C}^1[0,1]$.
(b) Show that there is a (unique) solution to this problem if X consists of all $f \in \mathbf{C}[0,1]$ such that (i) f'_+ and f'_- exist everywhere on $(0,1)$, and (ii) $\{t \in (0,1) : f'_+(t) \neq f'_-(t)\}$ is a finite set.

EXERCISE 52^{H} Let $[\mathbb{R}^2, \varphi, x_0, \tau, \mathcal{F}]$ be a simplest variational problem, where $\varphi : [0,\tau] \times \mathbb{R}^2 \to \mathbb{R}$ is defined by

$$\varphi(t,a,b) := U(t,a)b + V(t,a)$$

for some real maps $U, V \in \mathbf{C}^2([0,\tau] \times \mathbb{R})$. Prove:
(a) If (29) holds, then $\partial_1 U(t, f_*(t)) = \partial_2 V(t, f_*(t))$ for all $0 < t < \tau$.
*(b) If $\partial_1 U = \partial_2 V$, then every $f \in \mathcal{F}$ solves $[\mathbb{R}^2, \varphi, x_0, \tau, \mathcal{F}]$.

EXERCISE 53 (*The DuBois-Reymond Equation*) Let $[\Omega, \varphi, x_0, \tau, \mathcal{F}]$ be a simplest variational problem. Show that if

$$f_* \in \arg\max \left\{ \int_0^\tau \varphi(t, f, f')dt : f \in \mathcal{F} \cap \mathbf{C}^2[0,\tau] \right\},$$

with $(f_*, f'_*) \in int_{\mathbb{R}^2}(\Omega)$, then

$$\tfrac{d}{dt}\left(\partial_3 \varphi(t, f_*(t), f'_*(t)) f'_*(t) - \varphi(t, f_*(t), f'_*(t))\right) = \partial_1 \varphi(t, f_*(t), f'_*(t))$$

for any $0 < t < \tau$.

EXERCISE 54[H] (*Legendre's Necessary Condition*) Let $[\Omega, \varphi, x_0, \tau, \mathcal{F}]$ variational problem. Show that if (29) holds for some $f_* \in \mathcal{F}$ with $(f_*, f_*') \in int_{\mathbb{R}^2}(\Omega)$, then

$$\partial_{33}\varphi(t, f_*(t), f_*'(t)) \leq 0, \qquad 0 < t < \tau.$$

EXERCISE 55[H] (*A "Baby" Tonelli Existence Theorem*) Let $[\mathbb{R}^2, \varphi, x_0, \tau, \mathcal{F}]$ be a simplest variational problem such that

(i) $\varphi(t, a, \cdot)$ is concave for each $(t, a) \in [0, \tau] \times \mathbb{R}$;

(ii) $\mathcal{F}$ is a compact subset of $\mathbf{C}^1[0, \tau]$.[48]

Show that the map $\varsigma : \mathcal{F} \to \mathbb{R}$ defined by $\varsigma(f) := \int_0^\tau \varphi(t, f, f')dt$ is upper semicontinuous, and conclude that $\arg\max\{\varsigma(f) : f \in \mathcal{F}\} \neq \emptyset$.

While Theorems 1 and 2 concern only simplest variational problems, it is often not difficult to modify them to find the corresponding results for other kinds of variational problems. The following result, which is a reflection of Theorems 1 and 2 for free-end smooth variational problems, provides a case in point.

COROLLARY 5

Let $[\Omega, \varphi, x_0, \tau, \mathcal{F}]$ be a free-end smooth variational problem. If (29) *holds for some $f_* \in \mathcal{F}$ with $(f_*, f_*') \in int_{\mathbb{R}^2}(\Omega)$, then* (30) *holds, and we have*

$$\partial_3\varphi(\tau, f_*(\tau), f_*'(\tau)) = 0. \tag{32}$$

If $\varphi(t, \cdot, \cdot)$ is concave for each $0 \leq t \leq \tau$, then the converse is also true.

PROOF

If (29) holds, then

$$f_* \in \arg\max\left\{\int_0^\tau \varphi(t, f, f')dt : f \in \mathcal{F} \text{ and } f(\tau) = f_*(\tau)\right\},$$

[48] It is this hypothesis that takes away from the power of the existence theorem I'm about to state. $\mathcal{F}$ is rarely compact in applications, and the actual version of the Tonelli Existence Theorem replaces this requirement with much milder hypotheses (which control the "growth" of φ on its domain).

so, given that $(f_*, f_*') \in int_{\mathbb{R}^2}(\Omega)$, (30) follows from Theorem 1. Moreover, by Proposition 8, Lemma 3, and integrating by parts,

$$0 = \mathsf{D}_{\varphi,f_*}(g) = \int_0^\tau \partial_2\varphi(t,f,f')g(t)dt + \partial_3\varphi(\tau,f_*(\tau),f_*'(\tau))g(\tau) - \int_0^\tau \tfrac{d}{dt}\partial_3\varphi(t,f,f')g(t)dt$$

for all $g \in \mathbf{C}^1[0,\tau]$ with $g(0) = 0$. In view of (30), we thus obtain (32). The second assertion readily follows from the computation above and Proposition 10. ■

For example, consider the following optimization problem:

$$\text{Minimize} \int_0^1 \left(\tfrac{1}{2}f'(t)^2 + 12t^2 f(t)\right) dt \quad \text{such that}$$

$$f \in \mathbf{C}^1[0,1] \text{ and } f(0) = 0.$$

(We have solved this problem under the additional constraint $f(1) = 2$ in Example 12.[5], and identified the map $t \mapsto t + t^4$ as the solution.) By Corollary 5, $f \in \mathbf{C}^1[0,1]$ is a solution of this problem iff $f(0) = 0$, $f'(t) = f_+'(0) + 4t^3$, $0 < t < 1$, and $f_-'(1) = 0$, while this holds iff $f(t) = -4t + 4t^4$ for any $0 \leq t \leq 1$. (Draw a picture to compare the solutions to the present problem with and without the constraint $f(1) = 2$.)

EXERCISE 56^H

(a) For each $\alpha \in \mathbb{R}$, solve the following optimization problem:

$$\text{Minimize} \int_0^1 \left(\tfrac{1}{2}f'(t)^2 + t^\alpha f(t)\right) dt \quad \text{such that}$$

$$f \in \mathbf{C}^1[0,1] \text{ and } f(0) = 0.$$

(b) Solve the following optimization problem:

$$\text{Minimize} \int_0^1 \left(\tfrac{1}{2}f'(t)^2 + e^t f(t)\right) dt \quad \text{such that}$$

$$f \in \mathbf{C}^1[0,1] \text{ and } f(0) = 0.$$

(c) Solve the previous problem under the additional constraint $f(1) = e$.

EXERCISE 57[H] (*Calculus of Variations with Salvage Value*) Let $[\mathbb{R}^2, \varphi, x_0, \tau, \mathcal{F}]$ be a free-end smooth variational problem, and $W \in \mathbf{C}^3[0, \tau]$. Show that if

$$f_* \in \arg\max \left\{ \int_0^\tau \varphi(t, f, f')dt + W(f(\tau)) : f \in \mathcal{F} \right\},$$

and $(f_*, f_*') \in int_{\mathbb{R}^2}(\Omega)$, then (30) holds and

$$\partial_3\varphi(\tau, f_*(\tau), f_*'(\tau)) + W'(f_*(\tau)) = 0.$$

If $\varphi(t, \cdot, \cdot)$ is concave for each $0 \leq t \leq \tau$, then the converse is also true.

EXERCISE 58 Let $[\Omega, \varphi, x_0, \tau, \mathcal{F}]$ be a smooth variational problem, where $\mathcal{F} := \{f \in \mathbf{C}^1_{\Omega, x_0}[0, \tau] : f(\tau) \geq x_\tau\}$ for some $x_\tau \in \mathbb{R}$. Show that if (29) holds for some $f_* \in \mathcal{F}$ with $(f_*, f_*') \in int_{\mathbb{R}^2}(\Omega)$, then (30) holds, and $\partial_3\varphi(\tau, f_*(\tau), f_*'(\tau)) \leq 0$ while $\partial_3\varphi(\tau, f_*(\tau), f_*'(\tau)) = 0$ if $f_*(\tau) > x_\tau$. If $\varphi(t, \cdot, \cdot)$ is concave for each $0 \leq t \leq \tau$, then the converse is also true.

5.3* More on the Sufficiency of the Euler-Lagrange Equation

In this section we revisit the sufficiency part of Theorem 2 and attempt to improve on it a bit. Here is what we have in mind.

PROPOSITION 12
Let $[\Omega, \varphi, x_0, \tau, \mathcal{F}]$ be a simplest variational problem, and let $U, V \in \mathbf{C}^2([0, \tau] \times \mathbb{R})$ satisfy

(i) $\partial_1 U = \partial_2 V$;

(ii) *for any $0 \leq t \leq \tau$, the map*

$$(a, b) \mapsto \varphi(t, a, b) + U(t, a)b + V(t, a) \tag{33}$$

is concave on Ω.

If (30) holds for some $f_* \in \mathcal{F}$ with $(f_*, f_*') \in int_{\mathbb{R}^2}(\Omega)$, then (29) holds.

This is a genuine generalization of the first part of Theorem 2. Indeed, the latter result is obtained as a special case of Proposition 12 by setting $U = 0 = V$.

What is the idea behind Proposition 12? (If you have solved Exercise 52, then you know the answer.) All we are doing here is to transform our original problem to the following:

$$\text{Maximize} \int_0^1 \varphi(t,f,f')dt + \int_0^1 \big(U(t,f(t))f'(t) + V(t,f(t))\big)\, dt$$

such that $f \in \mathcal{F}$.

One can show that the condition $\partial_1 U = \partial_2 V$ implies that the second integral above is constant for *any* $f \in \mathcal{F}$ (Exercise 52.(b)). Thus an $f \in \mathcal{F}$ solves our original problem iff it solves the transformed problem. But, by hypothesis (ii) and Theorem 2, if (30) holds for some $f_* \in \mathcal{F}$ with $(f_*, f_*') \in int_{\mathbb{R}^2}(\Omega)$, then f_* must be a solution to the transformed problem. This f_*, then, solves our original problem as well.

The key step missing in this argument is supplied by the following lemma which we borrow from advanced calculus.

LEMMA 6
For any given $\tau > 0$ and $U, V \in \mathbf{C}^2([0,\tau]\times\mathbb{R})$ *with* $\partial_1 U = \partial_2 V$, *there exists a* $G \in \mathbf{C}^1([0,\tau]\times\mathbb{R})$ *such that* $\partial_1 G = V$ *and* $\partial_2 G = U$.

We shall omit the proof of this lemma, as we do not really need it.[49] But it is important to note the following consequence of Lemma 6: Given any simplest variational problem $[\Omega, \varphi, x_0, \tau, \mathcal{F}]$, if $U, V \in \mathbf{C}^2([0,\tau]\times\mathbb{R})$ satisfy $\partial_1 U = \partial_2 V$, then

$$\int_0^1 \big(U(t,f(t))f'(t) + V(t,f(t))\big)\, dt$$

$$= \int_0^1 \tfrac{d}{dt} G(t,f(t))dt = G(\tau,f(\tau)) - G(0,x_0),$$

[49] So, why is this lemma here? Well, I wish to use it only to tell you about the "big picture" here, nothing more. What I am really after is Corollary 6, and that result can be proved directly once the big picture is understood. (See Exercise 59.)

By the way, if you are good at calculus (and remember the Leibniz Rule, etc.), then Lemma 6 shouldn't be hard to prove. Just define your G by

$$G(t,a) := \tfrac{1}{2}\int_{-\infty}^{a} U(t,\omega)d\omega + \tfrac{1}{2}\int_0^t V(s,a)ds,$$

and check that this function is equal to the task.

where G is found by Lemma 6, whence

$$\int_0^1 \left(U(t,f(t))f'(t) + V(t,f(t))\right) dt = \text{constant} \quad \text{for any } f \in \mathcal{F}.$$

This observation hands us Proposition 12 on a silver platter.

PROOF OF PROPOSITION 12
Define $\psi \in \mathbf{C}^2([0,\tau] \times \Omega)$ by $\psi(t,a,b) := U(t,a)b + V(t,a)$, and let $\varphi_\bullet := \varphi + \psi$. By the preceding discussion, we have

$$\arg\max \left\{ \int_0^\tau \varphi_\bullet(t,f,f')dt : f \in \mathcal{F} \right\} = \arg\max \left\{ \int_0^\tau \varphi(t,f,f')dt : f \in \mathcal{F} \right\}.$$

Thus, applying Theorem 2 to the problem $[\Omega, \varphi_\bullet, x_0, \tau, \mathcal{F}]$ completes the proof. ■

But is Proposition 12 good for anything? That is, how are we supposed to come up with a (U, V) pair that would satisfy the requirements of this result? It is not possible to answer this query in general—the answer really depends on the kind of application one is working on—but the following corollary may provide some help. It is a nice improvement over Theorem 2.

COROLLARY 6
Let $[\Omega, \varphi, x_0, \tau, \mathcal{F}]$ be a simplest variational problem such that there exists a map $G \in \mathbf{C}^3([0,\tau] \times \mathbb{R})$ such that, for each $0 \le t \le \tau$, the map

$$(a,b) \mapsto \varphi(t,a,b) + \partial_2 G(t,a)b + \partial_1 G(t,a)$$

is concave on Ω. If (30) *holds for some $f_* \in \mathcal{F}$ with $(f_*, f_*') \in int_{\mathbb{R}^2}(\Omega)$, then* (29) *holds.*

PROOF
Define $U := \partial_2 G$ and $V := \partial_1 G$, and apply Proposition 12. ■

In the final section of this text we will be able to use this result in an instance where Theorem 2 does not provide immediate help.

EXERCISE 59 Modify the argument we gave for Proposition 12 to provide a direct proof for Corollary 6 that does not depend on Lemma 6.

5.4 Infinite-Horizon Variational Problems

In the majority of economic applications, the involved variational problems arise from a dynamic model that involves an infinite time horizon. Consequently, we now turn our attention to extending the theory of calculus of variations we have exposed so far to the case of infinite-horizon variational problems. Let x_0 be a real number and Ω a convex subset of $\mathbb{R}^2$ with $int_{\mathbb{R}^2}(\Omega) \neq \emptyset$. Denoting the statement "$(f(t),f'(t)) \in \Omega$ for all $t \geq 0$" as "$(f,f') \in \Omega$" for any $f \in \mathbf{C}^1(\mathbb{R}_+)$, we define

$$\mathbf{C}^1_{\Omega,x_0}(\mathbb{R}_+) := \{f \in \mathbf{C}^1(\mathbb{R}_+) : f(0) = x_0 \text{ and } (f,f') \in \Omega\}.$$

Now consider a nonempty set $\mathcal{F} \subseteq \mathbf{C}^1_{\Omega,x_0}(\mathbb{R}_+)$ and a map $\varphi \in \mathbf{C}(\mathbb{R}_+ \times \Omega)$ such that

(i) $\varphi \in \mathbf{C}^2(\mathbb{R}_+ \times \Omega)$;

(ii) $\int_0^\tau \varphi(t,f,f')dt < \infty$ for any $f \in \mathcal{F}$.[50]

We refer to the list $[\Omega, \varphi, x_0, \mathcal{F}]$ as an **∞-horizon smooth variational problem**, and understand that this list corresponds to the following optimization problem:

$$\text{Maximize} \int_0^\infty \varphi(t,f,f')dt \quad \text{such that} \quad f \in \mathcal{F}. \tag{34}$$

(With an innocent abuse of terminology, we refer to the problem (34) also as an *∞-horizon smooth variational problem.*) Of particular interest are those ∞-horizon smooth variational problems where

$$\mathcal{F} = \mathbf{C}^1_{\Omega,x_0}(\mathbb{R}_+). \tag{35}$$

The following is sometimes called the *Rounding-Off Corners Lemma*.

[50] This is not a crazy requirement. For instance, if $\varphi(t,a,b) = \eth(t)U(a,b)$, where $\eth \in \mathbf{C}^2(\mathbb{R}_+)$ satisfies $\int_0^\infty \eth(t)dt < \infty$, and U is a bounded function in $\mathbf{C}^2(\Omega)$, then these two hypotheses are fulfilled.

Lemma 7

Let $[\Omega, \varphi, x_0, \mathcal{F}]$ be an ∞-horizon smooth variational problem. Let $\mathcal{X}$ stand for the class of all $f \in \mathbf{C}(\mathbb{R}_+)$ such that

(i) $f(0) = 0$;

(ii) f'_+ *and* f'_- *exist everywhere on* $\mathbb{R}_{++}$;

(iii) *there exists a* $t_o > 0$ *such that* $f|_{\mathbb{R}_+\backslash\{t_o\}} \in \mathbf{C}^1(\mathbb{R}_+\backslash\{t_o\})$ *and* $(f(t), f'(t)) \in \Omega$ *for all* $t \in \mathbb{R}_+\backslash\{t_o\}$.

Then

$$\sup\left\{\int_0^\infty \varphi(t,f,f')dt : f \in \mathcal{F}\right\} = \sup\left\{\int_0^\infty \varphi(t,f,f')dt : f \in \mathcal{X}\right\}.$$

Exercise 60 Prove Lemma 7.

We can now easily derive the Euler-Lagrange equation for ∞-horizon smooth variational problems. The idea is simply to use Theorem 1 for *any* truncation of the problem at hand, and then utilize Lemma 7 to "patch up" what we find in these truncations.

Theorem 3

Let $[\Omega, \varphi, x_0, \mathcal{F}]$ be an ∞-horizon smooth variational problem, and let $f_* \in \mathcal{F}$ satisfy $(f_*, f_*') \in \mathrm{int}_{\mathbb{R}^2}(\Omega)$. If

$$f_* \in \arg\max\left\{\int_0^\infty \varphi(t,f,f')dt : f \in \mathcal{F}\right\}, \tag{36}$$

then,

$$\partial_2\varphi(t,f_*(t),f_*'(t)) = \tfrac{d}{dt}\partial_3\varphi(t,f_*(t),f_*'(t)) \quad \text{for all } t > 0,$$

while the converse also holds when $\varphi(t,\cdot,\cdot)$ is concave for each $0 \le t \le \tau$.

Proof

Assume that (36) holds. In view of Theorem 1, it is enough to show that, for any $\tau > 0$,

$$f_\tau|_{[0,\tau]} \in \arg\max\left\{\int_0^\tau \varphi(t,f,f')dt : f \in \mathcal{F}_\tau\right\},$$

where $f_\tau := f_*|_{[0,\tau]}$ and $\mathcal{F}_\tau := \{f \in \mathbf{C}^1_{\Omega,x_0}[0,\tau] : f(\tau) = f_*(\tau)\}$. (Right?) To derive a contradiction, suppose the latter statement is false, that is, there exist a $\tau > 0$ and an $h \in \mathcal{F}_\tau$ such that $\int_0^\tau \varphi(t,h,h')dt > \int_0^\tau \varphi(t,f_\tau,f_\tau')dt$. Define $g \in \mathbf{C}(\mathbb{R}_+)$ as

$$g(t) := \begin{cases} h(t), & \text{if } 0 \leq t \leq \tau \\ f_*(t), & \text{otherwise} \end{cases}.$$

Then, by Lemma 7, we have

$$\int_0^\infty \varphi(t,f_*,f_*')dt \geq \int_0^\infty \varphi(t,g,g')dt > \int_0^\infty \varphi(t,f_*,f_*')dt,$$

which is absurd. The proof of the second assertion is left to the reader. ■

EXERCISE 61[H] (Blot-Michel) Let $[\Omega,\varphi,x_0,\mathcal{F}]$ be an ∞-horizon smooth variational problem. Suppose that

$$f_* \in \arg\max\left\{\int_0^\tau \varphi(t,f,f')dt : f \in \mathcal{F} \cap \mathbf{C}^2(\mathbb{R}_+)\right\},$$

where $(f_*,f_*') \in int_{\mathbb{R}^2}(\Omega)$ and $\int_0^\infty \partial_1\varphi(t,f_*,f_*')dt < \infty$. Show that

$$\lim_{t\to\infty}\left(\partial_3\varphi(t,f_*,f_*')f_*'(t) - \varphi(t,f_*,f_*')\right)$$
$$= \partial_3\varphi(t,f_*(0),f_*'(0))f_*'(0) - \varphi(t,f_*(0),f_*'(0)) - \int_0^\infty \partial_1\varphi(t,f_*,f_*')dt.$$

EXERCISE 62 State and prove a Legendre-type necessary condition (Exercise 54) for ∞-horizon smooth variational problems.

*EXERCISE 63 Give an example of an ∞-horizon smooth variational problem $[\Omega,\varphi,x_0,\mathcal{F}]$ such that (36) holds for some $f_* \in \mathcal{F}$ with $(f_*,f_*') \in int_{\mathbb{R}^2}(\Omega)$, but

$$\lim_{\tau\to\infty} \partial_3\varphi(\tau,f_*(\tau),f_*'(\tau)) \neq 0.$$

5.5 Application: The Optimal Investment Problem

We consider the optimal investment problem introduced in Example 11 under the following assumptions:

ASSUMPTION (A)

(i) $P \in \mathbf{C}^2(\mathbb{R}_+)$ is a bounded function with $P(0) = 0$, $P' > 0$, $P'' \leq 0$ and $P'_+(0) = \infty$.

(ii) $C \in \mathbf{C}^2(\mathbb{R}_+)$, $C(0) = 0 = C'(0)$, $C|'_{\mathbb{R}_{++}} > 0$ and $C'' > 0$.

(iii) $\eth \in \mathbf{C}^2[0, \tau]$, $\eth(0) = 1$, $\eth > 0$ and $\eth' < 0$.

Let us first look into the case where the firm has a target level of capital stock at the end of the planning horizon, say $x_\tau > 0$, so the problem it faces is:

$$\text{Maximize} \int_0^\tau \eth(t)(P(f(t)) - C(f'(t)))dt$$

such that

$$f \in \mathbf{C}^1[0, \tau], \quad f(0) = x_0, \quad f(\tau) = x_\tau \quad \text{and} \quad f' \geq 0.$$

It can be shown that, under the premises of (A), this problem admits a solution $f \in \mathbf{C}^1[0, \tau]$ with $f, f' \gg 0$. (We omit the proof.) Since the map $(a, b) \mapsto P(a) - C(b)$ is strictly concave on $\mathbb{R}^2$, this solution is unique, and by Theorem 1, it is characterized by the Euler-Lagrange equation of the problem and its boundary conditions. It follows that the optimal capital accumulation plan f satisfies

$$P'(f(t)) = -\frac{\eth'(t)}{\eth(t)} C'(f'(t)) - C''(f'(t))f''(t), \qquad 0 < t < \tau.$$

We can interpret this differential equation as follows. $P'(f(t))$ is the time t marginal benefit of the firm that corresponds to the capital accumulation plan f. On the other hand, $-\frac{\eth'(t)}{\eth(t)}$ is the relative discount rate at time t, so $-\frac{\eth'(t)}{\eth(t)} C'(f(t))$ corresponds to forgone (subjective) interest on the money invested in capital. Finally, the term $C''(f'(t))f''(t)$ can be interpreted as the capital gains or losses that may have occurred through time. All in all, the Euler-Lagrange equation in the present case is a dynamic formulation of the principle that the firm has to invest in such a way that its marginal benefit and marginal cost at any given point in time are equal to each other.

Suppose we now omit the condition $f(\tau) = x_\tau$, and hence consider the problem of the firm as a free-end smooth variational problem. In that case, Corollary 5 tells us that we must have $C'(f'(\tau)) = 0$ in the optimum. Since $C'|_{\mathbb{R}_{++}} > 0$, this is possible iff $f'(\tau) = 0$. So we learn that additional investment at the final period τ should be zero in the optimum. (This makes sense, of course, as there is no time left to make use of an investment undertaken at τ.)

Incidentally, this observation shows that if we assumed $C'(0) > 0$ in (A.ii) (as would be the case if C was a linear function, for instance), then there would be no solution to the free-end variational problem of the firm.

EXERCISE 64[H] Solve the following optimization problem:

$$\text{Maximize} \quad \int_0^\tau e^{-rt}(pf(t) - f'(t)^2)dt$$

$$\text{such that} \quad f \in \mathbf{C}^1[0,\tau] \ \text{ and } \ f(0) = \tfrac{pe^{r\tau}}{2r^2},$$

where p and r are positive real numbers.

5.6 Application: The Optimal Growth Problem

We next consider the evolution of a one-sector economy through time, starting at initial time 0 and moving on to the indefinite future. The model parallels the one considered in Section E.4.4, but we now model time continuously. The initial capital stock in the economy is $x_0 > 0$, while the production function is given by a self-map F on $\mathbb{R}_+$. Thus the national level of production at time $t > 0$ is $F(x(t))$, where $x \in \mathbf{C}^1[0,\tau]$ describes the capital accumulation through time. The instantaneous change in the capital stock at time t, that is, $x'(t)$, is viewed as investment made at time t. Then the consumption level available to the population at time t is $F(x(t)) - x'(t)$. Therefore, if the (social) utility function is given by a map $u : \mathbb{R}_+ \to \mathbb{R}$, we find that the aggregate level of felicity in the economy at time t is $u(F(x(t)) - x'(t))$, if capital is accumulated according to the plan x. Finally, we posit that the social utility is discounted through time according to a discount function $\eth : \mathbb{R}_+ \to [1, 0)$. Thus, the aggregate felicity at time t is perceived as $\eth(t)u(F(x(t)) - x'(t))$ at time 0.

We maintain the following standard assumptions.

ASSUMPTION (B)

(i) $u \in \mathbf{C}^2(\mathbb{R}_+)$ is a bounded function with $u'_+(0) = \infty$, $u' > 0$ and $u'' < 0$.

(ii) $F \in \mathbf{C}^2(\mathbb{R}_+)$ is a bounded function with $F(0) = 0$, $F'_+(0) = \infty$, $F' > 0$ and $F'' < 0$.

(iii) $\eth \in \mathbf{C}^2(\mathbb{R}_+)$ while $\eth(0) = 1$, $\eth > 0$, $\eth' < 0$ and $\int_0^\infty \eth(t)dt < \infty$.

The list $(x_0, F, u, \eth)$ is called a **continuous-time optimal growth model**, provided that u, F, and $\eth$ satisfy (B). Given such a model, the problem of the planner is the following:

$$\text{Maximize} \int_0^\infty \eth(t)u(F(x(t)) - x'(t))dt$$

such that

$$x \in \mathbf{C}^1[0, \tau], \quad x \geq 0, \quad F(x) \geq x' \quad \text{and} \quad x(0) = x_0.$$

Obviously, this problem corresponds to the ∞-horizon smooth variational problem $[\Omega, \varphi, x_0, \mathcal{F}]$, where

$$\Omega := \{(a, b) \in \mathbb{R}^2 : a \geq 0 \text{ and } F(a) \geq b\},$$

φ is defined on $\varphi : \mathbb{R}_+ \times \Omega$ by $\varphi(t, a, b) := \eth(t)u(F(a) - b)$, and $\mathcal{F} := \mathbf{C}^1_{\Omega,x_0}(\mathbb{R}_+)$.[51]

Let us presume that $[\Omega, \varphi, x_0, \mathcal{F}]$ has a solution x_* such that $x_* > 0$ and $F(x_*) > x'_*$. Then, by Theorem 3, this solution satisfies the Euler-Lagrange equation of the problem, that is,

$$\eth u'(F(x_*) - x'_*)F'(x_*) = -\eth' u'(F(x_*) - x'_*) - \eth u''(F(x_*) - x'_*)(F'(x_*)x'_* - x''_*).$$

To simplify this expression a bit, let us define the map $c_* := F(x_*) - x'_*$, which is the (optimal) consumption plan for the society induced by the capital accumulation plan x_*. Then, we have

$$\eth u'(c_*)F'(x_*) = -\eth' u'(c_*) - \eth u''(c_*)c'_*,$$

51 A special case of this model was considered for the first time by Frank Ramsey in 1928. The present formulation is standard; it was studied first by Cass (1965) using control theory.

that is,

$$c_*' = -\frac{u'(c_*)}{u''(c_*)}F'(x_*) - \frac{\eth'}{\eth}\frac{u'(c_*)}{u''(c_*)},$$

or equivalently,

$$\frac{c_*'(t)}{c_*(t)} = \left(\frac{u'(c_*(t))}{u''(c_*(t))c_*(t)}\right)\left(\left(\frac{-\eth'(t)}{\eth(t)}\right) - F'(x_*(t))\right), \qquad t > 0. \tag{37}$$

Although solving this differential equation explicitly is impossible, we can still learn quite a bit by studying it in its implicit form. Begin by noting that $\frac{c_*'}{c_*}$ is the *growth rate of consumption* and $\frac{u''(c_*)c_*}{u'(c_*)}$ is the *elasticity of marginal utility of consumption*, which is negative (at any $t \geq 0$). It follows that whether or not the country has a positive growth rate of consumption at time $t > 0$ depends on whether or not the marginal productivity of capital exceeds the relative discount rate $\frac{-\eth'}{\eth}$ at time t:

$$\frac{c_*'(t)}{c_*(t)} \geq 0 \quad \text{iff} \quad F'(x_*(t)) \geq \frac{-\eth'(t)}{\eth(t)}.$$

To consider an example, let us specialize the model at hand by setting

$$u(c) := \frac{c^{1-\sigma}}{1-\sigma}, \quad c \geq 0, \qquad \text{and} \qquad \eth(t) := e^{-rt}, \quad t \geq 0,$$

where $0 < \sigma < 1$ and $r > 0$. (As you know, r is often viewed as the *interest rate* here.) In this case, (37) becomes

$$\frac{c_*'(t)}{c_*(t)} = \frac{1}{\sigma}\left(F'(x_*(t)) - r\right), \qquad t > 0.$$

Thus in the optimal capital accumulation plan, we have positive growth rate of consumption at all times iff the marginal productivity of capital exceeds the interest rate at all times. (It is also evident that the lower the interest rate, the higher the growth rate of consumption.)

Now define, still within the confines of this example, the number $x^G > 0$ by the equation $F'(x^G) = r$. (x^G is the *golden-rule level of investment* in this model.) Then, by concavity of F, it follows that

$$\frac{c_*'(t)}{c_*(t)} \geq 0 \quad \text{iff} \quad x^G \geq x_*(t)$$

for any $t \geq 0$.

A lot more can be deduced from the Euler-Lagrange equation in the case of this continuous-time optimal growth model, but this is not the place to go into the details of this matter. We thus stop here, but after we put on record the following, perhaps unexpected, characterization of the *exponential discounting hypothesis* that emanates from the present analysis.

Observation 4. Let $\{(x_0, F, u, \eth) : x_0 > 0\}$ be a class of continuous-time optimal growth models. Then $(x_0, F, u, \eth)$ has a constant (steady-state) solution for some $x_0 > 0$ if, and only if, there exists an $r > 0$ such that $\eth(t) := e^{-rt}$ for all $t \geq 0$.

Proof
The "if" part of the assertion is easily deduced by setting $x_0 = x^G$ and using Theorem 3. To prove the "only if" part, assume that x_* is a steady state solution to the optimization problem induced by $(x_0, F, u, \eth)$ for some $x_0 > 0$. Then $x_*(t) = x_0$ for every $t \geq 0$, and hence $c_*(t) = F(x_0)$ for all $t \geq 0$. By (37), therefore, $\frac{-\eth'}{\eth}$ is a map that equals a real number, say r, everywhere on $\mathbb{R}_{++}$. Then

$$-\int_0^t \tfrac{d}{dt}(\ln \eth(t))dt = \int_0^t \frac{-\eth'(t)}{\eth(t)}dt = \int_0^t rdt, \qquad t > 0,$$

so, for some $\beta \in \mathbb{R}$, we have $\ln \eth(t) = -rt + \beta$ for all $t > 0$. Since $\eth$ is a continuous function on $\mathbb{R}_+$ with $\eth(0) = 1$, therefore, we find that $\eth(t) := e^{-rt}$ for all $t \geq 0$, as we sought. □

5.7* Application: The Poincaré-Wirtinger Inequality

We pick our final application from real function theory; the present course being on real analysis and all, this seems rather appropriate. What we wish to prove—by means of the calculus of variations, of course—is a functional integral inequality of extraordinary caliber. (Really, it is almost too good to be true.) It goes by the following name:

The Poincaré-Wirtinger Inequality
For any $f \in \mathbf{C}^1[0,1]$ with $f(0) = 0 = f(1)$, we have

$$\int_0^1 f'(t)^2 dt \geq \pi^2 \int_0^1 f(t)^2 dt. \tag{38}$$

Proof

Fix an arbitrary number α with $0 \leq \alpha < \pi$, and consider the simplest variational problem $[\mathbb{R}^2, \varphi_\alpha, 0, 1, \mathcal{F}]$, where φ_α is defined on $[0,1] \times \mathbb{R}^2$ by

$$\varphi_\alpha(t, a, b) := \tfrac{1}{2}(b^2 - \alpha^2 a^2),$$

and

$$\mathcal{F} := \{f \in \mathbf{C}^1[0,1] : f(0) = 0 = f(1)\}.$$

The Euler-Lagrange equation of the problem yields

$$\alpha^2 f(t) = f''(t), \qquad 0 < t < 1. \tag{39}$$

Define the map $G \in \mathbf{C}^3([0,1] \times \mathbb{R})$ by

$$G(t, a) := \tfrac{\alpha}{2} \tan\left(\alpha\left(t - \tfrac{1}{2}\right)\right) a^2.$$

(This is the smart trick in the argument.[52]) For any $0 \leq t \leq 1$, you can check that the map

$$(a, b) \mapsto \varphi_\alpha(t, a, b) + \partial_2 G(t, a) b + \partial_1 G(t, a),$$

that is,

$$(a, b) \mapsto \tfrac{1}{2} b^2 + \alpha \tan\left(\alpha\left(t - \tfrac{1}{2}\right)\right) ab + \tfrac{\alpha^2}{2} \tan^2\left(\alpha\left(t - \tfrac{1}{2}\right)\right) a^2. \tag{40}$$

As this map is of quadratic form, it is easily seen to be convex.[53]

[52] *Reminder.* $x \mapsto \tan x$ is the map defined on the interval $(-\frac{\pi}{2}, \frac{\pi}{2})$ by $\tan x := \frac{\sin x}{\cos x}$. This map is strictly increasing and continuously differentiable (up to any order), and we have $\frac{d}{dx} \tan x = 1 + (\tan x)^2$.

[53] I can now tell you why I defined G above the way I did. All I'm after is to choose a $G \in \mathbf{C}^3([0,1] \times \mathbb{R})$ such that the map in (40) has a (positive-definite) quadratic form, and hence it is convex. And for this, it seems like a good starting point to define G as $G(t, a) := \frac{1}{2}\theta(t) a^2$, where $\theta \in \mathbf{C}^3[0,1]$ is a map yet to be specified. Then the map in (40) has the following form:

$$(a, b) \mapsto \tfrac{1}{2} b^2 + \theta(t) ab + \tfrac{1}{2}(\theta'(t) - \alpha^2) a^2.$$

Aha! Now you see that I should choose θ so as to "complete the square," but I should also make sure that $\theta'(t) \geq \alpha^2$ for every t. Well, to complete the square, I obviously need to have $\theta(t) = \sqrt{\theta'(t) - \alpha^2}$ for each t. (Yes?) It is not difficult to see that the general solution of this differential equation is

$$\theta(t) = \alpha \tan(\alpha(t - \beta)), \qquad 0 \leq t \leq 1,$$

where β is a constant of integration. (Notice that we have $\theta'(t) = \alpha^2(1 + (\tan(\alpha(t-\beta)))^2) \geq \alpha^2$, provided, of course, $\theta(t)$ is well-defined.)

Now, β should be chosen to ensure that θ is well-defined as such, that is, we need to have $-\frac{\pi}{2} < \alpha(t - \beta) < \frac{\pi}{2}$. Given that $\alpha < \pi$ and $0 \leq t \leq 1$, it is clear that we need to set $\beta = \frac{1}{2}$ for this purpose. Hence, we arrive at the specification $\theta(t) = \alpha \tan\left(\alpha\left(t - \frac{1}{2}\right)\right)$ for any $t \in [0,1]$, and the mystery behind the definition of G in the proof above is solved.

Now notice that the map that equals zero everywhere on $[0, 1]$ satisfies (39), so, by Corollary 6, it is a solution to $[\mathbb{R}^2, \varphi_\alpha, 0, 1, \mathcal{F}]$. Thus

$$\int_0^1 \left(f'(t)^2 - \alpha^2 f(t)^2\right) dt \geq 0 \qquad \text{for all } f \in \mathcal{F} \text{ and } 0 \leq \alpha < \pi.$$

From this observation follows (38). ∎

EXERCISE 65 Let $-\infty < a < b < \infty$. Show that

$$\int_a^b f'(t)^2 dt \geq \left(\tfrac{\pi}{b-a}\right)^2 \int_a^b f(t)^2 dt$$

for any $f \in \mathbf{C}^1[a, b]$ with $f(a) = 0 = f(b)$.

EXERCISE 66 Show that

$$\int_0^1 f'(t)^2 dt = \pi^2 \int_0^1 f(t)^2 dt$$

for some $f \in \mathbf{C}^1[0, 1]$ with $f(0) = 0 = f(1)$ iff there is an $\alpha \in \mathbb{R}$ such that $f(t) = \alpha \sin(\pi t)$ for all $0 \leq t \leq 1$.

EXERCISE 67 Show that

$$\int_0^1 f'(t)^2 dt \geq \tfrac{\pi^2}{4} \int_0^1 f(t)^2 dt$$

for any $f \in \mathbf{C}^1[0, 1]$ with $f(0) = 0$.

Needless to say, there is much more to calculus of variations in particular, and to infinite-dimensional optimization theory in general. But that story is better told some other time and place.

Hints for Selected Exercises

Chapter A

EXERCISE 7. Let me show that if R is transitive, then xP_Ry and yRz implies xP_Rz. Since $P_R \subseteq R$, it is plain that xRz holds in this case. Moreover, if zRx holds as well, then yRx, for R is transitive and yRz. But this contradicts xP_Ry.

EXERCISE 13. (c) Suppose $c_{\succsim}(S) = \emptyset$ (which is possible only if $|S| \geq 3$). Take any $x_1 \in S$. Since $c_{\succsim}(S) = \emptyset$, there is an $x_2 \in S\backslash\{x_1\}$ with $x_2 \succ x_1$. Similarly, there is an $x_3 \in S\backslash\{x_1, x_2\}$ with $x_3 \succ x_2$. Continuing this way, I find $S = \{x_1, \ldots, x_{|S|}\}$ with $x_{|S|} \succ \cdots \succ x_1$. Now find a contradiction to $\succsim$ being acyclic.

EXERCISE 14. Apply Sziplrajn's Theorem to the transitive closure of the relation $\succsim^* := \succsim \cup (\{x_*\} \times Y)$.

EXERCISE 16. (e) $\inf \mathcal{A} = \cap\mathcal{A}$ and $\sup \mathcal{A} = \cap\{B \in \mathcal{X} : \cup\mathcal{A} \subseteq B\}$ for any class $\mathcal{A} \subseteq \mathcal{X}$.

EXERCISE 18. Define the equivalence relation $\sim$ on X by $x \sim y$ iff $f(x) = f(y)$, let $Z := X/_\sim$, and let g be the associated quotient map.

EXERCISE 20. If f were such a surjection, we would have $f(x) = \{y \in X : y \notin f(y)\}$ for some $x \in X$.

EXERCISE 29. Show that $\inf S = -\sup\{-s \in \mathbb{R} : s \in S\}$.

EXERCISE 30. Consider first the case where $a \geq 0$. Apply Proposition 6.(b) (twice) to find some $a', b' \in \mathbb{Q}$ such that $0 \leq a < a' < b' < b$. Now $x \in (a', b')$ iff $0 < \frac{x-a'}{b'-a'} < 1$, while $\frac{1}{\sqrt{2}} \in (0,1)\backslash\mathbb{Q}$. (Why?)

EXERCISE 36. $\left|y_m - a\right| \leq z_m - x_m + |x_m - a|$ for each m.

EXERCISE 38. For every $m \in \mathbb{N}$, there exists an $x_m \in S$ with $x_m + \frac{1}{m} > \sup S$.

EXERCISE 40. (b) By part (a) and the Bolzano-Weierstrass Theorem, every real Cauchy sequence has a convergent subsequence.

EXERCISE 41. Note that

$$\left|y_k - y_{k'}\right| \leq \left|y_k - x_{kl}\right| + \left|x_{kl} - x_{k'l}\right| + \left|x_{k'l} - y_{k'}\right|$$

for all $k, k', l \in \mathbb{N}$. Use (i) and (ii) and a 3ε argument to establish that (y_k) is Cauchy. So, by the previous exercise, $y_k \to x$ for some $x \in \mathbb{R}$. Now use the inequality

$$\left|x_{kl} - x\right| \leq \left|x_{kl} - y_k\right| + \left|y_k - x\right|, \qquad k, l = 1, 2, \ldots$$

to conclude that $x_{kl} \to x$.

EXERCISE 43. How about $\left(1, -1, \frac{1}{2}, 1, -1, \frac{1}{2}, \ldots\right)$ and $\left(-1, 1, \frac{1}{2}, -1, 1, \frac{1}{2}, \ldots\right)$?

EXERCISE 45. (a) This follows from the fact that $\left|x_{k+1} + x_{k+2} + \cdots\right| = \left|x - \sum^k x_i\right|$ for all $k \in \mathbb{N}$. (By the way, the converse of (a) is true, too; just verify that $(\sum^m x_i)$ is a real Cauchy sequence (Exercise 40).)

EXERCISE 48. (a) Try the sequence $((-1)^m)$.
(b) Suppose that $\sum^\infty x_i = x \in \mathbb{R}$. Then $s_m \to x$, so, for any $-\infty < y < x$, there is an $M \in \mathbb{N}$ with $s_m \geq y$ for all $m \geq M$. But then

$$\frac{1}{m}\left(\sum_{i=1}^{M} s_i + \sum_{i=M+1}^{m} s_i\right) \geq \frac{1}{m}\sum_{i=1}^{M} s_i + \frac{m-M}{m} y, \qquad m \geq M,$$

so letting $m \to \infty$, and recalling that y is arbitrarily chosen in $(-\infty, x)$, I find $\liminf \frac{1}{m}\sum^m s_i \geq x$. Now argue similarly for $\limsup \frac{1}{m}\sum^m s_i$.

EXERCISE 53. Use the Bolzano-Weierstrass Theorem.

EXERCISE 54. Take any $y \in (f(a), f(b))$, and define $S := \{x \in [a, b] : f(x) < y\}$. Show that $f(\sup S) = y$.

EXERCISE 56. (a) Define $g \in \mathbf{C}[a, b]$ by $g(t) := f(t) - f(a) + \left(\frac{t-a}{b-a}\right)(f(b) - f(a))$, and apply Rolle's Theorem to g.

EXERCISE 57. Consider the case $\alpha = 1$, and write $\int_a^b f := \int_a^b f(t)dt$ and $\int_a^b g := \int_a^b g(t)dt$. Take any $\varepsilon > 0$, and show that there exists an $\mathbf{a} \in \mathcal{D}[a, b]$

with

$$\int_a^b f + \tfrac{\varepsilon}{4} > R_{\mathrm{a}}(f) \geq \int_a^b f > r_{\mathrm{a}}(f) - \tfrac{\varepsilon}{4} \quad \text{and}$$

$$\int_a^b g + \tfrac{\varepsilon}{4} > R_{\mathrm{a}}(g) \geq \int_a^b g > r_{\mathrm{a}}(g) - \tfrac{\varepsilon}{4}.$$

(Find one dissection for f that does the job, and another for g. Then combine these dissections by $\Cup$.) Now verify that $R_{\mathrm{a}}(f+g) \leq R_{\mathrm{a}}(f) + R_{\mathrm{a}}(g)$ and $r_{\mathrm{a}}(f+g) \geq r_{\mathrm{a}}(f) + r_{\mathrm{a}}(g)$. Put it all together to write $R_{\mathrm{a}}(f+g) - r_{\mathrm{a}}(f+g) < \varepsilon$ and $\left|R_{\mathrm{a}}(f+g) - (\int_a^b f + \int_a^b g)\right| < \varepsilon$.

EXERCISE 61. Let me show that $f(a) > 0$ would yield a contradiction, and you do the rest. Suppose $f(a) > 0$. Then, by continuity of f, we have $c := \inf\{x \in (a,b] : f(x) > \frac{f(a)}{2}\} > a$. (Yes?) So, by Exercise 58, $\int_a^b f(t)dt = \int_a^c f(t)dt + \int_c^b f(t)dt \geq \frac{f(a)}{2}(c-a) > 0$—a contradiction.

EXERCISE 68. See Example K.9.

CHAPTER B

EXERCISE 2. As simple as the claim is, you still have to invoke the Axiom of Choice to prove it. Let's see. Take any set S with $|S| = \infty$. I wish to find an injection g from $\mathbb{N}$ into S. Let $\mathcal{A} := 2^S \backslash \{\emptyset\}$. By the Axiom of Choice, there exists a map $f : \mathcal{A} \to \cup\mathcal{A}$ such that $f(A) \in A$ for each $A \in \mathcal{A}$. Now define $g : \mathbb{N} \to S$ recursively, as follows: $g(1) := f(S)$, $g(2) := f(S\backslash\{g(1)\})$, $g(3) := f(S\backslash\{g(1), g(2)\})$, and so on.

EXERCISE 5. (b) There is a bijection between $[0,1]$ and $\mathbb{N}^\infty$. You can prove this by "butterfly hunting." (If you get stuck, sneak a peek at the hint I gave for Exercise 15 below.)

EXERCISE 7. Let me do this in the case where A is countably infinite. Take any countably infinite subset C of B (Exercise 2). Since both C and $A \cup C$ are countably infinite, there exists a bijection g from C onto $A \cup C$. (Why?) Now define $f : B \to A \cup B$ by

$$f(b) := \begin{cases} g(b), & \text{if } b \in C \\ b, & \text{otherwise} \end{cases}.$$

This is a bijection, is it not?

EXERCISE 8. Use Proposition 3.

EXERCISE 9. Its concavity ensures that f is right-differentiable (Exercise A.68), and that f'_+ is decreasing (Example K.9). Now show that $x \in \mathbf{D}_f$ iff x is a point of discontinuity of f'_+, and use Exercise 8.

EXERCISE 10. (c) Recall Exercise A.20.

EXERCISE 13. Let $X := 2^S$, and consider $f \in X^S$ defined by $f(x) := \{y \in S : x \succsim y\}$.

EXERCISE 14. Adapt the proof of Proposition 5.

EXERCISE 15. (b) You know that $2^{\mathbb{N}} \sim_{\text{card}} \{0,1\}^\infty$, so it is enough to prove that $[0,1] \sim_{\text{card}} \{0,1\}^\infty$. You can do this by adopting the trick we used to prove the uncountability of $\mathbb{R}$. Construct the bijection f from $[0,1]$ onto $\{0,1\}^\infty$ as follows. For any given $x \in [0,1]$, the first term of $f(x)$ is 0 if $x \in [0,\frac{1}{2}]$ and 1 if $x \in (\frac{1}{2},1]$. In the former case, the second term of $f(x)$ is 0 if $x \in [0,\frac{1}{4}]$ and 1 if $x \in (\frac{1}{4},\frac{1}{2}]$. In the latter case, ...

EXERCISE 16. Recall Exercise 2.

EXERCISE 22. False.

EXERCISE 25. Adapt the proof of Proposition 9 for the "if" part. For the converse, let u represent $\succsim$, and consider the set of all closed intervals with distinct rational points. Assign to each such interval I one (if any) $y \in X$ such that $u(y)$ belongs to I. This procedure yields a countable set A in X. Define next $B := \{(x,y) \in (X \backslash A)^2 : x \succ y \text{ and } x \succ z \succ y \text{ for no } z \in A\}$, and check that B is countable. (The key is to observe that if $(x,y) \in B$, then $x \succ w \succ y$ for no $w \in X$.) Now verify that $Y := A \cup \{t \in X : \text{either } (t,x) \text{ or } (x,t) \text{ is contained in } B \text{ for some } x \in X\}$ works as desired.

EXERCISE 30. No.

EXERCISE 33. How about $\mathcal{U} := \{u, v\}$, where

$$u(x) := \begin{cases} \frac{x}{x+1}, & \text{if } x \in \mathbb{Q}_+ \\ \frac{-x}{x+1}, & \text{otherwise} \end{cases} \quad \text{and} \quad v(x) := \begin{cases} 1 - u(x), & \text{if } x \in \mathbb{Q}_+ \\ -1 - u(x), & \text{otherwise} \end{cases}$$

CHAPTER C

EXERCISE 1. The answer to the first question is no.

EXERCISE 7. No. In a metric space X, for any distinct $x, y \in X$ there exist two disjoint open subsets O and U of X such that $x \in O$ and $y \in U$.

EXERCISE 9. $bd_{\mathbb{R}}(\mathbb{Q}) = \mathbb{R}$.

EXERCISE 10. Observe that $int_X(S) \cap Y$ is an open subset of Y that is contained in $S \cap Y$. Conversely, if Y is open in X, then $int_Y(S \cap Y)$, being open in Y, must be open in X. Since this set is obviously contained in $S \cap Y$, we have

$$int_Y(S \cap Y) \subseteq int_X(S \cap Y) = int_X(S) \cap int_X(Y) = int_X(S) \cap Y.$$

EXERCISE 12. Try a suitable indiscrete space.

EXERCISE 14. $(a) \Rightarrow (b)$ The answer is hidden in Example 3.[4].
$(b) \Rightarrow (c)$ In this case there exists at least one x^m in $N_{\frac{1}{m},X}(x) \cap S$ for each $m \in \mathbb{N}$. What is $\lim x^m$?
$(c) \Rightarrow (a)$ Use Proposition 1.

EXERCISE 19. Pick any $x \in X$, and use the countability of X to find an $r > 0$ such that $d(x, y) \neq r$ for all $y \in X$ and $d(x, y) > r$ for some $y \in X$. Isn't $\{N_{r,X}(x), X \backslash N_{r,X}(x)\}$ a partition of X?

EXERCISE 22. For any $f, g \in \mathbf{C}[0, 1]$ with $f|_{\mathbb{Q}} = g|_{\mathbb{Q}}$, we have $f = g$.

EXERCISE 25. Take any $x, y \in X$ with $x \succ y$. Show that $U_{\succ}(y) \cup L_{\succ}(x)$ is clopen, so since X is connected, we have $U_{\succ}(y) \cup L_{\succ}(x) = X$. Now to derive a contradiction, assume that z and w are two points in X with $z \bowtie w$. Then either $z \in U_{\succ}(y)$ or $z \in L_{\succ}(x)$. Suppose former is the case; the latter case is analyzed similarly. Since $z \bowtie w$, we must then also have $w \in U_{\succ}(y)$. Now show that $O := L_{\succ}(z) \cap L_{\succ}(w) = L_{\succsim}(z) \cap L_{\succsim}(w)$. Conclude from this that O must be clopen in X, so $O = X$ by Proposition 2. But $z \notin O$.

EXERCISE 28. (a) Try the open cover $\left\{N_{\frac{1}{m},X}(x) : x \in X\right\}$ for each $m \in \mathbb{N}$.

EXERCISE 29. (b) Give a counter example by using a suitable discrete space.

EXERCISE 30. Take any $f \in X$ and let $S_m := \{x \in T : |f(x)| \geq m\}$ for each $m \in \mathbb{N}$. Now, each S_m is a compact in T, and $S_1 \supseteq S_2 \supseteq \cdots$. Then

$\{S_m : m = 1, 2, \ldots\}$ has the finite intersection property, while, if f was not bounded, we would have $\cap^{\infty} S_i = \emptyset$.

EXERCISE 35. The first statement easily follows from Theorem 2 and the definition of d_∞ and $d_{\infty,\infty}$. Use the "baby" Arzelà-Ascoli Theorem to prove the second one.

EXERCISE 40. $\ell^\infty = \mathbf{B}(\mathbb{N})$. In the case of $\mathbf{C}[0,1]$, proceed just as I did in Example 11.[5].

EXERCISE 41. Use Theorem 2.

EXERCISE 42. (a) Recall Exercise A.61.
(b) (i) Watch the metric! Watch the metric! (ii) What is the domain of d_1?

EXERCISE 43. (c) Let (f_m) be a Cauchy sequence in $\mathbf{C}^1[0,1]$. Then (f_m) is also Cauchy in $\mathbf{C}[0,1]$, so $d_\infty(f_m, f) \to 0$ for some $f \in \mathbf{C}[0,1]$. Similarly, $d_\infty(f'_m, g) \to 0$ for some $g \in \mathbf{C}[0,1]$. Now use the result you proved in part (b).

EXERCISE 44. To prove the "if" part (which is a tad harder than proving the "only if" part), pick a nonconvergent Cauchy sequence (x^m) in X, and define $S_m := \{x^k : k = m, m+1, \ldots\}$ for each $m \in \mathbb{N}$.

EXERCISE 46. (a) Recall Exercise 41.

EXERCISE 47. The argument is analogous to the one I gave to prove Theorem 3.

EXERCISE 50. (a) First draw a graph, and then try the map $t \mapsto 1 + \ln(1 + e^t)$ on $\mathbb{R}$.

EXERCISE 51. I wish to show that Φ is a contraction. First, observe that for any $m \in \mathbb{N}$ and $x, y \in X$,

$$d(\Phi(x), \Phi(y)) \leq d(\Phi(x), \Phi_m(x)) + d(\Phi_m(x), \Phi_m(y)) + d(\Phi_m(y), \Phi(y)).$$

Now pick any $\varepsilon > 0$. Since $\sup\{d(\Phi_m(x), \Phi(x)) : x \in X\} \to 0$, there exists an $M \in \mathbb{N}$ such that $d(\Phi(z), \Phi_m(z)) < \frac{\varepsilon}{2}$ for all $m \geq M$ and all $z \in X$. Thus, letting $K := \sup\{K_m : m = 1, 2, \ldots\}$, I get

$$d(\Phi(x), \Phi(y)) \leq \varepsilon + d(\Phi_M(x), \Phi_M(y)) \leq \varepsilon + Kd(x, y)$$

for all $x, y \in X$. Since $K < 1$ by hypothesis, it follows that Φ is a contraction.

The claim would not be true if all we had was $\lim d(\Phi_m(x), \Phi(x)) = 0$ for all $x \in X$. For instance, consider the case where $X = \mathbb{R}$ and $\Phi_m := (1 - \frac{1}{m})\mathrm{id}_{\mathbb{R}}$, $m = 1, 2, \ldots$.

EXERCISE 52. Metrize $\mathbb{R}^n$ by d_∞.

EXERCISE 53. Recall Exercise 48.

EXERCISE 56. $\ln\left(\frac{1+b}{1+a}\right) \leq b - a$ for all $0 \leq a \leq b$. (Study the map $t \mapsto (t-a) - \ln\left(\frac{1+t}{1+a}\right)$. Is this map monotonic on $[a, b]$? What is its minimum/maximum?)

EXERCISE 63. Completeness is not invariant under the equivalence of metrics.

EXERCISE 65. No. Consider, for instance, the case in which (X_i, d_i) is $\mathbb{R}$ and $O_i := \left(-\frac{1}{i}, \frac{1}{i}\right)$, $i = 1, 2, \ldots$.

EXERCISE 68. $\mathbf{c}_0$ is dense $\mathbb{R}^\infty$ relative to the product metric.

EXERCISE 69. Exercise 66 shows that any open set in $\mathsf{X}^\infty(X_i, d_i)$ can be written as a union of sets of the form of $\mathsf{X}^m O_i \times X_{m+1} \times X_{m+1} \times \cdots$, where O_i is an open subset of X_i, $i = 1, \ldots, m$. Apply this characterization to the definition of connectedness, and see what you get.

CHAPTER D

EXERCISE 2. (a) If $x \in X \backslash A$, then there exists an $\varepsilon > 0$ such that $N_{\varepsilon,X}(x) \subseteq X \backslash A$, so $d(x, A) \geq \varepsilon$.

EXERCISE 4. Yes.

EXERCISE 7. (c) Yes.

EXERCISE 8. For any $\varepsilon > 0$, there exists a $\delta > 0$ such that for all $x \in X$, we have $|f(x) - f(y)| < \frac{\varepsilon}{2K}$ and $|g(x) - g(y)| < \frac{\varepsilon}{2K}$ whenever $y \in N_{\delta,X}(x)$. Then

$$|f(x)g(x) - f(y)g(y)| \leq |f(x)|\,|g(x) - g(y)| + |g(y)|\,|f(x) - f(y)| < \varepsilon.$$

Without boundedness, the claim would not be true. For instance, consider the case $X = \mathbb{R}$ and $f = g = \mathrm{id}_X$.

EXERCISE 15. Use Theorem C.2 to show that if (x_m) is a sequence in X that converges to x, then the set $\{x, x^1, x^2, \ldots\}$ is compact in X.

EXERCISE 16. (b) $\mathbb{R}$ is complete, but $(0, 1)$ is not.
(c) Because it is nonexpansive.

EXERCISE 19. This is just Exercise 9 in disguise.

EXERCISE 20. (a) Here is the "only if" part. If $S = \emptyset$, the claim is trivial. So take any nonempty set $S \subseteq X$, and let $y \in f(cl_X(S))$. Then there exists an $x \in cl_X(S)$ such that $y = f(x)$. Clearly, there must exist an $(x^m) \in S^\infty$ such that $x^m \to x$ (Exercise C.14). By continuity of f, we then find $f(x^m) \to f(x)$. So $f(x) \in cl_X(f(S))$.

EXERCISE 24. $\mathbb{Q}$ is countable and $\mathbb{R}\backslash\mathbb{Q}$ is not. So, by the Intermediate Value Theorem . . .?

EXERCISE 25. Punch a hole in $\mathbb{R}^n$. Now do the same in $\mathbb{R}$. Do you get similar spaces?

EXERCISE 27. You can easily prove this by using Theorem C.2 and imitating the way we proved Proposition A.11. Let me suggest a more direct argument here. Take any $\varepsilon > 0$. Since f is continuous, for any $x \in X$ there exists a $\delta_x > 0$ such that $d_Y(f(x), f(y)) < \frac{\varepsilon}{2}$ for all $y \in X$ with $d(x, y) < \delta_x$. Now use the compactness of X to find finitely many $x^1, \ldots, x^m \in X$ with $X = \cup\{N_{\frac{1}{2}\delta_{x_i},X}(x^i) : i = 1, \ldots, m\}$. Let $\delta := \frac{1}{2}\min\{\delta_{x^1}, \ldots, \delta_{x^m}\}$, and observe that $d(x, y) < \delta$ implies $d(x^i, y) < \delta_{x^i}$ for some $i \in \{1, \ldots, m\}$ with $d(x, x^i) < \frac{1}{2}\delta_{x^i}$. (Good old triangle inequality!) But $d_Y(f(x), f(y)) \leq d_Y(f(x), f(x^i)) + d_Y(f(x^i), f(y))$. So?

EXERCISE 30. Suppose there exists a sequence $(x^m) \in T^\infty$ without a convergent subsequence. It is okay to assume that all terms of this sequence are distinct. Define $S := \{x^1, x^2, \ldots\}$. Show that S is closed, and define $\varphi \in \mathbb{R}^S$ by $\varphi(x^i) := i$. Is φ continuous on S? Can you extend φ to $\mathbb{R}^n$?

EXERCISE 31. Define $g \in \mathbb{R}^{[a,b]}$ by $g(t) := f(t) - \alpha t$, and show that g must have a minimum on $[a, b]$.

EXERCISE 37. (a) Since u is strictly increasing, $\{\alpha \in \mathbb{R}_+ : x \succsim (\alpha, \ldots, \alpha)\}$ is bounded from above, so its sup belongs to $\mathbb{R}$. Since u is lower semicontinuous, this set is closed, and hence contains its sup.

EXERCISE 40. (a) The idea is to show that in this case, f must be upper semicontinuous everywhere. Let's use Proposition 4 to this end. Take any $x \in X$, and let (x_m) be any real sequence with $x_m \to x$. Then

$$\begin{aligned}\limsup f(x_m) &= \limsup f((x_m - x + x^*) + (x - x^*)) \\ &= \limsup f(x_m - x + x^*) + f(x - x^*).\end{aligned}$$

Now define $(y_m) := (x_m - x + x^*)$, and notice that $y_m \to x^*$. So?

EXERCISE 41. (a) $f - f(0)$ satisfies Cauchy's functional equation.
(b) Show first that $f(\lambda x + (1-\lambda)y) = \lambda f(x) + (1-\lambda)f(y)$ for any $0 \leq x, y \leq 1$ and $\lambda \in \mathbb{R}$ such that $0 \leq \lambda x + (1-\lambda)y \leq 1$.
(*Note.* Section F.2.2 contains a generalization of this exercise.)

EXERCISE 48. Define $f_m \in \mathbf{C}[-1,1]$ by $f_m(t) := \sum^m \frac{t^i}{i^2}$, $m = 1, 2, \ldots$ and show that $d_\infty(f, f_m) \leq \sum_{i=m+1}^{\infty} \frac{1}{i^2}$ for each m. Since $\sum^m \frac{1}{i^2}$ converges—this is why I'm so sure that f_ms are well-defined—$\sum_{i=m+1}^{\infty} \frac{1}{i^2} \to 0$ as $m \to \infty$. Thus $f_m \to f$ uniformly.

EXERCISE 50. Define $\alpha_m := \sup\{|\varphi_m(x)| : x \in T\}$ for each m and $\alpha := \sup\{|\varphi(x)| : x \in T\}$. (Is α finite?) Since $d_\infty(\varphi_m, \varphi) \to 0$, there exists an $M \in \mathbb{N}$ such that $|\varphi(x) - \varphi_m(x)| < 1$ for all $x \in T$ and $m \geq M$. Then $K := \max\{\alpha_1, \ldots, \alpha_M, \alpha + 1\}$ should do it.

EXERCISE 55. Take $T = \mathbb{N}$ and recall that ℓ^∞ is not separable.

EXERCISE 58. Imitate the argument given in the proof of the Arzelà-Ascoli Theorem.

EXERCISE 62. This is just Exercise 30 (with a touch of the Tietze Extension Theorem).

EXERCISE 66. (a) Since T is not necessarily closed in X, the closedness of these sets in T does not imply their closedness in X. You should use uniform continuity.

EXERCISE 67. Write $\phi = (\phi_1, \phi_2, \ldots)$, and define $\phi_i^* \in \mathbb{R}^T$ by $\phi_i^*(x) := \inf\{\phi_i(w) + Kd(w,x)^\alpha : w \in T\}$ for each i. Now check that $\phi^* = (\phi_1^*, \phi_2^*, \ldots)$ does the job.

EXERCISE 68. (a) Recall the Homeomorphism Theorem (Section 3.1).

(b) Let X be $[0,1]$ with the discrete metric and Y be $[0,1]$ with the usual metric.

EXERCISE 69. Try the projection operator (Example 5).

EXERCISE 73. If f is a fixed point of Φ, then, for any $0 < t < 1$, we must have $f(t) = f(t^{m^2})$ for any $m \in \mathbb{N}$. Then $f|_{[0,1)} = 0$.

EXERCISE 77. Let $\mathbf{0}$ denote the n-vector of 0s, and write $\beta_f(x)$ for $(f_1(x), \ldots, f_n(x))$ for any $x \in B_\alpha^n$. If $\beta_f(x) \neq \mathbf{0}$ for all $x \in B_\alpha^n$, then I can define the map $\Phi : B_\alpha^n \to \mathbb{R}^n$ by $\Phi(x) := \frac{-\alpha}{d_2(\beta_f(x), \mathbf{0})}\beta_f(x)$. Would a fixed point of Φ satisfy the given boundary condition? Does Φ have a fixed point?

CHAPTER E

EXERCISE 1. $\sigma(\mathbb{R}_+) = [0, \frac{\pi}{2}) \cup \{\frac{\pi}{2}, \frac{5\pi}{2}, \frac{9\pi}{2}, \ldots\}$.

EXERCISE 2. $\Gamma(X) = S$.

EXERCISE 4. For the "only if" part, define f by $f(y) := x$, where x is any element of X with $y \in \Gamma(x)$. Is f well-defined?

EXERCISE 8. How about $\Gamma(0) = [-1, 1]$ and $\Gamma(t) := \{0\}$ for all $0 < t \leq 1$?

EXERCISE 10. Yes. To verify its upper hemicontinuity at any (p, ι) with $p_i = 0$ for some i, use the definition of upper hemicontinuity directly. By Example 2, this is the only problematic case.

EXERCISE 11. If $(y_m) \in \Gamma(S)^\infty$, then there exists an $(x_m) \in S^\infty$ such that $y_m \in \Gamma(x_m)$ for each m. Use Theorem C.2 to extract a subsequence of (x_m) that converges to some $x \in S$, and then use Proposition 2 to get your hands on a suitable subsequence of (y_m).

EXERCISE 12. (a) Suppose f is closed. Take any $y \in Y$ and let O be an open subset of X with $f^{-1}(y) \subseteq O$. Notice that $Y \backslash f(X \backslash O)$ is an open subset of Y that contains y. Moreover, $f^{-1}(Y \backslash f(X \backslash O)) \subseteq O$. Conversely, if S is a closed subset of X and $y \in Y \backslash f(S)$, then $X \backslash S$ is open in X and $f^{-1}(y) \subseteq X \backslash S$, so there exists a $\delta > 0$ such that $f^{-1}(N_{\delta,Y}(y)) \subseteq X \backslash S$. So?

EXERCISE 21. Use Proposition D.1 and the triangle inequality (twice) to verify that Γ has a closed graph. Then apply Proposition 3.(a).

EXERCISE 22. Let $X_0 := X$ and $X_{i+1} := \Gamma(X_i)$ for all $i \in \mathbb{Z}_+$. Recall Example C.8 to be able to conclude that $S := \cap^\infty X_i$ is a nonempty compact set. S is a fixed set of Γ.

EXERCISE 24. The answer to the last question here is no.

EXERCISE 27. Take any $(x^m) \in X^\infty$ and $(y_m) \in \mathbb{R}^\infty$ with $x^m \to x$ for some $x \in X$, and $0 \leq y_m \leq \varphi(x^m)$ for each m. Since φ is continuous, $(\varphi(x^m))$ converges. Deduce that (y_m) is a bounded sequence, and then use the Bolzano-Weierstrass Theorem and Proposition 2 to conclude that Γ is upper hemicontinuous.

To show that Γ is lower hemicontinuous, take any $(x, y) \in X \times \mathbb{R}$ and $(x^m) \in X^\infty$ such that $x^m \to x$ and $0 \leq y \leq \varphi(x)$. If $y = \varphi(x)$, then define $y_m := \varphi(x^m)$ for each m. If $y < \varphi(x)$, then there exists an integer M with $0 \leq y \leq \varphi(x^m)$ for each $m \geq M$, so pick y_m from $[0, \varphi(x^m)]$ arbitrarily for $m = 1, \ldots, M$, and let $y_m := y$ for each $m \geq M$. Either way, $y_m \to y$ and $y_m \in \Gamma(x_m)$ for each m. Apply Proposition 4.

EXERCISE 28. Γ is not upper hemicontinuous at any $x \in [0,1] \cap \mathbb{Q}$, but it is upper hemicontinuous at any $x \in [0,1]\backslash\mathbb{Q}$. To prove the latter fact, recall that $[0,1] \cap \mathbb{Q}$ is dense in $[0,1]$, so there is one and only one open subset of $[0,1]$ that contains $[0,1] \cap \mathbb{Q}$.

Γ is lower hemicontinuous. To see this, take any $x \in [0,1]$, $y \in \Gamma(x)$, and any $(x_m) \in [0,1]^\infty$ with $x_m \to x$. Let (z_m) be a sequence in $[0,1] \cap \mathbb{Q}$ with $z_m \to y$ and (w_m) a sequence in $[0,1]\backslash\mathbb{Q}$ with $w_m \to y$. Now define $y_m := z_m$ if x_m is irrational and $y_m := w_m$ if x_m is rational, $m = 1, 2, \ldots$. Then $(y_m) \in [0,1]^\infty$ satisfies $y_m \in \Gamma(x_m)$ for each m, and $y_m \to y$.

EXERCISE 29. Upper hemicontinuity is just the closed graph property here. For the lower hemicontinuity at any $v \in u(T)$, suppose there exists an open subset O of T such that $\Gamma(v) \cap O \neq \emptyset$, but for any $m \in \mathbb{N}$ there exists an $v_m \in N_{\frac{1}{m},\mathbb{R}}(v) \cap u(T)$ such that $\Gamma(v_m) \cap O = \emptyset$. Take any $x \in \Gamma(v) \cap O$, and show that $u(x) > v$ is impossible. If $u(x) = v$, then use the monotonicity of u.

EXERCISE 32. (a) If $a < 0 < b < 1$, then $d_H([0,1],[a,b]) = \max\{|a|, 1 - b\}$. (b) Yes.

EXERCISE 35. Take any Cauchy sequence (A_m) in $\mathbf{c}(Y)$ and define $B_m := \text{cl}_Y(A_m \cup A_{m+1} \cup \cdots)$ for each m. Show that B_1 (hence each B_m) is compact.

By the Cantor-Fréchet Intersection Theorem, $\cap^\infty B_i \in \mathbf{c}(Y)$. Show that $A_m \to \cap^\infty B_i$.

EXERCISE 36. Let $\mathcal{F} := \{f_1, \ldots, f_k\}$ for some $k \in \mathbb{N}$. Define the self-map Φ on $\mathbf{c}(Y)$ by $\Phi(A) := f_1(A) \cup \cdots \cup f_k(A)$, and check that Φ is a contraction, the contraction coefficient of which is smaller than the maximum of those of f_is. Now apply Exercise 35 and the Banach Fixed Point Theorem.

EXERCISE 42. Modify the argument given in the last paragraph of the proof of Maximum Theorem.

EXERCISE 43. Take any $(\theta^m) \in \Theta^\infty$ with $\theta^m \to \theta$ for some $\theta \in \Theta$, and suppose that $\varphi^*(\theta) < \limsup \varphi^*(\theta^m)$. Then there exist an $\varepsilon > 0$ and a strictly increasing $(m_k) \in \mathbb{N}^\infty$ such that, for some $x^{m_k} \in \Gamma(\theta^{m_k})$, $k = 1, 2, \ldots$, and $K \in \mathbb{N}$, we have $\varphi^*(\theta) + \varepsilon \leq \varphi(x^{m_k}, \theta^{m_k})$ for all $k \geq K$. (Why?) Now use the upper hemicontinuity of Γ to extract a subsequence of (x^{m_k}) that converges to a point in $\Gamma(\theta)$, and then derive a contradiction by using the upper semicontinuity of φ.

EXERCISE 44. (a) Let $\alpha := u(0, \ldots, 0)$ and $\beta := u(1, \ldots, 1)$. Define $\Gamma : [\alpha, \beta] \rightrightarrows T$ by $\Gamma(\upsilon) := u^{-1}([\upsilon, \beta])$. It is easy to check that Γ is upper hemicontinuous. It is in fact lower hemicontinuous as well. To see this, take any $\upsilon \in [\alpha, \beta]$, $x \in \Gamma(\upsilon)$, and any sequence $(\upsilon_m) \in [\alpha, \beta]^\infty$ with $\upsilon_m \to \upsilon$. Let $I := \{m : \upsilon_m \leq \upsilon\}$ and $J := \{m : \upsilon_m > \upsilon\}$. Define

$$S := T \cap \{\lambda x + (1 - \lambda)(1, \ldots, 1) : \lambda \in [0, 1]\}.$$

Clearly, $\beta \geq \upsilon_m \geq \upsilon$ for all $m \in J$. By the Intermediate Value Theorem, for each $m \in J$, there exists an $0 \leq \lambda_m \leq 1$ such that $u(\lambda_m x + (1 - \lambda_m)(1, \ldots, 1)) = \upsilon_m$. Now define $(x^m) \in T^\infty$ as $x^m := x$ if $m \in I$, and as $x^m := \lambda_m x + (1 - \lambda_m)(1, \ldots, 1)$ otherwise. Then $x^m \in \Gamma(\upsilon_m)$ for eah m and $x^m \to x$. (Why?) Conclusion: Γ is a continuous correspondence. The stage is now set for applying the Maximum Theorem.
(b) $\mathbf{e}_u(p, u^*(p, \iota)) \leq \iota$ follows from definitions. To show that $<$ cannot hold here, you will need the continuity of $\mathbf{e}_u$ and strict monotonicity of u^*.

EXERCISE 53. (a) Let me formulate the problem in terms of capital accumulation. In that case, letting $\gamma := 1 - \alpha$, the problem is to choose a real sequence (x_m) in order to

$$\text{Maximize } \sum_{i=0}^{\infty} \delta^i (pf(x_i) - (x_{i+1} - \gamma x_i))$$

such that

$$\gamma x_m \leq x_{m+1} \leq x_m + \theta, \; m = 0, 1, \ldots.$$

It is easily seen that f must have a unique positive fixed point, denote it by $\bar{x}$. Clearly, the firm will never operate with an input level that exceeds $\bar{x}$, so the solution of the problem must belong to X^∞, where $X := [0, \bar{x}]$. Define $\Gamma : X \rightrightarrows X$ by $\Gamma(x) := [\gamma x, x + \theta]$, and $\varphi : Gr(\Gamma) \to \mathbb{R}$ by $\varphi(a, b) := pf(a) - (b - \gamma a)$. Then the problem of the firm can be written as choosing $(x_m) \in X^\infty$ in order to

$$\text{Maximize } \sum_{i=0}^{\infty} \delta^i \varphi(x_i, x_{i+1}) \text{ such that } x_{m+1} \in \Gamma(x_m), \; m = 0, 1, \ldots.$$

(b) The optimal policy correspondence P for $\mathcal{D}(X, \Gamma, u, \delta)$ satisfies

$$P(x) = \arg\max\{\varphi(x, y) + \delta V(y) : \gamma x \leq y \leq x + \theta\},$$

where V is the value function of the problem. It is single-valued, so I can treat is as a function. Then, using the one-deviation property,

$$P(x) \in \arg\max\{pf(x) - (y - \gamma x) + \delta(pf(y) - (P^2(x) - \gamma y)) \\ + \delta^2 V(P^2(x)) : \gamma x \leq y \leq x + \theta\}.$$

By strict concavity of f, the solution must be interior, so the first-order condition of the problem shows that $-1 + \delta p f'(P(x)) + \delta\gamma = 0$. But there is only one x that satisfies this, no?

EXERCISE 57. Study the correspondence $\Gamma : T \rightrightarrows \mathbb{R}^n$ with $\Gamma(x) := \{y \in \mathbb{R}^n : (x, y) \in S\}$, where $T := \{x \in \mathbb{R}^n : (x, y) \in S \text{ for some } y \in S\}$.

EXERCISE 58. Study the correspondence $\Gamma : X \rightrightarrows X$ defined by $\Gamma(x) := \{z \in X : \varphi(x, z) \geq \max\{\varphi(x, y) : y \in X\}\}$.

EXERCISE 59. The "$\leq$" part of the claim is elementary. To prove the "$\geq$" part, define $\Gamma_1 : Y \rightrightarrows X$ by $\Gamma_1(y) := \arg\max f(\cdot, y)$ and $\Gamma_2 : X \rightrightarrows Y$ by $\Gamma_2(x) := \arg\min f(x, \cdot)$. Now define the self-correspondence Γ on

$X \times Y$ by $\Gamma(x,y) := \Gamma_1(y) \times \Gamma_2(x)$. Use Kakutani's Fixed Point Theorem to find an (x^*, y^*) such that $(x^*, y^*) \in \Gamma(x^*, y^*)$, and check that $\max_{x\in X} \min_{y\in Y} f(x,y) \geq f(x^*, y^*) \geq \min_{y\in Y} \max_{x\in X} f(x,y)$.

EXERCISE 60. Let $T := \{x \in X : x \in \Gamma(x)\}$, and note that $T \neq \emptyset$. Define $g := f|_T$, and show that g is a self-map on T. Moreover, if x is a fixed point of g, then it is a fixed point of both f and Γ.

EXERCISE 65. (a) Let $O_y := \{x \in X : y \in \Psi(x)\}$, $y \in \mathbb{R}^n$. Then $\{O_y : y \in \mathbb{R}^n\}$ is an open cover of X, so since X is compact, there are finitely many $y^1, \ldots, y^k \in \mathbb{R}^n$ such that $\{O(y^i) : i = 1, \ldots, k\}$ covers X. Now proceed as in the proof of the Approximate Selection Lemma.
(b) Apply part (a) and the Brouwer Fixed Point Theorem.

EXERCISE 66. (b) If the claim was false, then, by Carathéodory's Theorem, we could find $(x^m), (z^m), (y^{m,i}) \in X^\infty$ and $(\lambda_{m,i}) \in [0,1]^\infty$, $i = 1, \ldots, n+1$ such that $\sum^{n+1} \lambda_{m,i} = 1$, $\sum^{n+1} \lambda_{m,i} y^{m,i} \in \Gamma(z^m)$, $d_2(z^m, x^m) < \frac{1}{m}$, and $(x^m, \sum^{n+1} \lambda_{m,i} y^{m,i}) \notin N_{\varepsilon, X\times X}(Gr(\Gamma))$, for all $m \in \mathbb{N}$. Use the sequential compactness of X and the convex-valuedness and closed graph properties of Γ to derive a contradiction.

CHAPTER F

EXERCISE 2. $X = \{\mathbf{0}\}$ has to be the case.

EXERCISE 3. Sure. $((0,1), \oplus)$, where $x \oplus y := x + y - \frac{1}{2}$, is such a group.

EXERCISE 4. (c) Suppose there are $y, z \in X$ such that $y \in Y \backslash Z$ and $z \in Z \backslash Y$. If $(Y \cup Z, +)$ is a group, then $y + z \in Y \cup Z$, that is, either $y + z \in Y$ or $z \in Y \cup Z$. But if $y + z \in Y$, then $z = -y + y + z \in Y$, a contradiction.

EXERCISE 12. (b) Scalar multiplication does not play a role here; recall Exercise 4.(c).

EXERCISE 15. (a) $\mathbb{R}^2$.
(b) $\{(a,b) \in \mathbb{R}^2 : a \neq 0 \text{ and } b \neq 0\}$

EXERCISE 17. Let $m \geq n+2$, and pick any m vectors $x^1, \ldots, x^m$ in S. Since there can be at most $n+1$ affinely independent vectors in $\mathbb{R}^n$, $\{x^1, \ldots, x^m\}$ is affinely dependent. That is, there is a *nonzero* m-vector $(\alpha_1, \ldots, \alpha_m)$ such that $\sum^m \alpha_i x^i = \mathbf{0}$ and $\sum^m \alpha_i = 0$. Let $I := \{i : \alpha_i > 0\}$ and $J := \{i : \alpha_i < 0\}$.

Then $A := \{x^i : i \in I\}$ and $B := \{x^i : i \in J\}$ are equal to the task. Indeed, where $\beta_i := \alpha_i / \sum_{i \in I} \alpha_i$ for each $i \in I$, we have $\sum_{i \in I} \beta_i x^i \in co(A) \cap co(B)$.

EXERCISE 18. (a) The claim is obviously true if $|\mathcal{S}| = n+1$. Suppose that it is also true when $|\mathcal{S}| = m$, for an arbitrarily picked integer $m \geq n+1$. We wish to show that it would then hold in the case $|\mathcal{S}| = m+1$ as well. Indeed, if $|\mathcal{S}| = m+1$, the induction hypothesis implies that there is at least one vector x^i in $\cap(\mathcal{S} \backslash \{S_i\})$, for each $i = 1, \ldots, m+1$. We are done if $x^i = x^j$ for some $i \neq j$. Otherwise, the cardinality of the set $T := \{x^1, \ldots, x^{m+1}\}$ exceeds $n+2$, so by Radon's Lemma, there exist disjoints sets A and B in T such that $co(A) \cap co(B) \neq \emptyset$. Now show that if $x \in co(A) \cap co(B)$, then $x \in \cap \mathcal{S}$.
(b) By part (a), any nonempty finite subset of the class $\mathcal{S}$ has a nonempty intersection.

EXERCISE 23. Let S be a basis for X, and T for Y. Define $S' := \{(x, \mathbf{0}) : x \in S\}$ and $T' := \{(\mathbf{0}, y) : y \in T\}$. (Here the former $\mathbf{0}$ is the origin of X, and the latter that of Y.) Isn't $S' \cup T'$ a basis for $X \times Y$?

EXERCISE 26. (b) $dim(null(L)) = 2$.
(c) If $f \in \mathbf{P}(\mathbb{R})$ is defined by $f(t) := t^2$, then $L(\mathbb{R}^{2\times 2}) = span(\{\mathrm{id}_{\mathbb{R}}, f\})$.

EXERCISE 28. $KL(-x) = KL(x)$ for all $x \in X$.

EXERCISE 30. Start with a basis for $null(L)$ (which exists by Theorem 1), and then extend this to a basis for X. If S is the set of vectors you added in this extension, then $L(S)$ is a basis for $L(X)$.

EXERCISE 31. (b) Here is how to go from (i) to (iii). Take any $\alpha \in \mathbb{R}\backslash\{0\}$ and $x, x' \in X$. Suppose $z \in \Gamma(\alpha x)$. Then, by linearity of Γ, $\frac{1}{\alpha} z \in \Gamma(x)$, so $z \in \alpha\Gamma(z)$. Thus, $\Gamma(\alpha x) \subseteq \alpha\Gamma(x)$. Now take any $z \in \Gamma(x + x')$. Then, for any $w \in \Gamma(-x')$, we have

$$z + w \in \Gamma(x + x') + w \subseteq \Gamma(x + x') + \Gamma(-x') \subseteq \Gamma(x)$$

since Γ is linear. Thus $z \in \Gamma(x) - w \subseteq \Gamma(x) + \Gamma(x')$, since, by the first part of the proof, $-w \in \Gamma(x')$.
(c) If $y \in \Gamma(x)$, then, for any $z \in \Gamma(x)$, we have $z = y + z - y \in y + \Gamma(x) - \Gamma(x) \subseteq y + \Gamma(\mathbf{0})$.
(e) We have $P(\Gamma(\mathbf{0}) = \{\mathbf{0}\}$, so $P \circ \Gamma$ is single-valued by part (a). Take any $x \in X$ and let $y := P(\Gamma(x))$. Then there is a $z \in \Gamma(x)$ such that $y = P(z)$.

We have $z - P(z) \in null(P)$ (since P is idempotent), so $z - P(z) \in \Gamma(\mathbf{0})$ which means $z - P(z) + \Gamma(\mathbf{0}) = \Gamma(\mathbf{0})$ by part (c). But then, using part (c) again, $\Gamma(x) - y = z + \Gamma(\mathbf{0}) - y = z - P(z) + \Gamma(\mathbf{0}) = \Gamma(\mathbf{0})$. Since $\mathbf{0} \in \Gamma(\mathbf{0})$, it follows that $y \in \Gamma(x)$.

EXERCISE 32. (a) In $\mathbb{R}^2$ a carefully selected S with $|S| = 4$ would do the job. (b) Take any $\lambda \in \mathbb{R}$ and $x, y \in S$, and suppose $z := \lambda x + (1-\lambda)y \in S$. If $\lambda < 0$, since $\frac{1}{1-\lambda}z - \frac{\lambda}{1-\lambda}x = y$, we have $\frac{1}{1-\lambda}\varphi(z) - \frac{\lambda}{1-\lambda}\varphi(x) = \varphi(y)$ by pseudo-affineness of φ, so we get $\varphi(z) = \lambda\varphi(x) + (1-\lambda)\varphi(y)$. The case $\lambda > 1$ is settled by an analogous argument, and the case $0 \leq \lambda \leq 1$ is trivial. So, if φ is pseudo-affine, then $\varphi(\lambda x + (1-\lambda)y) = \lambda\varphi(x) + (1-\lambda)\varphi(y)$ for any $\lambda \in \mathbb{R}$ and $x, y \in S$ with $\lambda x + (1-\lambda)y \in S$. Now proceed with induction. It is at this point that you will need to use the convexity of S.

EXERCISE 34. The problem is to show that injectivity and surjectivity alone are able to entail the invertibility of L. Let $S := \{x^1, \ldots, x^m\}$ be a basis for X, $m \in \mathbb{N}$, and let $y^i := L(x^i)$ for each i. In both cases, the idea is to define L^{-1} as "the" linear operator in $\mathcal{L}(X, X)$ with $L^{-1}(y^i) = x^i$. If $\{y^1, \ldots, y^m\}$ is a basis for X, then we are fine. For in that case there is a unique $F \in \mathcal{L}(X, X)$ with $F(y^i) = x^i$, and this F is the inverse of L. (A linear operator is determined completely by its actions on the basis, right?) Well, if L is injective, then $L(S)$ is linearly independent in X, and if L is surjective, $L(S)$ spans $L(X) = X$. (Why?) So, in either case, $L(S)$ is a basis for X.

EXERCISE 35. (b) Let A be a basis for X and B a basis for Y. Suppose $|A| \leq |B|$, and let Σ be the set of all injections in B^A. For any $\sigma \in \Sigma$, define $L_\sigma \in \mathcal{L}(X, Y)$ by $L_\sigma(x) := \sigma(x)$ for all $x \in A$. (Is L_σ well-defined?) Now show that $\{L_\sigma : \sigma \in \Sigma\}$ is a basis for $\mathcal{L}(X, Y)$, while $|\Sigma| = |A|\,|B|$. A similar argument works for the case $|A| > |B|$ as well.

EXERCISE 37. $\mathbf{0} \notin H$ is needed only for the existence part.

EXERCISE 52. Define $x_i := \mathbf{v}(A_i \cup \{i\}) - \mathbf{v}(A_i)$, where $A_1 = \emptyset$ and $A_i := \{1, \ldots, i-1\}$ for each $i \geq 2$. Show that $x \in core(\mathbf{v})$.

EXERCISE 53. This is much easier than it looks. Use the Principle of Mathematical Induction to show that there is a unique potential function. To prove the second claim, fix any $(N, \mathbf{v}) \in \mathcal{G}$, and use Proposition 8 to write $\mathbf{v} = \sum_{A \in \mathcal{N}} \alpha_A^{\mathbf{v}} u_A$ for some $\alpha_A^{\mathbf{v}} \in \mathbb{R}$, $A \in \mathcal{N}$. Now set $P(N, \mathbf{v}) = \sum_{A \in \mathcal{N}} \frac{1}{|A|} \alpha_A^{\mathbf{v}}$. (Interpretation?) It remains to check that $\Delta_i P(N, \mathbf{v}) = L_i^{\mathrm{S}}(N, \mathbf{v})$ for each i.

CHAPTER G

EXERCISE 4. Let $H = \{x : L(x) = \alpha\}$, where $L \in \mathcal{L}(X, \mathbb{R})$ and $\alpha \in \mathbb{R}$. Use the convexity of S and the Intermediate Value Theorem to show that if $L(x) > \alpha > L(y)$ for some $x, y \in S$, then $L(z) = \alpha$ for some $z \in S$. (In fact, the hypothesis that X be Euclidean is redundant here. Can you see why?)

EXERCISE 8. (e) Let $\Gamma(x_0) = \{z\}$, and let $y, y' \in \Gamma(x)$ for some $x \in X$. Let $x' := 2x_0 - x$ (which belongs to the domain of Γ since X is a linear space), and observe that $\frac{1}{2}\{y, y'\} + \frac{1}{2}\Gamma(x') \subseteq \{z\}$ by convexity of Γ. Deduce that $y = y'$.

EXERCISE 9. (b) C is infinite-dimensional, because $\{(1,0,0,\ldots),(0,1,0,\ldots),\ldots\}$ is a basis for $\mathbf{c}^0$. Moreover, the sequences $(x_m) := (1,-1,0,0,\ldots)$ and $(y_m) := (-2,1,0,0,\ldots)$ belong to C, but $\frac{1}{2}(x_m) + \frac{1}{2}(y_m) \notin C \cup \{(0,0,\ldots)\}$.
(c) For any vector in $\mathbf{c}^0$, there exist two vectors in C such that that vector lies at the midpoint of those two vectors.

EXERCISE 12. The question is if $co(C)$ is a cone. (Yes it is!)

EXERCISE 13. The hardest part is to show that (ii) implies (iii). To show this, take any $x_0 \in B$ and let $Y := \mathit{aff}(B) - x_0$. Then Y is a linear subspace of X such that $x_0 \in X \setminus Y$. Take any basis S for Y, and show that there is a basis T for X with $S \cup \{x_0\} \subseteq T$. For all $x \in X$, there exists a unique $\lambda^x : T \to \mathbb{R}$ such that $x = \sum_{z \in T} \lambda^x(z) z$, where, of course, $\lambda^x(z) \neq 0$ for only finitely many $z \in T$. (Why?) Define $L \in \mathcal{L}(X, \mathbb{R})$ by $L(x) := \lambda^x(x_0)$; that should do the job.

EXERCISE 14. Recall that $span(X_+) = X_+ - X_+$ (Exercise 10).

EXERCISE 20. (a) Let $n := dim(\mathit{aff}(S))$. The case $n = 0$ is trivial, so assume $n \geq 1$, and pick any $x^1 \in S$. Then $span(S - x^1)$ is an n-dimensional linear space, so it contains n linearly independent vectors, say $x^2 - x^1, \ldots, x^{n+1} - x^1$. (This means that $\{x^1, \ldots, x^{n+1}\}$ is affinely independent in X.) Now show that $\sum^{n+1} \frac{1}{n+1} x^i \in ri(co\{x^1, \ldots, x^{n+1}\}) \subseteq ri(S)$.
(b) Use part (a) and Observation 1.

EXERCISE 22. No. Evaluate $al\text{-}int_{\mathbb{R}^2}(S)$ and $al\text{-}int_{\mathbb{R}^2}(al\text{-}int_{\mathbb{R}^2}(S))$ for the set S considered at the end of Example 4, for instance.

EXERCISE 25. $al\text{-}int_{\mathbf{C}[0,1]}(\mathbf{C}[0,1]_+) = \{f \in \mathbf{C}[0,1] : f \gg 0\}$.

EXERCISE 26. Let $x \in al\text{-}int_X(S)$, and pick any $y \in X$. By definition, there exist $\alpha_{x+y} > 0$ and $\beta_{x-y} > 0$ such that $x + \alpha y = (1-\alpha)x + \alpha(x+y) \in S$ and $x - \beta y = (1-\beta)x + \beta(x-y) \in S$ for all $0 \leq \alpha \leq \alpha_{x+y}$ and $0 \leq \beta \leq \beta_{x-y}$. Thus, defining $\varepsilon_y := \min\{\alpha_{x+y}, \beta_{x-y}\}$ establishes the "only if" part of the claim. To prove the converse, assume that x satisfies the stated property, pick any $y \in X$, and use the property with respect $x - y$.

EXERCISE 27. (a) Observe that $(1-\alpha)x \in al\text{-}int_X((1-\alpha)S)$ for any $0 \leq \alpha < 1$, and use Exercise 26.
(b) That $\mathit{aff}(S) = X$ is immediate from Observation 1. Let $T := al\text{-}int_X(S)$, and to derive a contradiction, suppose $\mathit{aff}(T) \subset X$. Then, by Observation 1, $al\text{-}int_X(T) = \emptyset$. Now use part (a) and the convexity of S to show that we cannot have $al\text{-}int_X(T) = \emptyset$.

EXERCISE 30. (a) $\mathbb{R}\backslash\mathbb{Q}$ is algebraically closed in $\mathbb{R}$, but $\mathbb{Q}$ is not algebraically open in $\mathbb{R}$.
(b) If x is in $S\backslash al\text{-}int_X(S)$, then, by convexity of S, there must exist a $y \in X\backslash S$ such that $(1-\alpha)x + \alpha y \in X\backslash S$ for all $0 < \alpha \leq 1$.

EXERCISE 33. Let $x \in \cap\{al\text{-}cl(A) : A \in \mathcal{A}\}$ and pick any $y \in \cap\{ri(A) : A \in \mathcal{A}\}$. Now use the Claim I proved in Example 10 to get $x \in al\text{-}cl(\cap\{ri(A) : A \in \mathcal{A}\})$.

EXERCISE 39. This is an easy extension problem—we don't need the power of the Hahn-Banach Extension Theorem 1 to settle it. Examine $L^* \in \mathbb{R}^{\ell^p}$ defined by $L^*((x_m)) := d_p((x_m), \mathbf{0})L\left(\frac{1}{d_p((x_m),\mathbf{0})}(x_m)\right)$.

EXERCISE 42. The argument underlying the Hahn-Banach Extension Theorem 1 will do fine here. Take any positive $L \in \mathcal{L}(Y, \mathbb{R})$ and $z \in X\backslash Y$, and try to find a positive $K \in \mathcal{L}(Z, \mathbb{R})$, where $Z := span(Y \cup \{z\})$. To this end, define $A := \{y \in Y : z \succsim_X y\}$ and $B := \{y \in Y : y \succsim_X z\}$, and check that these are nonempty sets. Show that there is an $\alpha \in \mathbb{R}$ with $\sup L(A) \leq \alpha \leq \inf L(B)$. This number will do the job of α I used in the proof of Hahn-Banach Extension Theorem 1.

EXERCISE 44. Pick any $x \in S$. Define $T := S - x$, $y := z - x$, and $Y := span(S - x)$. Then $al\text{-}int_Y(T) \neq \emptyset$ and $y \in Y\backslash T$. First apply Proposition 1 and then Corollary 2.

EXERCISE 45. False, even in $\mathbb{R}^2$!

EXERCISE 46. Separate $\{(x,t) : t \geq \varphi(x)\}$ and $\{(x,t) : t \leq \psi(x)\}$.

EXERCISE 59. The first possibility can be written as $\sum^m w_j^1 u^j + \sum^m w_j^2(-u^j) = v$ for some $w^1, w^2 \in \mathbb{R}_+^m$. Now use Farkas' Lemma.

EXERCISE 60. Define the $(n+1) \times n$ matrix $\mathbf{B} := [b_{ij}]$, where

$$b_{ij} := \begin{cases} a_{ij}, & \text{if } i = 1, \ldots, n \\ 1, & \text{if } i = n+1 \end{cases},$$

$j = 1, \ldots, n$. The claim is equivalent to the existence of an $x \in \mathbb{R}_+^n$ such that $\mathbf{B}x = (x_1, \ldots, x_n, 1)$. Use Farkas' Lemma.

EXERCISE 61. (a) Follow through the argument given before the statement of the result. That is, show that $A - B$ is a closed set, and then apply the argument outlined in the second paragraph of Proposition 4.
(b) Show that $cl_{\mathbb{R}^n}(A - B)$ is a closed and convex set, and use Proposition 4.

EXERCISE 62. Use the Minkowski Separating Hyperplane Theorem to find a nonzero $L \in \mathcal{L}(\mathbb{R}^n, \mathbb{R})$ and $\alpha \in \mathbb{R}$ such that $L(H) \geq \alpha \geq L(C)$. Clearly, $\alpha \geq 0$ whereas $span(H) = \mathbb{R}^n$. So?

EXERCISE 66. (b) Let S be a nonempty compact and convex subset of $\mathbb{R}^n$. Let's use induction on $dim(S)$ (Remark F.1). If $dim(S) = 0$, there is nothing to prove. Take any $k \in \mathbb{N}$, and suppose the claim is true for $dim(S) \leq k-1$. Now consider the case where $dim(S) = k$. Take any $x \in S$. If $x \in al\text{-}bd_{\mathbb{R}^n}(S)$, then there exists a hyperplane H that supports S at x. But then $dim(H \cap S) \leq k-1$ (yes?), so the induction hypothesis applies. If, on the other hand, $x \in al\text{-}int_{\mathbb{R}^n}(S)$, take any $y \in S \backslash \{x\}$ (which exists since $dim(S) > 0$) and observe that the line $\{\lambda x + (1-\lambda)y : \lambda \in \mathbb{R}\}$ intersects S in exactly two points in $al\text{–}bd_{\mathbb{R}^n}(S)$. Use what you learned in the previous case to express each of these points as convex combinations of the extreme points of S.

EXERCISE 73. By Example F.9, there is a $w \in \mathbb{R}^n$ such that $L(y) = wy$. But $L(y) = \omega y$ for all $y \in Y$ and $\omega \in w + Y^\perp$.

EXERCISE 76. By the Projection Theorem, any $x \in \mathbb{R}^n$ can be written as $x = y^x + z^x$ for some unique $(y^x, z^x) \in Y \times Y^\perp$. Define P by $P(x) := y^x$.

CHAPTER H

EXERCISE 1. (a) Pick any $s, t \in S$ with $s \succ t$, and any $r \in \textit{al-int}_Y(S)$. Show that there is an $0 < \alpha < 1$ small enough so that $s^* := (1-\alpha)r + \alpha s \in \textit{al-int}_Y(S)$ and $t^* := (1-\alpha)r + \alpha t \in \textit{al-int}_Y(S)$. Now take any $\sigma \in Y$ and use weak continuity and the fact that $s^* \in \textit{al-int}_Y(S)$ to find a $\theta_\sigma > 0$ small enough so that $S \ni (1-\theta)s^* + \theta(\sigma + t^*) \succ t^*$ for all $0 \leq \theta \leq \theta_\sigma$. Conclusion: $(1-\theta)(s^* - t^*) + \theta\sigma \in A$ for all $0 \leq \theta \leq \theta_\sigma$.
(b) $\textit{span}(C) \subseteq \textit{span}(S) = \{\lambda(s-t) : \lambda > 0 \text{ and } s, t \in S\}$.

EXERCISE 8. This is a showcase for the Minkowski Separating Hyperplane Theorem.

EXERCISE 9. Use Bewley's Expected Utility Theorem.

EXERCISE 11. (b) A variational preference need not satisfy the C-Independence Axiom*.

EXERCISE 17. Consider L first on Y, and then use the Hahn-Banach Extension Theorem 1.

EXERCISE 25. This axiom, SYM, C.INV, and IIA imply Axiom PO.

EXERCISE 27. (b) Roth (1980) argues strongly that the "right" solution here is $(\frac{1}{2}, \frac{1}{2}, 0)$. (But Aumann (1985) disagrees.)

EXERCISE 28. This requires you to find a clever way of using Kakutani's Fixed Point Theorem.

CHAPTER I

EXERCISE 8. For (b), use the continuity of vector addition to show that $A + x$ is open for any x. Now take the union over all x in B.

EXERCISE 9. $\mathbb{Q}^\infty$ doesn't work. Why?

EXERCISE 12. Let X be a metric linear space and $S_x := \{\lambda x : \lambda \in \mathbb{R}\}$ for each $x \in X$. Then $X = \cup\{S_x : x \in X\}$ and $\cap\{S_x : x \in X\} \neq \emptyset$. Moreover, for each x, S_x is the image of a connected set under a continuous map.

EXERCISE 13. The idea is that any number in $(0,1)$ can be approximated by iterating the process of taking midpoints. More precisely, let $A_0 := \{0, 1\}$, $A_1 := \{0, \frac{1}{2}, 1\}$, $A_2 := \{0, \frac{1}{4}, \frac{1}{2}, \frac{3}{4}, 1\}$, etc. (that is, $A_i := \{\frac{a}{2} + \frac{b}{2} : a, b \in A_{i-1}\}$,

$i = 1, 2, \ldots\}$). Let $A := A_0 \cup A_1 \cup \cdots$, and show that $cl_{[0,1]}(A) = [0, 1]$ while $\lambda\Gamma(x) + (1 - \lambda)\Gamma(y) \subseteq \Gamma(\lambda x + (1 - \lambda)y)$ for any $x, y \in S$ and $\lambda \in A$. But then, for any $0 < \alpha < 1$, there exists a $(\lambda_m) \in A^\infty$ such that $\lambda_m \to \alpha$ and $Gr(\Gamma)$ is λ_m-convex for each m. Now use the closed graph property of Γ and the continuity of scalar multiplication and vector addition.

EXERCISE 14. For each $m \in \mathbb{N}$, define $f_m \in \mathbf{C}^1[0, 1]$ by $f_m(t) := \frac{1}{m} \sin mt$. Check that $d_\infty(f_m, \mathbf{0}) \to 0$, whereas $d_\infty(f_m', \mathbf{0}) = 1$ for each m. Conclusion: D is a discontinuous linear operator.

EXERCISE 20. $x \mapsto |L(x)|$ is a continuous seminorm on X when L is continuous.

EXERCISE 21. (a) The "only if" part follows from the definition of continuity. To see the "if" part, pick any $K > 0$ with $|L(x)| \leq K$ for all $x \in O$. Take any $(x^m) \in X^\infty$ with $x^m \to \mathbf{0}$, and fix an arbitrary $\varepsilon > 0$. Since $\frac{\varepsilon}{K}O$ is an open set (why?), there exists an $M \in \mathbb{R}$ such that $x^m \in \frac{\varepsilon}{K}O$ for all $m \geq M$. Derive from this that $|L(x^m)| \leq \varepsilon$ for all $m \geq M$. Conclusion: L is continuous at $\mathbf{0}$.

EXERCISE 22. (a) Take any $\varepsilon > 0$ and $x \in X$. Then there exists a $\delta > 0$ with $\Gamma(N_{\delta,X}(\mathbf{0})) \subseteq N_{\varepsilon,Y}(\Gamma(\mathbf{0}))$. So, recalling Exercise F.31.(b),

$$\begin{aligned}\Gamma(N_{\delta,X}(x)) &= \Gamma(x + N_{\delta,X}(\mathbf{0}))\\ &= \Gamma(x) + \Gamma(N_{\delta,X}(\mathbf{0}))\\ &\subseteq \Gamma(x) + N_{\varepsilon,Y}(\Gamma(\mathbf{0}))\\ &= \Gamma(x) + \Gamma(\mathbf{0}) + N_{\varepsilon,Y}(\mathbf{0})\\ &= N_{\varepsilon,Y}(\Gamma(x)).\end{aligned}$$

(b) Take any $x, y \in X$ and $(x^m) \in X^\infty$ such that $y \in \Gamma(x)$ and $x^m \to x$. Then $x - x^m \to \mathbf{0}$ (why?), so there exists a sequence (z^m) with $z^m \in \Gamma(x - x^m)$ and $z^m \to \mathbf{0}$. Now consider the sequence $(y - z^m)$.

EXERCISE 23. (a) We have $\Gamma(x) = L(x) + \Gamma(\mathbf{0})$ for some $L \in \mathcal{L}(X, Y)$. (Recall Exercise F.31.(c).)
(b) Recall Exercise F.31.(e).

EXERCISE 26. (a) Take any $x \in O$. Given that O is open and convex, we can find a $y \in O$ and a $0 < \lambda < 1$ such that $x_0 = \lambda x + (1 - \lambda)y$. (Why?)

Then $\lambda\Gamma(x) + (1-\lambda)\Gamma(y) \subseteq \Gamma(x_0)$, so if I choose any $z \in \Gamma(y)$, I get $\Gamma(x) \subseteq \frac{1}{\lambda}(\Gamma(x_0) - (1-\lambda)z)$.

EXERCISE 30. You know that *cone*(S) must be algebraically closed from Theorem G.1.

EXERCISE 33. Note that $x^i + \theta N_{\varepsilon,X}(\mathbf{0}) = \theta N_{\varepsilon,X}(x_i)$ and imitate the argument I gave to prove Lemma C.1.

EXERCISE 38. (b) The closed unit ball B_X is a bounded set, obviously. Is $\mathrm{id}_X(B_X)$ compact when X is infinite-dimensional?

EXERCISE 39. (a) The image of a relatively compact set must be relatively compact under L.

EXERCISE 40. Start as in the hint I gave above for Exercise 13; $\frac{1}{2}$ was not really playing an essential role there.

EXERCISE 45. Observe that, by Proposition 11, it is enough to show that $co(S) \subseteq al\text{-}int_X(co(S))$ to prove the second claim. Let $x \in co(S)$. Then x can be written as a convex combination of finitely many members of S. Let's assume that two members of S are enough for this (since this is only a hint), that is, $x = \lambda y + (1-\lambda)z$ for some $y, z \in S$ and $0 \leq \lambda \leq 1$. Now take any $w \in X$. Since $y, z \in S$ and S is open, there exists a small enough $\alpha^* > 0$ such that $(1-\alpha^*)y + \alpha^* w \in S$ and $(1-\alpha^*)z + \alpha^* w \in S$. (Yes?) Thus,

$$(1-\alpha^*)x + \alpha^* w = \lambda((1-\alpha^*)y + \alpha^* w) + (1-\lambda)((1-\alpha^*)z + \alpha^* w) \in co(S).$$

Since $co(S)$ is convex, the line segment between $(1-\alpha^*)x + \alpha^* w$ and x is contained within $co(S)$, so $(1-\alpha^*)x + \alpha^* w \in co(S)$ for all $0 \leq \alpha \leq \alpha^*$.

EXERCISE 47. (c) Let me show that $cl_X(S) \subseteq cl_X(int_X(S))$. Let $y \in cl_X(S)$. Then there is a $(y^m) \in S^\infty$ with $y^m \to y$. Pick any $x \in int_X(S)$ and use Lemma 3 to conclude $\frac{1}{m}x + \left(1 - \frac{1}{m}\right) y^m \in int_X(S)$ for all $m = 1, 2, \ldots$. But $\frac{1}{m}x + \left(1 - \frac{1}{m}\right) y^m \to y$ as $m \to \infty$.

EXERCISE 48. The answer to the question is no.

EXERCISE 50. Either recall the proofs of Propositions G.2 and Theorem G.2, or use these results along with those of Section 5.2.

EXERCISE 51. Recall Example G.12.

CHAPTER J

EXERCISE 2. False. Recall $(\mathbb{R}^2, d_{1/2})$.

EXERCISE 6. Fix any $x, y \in X$ and $\beta \geq \alpha > 0$, and show that claimed inequality is equivalent to the following: $\left\|\frac{\beta}{\alpha}x + y\right\| - \left\|\left(\frac{\beta}{\alpha} - 1\right)x\right\| \leq \|x + y\|$. Now use the following change of variables: $z := \frac{\beta}{\alpha}x + y$ and $w := \left(\frac{\beta}{\alpha} - 1\right)x$.

EXERCISE 7. Two vectors x and y in X are linearly independent iff $\frac{1}{\|x\|}x$ and $\frac{1}{\|y\|}y$ are distinct.

EXERCISE 9. $\varphi^{-1}(0) = \mathbf{c}_0$.

EXERCISE 20. If you try to apply the local-to-global method, you will see that the only difficulty is to prove the following: If A and B are open subsets of S, and $\varphi|_A$ and $\varphi|_B$ are Lipschitz continuous, then so is $\varphi|_{A\cup B}$. To prove this, take arbitrary points x and y in A and B, respectively, and find $(z, w) \in A \times B$ with $z \neq x$ and $w \neq y$. Now define $T := co\{z, w\}$, and notice that $\varphi|_T$ is concave. Deduce from this that

$$\frac{\varphi(y) - \varphi(x)}{\|x - y\|} \leq \frac{\varphi(x) - \varphi(z)}{\|x - z\|} \leq K_A \quad \text{and} \quad \frac{\varphi(x) - \varphi(y)}{\|x - y\|} \leq \frac{\varphi(y) - \varphi(w)}{\|y - w\|} \leq K_B,$$

where K_A and K_B are the Lipschitz constants of $\varphi|_A$ and $\varphi|_B$, respectively. Letting $K := \max\{K_A, K_B\}$, therefore, we find $|\varphi(x) - \varphi(y)| \leq K\,\|x - y\|$.

EXERCISE 21. Recall Corollary I.1.

EXERCISE 22. Let (x^k) be a sequence (of sequences) in $\mathbf{c}_0$, and suppose $\|x^k - (x_m)\|_\infty \to 0$ for some $(x_m) \in \ell^\infty$. I wish to show that (x_m) converges to 0. Take any $\varepsilon > 0$. Since each $x^k \in \mathbf{c}_0$, for each $k \in \mathbb{N}$ there exists an $M_k \in \mathbb{R}$ such that $|x_m^k| < \frac{\varepsilon}{2}$ for all $m \geq M_k$. Thus $|x_m| < |x_m - x_m^k| + \frac{\varepsilon}{2}$ for all $k \geq 1$ and $m \geq M_k$. Now choose $K \in \mathbb{R}$ such that $\|x^k - (x_m)\|_\infty < \frac{\varepsilon}{2}$ for all $k \geq K$, and let $M := M_K$. Clearly, $|x_m| < \varepsilon$ for all $m \geq M$.

And yes, $\mathbf{c}$ is also a closed subspace of ℓ^∞.

EXERCISE 28. Recall Dirichlet's Rearrangement Theorem.

EXERCISE 31. (a) Pick any $z \in X\backslash Y$ and let $\beta := d_{\|\cdot\|}(z, Y) > 0$. For any (fixed) $0 < \alpha < 1$, there must exist a $y \in Y$ with $\beta \leq \|z - y\| \leq \frac{\beta}{\alpha}$. Define $x := \frac{1}{\|z-y\|}(z - y)$.

EXERCISE 32. Yes. No.

EXERCISE 34. Pick any open convex set O_1 with $\mathbf{0} \in O_1 \subseteq N_{1,X}(\mathbf{0})$, and say $d_1 := diam(O_1)$. Find $k_1 \in \mathbb{N}$ and $x^1(1), \ldots, x^{k_1}(1) \in S$ such that $S \subseteq \cup^{k_1} cl_X(N_{d_1,X}(x^i(1)))$. Now pick any open convex set O_2 with $\mathbf{0} \in O_2 \subseteq N_{d_1,X}(\mathbf{0})$, and say $d_2 := diam(O_2)$. Find $k_2 \in \mathbb{N}$ and $x^1(2), \ldots, x^{k_2}(2) \in S$ such that $S \subseteq \cup^{k_2} cl_X(N_{d_2,X}(x^i(2)))$. Proceed this way inductively to obtain a sequence (O_m) of open convex sets such that for each $m = 2, 3, \ldots$, (i) $\mathbf{0} \in O_m \subseteq N_{d_{m-1},X}(\mathbf{0})$; and (ii) there exist a $k_m \in \mathbb{N}$ and $x^1(m), \ldots, x^{k_m}(m) \in S$ with $S \subseteq \cup^{k_m} cl_X(N_{d_m,X}(x^i(m)))$. (Here $d_m := diam(O_m)$ for each m.) Now let $S(m) := co\{x^1(m), \ldots, x^{k_m}(m)\}$ and define the self-correspondence Γ_m on $S(m)$ by

$$\Gamma_m(x) := \left(\Gamma(x) + cl_X(N_{d_m,X}(\mathbf{0}))\right) \cap S(m),$$

$m = 1, 2, \ldots$. Proceed as in the proof of the Glicksberg-Fan Fixed Point Theorem.

EXERCISE 36. Recall how I proved Corollary E.2.

EXERCISE 38. (c) Since $cl_X(\Phi(S))$ is compact, there exists a subsequence of $(\Phi(y(m)))$ that converges in $cl_X(\Phi(S))$. The limit of this subsequence must be a fixed point of Φ.

EXERCISE 41. Let S be a nonempty, closed, bounded, and convex subset of X such that $\mathbf{0} \in int_X(S)$. (What if $int_X(S) = \emptyset$? What if $\mathbf{0} \notin int_X(S)$?) Define the retraction $r \in S^X$ by $r(x) := \begin{cases} x, & \text{if } x \in S \\ \frac{1}{\varphi_S(x)} x, & \text{otherwise} \end{cases}$, where φ_S is the Minkowski functional of S. Proceed as in the proof of Proposition D.10.

EXERCISE 42. (b) For any $f \in \mathcal{F}$, if $x \in Fix(f)$ and $g \in \mathcal{F}$, then $g(x) = g(f(x)) = f(g(x))$, so $g(x) \in Fix(f)$. That is, g is self-map on $Fix(f)$ for any $f, g \in \mathcal{F}$.

EXERCISE 46. Take an arbitrary $k > 0$, and let C_k stand for the class of all Lipschitz continuous real functions g on $[0, 1]$ such that $g(0) = 0$ and k is greater than or equal to the Lipschitz constant of g. Check that C_k is a convex set, and use the Arzelà-Ascoli Theorem to prove that C_k is a compact subset of $\mathbf{C}[0, 1]$. Now define the continuous self-map Φ on C_k by $\Phi(f)(x) := H(x, f(h(x)))$. Check that $\Phi(C_k) \subseteq C_k$ for any $k > \frac{p}{1-Kq}$, and apply the Schauder Fixed Point Theorem 1.

EXERCISE 47. Use Krasnoselsky's Theorem.

EXERCISE 50. No.

EXERCISE 51. Recall Lemma D.2, and show that L is linearly homogeneous.

EXERCISE 55. $\|L\|^* = \frac{\delta}{1-\delta}$.

EXERCISE 56. $\|L\|^* = \int_0^1 |g(t)|\, dt$.

EXERCISE 58. (This is quite similar to Example 6.[1].) Letting $M := \|x^*\|$, we have $|L(x)| \leq M \|x\|$ for any $x \in X$, by the Cauchy-Schwarz Inequality (of Exercise 12).

EXERCISE 60. For any $x \in X$, we have $\|K(L(x))\|_Z \leq \|K\|^* \|L(x)\|_Y \leq \|K\|^* \|L\|^* \|x\|$.

EXERCISE 63. (b) We have $\left\|f(x,\cdot)\right\|^* \leq |||f||| \, \|x\|$ for any $f \in \mathfrak{b}_{X,Y}$ and $x \in X$.

EXERCISE 68. (b) Since $\left|\|L_k\|^* - \|L_l\|^*\right| \leq \|L_k - L_l\|^* \to 0$ as $k, l \to \infty$, $(\|L_m\|^*)$ is a Cauchy sequence. Thus $\alpha := \sup\{\|L_m\|^* : m \in \mathbb{N}\} < \infty$, while $|L(x)| \leq \alpha \|x\|$ for all $x \in X$.
(c) Fix an arbitrarily small $\varepsilon > 0$. Since (L_m) is Cauchy, there is an $M \in \mathbb{R}$ such that $|L_k(x) - L_l(x)| < \varepsilon$ for all $x \in X$ and $k, l \geq M$. But then $\varepsilon \geq \lim_{l\to\infty} |L_k(x) - L_l(x)| = |L_k(x) - L(x)|$ for all $x \in X$ and $k \geq M$.

EXERCISE 70. Let $\mathcal{H}$ be the set of all closed half-spaces that contain S. Since $\cap\mathcal{H}$ is closed and convex, it follows that $\overline{co}_X(S) \subseteq \cap\mathcal{H}$. Conversely, if $x \notin \overline{co}_X(S)$, then, by Theorem 1, there exist an $\alpha \in \mathbb{R}$ and a nonzero $L \in X^*$ such that $\inf L(\overline{co}_X(S)) \geq \alpha > L(x)$. So $x \notin L^{-1}([\alpha, \infty)) \in \mathcal{H}$. (Could you use Proposition G.2 here?)

EXERCISE 75. Two.

EXERCISE 80. First apply Milman's Converse to $L(cl_X(ext(S)))$, then apply the Krein-Milman Theorem to S.

EXERCISE 81. Let $A := cl_X(\{L^k(x) : k = 0, 1, \ldots\})$ and $S := \overline{co}_X(A)$. Use Milman's Converse to show that $S \subseteq X_+ \backslash \{\mathbf{0}\}$, and then apply the Schauder Fixed Point Theorem 2 to $L|_S$.

EXERCISE 82. Let $Y := span\{x\}$ and define $L \in Y^*$ by $L(\lambda x) := \lambda \|x\|$ for any $\lambda \in \mathbb{R}$. Now use the Hahn-Banach Extension Theorem 2.

EXERCISE 83. We have $\|L_0\|^* = L_0(x)$.

EXERCISE 84. By Exercise 82, $\|x\| \leq \sup\{|L(x)| : L \in X^* \text{ and } \|L\|^* \leq 1\}$. Now invoke the definition of $\|\cdot\|^*$.

EXERCISE 86. There must exist a $z \in X$ with $d_{\|\cdot\|}(z, Y) > 0$. Let $Z := span(Y \cup \{z\}) = Y + \{\lambda z : \lambda \in \mathbb{R}\}$. Define $L_0 \in \mathbb{R}^Z$ by $L_0(y + \lambda z) := \lambda d_{\|\cdot\|}(z, Y)$ for all $y \in Y$ and $\lambda \in \mathbb{R}$, and check that $L_0 \in Z^*$. (In fact, $\|L_0\|^* = 1$.) Now apply the Hahn-Banach Extension Theorem 2.

EXERCISE 89. Suppose there is a subspace Y of X and an $L_0 \in Y^*$ such that $L_1|_Y = L_0 = L_2|_Y$ and $\|L_1\|^* = \|L_0\|^* = \|L_2\|^*$ for two distinct $L_1, L_2 \in X^*$. If $L := \frac{1}{2}L_1 + \frac{1}{2}L_2$, then $L \in X^*$ and $\|L\|^* = \|L_0\|^*$. Show that this contradicts the rotundity of X^*.

EXERCISE 92. For each $x \in S$, define $f_x : X^* \to \mathbb{R}$ by $f_x(L) := L(x)$. Since X^* is Banach (Proposition 6), we may apply the Uniform Boundedness Principle to $\{f_x : x \in S\}$. It follows that there is a real number $M > 0$ such that $\sup\{|L(x)| : x \in S\} \leq M$ for each $L \in X^*$. Now use the Duality Theorem of Exercise 84.

CHAPTER K

EXERCISE 6. If the claim is false, then we can find an $\varepsilon > 0$ and a sequence (x^m) in B_X such that $\left\|\mathbf{D}_{\Phi,x}(x^k) - \mathbf{D}_{\Phi,x}(x^l)\right\|_Y > 3\varepsilon$ for all distinct $k, l \in \mathbb{N}$. (Why?) Define the map $\mathbf{e}$ as in Proposition 2, and choose $\delta > 0$ such that $\|\mathbf{e}(\omega)\|_Y \leq \varepsilon \|x - \omega\|$ for all $\omega \in N_{\delta,X}(x)$. Use the definition of $\mathbf{e}$ to deduce that, for each $k \neq l$,

$$\begin{aligned}&\left\|\Phi(x+\delta x^k) - \Phi(x+\delta x^l)\right\|_Y \\ &\quad \geq \delta\left\|\mathbf{D}_{\Phi,x}(x^k - x^l)\right\|_Y - \left\|\mathbf{e}(x+\delta x^k)\right\|_Y - \left\|\mathbf{e}(x+\delta x^l)\right\|_Y \geq \delta\varepsilon.\end{aligned}$$

Conclude that $\{\Phi(x + \delta x^m) : m \in \mathbb{N}\}$ is not compact, contradicting our hypothesis.

EXERCISE 9. Assume $n = 2$ and $\mathbf{0} \in O$. It is without loss of generality to posit that $\varphi(\mathbf{0}) = \partial_1\varphi(\mathbf{0}) = \partial_2\varphi(\mathbf{0}) = 0$. (Otherwise, we would work with the map $\phi \in \mathbf{C}(O)$ defined by $\phi(x) := \varphi(x) - \alpha x_1 - \beta x_2 - \gamma$, with suitably chosen $\alpha, \beta, \gamma \in \mathbb{R}$.) We wish to show that $\mathbf{D}_{\varphi,\mathbf{0}}$ is the zero functional. Under the present assumptions, this means that $\lim_{\omega\to\mathbf{0}} \frac{\varphi(\omega)}{\|\omega\|_2} = 0$. Take any $\varepsilon > 0$, and choose $\delta > 0$ such that $|\partial_1\varphi(\omega)| < \varepsilon$ and $|\partial_2\varphi(\omega)| < \varepsilon$ for all $\omega \in N_{\delta,\mathbb{R}^2}(\mathbf{0})$. Let $y \in N_{\delta,\mathbb{R}^2}(\mathbf{0})$. By the Mean Value Theorem, there exist a $0 \leq z_1 \leq y_1$ and a $0 \leq z_2 \leq y_2$ such that

$$\varphi(y) = \varphi(0, y_2) + \partial_1\varphi(z_1, y_2)y_1 \quad \text{and} \quad \varphi(0, y_2) = \partial_2\varphi(0, z_2)y_2.$$

So, $\varphi(y) = \partial_1\varphi(z_1, y_2)y_1 + \partial_2\varphi(0, z_2)y_2$, and we thus find $|\varphi(y)| \leq 2\varepsilon \|y\|_2$. Since $\varepsilon > 0$ was arbitrary here, the claim follows from this observation.

EXERCISE 12. $\mathsf{D}_\varphi(f)(g) = 2f(0)g(0)$ for all $f, g \in \mathbf{C}[0,1]$.

EXERCISE 13. $\mathsf{D}_\varphi(f)(g) = \int_0^1 f(t)^2 g(t)dt$ for all $f, g \in \mathbf{C}[0,1]$.

EXERCISE 18. There are no such points.

EXERCISE 20. Recall Exercise 14.

EXERCISE 29. Fix any distinct $x, y \in S$, and define $F : (0,1) \to \mathbb{R}$ by $F(\lambda) := \varphi(\lambda x + (1-\lambda)y)$. Now apply the Second Mean Value Theorem, Lemma 1, and Example 7.

EXERCISE 32. Take any $x_o \in O$, and let $T := \{x \in O : \Phi(x) = \Phi(x_o)\}$. Show that T is a nonempty clopen subset of O, and hence $T = O$. (To show that T is open in O, use the Mean Value Inequality.)

EXERCISE 33. Let $\Psi := \Phi - \mathsf{D}_{\Phi,x_o}$. Show that $\mathsf{D}_{\Psi,w} = \mathsf{D}_{\Phi,w} - \mathsf{D}_{\Phi,x_o}$ and apply the Mean Value Inequality.

EXERCISE 35. $\vartheta(\omega) := \varphi(x) + \mathsf{D}_{\varphi,x}(\omega - x)$ for all $\omega \in X$.

EXERCISE 38. The task is to show that φ is Fréchet differentiable at x; once this is done, invoking Proposition 6 will prove the first assertion. To this end, show that $\mathsf{D}_{\varphi,x}(t_1, \ldots, t_n) = \sum^n \partial_i\varphi(x)t_i$ for any $(t_1, \ldots, t_n) \in \mathbb{R}^n$. The second assertion follows from Exercise 9 and Proposition 6.

EXERCISE 40. The only supergradient of f at 1 is the linear map $t \mapsto \frac{1}{2}t$ on $\mathbb{R}$. On the other hand, $\partial g(1)$ consists of the maps $t \mapsto \alpha t$, $\frac{1}{2} \leq \alpha \leq 1$, while $\partial h(1) = \emptyset$.

EXERCISE 42. To prove the nonemptiness assertion, let $S := \{(y,t) \in O \times \mathbb{R} : \varphi(y) \geq t\}$. Since φ is continuous (Proposition I.8), S is a closed and convex set with nonempty interior. Apply the Supporting Hyperplane Theorem.

To prove the closedness assertion, verify first that, for any $\alpha \in \mathbb{R}$ and $L_m, L \in X^*$, $m = 1, 2, \ldots$, if $\|L_m - L\|^* \to 0$ and $L_m(\omega) \geq \alpha$ for each m, then $L(\omega) \geq \alpha$.

EXERCISE 44. (a) If $L \in \partial\varphi(x)$, then

$$\varphi(\omega) \leq \varphi(x) + L(\omega - x) \qquad \text{for all } \omega \in O,$$

while

$$\varphi(\omega) = \varphi(x) + \mathsf{D}_{\varphi,x}(\omega - x) + \mathbf{e}(\omega) \qquad \text{for all } \omega \in O,$$

for some $\mathbf{e} \in \mathbb{R}^O$ with $\lim_{\omega \to x} \frac{\mathbf{e}(\omega)}{\|\omega - x\|} = 0$. Fix an arbitrary $\varepsilon > 0$, and find a $\delta > 0$ with $|\mathbf{e}(\omega)| < \varepsilon\delta$ for all $\omega \in N_{\delta,X}(x)$. Then $L(z) \geq \mathsf{D}_{\varphi,x}(z) - \varepsilon\delta$ for all $z \in N_{\delta,X}(\mathbf{0})$. Define $K := L - \mathsf{D}_{\varphi,x}$, and check that $|K(z)| \leq 2\varepsilon$ for all $z \in N_{1,X}(\mathbf{0})$. Conclude that K must be the zero functional.

EXERCISE 46. By Proposition 8 and Exercise 30, there exists an $\mathbf{r} \in \mathbb{R}^X$ such that

$$\varphi(x^* + z) - \varphi(x^*) = \tfrac{1}{2}\mathsf{D}^2_{\varphi,x^*}(z, z) + \mathbf{r}(z)$$

for any $z \in X$ with $x^* + z \in O$, and $\lim_{z \to \mathbf{0}} \frac{\mathbf{r}(z)}{\|z\|^2} = 0$. It follows that there exists $\delta > 0$ small enough so that $\mathsf{D}^2_{\varphi,x^*}(z, z) + 2\mathbf{r}(z) \leq 0$ for all $z \in N_{\delta,X}(\mathbf{0})$.

EXERCISE 48. φ is continuous (Proposition I.8).

EXERCISE 51. (b) The solution is $f_*(t) := \begin{cases} t, & \text{if } 0 \leq t \leq \frac{1}{2} \\ 1 - t, & \text{if } \frac{1}{2} < t \leq 1 \end{cases}$.

EXERCISE 52. (b) A complete answer requires you to prove Lemma 6.

EXERCISE 54. Use Exercise 46.

EXERCISE 55. Take any $f \in \mathcal{F}$ and any $(f_m) \in \mathcal{F}^\infty$ with $\|f_m - f\|_{\infty,\infty} \to 0$. Use hypothesis (i) to get

$$\begin{aligned} \varsigma(f_m) \leq & \int_0^\tau \varphi(t, f_m, f')dt + \int_0^\tau \partial_3\varphi(t, f, f')(f'_m - f')dt \\ & + \int_0^\tau \left(\partial_3\varphi(t, f_m, f') - \partial_3\varphi(t, f, f')\right)(f'_m - f')dt \end{aligned}$$

Since $f'_m \to f'$ uniformly, the second term on the right-hand side goes to 0. But hypothesis (i) ensures that $\sup\{\|f'_m - f'\|_\infty : m \in \mathbb{N}\} < \infty$, so, similarly, the third term on the right-hand side goes to 0. Then, since $f_m \to f$ uniformly, we find

$$\limsup \varsigma(f_m) \leq \limsup \int_0^\tau \varphi(t, f_m, f')dt = \int_0^\tau \varphi(t, f, f')dt.$$

Finally, recall Proposition D.5.

(*Note.* Here I assumed that you know the following fact: For any $-\infty < a < b < \infty$, if $h_m, h \in \mathbf{C}[a,b]$, $m = 1, 2, \ldots$, and $h_m \to h$ uniformly, then $\int_a^b h_m(t)dt \to \int_a^b h(t)dt$. Please prove this if you didn't see it before; it is not hard.)

EXERCISE 56. (b) $f(t) = e^t - et - 1$, $0 \leq t \leq 1$.
(c) $f(t) = e^t + t - 1$, $0 \leq t \leq 1$.

EXERCISE 57. $f_* \in \arg\max \left\{ \int_0^\tau \left(\varphi(t, f, f') + W'(f(t)) f'(t) \right) dt : f \in \mathcal{F} \right\}$.

EXERCISE 61. Use Theorem 3 to get

$$\tfrac{d}{dt} \left(\partial_3 \varphi(t, f_*, f_*') f_*' - \varphi(t, f_*, f_*') \right) = \partial_1 \varphi(t, f_*, f_*').$$

Then define $H : \mathbb{R}_+ \to \mathbb{R}$ by

$$H(t) := \partial_3 \varphi(t, f_*(t), f_*'(t)) f_*'(t) - \varphi(t, f_*(t), f_*'(t)),$$

and check that $\int_0^\tau \partial_1 \varphi(t, f_*, f_*') dt = H(\tau) - H(0)$ for any $\tau > 0$. Now let $\tau \to \infty$.

EXERCISE 64. The solution is $f(t) = \frac{p}{2r}(t + \frac{1}{r} e^{r(\tau - t)})$, $0 \leq t \leq \tau$.

References

Aczel, J. 1966. *Lectures on Functional Equations and Applications.* New York: Academic Press.

Agarwal, R., M. Meehan, and D. O'Regan. 2001. *Fixed Point Theory and Applications.* Cambridge: Cambridge University Press.

Aigner, M., and G. Ziegler. 1999. *Proofs from THE BOOK.* Berlin: Springer-Verlag.

Aliprantis C., and K. Border. 1999. *Infinite Dimensional Analysis: A Hitchhiker's Guide.* New York: Springer-Verlag.

Aliprantis, C., R. Tourky, and N. Yannelis. 2000. "Cone Conditions in General Equilibrium Theory." *Journal of Economic Theory* 92: 96–121.

Anscombe, F., and R. Aumann. 1963. "A Definition of Subjective Probability." *Annals of Mathematical Statistics* 34: 199–205.

Apostol, T. 1974. *Mathematical Analysis.* Reading, Mass.: Addison-Wesley.

Asplund, E. 1968. "Fréchet Differentiability of Convex Functions." *Acta Mathematica* 121: 31–48.

Aumann, R. 1962. "Utility Theory without the Completeness Axiom." *Econometrica,* 30: 445–62.

Aumann, R. 1985. "On the Nontransferable Utility Value: A Comment on the Roth-Shafer Examples." *Econometrica* 53: 667–77.

Aumann, R., and S. Hart, eds. 1992. *Handbook of Game Theory,* Vol. 1. Amsterdam: North-Holland.

Banks, J., and J. Duggan. 1999. "Existence of Nash Equilibria on Convex Sets." Mimeo, California Institute of Technology.

Basu, K., and T. Mitra. 2003. "Aggregating Infinite Utility Streams with Intergenerational Equity: The Impossibility of Being Paretian." *Econometrica* 71: 1557–63.

Baye, M., G. Tian, and J. Zhou. 1993. "Characterizations of the Existence of Equilibria in Games with Discontinuous and Non-Quasiconcave Payoffs." *Review of Economic Studies* 60: 935–48.

Beardon, A. 1992. "Debreu's Gap Theorem." *Economic Theory* 2: 150–52.

Beardon, A., J. Candeal, G. Herden, E. Induráin, and G. Mehta. 2002. "The Non-Existence of a Utility Function and the Structure of Non-Representable Preference Relations." *Journal of Mathematical Economics* 37: 17–38.

Becker, R., and J. Boyd III. 1997. *Capital Theory, Equilibrium Analysis and Recursive Utility.* Oxford: Blackwell.

Becker, R., and S. Chakrabarti. 2005. "Satisficing Behavior, Brouwer's Fixed Point Theorem and Nash Equilibrium." *Economic Theory* 26: 63–83.

Bellman, R. 1957. *Dynamic Programming*. Princeton: Princeton University Press.

Bellman, R. 1984. *Eye of the Hurricane*. Singapore: World Scientific.

Benacerraf, P., and H. Putnam, eds. 1983. *Philosophy of Mathematics*. Cambridge: Cambridge University Press.

Benoit, J-P., and E. A. Ok. 2007. "Delay Aversion." Forthcoming in *Theoretical Economics*.

Berge, C. 1963. *Topological Spaces*. New York: Macmillan.

Bertsekas, D. 1976. *Dynamic Programming and Stochastic Control*. New York: Academic Press.

Bewley, T. 1986. "Knightian Uncertainty Theory. Part I." Cowles Foundation Discussion Paper No. 807.

Bewley, T. 2002. "Knightian Uncertainty Theory. Part I." *Decisions in Economics and Finance* 25: 79–110.

Billot, A., A. Chateauneuf, I. Gilboa, and J-M. Tallon. 2000. "Sharing Beliefs: Between Agreeing and Disagreeing." *Econometrica* 68: 685–94.

Blackwell, D., and M. Girshick. 1954. *Theory of Games and Statistical Decisions*. New York: Dover.

Blot, J., and P. Michel. 1996. "First-Order Necessary Conditions for Infinite-Horizon Variational Problems." *Journal of Optimization and Applications* 88: 339–64.

Boel, S., T. Carlsen, and N. Hansen. 2001. "A Useful Generalization of the Stone-Weierstrass Theorem." *American Mathematical Monthly* 108: 642–43.

Bonnet, R., and M. Pouzet. 1982. "Linear Extensions of Ordered Sets." In *Ordered Sets*. I. Rival, ed. Dordrecht: Reidel.

Border, K. 1989. *Fixed Point Theorems in Economics*. Cambridge: Cambridge University Press.

Borwein, J. 1981. "Convex Relations in Analysis and Optimization." In *Generalized Convexity in Optimization and Economics*. S. Schaible and W. Ziemba, eds. New York: Academic Press.

Borwein, J., and A. Lewis. 2000. *Convex Analysis and Nonlinear Optimization*. New York: Springer-Verlag.

Boyd, D., and J. Wong. 1969. "On Nonlinear Contractions." *Proceedings of the American Mathematical Society* 20: 456–64.

Boyer, C., and U. Merzbach. 1989. *A History of Mathematics*. New York: Wiley.

Broida, J., and G. Williamson. 1989. *A Comprehensive Introduction to Linear Algebra*. New York: Addison-Wesley.

Browder, F. 1968. "The Fixed Point Theory of Multi-Valued Mappings in Topological Vector Spaces." *Mathematische Annalen* 177: 283–301.

Bröndsted, A. 1976. "Fixed Points and Partial Orders." *Proceedings of the American Mathematical Society* 60: 365–66.

Bröndsted, A., and R. T. Rockafellar. 1965. "On the Subdifferentiability of Convex Functions." *Proceedings of the American Mathematical Society* 16: 605–11.

Butazzo, G., M. Giaquinta, and S. Hildebrandt. 1998. *One-dimensional Variational Problems*. Oxford: Clarendon Press.

Cain, G. and M. Nashed. 1971. "Fixed Points and Stability for a Sum of Two Operators in Locally Convex Spaces." *Pacific Journal of Mathematics* 39: 581–92.

Camerer, C. 1995. "Individual Decision Making." In *The Handbook of Experimental Economics*. J. Kagel and A. Roth, eds. Princeton: Princeton University Press.

Caristi, J. 1976. "Fixed Point Theorems for Mappings Satisfying Inwardness Conditions." *Transactions of the American Mathematical Society* 215: 241–51.

Carothers, N. 2000. *Real Analysis*. Cambridge: Cambridge University Press.

Cartan, H. 1972. *Calculo Diferencial*. Barcelona: Omega.

Cass, D. 1965. "Optimum Growth in an Aggregative Model of Capital Accumulation." *Review of Economic Studies* 32: 233–40.

Cauty, R. 2001. "Solution du Problème de Point Fixe de Schauder." *Fundamenta Mathematicae* 170: 231–46.

Cellina, A. 1969. "Approximation of Set-Valued Functions and Fixed-Point Theorems." *Annali di Matematica Pura et Applicata* 82: 17–24.

Chae, S. 1995. *Lebesque Integration*. Berlin: Springer-Verlag.

Crémer, J. 1982. "A Simple Proof of Blackwell's "Comparison of Experiments" Theorem." *Journal of Economic Theory* 27: 439–43.

Corchón, L. 1996. *Theories of Imperfectly Competitive Markets*. Berlin: Springer-Verlag.

Cox, H. 1968. "A Proof of the Schröder-Bernstein Theorem." *American Mathematical Monthly* 75: 508.

Cubiotii, P. 1997. "Existence of Nash Equilibria for Generalized Games without Upper Semicontinuity." *International Journal of Game Theory* 26: 267–73.

Dacorogna, B. 2004. *Introduction to the Calculus of Variations*. London: Imperial College Press.

Daffer, P., H. Kaneko, and W. Li. 1996. "On a Conjecture of S. Reich." *Proceedings of the American Mathematical Society* 124: 3159–62.

Dasgupta, P., and E. Maskin. 1986. "The Existence of Equilibrium in Discontinuous Economic Games: Theory." *Review of Economic Studies* 53: 1–26.

Dauben, J. 1980. "The Development of Cantorian Set Theory." In *From the Calculus to Set Theory*: 1630–1910. I. Grattan-Guinness, ed. Princeton, N.J.: Princeton University Press.

Debreu, G. 1954. "Representation of a Preference Relation by a Numerical Function." In *Decision Process*. R. M. Thrall, C. H. Coombs, and R. L. Davis, eds. New York: Wiley.

Debreu, G. 1964. "Continuity Properties of Paretian Utility." *International Economic Review* 5: 285–93.

de la Fuente, A. 1999. *Mathematical Methods and Models for Economists*. Cambridge: Cambridge University Press.

Devlin, K. 1993. *The Joy of Sets*. New York: Springer-Verlag.

Dieudonné, J. 1969. *Foundations of Modern Analysis*. New York: Academic Press.

Dieudonné, J. 1981. *History of Functional Analysis*. Amsterdam: North-Holland.

Dow, J., and S. Werlang. 1992. "Uncertainty Aversion, Risk Aversion and the Optimal Choice of Portfolio." *Econometrica* 60: 197–204.

Dreyfus, S. 2000. "Richard Bellman on the Birth of Dynamic Programming." *Operations Research* 50: 48–51.

Dubey, P., A. Mas-Colell, and M. Shubik. 1980. "Efficiency Properties of Strategic Market Games." *Journal of Economic Theory* 22: 339–62.

Dubra, J., and F. Echenique. 2001. "Monotone Preferences over Information." *Topics in Theoretical Economics* 1: Article 1.

Dubra, J., and E. A. Ok. 2002. "A Model of Procedural Decision Making in the Presence of Risk." *International Economic Review* 43: 1053–80.

Dubra, J., F. Maccheroni, and E. A. Ok. 2004. "The Expected Utility Theorem without the Completeness Axiom." *Journal of Economic Theory* 115: 118–33.

Dudley, R. 2002. *Real Analysis and Probability*. Cambridge: Cambridge University Press.

Duffie, D. 1996. *Dynamic Asset Pricing Theory*. Princeton: Princeton University Press.

Dugundji, J., and A. Granas. 1982. *Fixed Point Theory*, Vol. 1. Warsaw: Polish Scientific Publishers.

Eliaz, K., and E. A. Ok. 2006. "Indifferent or Indecisive? Revealed Preference Foundations of Incomplete Preferences." *Games and Economic Behavior* 56: 61–86.

Enderton, H. 1977. *Elements of Set Theory*. Boston: Academic Press.

Enflo, P. 1973. "A Counterexample to the Approximation Problem in Banach Spaces." *Acta Mathematica* 130: 309–17.

Epstein, L., and T. Wang. 1994. "Intertemporal Asset Pricing under Knightian Uncertainty." *Econometrica* 62: 183–322.

Federer, H. 1996. *Geometric Measure Theory*. Berlin: Springer-Verlag.

Fishburn, P. 1970. *Utility Theory for Decision Making*. New York: Wiley.

Fishburn, P. 1991. "Nontransitive Preferences in Decision Theory." *Journal of Risk and Uncertainty* 4: 113–34.

Folland, G. 1999. *Real Analysis: Modern Techniques and Their Applications*. New York: Wiley.

Franklin, J. 1980. *Methods of Mathematical Economics: Linear and Nonlinear Programming, Fixed-Point Theorems*. Berlin: Springer-Verlag.

Friedman, J. 1990. *Game Theory with Applications to Economics*. New York: Oxford University Press.

Fudenberg, D., and J. Tirole. 1991. *Game Theory*. Cambridge: MIT Press.

Gamelin, T., and R. Greene. 1999. *Introduction to Topology*. New York: Dover.

Geanakoplos, J. 2003. "Nash and Walras Equilibrium." *Economic Theory* 21: 585–603.

Gelbaum, B., and J. Olmsted. 1991. *Theorems and Counterexamples in Mathematics*. New York: Springer-Verlag.

Gelfand, I., and S. Fomin. 1963. *Calculus of Variations*. New York: Prentice-Hall.

Gilboa, I., and D. Schmeidler. 1989. "Maxmin Expected Utility with Non-Unique Prior." *Journal of Mathematical Economics* 18: 141–53.

Gilboa, I., and D. Schmeidler. 1994. "Additive Representations of Non-Additive Measures and the Choquet Integral." *Annals of Operations Research* 52: 43–65.

Ghirardato, P., and J. Katz. 2006. "Indecision Theory: Explaining Selective Abstention in Multiple Elections." *Journal of Public Economic Theory* 8: 379–400.

Ghirardato, P., and M. Marinacci. 2001. "Risk, Ambiguity, and the Separation of Utility and Beliefs." *Mathematics of Operations Research* 26: 864–90.

Ghirardato, P., and M. Marinacci. 2002. "Ambiguity Made Precise: A Comparative Foundation." *Journal of Economic Theory* 102: 251–89.

Gleason, A. 1991. *Fundamentals of Abstract Analysis*. Boston: Jones and Bartlett.

Glicksberg, L. 1952. "A Further Generalization of the Kakutani Fixed Point Theorem with Application to Nash Equilibrium Points." *Proceedings of the American Mathematical Society* 38: 170–74.

Goebel, K., and W. Kirk. 1990. *Topics in Metric Fixed Point Theory*. Cambridge: Cambridge University Press.

Goldberg, S. 1966. *Unbounded Linear Operators*. New York: McGraw-Hill.

Goldstine, H. 1980. *A History of Calculus of Variations from the 17th through 19th Century*. New York: Addison-Wesley.

Grandmont, J.-M. 1972. "Continuity Properties of a von Neumann-Morgenstern Utility." *Journal of Economic Theory* 4: 45–57.

Haaser, N., and J. Sullivan. 1991. *Real Analysis*. New York: Dover.

Halkin, H. 1974. "Necessary Conditions for Optimal Control Problems with Infinite Horizon." *Econometrica* 42: 267–72.

Halmos, P. 1960. *Naive Set Theory*. New York: van Nostrand.

Harrison, J., and D. Kreps. 1979. "Martingales and Arbitrage in Multiperiod Securities Markets." *Journal of Economic Theory* 20: 381–408.

Harsanyi, J. 1955. "Cardinal Welfare, Individual Ethics, and Interpersonal Comparisons of Utility." *Journal of Political Economy* 63: 309–21.

Hart, S., and A. Mas-Colell. 1989. "Potential, Value, and Consistency." *Econometrica* 57: 589–614.

Hennefeld, J. 1980. "A Nontopological Proof of the Uniform Boundedness Theorem." *American Mathematical Monthly* 87: 217.

Herstein, I., and J. Milnor. 1953. "An Axiomatic Approach to Measurable Utility." *Econometrica* 21: 291–97.

Hewitt, E. 1960. "The Role of Compactness in Analysis." *American Mathematical Monthly* 67: 499–516.

Hewitt, E., and K. Stromberg. 1965. *Real and Abstract Analysis*. New York: Springer-Verlag.

Hildenbrand, W., and A. Kirman. 1988. *Equilibrium Analysis*. Amsterdam: North-Holland.

Hiriart-Urruty, J.-B., and C. Lemaréchal. 2000. *Fundamentals of Convex Analysis.* Berlin: Springer-Verlag.

Hoffman, K., and R. Kunze. 1971. *Linear Algebra.* Englewood Cliffs: Prentice Hall, 1971.

Holmes, R. 1975. *Geometric Functional Analysis and its Applications.* New York: Springer-Verlag.

Hörmander, L. 1994. *Notions of Convexity.* Boston: Birkhäuser.

Hu, T. 1967. "On a Fixed-Point Theorem for Metric Spaces." *American Mathematical Monthly* 74: 436–37.

Jachymski, J. 1995. "On Reich's Question Concerning Fixed Points of Multimaps." *Unione Matematica Italiana Bollettino* 9: 453–60.

Jachymski, J. 1998. "Caristi's Fixed Point Theorem and Selections of Set-Valued Contractions." *Journal of Mathematical Analysis and Applications* 227: 55–67.

Jachymski, J., B. Schroder, and J. Stein, Jr. 1999. "A Connection Between Fixed-Point Theorems and Tiling Problems." *Journal of Combinatorial Theory* 87: 273–86.

Jaffray, J.-Y. 1975. "Existence of a Continuous Utility Function: An Elementary Proof." *Econometrica* 43: 981–83.

James, I. 2002. *Remarkable Mathematicians: From Euler to von Neumann.* Cambridge: Cambridge University Press.

Kakutani, S. 1941. "A Generalization of Brouwer's Fixed Point Theorem." *Duke Mathematical Journal,* 8: 457–59.

Kannai, Y. 1981. "An Elementary Proof of the No-Retraction Theorem." *American Mathematical Monthly,* 88: 264–68.

Kaplansky, I. 1977. *Set Theory and Metric Spaces.* New York: Chelsea.

Karni, E., and D. Schmeidler. 1991. "Utility Theory with Uncertainty." In *Handbook of Mathematical Economics,* Vol. 4. W. Hildenbrand and H. Sonnenschein, eds. Amsterdam: North-Holland.

Kelley, J. 1955. *General Topology.* New York: van Nostrand.

Kirzbraun, M. 1934. "Über die Zusammenzichenden und Lipschitzschen Transformationen." *Fundamenta Mathematicae* 22: 77–108.

Klee, V. 1950. "Decomposition of an Infinite-Dimensional Linear System into Ubiquitous Convex Sets." *American Mathematical Monthly* 50: 540–41.

Klee, V. 1951. "Convex Sets in Linear Spaces I." *Duke Mathematical Journal* 18: 443–66.

Klee, V. 1955. "Some Topological Properties of Convex Sets." *Transactions of American Mathematical Society* 178: 30–45.

Klee, V. 1969. "Separation and Support Properties of Convex Sets." In *Control Theory and the Calculus of Variations.* A. Balakrishnan, ed. New York: Academic Press.

Klee, V. 1971. "What Is a Convex Set?" *American Mathematical Monthly* 78: 616–31.

Klein, E. 1973. *Mathematical Methods in Theoretical Economics.* New York: Academic Press.

Klein, E., and A. Thompson. 1984. *Theory of Correspondences.* New York: Wiley.

Knill, R. 1965. "Fixed Points of Uniform Contractions." *Journal of Mathematical Analysis and Applications* 12: 449–55.

Koçkesen, L., E. A. Ok, and R. Sethi. 2000. "The Strategic Advantage of Negatively Interdependent Preferences." *Journal of Economic Theory* 92: 274–99.

Kolmogorov, A., and S. Fomin. 1970. *Introductory Real Analysis*. New York: Dover.

Körner, T. 2004. *A Companion to Analysis: A Second First and First Second Course in Analysis*. Providence, R.I.: American Mathematical Society.

Köthe, G. 1969. *Topological Vector Spaces I*. New York: Springer-Verlag.

Krein, M., and M. Rutman. 1950. "Linear Operators Leaving Invariant a Cone in a Banach Space." *American Mathematical Society Translations* No. 6.

Kreps, D. 1988. *Notes on the Theory of Choice*. Boulder: Westview Press.

Kreyzig, E. 1978. *Introductory Functional Analysis with Applications*, New York: Wiley.

Laugwitz, D. 1999. *Bernhard Riemann, 1826–1865: Turning Points in the Conception of Mathematics*. Boston: Birkhäuser.

Lax, P. 1999. *Linear Algebra*. New York: Wiley.

Leininger, W. 1984. "A Generalisation of the 'Maximum Theorem'." *Economics Letters* 15: 309–13.

Lin, P., and Y. Sternfeld. 1985. "Convex Sets with Lipschitz Fixed Point Property Are Compact." *Proceedings of American Mathematical Society* 93: 633–39.

Ljungqvist, L., and T. Sargent. 2004. *Recursive Macroeconomic Theory*. Cambridge, Mass.: MIT Press.

Luenberger, D. 1969. *Optimization by Vector Space Methods*. New York: Wiley.

MacCluer, C. 2000. "The Many Proofs and Applications of Perron's Theorem." *SIAM Review* 42: 487–98.

Maccheroni, F., M. Marinacci, and A. Rustichini. 2005. "Ambiguity Aversion, Malevolent Nature, and the Variational Representation of the Preferences." Mimeo, University of Minnesota.

Machina, M. 1987. "Choice under Uncertainty: Problems Solved and Unsolved." *Journal of Economic Perspectives* 1: 121–54.

Maddox, I. 1988. *Elements of Functional Analysis*. Cambridge: Cambridge University Press.

Maddox, I. 1989. "The Norm of a Linear Functional." *American Mathematical Monthly* 96: 434–36.

Maligranda, A. 1995. "A Simple Proof of the Hölder and Minkowski Inequalities." *American Mathematical Monthly* 92: 256–59.

Mandler, M. 2005. "Incomplete Preferences and Rational Intransitivity of Choice." *Games and Economic Behavior* 50: 255–77.

Marek, W., and J. Mycielski. 2001. "Foundations of Mathematics in the Twentieth Century." *American Mathematical Monthly* 108: 449–68.

Marinacci, M., and L. Montrucchio. 2004. "Introduction to the Mathematics of Ambiguity." In *Uncertainty in Economic Theory: Essays in Honor of David Schmeidler's 65th Birthday*. I. Gilboa, ed. New York: Routledge.

Marsden, J., and M. Hoffman. 1993. *Elementary Classical Analysis*. San Francisco: W. H. Freeman.

Marshall, A., and I. Olkin. 1979. *Inequalities: Theory of Majorization and Its Applications*. San Diego: Academic Press.

Masatlioglu, Y., and E. A. Ok. 2005. "Rational Choice with Status Quo Bias." *Journal of Economic Theory* 121: 1–29.

Mas-Colell, A. 1989. *The Theory of General Economic Equilibrium: A Differentiable Approach*. Cambridge: Cambridge University Press.

Mas-Colell, A., M. Whinston, and J. Green. 1995. *Microeconomic Theory*. Oxford: Oxford University Press.

Matkowski, J. 1973. "On Lipschitzian Solutions of a Functional Equation." *Annales Polonici Mathematici* 28: 135–39.

Matkowski, J. 1975. "Integrable Solutions of Functional Equations." *Dissertationes Mathematicae* 127.

McShane, E. 1934. "Extension of Ranges of Functions." *Bulletin of American Mathematical Society* 40: 837–42.

McShane, E., and T. Botts. 1959. *Real Analysis*. New York: van Nostrand.

Merryfield, J., and J. Stein, Jr. 2002. "A Generalization of the Banach Contraction Principle." *Journal of Mathematical Analysis and Applications* 273: 112–20.

Megginson, R. 1998. *An Introduction to Banach Space Theory*. New York: Springer.

Michel, P. 1982. "On the Transversality Condition in Infinite-Horizon Problems." *Econometrica* 50: 975–85.

Milgrom, P., and J. Roberts. 1990. "Rationalizability, Learning and Equilibrium in Games with Strategic Complementarities." *Econometrica* 58: 1255–78.

Minty, G. 1970. "On the Extension of Lipschitz, Lipschitz-Hölder Continuous, and Monotonic Functions." *Bulletin of American Mathematical Society* 76: 334–39.

Mitra, T. 2000. "Introduction to Dynamic Optimization Theory." In *Optimization and Chaos*. M. Majumdar, T. Mitra, and K. Nishimura, eds. New York: Springer-Verlag.

Moulin, H. 2001. "Axiomatic Cost and Surplus-Sharing." In *Handbook of Social Choice and Welfare*. K. Arrow, A. Sen, and K. Suzumura, eds. Amsterdam: North-Holland.

Mukerji, S. 1998. "Ambiguity Aversion and Incompleteness of Contractual Form." *American Economic Review* 88: 1207–31.

Myerson, R. 1991. *Game Theory*. Cambridge: Harvard University Press.

Nadler, S. 1969. "Multi-valued Contraction Mappings." *Pacific Journal of Mathematics* 30: 475–87.

Nadler, S. 1978. *Hyperspaces of Sets*. New York: Marcel Dekker.

Nash, J. 1950. "The Bargaining Problem." *Econometrica* 28: 155–62.

Nash, J. 1951. "Noncooperative Games." *Annals of Mathematics* 54: 286–95.

Negishi, T. 1960. "Welfare Economics and Existence of an Equilibrium for a Competitive Economy." *Metroeconomica* 12: 92–7.

Nikodem, K. 1987. "On Midpoint Convex Set-Valued Functions." *Aequationes Mathematicae* 33: 46–56.

Nirenberg, L. 1974. *Functional Analysis.* Lecture notes, Courant Institute of Mathematical Sciences, New York University.

Ok, E. A. 2000. "Utility Representation of an Incomplete Preference Relation." *Journal of Economic Theory* 104: 429–49.

Ok, E. A. 2002. "Nonzero Fixed Points of Power-Bounded Linear Operators." *Proceedings of the American Mathematical Society* 131: 1539–51.

Ok, E. A. 2004. "Fixed Set Theory for Closed Correspondences." *Nonlinear Analysis* 56: 309–30.

Ok, E. A. 2005. "Functional Representation of Rotund-Valued Proper Multifunctions." Mimeo, Department of Economics: New York University.

Ok, E. A. 2006. "Fixed Set Theorems of Krasnoselsky Type." Mimeo, Department of Economics: New York University.

Ok, E. A. 2007. *Probability Theory with Economic Applications.* Princeton: Princeton University Press.

Ok, E. A, and L. Koçkesen. 2000. "Negatively Interdependent Preferences." *Social Choice and Welfare* 3: 533–58.

Ok, E. A., and Y. Masatlioglu. 2005. "A Theory of (Relative) Discounting," Mimeo, Department of Economics: New York University.

Osborne, M., and A. Rubinstein. 1994. *A Course in Game Theory.* Cambridge, Mass.: MIT Press.

Parthasarathy, T. 1971. *Selection Theorems and their Applications.* Berlin: Springer-Verlag.

Peleg, B. 1970. "Utility Functions for Partially Ordered Topological Spaces." *Econometrica* 38: 93–6.

Phelps, R. 1957. "Convex Sets and Nearest Points." *Proceedings of the American Mathematical Society* 8: 790–97.

Preiss, D., and Zajicek, L. 1984. "Fréchet Differentiation of Convex Functions in a Banach Space with a Separable Dual." *Proceedings of American Mathematical Society* 91: 202–4.

Rader, T. 1963. "The Existence of a Utility Function to Represent Preferences." *Review of Economic Studies* 30: 229–32.

Ramsey, F. 1928. "A Mathematical Theory of Saving." *Economic Journal* 38: 543–9.

Ray, D. 1995. *Dynamic Programming and Dynamic Games.* Lecture notes, Boston University.

Reich, S. 1972. "Fixed Points of Contractive Functions." *Unione Matematica Italiana Bollettino* 5: 26–42.

Reny, P. 1999. "On the Existence of Pure and Mixed Strategy Nash Equilibria." *Econometrica* 67: 1029–56.

Richter, M. 1966. "Revealed Preference Theory." *Econometrica* 34: 635–45.

Richter, M. 1980. "Continuous and Semi-Continuous Utility." *International Economic Review* 21: 293–9.

Riesz, F., and B. Sz.-Nagy. 1990. *Functional Analysis.* New York: Dover.

Rigotti, L., and C. Shannon. 2005. "Uncertainty and Risk in Financial Markets." *Econometrica* 73: 203–43.

Roberts, A., and D. Varberg. 1973. *Convex Functions*, New York: Academic Press.

Roberts, J. 1977. "A Compact Convex Set with No Extreme Points." *Studia Mathematica* 60: 255–66.

Robinson, S., and R. Day. 1974. "A Sufficient Condition for Continuity of Optimal Sets in Mathematical Programming." *Journal of Mathematical Analysis and Applications* 45: 506–11.

Rockefellar, T. 1976/2000. *Convex Analysis*. Princeton: Princeton University Press.

Rogers, C. 1980. "A Less Strange Version of Milnor's Proof of Brouwer's Fixed Point Theorem." *American Mathematical Monthly* 87: 525–7.

Rolewicz, S. 1985. *Metric Linear Spaces*. Dordrecht: Reidel Publishing.

Ross, S. 1978. "A Simple Approach to the Valuation of Risky Streams." *Journal of Business* 51: 453–75.

Rota, G. 1964. "Theory of Mobius Functions." *Zeitschrift für Wahrscheinlichkeitstheorie und Verwandte Gebiete* 2: 340–68.

Roth, A. 1980. "Values for Games without Side Payments: Some Difficulties with the Current Concepts." *Econometrica* 48: 457–65.

Royden, H. 1994. *Real Analysis*. New York: Macmillan.

Rubinstein, A. 1991. "Comments on the Interpretation of Game Theory." *Econometrica* 59: 909–24.

Rubinstein, A. 1998. *Lectures on Modeling Bounded Rationality*. Cambridge, Mass.: MIT Press.

Rucker, R. 1995. *Infinity and the Mind*. Princeton: Princeton University Press.

Rudin, W. 1976. *Introduction to Mathematical Analysis*. New York: McGraw-Hill.

Saaty, T., and J. Bram. 1964. *Nonlinear Mathematics*. New York: Dover.

Schechter, E. 1997. *Handbook of Analysis and its Foundations*. San Diego: Academic Press.

Schechter, M. 2002. *Principles of Functional Analysis*. Providence: American Mathematical Society.

Schmeidler, D. 1971. "A Condition for the Completeness of Partial Preference Relations." *Econometrica* 39: 403–4.

Schmeidler, D. 1986. "Integral Representation without Additivity." *Proceedings of the American Mathematical Society* 97: 255–61.

Schmeidler, D. 1989. "Subjective Probability and Expected Utility without Additivity." *Econometrica* 57: 571–87.

Segal, I. 1947. "Postulates of General Quantum Mechanics." *Annals of Mathematics* 48: 930–48.

Seierstad, A., and K. Sydsaeter. 1987. *Optimal Control Theory with Economic Applications*. Amsterdam: North Holland.

Sen, A. 1997. *On Economic Inequality*. Oxford: Clarendon Press.

Shapley, L. 1953. "A Value for *n*-Person Games." In *Contributions to the Theory of Games*. H. Kuhn and A. Tucker, eds. Princeton: Princeton University Press.

Simon, C., and L. Blume. 1994. *Mathematics for Economists.* New York: Norton.

Simon, L. 1987. "Games with Discontinuous Payoffs." *Review of Economic Studies* 54: 569–97.

Simon, L., and W. Zame. 1990. "Discontinuous Games and Endogenous Sharing Rules." *Econometrica,* 58: 861–72.

Smart, D. 1974. *Fixed Point Theorems.* Cambridge: Cambridge University Press.

Starmer, C. 2000. "Developments in Non-Expected Utility Theory: The Hunt for a Descriptive Theory of Choice under Risk." *Journal of Economic Literature* 38: 332–82.

Stokey, N., and R. Lucas. 1989. *Recursive Methods in Economic Dynamics.* Cambridge: Harvard University Press.

Stoll, R. 1963. *Set Theory and Logic.* New York: Dover.

Strang, G. 1988. *Linear Algebra and Its Applications.* Philadelphia: Saunders.

Sundaram, R. 1996. *A First Course in Optimization Theory.* Cambridge: Cambridge University Press.

Sutherland, W. 1975. *Introduction to Metric and Topological Spaces.* Oxford: Clarendon Press.

Tan, K.-K., Yu J., and X-Z. Yuan. 1995. "Existence Theorems of Nash Equilibria for Non-Cooperative *n*-Person Games." *International Journal of Game Theory* 24: 217–22.

Tarafdar, E. 1974. "An Approach to Fixed-Point Theorems on Uniform Spaces." *Transactions of the American Mathematical Society* 191: 209–25.

Taylor, A. 1982. "A Study of Maurice Fréchet: I. His Early Work on Point Set Theory and the Theory of Functionals." *Archive for History of Exact Sciences* 27: 233–95.

Thomson, W. 1994. "Cooperative Models of Bargaining." In *Handbook of Game Theory,* Vol. 2. R. Aumann and S. Hart, eds. New York: North Holland.

Thurston, H. 1994. "A Simple Proof That Every Sequence Has a Monotone Subsequence." *American Mathematical Monthly* 67: 344.

Topkis, D. 1998. *Supermodularity and Complementarity.* Princeton: Princeton University Press.

Tsing, N.-K. 1984. "Infinite-Dimensional Banach Spaces Must Have Uncountable Basis: An Elementary Proof." *American Mathematical Monthly* 91: 505–6.

Tukey, J. 1942. "Some Notes on the Separation of Convex Sets." *Portugaliae Mathematicae* 3: 95–102.

Vainberg, M. 1964. *Variational Methods for the Study of Nonlinear Operators.* San Francisco: Holden-Day.

van Brunt, B. 2004. *The Calculus of Variations.* New York: Springer.

Vives, X. 1990. "Nash Equilibrium with Strategic complementarities." *Journal of Mathematical Economics* 19: 305–21.

Walker, M. 1979. "A Generalization of the Maximum Theorem." *International Journal of Economics* 20: 260–72.

Weymark, J. 1991. "A Reconsideration of the Harsanyi-Sen Debate on Utilitarianism." In *Interpersonal Comparisons of Utility.* J. Elster and J. Roemer, eds. Cambridge: Cambridge University Press.

Wilansky, A. 1951. "The Bounded Additive Operation on Banach Space." *Proceedings of the American Mathematical Society* 2: 46.

Wilson, R. 1971. "Computing Equilibria of n-Person Games." *SIAM Journal of Applied Mathematics* 21: 80–87.

Wong, C. 1976. "On a Fixed Point Theorem of Contraction Type." *Proceedings of the American Mathematical Society* 57: 253–54.

Young, P. 1985. "Monotonic Solutions of Cooperative Games." *International Journal of Game Theory* 14: 65–72.

Yu-Qing, C. 1996. "On a Fixed Point Problem of Reich." *Proceedings of the American Mathematical Society* 124: 3085–88.

Zajicek, L. 1992. "An Elementary Proof of the One-Dimensional Rademacher Theorem." *Mathematica Bohemica* 117: 133–6.

Zhou, L. 1997. "Harsanyi's Utilitarianism Theorems." *Journal of Economic Theory* 72: 198–207.

Glossary of Selected Symbols

Set Theory

$A \subseteq B$	A is a subset of B	4
$A \subset B$	A is a proper subset of B	5
$\|A\|$	cardinality of the set A	5
$\emptyset$	empty set	5
2^S	power set of S	5
$\mathcal{P}(S)$	class of all nonempty finite subsets of S	6
$A \backslash B$	difference between A and B	6
$\cup\mathcal{A}$	union of sets that belong to the class $\mathcal{A}$	7
$\cap\mathcal{A}$	intersection of sets that belong to the class $\mathcal{A}$	7
$\mathbf{X}\mathcal{A}$	(Cartesian) product of sets that belong to the class $\mathcal{A}$	31
X^n	set of all n-vectors in X	28
X^∞	set of all sequences in X	27
$tr(R)$	transitive closure of R	11
$X \backslash \sim$	quotient set of X relative to $\sim$	12
$[x]_\sim$	equivalence class of x relative to $\sim$	12
$(X, \succsim)$	preordered set	15
$U_{\succsim}(x)$, $U_{\succ}(x)$	upper (strict upper) contour set of x relative to $\succsim$	102
$L_{\succsim}(x)$, $L_{\succ}(x)$	lower (strict lower) contour set of x relative to $\succsim$	102
$\succcurlyeq_{\text{card}}$	cardinality ordering	93
$f : X \to Y$	function from X into Y	20
Y^X	set of all functions from X into Y	21
$Gr(f)$	graph of f	21
id_X	identity function on X	23
$\mathbf{1}_S$	indicator function of S	23
$f\|_Z$	restriction of f to Z	23
$f(A)$	(direct) image of A under f	25
$f^{-1}(B)$	inverse image of B under f	25
f^{-1}	inverse of f	26
$g \circ f$	composition of f and g	24
$\Gamma : X \rightrightarrows Y$	correspondence from X into Y	284
$\Gamma(S)$	image of S under Γ	284
$\Gamma^{-1}(O)$	upper inverse image of O under Γ	289

$\Gamma_{-1}(O)$	lower inverse image of O under Γ	298
$\mathrm{Gr}(\Gamma)$	graph of Γ	294
arg max	set of maximizing points	286
$\mathbb{N}$	set of all positive integers	37
$\mathbb{Z}$	set of all integers	39
$\mathbb{Z}_+$	set of all nonnegative integers	39
$\mathbb{Q}$	set of all rational numbers	39
$\mathbb{Q}_+, \mathbb{Q}_{++}$	set of all nonnegative (positive) rational numbers	41
$\mathbb{R}$	set of all real numbers	41
$\mathbb{R}_+, \mathbb{R}_{++}$	set of all nonnegative (positive) real numbers	41
$\overline{\mathbb{R}}$	set of all extended real numbers	45
$\lvert t \rvert$	absolute value of t	46
$\sup S, \inf S$	supremum (infimum) of S	41
$\max S, \min S$	maximum (minimum) of S	45
$\mathbb{R}^n$	set of all real n-vectors	28
$\mathbb{R}^\infty$	set of all real sequences	46
$\mathbb{R}^T$	set of all real maps on T	62
$x_m \to x$	(x_m) converges to x	47
$x_m \nearrow x$	(x_m) is increasing and converges to x	51
$x_m \searrow x$	(x_m) is decreasing and converges to x	51
$\lim x_m$	limit of (x_m)	47
$\limsup x_m$	limit superior of (x_m)	54
$\liminf x_m$	limit inferior of (x_m)	54
$\sum^\infty x_i$	infinite series	57
$\prod^\infty x_i$	infinite product	61
$\lim_{t \to x} f(t)$	limit of f at x	64
f'_+	right-derivative of f	67
f'_-	left-derivative of f	67
$f', \frac{d}{dt} f$	derivative of f	67
f''	second-derivative of f	67
$\mathbf{C}[a, b]$	set of continuous real functions on $[a, b]$	65
$\mathbf{C}^k[a, b]$	set of k times continuously differentiable real functions on $[a, b]$	68
$\mathbf{P}[a, b]$	set of polynomials on $[a, b]$	66
$\int_a^b f(t)dt$	Riemann integral of f on $[a, b]$	71

Linear Spaces

$(X, +, \cdot)$	linear space	361
$\mathbf{0}$	origin	361
$A + B$	sum of A and B	362

$A+y, y+A$	sum of A and $\{y\}$	362
λA	$\{\lambda x : x \in A\}$	362
$\mathbf{c}$	space of convergent real sequences	365
$\mathbf{c}^0$	space of all real sequences with finite support	365
$\mathbf{c}_0$	space of all real sequences that converge to 0	608
$\sum^m x^i$	sum of the vectors $x^1, \ldots, x^m$	368
$span(S)$	span of S	369
$aff(S)$	affine hull of S	369
$\dim(X)$	dimension of the linear space X	376
$\dim(S)$	dimension of the set S	378
$X \times Y$	product of linear spaces X and Y	379
$null(L)$	null space of L	382
$\mathcal{L}(X, Y)$	set of linear operators from X into Y	382
$\mathfrak{L}_X$	set of all lotteries on X	396
δ_x	lottery that puts unit mass on x	396
$\mathsf{E}_p(u)$	expected value of u with respect to p	396
$\mathbf{v}$	capacity	410
$\mathcal{V}_N$	space of capacities on N	410
$co(S)$	convex hull of S	426
$cone(S)$	conical hull of S	430
$ext(S)$	set of extreme points of S	655
$(X, +, \cdot, \succsim)$	preordered linear space	432
X_+, X_{++}	positive (strictly positive) cone of X	433
$al\text{-}int_X(S)$	algebraic interior of S (in X)	437
$ri(S)$	relative interior of S	438
$al\text{-}cl(S)$	algebraic closure of S	448
$al\text{-}bd_X(S)$	algebraic boundary of S (in X)	448
$x \perp y$	x and y are orthogonal	492
$S^\perp$	orthogonal complement of S	492

Metric/Normed Spaces

$d(x, y)$	distance between x and y	119
(X, d)	metric space	119
$(X, \|\cdot\|)$	normed linear space	605
$\|\cdot\|, \|\cdot\|_Y$	norm (norm of Y)	605, 606
$d_p, \|\cdot\|_p$	p-metric (p-norm)	122, 608
$d_\infty, \|\cdot\|_\infty$	sup-metric (sup-norm)	123, 608
$d_{\infty,\infty}, \|\cdot\|_{\infty,\infty}$	sup-sup-metric (sup-sup-norm)	125, 609
d_H	Hausdorff metric	302
$d_{\|\cdot\|}$	metric induced by $\|\cdot\|$	611

$\mathbb{R}^{n,p}$	$(\mathbb{R}^n, d_p)$	121
$\mathbb{R}^n$	n-dimensional Euclidean space	121
ℓ^p	space of p-summable real sequences	122, 608
ℓ^∞	space of bounded real sequences	123, 608
$\mathbf{B}(T)$	space of bounded real maps on T	123, 608
$\mathbf{B}[a,b]$	space of bounded real maps on $[a,b]$	125
$\mathbf{C}(T), \mathbf{CB}(T)$	space of continuous (and bounded) real maps on T	249, 608
$\mathbf{C}(T, \mathbb{R}^n)$	space of continuous maps from T into $\mathbb{R}^n$	250
$\mathbf{C}^1[a,b]$	space of continuously differentiable real maps on $[a,b]$	125, 609
$N_{\varepsilon,X}(x)$	ε-neighborhood of x in X	127, 612
$int_X(S)$	interior of S (relative to X)	128
$cl_X(S)$	closure of S (relative to X)	128
$bd_X(S)$	boundary of S (relative to X)	128
$T^\diamond$	interior of T in $aff(T)$	695
$x^m \to x$	(x^m) converges to x	132
$\lim x^m$	limit of (x^m)	132
$\sum^\infty x^i$	infinite series in a normed linear space	618
$diam(S)$	diameter of S	169
ρ	product metric	193, 194
$\mathsf{X}^n(X_i, d_i)$	product of (X_i, d_i), $i = 1, \ldots, n$	193
$\mathsf{X}^\infty(X_i, d_i)$	product of (X_i, d_i), $i = 1, 2, \ldots$	194
$[a,b]^\infty$	Hilbert cube	195
$\mathsf{p}_S(x)$	projection of x on S	227
$\varphi^\bullet, \varphi_\bullet$	$\limsup \varphi$, $\liminf \varphi$	231
$\varphi_m \to \varphi$	(φ_m) converges to φ pointwise	253
$\varphi_m \nearrow \varphi$	$\varphi_1 \leq \varphi_2 \leq \cdots$ and $\varphi_m \to \varphi$	255
$\varphi_m \searrow \varphi$	$\varphi_1 \geq \varphi_2 \geq \cdots$ and $\varphi_m \to \varphi$	255
B^n_α	closed α-ball around $\mathbf{0}$ in $\mathbb{R}^n$	275
B^n	closed unit ball in $\mathbb{R}^n$	275
S^{n-1}	$n-1$ dimensional unit sphere	276
B_X	closed unit ball of X	612
$\mathcal{B}(X,Y)$	space of bounded linear operators from X into Y	639
X^*	space of bounded linear functionals on X	639
$\lVert\cdot\rVert^*$	operator norm	644
$\lim_{\omega \to x} \Phi(\omega)$	limit of Φ at x	671
D_Φ	Fréchet derivative of Φ	675, 696
$\mathsf{D}_{\Phi,x}$	Fréchet derivative of Φ at x	675, 696
$\mathsf{D}^2_{\Phi,x}$	second Fréchet derivative of Φ at x	91, 696
$\partial\varphi(x)$	superdifferential of φ at x	711

Index